SPECIATION

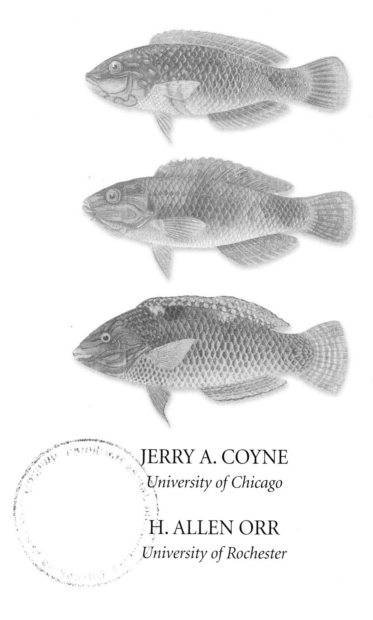

JERRY A. COYNE
University of Chicago

H. ALLEN ORR
University of Rochester

Sinauer Associates, Inc. • *Publishers*
Sunderland, Massachusetts U.S.A.

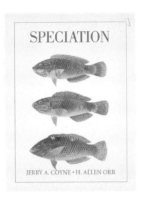

About the cover

Three congeneric species of Pacific wrasse; from top to bottom, *Halichoeres trimaculatus*, *H. margaritaceus*, and *H. hortulanus*. These paintings, by the Japanese artist Kako Morita, are reproduced from Plates 46 and 47 of *The Fishes of Samoa* by David Starr Jordan and Alvin Seale (1906, Bulletin of the United States Bureau of Fisheries 25:173–456). The illustrations were published with the help of Jordan's friend and fellow naturalist, President Theodore Roosevelt, who interceded when the government's committee on publication deemed the plates too expensive to print. Jordan (1851–1931) was an influential evolutionist, ichthyologist, and a staunch defender of Darwinism at a time when it was unpopular. A prolific author, his most notable contribution to the study of speciation was his emphasis on the importance of geographic barriers.

SPECIATION

Copyright©2004 by SINAUER ASSOCIATES, Inc. All rights reserved. This book may not be reprinted in whole or in part without permission from the publisher.

For information or to order, address:
SINAUER ASSOCIATES, Inc.
23 Plumtree Road/PO Box 407
Sunderland, MA 01375 U.S.A.

FAX: 413-549-1118

Email: publish@sinauer.com

www.sinauer.com

Library of Congress Cataloging-in-Publication Data
Coyne, Jerry A.
  Speciation / Jerry A. Coyne, H. Allen Orr.
    p. cm.
  Includes bibliographical references and index.
  ISBN 0-87893-091-4 (hardcover) -- ISBN 0-87893-089-2 (paperback)
  1. Species. I. Orr, H. Allen. II. Title.
QH380.C68 2004
576.8'6--dc22   2004009505

Printed in U.S.A.   5 4 3 2

To Anne and Lynne

Contents

Introduction 1

1 Species: Reality and Concepts 9

The Reality of Species 10
  Sexually reproducing eukaryotic taxa 12
  Groups with little or no sexual reproduction 17
  Conclusions 25

Species Concepts 25
  The biological species concept (BSC) 26
  Advantages of the BSC 38
  Problems with the BSC 39
  Other species concepts 48

Why Are There Species? 48

2 Studying Speciation 55

The Problem of Speciation 57

Identifying and Measuring Reproductive Isolation 61
  Absolute strength of isolating barriers 62
  Relative strength of isolating barriers 63
  Prezygotic versus postzygotic isolation 65
  Which isolating barriers caused speciation? 69

Comparative Studies of Isolating Barriers 72
  How fast does reproductive isolation appear? 72
  Which traits promote the evolution of reproductive isolation? 81

3 Allopatric and Parapatric Speciation 83

Allopatric Speciation 85
  Vicariant speciation 86
  Peripatric speciation 105
Parapatric Speciation 111
  Theory 112
  Experimental evidence 117
  Evidence from nature 118
Conclusions 123

4 Sympatric Speciation 125

Theory 127
  Disruptive sexual selection 128
  Disruptive natural selection 130
  Conclusions 136
Experimental Evidence 138
Evidence from Nature 141
  Evidence from habitat "islands" 143
  Evidence from host races and host-specific species 157
  Allochronic (temporal) isolation in sympatry 166
  Comparative studies of the biogeography of speciation 168
Conclusions 175

5 Ecological Isolation 179

Habitat Isolation 182
  Detecting and measuring habitat isolation 184
  The problem of allopatry 185
  Examples of habitat isolation 186
  Relative importance of habitat isolation 188
  The evolution of habitat isolation 188
  The genetics of habitat isolation 191
Pollinator (Floral) Isolation 193
  Detecting and measuring pollinator isolation 194
  Examples of pollinator isolation 195
  Relative importance of pollinator isolation 197
  The evolution of pollinator isolation 198
  The genetics of pollinator isolation 201
Temporal (Allochronic) Isolation 202
  Detecting and measuring temporal isolation 203
  Examples of temporal isolation 204
  Relative importance of temporal isolation 205
  The evolution of temporal isolation 206
  The genetics of temporal isolation 210
Conclusions 210

6 Behavioral and Nonecological Isolation 211

Mating System "Isolation" 211
Behavioral Isolation 213
  Detecting and measuring behavioral isolation 213
  Examples of behavioral isolation 214
  Relative importance of behavioral isolation 215
  The evolution of behavioral isolation 216
  The genetics of behavioral isolation 223
Mechanical Isolation 227
  Examples of mechanical isolation 228
  Relative importance of mechanical isolation 229
  The evolution of mechanical isolation 230
  The genetics of mechanical isolation 231
Gametic (Postmating, Prezygotic) Isolation 232
  Examples of gametic isolation 233
  Relative importance of gametic isolation 238
  The evolution of gametic isolation 241
  Conclusions 245

7 Postzygotic Isolation 247

Extrinsic Postzygotic Isolation 249
Intrinsic Postzygotic Isolation 253
The Frequency of Various Forms of Postzygotic Isolation 255
The Evolution of Extrinsic versus Intrinsic Postzygotic Isolation 255
Genetic Modes of Intrinsic Postzygotic Isolation 256
  Chromosomal speciation: theory 256
  Chromosomal speciation: data 259
  Genic incompatibilities 267
  The evolution of genic incompatibilities: the Dobzhansky–Muller model 269
  Mathematical models of genic speciation 272
  Wolbachia and cytoplasmic incompatibility 276
Conclusions 280

8 The Genetics of Postzygotic Isolation 283

Haldane's Rule 284
  The phenomenon 284
  The causes of Haldane's rule 286
  Conclusions 298
The Genetic Basis of Postzygotic Isolation 299
  How many genes cause postzygotic isolation? 299

Complexity of hybrid
  incompatibilities 307
Probability of hybrid
  incompatibilities 308
Where are the genes causing postzygotic
  isolation? 308
Developmental basis of postzygotic
  isolation 309
Are duplicate genes important? 312
Which genes cause postzygotic
  isolation? 313

9 Polyploidy and Hybrid Speciation 321

Polyploidy 321
  Classification 322
  Pathways to polyploidy 324
  Incidence 326
  Frequency of auto- versus
    allopolyploidy 328
  Ecology and persistence 330
  Why is polyploidy rarer in animals than
    in plants? 333

Recombinational Speciation 337
  What is recombinational speciation? 337
  Theory 338
  The data: frequency and artificial
    hybrids 342
  The data: natural recombinational
    speciation 344
  The data meet the theory 350

10 Reinforcement 353

The Data 354
  Selection experiments 355
  Evidence from nature: case studies 357
  Evidence from nature: comparative
    studies 362
  Reinforcement of postzygotic
    isolation 365

The Theory 366
  Early enthusiasm 366
  Objections to reinforcement 369

  The revival of reinforcement 372
Alternative Explanations 375
  Publication bias 375
  Differential fusion 376
  Direct ecological effects 377
  Ecological character displacement 377
  Runaway sexual selection 378
  Sympatric speciation 378
Distinguishing the Alternatives 379

11 Selection versus Drift 383

Speciation by Selection 383
  Natural selection 385
  Sexual selection 386
  Mathematical theories of selection-based speciation 387
Speciation by Drift 387
  Peak shift models 388

Theoretical Criticisms 394
Recent Peak Shift Models 396
The Data 398
  Evidence from the laboratory 398
  Evidence from nature 401
Conclusions 410

12 Speciation and Macroevolution 411

Rates of Speciation 411
  What is a speciation rate? 412
  Theory and speciation rates 413
  Calculating speciation intervals 416
  Extreme rates of speciation 425
  What is the effect of biogeography? 427
  Conclusions 428

Factors Affecting Speciation Rates 429
  Tests for the effects of key factors 431
  Distinguishing speciation from extinction 435
  The data 436
  Conclusions 441
Species Selection 442

Appendix: A Catalogue and Critique of Species Concepts 447

Genotypic Cluster Species Concept 447
Recognition Species Concept 451
Cohesion Species Concept 452

Evolutionary Species Concept 456
Ecological Species Concept 457
Phylogenetic Species Concepts 459

References 473

Author Index 523

Subject Index 533

Preface

Writing a book is an adventure. To begin with, it is a toy and an amusement; then it becomes a mistress, and then it becomes a master, and then a tyrant. The last phase is that just as you are about to be reconciled to your servitude, you kill the monster, and fling him out to the public.
—Winston Churchill

This book grew out of our long-standing interest in speciation, which began as a relationship between student and teacher, but quickly evolved into a collegial collaboration that has lasted more than fifteen years. Over these years, the study of speciation has expanded from a modest backwater of evolutionary biology into a large and vigorous discipline. The result is that the literature on speciation has grown explosively, along with the number of researchers and students working on the problem. Despite this, no recent book summarizes and critically reviews current research in the field. *Speciation* tries to fill this gap.

While this volume has only two authors, it could not have been written without extensive help and advice from many biologists. In a field as wide-ranging as speciation, no single person—or even two people—can hope to master the relevant literature. We are grateful to many colleagues for pointing us to important work outside our areas of expertise, and for helpful discussion and criticism. We extend special thanks to those who patiently answered our endless questions: Tim Barraclough, Spencer Barrett, Nick Barton, Stewart Berlocher, Fred Cohan, Jeff Feder, Dave Jablonski, Trevor Price, Steve Pruett-

Jones, Dolph Schluter, Doug Soltis, George Turner, and Victor Vacquier. The manuscript was greatly improved by critical comments from Brian Charlesworth, Doug Futuyma, Dan Howard, John Jaenike, Anne Magurran, Mohamed Noor, Daven Presgraves, Loren Rieseberg, Howard Rundle, Doug Schemske, Brit Smith, Michael Turelli, and Phil Ward.

We also received valuable assistance from Peter Abrams, Chip Aquadro, Daniel Barbash, Michael Bell, Stewart Berlocher, Andrew Berry, Andrea Betancourt, Dan Bolnick, Seth Bordenstein, Jim Bull, Deborah Charlesworth, Dale Clayton, Andrew Cohen, John Cooley, Cliff Cunningham, Rob DeSalle, Tim Dickinson, Susannah Elwyn, Ben Fitzpatrick, David Foote, Jim Fry, Geoffrey Fryer, Sergey Gavrilets, Laura Geyer, Todd Grantham, Russell Greenberg, Susan Harrison, Dan Hartl, Ralph Haygood, Andrew Hendry, Charles Henry, Allen Herre, Jody Hey, Richard Highton, Dave Hillis, Richard Hudson, Molly Hunter, Darryl Irwin, Fran James, Corbin Jones, Ada Kaliszewska, Bob Kimsey, Mark Kirkpatrick, Nancy Knowlton, Irv Kornfield, Russ Lande, Mathew Leibold, Michael Lynch, Mark Macnair, David Marshall, J. P. Masly, Natasha Mehdiabadi, Eldredge Moores, Leonie Moyle, Michael Nachman, Sally Otto, Steve Palumbi, Ole Pellmyr, Silvia Pihu, Claudia Ricci, Sievert Rohwer, Mike Ryan, Mike Sanderson, Tom Schoener, Jon Seger, Sylvia Sepp, Maria Servedio, Kerry Shaw, Christine Simon, Stevan Springer, Victor Springer, Eli Stahl, Yun Tao, John Thompson, James Thomson, Peter Van Dijk, Sara Via, Ron Wagner, Barbara Wakimoto, Jack Werren, Mike Whitlock, Jeannette Whitton, Mariana Wolfner, (the late) Tom Wood, and Tim Wootton. Theodore B. Coyne provided unstinting support and affection but, sadly, did not live to see this book published. It goes without saying that not all of these people agree with our conclusions—indeed, all of them surely disagree with some of our conclusions. We apologize to those who have helped us but whose names have been inadvertently omitted, and to those whose work we have accidentally overlooked in a vast literature.

Our work has relied heavily on the support of several foundations and institutions. These include the National Institutes of Health, which has continuously funded both of us (we are particularly grateful to Irene Eckstrand for her help over the years); the David and Lucile Packard Foundation, which funded HAO; the John Simon Guggenheim Foundation, whose fellowships allowed us extended periods to work on speciation; our home institutions, The University of Chicago and The University of Rochester, which gave us respite from our teaching duties; The University of Edinburgh, which granted JAC space, hospitality, and intellectual stimulation during a sabbatical leave, and The Rockefeller Foundation, whose residency program at Bellagio, Italy allowed us to complete this book. Our stay in Bellagio was immensely enhanced by the kindness and generosity of Gianna Celli. Kelly Dyer, Susannah Elwyn, and Shannon Irving kept our labs running when we were forced to forsake flies for writing. We are grateful to our (joint) undergraduate institution, the College of William and Mary—and especially to Professor Bruce

Grant—for a superb introduction to evolutionary biology. JAC extends special thanks to Dick Lewontin for his advice and inspiration over the years.

This book would never have reached fruition without the help of the good people at Sinauer Associates. We are grateful to Andy Sinauer, the editor and publisher, for supporting us from the beginning, when *Speciation* was only an idea. We also are indebted to Kathaleen Emerson, our project manager, for her many contributions and her remarkable ability to calm overwrought authors, and to Joan Gemme, Janice Holabird, Suzanne Lain, Michele Ruschhaupt, Marie Scavotto, and Christopher Small for their work in design, production, advertising, and editing.

Finally, we thank Anne Magurran and Lynne Orr, to whom this book is dedicated, for their love and support.

Introduction

One of the most striking developments in evolutionary biology during the last 20 years has been a resurgence of interest in the origin of species. One can hardly open a new issue of *Evolution*, *Genetics*, or *The Proceedings of the Royal Society* without finding several papers on speciation. This revival has produced immense amounts of data that, while sometimes supporting old ideas, have often overturned them.

Despite these developments, there has been no book-length treatment of speciation in several decades. Mayr's *Animal Species and Evolution* (1963), White's *Modes of Speciation* (1978), and Grant's *Plant Speciation* (1981)—the last overviews of this area—are badly dated. As a result, both the seasoned scholar and the newcomer to speciation find no ready guide to the recent literature—a body of work that is enormous, scattered, and increasingly technical. Although several excellent symposium volumes have appeared in the last 15 years (e.g., Otte and Endler 1989; Howard and Berlocher 1998), these collections do not give a unified and comprehensive overview of the field. We hope that *Speciation* will.

The recent burst of work on speciation constitutes what we see as the third phase of work on the problem. The first began with Darwin. The point has often been made that, despite its title, *The Origin of Species* (1859) had much more to say about change *within* species than about the origin of new species. Yet Darwin at least recognized that species not only evolve but also divide. Indeed, the only figure in *The Origin* depicts this splitting. But Darwin's own theory of how splitting occurs made little distinction between speciation and adaptation: he saw the origin of species as a direct consequence of the struggle between individuals for ecological elbow room.

Darwin's immediate successors considered his theory of speciation inadequate, but dwelt on two different problems, yielding two opposing schools of thought. Naturalists, well acquainted with biogeography, felt that Darwin unduly emphasized sympatric speciation (i.e., speciation occurring within a single interbreeding population). Observing that most closely related species are allopatric (geographically isolated), workers like Moritz Wagner (1889), Karl Jordan (1896), and David Starr Jordan (1905, 1908) suggested that geographic isolation played a pivotal role in speciation (Mayr, 1982, pp. 561–566). These workers agreed with Darwin, however, that natural selection was the most important force in speciation. In a remarkably prescient essay, Poulton (1908) not only defined species as reproductively isolated entities, but provided a list of isolating barriers and suggested how these barriers might arise as byproducts of natural selection.

The second school of thought—the mutationists—arose after the rediscovery in 1900 of Mendel's work. Rejecting Darwin's claim that speciation was gradual and driven by natural selection, biologists like De Vries (1906) and Bateson (1922) argued that speciation is divorced from such selection, and instead involves nonadaptive and macromutational leaps. This view may have derived from an inability to understand how a continuous process could create the discontinuous entities seen in nature. Bateson (1909, p. 99) expressed a typical mutationist view: "As Samuel Butler so truly said: 'To me it seems that the *Origin of Variation*, whatever it is, is the only true *Origin of Species.*'" The reviews of O. W. Richards and G. C. Robson (1926; Robson 1928) supported the mutationist view by claiming that species differences had little or nothing to do with natural selection. Until about 1935, there were prominent advocates of both gradualism and mutationism (see Provine 1971; Gould 2002). In the absence of decisive evidence, research on speciation languished.

The second phase of work on speciation began with the Modern Synthesis, when evolutionists decisively rejected mutationism and reconciled Mendelism, biogeography, and natural selection. We date the beginning of this era to 1935, when Dobzhansky published his paper "A critique of the species concept in biology." As a naturalist and geneticist, Dobzhansky recognized that speciation was both a novel and an unsolved problem. Most important, he realized that the solution to the conundrum involved understanding how a continuous evolutionary process—changes in allele frequencies—could produce genetically and morphologically discrete groups living in one habitat. He also saw that Darwin's notion of speciation was implausible, or at least incomplete, since ecologically distinct forms cannot coexist without barriers to gene exchange.

Dobzhansky thus stressed the importance of "reproductive isolating mechanisms," the diverse set of traits that prevent gene flow between taxa. In his view, the problem of speciation was not the occupation of new niches, but the origin of reproductive isolation. These ideas were enlarged and codified in what is often considered the seminal work of the Modern Synthesis, Dobzhansky's *Genetics and the Origin of Species* (1937). Dobzhansky's book inspired a flood of research on speciation. This included the first rigorous genetic studies of reproductive isolation, including Dobzhansky's own research, and that of Muller, Pontecorvo, Tan, and others. This work established an important point that is now taken for granted: reproductive isolating barriers, like adaptations within species, are usually based on changes in "ordinary" Mendelian genes.

This genetical work was complemented by important studies of the natural history and biogeography of speciation, first synthesized in Mayr's (1942) *Systematics and the Origin of Species*. Mayr's book made two important contributions. First, he codified the meaning of species. Mayr's "biological species concept" identified species as groups of interbreeding populations that are reproductively isolated from other groups, thus representing independent units of evolution. Mayr also argued that species arise only from populations that are allopatric. While neither of these ideas has survived completely unscathed, they stimulated a great deal of research and remain central parts of our current view of speciation. Although Dobzhansky and Mayr were both zoologists, Stebbins's (1950) *Variation and Evolution in Plants* showed that, with the addition of polyploidy, earlier theories could also explain speciation in plants.

Curiously, however, the Modern Synthesis neglected important aspects of speciation. Although evolutionists talked endlessly about reproductive isolating barriers, they paid almost no attention to how those barriers evolved. Thus, while the Modern Synthesis yielded a strong consensus that speciation is driven by natural selection, there were few attempts to describe exactly how selection produces reproductive isolation. Perhaps the most glaring omission involved ethological isolation—differences in sexual preferences between species that prevent them from mating. Although sexual selection was proposed by Darwin in 1871 and elaborated by Fisher in 1930, West-Eberhard (1983) was the first to emphasize what now seems an obvious link between sexual selection and speciation: populations experiencing divergent sexual selection could become behaviorally isolated. (Haskins and Haskins 1949 and Nei 1976 made this suggestion earlier but more briefly.) Moreover, although both Dobzhansky and Mayr were naturalists who recognized the importance of ecology in evolution, neither of them pursued, nor suggested, research programs connecting adaptation and speciation. Evolutionary ecologists, as exemplified by the Oxford school of ecological genetics, were instead preoccupied with demonstrating natural selection in the wild.

Moreover, the enormous contributions of theoretical population geneticists to the Modern Synthesis dealt almost exclusively with genetic change within species, not the origin of new ones. Haldane, for instance, was not even convinced that species were real. His treatment of speciation in *The Causes of Evolution* (1932) is merely a discourse on the genetics of morphological differences between taxa. Wright, who accepted both the biological species concept and allopatric speciation, nevertheless largely ignored the topic. Among theorists, only Fisher (1930) was seriously concerned with speciation, but even his discussion was brief and verbal.

The situation was different among experimental geneticists. Although early in the Modern Synthesis they made important contributions to understanding speciation—the origin of species represented a major, if not *the* major, problem confronting evolutionary genetics—by the end of the Synthesis they had largely abandoned the problem. Indeed, by the 1950s it is hard to find a paper by Dobzhansky, Muller, or their peers on the genetics of speciation. The cause of this neglect is clear. As the Modern Synthesis progressed, evolutionary geneticists grew increasingly obsessed with measuring and explaining genetic variation within species (Lewontin 1974). In retrospect, this shift in emphasis was unfortunate in at least one way: understanding speciation, it turns out, may *not* depend critically on how genetic variation is maintained.

Although interest in the genetics of speciation waned, progress continued on the biogeography and natural history of species and speciation. Mayr synthesized the relevant zoological work in *Animal Species and Evolution* (1963), best known for its summary of the evidence for allopatric speciation. Grant presented the botanical equivalent in his *Plant Speciation* (1981), notable for its analysis of polyploid speciation.

For two decades after the publication of Mayr's book, speciation again became a relative backwater of evolutionary biology. With the advent of protein gel electrophoresis and DNA sequencing, evolutionists' obsession with genetic variation grew even more intense. These seductive new technologies lured experimentalists into a lengthy affair with measuring heterozygosity in nearly every available species. The resulting data unleashed a flood of work in theoretical population genetics, accompanied by the rise of the neutral theory. Though obviously important, these developments pushed other aspects of evolutionary biology—including speciation—into the background.

The third phase of work on speciation began in the early 1980s. A detailed explanation of this renaissance is perhaps best left to historians of science. We can, however, offer a few ideas. First, some older evolutionary geneticists simply grew bored with the perpetual accumulation of molecular-genetic data. In addition, younger evolutionary biologists realized that new tools—involving molecular genetics, systematics, and comparative data—allowed novel and more powerful approaches to old and unresolved questions. Many of these questions concerned the origin of species. Molecular-genetic tools, for example, allowed not only finer scale genetic analysis of isolating barriers and species differences, but also led to the rise of molecular systematics, which in turn produced data essential for answering many questions about speciation. Finally, a small group of evolutionary ecologists realized that ecological approaches to speciation had been almost completely neglected during the Modern Synthesis, and that long-term field studies could not only identify reproductive barriers in nature, but also suggest how natural selection had shaped them.

As a result, more work on speciation has been performed over the last two decades than over the entire period from 1859 to 1980. This latest phase has involved reexamining nearly every conclusion about speciation reached during the Modern Synthesis. Debate about species concepts—virtually quashed by Mayr's forceful arguments in *Animal Species and Evolution*—was revived as biologists not only introduced dozens of new concepts, but even questioned whether species exist. Geneticists who accepted the importance of reproductive isolating barriers performed more rigorous genetic analyses of these barriers. Their work concentrated on counting the genes causing reproductive isolation, locating their positions on chromosomes, measuring their relative effects, and, ultimately, identifying their normal functions within species and the evolutionary forces that drove their divergence. The genetic work also focused on several patterns characterizing reproductive isolation in animals, such as Haldane's rule, the more frequent sterility or inviability of heterogametic (XY) than of homogametic (XX) hybrids. Moreover, the dominant view of the biogeography of speciation—the allopatric model—came under vigorous attack on both theoretical and empirical fronts. New theory suggested that species could arise in the face of limited or even pervasive gene flow (parapatric and sympatric speciation, respectively), and new data suggested that some species had in fact arisen without geographic isolation. Biologists also began to reassess reinforcement, the idea that natural selection can increase the strength of reproductive isolation between two taxa after secondary contact.

Recent work has also taken up the neglected question of the relative roles of natural selection and genetic drift in speciation. While both Mayr and Dobzhansky recognized the importance of selection, Mayr (1963) maintained that genetic drift played an important part in speciation. Although Mayr's view was embraced and extended by others, it was severely criticized in the 1980s. The resulting debate stimulated new experimental work that, in the main, failed to support an important role for drift. It thus seemed clear that natural and sexual selection are the main engines of speciation. But given our almost complete ignorance about how these forms of selection give rise to new species, this conclusion was based more on intuition than on data. Getting such data usually requires laborious, long-term fieldwork. In the 1980s and 1990s, evolutionary ecologists began this research, and we now have several good cases showing a clear link between selection and reproductive isolation. These links have been strengthened by recent molecular analysis showing that natural selection was almost certainly involved in genetic changes that cause reproductive isolation. Molecular work has also yielded more accurate phylogenies, which are essential for comparative studies and which can tell us about the biogeography of speciation and about what factors may promote the evolution of reproductive isolation. Finally, some biologists took up the neglected idea that new species can arise not only by the fission of existing ones, but also by their hybridization, a process that might involve either chromosomal evolution (polyploidy) or genic evolution (recombinational speciation).

In summary, the third phase of speciation involves at least five themes. The first is the continuation of genetic approaches that dominated the second phase. But the other four themes—mathematical theory, an emphasis on ecology, molecular analysis, and the use of comparative studies—are novel. It is these new approaches that stimulated the critical re-examination of virtually every accepted idea about speciation. This broad reexamination may not have produced a radical new paradigm, but it has led many researchers (including the present authors) to change their minds about fundamental questions, such as the plausibility of sympatric speciation and reinforcement.

Speciation summarizes and critically reviews current work and ideas in the field. Because of the huge literature that has accumulated since Mayr (1963) and Grant (1981), we obviously cannot claim expertise in all aspects of speciation, which range from molecular population genetics to ecological fieldwork. We are evolutionary biologists who have worked largely with *Drosophila*, using genetic approaches to understand the evolution of reproductive isolation. Although we have tried to cover the most important aspects of speciation in plants and animals, the content of this book surely reflects our interests and research backgrounds. Partly for this reason, we realize that readers will sometimes disagree with our arguments and conclusions. But we have not refrained from criticizing what we see as flawed work (including some of our own previous studies), or from taking stands on difficult and controversial topics. We have no doubt that future work will show that some of our conclusions are mistaken. But we believe that we will have done our job if we stimulate rigorous new research, even if that research is driven by a desire to prove us wrong.

As the title of our book suggests, we focus on the *process* of species formation, not on the end products of that process. We thus devote relatively little space to species concepts, the nature of species differences, and higher-level selection acting among existing species. Instead, most of the book deals with the evolution of various forms of reproductive isolation, and with the roles that biogeography and natural selection play in the origin of this isolation.

Several themes run through *Speciation*: the resurrection of important but forgotten literature; the emphasis on neglected, unsolved, and even unrecognized problems; and an insistence on hypotheses that are testable. Because interest in speciation has been episodic, a good deal of old but important literature has been largely forgotten. That literature focuses mainly on hybrid sterility and inviability, the subject of intensive work in the 1930s and 1940s. Some of these studies bear critically on modern arguments, but are rarely cited.

We also try to highlight important but unsolved problems. Some of these are well known (the role of geographic isolation); others less so (the identification of traits that facilitate speciation). Finally, if any single difficulty has impeded progress in this field, it is a preoccupation with vague and untestable ideas. There has been, for example, nearly endless discussion of species concepts. Although we will necessarily engage in some of this ourselves (after all, it seems wise to decide what species are before considering how they arise), this vast and stupefying literature has for the most part produced little new or interesting biology. In our view, recent progress in speciation largely reflects a shift from a fascination with nebulous and untestable ideas to empirically tractable ones. It is this shift, we believe, that has allowed the field to attain scientific maturity.

We begin our discussion with what we consider to be *the* problem of speciation: the origin of discrete groups of organisms living together in nature. This problem involves several related topics. The first is whether species are real entities or arbitrary constructs of the human mind. Several lines of evidence show that species are real, which leads to the question of how one can best conceptualize or characterize species. We argue that a modified version of Mayr's "biological species concept" proves most useful. Our choice of this concept, however, is dictated less by a priori philosophical considerations than by pragmatic ones: regardless of the philosophical merits of any species concept, it seems undeniable that nearly all recent progress on speciation has resulted from adopting some version of the biological species concept. We conclude Chapter 1 by considering why species exist at all: Why do organisms form discrete clusters instead of an organic continuum? This is, we believe, one of the most intriguing unsolved problems of evolutionary biology.

Chapter 2, "Studying Speciation," considers what is unique about speciation compared to other aspects of evolution. Identifying these novel qualities allows us to outline a research program for understanding the origin of species. We argue that identifying the current barriers to gene flow between species may mislead us about which barriers were *historically* important in speciation. We then suggest ways to identify these historically significant barriers.

Chapters 3 and 4 discuss the biogeography of speciation. Does speciation require complete geographic isolation between populations, or can it occur in the face of gene flow? The former idea, allopatric speciation, is well supported and uncontroversial, and we discuss it only briefly. The latter idea, comprising parapatric and sympatric speciation, is less well supported and more controversial. We conclude that while speciation with substantial gene flow is possible—and that several plausible cases exist—comparative work suggests that it is far less frequent than allopatric speciation.

Chapters 5–8 deal with the nature and origin of reproductive isolating barriers. We follow the tradition of dividing these barriers into those that act before fertilization ("prezygotic") and those that act after fertilization ("postzygotic"), and present them in this order. Chapters 5 and 6 deal with prezygotic isolation, which we separate into "ecological" and "nonecological" barriers. This partition corresponds roughly to whether natural or sexual selection was the force that drove the evolution of reproductive isolation. Chapter 7 considers various forms of postzygotic isolation. These include "intrinsic" postzygotic barriers, in which developmental problems cause hybrid sterility or inviability, and "extrinsic" postzygotic isolation, in which sterility and inviability depend on particular environments. Chapter 8 discusses detailed genetic and molecular analyses of postzygotic isolation, as well as the basis of Haldane's rule.

In Chapter 9 we consider two novel forms of speciation limited largely to plants. One of these, polyploidy, involves an increase in chromosome number and is nearly instantaneous. The other, recombinational speciation, does not involve change in chromosome number; instead, a diploid hybrid between two species undergoes recombination, giving rise to a population whose novel genome—a combination of genes from both species—renders it reproductively isolated from both ancestors.

Chapter 10 deals with reinforcement, the enhancement of prezygotic isolation by natural selection in response to maladaptive hybridization. This topic has been controversial, its popularity rising and falling over the years. We conclude that recent theory shows that reinforcement is formally possible and that recent empirical work reveals evolutionary patterns consistent with the process. But because other processes can also produce these patterns, we cannot conclude with confidence that reinforcement is common in nature. In the end, we suggest a new way to distinguish between reinforcement and these alternative possibilities.

We then take up a controversy that we believe is now settled: the relative roles of selection versus drift in speciation. While Chapters 5–8 deal in passing with the evolutionary origin of reproductive isolation, Chapter 11 summarizes and evaluates this scattered evidence. We conclude that, despite the perennial popularity of models based on genetic drift, there is little evidence that drift plays an important role in speciation. There is, in contrast, a growing body of evidence for the importance of natural and sexual selection.

Finally, Chapter 12 addresses several macroevolutionary problems connected with speciation. The first involves calculating speciation rates, a problem as difficult to frame as to solve. The second involves identifying which, if any, biological factors affect speciation rates. The development of comparative methods now allow us to infer those organismal traits that have increased or decreased biodiversity, an enterprise that may tell us which isolating barriers were important in speciation. This leads naturally to a consideration of "species selection"—the differential proliferation of traits due to their association with higher or lower speciation rates. While species selection is a controversial topic, we argue that comparative studies strongly support its action in nature.

In summary, we have tried to survey, analyze, and synthesize what is known about speciation, offering not only a critique of the field but some new (and hopefully fruitful) ideas for research. Although our attempt surely suffers from the problems afflicting any broad survey—too little detail about some issues, and an occasional factual error—we hope that it at least imposes some order on, and draws some nontrivial conclusions from, a vast literature. More important, we hope that this book will stimulate younger scientists to pursue their own work on speciation.

1

Species: Reality and Concepts

When on board the H.M.S. 'Beagle,' as naturalist, I was much struck with certain facts in the distribution of the inhabitants of South America, and in the geological relations of the present to the past inhabitants of that continent. These facts seemed to me to throw some light on the origin of species—that mystery of mysteries, as it has been called by one of our greatest philosophers (Darwin 1859).

So begins *The Origin of Species*, whose title and first paragraph imply that Darwin will have much to say about speciation. Yet his magnum opus remains largely silent on the "mystery of mysteries," and the little it does say about this mystery is seen by most modern evolutionists as muddled or wrong. The study of speciation is thus one of the few areas of evolutionary biology not overshadowed by Darwin's immense achievements. For years after publication of *The Origin*, biologists struggled, and failed, to reconcile the continuous process of evolution with the discrete entities, namely species, that it produces. Now, 120 years after Darwin's death, a reconciliation has been achieved: we have a reasonably complete picture of what species are and how they arise.

But we must start at the beginning—with the question of whether biological nature really is discontinuous. Do species exist as discrete, objective entities, or are they, as Darwin believed, purely arbitrary constructs? If species are not real, then the problem of speciation is moot and we need go no further.

Most biologists certainly *act* as if species are real: naturalists label their specimens, systematists reconstruct the history of life from species-specific traits, population geneticists measure DNA variation within species, and ecologists calculate species diversity. Yet a vocal group of biologists, including many botanists, dissent, claiming that species are subjective divisions of nature made for human convenience. This view is common enough to merit serious examination.

If species *are* real, a second question immediately arises: How do we define them? That is, how do we encapsulate in words the discrete groups that we see in nature? There have been endless arguments about the "right" species concept, and it is clear that one's favorite answer depends on what one wants to understand—how one views the "species problem." Systematists, whose task is unraveling the history of life, often prefer species concepts different from those used by evolutionists more interested in evolutionary processes. Accordingly, deriving a species concept is important because it frames one's entire research program on the origin of species.

Finally, if one assumes that species are real, one can ask a related question: *Why* do they exist? This query does not involve describing species, but rather determining what properties of organisms and their environments cause nature to be divided into discrete groups.

In this chapter, we consider three fundamental questions: Are species real? If so, what are they? Finally, why do they exist? We contend that species are in fact real, and that the species concept most useful for understanding their origin is a modified version of Ernst Mayr's "biological species concept." At the end, we offer some approaches to the question of why species exist—a badly neglected problem.

The Reality of Species

As Mayr (1982, p. 285) noted, "The so-called species problem can be reduced to a simple choice between two alternatives: Are species realities of nature or are they simply theoretical constructs of the human mind?"

Thus, when one inquires about the reality of species, one asks whether assemblages of individuals—populations—are partitioned into discrete units that are objective, not subjective. Determining whether such groups exist is best accomplished by studying organisms that live in the same area—in *sympatry*. Because entities widely considered to be species show geographical variation in traits (e.g., *Homo sapiens*), one can easily demonstrate morphological or genetic "gaps" between populations from different regions. If hybridization occurs, however, such gaps would often disappear were the populations to inhabit the same area. Such is the case in modern humans.

In *The Origin*, Darwin apparently felt that species were not real:

> From these remarks it will be seen that I look at the term species, as one arbitrarily given for the sake of convenience to a set of individuals closely resembling each other, and that it does not essentially differ from the term variety, which is given to less distinct and more fluctuating forms (Darwin 1859, p. 52).
>
> In short, we shall have to treat species in the same manner as those naturalists treat genera, who admit that genera are merely artificial combinations made for convenience. This may not be a cheering prospect; but we shall at least be freed from the vain search for the

undiscovered and undiscoverable essence of the term species (Darwin 1859, p. 485).

For Darwin, the origin of species was identical to the origin of adaptations within species—the production of different varieties. He therefore conflated the problem of change within a lineage with the problem of the origin of new lineages. Surprisingly, however, in his unpublished notebooks and post-*Origin* publications, Darwin sometimes took a different stance, tacitly accepting the idea of organic discontinuity and even suggesting (in the first quote given below) that this discontinuity might result from reproductive barriers:

> My definition [in wild] of species, has nothing to do with hybridity, is simply, an instinctive impulse to keep separate, which no doubt be overcome, but until it is the animals are distinct species (Notebook C, entry 616, Barrett et al. 1987; see also Kottler 1978).
>
> Independently of blending from intercrossing, the complete absence, in a well-investigated region, of varieties linking together any two closely-allied forms, is probably the most important of all the criterions of their specific distinctness... (Darwin 1871, p. 215).

It is unclear why these views did not find their way into *The Origin*.

A number of biologists have agreed with Darwin's published view that species are arbitrary constructs. Surprisingly, this group includes the evolutionist J. B. S. Haldane, who observed that "the concept of a species is a concession to our linguistic habits and neurological mechanisms.... a dispute as to the validity of a specific distinction is primarily a linguistic rather than a biological dispute" (1956, p. 96). Raven (1976), Mishler and Donoghue (1982), and Nelson (1989) have made similar arguments. Still others consider the gradual nature of speciation as evidence against the distinctness of its products: "Today, an essential species 'reality' strongly conflicts with our understanding of gradual speciation, and is no longer accepted at all generally..."(Mallet 2001, p. 887). We contend, however, that the process of speciation is likely to be short relative to the duration of well-demarcated species, and that brief transitions between long lasting and discrete entities do not make those entities unreal. The existence of puberty, for example, does not mean that one cannot distinguish between children and adults.

A different view, common among botanists, is that while *some* species are real, other groups are less discrete owing to extensive hybridization or the presence of uniparental reproduction (e.g., selfing or clonal reproduction). We find it puzzling, if not contradictory, that many evolutionists who doubt the reality of species nevertheless act as if species were real when doing their own research, using Linnaean names and treating members of one species as equivalents.

Because of the continuing debate about the reality of species, we will describe methods that can help determine whether species are subjective or objective, and will show the outcome when these methods are applied to real organisms. We treat sexually and asexually reproducing taxa separately, for a group's mode of reproduction may affect its propensity to form discrete taxa. We will conclude that species are indeed discrete in sexually reproducing organisms, probably discrete in asexually reproducing organisms, but often *not* discrete in organisms that reproduce both sexually and asexually.

Sexually reproducing eukaryotic taxa

Biologists have used three methods to determine whether species are real in sexually reproducing groups. We discuss these methods in order of increasing rigor, weighing their pros and cons.

1. Arguments from common sense. This method settles the question by fiat: species are real because everyone recognizes that they are real. This was the argument Dobzhansky used in *Genetics and the Origin of Species*: "Discrete groups are encountered among animals as well as plants, in those that are structurally simple as well as in those that are very complex. Formation of discrete groups is so nearly universal that it must be regarded as a fundamental characteristic of organic diversity" (Dobzhansky 1937a, p. 5).

Indeed, clusters in a given locality are often discrete to even the casual and nonscientific observer. This is especially true in well-studied groups such as birds: nobody, for example, claims that there is a continuum between eagles and crows. The value of field guides is proportional to the discreteness of the taxa they cover, and of course, many such guides are useful.

While we find these arguments intuitively convincing, they are not hard evidence. We must confront, for example, the argument that humans have a propensity to divide a continuous array of organisms into discrete units, just as we separate the rainbow's continuous spectrum of light into seven discrete colors. To investigate this claim, we can compare the way that people from very different cultures divide up the organisms living in one area.

2. Concordance between "folk" and "scientific" species. Scientific facts ultimately derive from the agreement of independent observers. One can apply this principle to the problem of species reality by determining whether different observers—particularly those not sharing obvious biases—see the same divisions in nature. Biologists and anthropologists alike have conducted these studies. Typically, they survey a region's indigenous people, who lack formal biological training, and ask them to list the types of animals or plants in their habitat. These groupings of organisms into "folk species" can then be compared to the "Linnaean species" recognized by modern taxonomists.

Such comparisons can yield three possible results. First, there can be a one-to-one correspondence between Linnaean and folk species, which is strong evidence that nature is partitioned into units consistently recognized by people of different backgrounds. Second, a folk species can be *underdifferentiated*, meaning that it includes two or more Linnaean species. Such a result might be considered evidence for the reality of species if the Linnaean species are very similar to each other, differing in traits cryptic to nonscientists. Finally, folk species can be *overdifferentiated*, with a Linnaean species in one area being described as two or more folk species. Consistent overdifferentiation would constitute evidence against the reality of species.

The results of these studies are consistent: there is a remarkable coincidence between folk species and Linnaean species. Moreover, of the exceptions that do exist, most involve under- rather than overdifferentiation. The first compelling evidence was Mayr's (1963) observation that tribesmen of the Arfak Mountains of New Guinea had 136 vernacular names for the 137 Linnaean species of birds they encountered. In a more thorough analysis, Diamond (1966) studied bird names used by the Fore people of New Guinea. Their habitat contained 120 Linnaean species, with roughly 80% of these showing a one-to-one correspondence with Fore names. Diamond notes (1966, p. 1103):

> To a zoologist, the ability of the Fore to distinguish between closely similar species is impressive... [In two species of warblers] the differences are sufficiently subtle that I was often in doubt about the identity of the species held in the hand. Nevertheless the Fore not only had different names for the two birds... but also could identify them correctly in the field at moderate distances without binoculars. In this case small differences in behavior and call-note had probably alerted them to the fact that more than one kind of bird was present.

Diamond also took a group of Fore to lower elevations and asked them to give names to bird species they had never encountered. Ninety percent of the 103 Linnaean species were recognized as distinct folk species. Diamond argues, "That the elements in these two dissimilar classificatory systems nevertheless usually show a one-to-one correspondence strikingly illustrates the objective reality of the species" (1966, p. 1104).

Majnep and Bulmer (1977) obtained similar results studying animals encountered by the Kalam people of New Guinea. Of 176 bird species recognized by Western zoologists, 123 had a one-to-one correspondence with the folk designation, while there were 24 cases of underdifferentiation. This yields a 70–80% correspondence between names. The concordance between Kalam and Linnaean species is about 80% for frogs and 95% for reptiles (Bulmer and Tyler 1968; Bulmer et al. 1975).

In view of the common claim that species are less "real" in plants than in animals, one might expect to find less correspondence between Linnaean and folk species of plants. However, the work of Berlin et al. (1974) among the Tzeltal of southern Mexico shows that this expectation is incorrect. The Tzeltal have 471 folk names for plants growing in their area; of these, 66% are identical to Linnaean species. (Some of the plants surveyed are not sympatric, and geographic variation within Linnaean species may have reduced this correspondence.)

Regardless of the group surveyed, then, there is a very strong, although not perfect, correspondence between folk and Linnaean species—a correspondence of around 70%. Given the subtle traits used to designate Linnaean species (which can cause underdifferentiation of folk names) and the geographic variation of some species, there can be no reasonable expectation of perfect correspondence. Strikingly, one sees few cases of overdifferentiation. These results strongly support the view that people of different backgrounds recognize similar units of natural diversity. This buttresses the claim that species are real.

It must be admitted that not everyone finds this evidence convincing. Ridley (1996, p. 421) notes, "[T]he fact that independently observing humans see much the same species in nature does not show that species are real rather than nominal categories. The most it shows is that all human brains are wired up with a similar perceptual cluster statistic." (See also Mishler and Donoghue 1982, p. 493.)

These views can be interpreted in two ways. The weaker claim is that humans have an evolved tendency to subdivide and categorize, even when presented with a continuum. Yet this hypothesis does not explain why, if species are not discrete, people of widely diverse backgrounds—geographical, cultural, and scientific—tend to recognize the same groups. Proponents of the view that species are illusory must then make the stronger claim that human neurological wiring somehow constrains us to divide continua *at the same boundaries*. They might argue, for instance, that the three types of cones in the human eye—with differential sensitivities to blue, red, and green light—cause all humans to divide the continuous spectrum of light into a largely identical set of colors. Indeed, as different societies incorporate colors into their vocabulary, the six "primary constituent colors" are added in a nearly identical sequence (Durham 1991).

However, while the "neurological wiring" argument might conceivably explain congruent divisions of *single* traits, the claim loses force when dealing with *groups* of traits. We must remember that the congruence of species names between folk and Western taxonomy reflects the assessment of multiple traits. And there is no reason why our neurological wiring for recognizing, say, size, would divide up a biological continuum into the same groups as would our wiring for shape, for color, and so on. Moreover, using one set of traits yields clusters identical to those recognized using a different set of traits. It is well known, for example, that morphological discontinuities almost always coincide with genetic discontinuities in DNA sequences. This consistent carving of nature at the same joints is a powerful argument for the reality of species.

One can make a related argument based not on humans but on other species. In animals, individuals recognize conspecifics but not heterospecifics during the breeding season—the same differences recognized by humans. A male robin courts only female robins, not birds that humans consider members of other species. Pollinator-specific insects also discriminate between plant species recognized as different by humans. Likewise, many host-specific herbivores and parasites are good "taxonomists," recognizing the same species as do biologists. Even if one accepts that all human brains are wired with the same "perceptual cluster statistic," it hardly seems reasonable to assume that this statistic is identical in other animals.

3. Statistical identification of clusters. Folk taxonomy is a form of cluster analysis, but one can use more sophisticated statistical tools to look for clustering. When we apply these advanced tools to various traits—morphological, behavioral, reproductive, and molecular—do we still see sympatric individuals falling into distinct clusters? While such methods are designed to determine whether clusters exist or to choose the characters that best discriminate groups designated a priori, it is important to realize that these methods cannot identify such groups if they do not exist.

Given persistent arguments about the reality of species, it is curious that these statistical studies are rare. Aiming to distinguish rare hybrids from parental species, Neff and Smith (1978), for example, used discriminant-function analysis of morphology in the sunfish *Lepomis macrochirus* vs. *L. cyanellus* and in the shiners *Notropis spliopterus* vs. *N. whippeli*. In both cases, sympatric species were well separated and hybrids morphologically intermediate. Humphries et al. (1981) used combinations of traits to discriminate sympatric species of the pupfish *Cyprinodon*, sympatric species of minnows (*Richardonsius* and *Rhinichthys*), and allopatric populations of ciscoes (*Coregonus*) that had been considered two species based on size and shape. While the sunfish and pupfish were completely distinguished, the populations of ciscoes could not be separated unambiguously. The authors conclude that the cisco "species" are probably only morphologically differentiated populations. This underscores the difficulty of distinguishing discrete taxa when samples are taken from different places.

Avise (2000) notes that there is usually a strong concordance between vertebrate species differentiated by morphological criteria and by mitochondrial DNA (mtDNA) sequence. He concludes (p. 309) that "this compatibility of outcomes probably reflects an underlying historical reality to many of the biotic discontinuities traditionally recognized as species." Especially well-studied groups, such as birds and *Drosophila*, show almost no cases of indistinct sympatric taxa except for rare, morphologically intermediate hybrids (Gupta et al. 1980; Grant and Grant 1992). The question remains, however, whether birds and *Drosophila* form more discrete clusters than do other groups, such as plants.

Mayr (1992) conducted a comprehensive study of discontinuities in vascular plants using the flora of Concord Woods, Massachusetts. While this represented Mayr's attempt to show that his "biological species concept" (based on interbreeding and reproductive isolation) applied to plants, he examined discontinuities not of reproductive compatibility but of morphology and chromosome number (Whittemore 1993). His investigation is thus a better test of the reality of species than of the usefulness of his species concept.

According to Mayr, of 838 plant species in this area recognized by previous workers, 616 fell into easily recognized morphological groups. Fifty-three others were either allopolyploids, autopolyploids, or sibling species, recognizable on genetic or chromosomal grounds. Thus, 669 taxa—80% of the total named species—were easily distinguished. Fifty previously named "species" were found to be only aberrant individuals, possibly chromosomal aneuploids or nongenetic, developmental variants. Only 72 of the 838 groups were truly problematic, including possible hybrid swarms, offspring of polyploids that had mated with their ancestors, clones, and variable entities that were not well studied. Overall, about 9% of named species did not correspond to locally well-demarcated groups. (See Whittemore 1993 for a critique of this study.)

The presence of hybrids does not necessarily refute the distinctness of species, for hybrids can be rare or sterile. Indeed, some studies suggest that the "fuzziness" of plant species boundaries caused by hybridization may be overstated. Ellstrand et al. (1996) estimated the frequency of plant hybrids in five regions: the British Isles, Scandinavia, the Great Plains and Great Basin of the United States, and the Hawaiian Islands. They found a low rate of hybridization: between only 6 and 16% of genera within an area contained one or more reported hybrids. Given the likelihood that some hybrids have not been observed, this is almost certainly an underestimate of the frequency of hybrid-producing genera. On the other hand, not every species in a genus forms hybrids, so the frequency of hybridizing *species* is certainly much lower than 6%.

Nevertheless, it is likely that some taxa in plants are less distinct than those in animals, for plants have a greater diversity of mating systems, including selfing and apomixis, that can blur species boundaries. Animals generally lack the "difficult" complexes, such as dandelions, that plague plant systematists (see below). We know of no systematic data supporting the claims of some botanists that plant species are not real or hybridize promiscuously (Stebbins 1950; Raven 1976). Diamond (1992) concludes that such claims derive from "anecdotal horror stories" of botanists who concentrate on difficult groups or hybridizing taxa.

It appears, then, that most sexually reproducing organisms (which form a sizeable majority of plants and animals) fall into discrete groups in sympatry, confirming the intuition of most biologists that species are real. Of course, most taxa have not been examined carefully, and it would be useful to do more cluster analyses of groups like angiosperms or insects living in one area. In light of existing evidence, however, it seems fair to ask those who deny the existence of species to support their claim with systematic surveys instead of anecdotes.

ARE "HIGHER" TAXA REAL?   Most biologists agree that species are real in a way that supraspecific taxa—including ranks like genera and families—are not. Higher-level groups often share a common ancestor (i.e., are *monophyletic*), and can even be distinguished as large morphological clusters. Yet because evolutionary trees can branch at any level, higher-level groupings are necessarily somewhat arbitrary. Indeed, systematists such as Griffiths (1976) have even suggested doing away with formal taxonomic ranks altogether.

We can demonstrate the difference between the reality of species and of higher taxa by examining the three species *Drosophila pseudoobscura*, *D. persimilis*, and *D. miranda*. All occur sympatrically in northwestern North America and can be clearly distinguished by breeding relationships, chromosome configuration, and morphology (Dobzhansky and Powell 1975). The first two are sister species, with *D. miranda* the outgroup. All are, in turn, more closely related to each other than to species in the European *D. obscura* group. What are the "real" higher taxa here? *D. pseudoobscura* + *D. persimilis* is one, but the combination of these species with *D. miranda* is another. Both of these two higher groups are monophyletic and thus share at least one derived trait that distinguishes them from other such groups. But there are many such groups, which overlap in a nested fashion. The "reality" of such groups thus consists only of their common ancestry and the traits that allow us to recognize it. Unlike species, such groups do not evolve as a unit nor are they homogenized by interbreeding. Nevertheless, their common ancestry—reflected in the possession of shared derived traits—might sometimes channel their fate in a collective fashion, leading to broad patterns of speciation and extinction (Chapter 12).

The fact that the "reality" of higher-level groupings reflects only ancestry means that these groups are less distinct to laymen than to professional systematists. This is shown by the breakdown of the correspondence between folk and Linnaean taxonomy when one goes above the species level. "Higher" folk taxa are often based on superficial morphological traits such as body size. Considering the Tzeltal's higher-level classification of plants, Berlin (1992, p. 167) notes that "life-form groupings do not generally represent biologically natural categories in the same sense that taxa of intermediate or generic rank do, in that they often cross-cut biologically natural groupings." The Tzeltal divide plants, for example, into "trees," "vines," "grasses," and "broad-leafed herbs," reflecting evolutionary convergences among distantly related groups. Some species, like bamboo and agave, are not included in *any* higher group. The Rofaifo of New Guinea have five higher groupings of animals (Berlin 1973), including *Hefa* (eels, cassowaries, large marsupials, and rodents), *Huneme* (small marsupials and rodents), *Nema* (bats, and all birds except cassowaries), *Hoiafa* (lizards, snakes, fish, molluscs, worms, and centipedes), and *Hera* (frogs other than those of three genera). Such groupings do not seem to reflect functional or economic considerations, such as our distinction between farm animals and wild animals.

Groups with little or no sexual reproduction

While species may be real in sexually reproducing groups, species are often said to be either absent or much harder to distinguish in partly or fully asexual groups. In sexual groups, interbreeding among individuals can enforce genetic and phenotypic homogeneity. As interbreeding diminishes, however, individuals can diverge more extensively, effacing genetic and morphological gaps between clusters. This leads to a simple prediction for sympatric organisms: as sexual reproduction becomes less important, one should see less differentiation between species and greater differentiation among individuals within species. This tendency will make species less distinguishable in groups having limited sexual reproduction. Does this prediction hold? As we shall see, the answer depends on how much sex there is.

UNIPARENTALLY REPRODUCING EUKARYOTIC TAXA. Eukaryotes have several forms of uniparental reproduction, including vegetative reproduction, self-fertilization, and the production of seeds or fertile eggs without sexual reproduction (apomixis or agamospermy). Although such modes of reproduction are more common in plants than in animals, there is nonetheless a surprisingly large number of unisexually reproducing vertebrates (Vrijenhoek et al. 1989). Strictly uniparental groups are expected to form clones, which in theory could be recognized as distinct entities whose members are genetically identical to each other and distinct from other groups. If one is willing to regard *completely* asexual clones as distinct units, then one can indeed define and group them into "species." However, as more complete DNA sequences become available, such species will break down, for one must then delimit species based on differences at single nucleotide sites. Such a practice makes each individual, with its own unique mutations, a distinct species.

Complete uniparental reproduction is, however, quite rare. For example, most selfing plant species engage in at least some form of outcrossing. Recent work on some species thought to be completely asexual suggests that they might occasionally engage in covert sex (Walliker et al. 1987; Hurst et al. 1992; Pernin et al. 1992). Only among the bdelloid rotifers is there evidence for nearly complete asexuality over very long periods of time (Welch and Meselson 2000).

Reviewing the origin and evolutionary fate of unisexual species in vertebrates, Vrijenhoek et al. (1989) and Avise et al. (1992) show that nearly all of the 70-odd uniparental "species" are generated by hybridization of sexual species, with the hybrids reproducing via parthenogenesis. These "species" are well demarcated phenotypically, but can be *polyphyletic* (i.e., having multiple independent origins from repeated hybridizations between parental species). These hybridizations produce multiple clones, which, as in the hybrid fish *Poeciliopsis monacha-lucida*, can be ecologically diverse. The question then becomes whether to recognize as a cluster the entire group of hybrids, an ecologically similar group of clones, or a single genetically distinguishable clone. If one adopts the first solution, different unisexual vertebrates cluster simply because they are hybrids between distinct pairs of sexually reproducing species. That is, asexual taxa are distinct because their sexual ancestors are distinct. This may also be true for many uniparental plant "species" that are repeatedly derived from sexual ancestors.

Some plants form important but taxonomically confusing "agamic complexes" that harbor a core of diploid species with obligate sexual reproduction. The sexual species hybridize to form polyploids that may themselves reproduce sexually, but more often reproduce through agamospermy. Repeated bouts of interspecific hybridization, polyploid formation, and occasional sexual reproduction of the agamospermous forms can produce a continuum of variation between the sexual forms (Figure 1.1). The dandelion genus, *Taraxacum*, for example, contains 26 diploid sexual core species, all of which can be crossed in the greenhouse, producing polyploids that can be either obligately or facultatively apomictic (Grant 1981). The polyploids and their derivates constitute the nearly 2000 described "microspecies" of *Taraxacum* (Richards 1973). Similar agamic complexes include *Alchemilla* (lady's-mantle), *Crataegus*

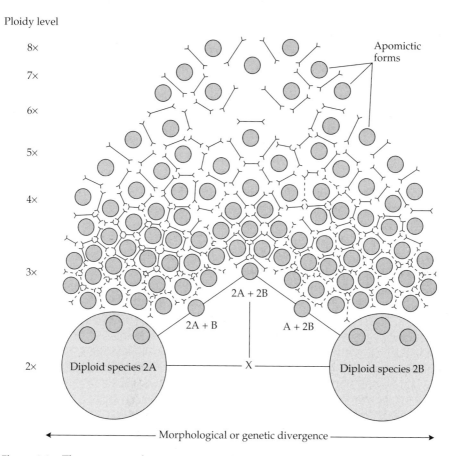

Figure 1.1 The structure of an agamic complex such as *Taraxacum* (dandelions). The complex includes core diploid sexual species (2A and 2B), whose F_1 hybridization can produce triploids (2A + B; A + 2B) and allotetraploids (2A + 2B). Individuals occupy all levels of ploidy from diploid (2X) to octaploid (8X), yielding a variety of apomictic forms that, as shown by the barriers, are largely reproductively isolated from each other (dashed lines between nondiploid taxa indicate some possibility of gene exchange). (From Grant 1981, after Babcock and Stebbins 1938.)

(hawthorn), *Hieraceum* (hawkweed), and *Rubus* (Grant 1981; Richards 1997; Sepp and Paal 1998).

One might expect groups with such diverse modes of reproduction to form clusters less distinct than those seen in sexual groups. Unconstrained by the cohesion of sexual reproduction, uniparental taxa are free to fill up the genetic, ecological, and morphological gaps that exist between sexual species. One might also expect a similar lack of distinctness in selfing groups. Yet even experienced botanists disagree about whether agamic groups comprise distinct clusters:

> When examined closely, species in these predominantly asexual genera [*Taraxacum, Hieracium,* and *Rubus*] are every bit as distinct (morphologically, geographically, and ecologically) as species in large, complex, exclusively sexual genera such as *Carex* or *Senecio* (Mishler 1990, p. 95).

Versus:

> The delineation of morphologically-based species [in obligate apomicts] becomes an arbitrary matter (Baker 1959, p. 188).

The genera *Rubus, Crataegus, Taraxacum,* and *Hieracium* have each been divided into hundreds or thousands of "microspecies" or "agamospecies" based on minute morphological differences, many of which might reflect only developmental plasticity in different habitats. There is little agreement among systematists on how many groups should be recognized. Camp (1951) recounts how three botanists, working independently, divided North American *Rubus* into 24, 205, and 494 species, respectively. Even single bushes have been designated as species (Asker and Jerling 1992).

The heavily studied group *Taraxacum* best exemplifies this confusion. Richards (1972) lists 132 dandelion species in the British Isles, and notes (p. 2) that he "became increasingly convinced that the microspecies obeyed all the dictates of 'good' species, being well-defined by constantly correlated characters, each microspecies with its own diagnostic geographical, ecological and genetic behavior." Yet Richards adds that when he gave his key to other botanists, they correctly identified only 40% of all species. Even a cursory survey of this literature validates Stebbins's (1950, p. 409) assessment of agamic groups: systematists "have not been able to agree on the boundaries of species. ... [I]n attempting to set up species like those found in sexual entities, they are looking for entities which in the biological sense are not there."

The best way to test whether such groups form discrete taxa is to perform cluster analysis on many sympatric individuals. Unfortunately, only one such study exists: that of Sepp and Paal (1998) on *Alchemilla*. This genus has been divided into more than 1000 microspecies, with some biologists claiming that species are discrete and easily recognized (Walters 1986). Sepp and Paal analyzed 23 named species from Estonia, scoring herbarium specimens for 43 morphological traits. A principal component analysis (Figure 1.2) showed that the specimens largely formed a morphological continuum containing almost no distinct clusters. They conclude that only three species (one of them represented by the filled dots in Figure 1.2) constitute discrete entities in morphospace, and that the

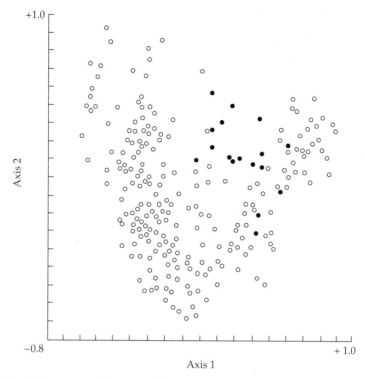

Figure 1.2 Principal coordinates analysis of 373 individuals (dots) in 23 named "species" of the agamic complex *Alchemilla*. Only the first two axes are shown. Filled dots represent individuals of one of the three species that the authors consider truly distinct (*A. plicata*). (From Sepp and Paal 1998.)

remaining 20 species form a continuum. They further note (p. 531) that "most of the species cannot be clearly distinguished, and surprisingly some pairs of species that are considered by several authors to be quite different (i.e. belonging to different series or sections), cannot be distinguished from a statistical point of view." In light of these results, one should be cautious about claims that agamic species are discrete when these assertions lack statistical support.

Although some researchers (e.g., Grant 1981; Dickinson 1998) have proposed ways to designate species in agamic complexes, most of these methods are more or less arbitrary, serving, as Mayr notes (1992, p. 411), "simply as a means of bringing some type of order to a situation which from the biological point of view is incapable of resolution."

Turning to plants that are largely selfing (autogamous), we are unable to judge from the literature whether they usually include distinct taxa. Some outcrossing species in genera such as *Linanthus* and *Mimulus* can repeatedly produce derivates that are largely selfing (e.g., Macnair and Gardner 1998; Goodwillie 1999). These selfers may form small allopatric populations adapted to restricted ecological conditions. Because of their homozygosity and allopatry, such selfers are indeed distinct.

CLUSTERING IN UNIPARENTAL VERSUS SEXUALLY REPRODUCING EUKARYOTES. There have been only a few tests of the relative distinctness of groups in sexual versus asexual taxa, none of which is completely satisfactory. Holman (1987) compared bdelloid rotifers (which include roughly 370 named species that appear to have reproduced without sex for at least 40 million years) with monogonont rotifers (consisting of about 1450 named species capable of sexual reproduction). Using three taxonomic monographs to estimate the stability of nomenclature in these groups, Holman quantified stability by first counting the number of "synonymous" genus and species names, and then calculating the ratio of synonymous species names to synonymous genus names. ("Synonymy" is a sign of taxonomic uncertainty, with greater synonymy reflecting greater difficulty in classifying individuals.)

Noting that the ratio was actually *higher* for monogonts than for bdelloids, Holman concluded that "bdelloid species are more consistently recognized than monogonont species" (1987, p. 384). If true—and this is not entirely clear given the use of synonymous genera in the statistic—this difference in recognizability seems to show that, contrary to expectation, sexually reproducing rotifers are *less* distinct than those that breed asexually. But it is questionable whether the distinctness of taxa can be judged from the stability of nomenclature. In fact, systematists have paid far more attention to monogonts than to bdelloids (C. Ricci, pers. comm.), resulting in more taxonomic revisions of the former than of the latter. As all systematists know, the number of synonyms increases with the number of revisions. Holman's results may thus be an artifact of different amounts of taxonomic work in different groups.

Deploying a different strategy, Baker (1953) compared morphological variation among populations of an obligatorily outcrossing and a partially selfing subspecies of the plant *Armeria maritima*. The homogenizing effect of gene flow leads one to expect less variation among populations of sexual than of asexual species. In fact, Baker observed just the opposite. He concludes that, when taking into account geographic variation, sexually reproducing species are not more discrete than those that breed asexually. But this conclusion is questionable. The proper test for species distinctness is not the relative amount of geographic variation, but the recognizability of taxa *in a single location*. If, for example, asexual taxa form clones that are widely distributed, one may see less spatial variation among asexual than among sexual taxa.

None of these studies answers the question of whether clusters are as distinct in asexual as in sexual taxa. Arriving at an answer requires measuring a wide variety of traits and/or genotypes in many individuals from a single locality, regardless of species, and subjecting these data to multivariate analysis. One can then compare the discreteness of sexual versus asexual clusters in morphospace or "genospace." Such studies are badly needed.

PROKARYOTES. It is a mistake to regard prokaryotes—or at least Eubacteria—as asexual. Transduction, transformation, and conjugation can cause rare gene transfer and recombination ("homologous exchange") between individuals of the same taxon, or between different taxa that are not too distantly related. In fact, bacteria even show a form of reproductive isolation: homologous recombination rates decay exponentially with increasing sequence divergence between taxa. It is difficult to recombine genomes between strains whose DNA sequences differ by more than 20% (Vulic et al. 1997; Majewski and Cohan 1998). However, genes can also be acquired from more distant relatives by "horizontal transfer," probably involving the uptake of naked DNA from the environment. This process can have important effects on bacterial genomes. For example, about 18% of the loci in *Escherichia coli* were acquired by horizontal transfer from different lineages over a period of nearly 200 million years, with many of these genes coding for ecological adaptations, antibiotic resistance, and pathogenicity (Lawrence and Ochman 1998).

Given that bacteria can exchange genes with both close and distant relatives, and that they are largely asexual, one might expect them to resemble agamic complexes of plants, in which clusters are difficult to discern. Cohan (2001 p. 515) notes, "One could imagine that an asexual or rarely sexual species might have no cap on its divergence, such that a closely related group of bacteria would grow indefinitely in its diversity of sequences and phenotypes." However, most microbiologists (e.g., Barrett and Sneath 1994; Roberts and Cohan 1995; Ochman et al. 2000; Cohan 2001, 2002 and references therein) agree that many bacteria do form discrete clusters, a view supported by evidence from bacterial phenotypes and DNA sequences. Although most studies involve bacteria taken from different hosts or localities (multiple samples of bacteria are rarely taken in sympatry), distinct taxa of *Bacillus* have been found in soil samples retrieved from a single location (Roberts and Cohan 1995; Roberts et al. 1996). In addition, analysis of multiple genes has revealed discrete clonal complexes within the named species *Neisseria meningitidis* and *Streptococcus pneumoniae* (Feil et al. 1999, 2000).

The most thorough study included 315 strains of the endoparasitic bacterial genus *Neisseria*, involving analysis of 155 phenotypic traits (Barrett and Sneath 1994). Although infecting many species, *Neisseria* has been especially well studied in humans because some "species" cause gonorrhea and meningitis. While some strains could not be placed in distinct groups, cluster analysis distinguished 31 fairly discrete groups in phenotypic space ("phenons") (Figure 1.3). Surprisingly, some clusters included strains from hosts as diverse as guinea pigs, fur seals, deer, and rabbits. Distinct clusters were also seen within the same host, but the clusters most difficult to distinguish occurred sympatrically—in the human nasopharynx. These strains, however, generally fell into clusters when DNA sequences were examined (Smith et al. 1999).

Because only a few groups have been examined and many of the strains are from different locations, the general degree of distinctness in bacteria is unclear. Moreover, one might often *expect* pathogenic asexual taxa to cluster by hosts (if transmission to other host species is rare) or for allopatric strains to cluster by location because of episodes of "periodic selection" that purge

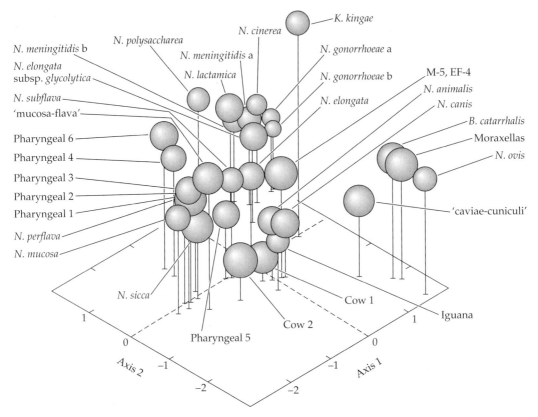

Figure 1.3 Phenotypic clusters of bacterial clones in the genus *Neisseria*. 315 strains were categorized using 155 phenotypic traits. Figure shows the centroids of the 31 named phenotypic clusters (phenons), sorted on the first three axes of a principal coordinate analysis (vertical direction is Axis 3). The degree of overlap between phenons is shown by the proximity of clusters. (After Barrett and Sneath 1994.)

genetic variation within one area. Hence, answering the question of distinctness of *sympatric* bacterial taxa is difficult. Ideally, one should examine samples of many isolates taken from a single substrate (such as soil) in a single area, as did Roberts and Cohan (1995). But most bacteria remain unknown. Although more than 8000 bacterial "species" have been named, there may be as many as a billion ecologically distinct taxa, most impossible to culture and study (Dykhuizen 1998).

Preliminary observations that bacterial taxa appear discrete may seem somewhat surprising. Recent work of Cohan and his colleagues (Majewski and Cohan 1999; Cohan 2001, 2004), however, suggest that episodic natural selection, coupled with a diversity of ecological niches, can produce distinct clusters of bacteria in sympatry. We discuss this process of bacterial "speciation" in greater detail below.

Conclusions

Although most biologists agree that species are real, we lack the rigorous studies needed to convince skeptics that nature is discontinuous. Discrete clusters appear to characterize sympatric, sexually reproducing eukaryotes and perhaps many prokaryotes. However, clusters seem less distinct in groups with mixed mating systems, such as agamic complexes in plants. It may seem odd that taxa appear most distinct in groups that are either completely sexual or nearly completely asexual, and less distinct in groups having both forms of reproduction. Such a result, however, can be understood if one considers how clusters form. We return to this problem at the end of the chapter.

Species Concepts

> The essence of the "species problem" is the fact that, while many different authorities have very different ideas of what species are, there is no set of experiments or observations that can be imagined that can resolve which of these views is the right one. This being so, the "species problem" is not a scientific problem at all, merely one about choosing and consistently applying a convention about how we use a word. So, we should settle on our favorite definition, use it, and get on with the science (Brookfield 2002, p. 107).

Most biologists agree that discrete clusters exist among sexually reproducing organisms, and behave in their own research as if these groups were real. However, evolutionists disagree about whether these groups constitute "species," and, if so, how to best define them. As we have seen, the species concept is one of the most hotly debated issues in speciation. While much of the debate seems more philosophical than scientific, the issue is important, for we cannot study how species form until we determine what they are.

Mayr (1942, 1982) reviews the history of species concepts up to about 1980. During the Modern Synthesis, only a few concepts competed for the allegiance of biologists, most prominently Mayr's own "biological species concept" or those based on morphological difference ("typological" concepts). However, in the last twenty years the debate has intensified. New species concepts appear yearly, and there are now entire books devoted to the problem (e.g., Ereshefsky 1992; Claridge et al. 1997; Wilson 1999; Wheeler and Meier 2000; Hey 2001). We count at least 25 concepts, by no means an exhaustive list. It is somewhat depressing that evolutionary biologists continue to spend so much time arguing about what constitutes a species when, as noted by Brookfield (2002), the debate cannot be resolved by normal scientific methods.

There are several reasons why these debates persist. First, there is no concept that, when applied to nature, is free from ambiguities (Hey 2001). Some ambiguities derive from evolution itself: species arise from other species, and during this process there will be unclear cases, no matter how one defines species. Moreover, any strict concept fails in some situations, and different concepts fail in different situations. If one sees species as groups separated from other groups by reproductive barriers, what does one do upon finding a single fertile hybrid among a million individuals? Or, if one defines species as groups possessing at least one unique, diagnosable trait, does a single nucleotide in the genome suffice? How does one deal with geographically isolated populations that are genetically or morphologically divergent? Further problems arise from the diverse ways in which organisms reproduce. Evolutionists now appreciate that no single species concept can encompass sexual taxa, asexual taxa, and taxa having mixed modes of reproduction. As Kitcher (1984, p. 309) notes: "There is no unique relation which is privileged in that the species taxa it generates will answer to the needs of all biologists and will be applicable to all groups of organisms."

Moreover, biologists want species concepts to be useful for some purpose (i.e., be "operational"), but differ in what that purpose should be. We can think of at least five such goals. Species can be defined in a way that

1. helps us classify them in a systematic manner;
2. corresponds to the discrete entities that we see in nature;
3. helps us understand how discrete entities *arise* in nature;
4. represents the evolutionary history of organisms; and
5. applies to the largest possible number of organisms.

No species concept will accomplish even most of these purposes. We therefore feel that, when deciding on a species concept, one should first identify the nature of one's "species problem," and then choose the concept best at solving that problem.

The biological species concept (BSC)

Our own species concept is one that comes closest to deciphering what we (and many of our predecessors) consider the most important "species problem," namely, why do sympatric, sexually reproducing organisms fall into discrete clusters? This view of the species problem antedates the Modern Synthesis, going back to Bateson (1894). In our opinion, the discontinuities of nature are best encapsulated, and their origin best understood, using a modified version of the biological species concept (BSC; Table 1.1). We do not wish to describe and evaluate here every species concept ever proposed. Table 1.1 also lists eight of the most popular alternatives to the BSC, which we explain and evaluate in the Appendix. Here we describe our version of the BSC and consider its advantages and disadvantages.

To an evolutionary geneticist, the observation of discrete, sexually reproducing groups in sympatry immediately suggests a species concept based on interbreeding and its absence. As Dobzhansky (1937c, p. 281) recognized:

Table 1.1 *The biological species concept and some recently proposed alternatives*[a]

Basis of concept	Concept	Definition
1. Interbreeding	Biological Species Concept (BSC)	Species are groups of interbreeding natural populations that are reproductively isolated from other such groups (Mayr 1995).
2. Genetic or phenotypic cohesion	Genotypic Cluster Species Concept (GCSC)	A species is a [morphologically or genetically] distinguishable group of individuals that has few or no intermediates when in contact with other such clusters (Mallet 1995).
	Recognition Species Concept (RSC)	A species is that most inclusive population of individual biparental organisms which shares a commmon fertilization system (Paterson 1985).
	Cohesion Species Concept (CSC)	A species is the most inclusive population of individuals having the potential for phenotypic cohesion through intrinsic cohesion mechanisms (Templeton 1989).
3. Evolutionary cohesion	Ecological Species Concept (EcSC)	A species is a lineage (or a closely related set of lineages) which occupies an adaptive zone minimally different from that of any other lineage in its range and which evolves separately from all lineages outside its range (Van Valen 1976).
	Evolutionary Species Concept (EvSC)	A species is a single lineage of ancestral descendant populations or organisms which maintains its identity from other such lineages and which has its own evolutionary tendencies and historical fate (Wiley 1978, modified from Simpson, 1961).
4. Evolutionary history	Phylogenetic Species Concept 1 (PSC1)	A phylogenetic species is an irreducible (basal) cluster of organisms that is diagnosably distinct from other such clusters, and within which there is a paternal pattern of ancestry and descent (Cracraft 1989).
	Phylogenetic Species Concept 2 (PSC2)	A species is the smallest [exclusive] monophyletic group of common ancestry (de Queiroz and Donoghue 1988).
	Phylogenetic Species Concept 3 (PSC3) or Genealogical Species Concept (GSC)	A species is a basal, exclusive group of organisms all of whose genes coalesce more recently with each other than with those of any organisms outside the group, and that contains no exclusive group within it (Baum and Donoghue 1995; Shaw 1998).

[a]The Appendix discusses and evaluates all of these concepts except the BSC.

Any discussion of these problems [of discontinuities in the living world] should have as its logical starting point a consideration of the fact that no discrete groups of organisms differing in more than a single gene can maintain their identity unless they are prevented from interbreeding with other groups . . . Hence, the existence of discrete groups of any size constitutes evidence that some mechanisms prevent their interbreeding, and thus isolate them.

Dobzhansky (1935, p. 353) proposed that "a species is a group of individuals fully fertile inter se, but barred from interbreeding with other similar groups by its physiological properties (producing either incompatibility of parents, or sterility of hybrids, or both)." (Among "physiological properties" Dobzhansky also included genetic barriers acting *before* fertilization, such as the unwillingness to mate with dissimilar individuals.) This is close to the definition that we adopt. However, Dobzhansky's implication that different species must exchange *no* genes seems too extreme, and has promoted both confusion in the field and suggestions that the BSC be rejected.

The BSC is, however, most closely associated with Ernst Mayr, who not only provided its most famous formulation—"Species are groups of actually

Table 1.2 *Classification of reproductive isolating barriers*

I. **Premating isolating barriers.** Isolating barriers that impede gene flow before transfer of sperm or pollen to members of other species.

  A. **Behavioral isolation** (also called "ethological" or "sexual" isolation). Includes all differences that lead to a lack of cross-attraction between members of different species, preventing them from initiating courtship or copulation.

  B. **Ecological isolation.** Isolating barriers based primarily on differences in species' ecology, i.e., barriers that are direct byproducts of adaptation to the local environment.

   1. **Habitat isolation.** Species have genetic or biological propensities to occupy different habitats when they occur in same general area, thus preventing or limiting gene exchange through spatial separation during the breeding season. This isolation can be caused by differential adaptation, differential preference, competition, or combinations of these factors.

   2. **Temporal (allochronic) isolation.** Gene flow between sympatric taxa is impeded because they breed at different times.

   3. **Pollinator isolation.** Gene flow between angiosperm species is reduced by their differential interactions with pollinators. This can occur via pollination by different species, or by pollen transfer involving different body parts of a single pollinator species.

  C. **Mechanical isolation.** Inhibition of normal copulation or pollination between two species due to incompatibility of their reproductive structures. This incompatibility can result from lack of mechanical fit between male and female genitalia (structural isolation) or the failure of heterospecific genitalia to provide proper stimulation for mating (tactile isolation).

  D. **Mating system "isolation."** The evolution of partial or complete self-fertilization (autogamy) or the asexual production of offspring (apomixis) that can result in the creation of a new taxon or set of lineages. As noted in Chapter 6, this is not an isolating barrier in the same sense as the others in this list.

or potentially interbreeding natural populations, which are reproductively isolated from other such groups"(1942, p. 120)—but also worked out the implications of this definition and defended it against critics (Mayr 1963, 1969).

Dobzhansky's later contributions to the BSC included compiling a list of various barriers to gene flow, which he called "isolating mechanisms" (1937a, 1951). To some, the word "mechanism" paints a misleading picture of speciation, implying that selection builds reproductive barriers to keep species distinct. But this process occurs only during reinforcement and some types of sympatric speciation, while the rest of the time species are not direct objects of natural selection, but accidental byproducts of evolutionary divergence. When referring to forms of reproductive isolation, we therefore use the less misleading term *isolating barriers*, which we define as *those biological features of organisms that impede the exchange of genes with members of other populations*. These barriers are usually, but not invariably, based on genetic differences between populations; we describe a few exceptions below. Table 1.2 describes and defines the

Table 1.2 *Classification of reproductive isolating barriers* (continued)

- **II. Postmating, prezygotic isolating barriers.** Isolating barriers that act after sperm or pollen transfer but before fertilization.
    - **A. Copulatory behavioral isolation.** Behavior of an individual during copulation is insufficient to allow normal fertilization.
    - **B. Gametic isolation.** Transferred gametes cannot effect fertilization.
        1. **Noncompetitive gametic isolation.** Intrinsic problems with transfer, storage, or fertilization of heterospecific gametes in single fertilizations between members of different species.
        2. **Competitive gametic isolation.** (conspecific sperm or pollen preference) Heterospecific gametes are not properly transferred, stored, or used in fertilization only when competing with conspecific gametes.
- **III. Postzygotic isolating barriers (hybrid sterility and inviability)**
    - **A. Extrinsic.** Postzygotic isolation depends on the environment, either biotic or abiotic.
        1. **Ecological inviability.** Hybrids develop normally but suffer lower viability because they cannot find an appropriate ecological niche.
        2. **Behavioral sterility.** Hybrids have normal gametogenesis but are less fertile than parental species because they cannot obtain mates. Most often, hybrids have intermediate phenotypes or courtship behaviors that make them unattractive.
    - **B. Intrinsic.** Postzygotic isolation reflects a developmental problem in hybrids that is relatively independent of the environment.
        1. **Hybrid inviability.** Hybrids suffer developmental difficulties causing full or partial lethality.
        2. **Hybrid sterility.**
            a. **Physiological sterility.** Hybrids suffer problems in the development of the reproductive system or gametes.
            b. **Behavioral sterility.** Hybrids suffer neurological or physiological lesions that render them incapable of successful courtship.

diverse forms of isolating barriers. Our list is indebted to Dozhansky's but is updated in light of recent work.

Because of the difficulty of determining the species status of allopatric taxa, Mayr later struck the word "potentially" from his definition and suggested the following version of the BSC, which we adopt with a few caveats:

> Species are groups of interbreeding natural populations that are reproductively isolated from other such groups (Mayr 1995, p. 5).

Groups of populations thus constitute different species under two conditions: (1) their genetic differences preclude them from living in the same area, or (2) they inhabit the same area but their genetic differences make them unable to produce fertile hybrids.

In our view, distinct species are characterized by *substantial but not necessarily complete reproductive isolation*. We thus depart from the "hard line" BSC by recognizing species that have limited gene exchange with sympatric relatives. But we feel that it is less important to worry about species status than to recognize that the *process* of speciation involves acquiring reproductive barriers, and that this process yields intermediate stages when species status is more or less irresolvable.

The reader may have noticed an apparent discrepancy between the way we recognize species and the way we define them. If we *distinguish* species as discrete morphological and genetic units coexisting in sympatry, why do we not *define* them as such, considering speciation to be the acquisition of diagnostic traits and genes? Indeed, one species concept—the "genotypic cluster species concept" (GCSC)—does exactly that (see Appendix). Schilthuizen (2000, p. 1135) emphasizes this discrepancy between recognition and definition:

> In Mayr's writings, two views on species appear. The first is that all individuals of a species share the same well-integrated complex of epistatically and pleiotropically interacting genes. This is the species *concept*, and Mayr [1963] writes that the evolution of two well-integrated gene complexes from a single ancestral one is "the essence of speciation." At the same time, however, the biological species definition makes no mention of gene complexes, but rather of devices for reproductive isolation. Consequently Mayr [1963] can also be found writing that 'speciation is characterized by the acquisition of these devices.'

Schilthuizen's point is clear: If distinctness in sympatry is all that matters, then the BSC is problematic, for he believes that populations can remain distinct in sympatry for reasons other than reproductive isolation. Schilthuizen and others (e.g., Mallet 1995) suggest several ways this can happen.

The first involves disruptive selection in one area. Selection favoring individuals at two extremes of habitat or resource use, for example, can create and maintain groups that differ in genes causing local adaptation. If this selection is strong, it can create groups that remain distinct at several to many loci, although genes not subject to selection will be freely exchanged. Schilthuizen notes that such groups include "host races," such as the apple and hawthorn races of the apple maggot fly *Rhagoletis pomonella* (Chapter 4). Hybrid zones, in which two forms with contiguous ranges hybridize where they meet but remain distinct, are not uncommon (Barton and Hewitt 1985).

Schilthuizen (2000, p. 1136) argues that these cases show that "the BSC with its reproductive-isolation criterion does not automatically follow from a concept of species as a coadapted gene complex, because the latter can persist in spite of the absence of reproductive barriers." But this contention is incorrect. In sexually reproducing organisms, *the stable coexistence of genetically distinct groups in sympatry requires reproductive barriers between them*. (By "genetically distinct," we mean groups differing at several loci, not discontinuities caused by simple Mendelian polymorphisms.) Without reproductive barriers, the groups would fuse. In many cases, such as strong disruptive selection that causes speciation, the barriers involve *extrinsic hybrid inviability* (see Table 1.2): intermediate forms are ecologically unfit. Such inviability preserves the distinctness of loci affecting the selected traits. Part of the confusion comes from the rather artificial distinction between "selection" and "reproductive isolation." If disruptive selection causes speciation, it does so by creating reproductive isolation. Indeed, much work on the host races of *Rhagoletis pomonella* has involved identifying barriers to gene exchange (Feder et al. 1994, 1997a, b). In many hybrid zones, intermediate forms are unfit, being relatively inviable or sterile (Barton and Hewitt 1985; Howard et al. 1997; Presgraves 2002). We are not claiming that reproductive barriers must exist before selection can create evolutionary divergence. This neoDarwinian view is obviously wrong. Rather, we maintain that disruptive selection and reproductive isolation are two sides of the same coin.

During sympatric speciation and reinforcement, the point at which sympatric taxa should be called "species" is arbitrary. In fact, one could consider speciation as the conversion of "genotypic cluster" species into "biological" species, a process that is continuous, yielding ever-increasing barriers to gene flow. In such situations we prefer to apply our version of the BSC, for under this concept one can view the entire process of speciation as the evolution of reproductive isolation. Arguments about the exact relationship between gene flow and species status have obscured the more important fact that reproductive barriers are essential for producing and maintaining distinct groups in sympatry.

Our view that reproductive barriers are the currency of speciation derives from our belief that understanding how these barriers arise is the solution to the species problem. This does not mean that selection can or should be ignored. Indeed, as we show in Chapter 11, most reproductive barriers probably result from natural selection. Yet before one can understand which forms of selection keep clusters distinct, one must understand which *barriers* keep clusters distinct.

Our version of the BSC differs from the GCSC in two respects. First, we do not consider clusters to be species if they are distinct at only a few loci but freely exchange genes in the rest of the genome. We view such clusters as races or incipient species. Indeed, even biologists of the "cluster" school appreciate the importance of isolating barriers, and recognize their evolution as a part of speciation (e.g., Mallet et al. 1998). For example, some advocates of the GCSC believe that sympatric speciation is common. However, those who model sympatric speciation consider that it is complete only when isolating barriers reduce gene flow to nearly zero (Rice 1984a; Dieckmann and Doebeli 1999; Kondrashov and Kondrashov 1999). Second, we consider the BSC better than the GCSC at stimulating research. Defining species simply as clusters offers no insight into how these clusters arise and are maintained.

Finally, we argue that the traits used to *recognize* groups need not be identical to the traits used to define or conceptualize them. This point was best made by Simpson (1961) using the example of identical twins. These twins are recognized by their extreme morphological similarity, but are defined as two individuals derived from a single fertilized egg. The latter concept seems more useful because it accounts for the morphological similarity. Likewise, reproductive isolation accounts for the existence of discrete clusters in sympatry.

In our view, then, reproductive isolation is the proper focus for the study of speciation. In fact, we can hardly imagine writing a substantive book on speciation using any concept other than the BSC. The recent explosion of work on speciation concentrates almost entirely on reproductive isolation.

Our acceptance of isolating barriers as the key to speciation does not mean, of course, that we adhere to every idea espoused by Dobzhansky, Mayr, and other proponents of the BSC. As noted above, for instance, we do not believe that evolutionary divergence in sympatry requires the prior evolution of reproductive isolation.

Moreover, we do not agree that species always form "integrated, coadapted gene complexes." This view was common during the Modern Synthesis, with some holding the almost teleological view that selection erects isolating barriers to protect such complexes:

> The division of the total genetic variability of nature into discrete packages, the so-called species, which are separated from each other by reproductive barriers, prevents the production of too great a number of disharmonious incompatible gene combinations. This is the basic biological meaning of species, and this is the reason why there are discontinuities between sympatric species (Mayr 1969, p. 316).

> Hence maintenance of life is possible only if the gene patterns whose coherence is tested by natural selection are prevented from disintegration due to unlimited hybridization. It follows that there must exist discrete groups of forms, species, which consist of individuals breeding inter se, but prevented from interbreeding with individuals belonging to other groups of similar nature (Dobzhansky 1937a, p. 405).

Although reproductively isolated groups will eventually acquire sets of harmoniously acting genes—the so-called coadapted complexes—newly formed species need not differ in any traits beyond those causing reproductive isolation. Some species can arise via changes in only one or a few genes, and some cases of speciation may involve no genetic change at all.

To prevent confusion, we deal with several questions that arise about our version of the BSC.

MUST REPRODUCTIVE ISOLATION BE COMPLETE BEFORE TAXA ARE CONSIDERED SPECIES UNDER THE BSC? The BSC is usually seen as requiring absolute barriers to gene flow between taxa. For example, Barton and Hewitt (1985, p. 114) argue that "if two populations are to belong to different biological species, reproductive isolation must be complete: no fertile hybrids can be formed." This strict construction has bothered biologists who consider "good species" to be those that maintain their distinctness in sympatry even if they occasionally hybridize with others. Indeed, molecular studies have shown that hybridization may be far more common than previously suspected.

Historically, one of the most common criticisms of the BSC has been that related species rarely show complete reproductive isolation. Mayr himself wavered about whether the BSC should be modified to deal with this problem. He often took the hard line of "no gene flow permitted," as when asserting that "species level is reached when the process of speciation has become irreversible, even if some of the (component) isolating mechanisms have not yet reached perfection" (Mayr 1963, p. 26). But he argued elsewhere that some hybridization is permissible between biological species so long as they maintain their distinctness. Referring to sympatric taxa of ducks, for example, he noted that "occasional hybrids occur, but at such a low rate that the elimination of the introgressing genes is not too severe a burden on the parental species" Mayr (1963, p. 552). Considering fish of the genus *Gila*, he observed that "the characters of a few specimens indicated the possibility of introgression, yet there was no blurring of the species border" (1963, p. 116). Such contradictory statements obviously reflect confusion about whether morphological distinctness requires absolute bars to hybridization.

Other contributors to the Modern Synthesis believed that good species could show limited hybridization:

> Two or more Mendelian populations can be sympatric, i.e., can coexist indefinitely in the same territory, only if they are reproductively isolated, *at least to the extent that the gene exchange between them can be kept under control by natural selection* (Dobzhansky 1951, p. 264, our italics).

> Natural hybridization and gene flow can take place between biological species, even though they are highly intersterile or isolated in other ways, as long as the breeding barriers are less than 100% effective. . . Some of these results of hybridization do not affect the distinctness of the species involved, and hence do not concern us now (Grant 1971, p. 51).

> Even if there is evidence of backcrossing but the intergrading types remain relatively uncommon in comparison with sharply distinct parent types, it may be presumed that there is so much selection against

the hybrids that they do not destroy the integrity of the two species" (Wright 1978, p. 5).

Although Dobzhansky, Grant, and Wright all adhered to the BSC, they obviously did not take a hard line on gene flow.

Our notion of species status, then, involves a sliding scale. We do not consider taxa having substantial gene flow despite morphological distinctness to be species. As reproductive barriers become stronger, taxa become more and more "species-like," and when reproductive isolation is complete we consider taxa to be "good species." This view obviously requires some subjective decisions about species status. But this is not unique to the BSC. As we show in the Appendix, *all* species concepts require some subjective judgments.

Some evolutionists have suggested guidelines for gene flow that would allow the BSC to appear more objective. Schemske (2000, p. 1070), for example, proposes that "as a gross yardstick, if the probability of successful hybrid formation is less than the mutation rate, then populations meet the criterion of good biological species." The rationale appears to be that species status is attained when the variation produced by mutation exceeds that introduced by introgression. Yet even this criterion is arbitrary. "Hybrid formation" is not equivalent to introgression, and most mutations are unconditionally deleterious. Moreover, the criterion is an operational nightmare: to determine species status, one would have to measure mutation and hybridization rates, usually impossible tasks. A further implication of Schemske's thesis is that, when introgression exceeds mutation, species borders blur. This is almost certainly untrue, as it ignores the fact that selection can eliminate introgressed genes.

Determining BSC status using a sliding scale is of course also difficult in its own right: groups can appear quite distinct while still exchanging many genes. This occurs, for instance, in sympatric morphs of the butterfly *Papilio dardanus* that are Batesian mimics of different species (Clarke and Sheppard 1963). Mimetic forms differ by several genes that are apparently closely linked in clusters of "supergenes," but appear to interbreed freely. Even taxa with substantial reproductive isolation can show rare gene exchange. Everyone considers *Drosophila pseudoobscura* and *D. persimilis* (two sympatric taxa that are classic subjects of evolutionary genetics) as distinct species. Nevertheless, they hybridize at a low rate: roughly one out of 10,000 females examined is a hybrid (Powell 1983), and hybrid females are fertile. The pattern of molecular variation in these two species also suggests some introgression after evolutionary divergence (Machado et al. 2002).

Adopting a species concept that allows some introgression does not trouble us. Indeed, throughout this book we use the term "species" even when a group exchanges some genes with sympatric relatives. We largely agree with McPhail (1994, p. 400) that "the goal of speciation studies is to understand how coexisting populations come into being, and it is unimportant whether or not systematists consider such divergent populations as species." However, we also recognize that systematists need a yardstick for delimiting species, and we are usually happy to recognize the groups that most biologists call species, even though many of these may not conform to our notion of "good" species because they exchange genes with other groups.

Finally, we emphasize that we do not regard our species concept as perfect, and discuss some of its problems later in this chapter.

WHY ISN'T ECOLOGICAL DIFFERENTIATION PART OF THE BSC? We have framed the species problem as the sympatric coexistence of discrete groups, thereby raising issues of ecology. Most ecologists believe that species can coexist only if they show a minimal degree of ecological divergence. Why, then, do we not define species as "reproductively isolated entities having sufficient ecological divergence to permit their coexistence"? Indeed, Mayr (1982, p. 273) amended the BSC to take ecology into account: "A species is a reproductive community of populations (reproductively isolated from others) that occupies a specific niche in nature." Van Valen's (1976) ecological species concept also requires that a species occupy a distinct "adaptive zone." These views are closely connected with Sewall Wright's idea that species sit atop peaks in the adaptive landscape, with each peak representing a discrete niche (e.g., Wright, 1982).

Not all ecologists, however, agree that extreme ecological similarity prevents the coexistence of species. The "limiting similarity" principle has its own large and controversial literature (e.g., Abrams 1983; Chesson 1991; Hubbell 2001). Coexistence of nearly identical species can be maintained by spatial and temporal fluctuation in resources, or by subtle and virtually undetectable differences in ecology, such as a difference in the shape of the relationship between resource abundance and consumption rate (Armstrong and McGehee 1980). Thus, the hypothesis that species coexistence requires ecological difference seems theoretically plausible but empirically untestable: if one cannot find ecological differences between sympatric species, one may have missed undetectable but important aspects of resource use. Nevertheless, there is much evidence for competition between closely related sympatric species (Schluter 2000), and so we assume that such species usually have some ecological difference.

Nevertheless, we see niche differences as more relevant to the *persistence* of, rather than to the definition of, species, for there is no necessary correlation between reproductive isolation and ecological differentiation. In fact, most biologists implicitly recognize that permanent coexistence is *not* a criterion for species status. This is shown by the number of cases in which one species outcompetes or replaces a close relative in nature. The Chinese parasitoid wasp *Aphytis lingnanensis* has displaced its Mediterranean relative *A. chrysomphali* in Southern California (DeBach and Sundby 1963), but their putative ecological similarity does not affect their acknowledged status as distinct species. Conversely, ecologically differentiated taxa lacking reproductive isolation can fuse in sympatry. Of course, ecological differences are clearly important in speciation. Such differences can themselves constitute barriers to gene flow, as with habitat isolation, or create selective pressures that promote the evolution of other isolating barriers. In many cases there will be considerable overlap between the factors that prevent gene flow between sympatric species and the factors that allow them to coexist.

This overlap between reproductive isolation and coexistence is especially important in three circumstances. First, divergent natural selection may produce adaptations that simultaneously reduce gene flow *and* allow species to coexist. This is true for habitat isolation (in which adaptation to different niches within one area spatially restricts hybridization), and extrinsic postzygotic isolation (in which two species occupying different niches produce hybrids ecologically inferior to either parent). Second, ecological differences allowing coexistence can promote the evolution of further barriers to gene flow. The ecological inferiority of hybrids, for example, may lead to the evolution of increased mating discrimination, an important part of sympatric speciation and reinforcement. Finally, the creation of a new polyploid plant species must often involve ecological changes that allow it to coexist with its ancestors (see Chapter 9).

MUST REPRODUCTIVE ISOLATION BE GENETIC? Dobzhansky (1937c) initially considered geographic isolation between populations as a form of reproductive isolation, although he later abandoned this view. While geographic barriers impede gene flow and are instrumental in allopatric speciation, we do not consider them isolating barriers, for they neither involve biological differences between taxa nor prevent gene flow between sympatric species.

While nearly all isolating barriers are genetic, there are some exceptions. Nongenetic barriers include "infectious speciation" caused by microorganisms that produce hybrid inviability between their hosts (Chapter 7), "cultural speciation" based on the imprinting of brood-parasitic birds on their hosts (Chapter 6), and "nongenetic allochronic speciation," as may have occurred in periodical cicadas and pink salmon (Chapters 4 and 5). New autopolyploid species are formed by differences in chromosome number, not gene sequence. Because all of these factors prevent gene flow in sympatry and are byproducts of the biology of organisms, we consider them genuine isolating barriers that are distinct from geographic barriers.

CAN ONE DETERMINE WHETHER SYMPATRIC SPECIES ARE REPRODUCTIVELY ISOLATED? Some critics have argued that it is impossible to apply the BSC in nature because one simply cannot perform the many hybridizations needed to determine the number of biological species in one area (Sokal and Crovello 1970). However, in reality this exercise is unnecessary, for reproductive isolation can be *inferred* from morphological, chromosomal, or molecular traits. Thus it is not necessary to identify the barriers to gene exchange to apply our version of the BSC; one need only show that two populations are reproductively isolated. This has traditionally been done (with great success) by analyzing the distribution of several morphological characters, such as bristles and genitalic traits in many insects. Fixed differences in chromosome inversions or molecular markers can serve equally well. Knowlton (1993) enumerates sympatric *sibling species* (related species showing only slight differences in morphology) in marine organisms. In nearly every case, species diagnosis is based not on reproductive isolation but on fixed differences in morphological, ecological, or molecular traits. Finally, one can show that reproductive isolation in the laboratory (such as hybrid sterility or inviability) invariably accompanies morphological or chromosomal differences seen in nature.

DOES THE BSC MAKE SPECIATION "CAPRICIOUS?" In allopatric speciation, reproductive isolation is a byproduct of evolutionary change in isolated populations, and thus can be considered an evolutionary accident. This accidental aspect of speciation violates the notion that species must be the direct object of natural selection—that selection favors isolating barriers *because they cause isolation*. This view of "adaptive speciation" probably derives from Darwin (1859, p. 112), who felt that species arose to pack available niches as fully as possible.

The idea that selection operates to increase isolation was refined by the founders of the Modern Synthesis, who, as we note, saw isolating "mechanisms" as nature's way of protecting coadapted gene complexes. Dobzhansky (1935, p. 349), for example, seemed reluctant to accept isolating barriers as mere byproducts of evolution:

> This diversity of isolating mechanisms is itself remarkable and difficult to explain. It is unclear how such mechanisms can be created at all by natural selection, that is, what use the organism derives directly from their development. We are almost forced to conjecture that the isolating mechanisms are merely by-products of some other differences between the organisms in question, these latter differences having some adaptive value and consequently being subject to natural selection.

This may be why Dobzhansky believed that reinforcement, in which selection acts directly to increase reproductive isolation, is a nearly ubiquitous final step in speciation.

IS SPECIATION REVERSIBLE? The BSC is sometimes described as a "prospective" concept because it characterizes species by their evolutionary potential—their ability to evolve independently without contamination by genes from other species. If reproductive isolation is complete and irreversible, this claim is true. Nevertheless, the BSC is concerned only with isolating barriers operating *at present* and makes no claims about their permanence. Obviously, many barriers can be reversed during speciation, fusing two "good" species back into one. Habitat, temporal, sexual, and extrinsic postzygotic isolation can disappear with a change in environment. The formation of hybrid swarms through human disturbance of the habitat has occurred in *Iris* (Riley 1938; Anderson 1949) and perhaps in Lake Victoria cichlids (Seehausen et al. 1997). Rhymer and Simberloff (1996) describe many other cases of "extinction through hybridization," all involving either human disturbance or artificial introduction. Many similar fusions must have occurred in the absence of humans.

Intrinsic postzygotic isolation, however, is quite efficient at preventing fusion. As species adapt and diverge, their developmental pathways become less compatible in hybrids, yielding hybrid sterility and inviability. The key point, as we elaborate in the next chapter, is that intrinsic incompatibilities are difficult to undo (Muller 1939). Moreover, the expected number of genetic incompatibilities between two taxa grows at least as fast as the square of the time since they diverged (Orr 1995). Thus, as time passes, the probabilities of reversing all of these incompatibilities quickly approaches zero. At this point speciation *has* become irreversible.

Fusion of species through hybridization contradicts Mayr's view that speciation is not complete until it is irreversible. We cannot predict whether future environmental or genetic changes will undo reproductive isolation that is now "complete." If humans had not disturbed the habitat of *Iris fulva* and *I. hexagona*, we would still consider them good species. The BSC, then, is best viewed as a static and not a prospective species concept.

Advantages of the BSC

In promoting the BSC, Mayr (1942) emphasized what he viewed as its advantages over its competitors. At the time, the strongest competitors were typological species concepts based on morphological difference. In those pre-molecular days, the BSC was superior in diagnosing sibling species showing little or no morphological difference. In the past two decades, however, new species concepts have arisen, many of them similar to the BSC. In fact, most of the concepts listed in Table 1.2 pick out nearly identical sets of sexually reproducing groups occurring in sympatry.

While we concede that our version of the BSC has its own problems (described in the next section), it nonetheless has a major advantage over other concepts: it alone helps solve the species problem—the existence of discontinuities among sexually reproducing organisms living in one area. Other concepts can help recognize and diagnose these entities: a species can be seen as a genotypically distinct cluster, as a group that evolves largely as a unit, or as a group whose genes are more closely related to each other than to genes from other groups. Yet none of these concepts helps us understand why populations fall into discrete groups. "Phylogenetic species," for example, can be recognized as the discrete tips of phylogenies, but phylogenetic species concepts do not tell us *why* the tips are distinct. Likewise, sympatric species can be diagnosed as morphological or genetic clusters, but one cannot understand how these clusters arise and persist without knowing what prevents them from fusing. Of course the BSC does not solve every aspect of the species problem. For instance, it cannot tell us why or how reproductive isolation develops in the first place. But only the BSC leads us to these other problems.

Other concepts, however, can sometimes be more useful in naming species. For example, the BSC cannot resolve the status of completely allopatric populations that produce viable and fertile hybrids. Using one version of the phylogenetic species concept, however, (PSC1, see Table 1.1), once can diagnose such populations as different species if they differ by as little as one trait, even a single nucleotide. Yet is this a substantial advantage? Such a practice enables one to name new species—many more than currently recognized—but forfeits any insight into the origin of distinct sympatric taxa.

Perhaps the most important advantage of the BSC is that it immediately suggests a research program to explain the existence of the entities it defines. Under the BSC, the nebulous problem of "the origin of species" is instantly reduced to the more tractable problem of the evolution of isolating barriers. While some evolutionists argue that the choice of a species concept should not include its pragmatic value, we feel that the best species concepts produce the richest research programs.

Indeed, this very book reflects the increased understanding of nature derived from using the BSC. It is a testament to the BSC that the study of reproductive isolation has become a major enterprise in evolutionary biology. When it comes to actually studying speciation, even severe critics of the BSC concentrate on reproductive isolation, working on barriers such as assortative mating and extrinsic postzygotic isolation. Virtually every recent paper on the origin of species, theoretical or experimental, deals with the origin of isolating barriers. This rich literature stands in vivid contrast to the paucity of research inspired by other species concepts.

Problems with the BSC

Problems with the BSC, including ambiguities of species status and the existence of groups to which the concept cannot be applied, have been extensively discussed by Mayr and others (e.g., Mayr 1963, 1982, 1992; Ereshefsky 1992; Claridge et al. 1997; Wilson 1999; Wheeler and Meier 2000). Rather than retread this familiar ground, we will briefly discuss a few of the most serious concerns.

ALLOPATRIC TAXA. Biological species are best diagnosed in sympatry, and yet some taxa include geographically isolated and morphologically differentiated populations. The European red deer and the North American elk, for example, are both placed in the species *Cervus elaphus*, but are allopatric and differ in traits such as size and color. Such populations are difficult to categorize using the BSC. We do not know whether their differences—assuming they are genetic—would allow them to coexist in sympatry without exchanging genes. In some groups this problem is severe. In the African rift lakes, for example, dozens of allopatric cichlid populations have been diagnosed as species because of differences in male breeding color (Turner et al. 2001). We cannot be sure whether such differences would prevent hybridization in sympatry. Yet the problem of allopatry is not limited to the BSC: all species concepts, save those based on phylogenetics, have problems with allopatric populations.

Nevertheless, the BSC is not completely powerless in this situation. Many "allopatric" populations are not completely isolated, but exchange migrants. The ability of these migrants to interbreed with local individuals can help resolve their species status. This is why all human populations belong to a single biological species. In addition, some allopatric populations with little or no migration can be unambiguously diagnosed as *different* biological species. This is possible when interpopulation crosses in the greenhouse or laboratory yield hybrids that are completely sterile or inviable due to intrinsic developmental problems. Such problems reflect genomic incompatibilities that would also act in nature. We know of no cases in which hybrids that are intrinsically sterile or inviable in the laboratory are fertile or viable in nature. Allopatric populations can also be considered different species if they show some forms of postmating, prezygotic isolation, such as the failure of pollen to germinate on foreign stigmas.

When experimental studies of allopatric taxa demonstrate that no single isolating barrier is complete, one can only make reasonable guesses about biological species status. These guesses, however, can be informed by measuring reproductive isolation in the laboratory. Coyne and Orr (1989a, 1997) compared estimates of premating and postmating isolation between allopatric *Drosophila* taxa with similar estimates from sympatric species. This comparison allowed judgments about whether allopatric taxa would probably be reproductively isolated if they became sympatric. Similar decisions can be made using morphological or genetic-distance criteria (e.g., Highton 1991), but this is riskier.

Determining whether allopatric populations are biological species is thus a one-way test. Artificial hybridizations can demonstrate that such populations are members of different biological species, but cannot determine whether they belong to the same biological species, since many taxa that produce fertile and viable offspring in the laboratory or greenhouse do not hybridize in nature. The lion *(Panthera leo)* is sympatric with the leopard *(Panthera pardus)* in Africa. Hybrids have not been reported from the wild, but these "leopons" can be produced in zoos, and females are fertile. Obviously, premating barriers break down under the artificial conditions of confinement. Similarly, many orchids that occur sympatrically without hybridization are easily crossed in the greenhouse.

HYBRIDIZATION AND INTROGRESSION. Many critics argue that the BSC fails to deal with gene flow between sympatric taxa. As Grant (1957, p. 75) wrote, "The most important single cause of a species problem in plants is natural hybridization." Indeed, hybridization would be a serious problem for the BSC under two conditions: (1) if one adhered to the strict construction of the BSC in which *no* exchange can occur between species, or (2) if gene exchange were *widespread and substantial* between sympatric taxa. Our version of the BSC does not demand complete reproductive isolation, so a low frequency of gene exchange is not a problem. This concept would thus be inapplicable only if nature formed a *syngameon* (a morphological or genetic continuum), so that distinct groups were rarely distinguishable, or if distinct groups seen in sympatry usually differed at only a few loci but exchanged genes freely throughout the rest of the genome. Whether or not recognized sympatric "species" exchange genes promiscuously is a matter for empirical work. In groups like *Drosophila*, in which morphologically distinct taxa have also been thoroughly scrutinized for genetic traits such as chromosome structure and DNA sequence, we find strong concordance between the ability to interbreed and the degree of morphological and genetic similarity. In this genus, pervasive introgression is not a problem. For most groups, however, such information does not exist. Our guess is that morphologically distinct taxa showing rampant gene exchange at many loci will be rare. Syngameons appear to be uncommon except among agamic complexes of plants.

Nevertheless, recent work shows that hybridization and introgression are more frequent than imagined by earlier evolutionists such as Mayr and Dobzhansky. But three recent surveys suggest that such hybridization is not rampant. In birds, 895 out of 9672 described species (9.2%) are known to have produced at least one hybrid with another species in nature (Grant and Grant 1992). Among the roughly 2000 described species of *Drosophila*, there are only 10 examples of naturally formed interspecific hybrids (Gupta et al. 1980; Powell 1983; Lachaise et al. 2000). Some *Drosophila* hybrids have undoubtedly gone undetected, but given the amount of work on this genus it is reasonable to conclude that interspecific hybridization is rare. As noted above, Ellstrand et al. (1996) reviewed the frequency of hybridization in plants, estimating that 6–16% of *genera* contain at least one species that forms hybrids, probably a substantial overestimate of the fraction of species that hybridize. Moreover, in each geographic area hybridization was limited to relatively few groups. Ellstrand et al. conclude (1996; p. 5093) that in plants spontaneous hybridization "is not as ubiquitous as is frequently believed" and is "not universal, but concentrated in a small fraction of families and an even smaller fraction of genera."

Studies of hybridization based on the appearance of morphological or genetic intermediates can either underestimate or overestimate the true amount of gene flow between taxa. Some hybrids, for example, have simply been overlooked. In plants, many hybrids have been collected only once or twice from a single location. In addition, hybrids are usually recognized by morphological intermediacy. This can seriously underestimate the amount of intercrossing if some hybrids, such as individuals from backcrosses, resemble individuals of pure species but still carry foreign genes.

Cryptic introgression can be inferred if phylogenies based on different loci are not concordant (Hey 2001); that is, many or most genes might be highly diverged between taxa, while others are nearly identical. Unfortunately, this observation cannot always distinguish between gene exchange that occurred in the past (before reproductive isolation was substantial), gene exchange occurring now (Machado et al. 2002), or simply the persistence of ancestral polymorphisms. Yet, one observation can provide indisputable evidence for *current* hybridization: alleles are shared between taxa where they are sympatric but not where they are allopatric. Whittemore and Schaal (1991) describe such a pattern in oaks.

Observing hybrids may also *overestimate* gene exchange because hybridization (the production of individuals from an inter-taxon cross) is not identical to introgression (the infiltration of genes between taxa through the bridge of F_1 hybrids). Among the ten naturally occurring hybridizations in *Drosophila*, three produce completely sterile or inviable offspring, and four produce sterile males. Sterile interspecific hybrids are common in the frogs of the genus *Rana* (Hillis 1988), in Lepidoptera (Presgraves 2002) and in the sedge genus *Carex* (Cayouette and Catling 1992). In the area where the black-capped and Carolina chickadees (*Poecile atricapilla* and *P. carolinensis*) are sympatric, hybridization is pervasive, but introgression is restricted because hybrids show strong intrinsic postzygotic isolation (Sattler and Braun 2000; Bronson et al. 2003). Vollmer and Palumbi (2002) describe a widespread coral "species" composed entirely of hybrids, but these are effectively sterile.

Unfortunately, we lack information about intrinsic and extrinsic postzygotic isolation in nearly all of the bird and plant hybrids described by Grant and Grant (1992) and Ellstrand et al. (1996). The survey of Price and Bouvier (2002) suggests that bird hybrids are unlikely to suffer intrinsic sterility or inviability, but introgression in at least some groups, such as Galápagos finches, is prevented by *extrinsic* hybrid sterility involving differences in ecology or mating behavior (Grant and Grant 1997). The continued persistence of distinct taxa that hybridize surely implies some form of postmating isolation.

Evidence from hybrid zones also suggests that the mere presence of hybrids need not imply massive gene exchange. As we discuss in the next chapter, in many such zones, hybrids are unfit. Estimates of the number of genes involved in this loss of fitness can be large, suggesting that much of the genome cannot move between species because it is linked to divergently selected alleles. This lack of introgression can be seen in clines of allozyme alleles that are diagnostic for hybridizing species. Frequencies of such alleles often go from 0% to 100% as one moves across a hybrid zone, suggesting little introgression outside of the area of contact (e.g., Kocher and Sage 1986; Szymura and Barton 1986).

Several other factors should be considered before concluding that hybrids pose a severe problem for our version of the BSC.

1. Much current hybridization probably results from human disturbance of the habitat—disturbance that is likely to be less common under natural conditions. Cayouette and Catling (1992, pp. 371–372) note that 252 different hybrids have been reported among species in the sedge genus *Carex*, but add that "sedge hybrids vary a great deal in practically all of their characteristics, but the one thing that they almost all have in common is disturbed

site ecology. It is quite possible that sedge hybrids were formerly rare, but have increased dramatically as a consequence of disturbance resulting from human activity." Rieseberg and Gerber (1995) suggest that some hybrids between Hawaiian plants described in the survey of Ellstrand et al. (1996) may have resulted from human disturbance. This situation may be common in plants given the tendency of some botanists to collect along roadsides.

2. "Hybridization" may be a transient phase of evolution. During sympatric speciation and reinforcement, individuals may appear that are intermediate between two well-demarcated forms, but these intermediates disappear when reproductive isolation becomes complete. Alternatively, hybridizing taxa might be in the process of fusing into a single species.

3. What appear to be hybrids might be only geographic variants for one or a few traits, or nongenetic variants produced by local conditions. This possibility has received little attention despite the ubiquity of developmental plasticity and geographic variation. Plant morphology, for example, can be dramatically altered by environmental differences (Sultan 2000).

OAKS: THE WORST-CASE SCENARIO. The classic example of the supposed failure of the BSC to deal with hybridization is the oak genus *Quercus* in North America and Europe. Oaks thus constitute a good case for testing the validity of the BSC.

Quercus is variously described as either a rampantly hybridizing complex in which distinct taxa cannot be seen, or as a group of fairly well-differentiated entities that sometimes hybridize (Burger 1975; Van Valen 1976). On the other hand, some botanists claim that genuine hybrids are infrequent and that most recognized "hybrids" are actually trivial intraspecific variants (Muller 1952; Jones 1959).

Stebbins (1950, pp. 61–66) reviews the problems in this genus, and Whittemore and Schaal (1991) and Howard et al. (1997) discuss more recent data. There are 16 species of white oaks (subgenus *Quercus*) in eastern North America, distinguished largely by the morphology of leaves and acorns. Fourteen of these are known to hybridize with other species (Hardin 1975). Most botanists who work on *Quercus* describe hybrids as being uncommon, rarely obscuring the morphological boundaries of species (Palmer 1948; Jones 1959), although in some localities hybrid swarms have been described. The situation is complicated by the tendency of some species to hybridize at some locations but not others. Moreover, what are described as "hybrids" may actually be localized genetic ecotypes or even environmental variants having no genetic basis (Jones 1959). Thus, the distinctness of oak species could reflect two possibilities: the species might maintain differences in a few diagnostic traits despite extensive introgression, or they might represent truly distinct gene pools whose hybrids are unfit. Recent molecular work has begun to clarify the situation.

Using both chloroplast DNA (cpDNA) and nuclear DNA, Whittemore and Schaal (1991) studied gene flow among five species of white oaks in the central United States. Despite no morphological evidence for hybridization, there was extensive interspecific exchange of cpDNA among sympatric species. In fact, phylogenies based on cpDNA showed that different species living in the same place are genetically more similar than are members of the *same* species inhabiting different places. However, one nuclear marker was species specific, and Whittemore and Schaal note (p. 2543) that "the five species studied here are well differentiated with respect to many morphological characters, allozyme loci (Guttman and Weigt 1989), and probably, judging from their different ecological and geographic range, many physiological traits." In a similar study, Martinesen et al. (2001) describe much more exchange of cpDNA and mtDNA than of nuclear DNA between two species of cottonwood (*Populus*). It is likely that oaks and cottonwoods, like other plant and animal species, show more extensive introgression of organelle DNA than of nuclear DNA (see Appendix). It is thus risky to assume extensive hybridization based on observations of organelle DNA alone.

The situation in *Populus* is mirrored by two species of European oaks, *Quercus robur* and *Q. petraea*, which are sympatric in many places and have been described as hybridizing freely. However, a study of 20 nuclear microsatellite loci from five locations showed that the species were well demarcated from each other, forming two well-separated clusters in all locations (Muir et al. 2000). The authors raise the question of "how the species differences are maintained despite the high levels of interspecific gene flow" (p. 1016). But the observation that the species differ at many loci suggests that gene flow is *not* high.

Likewise, Howard et al. (1997) reported limited introgression between *Quercus gambelii* and *Q. grisea*, whose ranges overlap in the southwestern United States. Although the species are segregated by altitude, in the area of sympatry they form a "mosaic hybrid zone" in which the transition between the species' ranges is not smooth but patchy. Many individuals within this zone appear to be morphologically pure species, but nevertheless carry some foreign genes. However, the extent of introgression drops rapidly outside the area of overlap. Only two kilometers away, one finds few individuals of *Q. gambelii* that carry genes from *Q. grisea*. The authors suggest that "the abrupt genetic and morphological discontinuity between *Q. gambelii* and *Q. grisea*, despite areas of hybridization, indicates that selection acts to maintain coadapted complexes of alleles in the two species" (p. 754).

In California, the genotypes of the few morphological intermediates between *Q. lobata* and *Q. douglasii* show them to be pure-species individuals rather than hybrids (Craft et al. 2002). The authors propose that morphological intermediacy reflects not hybridization but phenotypic plasticity. Nason et al. (1992) found that morphological intermediates between sympatric *Q. kelloggii* and *Q. wislizenii* var. *frutescens* in Southern California were almost all first-generation (F_1) hybrids. They suggest that the absence of backcross or later-generation hybrids reflects their inability to compete with the parental species (i.e., there is extrinsic postzygotic isolation).

The data thus suggest that in many cases nuclear gene flow between oak species is restricted by unknown forms of selection against hybrids. Williams et al. (2001) identified one reproductive barrier between *Q. gambelii* and *Q. grisea*: fruit set was significantly higher in conspecific than in heterospecific pollinations. This reproductive isolation, which reduced gene flow by about 60%, was caused by the inviability of hybrid embryos.

The situation in oaks is complex, and it is clear that named species do not always correspond to good biological species free from introgression. However, genetic studies also show that oak species are not rampantly hybridizing, and are not differentiated by only a few morphological or genetic traits. This implies that, as in *Q. gambelii* and *Q. grisea*, the distinctness of oak species in sympatry reflects disruptive selection causing intrinsic and extrinsic postzygotic isolation.

In summary, the boundaries between oak species may not be as porous as commonly thought. As Howard et al. (1997, p. 754) remark, "Oaks may not represent a greater challenge to traditional concepts of species than many other plant and animal taxa that form hybrid zones with close relatives." Although this group has been considered a problem for the BSC, detailed scrutiny suggests that the difficulties are exaggerated.

The intense interest that botanists have paid to hybridizing species might well overstate the challenge that plants—and other species—pose to our version of the BSC. To determine whether the BSC is *generally* inapplicable, one must extend the work on oaks to random samples of species in a wide variety of taxa.

TAXA WITH WHOLLY OR PARTIALLY UNIPARENTAL REPRODUCTION. The BSC obviously cannot deal with groups in which sexual reproduction is very rare. To the extent that such groups form distinct clusters in sympatry, we recommend using a species concept that addresses the origin and maintenance of such clusters. As described below, recent theories suggest that discrete bacterial groups can arise as a result of natural selection acting on ecologically equivalent clones, coupled with the occurrence of mutations that permit the occupation of new niches. If this is the case, both the cohesion and ecological species concepts (Table 1.2) seem appropriate for dealing with bacterial taxa. There is less evidence that agamic complexes in plants, with their combination of sexual and asexual reproduction, form discrete clusters, and we would be happy to adopt any species concept that helps us understand the evolution of such groups.

DELINEATING "SPECIES" IN A SINGLE EVOLVING LINEAGE, OR IN FOSSILS, OR PRESERVED MATERIAL. Mayr (1963) dealt extensively with the problems these matters pose for the BSC, and we can add little to his arguments. Under every species concept, the division of a continuously evolving lineage into named species is a purely subjective exercise, although one that may be necessary for scientific communication.

Diagnosing species in fossilized or preserved material from a single location is less arbitrary, as one can make reasonable guesses about the likelihood of reproductive isolation from discontinuities between phenotypes. If one has allopatric samples that show some phenotypic variation, one can search for material from intervening areas to see if the phenotype changes gradually over space (suggesting conspecific status) or if there is an abrupt geographic discontinuity (suggesting two species). If no such material can be found, one can tentatively diagnose allopatric taxa as species if they differ as much or more than bona fide species existing in sympatry. Ward (2001, pp. 591–592) shows how this can be done in ants.

DISTORTING EVOLUTIONARY HISTORY. The BSC has been severely criticized by systematists because species identified using interbreeding and reproductive isolation may distort evolutionary history (Mishler and Donoghue 1982; Cracraft 1989). The most frequent criticism is that populations of a single biological species can be less closely related to each other than populations belonging to different biological species.

Imagine, for example, that species A, consisting of three populations (A_1, A_2, and A_3), occupies a continent, and that migrants from population A_3 colonize an isolated island. The descendants of these colonists experience strong selective pressure and rapidly evolve into species B, whose members have isolating barriers strong enough to prevent hybridization with all populations of species A were they to re-invade the continent. In such a situation, genetic analysis might yield the phylogeny depicted in Figure 1.4. Here, individuals of population A_3 appear more closely related to individuals of species B than to those of populations A_1 and A_2. To use the terminology of modern systematics, species A is *paraphyletic* relative to species B, and the reproductive relationships do not mirror genealogical history. Avoiding use of the BSC because of this possibility has been called "fear of paraphyly" (Harrison 1998).

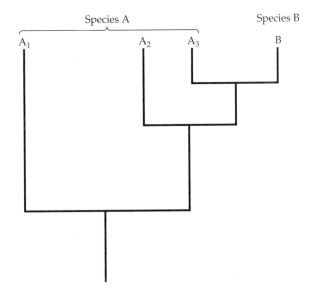

Figure 1.4 The phylogeny of two reproductively isolated species, A and B, in which reproductive relationships do not coincide with ancestry at some genetic locus or loci. Species A consists of three populations (A_1, A_2, and A_3), one of which (A_3) gave rise to species B. Phylogenetic analysis might show that individuals in population A_3 are more closely related to individuals in species B than to conspecific individuals in populations A_1 and A_2. In such cases, species A is considered "paraphyletic" with respect to species B. (After Harrison 1998.)

Our response to this critique is similar to that of Harrison (1998, p. 26): "If we accept that species are defined by isolation and/or cohesion and do not start with the assumption that they must be exclusive groups and the units of phylogeny, then including paraphyletic assemblages as species does not misrepresent history." Nevertheless, if situations such as that shown in Figure 1.4 are common, we would like to know about them: such a phylogeny might, for example, identify the source of an island endemic.

But is it really possible to reconstruct such histories using genetically based phylogenies? Since populations A_1, A_2, and A_3 are conspecific, gene flow will eventually homogenize them, destroying the phylogeny that shows a close relationship between population A_3 and species B. One then obtains a bifurcating phylogeny with species $[A_1 + A_2 + A_3]$ as one branch and species B as the other. Since evolutionary history is not seen directly but must be reconstructed, the history given by the latter phylogeny is indeed congruent with the BSC. The genetically distinct populations A_1, A_2, and A_3 are transitory entities, and it is not a gross distortion to conclude that species B derives from an indefinable group contained within species A.

Interbreeding among individuals of a biological species thus quickly eliminates our ability to detect paraphyly. This problem is especially serious because interbreeding is likely to erase the history of populations much faster than reproductive isolation can evolve in an isolate. Thus, reconstructing the history of populations is feasible only when these populations are fairly discrete and exchange genes only rarely. If these conditions do not obtain, constructing a bifurcating evolutionary tree will not yield an accurate history of populations. Such a history is complex, and is better represented by a reticulating network than by a tree. An example involves the species *Drosophila sechellia* and *D. mauritiana*, endemic to the Seychelles archipelago and the island of Mauritius, respectively. These species are closely related to *D. simulans*, which is widespread in east Africa. *D. mauritiana* and *D. sechellia* presumably arose after colonization of the islands by a *D. simulans*-like ancestor. Analysis of multiple loci, however, has shown that it is impossible to identify a contemporary population of *D. simulans* that was the source of these colonists (Kliman et al. 2000).

This raises the most serious problem facing those who claim that the BSC often distorts evolutionary history. It is important to recognize that advocates of this view take "evolutionary history" to be the branching sequence of the taxa themselves. (These taxa can be either populations or reproductively isolated species.) The problem, implied above, is that the history of taxa cannot be seen directly, but must be reconstructed from the history of genes. As we show in the Appendix, there are several reasons why these histories can differ. The most important is that each gene has its own evolutionary history that is not necessarily congruent with the history of other genes, or of the populations themselves.

There are thus two causes of a discrepancy between reproductive relationships and gene-based phylogenies. The first is that the reproductive relationships between taxa do indeed distort their true evolutionary history. The second is that phylogenies, while providing an accurate history of some genes, may give an inaccurate history of the taxa containing those genes. In this case it is the *phylogenies* that distort evolutionary history. It is hard to decide which of these two causes explains an incongruity between phylogeny and reproductive compatibility, especially when species are closely related. In such cases, reconstructing evolutionary history requires *congruent phylogenies of many different genes*, which is difficult to achieve when ancestral polymorphisms persist in descendant taxa, or when gene flow destroys the history of populations. As we note in the Appendix, most cases of "paraphyletic" species have been diagnosed using what is in effect a single gene: mitochondrial DNA. For many reasons, mtDNA behaves differently from nuclear genes, and paraphyly diagnosed using only mtDNA may not reflect the situation in the rest of the genome (Hudson and Coyne 2002; Shaw 2002).

In view of these problems, one can rarely assert with confidence that reproductive relationships distort evolutionary history. We know of only a few such cases, which we discuss in the Appendix. Thus, the seriousness of the "paraphyly problem," and of other cases in which the BSC seems to conflict with the history of taxa, may well have been exaggerated or misunderstood. Nevertheless, it is likely that some multi-gene phylogenies may show biological species to be truly paraphyletic, and that the relatedness of populations and individuals may not always be concordant with their assignment to biological species. Nevertheless, so long as one keeps these possibilities in mind, we see no compelling reason to abandon the BSC.

Other species concepts

As noted earlier, the Appendix considers the eight most popular rivals of the BSC (see Table 1.1). There we explain why these concepts were proposed as alternatives to the BSC, discuss their advantages and disadvantages, show how they compare to the BSC in dealing with difficult cases, and describe how they define the process of speciation.

These concepts fall into two classes. The first, comprising groups 2 and 3 in Table 1.1, follows the BSC by assuming the species problem to be the origin of organic discontinuities, but considers the BSC an inadequate solution to this problem. Phylogenetic species concepts, on the other hand, take as the species problem the reconstruction of evolutionary history and the assessment of evolutionary relatedness among individuals and groups. These purposes often overlap, for individuals within discrete species must usually share an evolutionary history. Moreover, most species concepts will diagnose the same species in sympatry. Where they differ is how they treat allopatric or hybridizing taxa.

Why Are There Species?

Studying speciation may reveal the origin of discontinuities between sympatric groups, but does not explain why such discontinuities are inevitable. What properties of sexually reproducing organisms and their environments inevitably lead to the evolution of discrete species? Why are organisms apportioned into clusters separated by gaps? Dawkins (1982) argues that natural selection is an inevitable consequence of any type of life; in fact, he defines "life" as the property that allows its bearers to experience natural selection. Can we also conclude that species are the inevitable results of life—at least life that reproduces sexually?

Dobzhansky (1935, p. 347) found this question intractable: "The manifest tendency of life toward formation of discrete arrays is not deducible from any a priori considerations. It is simply a fact to be reckoned with." Perhaps we cannot deduce such arrays from a priori considerations, but the inevitability of species might still be understandable a posteriori. Here we consider why discrete clusters might be inevitable in both sexual and uniparental organisms.

In some ways, this topic is more difficult than understanding the origin of species because it is more abstract. Nevertheless, we regard it as one of the most important unanswered questions in evolutionary biology—perhaps *the* most important question about speciation. Yet despite its importance, it has been almost completely neglected: the only extensive discussion is by Maynard Smith and Szathmáry (1995, pp. 163–167). These authors suggest several hypotheses, which we discuss below. An additional explanation is that evolutionary history itself can create clusters: splitting and extinction of lineages will ultimately create groups of genetically and morphologically similar organisms separated by gaps from other clusters—groups such as mammals, fish, and conifers. (See Raup and Gould 1974 for a model of clustering based on random branching and extinction). Nevertheless, while history can create discrete clusters containing *groups* of species, we do not see how it can produce species themselves, at least in sexually reproducing organisms.

Maynard Smith and Szathmáry consider three other explanations:

1. Species exist because they are discrete "stable states" formed by the self-organizing properties of biological matter. This view is closely connected with the "structuralist" school of biology, which claims that many adaptations and aspects of development result not from natural selection acting on genes, but from the self-organizing properties of biological molecules (Ho and Saunders 1984). This view of species seems untenable for several reasons. First, it lacks any mechanism that explains the origin of such states. Second, it does not explain the origin of *new* "stable states" (species), which must arise after some unspecified and temporary instability—an "adaptive valley" of molecular organization. Finally, as Maynard Smith and Szathmáry (1995) argue, the pervasive geographic variation of morphological, physiological, and ecological traits within species casts severe doubt on the inherent stability of species.

2. Species exist because they fill discrete ecological niches. This explanation sees clusters as resulting from intrinsic discreteness in ways of using resources. For example, the mechanisms by which microorganisms use alternative carbon sources or capture energy might impose distinct phenotypic solutions on the organism, in the same way that different jaw morphologies are needed to efficiently handle different prey. This effect accumulates as one goes from lower to higher trophic levels, because clusters at lower levels provide discrete niches for organisms at higher levels. This ecological explanation also rests on the inevitability of tradeoffs: being suited for one way of life makes one less suited for another. Such tradeoffs create disruptive selection, with hybrids that fall between niches being unfit. Note that this explanation is not independent of reproductive isolation because it depends on a reproductive barrier: extrinsic postzygotic isolation.

Historically, the ecological explanation is closely wedded to Sewall Wright's view of the adaptive landscape. Dobzhansky (1951, pp. 9–10) emphasized this connection:

> The enormous diversity of organisms may be envisaged as correlated with the immense variety of environments and of ecological niches which exist on earth. But the variety of ecological niches is not only immense, it is also discontinuous... Hence, the living world is not a formless mass of randomly combining genes and traits, but a great array of families of related gene combinations, which are clustered on a large but finite number of adaptive peaks. Each living species may be thought of as occupying one of the available peaks in the field of gene combinations.

This view does not require that the environment present a discrete array of niches that antedate the evolution of organisms—a difficulty given that organisms create new niches through their own evolution and that the environment itself includes organisms. The ecological explanation merely requires tradeoffs: there is a finite number of ways to make a living in nature, and organisms adopting one way sacrifice their ability to adopt others.

3. Species exist because reproductive isolation is an inevitable result of evolutionary divergence. This explanation, which is limited to sexually reproducing groups, relies on the fact that divergent evolution is likely (and given enough time, certain) to yield reproductive isolation. Such isolation allows both the permanent coexistence of taxa in sympatry and future evolutionary divergence without gene flow, factors that both contribute to discreteness. This explanation is also related to the existence of ecological niches, for divergent adaptation to such niches can impede gene flow by producing reproductive isolation as a byproduct. (Plants, for example, can develop reproductive barriers by adapting to different soil types or pollinators.) There are also "developmental niches" that arise because development requires the joint action of many coadapted genes. Sufficiently diverged developmental systems cannot work properly in hybrids, yielding intrinsic hybrid sterility and inviability. Finally, sexual reproduction itself leads to the evolution of anisogamy (disparate sizes of male and female gametes), which in turn creates the possibility of sexual selection. Divergent sexual selection will almost inevitably lead to behavioral or gametic isolation.

The "ecological" and "reproductive-isolation" explanations of species are not mutually exclusive. Indeed, they are intimately connected. Although Dobzhansky leaned more toward the ecological explanation, he also saw a role for reproductive isolation (1951, p. 255):

> The patterns with the superior adaptive values [i.e., species] form the "adaptive peak"; the peaks are separated by the "adaptive valleys" which symbolize the gene combinations that are unfit for survival and perpetuation. The reproductive isolating mechanisms, as well as the geographic isolation, interdict promiscuous formation of the gene combinations corresponding to the adaptive valleys, and keep the existing genotypes more or less limited to the adaptive peaks.

In sexually reproducing species, the ecological and reproductive-isolation explanations are intertwined because adaptive valleys between niches imply some reproductive isolation, and isolating barriers may result largely from adaptation to distinct niches. Is it possible to assess the relative importance of these explanations?

One possibility is to see what happens when one leaves niches intact but removes reproductive isolation. If the "ecological" explanation were correct, one would still see distinct clusters in sympatry. This could be addressed by looking at organisms that are almost completely uniparental, thus lacking the possibility of reproductive isolation. As noted above, the jury is out on whether uniparentally reproducing eukaryotes form discrete clusters in sympatry, but there is some evidence for clustering in bacteria.

Recent theory suggests that one can explain the existence of uniparental clusters by considering the invasion of new niches (Cohan 2001, 2004; Barraclough et al. 2003). One might naively expect uniparental organisms to continuously accumulate mutations, producing an infinite variety of clones, each adapted to a slightly different habitat. Cohan (1984, 2001), however, suggested a type of bacterial "speciation" that produces distinct clusters. A lineage of bacteria may indeed accumulate new mutations and begin to fill up ecospace with a panoply of clones. Periodically, however, an individual experiences a new mutation that is generally adaptive. The clone containing this mutation will replace all other clones with which it is ecologically equivalent. The genetic variation within the group of clones then collapses to the genotype of the single mutant clone. These recurrent episodes of "periodic selection" limit the degree to which asexual groups can diverge to form microspecies.

In this theory, a new bacterial "species" arises when a mutation gives an individual the ability to invade a new ecological niche, rendering it and its descendants immune from extinction during episodes of periodic selection. (Such mutations may be relatively common in bacteria because of their ability to incorporate genes from distantly related taxa.) If recombination in bacteria is rare and periodic selection common, the new "species" will form a distinct cluster that could coexist with its ancestor. Such speciation could occur either allopatrically, when a migrant individual lands in a novel habitat, or sympatrically. In the allopatric case, mutations of large effect are not required, for adaptation to a new niche can be built up gradually. In sympatry, the new bacterial species will persist if the "macromutation" allowing occupation of a new niche has a selective advantage higher than that of subsequent mutations causing periodic selection in the ancestral species.

Bacterial "speciation" thus involves occupying new ecological niches, and a bacterial species can be defined as an "ecological population, [which is] the domain of competitive superiority of an adaptive mutant" (Palys et al. 1997, p. 1145). This is closely related to Templeton's (1989) cohesion species concept, which incorporates demographic exchangeability as one of the "cohesion mechanisms" that defines species (see Appendix).

The importance of niche differentiation in understanding asexual clustering suggests that ecology might form the basis of an asexual species concept. Just as reproductive isolation suggests why sexual organisms remain discrete, so the occupation of distinct niches by demographically nonexchangeable clones suggests why asexual clusters remain discrete. This idea also yields a research program for bacterial speciation. Sympatric clusters ("species") of bacteria should always occupy different ecological niches, and should remain distinct when periodic selection occurs in any of them. Moreover, bacteria showing greater gene exchange should form clusters that are less distinct than those seen in more-clonal species. Finally, different sympatric clones within a single bacterial "species" should not be strongly adapted to their local habitat, because such adaptation would prevent periodic selection that homogenizes each cluster. Belotte et al. (2003) support this prediction in a study of *Bacillus mycoides* from a Canadian forest.

More recent theories consider other explanations for clustering besides periodic selection and macromutations. These theories see asexual clusters as simple adaptive responses to resource gradients in either sympatry or parapatry (Dieckmann and Doebeli 1999; Doebeli and Dieckmann 2003). However, in both geographic situations, the clustering appears to be either an artifact of the models' assumptions, or a temporary phenomenon that disappears when resource space eventually becomes filled with a continuum of asexual organisms (Polechova and Barton 2004).

A third explanation for clustering in asexual organisms is that clonal reproduction, coupled with occasional mutations affecting morphology or DNA sequence, will eventually produce clumps as a simple artifact of history. Barraclough et al. (2003) show that this can occur in both sympatric and allopatric populations. But unless this clustering is accompanied by ecological diversification, it will disappear in sympatry—the only place where clusters are truly discernible—through either periodic selection or the relentless accumulation of alleles adapting clones to new microhabitats.

Because we rejected ecological differentiation as part of the BSC in sexually reproducing groups, we obviously endorse the use of different species concepts in different groups. We do not consider this pluralism to be a weakness of the BSC. Because the causes of discreteness may well differ among taxa, so may the concepts appropriate to addressing the species problem. If groups without sex form distinct clusters, and the explanation for such clusters resembles Cohan's theory of bacterial speciation, then the answer to "Why are there species?" in such taxa seems to be "Because there are discrete ways of making a living."

While this may explain species in asexual groups, it will not suffice for sexual groups. For example, in taxa having a mixture of sexual and uniparental reproduction, as in agamic plants, periodic selection cannot eliminate all genetic variation within a group of "demographically exchangeable" individuals: as the new adaptive mutation spreads, recombination will separate it from the genome in which it arose. Moreover, the occurrence of macromutations that create new species by allowing invasion of a new niche must be rare in eukaryotes, which almost never experience the wide gene transfer that causes adaptive leaps in bacteria.

In fact—although this conclusion is tentative—taxa with some sexual reproduction, such as agamic complexes, seem to form clusters that are *less distinct* than those seen in taxa with largely asexual reproduction. If adding a little bit of sex erodes the discreteness of groups, then ecology cannot be the only explanation for discreteness.

When one moves to fully sexual groups, one again finds discrete clusters of genes and traits. This is a clue that sexual reproduction itself must play a role in distinctness. In fact, we suggest that in sexually reproducing groups it is reproduction itself, combined with differential adaptation and the existence of tradeoffs, that ineluctably produces species. This idea derives from understanding how clusters are formed.

Recent theoretical models (Chapter 4) suggest that in sexual groups the ecological explanation is at least partly necessary for the existence of species that arise sympatrically, as the initial steps in sympatric speciation often involve adaptation to discrete resources. Yet, these same models show that clusters will exist only for those traits involved in resource use, and that differentiation of other traits requires the evolution of further isolating barriers such as behavioral isolation. In fact, it is sexual reproduction that allows the coupling of resource use to other isolating barriers, a coupling that is necessary to complete speciation (Dieckmann and Doebeli 1999; Kondrashov and Kondrashov 1999).

However, evidence adduced in Chapter 4 suggests that most speciation is allopatric. Although discrete niches might be necessary to explain the sympatric *coexistence* of allopatrically formed species, such niches are not required for the *formation* of distinct and recognizable species in allopatry. Following geographic isolation, good biological species can arise via nonecological processes (such as sexual selection) that yield behavioral, mechanical, gametic, or intrinsic postzygotic isolation (Chapter 6). Alternatively, when ecology is involved in the allopatric evolution of reproductive barriers, it need not produce a difference in niches. Identical environments, for example, can select for identical traits having different genetic bases, yielding developmental incompatibilities in hybrids.

We suggest, then, that there are different reasons for discreteness in different groups. Clustering in completely asexual or uniparental taxa may rest largely on the ecological explanation, while clustering in sexual taxa rests on a combination of ecology and reproductive isolation. Since there may be different causes for clustering in asexual versus sexual taxa, should one use "species" as the term for asexual clusters and "speciation" for the processes by which they form? We see no problem with this so long as one recognizes that these words mean different things in different taxa.

We predict, then, that statistical analyses of groups having both sexual and asexual reproduction will show that they form clusters less distinct than those seen in either completely sexual or completely asexual groups. Groups with mixed modes of reproduction have too much sex to permit the homogenizing effects of periodic selection, but too little sex to homogenize members of diverging "microspecies." Intermediate levels of sexual reproduction are not conducive to forming discrete taxa.

2

Studying Speciation

It is sometimes argued that speciation is not a distinct field of research. After all, species are largely byproducts of evolution within lineages, a process that has always been the purview of evolutionary genetics. Julian Huxley, for example, declared that

> The formation of many geographically isolated and most genetically isolated species is thus without any bearing upon the main processes of evolution. . . . Species-formation constitutes one aspect of evolution; but a large fraction of it is in a sense an accident, a biological luxury, without bearing upon the major and continuing trends of the evolutionary process (Huxley 1942, p. 389).

But while *anagenesis* (evolutionary change within a lineage) is the underpinning of *cladogenesis* (the creation of new lineages by splitting), these two processes are analyzed with different methods. In this chapter we explain why speciation is unique, and suggest ways to study it.

As Mayr has emphasized, a key aspect of species is that they can be defined only relative to other species. Unlike anagenesis, then, speciation involves the joint evolution of two or more groups:

> The word species thus became a word expressing relationship, just like the word brother, which does not describe any intrinsic characteristics of an individual but only that of relationship to other individuals; that is, to other offspring of the same parents (Mayr 1992, p. 223).

Critics consider this relativistic aspect of the BSC a weakness. The BSC, however, is not the only relativistic species concept: *every* concept requires comparing different groups of individuals, whether this comparison involves reproductive isolation, morphological distinctness, or phylogenetic relationship.

Under the BSC, however, the origin of species does involve a unique character that is a joint property of two species: reproductive isolation. The evolution of traits in a single lineage is relevant to speciation only insofar as it contributes to reproductive isolation from other lineages. Thus, one essential feature of speciation is *pleiotropy*: characters that evolve within a group have the side effect of producing isolating barriers. In some cases pleiotropy is direct: adaptation to a new habitat may automatically guarantee spatial isolation from a relative. In other cases pleiotropy is indirect: divergent selection for adaptive differences between two taxa, for example, can yield intrinsic hybrid inviability or sterility as a byproduct.

Another novel (but not ubiquitous) feature of speciation is between-lineage *epistasis*: the phenomenon in which genes that evolve in one group produce reproductive isolation by interacting with genes evolving in another group. Like pleiotropy, epistasis is usually considered an intraspecific phenomenon. In intraspecific epistasis, the effects of an allele depend on other genes in the genome. In contrast, interspecific epistasis acts between loci from *diverged* genomes, causing reproductive isolation. Epistasis is almost always involved in intrinsic postzygotic isolation, whereby hybrids are sterile or inviable because of incompatible developmental systems, ecological disadvantage, or inappropriate mating behavior. Such fitness-reducing interactions almost always require at least two genetic changes. Some types of postmating, prezygotic isolation also depend on disrupted interactions between gametes from different species.

Epistasis can also cause premating isolation. Behavioral isolation, for example, usually results from coevolution of male traits and female preferences occurring in each of two lineages. The effect of a new male trait on reproductive isolation thus depends on genetic changes that have occurred in females of the other lineage, and vice versa. Interspecific epistasis also occurs in mechanical or pollinator isolation, when changes in genitalia or the placement of pollinia on insects reduces fertilization in crosses between populations.

Pleiotropy and (to a lesser extent) epistasis are key features of speciation. While they can also be important during anagenesis, they are not necessary for evolutionary change and may be largely irrelevant to many evolutionary phenomena, such as the origin of cryptic coloration. During speciation, however, the complex interactions between the genomes of two taxa guarantee that both the mathematical models and genetic analysis of speciation will differ from those used within species. This implies that speciation may show emergent properties not seen in traditional population-genetic models. Indeed, such properties (e.g., the "snowball effect" and the asymmetry of postzygotic isolation discussed in Chapters 7 and 8) have already been seen. Recognizing the importance of epistasis leads to other insights about speciation. For instance, it is sometimes assumed that because hybrids between two species are unfit, a population undergoing speciation must itself experience a loss of fitness (Wade and Goodnight 1998). Wright (1982a) suggested that genetic drift is required to help such incipient species cross an "adaptive valley." Models of postzygotic isolation that include epistasis (Chapter 7), however, clearly show that speciation can occur without any loss of fitness and without any need for genetic drift.

The Problem of Speciation

If one accepts some version of the biological species concept, then the central problem of speciation is understanding the origin of those isolating barriers that actually or potentially prevent gene flow in sympatry. This problem is conceptually simple, but its solution involves two formidable tasks. First, one must determine which reproductive barriers were involved in the *initial* reduction of gene flow between populations; we outline some ways to address this problem immediately below. Second, one must understand which evolutionary forces produced these barriers; we deal with this issue in Chapters 5–11.

Several difficulties arise when trying to determine which isolating barriers were important in a given speciation event. First, speciation involves the study of isolating barriers only up to the point at which gene flow between taxa is close to zero, but such barriers continue to accumulate thereafter. Presgraves (2003), for example, estimates that about 200 genes are involved in the inviability of hybrids between the sympatric species *Drosophila melanogaster* and *D. simulans*. Each of these genes, however, causes nearly complete inviability in a hybrid genetic background, producing a virtually complete barrier to gene flow. Nevertheless, these genes surely evolved one after the other, so most could have had nothing to do with the completion of hybrid inviability.

Second, several isolating barriers may act together to prevent gene flow, and their *present* importance may distort their *historical* importance during speciation. This problem is particularly acute for species that are at least partially sympatric. Isolating barriers are best studied in sympatry, where they actually (rather than potentially) prevent gene flow. Indeed, some barriers, such as gametic isolation or habitat isolation enforced by competition, may not even be detectable in allopatric taxa. But in fully isolated sympatric species, additional forms of reproductive isolation may have evolved after speciation was complete. Thus, by the time allopatrically formed species become sympatric, they may carry little information about the evolutionary sequence of isolating barriers.

One can appreciate this problem by looking at sympatric taxa in which several isolating barriers have been identified and measured. Perhaps the best of these studies is that of Ramsey et al. (2003) on two sister species of monkeyflower from western North America, *Mimulus lewisii* and *M. cardinalis*. Although substantial habitat isolation results from altitudinal segregation, the two species co-occur in some regions of the Sierra Nevada Mountains. In sympatry, the species are further isolated by having nearly nonoverlapping groups of pollinators, strong conspecific pollen precedence, and partial inviability and sterility of F_1 hybrids. Table 2.1 (middle column) shows the absolute strength of each isolating barrier as calculated by Ramsey et al. (2003). These strengths

Table 2.1 *Reproductive isolating barriers between* Mimulus lewisii *and* M. cardinalis[a]

Isolating barrier	Absolute strength of isolating barrier[b]	Proportional contribution to total isolation
Habitat isolation	0.587	0.587
Pollinator isolation	0.976	0.403
Interspecific pollen competition	0.833	0.0083
F_1 hybrid inviability (seed germination)	0.125	0.0002
F_1 hybrid sterility[c]	0.609	0.0009
Total isolation		0.9986

Source: Ramsey et al. 2003
[a]Figures are averages of values from each of the two species, some recalculated from data of Ramsey et al. Values for the intensity of each isolating barrier range from 0 (interspecific gene flow equal to intraspecific gene flow) to 1 (complete reproductive isolation between species).
[b]Strength of barrier considered in absence of other barriers; 0 = free gene flow, 1 = no gene flow.
[c]Average of male (pollen viability) and female (seed mass) components of sterility.

are measured on a scale ranging from 0 to 1, with 0 representing free gene flow between taxa and 1 representing complete reproductive isolation. Each barrier has a significant effect on gene flow, and, acting together, all reduce interspecific gene flow to about 0.1% of that seen between conspecific individuals. The important point is that these data cannot tell us which barriers evolved before others. *Any* combination of three of the five barriers reduces gene flow to nearly zero.

Third, isolating barriers act sequentially over the life cycle of the organism, beginning (in adults) with habitat and temporal isolation and ending with hybrid sterility and inviability. Each barrier reduces only the gene flow that has escaped barriers acting previously. If, for example, there are n isolating barriers acting sequentially over the life cycle (I_1 through I_n, where each I, the strength of isolation, falls between 0 and 1), then the total gene flow between a pair of taxa, G, is

$$(1-I_1)(1-I_2)(1-I_3)\ldots(1-I_n), \text{ or } \prod_{i=1}^{n}(1-I_i)$$

It is obvious that earlier-acting reproductive barriers will reduce gene flow proportionally more than will later-acting barriers, even if the later barriers are stronger in an absolute sense. For example, hybrids between *M. cardinalis* and *M. lewisii* have on average only 60.9% the fertility of pure species (I = 0.609). In absolute terms, hybrid fertility is thus a more effective isolating barrier than habitat isolation (I = 0.587). Yet the proportion of total gene flow reduced by hybrid sterility is only 0.0009, while that of the earlier-acting habitat isolation is 0.587. The reduced importance of later-acting isolating barriers can be seen in the right-hand column of Table 2.1.

It is easy to find the conditions under which a later-acting barrier to gene flow, although absolutely stronger than an earlier one, has a smaller effect on total reproductive isolation. Assume that there are two isolating barriers, a prezygotic barrier of absolute strength I_1 and a stronger postzygotic barrier of absolute strength I_2. When acting together, the reduction of gene flow by the earlier-acting barrier is simply I_1. The effect of the second, however, is $I_2(1 - I_1)$. The proportions of total gene flow impeded by the first (p_1) and second (p_2) barriers are

$$p_1 = I_1 / [I_1 + I_2(1-I_1)], \text{ and } p_2 = [I_2(1-I_1)] / [I_1 + I_2(1-I_1)]$$

It follows that if the postzygotic barrier is stronger in absolute terms than the prezygotic barrier ($I_2 > I_1$), the prezygotic barrier will have a greater proportional effect ($p_1 > p_2$) when

$$I_2 \times I_1 > I_2 - I_1$$

that is, when the product of the individual strengths of isolation exceeds their difference. Figure 2.1 shows the region of parameter space in which the earlier-acting barrier has a greater proportional effect on gene flow even though the later-acting barrier is stronger. If an even stronger yet still later-acting bar-

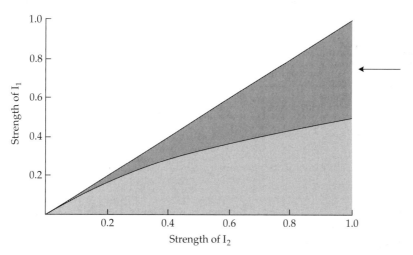

Figure 2.1 The relative contributions of two sequentially acting isolating barriers (I_1 acts earlier than I_2) to total reproductive isolation. The shaded space below the line of equality includes all values in which the later-acting barrier is stronger than the earlier-acting barrier ($I_2 > I_1$). The darker-shaded subset of this area (indicated with an arrow) is the region in which the barrier I_1 reduces gene flow *proportionately* more than barrier I_2, even though barrier I_2 is stronger in an absolute sense.

rier, I_3, were added to the life cycle, there would be an even wider range of parameters for which I_3 would have a weaker proportional effect on gene flow.

One implication of these calculations is that the isolating barriers currently most important in restricting gene flow are not necessarily those *historically* most important during speciation. Over time, most isolating barriers will approach completion. Considering the inequality above, one sees that as both I_2 and I_1 approach their limits of 1, the right-hand side will always approach zero and so the inequality will be satisfied. With time, then, earlier-acting barriers invariably appear proportionately more important than later-acting ones. Moreover, when postzygotic isolation is *complete* ($I_2 = 1$), prezygotic isolation is a more important impediment to gene flow when it is above only 0.5 (Fig 2.1).

In *Drosophila* and Lepidoptera, for example, we know that hybrid sterility evolves faster than hybrid inviability (Wu 1992; Presgraves 2002). Yet as divergence proceeds, both barriers will eventually become complete, and hybrids will be inviable. In such cases, hybrid inviability will seem to be of paramount importance in speciation—no hybrids will appear and thus it will be impossible to know if they are sterile. Similarly, polyploidy yields reproductive isolation via hybrid sterility and prezygotic ecological isolation within only a few generations (Chapter 9). These are not only the first isolating barriers to arise between polyploids and their diploid ancestors, but must be nearly *complete* barriers if the polyploid species is to persist. Nevertheless, other pre- and postzygotic barriers can evolve later and may currently act as more important impediments to gene flow. Polyploidy is nearly unique in that we have a good idea of the order in which reproductive barriers arose. This makes even more explicit the error of assuming that one can infer how speciation occurred by studying current barriers to gene flow. This crucial point is, however, usually ignored in practice owing to the impossibility of reconstructing the evolutionary sequence of reproductive isolation. We suggest some solutions to this dilemma later in the chapter.

The difficulty of reconstructing speciation in strongly isolated sympatric species suggests two approaches: (1) studying taxa in the process of splitting by sympatric speciation (and thus showing incipient reproductive isolation), or (2) examining allopatric taxa that have evolved some, but not complete, reproductive isolation. Yet even these alternatives have their problems. In neither situation do we know for certain whether speciation will be completed, and thus cannot tell whether the results can be extrapolated to "good" species.

Likewise, reproductive barriers between allopatric taxa must be inferred from observations in nature or tests in the laboratory. The problem here is that, with few exceptions, we are unable to determine whether characters that *seem* able to prevent gene flow in sympatry would actually do so. Would differences between allopatric populations in mating behavior, pollinator species, or habitat preference substantially reduce introgression if the taxa became sympatric?

Sometimes habitat isolation between allopatric taxa is strong: *Drosophila sechellia* (endemic to the Seychelles islands) oviposits, feeds, and breeds only on fruit of *Morinda citrifolia*, which is completely toxic to adults and larvae of its closest relatives (Legal et al. 1992; Legal and Plawecki 1995). Clearly such a barrier would act in sympatry. Nevertheless, differences between allopatric taxa in habitats or pollinator array might reflect only biogeographic history and not constitute isolating barriers in sympatry. Differences among allopatric populations may also reflect plastic responses to local environmental conditions and not genetically based isolating barriers. Thus, the completely asynchronous spawning cycles of the sea urchins *Diadema antillarum* and *D. mexicanum*, which live on opposite sides of the Isthmus of Panama, might rest not on genetic differences but on different ecological conditions on the two sides of the isthmus (Lessios 1984). Transplant experiments that could settle the issue are illegal.

However, if allopatric taxa show strong behavioral isolation when tested in the laboratory, as do many species of *Drosophila* (Coyne and Orr 1997), one can reasonably conclude that these factors would operate at least as strongly in sympatry. This is also true for mechanical isolation, postmating, prezygotic isolation, and intrinsic developmentally based hybrid sterility and inviability, few of whose effects are likely to be strongly altered in new environments. For example, laboratory studies of two species of crickets in the genus *Allonemobius* show strong conspecific sperm precedence: females doubly inseminated by both conspecific and heterospecific sperm produce few hybrid offspring (Gregory and Howard 1994). Direct observations in the field confirm the strength of this mechanism (Howard et al. 1998a). But *extrinsic* forms of postzygotic isolation, except for those involving abnormal sexual behavior of hybrids, are often dependent on particular environments, and thus difficult to study in geographically isolated populations.

The problems surrounding the study of individual pairs of species suggest a different strategy, namely the use of comparative methods to search for global patterns in the evolution of isolating barriers. We describe the results of these studies at the end of this chapter.

Identifying and Measuring Reproductive Isolation

There are several ways that an isolating barrier can be considered "important." First, it can be important in an *absolute* sense—that is, it presents a strong impediment to gene flow when it acts alone. Table 2.1 shows several barriers in *Mimulus* that are significant in this sense. Determining the absolute strength of a barrier sometimes depends on circumventing others, such as estimating hybrid sterility when sexual isolation is very strong. While knowing the strength of a barrier tells us neither its *relative* contribution to currently restricting gene flow nor its historical importance in speciation, a barrier cannot have played a major role in speciation unless it currently has a large absolute effect. Moreover, the strength of a barrier determines how useful that barrier is for genetic studies, because only strong (but not complete) barriers can be genetically dissected.

Second, a barrier can be considered important in its *relative* contribution to current reproductive isolation. To estimate this contribution, one must measure all possible forms of isolation and the sequence in which they act. When reproductive isolation is incomplete, the relative contributions of these barriers can be interpreted as the relative importance of barriers that prevent species from fusing via hybridization. These contributions, however, cannot tell us which barriers were critical in speciation.

Finally, a barrier can be important because it *caused* speciation; that is, it substantially reduced gene flow before reproductive isolation became complete. Identifying these barriers requires reconstructing the order in which different forms of reproductive isolation evolved, as well as knowing the relative strengths of each barrier up to the time when gene flow effectively ceased. This is the real question in understanding how speciation proceeds, and it is the most difficult to answer.

We will discuss each of these three approaches. In doing so, we clarify a debate about the importance of prezygotic versus postzygotic barriers.

Absolute strength of isolating barriers

Measuring the absolute strength of an isolating barrier is conceptually straightforward but often technically demanding (e.g., Chari and Wilson 2001). One estimates the amount of gene flow impeded by the barrier relative to the amount of gene flow occurring between conspecific individuals. Consider two marine invertebrates that spawn for the same length of time but at different periods. If the plots of the frequency of spawning versus date for the two species have a relative overlap of 0.3, then the strength of temporal isolation is 0.7, regardless of whether there are other isolating barriers. In *Drosophila* there have been many estimates of the strength of sexual and intrinsic postzygotic isolation (Coyne and Orr 1997).

Of all forms of reproductive isolation, those involving ecological differences are the most difficult to study and measure, and yet they must often be important in both plants and animals (Schemske 2000). Because they are the first barriers to act during the life cycle, they also have a disproportionate effect on gene flow. Habitat isolation and extrinsic postzygotic isolation are especially problematic. Consider a common situation: habitat isolation evolves by divergent adaptation of two allopatric populations to their environments. After speciation, the ranges of the two species expand and overlap, but the overlap is not complete. Moreover, in the area of sympatry, members of the two taxa still show microhabitat segregation within the "cruising range" of individuals or gametes. How do we estimate the strength of habitat isolation? Is it the proportion of the range of each taxon that is allopatric to the range of the other (as calculated by Ramsey et al. 2003), or is it the degree to which gene flow is restricted by habitat segregation in sympatry? In fact, both phenomena are potentially important, but the correct answer requires knowing the extent to which allopatry is the result of differential adaptation rather than historical circumstances. Transplant experiments can help settle this issue (Chapter 5).

Relative strength of isolating barriers

There are several reasons to expect that gene flow between closely related sympatric species must usually be restricted by several isolating barriers. First, many single barriers are "leaky," allowing some gene flow, yet the absence of hybrids implies that other barriers must operate. For example, four closely related species of the sea urchin *Echinometra* are sympatric off some Pacific islands, and they show some habitat divergence that leads to spatial restriction of gene flow. Yet habitat isolation is incomplete because members of different species live in close proximity on the reefs. There is also temporal isolation caused by differences in spawning time, but this is also incomplete. Hybrids, however, are extremely rare, almost certainly because strong gametic incompatibility complements the other barriers (Palumbi and Metz 1991; Metz et al. 1994). Leakiness of some barriers is also revealed by ongoing hybridization between species in well-studied taxa like birds and butterflies (Grant 1992; Presgraves 2002). The rarity of hybrids probably means that there is incomplete prezygotic isolation, but the distinctness of species suggests that there is also intrinsic or extrinsic postzygotic isolation that makes hybrids unfit. Direct observation often supports this conclusion. Schluter (1998), for example, describes congeneric species of stickleback fish and Galápagos finches that sometimes hybridize. The species remain distinct, however, because hybrids are ecologically unfit.

Multiple barriers are also likely simply because speciation is usually a long process (Chapter 12). During this time, many isolating barriers can arise as byproducts of selection, and it is reasonable to expect that several of these will be present when reproductive isolation is nearly complete. The comparative work of Coyne and Orr (1989a, 1997; see below) showed that sexual isolation and intrinsic postzygotic isolation evolve at roughly similar rates between allopatric species of *Drosophila*. Using a molecular clock, Coyne and Orr found that when populations have been geographically isolated for 1–2 million years, sexual and intrinsic postzygotic isolation alone have become strong enough to nearly stop gene flow in sympatry. Moreover, most plausible models of sympatric speciation, which in principle can occur quickly, involve at least two isolating barriers, typically assortative mating and extrinsic postzygotic isolation based on differential resource use (Chapter 4). Finally, and most important, careful studies of individual species pairs have shown that gene flow is blocked by the joint action of several isolating barriers. In fact, with the possible exception of coiling changes in snails (Chapters 3 and 11), we know of no case of speciation in which only one isolating barrier is involved.

Few studies have tried to identify and measure the diverse reproductive barriers that separate related species, yet there is really no other way to study individual speciation events. Recent examples in plants include studies of the Australian shrub *Conospermum* (Morrison et al. 1994), the annual *Linanthus* (Schemske and Goodwillie 1996), the monkeyflower *Mimulus* (Ramsey et al. 2003), and the fireweed *Chamerion angustifolium* and its autopolyploid derivative (Husband and Sabara 2004). All four cases show substantial habitat isolation, gametic isolation, and hybrid sterility between sister species. These and other barriers reduce gene flow between sympatric relatives to nearly zero.

The most extensive studies in animals have been on the genus *Drosophila*, which, although ideal for laboratory studies of behavioral, postmating prezygotic, and intrinsic postzygotic isolation, is not as well suited for studies of ecologically based isolation. A typical pair of allopatric taxa in the early stages of speciation is the pan-African *D. simulans* and its sister species *D. mauritiana*, endemic to the island of Mauritius. These species diverged about 250,000 years ago (Kliman et al. 2000). Strong behavioral isolation reduces gene flow by about 60% (Watanabe and Kawanishi 1979), shortened copulation in the other direction of the cross by 62% (Coyne 1993), and complete sterility of hybrid males by 50%. Other barriers include partial sterility of backcross females, strong conspecific sperm precedence, and other forms of postmating prezygotic isolation (Hollocher and Wu 1996; Price et al. 2000). If these mechanisms worked as strongly in sympatry as they do in the laboratory, reproductive isolation would be at least 0.98, even in the absence of ecological isolation. The two sympatric host races of *Rhagoletis pomonella* also show multiple forms of isolation including differences in female acceptance of fruits for oviposition, homing to the appropriate hosts, genetically based differences in pupal diapause, and duration and survivorship at extreme temperature (Feder 1998). However, in a very few cases (described below), there appear to be only two significant barriers to gene flow.

The "folk wisdom" of biologists holds that in some groups only a single isolating barrier is important. When examined closely, however, these generalizations fall apart. Pollinator isolation, for example, is often considered the sole—or at least the most important—isolating barrier in orchids. This view comes from observing that most of the estimated 25,000 species of orchids are pollinated by insects, and that there is often a strict one-to-one relationship between orchid and pollinator species (Paulus and Gack 1990). Careful study, however, has revealed other barriers to gene flow including differences in flowering time, habitat isolation, and extrinsic postzygotic isolation based on the failure of hybrids to attract pollinators or to transfer pollen because of aberrant reproductive morphology (Sanford 1968; Dressler 1981). There have been few attempts to measure the relative strength of these barriers.

The most careful study involved the sympatric, moth-pollinated species *Plantathera bifolia* and *P. chlorantha* from Scandinavia (Nilsson 1983). Here, several mechanisms prevent gene flow. *P. bifolia* is pollinated largely by sphingid moths and *P. chlorantha* by noctuid moths (this specificity rests largely on different floral odors), but the species do share some pollinators. There is also temporal isolation caused by differences in flowering time. The strongest barrier, however, is mechanical, acting both pre- and postzygotically. *P. bifolia* places its pollinia on the base of the moth's proboscis, and *P. chlorantha* on the moth's eyes. Cross-pollination occurs only rarely because stigmas are positioned to receive only conspecific pollen. Hybrid orchids produce viable pollen and ovules but suffer extrinsic postzygotic isolation for two reasons: (1) their intermediate morphology causes the pollinia to stick to the moth's hairy palps, where it often falls off before transfer to another orchid, and (2) the hybrids' intermediate fragrance is not attractive to insects. Although hybrids are rare, their extrinsic sterility—which must have appeared concurrently with the evolution of differences in fragrance and reproductive structures—is strong and probably important in preventing the formation of hybrid swarms. The situation in orchids is surely more complicated than the simple view that pollinator specificity is the preeminent isolating barrier.

The animal equivalent of orchids are ducks, often considered a paradigmatic case of speciation involving a single reproductive barrier: behavioral isolation. This conclusion rests on the strong sexual dimorphism for plumage color and ornamentation seen in many species. If this dimorphism reflects female preference, it seems obvious that the marked differences in plumage between males of closely related species would prevent hybridization. Yet the Anseriformes (ducks and geese) do hybridize in nature: 42% of all studied species (67/161) have produced natural hybrids, most of these among ducks (Grant and Grant 1992; T. Price, pers. comm). The observations that species remain distinct and do not form hybrid swarms suggest, however, that introgression is limited by *postzygotic* isolation. Based on the absence of intrinsic sterility and inviability in many duck hybrids (Price and Bouvier 2002), it is likely that such isolation is extrinsic, based on ecological intermediacy of hybrids or behavioral sterility caused by their intermediate morphology or sexual preferences. Such barriers may be as strong as behavioral isolation between pure species.

The multiple forms of isolation described in well-studied species suggest that we have underestimated the diversity of barriers preventing introgression between related taxa. Researchers rarely look for postmating, prezygotic barriers (pollen viability on foreign stigmas is an exception), rarely measure habitat isolation, and often neglect the possibility of environment-dependent postzygotic isolation, especially in animals.

Since gene flow between closely related species is often blocked by several strong isolating barriers, we need other methods to reconstruct the historical order in which those barriers arose. We describe these methods below.

Prezygotic versus postzygotic isolation

Prezygotic isolation has received considerable attention and is often cited as a more important, critical, or effective barrier to gene flow than postzygotic (or postmating prezygotic) isolation. Jiggins et al. (2001a, p. 1636), note, "As in hybrid zones, premating isolation is likely to be more effective than hybrid incompatibility in maintaining species differences despite gene flow." This sentiment is echoed by Kirkpatrick and Ravigne (2002, p. S23): "Prezygotic isolation is by far the most important factor keeping populations separate."

Such claims may well be true. Given that isolating barriers act sequentially, prezygotic barriers will almost surely be the strongest current impediments to gene exchange. But this sidesteps the issue of whether such isolation was important during speciation. Current barriers are sometimes seen, erroneously, as answering this question. Discussing hybrid sterility and inviability in fish, for example, Russell (2003, p. 325), argues that "... because sexual isolation would act first to isolate taxa, the general relevance of post-zygotic isolation to speciation remains somewhat controversial." Mallet et al. (1998, Figure 30.5) present a flowchart showing the course of speciation in *Heliconius* butterflies. In the sequence of evolutionary change, "speciation" occurs after the evolution of "ecological divergence" and "mate choice," but well before the appearance of "hybrid unfitness." However, three of the four interspecific crosses described in that paper show postzygotic isolation, including the sister species *H. melpomene* and *H. cydno*, which hybridize in nature (Naisbit et al. 2001). Moreover, intrinsic hybrid sterility has been documented between geographic *populations* of the subspecies *Heliconius melpomene melpomene* (Jiggins et al. 2001a).

Thus, although prezygotic barriers may well have been more important than postzygotic barriers in speciation, we lack decisive evidence for this view. The fact that postzygotic barriers act late in the life cycle does not mean that they were insignificant during speciation. Indeed, there are several reasons to think that they may be important:

1. There are ascertainment biases in favor of detecting premating isolation. If prezygotic isolation is essentially complete, then hybrids can never be formed and we will know nothing about the strength of postzygotic isolation. In some plants, hybrid embryos are viable if dissected from the seed and reared artificially, but the seeds will not sprout because interaction with the endosperm makes the embryo inviable (Stebbins 1950; Arisumi 1985). Moreover, if two species produce intrinsically sterile or inviable hybrids, but lack prezygotic isolation, the rarer species will go extinct (Liou and Price 1994). This bias makes us unable to observe secondarily sympatric species that originally had high postzygotic but low prezygotic isolation.

   A similar bias holds for extrinsic postzygotic isolation (Chapter 7), which, though potentially important in speciation, may eventually become *undetectable*. Imagine, for example, two allopatric populations of birds that have evolved different male colors and corresponding female preferences. Upon secondary contact, some hybrids may form, but male hybrids with intermediate coloration may find it hard to secure mates. Such behavioral sterility can promote the evolution of reinforcement that drives sexual isolation to completion (Chapter 10). In such cases, hybrids will no longer appear and one would not know that behavioral sterility of these hybrids was critical in driving speciation. Reinforcement always replaces postzygotic barriers with earlier-acting barriers.

2. Postzygotic isolation via hybrid sterility is a critically important factor in restricting gene flow between ancestral species and their newly formed polyploid or diploid hybrid descendants. (As we show in Chapter 9, this hybrid sterility must be accompanied by ecologically based prezygotic barriers, which allow the new polyploid to coexist with its ancestors and protect it from extinction caused by mating with them). Polyploidy has been frequent

in the history of plants; diploid hybrid speciation less so. Moreover, the observation that the sterility of hybrid diploid or triploid intermediates disappears when they form tetraploids shows that the parental species are isolated by chromosomal differences that cause hybrid sterility (Chapter 7).

3. *Extrinsic* hybrid sterility and hybrid inviability (the latter can be considered "ecological isolation") are varieties of postzygotic isolation. Although these isolating barriers are only now beginning to receive attention, we suspect that they are common.

4. Extrinsic postzygotic isolation may involve the same ecological differences or the same differences in sexual behavior that cause prezygotic isolation. Hybrids may have difficulty securing mates because they show traits and preferences intermediate to those of parental species, or may be relatively unfit because they fall between the niches of two well-adapted parental species. As sexual or disruptive selection causes taxa to diverge, hybrids will become unfit as a result of this divergence. Thus, many forms of prezygotic and postzygotic isolation must evolve concurrently.

5. The persistence of distinct taxa that form hybrids or hybrid zones implies that the hybrids suffer a disadvantage. In a review of 20 well-studied hybrid zones, Barton and Hewitt (1985) found evidence for low hybrid fitness in 11, and, in most of the remaining 9 zones, either no evidence was gathered or the data were scanty. Presgraves (2002) shows that 19% of all Lepidopteran species are known to hybridize in nature. Among these, at least 67% show intrinsic sterility or inviability of offspring. Surely many of these hybridizations also yield extrinsic postzygotic isolation. Hybrids between mimetic species of *Heliconius* butterflies, for example, are intermediate in color and behavior, an intermediacy that makes them less able to secure mates and presumably more vulnerable to predators (Jiggins et al. 2001b; Naisbit et al. 2001).

   As noted above, there are many pairs of bird species that hybridize but still remain distinct despite the absence of measurable intrinsic postzygotic isolation in F_1 hybrids (Price and Bouvier 2002). Hybrid unfitness might be extrinsic or appear only in generations beyond the F_1.

6. For reinforcement to increase sexual isolation between hybridizing taxa, there must be some postzygotic isolation that makes hybridization disadvantageous (see Chapter 10).

7. Unlike many forms of premating isolation and extrinsic postzygotic isolation, *intrinsic* postzygotic isolation is likely to be permanent. H. J. Muller (1939, 1942) particularly stressed this point:

   > As for the genetic [forms of prezygotic isolation] . . . the results of contrary selection as well as of the gradual infiltration of genes from the other group must usually bring about a phenotypic reversal which, although not actually reconstituting the original genotype, will undo the bars to crossing in so far as these merely involve differences to ecological circumstances. The above mechanism, or set of mechanisms,

will therefore be insufficient, in itself, to give rise to separated species capable of permanently inhabiting the same area without fusing ... We may at this point turn to a consideration of [intrinsic postzygotic isolation], the incapacitation of hybrids, for among the principles there involved are to be found those which operate to effect permanent genetic isolation in general (Muller 1942, p. 83).

As Muller explained, it is fairly easy to reverse an isolating barrier based on phenotypic divergence, as such a reversal can involve genes different from those producing the original barrier. So, for example, assortative mating between sympatric taxa of threespine sticklebacks (*Gasterosteus aculeatus*) is based on differences in body size, which in turn results from different feeding regimes (Schluter and Nagel 1995). This form of reproductive isolation would quickly vanish were the lake environment to change so that only one feeding niche was available. Other forms of premating isolation are equally susceptible to such reversal.

In contrast, it is much more difficult to undo hybrid sterility and inviability based on genetic incompatibilities between alleles in different taxa (Chapter 8). To effect this reversal, selection would have to act on *exactly the same genes* that diverged initially, restoring the original alleles. (One exception involves "hybrid rescue mutations," discussed in Chapter 8). Moreover, once *multiple* genetic incompatibilities have evolved, the probability of reversing intrinsic postzygotic isolation is close to zero.

If reproductive isolation between taxa is essentially complete, they are biological species regardless of what happens in the future. But as Mayr (1963) and Futuyma (1987) have emphasized, most populations that evolve some isolation will not become biological species: they will either become extinct, fuse with other populations because of changes in the environment or in the amount of migration, or maintain an intermediate level of reproductive isolation through a balance between selection and gene flow. Futuyma (1987) argues that by securing divergence between gene pools, reproductive isolation acts as a piton that allows species to scale different adaptive peaks. If this is true, then intrinsic postzygotic isolation is the strongest piton.

8. In two studies involving *Drosophila* and the sunflower *Helianthus*, molecular genetic analysis has shown that the parts of the genome that have not introgressed through natural hybrid zones are those that cause intrinsic hybrid sterility (Rieseberg et al. 1999; Machado et al. 2002). These genes are also associated with chromosome rearrangements, an association that would not exist had hybrid sterility evolved after, rather than during, speciation (Rieseberg 2001; Noor et al. 2001b; see below). Moreover, sex chromosomes, known to play a large role in hybrid sterility and inviability sometimes move through hybrid zones much less readily than do autosomes (Chapter 8). This observation is *prima facie* evidence that postzygotic isolation operated in the past—and continues to operate—to prevent introgression.

9. Finally, there is an asymmetry between how pre- and postzygotic isolation behave in hybrid generations beyond the F_1. Prezygotic and extrinsic postzy-

gotic isolation become weaker in F_2 and backcross generations because many hybrids become more like pure species. However, if genes causing intrinsic postzygotic isolation are partially recessive in hybrids, as they seem to be (Chapter 8), then F_2 and backcross hybrids will possess genetic incompatibilities that *increase* their sterility and inviability. In later generations of hybridization, then, postzygotic isolation might form an increasingly stronger barrier to gene flow.

Which isolating barriers caused speciation?

How can we tell which isolating barriers actually caused speciation instead of having evolved after speciation was complete? An obvious suggestion is to assay geographic populations of single species, determining which barriers exist in incipient form. This is the strategy that Mayr (1942, 1963) used to support the importance of allopatric speciation; but this strategy is problematic. Most such populations do not form species, and even if a single isolating barrier (such as behavioral isolation) is the first to arise, we do not know whether that barrier would be important in those populations that later become good species. However, comparative analysis of taxa at different stages of evolutionary divergence, ranging from populations through full species, may show which barriers persist throughout this transition.

There are several ways to determine those barriers important in speciation:

1. Find sympatric species having only one or two isolating barriers, or find allopatric taxa in which isolation in sympatry would almost certainly result from only one or two barriers.
2. Determine what sorts of ecological disturbance would cause two sympatric species to lose genetic isolation and form a hybrid swarm.
3. Observe those isolating barriers that are based on genes associated with chromosome rearrangements.
4. Conduct comparative studies of many species that allow one to estimate the relative rates at which isolating barriers arise.
5. Conduct comparative studies among higher-level taxa that tell us what organismal traits are associated with increased rates of speciation.

We examine these approaches in turn.

SPECIES WITH FEW ISOLATING BARRIERS. The preeminent case of speciation resulting from a few isolating barriers is polyploidy. In all cases of polyploidy, postzygotic isolation, in the form of chromosomally based hybrid sterility, is a major reproductive barrier. However, a single polyploid individual is not a new species: biological species comprise *populations*. Since a new population of polyploids is necessarily sympatric or parapatric with its ancestors, it can persist only if it has acquired sufficient ecological difference to coexist with and reduce gene flow from its diploid ancestors (Ramsey and Schemske 2002). As we note in Chapter 9, some of these differences may be an automatic byproduct of genome duplication. The relative roles of ecological and postzygotic barriers in the origin of polyploid species is an unanswered question.

There are several animal species in which gene flow is apparently prevented by only a few barriers. The sympatric ladybird beetles *Epilachna niponica* and *E. yasutomii*, for instance, appear isolated by only two barriers: preference for, and ability to survive on, different hosts (Katakura et al. 1989; Katakura and Hosogai 1994). Laboratory experiments show no behavioral isolation, conspecific sperm precedence, or intrinsic hybrid sterility and inviability. Unfortunately, these conclusions do not hold for other closely related congeneric species, which also show behavioral isolation, conspecific sperm precedence, and intrinsic hybrid inviability (Katakura 1997).

In four sea urchin species in the genus *Echinometra*, speciation is likely to have occurred largely through gametic incompatibility. Despite weak ecological isolation, members of different species often live close together on Pacific reefs. In Hawaii, for example, 30–60% of individuals have a heterospecific individual living within one meter, and spawning apparently occurs synchronously. Yet hybrids constitute only 0.05% of individuals in the field, almost certainly because the strength of gametic isolation—the failure of heterospecific fertilization—is extremely strong. There is some evidence, however, that hybrids also suffer slight inviability and sterility (Palumbi and Metz 1991; S. Palumbi, pers. comm.). Likewise, Howard et al. (1998a) found gametic isolation to be the only obvious isolating barrier between the sympatric crickets *Allonemobius fasciatus* and *A. socius*. Closely related sympatric species of African lake cichlids often appear to be behaviorally isolated. This conclusion comes from observing not only strong mating discrimination (Dominey 1984; Seehausen et al. 1998), but also high viability and fertility of artificially generated hybrids. Moreover, there seems to be little ecological difference between some closely related species (Genner et al. 1999a,b; Seehausen and van Alphen 1999). While ecological studies of these cichlids have been limited, the strong role of behavioral isolation is supported by studies in which species fusion has apparently followed habitat disturbance (see below).

Although such studies can tell us which reproductive barriers were important in individual speciation events, they cannot necessarily be extrapolated to other species, even within the same group. As we noted above, studies of *Epilachna* show that congeneric species can differ profoundly in the nature of their reproductive barriers. To determine whether single isolating barriers are *consistent* causes of speciation in a group, one must do comparative studies.

HYBRIDIZATION FOLLOWING ECOLOGICAL DISTURBANCE. Another way to judge the importance of particular isolating barriers in speciation is to determine whether and how the integrity of species breaks down when habitats are disturbed. If such a disturbance destroys an isolating barrier, causing two previously isolated species to fuse into one (or into a hybrid swarm), one can infer that that barrier was important in speciation. This follows because all other barriers acting together cannot prevent fusion.

Here we must add two caveats. First, one must know exactly which isolating barriers are affected by ecological disturbance. In plants, environmental changes may alter several forms of isolation—habitat, pollinator, or temporal. Extensive hybridization between two species of the Australian flower *Banksia*, for example, occurs only in disturbed habitats. This is due to the loss not of habitat isolation, but of temporal isolation: the species flower asynchronously in undisturbed habitats, but synchronously in disturbed areas (Lamont et al. 2003).

Second, a failure to observe fusion despite the removal of an isolating barrier does not mean that this barrier was unimportant in speciation. For example, the environment may change so that two plant species that previously flowered at different times now flower synchronously. If the species still maintain their integrity, this does not imply that temporal isolation was unimportant in speciation. It tells us only that other barriers that could have evolved later are now sufficient to prevent hybridization.

The haplochromine cichlids of Africa provide a good example of how to infer important isolating barriers from habitat disturbance. In Lake Victoria, Seehausen et al. (1997) found a positive correlation between the number of cichlid species (or color morphs within species) and water clarity: areas that had recently become turbid supported fewer species and color morphs. The authors conclude that the reduced diversity reflects the breakdown of behavioral isolation— isolation based on male breeding colors visible only in clear water. Laboratory experiments showed that species are more likely to hybridize when color differences are less visible. Although one cannot rule out the possibility of environmentally caused extinction, the loss of species may well reflect hybridization, implying that behavioral isolation was important in speciation.

Hybrid swarms of plants often form after habitat disturbance (Stebbins 1950). The most famous involves *Iris fulva* and *I. hexagona*, which occur without hybridizing in undisturbed habitats along the Mississippi delta. In such areas the species show habitat isolation: *I. hexagona* lives in wet soil in sunlit areas of marshes, while *I. fulva* is restricted to drier and more shaded understory along riverbanks. Massive hybrid swarms occur where human agriculture has disturbed these areas, suggesting that habitat isolation was important in speciation (Riley 1938; Anderson 1949; Arnold and Bennet 1993; and Chapter 9).

ISOLATING BARRIERS ASSOCIATED WITH CHROMOSOME REARRANGEMENTS. Noor et al. (2001b) and Rieseberg (2001) suggest that if genes causing reproductive isolation between two allopatric taxa reside within chromosome regions that are rearranged between those taxa (e.g., inversions), the association between genes and chromosome rearrangements will persist when the species come into secondary contact. This is because heterozygosity for such rearrangements prevents recombination between genes within them. The reduced recombination will in turn hinder the elimination of these genes by natural selection against hybrids, delaying the fusion of species (see Chapter 8 for details). This delay might, in turn, allow other isolating barriers to accumulate by reinforcement, completing speciation. In other words, chromosome arrangements promote speciation by allowing reinforcement to evolve before species can fuse. This theory does not require that genes for reproductive isolation *accumulate* in rearranged regions of chromosomes, only that genes so positioned will tend to survive in sympatry. The theory thus predicts that the genes for reproductive isolation between sympatric species will tend to be found in rearranged regions of the genome.

Observed associations between chromosomal rearrangements and genes for reproductive isolation thus suggest two conclusions. First, speciation was not completely allopatric, but was completed in the presence of gene flow. Second, isolating barriers associated with chromosomal rearrangements were probably important in speciation, since these barriers must have evolved before reproductive isolation was complete.

Comparative Studies of Isolating Barriers

Students of speciation want more than just a short list of species in which one or two isolating barriers predominate. Are some barriers consistently more important than others in certain groups? In animals, for example, does behavioral isolation evolve faster than hybrid sterility? In animal-pollinated plants, does pollinator isolation evolve before temporal isolation? Does postmating, prezygotic isolation evolve faster than behavioral isolation, even though both are probably driven by sexual selection?

How fast does reproductive isolation appear?

This question is best answered through comparative analysis. Such analysis requires a group of taxa for which one has two kinds of information: the absolute strength of various isolating barriers, and the time of divergence between taxa. In the absence of a fossil record, relative divergence time is best estimated by molecular clocks, usually based on divergence of DNA or protein sequence (Graur and Li 2000). With such information, one can reconstruct the time course of speciation, plotting the strength of isolating barriers between pairs of taxa against their divergence time. Such plots, made separately for each isolating barrier, might show regularities: one form of isolation, for example, might consistently evolve faster than others. If the molecular clock can be calibrated from fossil or biogeographic data, one can also estimate the absolute rate of evolutionary changes contributing to speciation.

The first such study compared very diverse taxa. Wilson et al. (1974) measured immunological distances between albumins (an index of protein-sequence divergence) in 31 pairs of mammalian species and 50 pairs of frog species, with members of each pair capable of producing viable hybrids. The authors found a large disparity between these taxa: mammals capable of producing hybrids showed an immunological distance equilavent to 0.6% divergence of albumin sequence, while pairs of frog species could still produce hybrids after much larger divergence (an average of 7.4%). Because the albumin clock appears to tick at the same rate in both groups (Wilson 1975), these results imply that hybrid inviability evolves faster in mammals than in frogs. This work was later extended by Prager and Wilson (1975) to 36 pairs of hybridizable birds. They found an albumin divergence similar to that seen among frogs. Converting molecular divergence to absolute time, the average divergence time between mammals, frogs, and birds still capable of producing viable hybrids is 2–3 million years, 21 million years, and 22 million years, respectively.

While it is reasonable to conclude that hybrid inviability evolves faster in mammals than in frogs or birds, these studies suffer from some problems. First, the mammalian data were based largely on primates (19 out of 31 mammalian pairs). Second, many of the species pairs were not phylogenetically independent; that is, most species were involved in more than one hybridization. More important, a faster loss of the ability to produce viable hybrids does not necessarily mean a faster evolution of hybrid inviability. The failure of some mammals to produce hybrids, for example, could reflect a faster evolution not of developmental incompatibilities, but of behavioral or of postmating, prezygotic isolation. (Taxa that failed to produce hybrids were not examined for insemination.) Nevertheless, Wilson et al. are probably correct in their main conclusion: although some distantly related species of birds can produce viable hybrids despite more than 15 million years of divergence (Price and Bouvier 2002), it is absurd to suppose that equally old mammalian species (e.g., humans vs. gibbons), could yield the same result. Moreover, the faster rate of postzygotic isolation in frogs than in birds is supported by comparative data discussed below.

The most complete data on relative rates of reproductive isolation come from studies of single groups. Coyne and Orr's (1989a, 1997) analysis of *Drosophila* involved 171 pairs of species for which there was information about at least one of three forms of reproductive isolation (behavioral isolation, hybrid sterility, and hybrid inviability), as well as allozyme-based estimates of genetic divergence. Molecular divergence was calculated as Nei's (1972) genetic distance, D. Nei's D appears to increase roughly linearly with time, so it can serve as a molecular clock, and ranges from 0 (all allele frequencies are identical between compared taxa) to ∞ (taxa have no alleles in common). All pairs of crossable species had D values between 0 and 2. Using data from Hawaiian *Drosophila*, Carson (1976) estimated that a D of 1 corresponds to roughly 2 million years of divergence.

Behavioral and postzygotic isolation (intrinsic hybrid sterility and inviability) were estimated from published laboratory experiments. Each form of isolation was scaled from 0 (no isolation) to 1 (complete isolation). Because single species were often used in multiple comparisons of reproductive isolation, the data were corrected to produce a set of phylogenetically independent pairs of taxa. Figure 2.2 shows the relationship between genetic distance and the strength

Figure 2.2 Time course for the evolution of prezygotic and postzygotic isolation among species of *Drosophila*. (A) Prezygotic (behavioral) isolation, and (B) Postzygotic isolation (hybrid sterility and inviability). Each point represents a phylogenetically independent pair of taxa. (From Coyne and Orr 1997.)

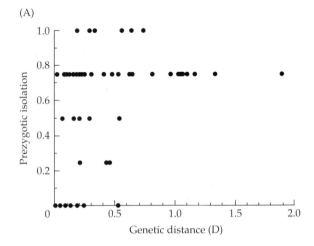

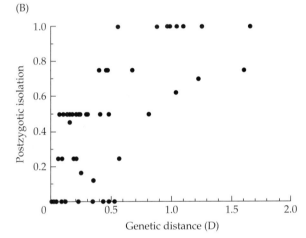

of reproductive isolation in these corrected data. "Postzygotic isolation" is calculated as the total restriction of gene flow due to the combination of hybrid sterility and inviability.

As expected, both behavioral and postzygotic isolation between species increase with time, although there is considerable scatter. Comparing only pairs of taxa that diverged recently ($D < 0.5$), one finds behavioral isolation to be significantly higher than hybrid sterility and inviability. This might lead one to conclude that behavioral isolation evolves faster than postzygotic isolation in *Drosophila*. This disparity, however, is attributable entirely to biogeography. Separating the taxa in Figure 2.2A into those showing some range overlap (sympatric) versus those with no overlap (allopatric), one finds that sympatric taxa develop behavioral isolation significantly earlier than do allopatric taxa of the same age. This difference may imply reinforcement in sympatry (see Chapter 10). Among allopatric taxa, there is no difference in the evolutionary rates of behavioral and intrinsic postzygotic isolation.

Comparing hybrid inviability and sterility, Coyne and Orr (1989a) originally concluded that these barriers arose at similar rates. However, this conclusion seems doubtful as the test for rate differences is insensitive (Wu 1992). In fact, other genetic evidence from *Drosophila* suggests that hybrid sterility generally evolves faster than hybrid inviability (Chapter 8). One conclusion that does seem reliable is that intrinsic postzygotic isolation evolves faster in males than in females, showing that Haldane's rule in *Drosophila* reflects the early evolution of male-specific incompatibilities (Chapter 8).

Finally, we can combine behavioral and intrinsic postzygotic isolation to estimate the evolutionary rate of "total" reproductive isolation: the reduction in interspecific gene flow caused by the joint action of both barriers (Figure 2.3). Total isolation increases rapidly, so that most species separated by a genetic distance of 0.5 or more have levels of pre- and postzygotic isolation sufficient to reduce gene flow to nearly zero in sympatry.

One can also use these data to determine the time at which taxa become good species. We assume that the degree of reproductive isolation between sympatric *Drosophila* taxa is sufficient to render them good species. Among such taxa, the mean level of total isolation is 0.94, and the lower bound of the 95% confidence interval around this mean is 0.90. Thus, total reproductive isolation of 0.90 or more can probably prevent fusion of sympatric taxa. Using this value and Carson's (1976) calibration of the *Drosophila* molecular clock, one can calculate that total isolation of 0.90 is attained at a divergence time of about 1.1 million years for allopatric taxa, but only 0.1 million years for sympatric taxa. This big difference in the rate of speciation between sympatric and allopatric groups is due to the much higher level of behavioral isolation seen in taxa with overlapping ranges.

These time estimates are of course crude and should not be taken too seriously given the variability of the data and possible errors in calibrating the molecular clock. Moreover, measures of "total" reproductive isolation certainly underestimate the true values in nature. Behavioral isolation is only one of several forms of prezygotic isolation (we know little about the ecology of most

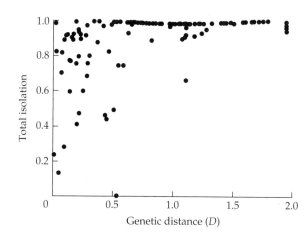

Figure 2.3 Time course for the evolution of "total" reproductive isolation (behavioral + postzygotic isolation) between pairs of *Drosophila* species. (From Coyne and Orr 1997.)

Drosophila species), there is little information about postmating, prezygotic isolation (which is known to occur in some hybridizations), and we lack data on extrinsic postzygotic isolation. However, while comparison of rates of different forms of reproductive isolation must be regarded as tentative, some conclusions—such as the faster evolution of male than female sterility and the rapid evolution of behavioral isolation in sympatric taxa—seem secure.

The only comparable study in animals is Mendelson's (2003) analysis of sexual isolation and hybrid inviability in freshwater darters of the genus *Etheostoma*. Using methods similar to those of Coyne and Orr (1989a), Mendelson measured sexual isolation (spawning success) in pairings between nine allopatric and phylogenetically independent pairs of darters, and one component of hybrid inviability (egg hatchability) in eight independent pairs of allopatric species. As Figure 2.4 shows, while hybrid inviability was appreciable, at a given

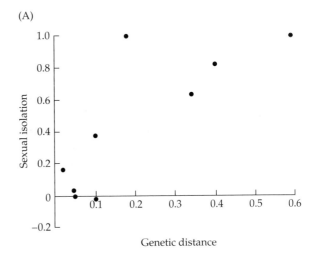

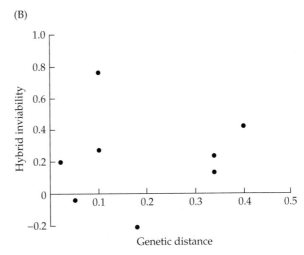

Figure 2.4 Time course for the evolution of prezygotic (behavioral) isolation (A), and postzygotic isolation (egg hatchability) (B) for pairs of darter species in the genus *Etheostoma*. (From Mendelson 2003.)

genetic distance it is almost always smaller than sexual isolation: in fact, there is no correlation between hatchability and genetic distance. Thus, sexual isolation appears to evolve more rapidly than hybrid inviability in allopatric populations. While this seems to differ from the conclusion of Coyne and Orr (1989a), the studies are not comparable: Coyne and Orr measured postzygotic isolation as the *combination* of hybrid sterility and hybrid inviability, while Mendelson measured inviability alone. In other teleost fish, as well as in *Drosophila* and Lepidoptera, sterility evolves faster than inviability (Wu 1992; Presgraves 2002; Russell 2003), and some crosses between darters do show hybrid sterility (Mendelson 2003). Thus, we cannot yet conclude that sexual isolation evolves faster than intrinsic postzygotic isolation in darters. However, these fish do resemble other taxa in conforming to Haldane's rule: in crosses showing hybrid sterility, the affected sex is male. This may be a byproduct of strong sexual selection (Chapter 8), for darters show extreme sexual dimorphism in the breeding season.

There have been several similar studies of animal taxa, but all are limited to postzygotic isolation (Sasa et al. 1998; Presgraves 2002; Price and Bouvier 2002; Russell 2003). Figure 2.5 shows the relationship in four groups (*Drosophila*, frogs, birds, and Lepidoptera) between Nei's genetic distance (or, in birds, sequence divergence of cytochrome *b*) and total postzygotic isolation.

Sasa et al. (1998) examined the relationship between genetic distance and hybrid sterility or inviability in 116 hybridizations among 46 species of frogs. Inviability was measured by the success of in vitro fertilization, and sterility by the condition of adult gonads. The results (Figure 2.5B) show a time-dependent increase in total postzygotic isolation strikingly similar to that seen by Coyne and Orr (1989a, 1997) (Figure 2.5A). In both groups, nearly complete postzygotic isolation does not appear until D is between 0.3 to 0.5, while intermediate levels of postzygotic isolation—usually representing only one sex that is sterile or inviable in hybrids—appear at much lower genetic distances. These latter cases may represent, as in *Drosophila*, the rapid evolution of postzygotic isolation in heterogametic hybrids. Depending on the species of frog, either males or females can be heterogametic, but the few cases of sex-limited postzygotic isolation obey Haldane's rule.

While the relationship between allozyme divergence and postzygotic isolation appears similar in frogs and *Drosophila*, Sasa et al.'s (1998) conclusion that the *evolutionary rates* of postzygotic isolation are similar in these taxa may be premature, as the molecular clock may tick at different rates in the two groups. While $D = 1$ corresponds to roughly 2–5 million years of divergence in *Drosophila*, Maxson and Maxson (1979) estimate that in salamanders it represents about 14 million years of divergence. Sasa et al. (1998) also note that in frogs, as in *Drosophila*, hybrid sterility seems to evolve faster than hybrid inviability. They present no data for prezygotic isolation, but some work on frogs suggests that temporal, ecological, and habitat isolation may precede the evolution of postzygotic isolation (see references in Hillis 1981).

Presgraves (2002) compiled data from 212 hybridizations among 182 species of Lepidoptera, correlating postzygotic isolation with genetic distance estimated from allozyme data or DNA sequences (Figure 2.5C). Compared to frogs

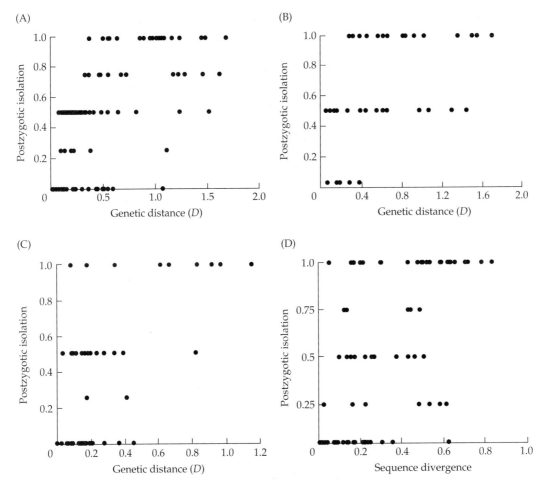

Figure 2.5 Time course for the evolution of total postzygotic isolation in four taxa; data includes all species pairs. Graphs should not be compared directly because genetic distance probably accumulates at different rates in different groups. All studies except that in birds use Nei's (1972) genetic distance as an index of divergence time. The index of molecular divergence in birds is the difference in DNA sequence of mitochondrial cytochrome oxidase. (A) *Drosophila*, (B) Frogs in seven genera, (C) Lepidoptera, (D) Birds. (A from Coyne and Orr 1997; B from Sasa et al. 1998; C from Presgraves 2002; D from Price and Bouvier 2002.)

and *Drosophila*, postzygotic isolation arises at lower genetic distances. (There is no calibrated molecular clock in Lepidoptera.) Hybrid sterility evolves faster than hybrid inviability. As in *Drosophila*, Haldane's rule is nearly ubiquitous. Of 114 crosses in which only one sex among hybrids is sterile or inviable, that sex is female (heterogametic) in 110. Haldane's rule also appears early in speciation and, as in *Drosophila*, appears to be a nearly ubiquitous phase in the progression to complete postzygotic isolation.

Price and Bouvier (2002) reviewed 254 bird hybridizations, 108 of which could be assigned relative divergence times using sequences of mitochondrial cytochrome *b*. Figure 2.5D shows the relationship between sequence divergence and postzygotic isolation. A regression line forced through the point of zero postzygotic isolation at zero divergence time suggests that complete postzygotic isolation is attained at roughly 21% sequence divergence. Using the authors' calibration of 2% sequence divergence per million years, the average time to complete postzygotic isolation is roughly 10.5 million years. This is slower than in *Drosophila*, where complete postzygotic isolation appears between 0.2 and 2.7 million years. Bird species capable of producing viable hybrids can be quite distantly related. In the most extreme case, viable (though aberrant) hybrids are obtained from the cross between African helmeted guineafowl (*Numida meleagris*) and domestic chicken (*Gallus gallus*)—species that diverged at least 55 million years ago. This is equivalent to obtaining a viable hybrid between humans and lemurs.

Hybrid sterility in birds appears to evolve faster than hybrid inviability, and Haldane's rule is pervasive: in 98% of hybridizations yielding only one sterile or inviable sex, that sex is female (heterogametic). Combining the crossing data with observations from nature, Price and Bouvier (2002) conclude that many species able to coexist in sympatry show little or no intrinsic postzygotic isolation. Since hybridization between sympatric bird species is not rare, Price and Bouvier suggest that species integrity is maintained by other forms of postzygotic isolation, including ecological or behavioral intermediacy of hybrids, or intrinsic sterility and inviability that appear in generations beyond the F_1. A recent study of hybrid inviability and genetic divergence in pigeons and doves (Lijtmaer et. al. 2003) confirms the pervasiveness of Haldane's rule, its ubiquity as an early stage of postzygotic isolation, and the slower evolution of postzygotic isolation in birds than in *Drosophila*.

Finally, in a study of 37 fish hybridizations, Russell (2003) found that both intrinsic hybrid sterility and inviability increased gradually with time, but that hybrid sterility appeared significantly earlier than hybrid inviability.

In the sole existing study of plants, Moyle et al. (2004) compared the evolution of prezygotic with postzygotic isolation in three genera of angiosperms: *Glycine*, *Silene*, and *Streptanthus*. Their measure of "post-pollination" prezygotic isolation in two genera (no data were available for *Streptanthus*) involved the relative proportion of unsuccessful interspecific pollinations compared to intraspecific pollinations. (This measure, however, could also include early inviability or abortion of zygotes before seed or fruit set—forms of *postzygotic* isolation.) Moyle et al. (2004) estimated postzygotic isolation as the proportion of sterile pollen in interspecific F_1 hybrids. Thus, measurements of sterility were limited to the male function. Like Mendelson (2003), Moyle et al. (2004) measured only one form of postzygotic isolation in each genus and thus could not compare the evolutionary rates of sterility and inviability.

This study yields four conclusions. First, both pre- and postzygotic isolation increase with divergence time between plant species. However, the genetic distances associated with "total" reproductive isolation (as defined by Coyne and Orr 1989a) were much lower in plants than in animals. Such isolation was attained in plants at genetic distances between 0.02 and 0.1, compared to about 0.5 in *Drosophila*. This difference between *Drosophila* and plants, however, cannot be interpreted as a difference in rate of speciation, as there is no calibration for the molecular clock in angiosperms. Second, in *Glycine* postzygotic isolation evolved more rapidly than prezygotic isolation; however, these rates were not obviously different in *Silene*. Third, unlike *Drosophila*, whether species were sympatric or allopatric had no effect on the relative rates of pre- and postzygotic isolation. Moyle et al. (2004) conclude that reinforcement does not occur in these plants. However, as the authors note, this conclusion applies only to reinforcement of gametic or early-acting postzygotic isolation, not to other barriers (such as pollinator or temporal isolation) that can evolve via reinforcement. Finally, Moyle et al. suggest that their results may have been complicated by polyploidy. Indeed, autopolyploids can appear instantaneously, producing reproductively isolated species having *no* genetic divergence from their ancestors.

Given these difficulties with the measurements of reproductive isolation, and the low number of phylogenetically independent species pairs, Moyle et al.'s conclusions need further confirmation. As we suggest below, however, plants may be the best material for this sort of comparative work.

These studies are merely a beginning. Only in *Drosophila*, plants, and darters, for example, are there comparisons between the evolutionary rates of pre- and postzygotic isolation. In every group, potentially important isolating barriers were not measured. Nevertheless, the results suggest a few conclusions:

1. Intrinsic hybrid sterility and inviability evolve gradually: the strength of postzygotic isolation rises slowly with time, and there are almost no recently diverged taxa whose hybrids are completely sterile or inviable. This rules out forms of "instantaneous speciation" (such as infection with *Wolbachia*; Chapter 7) as general causes of postzygotic isolation. In polyploids, of course, postzygotic isolation appears almost instantly.

2. Hybrid sterility evolves faster than hybrid inviability. We suspect that this will be a general pattern in animals, for sterility might often be a byproduct of genes that evolve rapidly via sexual selection. This hypothesis is supported by observations that genes related to sexual reproduction evolve more rapidly than genes having other functions (Chapter 6).

3. Haldane's rule for intrinsic postzygotic isolation is obeyed almost universally, regardless of whether the heterogametic sex is male or female. The early appearance of heterogametic sterility and inviability, as seen in birds and *Drosophila*, suggests that Haldane's rule is a nearly mandatory step in the evolution of intrinsic postzygotic isolation (Chapter 8).

4. Intrinsic postzygotic isolation evolves at different rates in different groups (see also Edmands 2002). This is not so much a pattern as the absence of one. Lijtmaer et al. (2003), for example, calculate that hybrid inviability evolves

fastest in frogs and *Drosophila*, somewhat more slowly in Lepidoptera, and slowest in birds. Inviability may evolve even more slowly in plants given the high fertility of hybrids between long-diverged species (Chapter 12). These disparities are no surprise given the biological differences between groups and the disparate evolutionary forces acting on them. Pending more accurate calibration of molecular clocks, however, conclusions about absolute rates of postzygotic isolation must be considered tentative.

Unfortunately, the lack of information about sexual isolation in groups outside *Drosophila* prevents us from testing the generality of Coyne and Orr's (1989a, 1997) conclusion that behavioral and intrinsic postzygotic isolation evolve at roughly the same rate in allopatric populations, but that behavioral isolation evolves much faster in sympatry. The observation that speciation in birds is completed faster than the evolution of intrinsic postzygotic isolation (roughly 2.6 million years [Klicka and Zink 1997] vs. 14 million years, respectively), suggests that avian speciation involves mostly prezygotic or extrinsic postzygotic isolation, with intrinsic sterility and inviability arising much later. This agrees with other direct and comparative evidence in birds (Price 1999). Nevertheless, the persistence of bird and butterfly species despite hybridization suggests that postzygotic isolation, even if not the *primary* cause of speciation, is important in preventing species fusion.

Given the limitations of these studies—especially their neglect of isolating barriers that may have been pivotal in speciation—we obviously need more complete analysis of these groups and more studies of other groups. As emphasized by Moyle et al. (2004), plants may be the most promising subject for such work. Ecological differences between species can be studied in the greenhouse and in the field via transplant experiments, pollinator isolation can be studied in the field, and intrinsic and extrinsic postzygotic isolation can be estimated through laboratory hybridization and transplantation of hybrids. There already exist considerable data on pollen germination, hybrid viability, and hybrid sterility in crosses within genera such as *Gilia*, *Phlox*, and *Mimulus* (Levin 1976, 1978; Vickery 1978; Grant 1981). Comparative studies of plants, however, await the production of good species-level phylogenies and a calibrated molecular clock.

Which traits promote the evolution of reproductive isolation?

In Chapter 12, we describe comparative studies in which some biological features of organisms (sexual dimorphism, body size, phytophagy, etc.) are significantly associated with diversification rates—the rate at which currently existing lineages arose. Although diversification rates involve both speciation and extinction, we will suggest a way to infer those features of organisms that enhance the rate of speciation per se. We find two such features: animal pollination of plants, and traits associated with sexual selection in birds and insects. These observations suggest that behavioral and pollinator isolation are associated with speciation. Such associations further imply that *these isolating barriers were primary factors in speciation.* If these barriers arose after reproductive isolation was complete, they would not be associated with higher rates of speciation. This may be the most powerful way to infer those isolating barriers important in speciation. Nevertheless, such inferences must be checked using direct measurements of reproductive barriers between taxa in different stages of evolutionary divergence.

Identifying the first isolating barriers to evolve is not a glamorous task, but it is essential for many studies of speciation. If, as we suspect, some barriers are consistently more important than others, the best way to show this is through comparative analysis.

3

Allopatric and Parapatric Speciation

Among all scientifically tractable questions about speciation, the most hotly contested concerns its biogeography. Does speciation require complete geographic isolation (allopatric speciation), or can it occur without such isolation—in populations that exchange genes either to a limited extent (parapatric speciation) or freely (sympatric speciation)? This controversy began, in fact, in the mind of Darwin, who struggled with the problem well before publishing *The Origin*. His early unpublished writings show deep ambivalence about not only the nature of species, but the biogeography of their origin (Sulloway 1979; Mayr 1982; Gould 2002). For example, in his "big species book," abandoned when he rushed *The Origin* into print, Darwin clearly thought that species could arise either allopatrically, parapatrically, or sympatrically (note that in the following quote he also emphasizes reproductive isolation, including temporal, habitat, and behavioral barriers):

> Finally, then, I suppose that a large number of closely allied or representative species, now inhabiting open & continuous areas, were originally formed in parts formerly isolated; or that varieties became in fact isolated from haunting different stations, disliking each other, breeding at different times &c, so as not to cross (Darwin, p. 269 in Stauffer 1975).

But Darwin's view of the importance of geographic isolation is not as clear in *The Origin*, where, according to Sulloway (1979, p. 48), Darwin's "ideas on this subject became so condensed as to be misleading." Those who were misled included many later biologists, including Moritz Wagner and David Starr Jordan, who believed that Darwin saw sympatric speciation as the main—if not the only—engine of diversity. In response, both men argued strenuously for the importance of allopatric speciation, citing abundant evidence from nature (Wagner 1873, 1889; Jordan 1905, 1908). This controversy continued up to and through the Modern Synthesis. Mayr, a strong proponent of allopatric speciation, summarizes this history in *Animal Species and Evolution* (1963, pp. 482–488). His forceful—some might say dogmatic—defense of allopatric speciation only aggravated the debate, as researchers like Bush (1969) and White (1978) became equally forceful proponents of sympatric speciation. Until the last decade, though, most evolutionists saw allopatric speciation as the norm.

Recently, however, sympatric and parapatric speciation have again gained credibility. This change resulted from new mathematical models demonstrating that species can form while exchanging genes, and from new data suggesting that speciation has occurred without geographic barriers. While most evolutionists still accept allopatric speciation as the most common mode, others claim that sympatric speciation may be nearly as frequent, or, in some groups, even more frequent. The controversy persists for several reasons:

1. It is relatively easy to demonstrate allopatric speciation by ruling out the possibility of gene flow, as, for example, in instances of speciation after colonization of distant islands. One can argue that allopatric speciation should be considered the "default" mode of speciation because it is supported by substantial evidence and occurs under a wider range of conditions than do other modes. In a given case, then, strong support for parapatric or sympatric speciation requires excluding the possibility of allopatry. Only rarely can this be done. Consequently, arguments about modes of speciation often degenerate into unproductive debates about plausibility.

2. Both theoretical and biological considerations imply that parapatric speciation is more common than sympatric speciation, yet it is harder to rule out allopatry in the former case. Oddly, then, while we can point to several plausible cases of sympatric speciation, we know of no well-established cases of parapatric speciation.

Taken together, these two problems mean that it is impossible to determine, by counting examples, the *proportion* of speciation events that occurred by sympatric or parapatric speciation. Fortunately, comparative methods described at the end of Chapter 4 have the potential to answer this question.

3. Evolutionists' opinions about the biogeography of speciation have undoubtedly been conditioned by their own scientific histories. Sympatric speciation, for example, may be far more common in phytophagous insects than in fish. This means that more entomologists than ichthyologists view sympatric speciation as common *in many taxa*.

4. Evolutionists have different views about the ubiquity of geographic barriers. Have these barriers really been common enough to create millions of species allopatrically? Even though sympatric speciation requires more stringent genetic and ecological requirements than does allopatric speciation, the opportunities for allopatry might be far less common. Sauer (1990, p. 2), for example, argues that allopatric speciation cannot explain the diversity of angiosperms, noting that this diversity "is too much to expect from a

process depending on vast time and widely empty space." Judging such claims requires historical information that is very hard to acquire.

In this chapter we describe and evaluate two major modes of speciation involving restricted gene flow: allopatric and parapatric speciation. In the next chapter we tackle the contentious issue of sympatric speciation. For each mode of speciation, we describe theory, laboratory experiments, and data from nature bearing on its plausibility. Because allopatric speciation has been extensively discussed in earlier literature, we devote most of our attention in these two chapters to speciation with gene flow, an area in which there has been much recent progress. At the end of Chapter 4, we review and analyze recent attempts to estimate the relative frequency of different biogeographic modes of speciation.

Admittedly, biogeography is not the only way to classify modes of speciation. One can categorize speciation by its genetic basis or by the evolutionary forces producing reproductive isolation (e.g., Kirkpatrick and Ravigné 2002). Why, then, do we concentrate so heavily on biogeography? Our rationale is that biogeography can limit the nature and strength of evolutionary forces potentially causing reproductive isolation. This is particularly obvious in speciation involving genetic incompatibilities that cause intrinsic hybrid sterility and inviability. In allopatry, the divergent evolution of populations can easily produce such incompatibilities, as the genes responsible never co-occur until isolated populations attain secondary contact and hybridize. In sympatry, however, these incompatibilities *do* co-occur in populations, are always deleterious, and will accumulate only if the genes causing them have other advantages that outweigh the disadvantages. In parapatry, such incompatibilities can accumulate only if gene flow between populations is very low.

Moreover, parapatric and sympatric speciation require that the evolutionary forces causing populations to diverge must be stronger than the gene flow causing them to fuse (Slatkin 1985). Allopatric speciation, on the other hand, necessitates no such balancing act: virtually *any* force causing divergence can eventually yield speciation. This implies that the biological conditions allowing parapatric and sympatric speciation are more restrictive than are those allowing allopatric speciation.

Moreover, classifying speciation by evolutionary forces or genetics still requires considering biogeography. For example, sexual selection can cause allopatric speciation under a wide variety of conditions, but can cause sympatric speciation only under far more restrictive—indeed, nearly prohibitive—conditions. Finally, as we show in Chapters 5 and 6, while it is hard to unravel the evolutionary forces and genetic changes causing speciation, the biogeographic mode of speciation is a more soluble question.

Allopatric Speciation

As Futuyma and Mayer (1980, p. 255) explain, "Whether speciation is accomplished by allopatric or non-allopatric means is really only a question of whether the initial reduction of m [the fraction of breeding individuals in a population that are immigrants from other populations] is accomplished by a physical barrier extrinsic to the organism (in which case the populations are considered allopatric) or by the biological characters of the organisms themselves." In population-genetic terms, allopatric speciation occurs when m is effectively zero from the beginning.

Allopatric speciation, in turn, is often subdivided into two modes: *vicariant speciation* (also known as the "dumbbell model"), in which reproductive isolation evolves after the geographic range of a species splits into two or more reasonably large, isolated populations; and *peripatric speciation*, in which reproductive isolation evolves after either an isolated habitat is colonized by a few individuals, or a small population becomes geographically isolated. Vicariant and peripatric speciation thus differ in the relative sizes of the populations involved.

Current support for allopatric speciation is based largely on evidence and arguments marshaled by Jordan (1905, 1908), Mayr (1942, 1963), and Dobzhansky (1937b, 1940). Jordan and Mayr relied largely on observations in nature, while Dobzhansky deduced the importance of allopatry from its ability to allow the evolution of "complementary genes" known to cause intrinsic postzygotic isolation.

Vicariant speciation

THEORY. For many years, the theory of vicariant allopatric speciation was purely verbal, based on the well known population-genetic conclusion that genetic divergence between populations, whether caused by selection or genetic drift, is hampered by gene flow (Haldane 1930). Recent mathematical models, while producing some counterintuitive results, have not overturned these earlier theories.

The vicariant model is simple and intuitive. Geographic separation of previously interbreeding populations can occur through diverse climatic or geological events, including elevation of land bridges, glaciation, formation of mountains, continental drift, climatic change, or extinction of intermediate populations. Geographic isolation allows allopatric populations to diverge unimpeded by hybridization. Such differentiation is nearly inevitable: populations usually occupy ecologically different habitats that impose different forms of selection, different mutations arise in different populations (so that divergence can occur even when habitats are identical), genetic drift fixes different genes in different isolates, and initial evolutionary divergence can be amplified by later selection. (Selection for ovipositing only on a toxic plant, for example, cannot evolve until larvae have developed the ability to survive on it.) Eventually, divergence will produce isolating barriers that allow species to remain distinct when they experience secondary contact.

Despite Darwin's reputation as an advocate of sympatric speciation, he describes both sympatric and allopatric speciation in *The Origin*, where he envisions divergence occurring after a large area of land becomes submerged, forming an archipelago:

> I conclude, looking to the future, that for terrestrial productions a large continental area, which will probably undergo many oscillations of level, and which consequently will exist for long periods in a broken condition, will be the most favourable for the production of many new forms of life, likely to endure long and to spread widely.... When converted by subsidence into large separate islands, there will still exist many individuals of the same species on each island: intercrossing on the confines of the range of each species will thus be checked: after physical changes of any kind, immigration will be prevented, so that new places in the polity of each island will have to be filled up by modification of the old inhabitants; and time will be allowed for the varieties in each to become well modified and perfected (Darwin 1859, pp. 107–108).

Mathematical models of such scenarios, which include explicit assumptions about the genetic basis of reproductive isolation, have appeared only in the last 25 years. These include models of pre- and postzygotic isolation caused by either drift alone (Nei et al. 1983) or a combination of drift and selection (Wills 1977), models of prezygotic isolation caused by sexual selection (Pomiankowski and Iwasa 1998), and models of postzygotic isolation caused by drift or selection (Orr 1995; Orr and Orr 1996; Orr and Turelli 2001). These models, discussed in later chapters, confirm the conclusion that geographic isolation can lead to complete reproductive isolation, that a small amount of gene flow can retard differentiation caused by drift (Nei et al. 1983), and that genetic drift usually causes speciation more slowly than does either natural selection alone or a combination of drift and selection. In Chapter 11 we consider the relative roles of selection and drift in the evolution of reproductive isolation.

EXPERIMENTAL EVIDENCE. Laboratory experiments on vicariant speciation can show whether divergent artificial selection or drift in isolated populations causes the rapid evolution of reproductive isolation. Such work can also identify which forms of selection are most effective, and whether prezygotic isolation evolves faster than postzygotic isolation. Given the rates of speciation estimated in nature (Chapter 12), we do not expect even long-term laboratory studies—those lasting 10–100 generations—to yield substantial reproductive isolation. The one exception appears to be Dobzhansky and Pavlovsky's (1966) observation that a strain of *Drosophila paulistorum* raised in the laboratory for five years without selection showed hybrid male sterility in crosses with other conspecific strains. But this result may have been due to contamination of cultures by other subspecies.

Table 3.1, a modified and updated version of tables compiled by Ringo et al. (1985) and Rice and Hostert (1993), lists laboratory studies of the effect of divergent selection on reproductive isolation between isolated lines. In Chapter 11 we discuss experiments that involve drift but no deliberate selection. Genetic drift,

Table 3.1 *Laboratory experiments modelling allopatric speciation by divergent selection*

Organism and trait	Generations of selection	Prezygotic isolation	Postzygotic isolation	Control for drift	Reference
Drosophila melanogaster abdominal bristle number	20–31	N	—	Y	Koref-Santibañez and Waddington 1958
D. melanogaster sternopleural bristle number	32	N	—	N	Barker and Cummins 1969
D. melanogaster escape reaction	18	Y	—	Y	Grant and Mettler 1969
D. melanogaster locomotor activity	112	Y	—	N	Burnet and Connolly 1974
D. melanogaster locomotor activity	45	N	—	N	van Dijken and Scharloo 1979
D. melanogaster phototaxis and geotaxis	20	N	—	N	Markow 1981
D. melanogaster temperature and humidity	~70–130	Y	N	Y	Kilias et al. 1980
D. melanogaster adaptation to EDTA	~20	N[a]	Y	N	Robertson 1966a,b
D. melanogaster adaptation to DDT	~600[b]	Y	N	N	Boake et al. 2003
D. simulans four groups of traits (development/ morphology, physiology, sexual behavior, nonsexual behavior)	~12–72	N	?[c]	N	Ringo et al. 1985
D. willistoni adaptation to pH	34–122	Y	?[c]	N	de Oliveira and Cordeiro 1980
D. pseudoobscura carbohydrate source	12	Y	—	Y	Dodd 1989
D. pseudoobscura temperature adaptation	25–60	?[d]	—	N	Ehrman 1964,1969
D. pseudoobscura photo- and geotaxis	5–11	Y	—	N	del Solar 1966
D. pseudoobscura temperature, photoperiod, food	37	N	—	Y	Rundle 2003

Table 3.1 *Laboratory experiments modelling allopatric speciation by divergent selection[a] (continued)*

Organism and trait	Generations of selection	Prezygotic isolation	Postzygotic isolation I	Control for drift	Reference
D. mojavensis development time	13	N[e]	—	Y	Etges 1998
Musca domestica geotaxis	38	Y	—	N	Soans et al. 1974
M. domestica geotaxis	16	Y	—	N	Hurd and Eisenberg 1975
Bactrocera cucurbitae (melon fly) development time	40–51	Y	—	Y	Miyatake and Shimizu 1999

Source: Updated and modified from Ringo et al. (1985) and Rice and Hostert (1993), removing experiments in which selection was practiced directly on prezygotic isolation.

Y = yes, N = no, Y/N = results variable, ? = results equivocal, "—" = not examined. Except for the study of Boake et al. (2003), reproductive isolation was examined between lines selected in opposite directions.

[a]After 20 generations of "allopatric" adaptation to EDTA, flies were put in cages that were connected by a small glass tube to other cages containing non-adapted flies. There was thus some gene flow between cages, although the authors note that it was restricted. After 20 generations of gene flow, flies from three pairs of such cages were tested for mate discrimination (none was observed).

[b]Selection for DDT resistance was practiced on one line for about 25 years and then relaxed for more than 15 years. The selected line (which maintained its resistance) was then tested against the original unselected control line.

[c]"Postzygotic isolation" was quantified as the average number of offspring produced per individual within vials containing either five or ten pairs of flies. Although both studies showed some sign of postzygotic isolation (i.e., a reduced number of hybrid offspring), under some conditions or in some tests this may have resulted at least partially from sexual isolation or a reduced propensity to remate, lowering the number of successful matings in some hybrid crosses.

[d]Results were not consistent among tests of identical populations.

[e]Selection for development time was performed on two geographic populations of *D. mojavensis* that initially showed moderate premating isolation. Divergent selection (one population selected in each direction) either *decreased* or did not change the level of premating isolation shown by controls. In addition, identical selection regimes practiced on the two populations led to a general *decrease* in sexual isolation between the selected lines.

which always occurs in selection experiments since relatively few individuals are chosen to start each generation, can also promote reproductive isolation. We thus note which experiments were accompanied by "drift controls": tests of whether restricted population size alone caused reproductive isolation.

All 19 experiments involved Diptera: house flies, melon flies, and *Drosophila*. All but one study incorporated post-selection tests for sexual isolation. Among all studies, ten resulted in behavioral isolation (positive assortative mating between lines), eight showed no behavioral isolation, and one gave equivocal results. In a few cases this isolation was strong: Hurd and Eisenberg's (1975) study showed assortative mating of 0.73 on a scale in which 0 represents random mating and 1 complete sexual isolation. In most studies, however, sexual isolation was much lower: the mean among nine studies analyzed by Ringo et al (1985) was 0.12. In only three studies was postzygotic isolation examined properly—by measuring the viability or fertility of offspring from F_1 or F_2 crosses between lines. Among these, only one experiment found such isolation: Robertson (1966a) observed complete sterility of a single hybrid genotype in the F_2 generation (females with homozygous "control" third chromosomes in a genetic background from the selected line).

Although the number of studies is small, and limited to one order of insects, we conclude that reproductive isolation—especially behavioral isolation—frequently evolves as a byproduct of selection. Moreover, such isolation develops quickly: in del Solar's (1966) experiment, appreciable sexual isolation appeared after only five generations of selection. In several experiments, however, sexual isolation emerged erratically, waxing and waning over the course of the study (Florin and Ödeen 2002). Although there are few tests of postzygotic isolation, in Robertson's (1966b) experiment it also evolved quickly, with intrinsic hybrid sterility appearing after about 20 generations of selection. (This result was, however, unreplicated.)

In these studies, reproductive isolation may have resulted from either pleiotropic effects of genes responding to selection, the effects of genes *linked* to selected genes ("hitchhiking"), or genetic drift. How can we distinguish among these alternatives? Only four of the fourteen experiments had controls for drift. In three of these, reproductive isolation was significantly higher in selected than in the drift control lines, suggesting that drift was not a major factor. Drift also seems unlikely in view of experiments showing little effect of extreme inbreeding on reproductive isolation (Chapter 11).

In four studies we can rule out hitchhiking as well as drift (Kilias et al. 1980; Dodd 1989; Etges 1998; Miyatake and Shimizu 1999) because each selection regime was replicated several times. In these experiments, behavioral isolation evolved among lines subject to different forms of selection, but not among lines subject to the same form of selection. Dodd (1989), for example, found that lines of *D. pseudoobscura* adapted to maltose were sexually isolated from lines adapted to starch, but that there was no isolation between replicate lines selected on either medium. It is unlikely that, in independent tests, genes responding to the same selective factor would be linked to alleles having similar effects on mating. These results imply that behavioral isolation was a byproduct of the genes that responded to selection.

Unfortunately, in most of these studies we know nothing about the *phenotypes* causing behavioral or postzygotic isolation, although adaptation to new environments and selection for altered movement seem to yield sexual isolation most readily. In one exceptional case, melon flies (*Bactrocera cucurbitae*) selected for long and short development time became sexually isolated because the response to selection also changed the diurnal rhythm of mating (Miyatake and Shimuzu 1999).

Collectively, these studies show that reproductive isolation in allopatry can evolve as a byproduct of divergent selection. More than half of the published experiments showed either pre- or postzygotic isolation. Although this proportion may reflect a publication bias, obtaining isolation as a byproduct of selection cannot be rare. More surprisingly, such isolation can evolve within 10–100 generations, a blink of the eye in evolutionary time. Although in no case was reproductive isolation complete, one does not expect full speciation to occur in so few generations. Surprisingly, no study examined whether adaptation to new environments also yielded a behavioral *preference* for those environments, a correlation that would promote sympatric speciation (Chapter 4).

EVIDENCE FROM NATURE.   After noting that closely related species almost invariably occupied opposite sides of geographical barriers, Moritz Wagner, David Starr Jordan and others suggested that such barriers were *required* for speciation:

> Rivers, mountain chains, and seas form distinct lines of demarcation in the distribution of many varieties, species and genera . . . On both sides of the boundaries endemic species will appear as so-called vicarious forms (or forms similar to neighbouring ones) separated from each other by these barriers. Such species show, as a general rule, a yet nearer typical relationship to each other than to species of the same genus formed at greater distances . . . The formation and continuance of a race will always be endangered where numerous individuals of the original stock become mixed up with, and by frequent crossing disturb, or entirely suppress it. Unless the separation of the colonists from the kindred continue for a long time, the formation of a new race cannot according to my conviction succeed, nor natural selection become active (Wagner 1873, pp. 24–25, 30–31).

> It is not claimed that species are occasionally associated with physical barriers, which determine their range, and which have been factors in their formation. We claim that such conditions *are virtually universal* among species as they exist in nature (Jordan and Kellogg 1908, p. 123; our italics).

Allen (1907, p. 654) called this pattern of geographically isolated close relatives "Jordan's law," which he codified as follows: "Given any species in any region, the nearest related species is not likely to be found in the same region, but in a neighboring district separated from the first by a barrier of some sort."

Mayr (1942, 1963), however, was the first to summarize the diverse and widespread evidence for vicariant allopatric speciation. Much of this evidence remains valid, but since Mayr's 1963 compendium better examples have appeared, many of them strengthened by molecular phylogenies. Moreover, some of Mayr's original evidence now seems weak. For example, his section on "proof for geographic speciation" in *Systematics and the Origin of Species* (1942, pp. 162–184) begins with two observations. First, the same traits that differ among species also differ to a lesser extent among populations of single species. Second, some species show reproductive barriers among *populations*, and these interpopulation barriers range from very weak to nearly complete. Although Mayr concluded from these data that geographically differentiated populations represent the earliest stages of vicariant speciation, this logic is flawed. Populations of a species are usually connected by migration, so that differentiation has occurred in the face of gene flow—and thus is not allopatric. Moreover, while the evolution of reproductive isolation between interbreeding populations suggests that selection or drift can counteract gene flow, it does not show that speciation can be *completed* in allopatry. It is entirely possible that such populations could never become good species without geographic barriers that completely block gene exchange. Hence, while this line of evidence is consistent with allopatric speciation, it is also consistent with parapatric speciation. The "all-degrees-of-isolation" argument becomes even more problematic when used to support parapatric and sympatric speciation (see below, and Chapter 4). But these arguments are unnecessary, as there is far better evidence for allopatric speciation.

In gathering such evidence, however, we face a problem. While biological species are best *recognized* in sympatry, how can we show that such species may have arisen in allopatry? Several forms of evidence have been used to support vicariant speciation. We discuss these in order from strongest to weakest:

1. Geographic concordance of species borders with existing geographic or climatic barriers. This type of evidence for allopatric speciation has a long history, beginning with Wagner. After observing that, in fish, many apparent "sister species" (i.e., species more closely related to each other than to any other species) occurred on opposite sides of the Isthmus of Panama, Jordan (1908) named such geographically isolated relatives "geminate species pairs." Their geographic distribution implied that the terrestrial barrier promoted speciation. Mayr (1954b) gave similar data for echinoids. There are, however, several problems with these early studies. First, species were defined by their morphological difference, and there was little evidence for reproductive isolation between members of a geminate pair. Second, there were no phylogenetic data to show that members of a pair were indeed sister species. Unless they are, one cannot assume that their common ancestor was divided by the geographic barrier. Finally, if populations of different species began diverging after the formation of a geographic barrier, different geminate pairs should often be of roughly equal age—the age of the barrier. (Actually, strict age-equality is not required, as the times when the barrier interrupts gene flow may differ among ecologically different species [e.g., Knowlton et al. 1993].) Without a molecular clock, earlier workers could not estimate divergence times.

An entire field, "vicariance biogeography," is now devoted to addressing these problems (Cracraft 1982; Brown and Lomolino 1998). As a result, we now have many examples of concordant species distributions that provide strong evidence for vicariant speciation.

The Isthmus of Panama is still the site of many such cases. The current Isthmus rose about 3 million years ago, cleaving the ranges of marine species. Populations on either side of the land bridge experience different biota, tidal fluctuations, and water temperature, clarity, and salinity (Rubinoff 1968). Lessios (1998) reviews the history of work on speciation around the Isthmus and describes several geminate pairs that appear to be reproductively isolated sister taxa.

Perhaps the best study involves snapping shrimp in the genus *Alpheus* (Knowlton et al. 1993). Molecular phylogenies reveal seven geminate pairs in this genus, each comprising sister species from the Caribbean Sea and Pacific Ocean. Laboratory work shows strong reproductive isolation between members of each pair: on average, only 1% of interspecific matings yield fertile clutches, compared to more than 60% success in conspecific matings. Molecular-based dating of 13 pairs show that they did not diverge at the same time, but became isolated over a period ranging from 3 million to 10 million years ago (Knowlton and Weigt 1998). This difference, however, is not strong evidence against vicariant speciation: as the authors note, the Isthmus closed gradually and different shrimp species live at different depths. Indeed, the longest-diverged pairs inhabit deeper or less heavily sedimented waters, and thus would be the first to be divided by a rising land bridge.

One sees a similar pattern in angiosperms from eastern Asia and eastern North America. All botanists know that these two areas harbor related floras, having many geminate pairs absent from intervening areas in western North America (Xiang et al. 1998; Wen 1999). Fossil evidence suggests that these disjunct distributions resulted from the onset of a cooler and dryer climate during the late Miocene. This change replaced much forest with grassland or other habitat, and restricted many plants to the eastern sections of both continents. Xiang et al. (2000) studied eleven closely related pairs of species, each pair including one North American and one Asian species. All but one of the pairs belong to genera containing only two sister species. Molecular clocks based on chloroplast DNA reveal a remarkable concordance of divergence times among pairs (Figure 3.1). The one exception involves the vines *Menispermum dauricum* and *M. canadensis*, which diverged more recently, perhaps through long-distance colonization. The mean divergence time between species in the other ten pairs is 5.4 ± 2.8 (standard error) million years, times that all fall within the late Miocene or Pliocene. A potential problem is that we do not know whether members of a pair, though morphologically different, are reproductively isolated. *Liriodendron tulipifera* and *L. chinese*, for example, produce vigorous and fertile hybrids despite more than six million years of divergence (Parks et al. 1983; Xiang et al. 2000), but the possibility of partial hybrid sterility or other isolating barriers has yet to be examined.

A familiar example of vicariant speciation is the pupfish (*Cyprinodon*), endemic to Death Valley and surrounding parts of California. A remnant of a system of lakes and rivers that dried up in the late Pleistocene, this region now harbors at least three allopatric species and as many as seven allopatric subspecies of pupfish, all members of a monophyletic group (Duvernell and Turner 1998, 1999). Each species or subspecies inhabits a different spring; all are morphologically different. Although there is some evidence of hybrid sterility (Miller 1950), we know nothing about other isolating barriers. The extreme genetic similarity of species supports the idea that differentiation occurred recently (Turner 1974; Echelle and Dowling 1992).

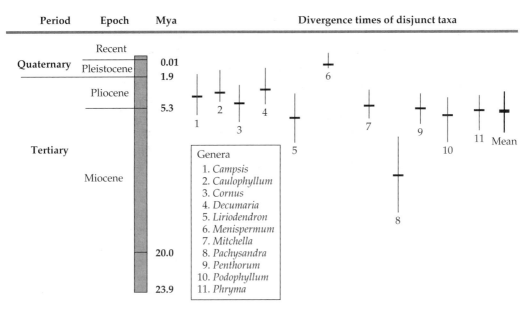

Figure 3.1 Estimated divergence times between geminate pairs of flowering plant species (one from eastern Asia and the other from eastern North America) showing general concordance among genera. Vertical lines give standard errors of divergence times. (From Xiang et al. 2000.)

2. Allopatry of young sister species. If sister species within a group almost always occur in allopatry, even if no geographic barrier is known, this constitutes fairly good evidence for allopatric speciation. One example involves marine bonefishes of genus *Albula* (Colborn et al. 2001). Molecular work shows that the species once recognized as *A. vulpes* may include as many as eight sibling species. Each species is allopatric to its sister species except for one pair that occurs near the Sunda shelf, a former terrestrial barrier between the Pacific and Indian Oceans.

Such evidence becomes stronger when allopatry is more pronounced between younger than between older pairs of species. Younger species are more likely to reflect the biogeography of speciation, which is obscured in older pairs when their ranges expand. In the tropical sea urchin *Diadema*, the youngest pairs of sister taxa are invariably allopatric, with geographic overlap seen only between members of older pairs (Lessios et al. 2001). This result is particularly interesting because urchins are broadcast spawners, and it is not clear how barriers to gene flow operate in the open ocean.

3. Geographic coincidence of species borders or hybrid zones among different taxa. If several pairs of related species have abutting distributions in the same location, one may infer that they speciated in allopatry and that the present distribution reflects secondary contact after a geographic barrier disappeared. But this observation can be considered strong evidence for vicariance only if we know that a geographic barrier existed earlier and if there is no boundary between ecologically different areas (an "ecotone") where the species' ranges meet.

With this type of evidence, however, one expects less concordance of divergence times among members of geminate pairs, for any hybridization between members of geminate pairs will make them more genetically similar, reducing estimates of divergence times based on molecular clocks. This is a special problem for clocks based on DNA from mitochondria or chloroplasts, organelles that often move freely between taxa (see Appendix). Thus, different amounts of gene flow between different pairs of species may obscure the fact that they diverged at the same time.

There are many places in South America where the ranges of related species coincide, often involving diverse groups such as birds and butterflies. This concordance is often explained by the previous presence of *refugia*: small patches of forest created during the Pleistocene when glacial advances reduced precipitation and temperature. Different taxa forced into such refugia would speciate in the same areas, expanding their distributions concurrently when the climate became wetter and the refugia merged back into continuous forest.

Originally proposed to explain the distribution of South American birds (Haffer 1969), the refugia theory has since been applied to other groups, including plants, mammals, and Lepidoptera (Prance 1982; Hall and Harvey 2002). But the theory is controversial because: (1) there is little independent evidence for refugia, (2) some related species clearly diverged well before the Pleistocene, (3) different groups sometimes suggest different patterns of refugia, and (4) concordant species boundaries might be explained by continuous ecological gradients—and thus parapatric speciation—instead of by geographical barriers (Endler 1982a; Kricher 1997; Knapp and Mallet 2003). The theory of parapatric speciation, however, has its own problems. If one invokes speciation occurring across existing ecotones, one must explain how those ecotones remained stable over the long periods necessary to permit the evolution of reproductive isolation, despite large changes in climate and environment. It is often suggested that the large area of Amazonia promoted parapatric speciation by severely reducing gene flow between distant populations (Bush 1994; Knapp and Mallet 2003). But although this suggestion involves speciation with gene flow, that flow would be only a mere trickle between those remote populations that became reproductively isolated. This scenario is an example of what we call *para-allopatric speciation* (see below).

However, some data do support allopatric speciation in isolated areas of the Neotropics. Hall and Harvey (2002) correlated the phylogeny and biogeography of 19 species of South American butterflies in the monophyletic *Charis cleonus* group, producing an "area cladogram" (Figure 3.2). Within this group, all sister species (and many higher-level sister taxa) are either parapatric or allopatric. The authors suspect that rivers may have been important geographic barriers for these butterflies. The *Charis* cladogram suggests seven areas of endemism, which are nearly identical to areas of endemism derived from phy-

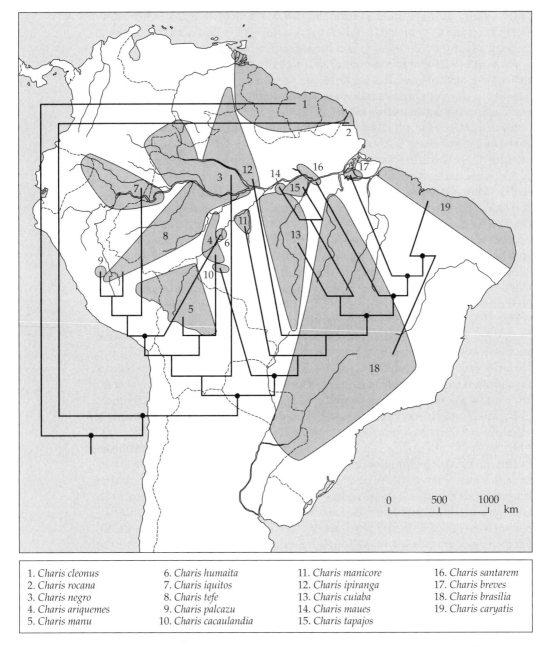

Figure 3.2 An area cladogram superimposing the phylogenetic relationships of 19 species of South American butterflies in the genus *Charis* with their ranges. The phylogeny is based on 36 morphological traits, and cladogram nodes that are well supported (bootstrap values exceeding 70) are indicated with black dots. All sister species have either allopatric or parapatric ranges. (From Hall and Harvey 2002.)

1. *Charis cleonus*
2. *Charis rocana*
3. *Charis negro*
4. *Charis ariquemes*
5. *Charis manu*
6. *Charis humaita*
7. *Charis iquitos*
8. *Charis tefe*
9. *Charis palcazu*
10. *Charis cacaulandia*
11. *Charis manicore*
12. *Charis ipiranga*
13. *Charis cuiaba*
14. *Charis maues*
15. *Charis tapajos*
16. *Charis santarem*
17. *Charis breves*
18. *Charis brasilia*
19. *Charis caryatis*

logenies of reptiles, primates, rodents, marsupials, birds, and other butterflies (Hall and Harvey 2002). The authors conclude (p. 1494) that "such corroborative pieces of independent evidence strongly suggest a common history of vicariant isolation events (Platnick and Nelson 1978) and argue against parapatric speciation models (Cracraft and Prum 1988) and those invoking ecological conditions of taxonomically narrow relevance (Tuomisto et al. 1995; Tuomisto and Ruokolainen 1997)."

We will probably never have strong evidence for refugia that is independent of such biogeographic patterns, and we obviously need molecular data showing that sister species of butterflies diverged at about the same time as sister species in other groups sharing proposed refugia. Speciation in the Amazon clearly involved factors other than refugia, including the formation of mountains and rivers, and perhaps even fragmentation by arms of the sea (Räsänen et al. 1995). And even if there were refugia, they may have formed and disappeared over long periods of time (Bush 1994; Hall and Harvey 2002). Nevertheless, much of the work on Amazonian fauna suggests that geographic isolation was important in speciation. It is likely that many cases of speciation in the neotropics occurred before the Pleistocene, although current species distributions may well reflect changes in habitat that occurred during the Pleistocene.

Geographic congruence of hybrid zones between diverse taxa gives similar evidence for vicariant speciation, the only difference being that in such cases related taxa show substantial gene exchange where their ranges meet. Although parapatric speciation might also explain this pattern, this mode of speciation appears less parsimonious than allopatry when hybrid zones coincide with known geographic barriers such as glacial refugia (Harrison 1990, 1992). Moreover, the congruence of clines for diverse genes across a hybrid zone suggests an earlier period of allopatry (Barton and Hewitt 1985).

Some of the best evidence in this category comes from Remington's (1968) analysis of six major "suture zones" in North America. Each zone is a fairly narrow area in which one finds hybrids in many groups, including plants, insects, mammals, amphibians, birds, and reptiles (Figure 3.3A). Most of these zones correspond roughly to the site of former geographic barriers, whose disappearance allowed incipient species to meet and hybridize. In some cases, hybrids are known to have low fitness, demonstrating reproductive isolation.

Moore and Price (1993) describe ten hybrid zones between species of North American birds. Six of these areas fall within Remington's "suture zone #4," the "Rocky Mountain-Eastern" area. Fig. 3.3B–E shows the coincidence of four hybrid zones in closely related species or subspecies of grosbeaks, orioles, buntings, and flickers. The members of three pairs show similar divergence times: 2.2 million years for grosbeaks, 2.4 million years for orioles, and 3.3 million years for buntings, respectively (Klicka and Zink 1997). The subspecies of flickers, however, are nearly identical in both mtDNA and allozymes (Grudzien et al. 1987; Moore et al. 1991), and although no divergence time has been calculated, it would presumably be close to zero. This genetic similarity surely reflects the rampant hybridization of flickers in the zone of over-

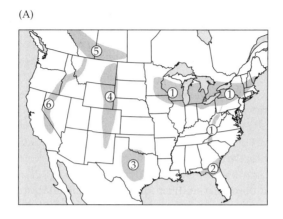
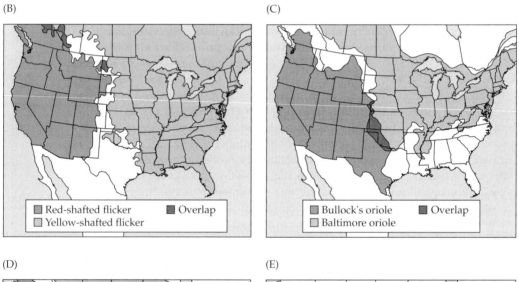
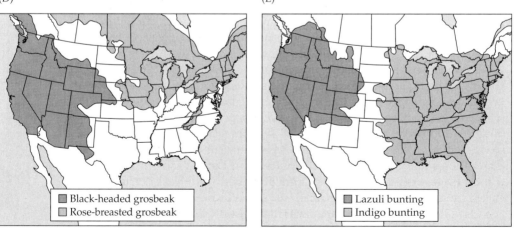

◀ **Figure 3.3 Evidence for allopatric speciation from congruent hybrid zones.** (A) Six major "suture zones" in North America as delineated by Remington (1968). Each zone is a region where the ranges of close relatives in many groups meet and the relatives hybridize. (B–E) Coincidence of bird hybrid zones in Remington's (1968) "Rocky Mountain Eastern" area hybrid zone (zone 4 in [A]). Distributions are shown as disjunct because they are ranges of "pure" parental species, but all four pair of taxa meet and hybridize. (B) Subspecies of the northern flicker: red-shafted flicker (*Colaptes auratus cafer*) and yellow-shafted flicker (*C. auratus auratus*) (C) Bullock's oriole (*Icterus bullockii*) and Baltimore oriole (*I. galbula*). (D) Black-headed grosbeak (*Pheucticus melanocephalus*) and rose-breasted grosbeak (*P. ludovicianus*). (E) Lazuli bunting (*Passerina amoena*) and indigo bunting (*P. cyanea*). (From Moore and Price 1993.)

lap (Moore 1987), and illustrates the difficulty of estimating divergence times between hybridizing groups.

Geological evidence suggests that these coincident zones reflect secondary contact of previously allopatric populations after the Pleistocene glaciers retreated around 15,000 years ago (Moore and Buchanan 1985). A complication is that the hybrid zones lie in the Great Plains/Rocky Mountains ecotone, which divides areas having different precipitation and vegetation (Moore and Price 1993). While the occurrence of hybrid zones at an ecotone might suggest parapatric rather than allopatric speciation, this possibility seems unlikely since the ecotone did not exist over most of the period when the species diverged. Thus, although the ecological transition may be responsible for *maintaining* the hybrid zones, the origin of these taxa almost certainly involved divergence in allopatry. Newton (2003) lists an additional eighteen pairs of avian sister taxa or closely related taxa in North America, each pair comprising an eastern and western form. Again, these ranges presumably reflect the distribution of ancient forests.

4. The absence of sister species in areas where geographic isolation was unlikely. If species are formed in areas containing geographic barriers, but not in areas lacking them, one can conclude that most speciation was probably allopatric (and perhaps vicariant). Such tests are difficult because we rarely know what would constitute a geographic barrier for a group, or even whether such barriers ever existed. One test, by Coyne and Price (2000), involved endemic birds on 51 remote oceanic islands (i.e., islands never connected to the mainland). Since birds are fairly mobile, one would expect to see sister species on these islands only if speciation were sympatric or parapatric. However, Coyne and Price found no such evidence, even on large islands, implying that bird speciation is allopatric. The absence of sympatric speciation in birds is particularly striking because related species often show the sort of niche differentiation associated with resource-mediated disruptive selection that can cause sympatric speciation (Dieckmann and Doebeli 1999). Similarly, Losos and Schluter (2000) found no evidence for sympatric speciation in *Anolis* lizards on Caribbean islands. Although intra-island speciation occurred on very large islands, no speciation was observed on islands smaller than 3000 km². Unfortunately, these tests can be applied only to fairly mobile organisms, as sedentary species like flightless insects can speciate "microallopatrically" in small areas.

5. *Concordance between present or past geographic barriers and genetic discontinuities within species.* This line of evidence involves abrupt changes in allele frequency occurring within species—breaks that occur in similar locations among diverse species. Avise (1994, 2000) and his colleagues provide many examples, usually involving mtDNA. The study of genetic patterns over space has been called "phylogeography" (Avise et al. 1987). Figure 3.4 shows one of the most striking patterns. Nine species from diverse different groups are distributed along coastal marshes and estuaries from the northeastern United States around the Florida peninsula to the Gulf of Mexico. These species show sharp breaks in mtDNA haplotype frequencies at the same location: the Florida peninsula. (This area is also a boundary for several subspecies and sister species.) Avise (1994, 2000) suggests that these species formerly had a continuous distribution

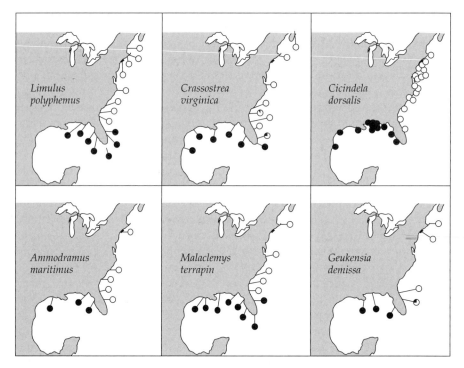

Figure 3.4 Pie diagrams showing the frequencies of diagnostic mtDNA haplotypes or allozyme alleles within six species inhabiting both the East Coast and the Gulf Coast of North America. *Limulus polyphemus* is the horseshoe crab; *Crassostrea virginica* is the American oyster; *Cicindela dorsalis* is a beach-dwelling tiger beetle; *Ammodramus maritimus* is the seaside sparrow; *Malaclemys terrapin* is the diamondback terrapin; and *Geukensia demissa* is the ribbed mussel. (From Avise 2000.)

but were divided when glaciation reduced sea levels, enlarging the peninsula and interposing a tropical barrier between Atlantic and Gulf Coast populations. After the populations diverged genetically, they reconnected when the glaciers receded, but the traces of history remain, perhaps maintained by current differences in temperature and habitat. Many of the mtDNA differences within species date to a similar period, around 2 million years ago. There are many examples of such "genealogical concordance" in other areas (Avise and Ball 1990; Cunningham and Collins 1998; Avise 2000; Colborn et al. 2001; Bernardi et al. 2003), although in some of these the geographic barrier can only be inferred.

The problem with such examples is that they show that allopatry promotes not speciation, but genetic differentiation. Increased divergence in the absence of gene flow is no surprise to anyone who has studied population genetics. Nevertheless, because the genetic differentiation occurs around geographic barriers, and because genetic differentiation is required for speciation, the patterns are at least consistent with allopatric speciation.

6. The increase of reproductive isolation with geographic distance between populations. If gene flow decreases with geographic distance, remote populations of a wide-ranging species may exchange genes only rarely, allowing them to evolve fairly independently. An increase of reproductive isolation with distance between populations—a pattern we call "clinal isolation"—could be considered weak evidence for vicariant speciation.

One complication of this argument is that if there is geographic variation in ecology, the same observation could reflect not attenuated gene flow but local differences in selection, which could produce reproductive isolation as a byproduct. This possibility has not been adequately addressed in the few known cases of clinal isolation. One way to distinguish these causes is to determine whether reproductive isolation correlates more strongly with distance than with local ecology. This may be impossible, however, if distance and ecological difference are themselves correlated (i.e., the environment changes clinally). Such a correlation is often expected in nature, so that isolation and selection will jointly cause divergence. Yet, reproductive isolation can also result from *identical* selection pressures that fix different genes in different places, making distance more important than ecological difference.

We know of only four cases of clinal isolation. In crosses between 11 populations of the plant *Strepthanthus glandulosus* subsp. *secundus*, which occupies small, disjunct outcrops of serpentine soils, Kruckeberg (1957) observed a negative correlation between pollen fertility of F_1 hybrids and geographic distance between parents. Tilley et al. (1990) found that behavioral isolation between populations of the salamander *Desmognathus ochrophaeus* in the Appalachian Mountains of the United States increased with geographic distance, with distant populations showing substantial isolation. (Tilley and Mahoney [1996] later reclassified some of these populations as new species based on allozyme data, but found that the correlation between isolation and distance remained both within and among species.) Edmands (1999) showed that some components of F_2 hybrid fitness decreased with increasing genetic distance between populations of the copepod *Tigriopus californicus*. Finally, sexual isolation in the stick insect *Timema cristinae* is positively correlated with geographic distance between parental populations (Nosil et al. 2002).

This pattern, however, is not universal: Edmands (2002) gives several exceptions. In many cases gene flow will be too high, or population extinction too frequent, to allow the evolution of a correlation between distance and reproductive isolation. Other factors may also affect this correlation. In some plants, one often finds the strongest isolation between the closest and the most distant populations, with populations at intermediate distances showing less isolation (Waser 1993). The low fitness of hybrids between nearby plants may be caused by inbreeding depression, and that between distant plants by differential adaptation ("outbreeding depression").

A related phenomenon involves "ring species," which Mayr (1942, 1963) considered strong evidence for allopatric speciation. A ring species encircles a geographic barrier, such as a mountain range or plateau. Populations around the ring appear to show free exchange of genes except at a single location, where adjacent populations are reproductively isolated. Ring species are said to form when a single ancestral population gradually expands its range around the barrier. During this expansion, populations exchange genes with their neighbors, but gene flow is low between more distant populations, allowing them to differentiate more extensively. The two populations that finally meet when the ring is closed have been genetically isolated for so long that they have achieved strong reproductive isolation. (Note that it is unclear whether these terminal populations can be considered "good" species. Even though they may show *complete* reproductive isolation, they can still exchange genes: a universally favorable mutation arising in one population can find its way into the other by moving the long way around the ring.)

The problem with Mayr's argument is that ring species do not demonstrate allopatric speciation, but speciation occurring through the attenuation of gene flow with distance. Mayr felt that, in a ring species, substantial distance between populations was equivalent to a geographic barrier. While this view may be justified, one could also argue that ring species provide better evidence for parapatric than for allopatric speciation. But this is not parapatric speciation in the conventional sense. Although the terminal populations are themselves parapatric, reproductive isolation evolved not through divergent selection in situ, but through the gradual accumulation of genetic differences as populations expanded around the ring.

Nevertheless, ring species are more convincing than cases of clinal isolation for showing that gene flow hampers the evolution of reproductive isolation. In clinal isolation, one can argue that reproductive isolation was caused by environmental differences that increase with distance between populations. One cannot make a similar argument for ring species because the most reproductively isolated populations occur in the *same* habitat.

But do ring species really exist? A convincing case must meet several criteria. First, there should be historical information that the ring was founded by one population rather than several genetically distinct populations, and that all individuals around the ring descend from this ancestral population. At least one of the terminal populations must represent the most recent range expansion. There must be no geographic barriers that interrupt gene flow around the ring, and such barriers must not have existed in the past. Finally, one must be able to show that gene flow is fairly extensive between adjacent populations, but decreases with distance around the ring.

Unfortunately, every proposed ring species fails to meet one or more of these criteria (Irwin et al. 2001b; Irwin and Irwin 2002). While molecular data can demonstrate the decrease of gene flow with distance, and reproductive isolation can be observed in the field, it is difficult to show that the ring expanded from one ancestral population and was never interrupted by geographic barriers.

These problems afflict even the most famous example of a ring species, the forest-dwelling plethodontid salamander *Ensatina eschscholtzii* of western North America. First described by Stebbins (1949), this case has been studied intensively by Wake and colleagues (e.g., Wake and Yanev 1986; Wake et al. 1986, 1989; Moritz et al. 1992; Jackman and Wake 1994). Stebbins (1949) divided *E. eschscholtzii* into seven subspecies based on morphological differences, primarily color pattern. Stebbins proposed that an ancestral population in Oregon or Northern California divided into two subpopulations, one expanding down the Coastal Range of California and the other down the western slopes of the Sierra Nevada Mountains. These expansions created a ring encircling the Central Valley of California. Geographically adjacent subspecies hybridize around this ring, but in southern California two subspecies exist in some areas without hybridizing (Brown 1974). The entire complex is considered one species despite complete reproductive isolation of the terminal populations. Studies of allozymes and mtDNA (Wake and Yanev 1986; Moritz et al. 1992) confirm that genetic distance is higher between the terminal populations than between most adjacent populations. Within each of several subspecies, one finds a positive correlation between genetic and geographic distance (Jackman and Wake 1994), showing that distance is indeed a barrier to gene flow.

Reanalyzing the genetic data, however, Highton (1998) concluded that *E. eschscholtzii* is not a true ring species but a "superspecies": a complex of roughly eleven distinct species whose divergence began in allopatry about 7–10 million years ago. (See Wake and Schneider [1998] for a response.) Some adjacent subspecies show exceptionally large genetic divergence, nearly three times higher than those characterizing sympatric sister species of other plethodontids. Moreover, the subspecies do not intergrade gradually, but form fairly narrow hybrid zones across which there is abrupt and substantial differentiation of allozymes and morphology. This coordinated change of characters suggests secondary contact after allopatric differentiation. Finally, it is not clear whether the required topography for ring speciation (two unbroken forest corridors meeting in Southern California) existed 5-10 million years ago when the salamanders began to diverge (Wakabayashi and Sawyer 2001). As Wake (1997, p. 7767) noted, "The history of this complex has probably featured substantial [geographic] isolation, differentiation, and multiple recontacts." In light of these problems, it seems premature to consider *Ensatina eschscholtzii* as a good example of a ring species.

Perhaps the most plausible ring species is the greenish warbler complex (*Phylloscopus trochiloides*) in Asia (Irwin et al. 2001a). Plate 1 shows the distribution of this species around the Tibetan Plateau, along with a mtDNA-based phylogeny of populations and sonograms of male songs from different populations. A presumed ancestral population in the Himalayas appears to have expanded northward about 2.5 million years ago, circling the plateau in two directions. The terminal populations encountered each other in central Siberia, overlapping in the hatched area of Plate 1 (the gap in the ring in northeast China probably reflects recent deforestation). As the expansion proceeded, populations diversified both in plumage and in male song, which became longer and more complex.

Playback experiments show that although males of adjacent populations respond to each other's songs, this recognition decreases with distance. Males from the terminal populations in Siberia do not respond to each other's songs, suggesting that they regard each other as members of different species. Irwin et al. (2001a) argue that the difference in songs reflects parallel sexual selection for more complex songs, which yielded divergent responses since songs can become complex in different ways. Reproductive isolation of the "terminal" populations is supported by the near-absence of introgression for mtDNA. Finally, molecular data provide the critical evidence that genetic exchange between populations decreases with distance. However, given the tendency of mtDNA to introgress more readily than nuclear DNA, this conclusion needs confirmation using nuclear genes.

The one problem with this story is that the mtDNA phylogeny in Plate 1 shows a deep bifurcation (about 6% sequence divergence) between populations east and west of Kashmir, a discontinuity that can be interpreted in two ways. First, it may represent a polymorphism in the ancestor that became differentially sorted by selection or drift into the two expanding lineages. Computer simulations show this can happen, albeit rarely (Irwin 2002). It is more likely, however, that the break is the remnant of two originally allopatric taxa that separately gave rise to eastern and western populations. This suggests that, as in *Ensatina*, genetic divergence along the ring was enhanced by geographic isolation. Although this would spoil the idea that reproductive isolation occurred without allopatry, it does not affect the conclusion that reproductive isolation was enhanced by reduced gene flow around the ring.

Realizing that many of the ring species he first described in 1943 probably involved periods of allopatry, Mayr later acknowledged (1963, p. 512) that "not a single case has been proved unequivocally." Molecular analyses of other putative ring species, such as the great tit (*Parus major*) and the herring gull (*Larus argentatus* complex), have also suggested the involvement of allopatry (Kvist et al. 2003; Liebers et al. 2004). We still have no airtight example of a ring species, and it may be too much to hope for one given the difficulty of showing that a ring was never broken. Nevertheless, the *Phylloscopus* example still supports a pivotal role for reduced gene flow in speciation. The real importance of ring species is not that they provide support for allopatric speciation—which is amply documented by other evidence—but that they show in a novel way that reducing gene flow promotes speciation.

There are many convincing examples of single vicariant speciation events, as seen in the numerous sister taxa occupying areas separated by continental drift or rising land masses. It is harder, however, to find evidence that vicariant speciation is *pervasive*. This would require good phylogenetic data for the existence of sister species, accurate molecular clocks, and strong evidence for geographic barriers in many groups. These requirements are, however, often met in cases of peripatric speciation.

Peripatric speciation

The only difference between vicariant and peripatric speciation is that in the latter case one population is very small. Two crucial consequences follow from this distinction. First, peripatric speciation may often involve the invasion of novel habitats that exert strong selection. Second, genetic drift becomes important in small populations, and could play a role in speciation.

The unequal division of an ancestral range can occur in three ways. First, in what is called a "founder event," one or a few individuals colonize a distant habitat, such as an isolated island or lake. Colonization is presumably rare enough so that there is no gene flow between ancestral and colonizing populations. Second, a small peripheral population may become isolated from the rest of the species by a geographic barrier. Finally, a non-peripheral population within a species' range could become isolated long enough to achieve reproductive isolation. The first two variants, particularly the island model, are the most credible, for it is hard to imagine that a central population could remain geographically isolated long enough to become a new species. The number of founders or isolated individuals is not specified in these models, but the generally accepted range is between 1 and 100.

THEORY. Models of peripatric speciation are usually verbal. Except for the possibility of strong drift after colonization, the forces causing speciation are thought to be identical to those causing vicariant speciation. However, colonization events have special features that may lead to especially rapid speciation. For example, colonization is often followed by rapid adaptation to novel habitats, which may have completely different biota or climates, as in the Galápagos archipelago vs. mainland South America. Such strong selection can cause rapid speciation as a byproduct. Moreover, colonists carry a restricted sample of the genetic variation present in the ancestral population. Thus, even if selection is identical in the two habitats, the alleles that respond may differ, and these genetic differences can form the basis for further divergence. Finally, genetic drift in a colonizing population can lead to the fixation of novel alleles, which, although deleterious or neutral, can promote the evolution of incompatibilities between ancestor and descendant.

There are, however, many mathematical and verbal models arguing that peripatric speciation can result from the interaction between selection and drift. Such interactions are called *founder effects*, and should not be confused with founder *events*, which are simply instances of colonization. Founder events are uncontroversial, but founder effects are not. Because founder-effect speciation involves genetic drift, we defer its discussion until Chapter 11, where we conclude that it is unlikely.

EXPERIMENTAL EVIDENCE. Chapter 11 also discusses laboratory work on founder-effect speciation, all of which involves repeatedly reducing the size of laboratory populations ("bottlenecking") and determining whether this procedure causes reproductive isolation. Here we consider only laboratory studies involving a *single* bottleneck, which seems a more realistic model of colonization. Very few experiments have used this single-bottleneck design. The best known is that of Powell (1978) and Dodd and Powell (1985) tested sexual isolation between 12 lines of *Drosophila pseudoobscura* put through four single-pair bottlenecks, each followed by a "flush" to large population size. While they observed weak sexual isolation between lines after a *series* of bottlenecks (see Chapter 11), only one pair of lines showed isolation after a single bottleneck. Moreover, this pair did not involve an ancestral and a founder population—as predicted by theories of founder-effect speciation—but two founder populations.

The artificial-selection experiments described in Table 3.1 may be considered models of peripatric speciation, as they involve not only small populations (selection experiments invariably begin with few individuals), but also divergent selection. The rapid evolution of behavioral isolation among these lines suggests that novel and strong selection pressures in small populations can promote speciation, but does not tell us if population size is important.

EVIDENCE FROM NATURE: SPECIATION ON ISLANDS. As Darwin (1859) first recognized, novel, species-rich groups endemic to isolated islands and archipelagoes prove the occurrence of peripatric speciation after colonization. However, the presence of such adaptive radiations, as in Hawaiian *Drosophila* and lobeliads (*Cyanea*), does not show that all or even most of the species originated peripatrically. The presence of, say, 20 species in a monophyletic group on an isolated archipelago does not imply 20 instances of colonization—there might have been only a few such colonizations followed by sympatric or parapatric speciation on single islands. In some cases, though, we can get a rough idea of the *proportion* of speciation events on islands and archipelagoes that involved colonization.

1. Peripatric speciation on single oceanic islands. The presence of endemic species on oceanic islands whose closest relatives inhabit a nearby continent is proof of peripatric speciation. However, many such isolates, particularly in plants, are designated as species based on morphological difference, with little or no information about reproductive isolation. Such cases become more convincing if the island species is also the sole representative of a genus, which indicates more pronounced morphological difference and a greater chance of reproductive isolation. Carlquist (1965) gives many examples of such species in plants, including the whitewood (*Petrobium arboreum*) on St. Helena and the bizarre "cucumber tree," (*Dendrosicyos socotrana*) on the Arabian Sea island of Socotra, which harbors ten endemic plant genera. (Some of these species, however, may not have evolved on the island but rather constitute remnants of formerly widespread distributions: *D. socotrana* is recorded from Djibouti and may have been widespread in Africa, to which Socotra was anciently connected [Evans 2001]). Stattersfield et al. (1998) list island endemics in birds, of which the classic example is the Cocos Island finch, *Pinaroloxias inornata*—the only Darwin's finch inhabiting this remote island.

In some cases, we know that divergent island endemics have achieved reproductive isolation. *Drosophila sechellia*, for example, is the only endemic species of the *D. melanogaster* subgroup on the Seychelles archipelago, and shows complete ecological isolation (Legal et al. 1994; Jones 1998, 2004), as well as substantial behavioral and postzygotic isolation, from its closest relatives (Coyne and Orr 1989a).

Occasionally a *pair* of endemic, related species is sympatric on an oceanic island. When these remain distinct, there is no doubt that they are reproductively isolated. While this might suggest parapatric or sympatric speciation on the island, it can also result from "double invasions," in which a mainland ancestor colonizes an island at two widely spaced times, with each colonist evolving into a new species. The classic evidence for double invasion is that one island species is much more similar morphologically to a mainland relative than is the other. Mayr (1942) and Williamson (1981) give many possible examples of double colonizations, mostly in birds. More recently, molecular/phylogenetic studies of such pairs indicate that they have indeed resulted from double colonization rather than intra-island speciation (Coyne and Price 2000). Each such observation represents two cases of peripatric speciation.

2. Peripatric speciation on archipelagos. Understanding the biogeography of speciation becomes more complicated for species on archipelagos. However, phylogenetic analysis may allow us to estimate how much speciation was truly peripatric (i.e., involving inter-island colonization).

The Hawaiian *Drosophila* is probably the most thoroughly studied—and indeed, a paradigmatic—example of adaptive radiation. Biologists have accumulated extensive data on the biogeography, phylogeny, and degree of reproductive isolation of many species. The archipelago harbors around 400 species, but there are probably over 800—fully a third of the world's *Drosophila* fauna (Carson and Kaneshiro 1976; Carson and Templeton 1984). Molecular analysis shows that all species in the group descended from a single ancestor

(DeSalle 1995). Nearly all species are endemic to single islands or to the three-island complex of Maui, Molokai, and Lanai that was a single island until about 0.3–0.4 million years ago (Carson and Clague 1995). Of 101 species in the well-studied "picture-winged group," for example, only two occur on more than one island. Closely related species, which are often allopatric, frequently show large ecological differences, as well as strong behavioral or postzygotic isolation in the laboratory (Carson and Kaneshiro 1976; Kaneshiro 1989). Surprisingly, molecular data suggest that the divergence of some lineages occurred before the oldest existing island was formed (5.1 million years ago), implying that colonization occurred on ancient islands now eroded below sea level (Beverley and Wilson 1985; DeSalle 1995).

One can use phylogenetic data, based on either chromosome banding patterns or molecular divergence, to estimate the frequency of colonization events (Carson 1970, 1983; Kaneshiro et al. 1995; DeSalle 1995). Molecular studies show that colonization nearly always involved movement from an older to a younger island (northwest to southeast), implying that species are formed more easily on islands with unoccupied niches. Using this assumption and the pattern of chromosome arrangements, Carson (1983) inferred that at least 45 inter-island colonizations are needed to explain the formation of 97 picture-winged species (Figure 3.5). Combining this result with likely cases of colonization *within* islands, one can conclude that more than 50% of species originated peripatrically. On more recently formed islands, where the chromosomal data are more reliable, this percentage is even higher.

Studies of flightless Hawaiian crickets in the genus *Laupala* also suggest that much speciation was peripatric (Otte 1994; Shaw 2002). Sister species of crickets usually inhabit different islands, and virtually no species occurs on more than one island. Shaw (2002) and Mendelson and Shaw (2004) estimate that a minimum of 17 out of 36 well-studied cases of speciation were peripatric, a ratio similar to that of *Drosophila*. This similarity is surprising as the crickets are flightless, which should facilitate intra-island speciation. Frequent peripatric speciation is also inferred in Hawaiian spiders of the genus *Tetragnatha*: sister species invariably occur on different islands (Gillespie and Croom 1995).

Givnish et al. (1995, 1998) estimated the frequency of peripatric speciation among monophyletic Hawaiian lobeliads of the genus *Cyanea*. This group contains 54 living and 10 extinct species that form a remarkable adaptive radiation on Hawaii, taking forms as diverse as trees, shrubs, herbs, and succulents. Many taxa are rare and have allopatric distributions, precluding observation of isolating barriers. In some areas, however, up to five species coexist without hybridizing (Givnish, pers. comm.). Molecular and biogeographic data (see Figure 3.5) suggest that at least 27% of existing species arose after colonizing a new island (Givnish 1998). This is probably a severe underestimate of the frequency of peripatric speciation given the highly localized distribution of many species and the intra-island barriers formed by volcanoes and erosion. As in the Hawaiian *Drosophila*, colonization in *Cyanea* has largely been from older to younger islands.

Finally, when islands of an archipelago are formed successively over time, as in Hawaii, peripatric speciation can be deduced from correlations between

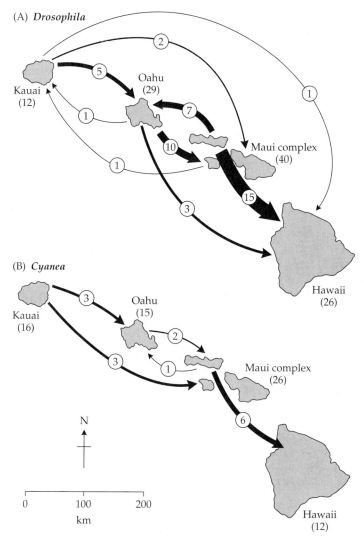

Figure 3.5 Minimum number of interisland colonization events ("founder events") needed to explain the present distribution of species in two groups. (A) Hawaiian *Drosophila* (97 picture-winged species); (B) Hawaiian *Cyanea* (54 living and 10 extinct lobeliad species). Arrows indicate directions of colonization, which in both cases are predominantly from the older (NE) to younger (SW) islands; thickness of arrows is proportional to the number of colonization events. The number of species in each group inhabiting each island is given in parentheses; these include all species and not just single-island endemics. (From Carson 1983 and Givnish 1998.)

the appearance of islands and of species. Grant and Grant (1996) describe a striking correspondence between the number of species of Darwin's finches and the number of islands existing at a given time (Figure 3.6), implying that island number was a rate-limiting factor in speciation. While such a coincidence could result from a combination of peripatric *and* parapatric or sympatric

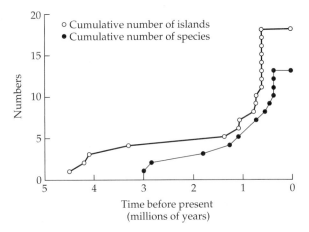

Figure 3.6 Coordinated increase over time in the number of islands in the Galápagos archipelago and the number of species of Darwin's finches on the archipelago. (From Grant and Grant 1998a.)

speciation, there is no evidence for the latter two processes in island birds (Coyne and Price 2000).

3. Peripatric speciation in peripheral isolates. It is much harder to detect peripatric speciation on continents than on islands or archipelagos. The existence of sister species having abutting distributions but grossly unequal range sizes may result not from peripatric speciation, but from vicariant speciation followed by reductions in range size. But if there is no evidence of a geographic barrier other than distance, it becomes more likely that colonization itself created the barrier to gene flow.

The best example of mainland speciation that probably involved an isolated population is a pair of plant species in central California, *Clarkia biloba* and *C. lingulata* (Lewis and Roberts 1956). The range of *C. biloba* extends down the canyon of the Merced River; all individuals have eight pair of chromosomes. *C. lingulata*, on the other hand, occupies only two restricted sites near the southern end of this distribution. The ranges of the two species are separated by roughly 100 yards, and there is no sign of a barrier to gene flow. Individuals of *C. lingulata* have nine pair of chromosomes, with the karyotype also differing from that of *C. biloba* by a translocation and a paracentric inversion. These chromosomal differences appear to cause meiotic problems in artificially produced hybrids, which are nearly sterile. No such hybrids are found in nature. Lewis and Roberts (1956) suggested that a colonizing population of *C. biloba*, genetically isolated because of limited seed dispersal and some self-compatibility, speciated rapidly through the accumulation of chromosomal rearrangements.

The pattern of allozyme variation supports this scenario. The genetic similarity of the species shows that they are close relatives, probably sister species. More important, *C. lingulata* carries only a subset of the alleles present in *C. biloba*, and has a heterozygosity that is 30% lower (Gottlieb 1984a). Finally, *C. lingulata* showed significantly less variation among individuals for nine morphological traits studied in the greenhouse (Lewis 1973). The biogeographic and genetic data thus suggest that *C. lingulata* arose as a small peripheral isolate of a *biloba*-like ancestor.

Another plausible case of peripatric speciation involves changes in chirality (direction of coiling) in several species of land snails (Gittenberger 1988; Orr 1991b; Van Batenburg and Gittenberger 1996; Ueshima and Asami 2003). Right-handed and left-handed coiling morphs, which differ by a single gene, are behaviorally isolated because they cannot copulate properly. Yet the transition from one form to another, producing new species, has occurred repeatedly. Because such a transition would be opposed by natural selection (new coiling morphs are at a reproductive disadvantage), it is likely that speciation occurred peripatrically, when isolated populations of snails experienced strong genetic drift that overcame selection. Crossing this adaptive valley requires a concatenation of rare events: single-locus determination of coiling, maternal inheritance, and extremely small population size (fewer than 20 individuals). This may be the only plausible example of founder-effect speciation involving an interaction between drift and selection (Chapter 11). Indeed, a single species of snail, *Euhadra aomoriensis*, may have originated several times independently by the repeated fixation of new coiling genes in different populations of a common ancestor (Ueshima and Asami 2003). Finally, Barraclough and Vogler's (2000) study correlating the biogeography of sister taxa with their divergence time (Chapter 4) shows that some recently formed allopatric taxa had very asymmetric range sizes—significantly more asymmetric than produced by null models involving random divisions of an ancestral range. This asymmetry is consistent with peripatric speciation, although, as Barraclough and Vogler note, other interpretations are possible.

Parapatric Speciation

Falling midway between allopatric and sympatric speciation, parapatric speciation involves the evolution of reproductive isolation between two populations that exchange genes, but not freely. In population-genetic terms, such speciation occurs when the fraction of genes *initially* exchanged between taxa lies between 0 (allopatric speciation) and 0.5 (sympatric speciation).

Models of parapatric speciation fall into two groups. "Clinal models" envision a single species continuously distributed across a variable environment. Subpopulations become adapted to their local habitats, although this adaptation is hindered by gene flow from adjacent, ecologically different populations. Once localities have differentiated, a form of reinforcement might operate, with prezygotic isolation evolving as a response to maladaptive hybridization between populations. It is not often remembered that, at least in his "big species book," Darwin strongly advocated this form of speciation:

> Although I believe the former broken & isolated state of parts of now continuous areas, & in a lesser degree the voluntary separation of the varieties of the higher animals, have played a very important part in the formation of species since become commingled, or just meeting in a border territory, I do not doubt that many species have been formed at

different points of an absolutely continuous area, of which the physical conditions graduate from one point to another in the most insensible manner (Darwin, p. 266, in Stauffer 1975).

Darwin's scenario involved gradual adaptation to the environment that somehow led to the formation of several distinct species along a gradient. The intermediate forms were driven to extinction by species at the end of the gradient, which then expanded their ranges until they met in the middle. (Darwin believed that the fittest species were those adapted to the most extreme conditions.) Fisher (1930) proposed a related model that we discuss below.

The second form of parapatric speciation is the "stepping-stone model," which posits discrete populations having restricted gene exchange. This restriction makes it is easier for selection or drift to override gene flow and produce reproductive barriers between populations as a byproduct. Given that marginal populations experience the least gene flow, it is often suggested that parapatric speciation is most likely near the edge of a species' range. But some models, like White's (1978) theory of "stasipatric speciation," propose that populations embedded well within a species' range can experience sufficiently low gene flow that they become new species. Obviously, there can be situations with features of both clinal and stepping-stone models. Many species do not comprise discrete populations, but show large differences in density among localities, leading to variable amounts of gene flow over space. But with the exception of stasipatric speciation, different forms of parapatric speciation leave identical biogeographic signatures: newly formed sister species that have abutting ranges.

Barton and colleagues (Barton and Hewitt 1981a, 1989; Barton 1988) proposed a hybrid between parapatric and allopatric speciation, which we call "para-allopatric speciation." Here, populations initially differentiate in parapatry but attain substantial reproductive isolation only after a subsequent period of allopatry (see below). Conversely, speciation via reinforcement can be considered *allo-parapatric* or *allo-sympatric*, whereby populations that diverged during allopatry evolve into full species only after they experience secondary contact. In these "hybrid" models, genetic variation fixed during allopatry is essential for completing speciation.

After reviewing mathematical theory and the data from nature and the laboratory given below we conclude that parapatric speciation is not only theoretically plausible, but probably occurs with reasonable frequency in nature. The difficulty lies in convincingly *demonstrating* its occurrence, for other modes of speciation can produce similar biogeographic patterns. This places us in the awkward position of being unable to document what might well be a common mode of speciation.

Theory

Parapatric speciation is fairly uncontroversial, at least as a theoretical possibility. Population genetics tells us that populations can diverge by adapting to local conditions even while exchanging genes—especially when gene exchange is limited (Haldane 1948; Slatkin 1973). It thus seems reasonable that reproductive isolation, especially of the ecological variety, may evolve between populations that exchange migrants. Indeed, for any level of hybridization, there is a threshold value of selection or drift that can override this gene flow and produce speciation. Moreover, unlike allopatric speciation, parapatric speciation does not require impermeable geographic barriers, but merely restricted gene flow and environmental differences between populations. We can already guess, then, that it is relatively easy to get parapatric speciation to work in theory. But whether such models are applicable to nature depends, as usual, on whether their assumptions are biologically realistic. In the case of parapatric speciation, these assumptions involve rates of gene flow, population sizes, and, most important, the genetic basis of reproductive isolation.

CLINAL SPECIATION.   Fisher (1930, pp. 125–129) was the first to propose a full theory of clinal speciation—a verbal model involving populations locally adapted to an environmental gradient. This geographic differentiation, argued Fisher, would favor selection for reduced migration, preventing the movement of individuals into areas where their genes were unfit. While Balkau and Feldman (1973) confirmed this conclusion mathematically, reduced migration per se will not produce biological species, but only strong genetic differentiation over space. In the limit, when migration becomes zero, one is left with a series of contiguous, differentiated populations that are not reproductively isolated. But Fisher, adhering to a morphological species concept, considered this differentiation to be speciation. As an alternative, he proposed that assortative mating would evolve among differentiated populations to prevent maladaptive hybridization. Although this scenario does yield biological species, Fisher seemed less interested in reproductive isolation than in morphological divergence. In a mathematical model, Lande (1982) extended Fisher's theory by incorporating sexual selection acting along an environmental gradient. Adaptive geographic differentiation in male traits can be amplified by the evolution of female preferences for those traits, yielding substantial but incomplete assortative mating between populations.

While other models have shown that clinal differentiation can lead to complete reproductive isolation, most have made assumptions that are either restrictive or biologically unrealistic. The first step in this process, the accumulation of genetic differences along a geographic gradient, is straightforward. Several studies have shown that genetic differences at one locus can accumulate over distances greater than $\sigma/\sqrt{2s}$, where σ is the standard deviation of dispersal distance per generation (the distance between birthplaces of parent and offspring), and s is a measure of selection acting on an allele at different positions in a cline (Slatkin 1973; Endler 1977). Clines in allele frequency need not be smooth but can form "steps": regions over which allele frequency changes abruptly. However, clines for alleles at different loci will not coincide unless these genes are affected by identical environmental factors.

Difficulties emerge when we try to understand how clinally varying populations become reproductively isolated. Two models have dealt with this problem. The first is related to Fisher's idea of selection for decreased migration. In this case, breeding times can diverge among subpopulations, causing temporal isolation. Caisse and Antonovics (1978) used computer simulation to show that reproductive isolation could evolve in this way. Their model, however, required a very simple genetic basis (a single locus under environmental selection and another affecting flowering time), large allelic effects, strong selection on the first gene, complete dominance of alleles at the second, and close linkage between the two genes. Finally, reproductive isolation was complete only between individuals at the ends of the cline, as there was gene flow along the entire gradient. Hence, the model yields clinal isolation, rather like a linearized ring species, instead of parapatric speciation. However, extinction of the intermediate populations could create two species.

Endler (1977) modeled Fisher's suggestion that locally adapted individuals might prefer to mate with individuals of similar phenotype, leading to the evolution of behavioral isolation. Endler's four-locus model involves one gene divergently selected along the cline, two other loci modifying the fitnesses of genotypes at this locus, and a fourth gene causing assortative mating of "ecological" genotypes. Although this model sometimes yields parapatric speciation in one part of the cline, the conditions are stringent: fitness modifiers must act in different directions on different genotypes, and alleles causing assortative mating must be neutral, with one completely dominant in heterozygotes. If such assumptions were relaxed (if, for example, alleles causing assortative mating were under selection), parapatric speciation might occur less often.

Doebeli and Dieckmann (2003) made a simulation model of parapatric speciation related to their earlier model of sympatric speciation (Dieckmann and Doebeli 1999; see Chapter 4). They assume a resource gradient along which individuals experience frequency- and density-dependent competition. This competition produces disruptive selection that drives phenotypes further apart. In both sexual and asexual taxa, the phenotypes of individuals eventually assume a bimodal distribution along the cline. In the sexual models, this bimodality is said to be caused by the evolution of assortative mating, triggering the development of reproductive isolation between emerging clusters. Such speciation occurs most readily when the cline is moderately steep and the movement of individuals restricted. Polechova and Barton (2004), however, showed that, in the asexual model, clusters still arise even when there is *no* competition between genotypes (i.e., no frequency-dependent selection based on the environment). In this case, parapatric speciation appears to be an artifact of the model's assumptions, with clusters forming as products of edge effects (population density becomes higher at the ends of the cline because there is less immigration of maladapted individuals). The importance of edge effects is shown by the failure of clusters to appear when the cline is

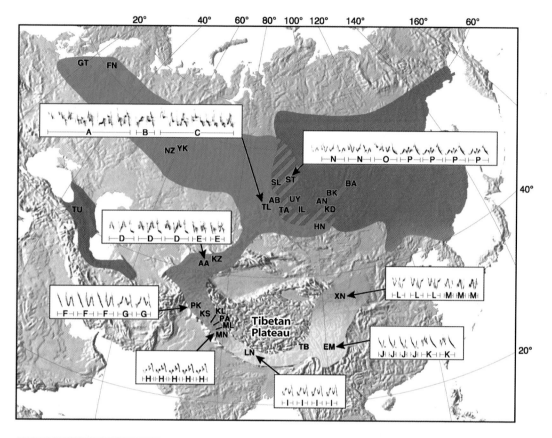
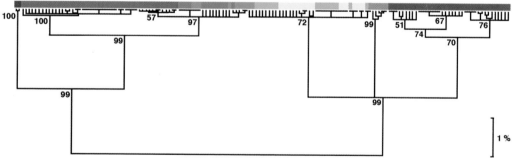

PLATE 1 A possible ring species: the distribution, song types, and mtDNA phylogeny of the greenish warbler complex (*Phylloscopus trochiloides*) in Asia. Ranges of the six recognized subspecies are shown in color: *nitidus* (purple); *viridanus* (blue); *ludlowi* (green); *trochiloides* (yellow); *obscuratus* (orange); and *plumbeitarsus* (red). The striped red/blue area is the presumed terminal meeting point of the ring, where *viridanus* and *plumbeitarsus* meet but do not hybridize. Pairs of letters show sampled populations, and the phylogeny of mtDNA haplotypes found in each population is given at the bottom of the map. Scale bar gives the substitution rate, and numbers at the nodes of the phylogeny give the support values (in percent) for major clades. (Note large bifurcation between populations east and west of Kashmir.) Sonograms are shown for representative songs of each subspecies: the horizontal axis is time, height of traces indicates frequency, and darkness of pattern is volume. Letters in sonograms indicate distinct song units that are repeated. The southern subspecies, with simple songs, are the presumed ancestors, and song becomes more complicated as one moves northward in both directions around the ring. (From Irwin et al. 2001, redrawn by Irwin.)

PLATE 2 *Mimulus lewisii* and its sister species *M. cardinalis*, an example of strong pollinator isolation (see Chapter 5). Top panel: *M. lewisii* (left), with broad pink flowers, recessed anthers, and low nectar volume, is pollinated almost exclusively by bees (here by the bumblebee *Bombus vosneskenskii*). In contrast, *M. cardinalis* (right), with red, tubular flowers, excerted anthers, and high nectar volume, is pollinated almost exclusively by hummingbirds (here by the rufous hummingbird, *Selasphorus rufus*). Bottom panel: The two pure species and various hybrids. Top row: *M. lewisii* (left), F_1 hybrid (middle), and *M. cardinalis* (right). Rows 2 and 3: a selection of F_2 hybrids showing the variation in color and morphology used for genetically mapping the species differences. (Photos courtesy of D. Schemske and T. Bradshaw; bottom panel from Schemske and Bradshaw 1999.)

placed on a topological figure lacking edges instead of on a linear gradient. Further analysis is required to determine whether speciation in the sexual model might also be an artifact rather than a result of competition and assortive mating.

"STEPPING-STONE" SPECIATION.   One might guess that a species divided into discrete populations would undergo parapatric speciation much more easily than one distributed along a cline. This appears to be true. The first such model was Felsenstein's (1981), an analysis of parapatric speciation between two hybridizing populations. His model involves three loci in a haploid species. Two diallelic loci (B and C) are involved in local adaptation. At a third diallelic "assortative-mating" locus, carriers of each allele mate preferentially with others of the same genotype (A with A, and a with a). To model parapatric speciation, Felsenstein assumed that populations exchange 10% of their genes each generation. If one assumes that assortative mating is strong (roughly of 40% or more), that there is tight linkage between the mating locus and one "ecological" locus, and that selection on the two ecological loci is very strong, the two populations become fixed for genotypes ABC and abc (or aBC and Abc) respectively. This yields populations showing substantial reproductive isolation, but these are not species unless assortative mating is nearly complete. Speciation could be completed by the evolution of modifier loci that increase assortative mating, and Felsenstein (1981) shows one computer simulation yielding this result. But the conditions for parapatric speciation are quite restrictive.

A related model by Gavrilets (2000b) involves speciation via local adaptation in a two-dimensional grid of populations that exchange migrants. In this case, alleles conferring adaptation yield behavioral or intrinsic postzygotic isolation as byproducts. Parapatric speciation was achieved with certain combinations of selection, mutation, and migration rates, but could be very slow. This raises the question of whether environmental grids in nature could persist long enough to allow speciation.

Gavrilets et al. (1998) also made non-ecological models of parapatric speciation occurring on a similar grid of populations. Here, alleles do not confer local adaptation but are slightly deleterious, and must be fixed by genetic drift. Parapatric speciation can occur if alleles from different populations are incompatible, causing behavioral or intrinsic postzygotic isolation. Again, the conditions are restrictive: populations must be small enough to allow fixation of deleterious alleles, and mutation rates to new alleles must be relatively large. Moreover, the rate of speciation is very slow because it depends on the rare fixation of deleterious mutations.

In Barton and Hewitt's (1981a, 1985) verbal models of para-allopatric speciation, genetic divergence occurs within a continuous species range, but only between distant rather than adjacent populations. In this model, the mechanisms of divergence that cause allopatric speciation can also yield "parapatric" speciation if ranges are sufficiently large. Like the terminal populations of ring species, distant populations can accumulate, through divergent selection, genetic differences that cause substantial reproductive isolation. Alternatively, divergence can produce clines at different loci that criss-cross in a complicated pattern. The species range then becomes a mosaic of genetically differentiated regions. If some populations become extinct and the remaining populations come into secondary contact by recolonizing empty habitats, the combination of alleles from some populations might be incompatible on the genetic background of others, producing postzygotic isolation. This process will be accelerated if species' ranges repeatedly expand and contract (Hewitt 1989).

Such changes in range are not only plausible, but probable. Given fluctuations in climate and geology over time, the present ranges of most species probably bear little relation to their ranges in the past. In the northern hemisphere, for instance, glacial cycles have led to repeated shrinking and expansion of species ranges. Many related taxa may thus have experienced multiple episodes of parapatry, sympatry, and allopatry. Consequently, models of para-allopatric speciation do not seem "parapatric" in the conventional sense (i.e., speciation that occurs between adjacent, hybridizing populations). Rather, incipient species are formed at a distance, with gene flow between them very restricted. The border between species formed para-allopatrically reflects not where they arose, but where they achieved secondary contact. Neither are these models "allopatric" in the conventional sense, for they produce reproductive isolation without geographic barriers.

In clinal and stepping-stone models, then, parapatric speciation is theoretically possible, but its plausibility depends on a balance among biological parameters that are hard to measure in nature. While the para-allopatric scenarios of Barton and Hewitt (1981a, 1985) seem more realistic, in nature they would be hard to distinguish from either pure allopatric speciation or allopatric differentiation followed by reinforcement.

STASIPATRIC SPECIATION AND OTHER CHROMOSOMAL MODELS. Stasipatric speciation is a form of parapatric speciation occurring in a large species subdivided into small populations. As White (1978, p. 177) characterizes it,

> Essentially, the stasipatric model envisages a widespread species generating within its range daughter species characterized by chromosomal rearrangements that play a primary role in speciation because of the diminished fecundity or viability of the heterozygotes. The daughter species are assumed to gradually extend their range at the expense of the parent species, maintaining a narrow parapatric zone of overlap at the periphery of their distribution within which hybridization leads to the production of genetically inferior individuals (usually inferior because of irregularities at meiosis).

In Chapter 7 we describe the problems that plague most models of speciation based on chromosomal differences. The main difficulty is the near impossibility of fixing chromosome rearrangements that are sufficiently unfit as heterozygotes to cause substantial reproductive isolation. In parapatry, where gene flow constantly opposes the fixation of such underdominant rearrangements, the probability of such speciation is vanishingly small. Indeed, this problem is so serious that the stasipatric model no longer seems viable (Futuyma and Mayer 1980; Barton and Hewitt 1981a).

Navarro and Barton (2003) suggest a somewhat more plausible model of chromosomally based parapatric speciation, a variant of the allo-parapatric model of Noor et al. (2001b) and Rieseberg (2001) described in Chapter 2. Navarro and Barton propose that different chromosome arrangements initially become fixed in two adjacent populations because they contain genes that are adapted to each habitat. After this differentiation, the arrangements can accumulate additional alleles that are adaptive in each locality but that are incompatible with alleles from the other locality. (Because such alleles would be eliminated by migration and recombination, it is hard for them to accumulate outside of rearrangements.) Thus, the rearranged part of the genome becomes an increasingly strong isolating barrier. By delaying the exchange of genes between two parapatric populations, chromosomal rearrangements can then allow reinforcement to complete the evolution of reproductive isolation before populations become genetically homogenized.

While this parapatric model seems plausible, it is not clear how efficiently it would operate in what Navarro and Barton (2003, p. 455) call "the more realistic situation of a continuous population." Nevertheless, the process they suggest might be involved in *allopatric* speciation and reinforcement (see Chapter 10). Regardless of the biogeographic scenario, this model predicts that in fully or partly sympatric taxa, alleles causing reproductive isolation will be located within rearranged areas of the genome, a prediction supported by some evidence (Noor et al. 2001a).

Experimental evidence

We are aware of only one experiment deliberately modeling parapatric speciation. Soans et al. (1974) studied reproductive isolation between two populations (*Musca domestica*), one selected for positive geotaxis and the other for negative geotaxis. Migration between these lines occurred at a rate of 30% per generation. After 38 generations of selection, significant assortative mating was found between individuals from the two lines. Moreover, the strength of assortative mating was roughly equal to that occurring between "allopatric" lines selected *without* gene flow.

However, successful studies of *sympatric* speciation, in which reproductive isolation develops between laboratory populations selected in different directions but allowed to exchange genes, also provide support for parapatric speciation. After all, it should be easier to evolve reproductive isolation with limited gene flow than with initially free gene flow. We describe these studies in the next chapter.

Evidence from nature

It is almost impossible to demonstrate parapatric speciation in nature, a problem that distinguishes this process from allopatric and sympatric speciation, which can be deemed likely or unlikely from observations in the field. For example, the existence of two sister species, one on an oceanic island and the other on the mainland, strongly suggests allopatric speciation. Similarly, the observation of a monophyletic group of species in a restricted area that is unlikely to have experienced geographic barriers (e.g., parasites on a single host species) strongly suggests sympatric speciation. However, sister species with adjacent ranges do not necessarily imply parapatric speciation: plausible alternative explanations are almost always possible, typically involving secondary contact between groups that diverged in allopatry, or competition between related species that prevent them from occupying the same area. We discuss several lines of evidence that have been or could be used to support parapatric speciation.

1. A pair of closely related species with abutting distributions. This observation provides only weak evidence for parapatric speciation for several reasons, chief among them being that species ranges must often change much faster than the evolution of reproductive isolation. Ranges that we see today, even in the tropics, may be so jumbled that they provide no biogeographic information about the history of speciation.

The possibility of range changes constitutes a special problem for parapatric speciation because it leaves allopatric speciation as a plausible alternative. The ranges of spatially isolated species may expand after geographic barriers disappear. There can then be secondary contact between good species, or between incipient species that subsequently attain full reproductive isolation via reinforcement. This secondary contact might produce abutting ranges for several reasons. First, competition between species might prevent them from occupying the same habitat. This is, of course, a special problem for closely related species, which often use similar resources. (One might posit that sympatric speciation could produce the same geographic pattern. But as we explain in Chapter 4, most theories of sympatric speciation involve evolution that reduces or eliminates competition, so that newly formed species need not be forced into parapatry.) In Chapter 5, we give several examples of parapatric distributions that are almost certainly due to competition: when one species is removed, another moves in to fill the vacuum. Second, species may have evolved habitat isolation in allopatry, in which case their range expansion will stop at an ecological boundary between two habitats. Finally, even without habitat differences, species experiencing secondary contact can maintain permanently adjacent distributions separated by narrow hybrid zones.

These hybrid zones, in which two "species" have parapatric distributions and hybridize at the border, are a classic bone of contention. They can be seen as either *primary hybrid zones* arising via parapatric speciation (e.g., Endler 1977) or as *secondary hybrid zones* arising from secondary contact occurring between previously allopatric populations (e.g., Barton and Hewitt 1985). There is wide agreement with Endler's (1982b) view that it is hard to discriminate between these two explanations.

Barton and Hewitt (1985) list 106 hybrid zones in diverse groups. About 40% of these zones show evidence of earlier geographic fragmentation, suggesting that they reflect secondary contact. For example, the well known hybrid zone between the fire-bellied toads *Bombina bombina* and *B. variegata* occurs in a part of eastern Europe known to have been glaciated (Szymura and Barton 1986).

Genetic data gives other evidence for secondary contact. Frequently, clines for diverse characters show near-identical changes in frequency as one moves across a hybrid zone (Hewitt 1989). In the *Bombina* hybrid zone in Poland, for example, these congruent clines involve allozymes, mtDNA, color patterns, skin thickness, fecundity, development time, and mating calls. Although parapatric speciation can occasionally produce congruent clines for a few traits (Endler 1977), it is unlikely to produce congruent clines for *many* traits (Hewitt 1989): unless alleles at all loci respond to the environment in the same way, they will tend to form clines at different locations. Moreover, neutral alleles will not change congruently with selected ones.

How, then, do congruent clines arise? The most plausible explanation involves a combination of two factors: secondary contact after allopatry, and epistasis between genes that cause reproductive isolation. Assume, for example, that intrinsic postzygotic isolation evolves between geographically isolated populations. This will produce epistasis for fitness in hybrids: alleles that work well together in a pure-species background are unfit in a hybrid genome. In a hybrid zone, this epistasis will cause groups of genes to be selected as a unit, so that their frequencies diminish in concert as they penetrate farther into the range of the other species. Neutral alleles linked to selected ones will behave in the same way. Such hybrid zones are called *tension zones* because they reflect a balance between the formation and dispersal of hybrids on the one hand, and selection against the genes carried by hybrids on the other. Indeed, many hybrid zones show evidence of postzygotic isolation in the zone of contact, sometimes based on many loci (Barton and Gale 1993; Bronson et al. 2003).

Congruent clines might reflect not *complete* allopatry, however, but secondary contact between populations that once lay far apart within a large continuous range and which had accumulated many genetic differences. These differences could produce epistasis for multiple loci in the same manner as evolution in allopatry, and could yield congruent clines if diverged populations meet after intermediate populations become extinct. In such cases, the hybrid zones would not be "primary": that is, they would result not from speciation in adjacent populations, but from secondary contact between once-distant populations. What is clear is that congruent clines for both selected and neutral genes imply range changes and not speciation in situ.

Without evidence for geographic barriers, it is hard to distinguish between the allopatric and para-allopatric explanations for congruent clines. If many genes are involved in reproductive isolation, it seems more likely that true allopatry is involved, as it is hard to imagine that a large number of alleles causing incompatibilities could arise between populations that exchange genes, however rarely. At least 150 loci appear to cause hybrid unfitness in the zone between two races of *Podisma pedestris* (Barton and Hewitt 1981b) and 500 loci between *Bombina bombina* and *B. variegata* (Barton and Hewitt 1985). Given that these estimates are very rough, we badly need similar data from other hybrid zones.

Although models of parapatric speciation predict that hybrid zones will often lie in areas of ecological transition, this pattern can also arise after allopatric speciation. For example, tension zones tend to settle in areas of low population density (Barton and Hewitt 1985), which may occur at ecotones. Moreover, if species evolved different ecological adaptations in allopatry, the species could meet—and remain permanently disjunct—at an interface between their respective habitats. This is particularly apparent in mosaic hybrid zones, where the transition between species involves a changing array of discrete environmental patches. In such hybrid zones, "pure types" predominate in their respective patches, while hybrids form on the borders between patches (Harrison and Rand 1989). As Harrison and Rand note, such patterns suggest—but do not prove—that mosaic zones result from secondary contact. As in "smooth" hybrid zones, many concordant traits are associated with patch types, implying coevolution in allopatry or distant parapatry. In some cases, hybridization at patch boundaries produces concordant clines among diverse genes, as in the crickets *Gryllus firmus* and *G. pennsylvanicus* (Ross and Harrison 2002). Mosaic zones like those in *Gryllus* often occur over wide areas that include habitats unsuitable for either pure species or hybrids. In such cases, it seems more parsimonious to invoke multiple secondary contacts in suitable habitats than to imagine repeated cases of parapatric speciation producing identical species.

To summarize, in some cases genetic and environmental patterns suggest that parapatric speciation is less likely than secondary contact. But this still leaves many hybrid zones that might have resulted from parapatric speciation. To demonstrate such speciation with confidence, one requires evidence that *no* period of allopatry ever occurred during the origin of the relevant species. We know of no cases that meet this criterion.

Plants are likely to provide the best material for demonstrating parapatric speciation. Many "edaphic endemic" species of plants are restricted to unusual and localized patches of habitat, such as serpentine soils. Yet despite extensive work on such endemics (e.g., Krukeberg 1984; Macnair and Gardner 1998), there has been little interest in their phylogeography. If, for example, one determined that the sister species of a serpentine endemic was an adjacent species living on non-serpentine soil, this would provide convincing evidence for parapatric speciation.

2. Multiple pairs of related species with abutting distributions. When many pairs of sister species show hybrid zones at the same location, particularly at an ecotone, parapatric speciation is a possibility. This observation might suggest that the ecotone itself was responsible for repeated speciation events. However, secondary contact following allopatry can also yield multiple hybrid zones at an ecotone, as in the North American birds described above (Moore and Price 1993). Species in different groups might easily evolve adaptations to the same pair of habitats in allopatry, and experience secondary contact where those habitats eventually meet. Parapatric speciation becomes more likely—but not certain—if one can show with a molecular clock that the hybridizing pairs are of very different ages. But disparate "ages" might reflect only different levels of gene flow that distort the divergence times between species. While we do find multiple hybrid zones in some areas, arguments continue as to whether they reflect allopatry or parapatry. The congruent parapatric distributions of South American taxa described above, for example, may have resulted from the coordinated spread of taxa from allopatric refugia (Hall and Harvey 2002) or from repeated parapatric speciation across environmental gradients (Endler 1982a,b), assuming that these gradients were stable over times required for speciation. Such situations could be resolved by a greater understanding of the geological history of such areas, the divergence times of species pairs, and the ecology of the species and of the areas where they meet.

3. Morphological or genetic discontinuities at ecotones. Single species often show morphological or genetic changes at transitions between habitats. These clines can be gradual or abrupt, and have been described in the little grenbul (*Andropadus virens*) in Africa (Smith et al. 1997) and the lizard *Anolis roquet* on Martinique (Ogden and Thorpe 2002). In both cases, control transects within a single habitat showed that differentiation among populations was not due to genetic drift. The authors consider these cases as evidence for parapatric speciation, although they actually provide evidence only for habitat-specific selection. There is no controversy about whether such selection can cause populations to differ: evidence for this can be found in any evolution textbook. The decisive question is whether such parapatric differentiation is the first step in the evolution of reproductive isolation. Neither of the studies cited above give evidence for such isolation. Such work thus verifies a necessary but not sufficient step in parapatric speciation. But surely most geographically differentiated populations do not become full species (Futuyma 1987).

4. Observations of all stages of parapatric speciation in nature. Jiggins and Mallet (2000) survey parapatric populations in several species, finding among them examples of every level of reproductive isolation, ranging from hybrid zones in which different populations mate randomly, to species boundaries maintained by strict reproductive isolation. They argue that this continuum supports parapatric speciation: if all phases of the process can be seen in nature, then it is reasonable to conclude that the entire process can occur in a single taxon. While we concur that this continuum supports the possibility of parapatric speciation, we cannot agree that it supports its likelihood. The data are not conclusive because one would observe this pattern even if all speciation were allopatric. If the parapatric populations surveyed by Jiggins and Mallet resulted from secondary contact between previously allopatric taxa, one could see all degrees of gene flow depending on the amount of reproductive isolation that had evolved in allopatry.

5. Biogeographic and phylogenetic patterns implying repeated parapatric speciation in a small clade. Members of a monophyletic group of species (a clade) might all speciate parapatrically, giving rise to a pattern in which closely related species would have parapatric distributions. Imagine, for example three species arrayed along a cline in the linear order A, B, and C, with little or no overlap between the ranges of any species. If a phylogeny showed that all three species arose at the same time, one might infer that a single ancestral species split parapatrically into three taxa (Cracraft 1982). One might also reach this conclusion if young sister species always had parapatric ranges. However, these patterns could also be explained by allopatric speciation followed by range expansion that was halted by habitat difference or interspecific competition. Moreover, hybridization between parapatric species would homogenize them genetically, giving the false impression that they diverged recently. This artifact could yield a misleading pattern of "young" parapatric taxa. Barraclough and Vogler (2000) and Fitzpatrick and Turelli (2004) give examples of apparently young taxa with abutting ranges.

6. Historical observation of speciation in parapatry. Can one actually observe the parapatric origin of reproductive isolation? This might seem impossible given that even rapid parapatric speciation surely requires hundreds of generations. There is, however, one set of examples—even though these exhibit only weak reproductive isolation and resulted from human disturbance of the habitat. These well-known examples involve the evolution of heavy-metal tolerance in plants.

The best-known studies are of two grasses growing on soil from abandoned Welsh mines: *Agrostis tenuis* on a copper mine, and *Anthoxanthum odoratum* on a lead and zinc mine. The transition between metal-containing mine soils through contaminated pasture to uncontaminated pasture is remarkably abrupt: about 5–10 meters. Yet plants growing on mines have evolved tolerance to high levels of metal on the soil, while adjacent populations are non-tolerant (McNeilly 1968, McNeilly and Antonovics 1968). Although the mines were first worked in the thirteenth century, McNeilly (1968) estimates that the current ecotones are 100–150 years old.

Although metal-tolerant plants do not thrive as well as susceptible plants on uncontaminated pasture, no intrinsic incompatibilities between tolerant and susceptible plants are known. Rather, in both *A. tenuis* and *A. odoratum* there has been parapatric divergence of flowering time, with plants on mine soils flowering earlier than those individuals in adjacent pasture. Common garden experiments show that this temporal isolation is genetically based (McNeilly and Antonovics 1968). However, ample gene flow is still possible between ecotypes: roughly 75% of the plants in each habitat flower synchronously with those in the other. This is arguably our strongest example of reproductive isolation evolving in parapatry. Nevertheless, the isolation is too weak to illustrate speciation.

Unitl recently, similar evidence for the evolution of reproductive isolation in parapatry appeared to come from *Drosophila melanogaster* inhabiting "Evolution Canyon" in Israel (Korol et al. 2000). The north- and south-facing slopes of this narrow canyon (100–400 m across) differ profoundly in temperature, moisture, and vegetation. Flies collected from the two slopes show divergent genetic adaptation to their habitats: those from the hotter south-facing slope, for example, show greater resistance to desiccation and higher survival after heat stress (Nevo et al. 1998; Michalak et al. 2001). Moreover, laboratory tests reveal appreciable sexual isolation between strains from opposite slopes (Korol et al. 2000). The evolution of microgeographic adaptation and assortative mating is surprising because *D. melanogaster* is a very mobile human commensal (Coyne and Milstead 1987). The Evolution Canyon example now seems doubtful, however, as recent experiments have failed to replicate both the DNA differences and the sexual isolation between populations on different slopes (Schlötterer and Agis 2002; Panhuis et al. 2004).

Conclusions

Allopatric speciation appears so plausible that it hardly seems worth documenting. Given enough time, and barring extinction, any pair of geographically isolated taxa is likely to evolve reproductive barriers. But one must be somewhat wary of accepting these verbal scenarios: after all, sympatric speciation seemed just as plausible to Darwin and many of his successors. Will geographically isolated populations inevitably become species if they remain allopatric? This seems probable: although some plant species that diverged around 5 million years ago can still produce viable and fertile hybrids (Wen 1999), comparative studies show that, in angiosperms and five animal taxa, reproductive isolation increases with divergence time, eventually becoming complete (Chapter 2).

Laboratory selection experiments (unfortunately limited to Diptera) support the basic premise of allopatric speciation—that divergent selection indirectly produces isolating barriers. Reproductive isolation in such studies arises surprisingly fast, and usually involves behavioral isolation. This isolation is probably a byproduct of the same genes that underlie the response to selection, but exactly how this happens remains unknown.

Earlier evidence for allopatric speciation involved geminate species pairs flanking geographic barriers (supporting vicariant speciation) and endemic species on oceanic islands (supporting peripatric speciation). These observations suffered from two problems: there was little evidence that allopatric taxa were indeed sister species, and estimating divergence times was difficult. To a large extent, the introduction of molecular systematics and molecular clocks has solved these problems. We can now confirm that geminate pairs are often sister species and that disjunct pairs often show similar divergence times. The current evidence for vicariant speciation is thus on a much stronger footing than when the process was first championed by Wagner and Jordan.

A remaining problem in allopatric speciation is the relative frequency of vicariant versus peripatric modes. This question produced the warring "vicariant" and "dispersal" schools of biogeography, each claiming that their favored mode of speciation was ubiquitous (Brown and Lomolino 1998). But it is clear that both modes operate. Vicariance cannot explain endemic species on oceanic islands, and dispersal cannot explain the distribution of marine species around the Isthmus of Panama. Nonetheless, it is still hard to judge the relative importance of these two modes. On archipelagos, we can use a combination of systematics and biogeography to infer the minimum percentage of speciation events involving peripatry. This appears to be around 50% for several groups on Hawaii, but has yet to be estimated for other archipelagos. But on large land masses, the possibility of changes in range size and the ephemeral nature of some geographic barriers make it difficult to distinguish between peripatric and vicariant speciation.

Parapatric speciation is intuitively reasonable, supported by theory, and has almost certainly occurred in nature. The problem is determining whether it has occurred with appreciable frequency. Although many closely related species have abutting ranges, one cannot consider this as evidence for parapatric speciation because the same pattern can result from allopatric speciation followed by range expansion.

There are two types of observations that can be explained more parsimoniously by parapatric than by allopatric speciation. The first involves sister species with contiguous ranges that meet at an ecotone. If there is little evidence that these species were ever separated by a geographic barrier, and if clines for different loci are not congruent, it is reasonable to infer that the species originated in situ via ecologically divergent selection. The evidence becomes stronger if this pattern is obeyed by sister species in several diverse groups, and if divergence times between members of different sister pairs are not congruent, suggesting that their ancestors were not divided by a geographic barrier at the same time. Such a conclusion will require accurate phylogenies and good estimates of divergence times, and we know of no study of parapatric speciation that satisfies these conditions.

The second type of evidence is both more compelling, and, in principle, easy to gather. If an edaphic endemic species of plant—one restricted to a small patch of aberrant habitat—has its sister species in an adjacent area, one can conclude that speciation was parapatric. There are many such endemics, but so far no serious studies of their phylogeography. Such habitat islands offer our best chance to demonstrate parapatric speciation. In the absence of compelling data, however, we can offer only the unsatisfying conclusion that parapatric speciation is probably more common than sympatric speciation, but we do not know whether it is more common than allopatric speciation. In the next chapter we describe comparative studies of species distributions that can potentially resolve these issues.

4

Sympatric Speciation

No aspect of speciation is as controversial as the view that new species can begin to form sympatrically, that is, within a freely interbreeding population. The idea of sympatric speciation was, of course, Darwin's. In *The Origin* (1859, p. 112), he saw this process as an important (but not exclusive) engine of biological diversity, arguing that new species arose in sympatry to fill empty niches in "the polity of nature." This formed the basis of his "principle of divergence," in which competition drives the formation of new species, with species that had diverged most strongly from their ancestors outcompeting the others:

> It has been experimentally proved, that if a plot of ground be sown with one species of grass, and a similar plot be sown with several distinct genera of grasses, a greater number of plants and a greater weight of dry herbage can thus be raised ... Hence, if any one species of grass were to go on varying and the varieties were continually selected which differed from each other in at all the same manner as distinct species and genera of grasses differ from each other, a greater number of individual plants of this species of grass, including its modified descendants, would succeed in living on the same piece of ground. And we well know that each species and each variety of grass is annually sowing almost countless seeds; and thus, as it may be said, is striving to the utmost to increase its numbers. Consequently, I cannot doubt that in the course of many thousands of generations, the most distinct varieties of any one species of grass would always have the best chance of succeeding and increasing in numbers, and thus of supplanting the less distinct varieties; and varieties, when rendered very distinct from each other, take the rank of species (Darwin 1859, pp. 113–114).

---

because the results of virtually all models of sympatric speciation depend heavily on assumptions about the strength of selection, the number of genes involved, the way that these genes operate, the rate of mutation, and so on. Unfortunately, these assumptions often seem either biologically unrealistic, artificial, or untestable. Moreover, few theories have explored whether the likelihood of speciation changes drastically if one alters a model's assumptions. It is only by scrutinizing these details, then, that we can know whether the models in question tell us anything about the likelihood of sympatric speciation. Our review of the theory is intended as a brief summary; more complete analyses can be found in Johnson and Gullberg (1998), Kirkpatrick and Ravigné (2002), and Gavrilets (2003).

Disruptive sexual selection

In many evolutionary radiations, species appear to differ more strongly in secondary sexual traits than in ecological adaptations. The most striking case, discussed below, involves cichlid fish in African great lakes. Such observations have inspired models in which sympatric speciation is driven by sexual selection.

Most sexual-selection models involve the initial evolution of two groups of females within a population, each group preferring males with a different trait. Turner and Burrows (1995), for example, propose a polygenic model of sexual selection that yields two reproductively isolated groups in sympatry, but their model requires assumptions that seem unrealistic, such as complete dominance of the genes for female preference and a "best of n" mating rule for females, which requires them to inspect many males before mating.

This raises the main problem of all models of sympatric speciation via disruptive sexual selection: they provide no mechanism for the coexistence of the new species in their habitat. Because they are isolated solely through mate discrimination, these species constitute what many ecologists have long deemed impossible: two taxa that are ecologically identical. At first it would seem that competition would drive one species extinct, but such species do not differ in the intensity of competition. Nevertheless, extinction can occur through "species drift": fluctuations in population size or difficulties of the rarer species in finding mates. Yet this elimination can take hundreds or thousands of generations, possibly allowing the evolution of enough ecological divergence to allow stable coexistence (Hubbell and Foster 1986).

A related model by Higashi et al. (1999) begins at a symmetric equilibrium: males vary substantially in their traits, and females in their preferences for these traits. The model is constructed so that the average female preference is zero for the mean male trait. This equilibrium would be unstable in the absence of natural selection, as any perturbation would trigger sexual selection that would change both male trait and female preference in one direction. But when one adds stabilizing selection on the male trait (intermediate trait values are fitter), the initial equilibrium becomes stable. When natural selection is relaxed,

The dominance of Darwin's sympatric theory, lasting well into the twentieth century, rested on several factors. First, of course, was the popularity of his species concept, which involved morphological difference. Without the corpus of population genetics (or, in the case of Darwin, *any* genetics), it was impossible to see that interbreeding might prevent the formation of discrete taxa within a small patch of habitat. Sympatric speciation also appealed to those who shared Darwin's belief in the omnipotence and creativity of natural selection. Although allopatrically formed species are accidental byproducts of genetic divergence, natural selection *directly favors* the multiplication of species during ecologically based sympatric speciation. If new adaptations can be the direct target of selection, why not new species?

Mayr (1963, Chapter 15) describes the history of work on sympatric speciation in the first century after *The Origin*. This chapter, however, is best known for its detailed analysis of sympatric speciation, one of the most famous critiques in modern evolutionary biology. Using arguments from natural history and genetics, Mayr diagnosed grave weaknesses in verbal theories of sympatric speciation and revealed flaws in their assumptions about the biology of speciation. Mayr's critique had an immense effect on the field: evolutionists no longer considered sympatric speciation the norm, or even very common. But theorists immediately set to work making more rigorous mathematical models of sympatric speciation (e.g., Maynard Smith 1966), experimentalists tried to duplicate the process in the laboratory, and field workers began to focus on possible examples in nature (e.g., Bush 1969).

In the last fifteen years, sympatric speciation has again become popular, reflecting a resurgence of what might be called "paleo-Darwinism"—the claim that Darwin's original views of adaptation and speciation have been unduly neglected (e.g., Schilthuizen 2000; Mallet 2001).

Here we examine the theory, experiments, and data that have fuelled the resurrection of sympatric speciation. Throughout our discussion, we use Futuyma and Mayer's (1980, p. 255) definition of the process as "the origin of an isolating mechanism (i.e., the evolution of a barrier to gene flow) among the members of an interbreeding population." As noted in Chapter 3, this can be interpreted as speciation occurring between two populations that show free migration (i.e., $m = 0.5$).

As characterized by Mayr (1963), sympatric speciation involves the evolution of reproductive isolation within the average dispersal distance ("cruising range") of a single individual. Sympatric speciation thus differs from parapatric speciation, in which interbreeding is initially restricted by distance. Futuyma and Mayer (1980) observe that sympatric speciation is also unique because the initial restriction of gene flow is caused not by geography or distance, but by biological features of organisms. A popular theory of sympatric speciation, for example, involves host specialization by phytophagous insects. An ancestral population uses one host but contains mutant individuals attracted to other hosts. If those mutants mate and feed on the new host, they could evolve into a new "host race" and ultimately a new species. Although the host races eventually become spatially isolated by choosing different microhabitats, this is not allopatric speciation since the evolution of isolating barriers was initially promoted not by distance, but by genetic variation.

Our analysis below leads us to conclude that, while sympatric speciation is theoretically plausible and supported by both laboratory work and observations from nature, there is little evidence that it is common. In fact, there are only a handful of credible examples. We therefore feel that the data, though certainly not ruling out sympatric speciation, fail to support the current enthusiasm for the process. An important exception is polyploid speciation in plants (Chapter 9), which usually occurs in sympatry.

Theory

All models of sympatric speciation face two fundamental problems—problems that do not arise with allopatric speciation. The first is an antagonism between selection and recombination (Felsenstein 1981). As selection tries to split a population into two parts, it is counteracted by interbreeding that continually breaks up the evolving gene complexes that produce reproductive isolation. If, for example, speciation in animals involves habitat isolation, recombination will break down the association between those alleles contributing to fitness in each habitat and those alleles involved in recognizing and preferring that habitat. This yields maladapted genotypes that prefer habitats in which they are unfit. Likewise, recombination will destroy the association between alleles of different genes that contribute to the *same* trait, such as multiple loci producing habitat preference. Theorists usually deal with the problem of recombination by invoking either close linkage between genes involved in reproductive isolation, or the evolution of assortative mating, both of which reduce recombination. Models of allopatric speciation do not face this difficulty because geographic isolation prevents recombination.

The second problem is coexistence. Sympatric speciation requires that populations develop sufficient ecological difference to coexist during and after the evolution of reproductive barriers. Most models deal with this issue by assuming that speciation is driven by selection for ecological divergence. Models based on disruptive *sexual* selection, however, do not easily solve the problem of coexistence. In sympatric speciation, then, reproductive isolation and the ability to coexist must evolve simultaneously. This is not required for allopatric speciation: ecological differences that allow coexistence must evolve only before species attain secondary contact, which could occur long after reproductive isolation is complete. All theoretical work on sympatric speciation can thus be seen as an attempt to overcome recombination and competitive exclusion.

Most recent theory involves two types of models: those based on disruptive sexual selection, or on disruptive natural selection involving competition for resources. We discuss these models in turn, examining the assumptions of a few models in detail. While such scrutiny may seem tedious, it is important as might happen when a population invades a new environment, disruptive sexual selection can then cleave the population into two species.

This model is problematic because its starting condition invokes an unrealistic symmetry. In general, one expects male traits and female preferences to reflect a compromise between natural and sexual selection. To produce their symmetrical initial condition, Higashi et al. assume that the male trait value most preferred by natural selection is precisely the trait for which females have no net preference. If there were any difference between these two "optima," sexual selection might lead not to speciation, but to evolutionary change in a single lineage. Moreover, the model does not address the issue of coexistence.

The problem of coexistence also besets Gavrilets and Waxman's (2002) model, which incorporates recent ideas about sexual conflict. When the reproductive interests of males and females differ, females may evolve traits that lessen their susceptibility to the deleterious effects of mating. This produces perpetual antagonistic coevolution of the sexes: males evolve to overcome the increasing resistance of females, and females evolve to counteract these male adaptations (Holland and Rice 1998). Gavrilets and Waxman show that this "chase-away" sexual selection can fragment one population into two behaviorally isolated groups, each evolving in a different direction. But as the authors note (p. 10538), "subsequent ecological differentiation or some spatial segregation are required for stable coexistence of these species."

Theorists have addressed the coexistence problem by adding other assumptions. Payne and Krakauer (1997), for example, propose that during disruptive selection, males disperse from areas in which they cannot find mates. Their model yields two parapatric taxa differentiated by male traits and female preferences, but showing incomplete behavioral isolation. As the authors state (p. 7), they "have only modeled the evolution of character divergence and not reproductive isolation." Moreover, the incipient species are segregated spatially but not ecologically, so coexistence remains a problem.

In a model designed to mimic the explosive speciation of African lake cichlids, van Doorn et al. (1998) showed that sympatric speciation can result from a combination of sexual selection and disruptive natural selection based on discrete niches. The model depends on many untested assumptions about mate preference and genetics, but at least incorporates ecology in a way that promotes coexistence. It also calls attention to how a change in environment can trigger sexual selection. In their example, an increase in water clarity promotes sexual selection and speciation by making male colors more visible, allowing females to better discriminate among different colors.

Among all models of sympatric speciation, those involving only sexual selection (or beginning with sexual selection) seem the least realistic. Unless accompanied or followed by the evolution of ecological divergence, the new species cannot coexist, at least stably. Moreover, models of disruptive sexual selection that lead to speciation frequently involve unrealistic assumptions. Reviewing these models, Arnegard and Kondrashov (2004) note that the evolution of strong reproductive isolation is unlikely unless the variation in male traits and female preferences is based on only one or two loci, or, if more genes are involved, when the initial distribution of traits and preferences are very symmetrical and disruptive sexual selection is quite strong. They conclude (p. 230) that "disruptive sexual selection by female mate choice can only drive sympatric speciation under extremely restrictive initial conditions." In contrast, models that start with ecological divergence *can* allow long-term coexistence of sympatrically formed taxa, permitting ample time for the evolution of other isolating barriers.

Disruptive natural selection

Models of sympatric speciation involving disruptive natural selection come in two varieties: those based on adaptation to discrete niches, and those based on subdivision of a continuous distribution of resources. In the past, discrete-habitat models were the most popular. Recently, however, continuous-resource models have attracted attention because they seem more relevant to some proposed cases of sympatric speciation.

DISCRETE-HABITAT MODELS. These models yield sympatric species that seek out different niches, are most fit in their own niches, and preferentially mate with members of their own species. One thus requires the evolution of three traits: niche preference, niche adaptation, and assortative mating.

Niche preference. Incipient species must differ genetically in their preference for niches. Such preferences may involve feeding, oviposition, or mating. Most models assume at least two such behaviors, which seems realistic. Many insects, for example, oviposit and mate on their hosts.

Niche adaptation. Habitat races must differ genetically in their ability to survive in niches. The usual assumption is a tradeoff in fitness: alleles improving adaptation to one niche reduce adaptation to others.

The evolutionary cause of niche preference and adaptation is competition between genotypes. Alleles that allow individuals to find and occupy new niches can enjoy a selective advantage over alleles conferring adaptation to old niches. Most models assume that the evolution of preference precedes that of adaptation (one cannot adapt to a niche unless one inhabits it), and that both traits precede the evolution of assortative mating.

Assortative mating. To complete speciation, there must be substantial assortative mating within groups: inhabitants of each niche must mate preferentially with others in that niche. While many models easily yield genotypes differing in habitat preference, it is harder to obtain assortative mating. Demonstrating the evolution of behavioral isolation is thus the critical step in modeling sympatric speciation.

Different models have assumed very different forms of assortative mating. Unfortunately, the forms that evolve most easily are also the most biologically unrealistic. In order of increasing realism, they include the following:

1. "No-gene models." Assortative mating is a *pleiotropic effect* of genes responding to selection for habitat preference or habitat adaptation (Kondrashov 1986; Rice 1987; Doebeli 1996). For instance, if adaptation to different niches produces differences in body size, as in threespine sticklebacks, the inherent tendency of these fish to mate with those of similar size yields assortative mating as a byproduct of adaptation (Nagel and Schluter 1998). In these models, there are no genes that directly affect mate preference.

2. "One-locus, one-allele" models (Felsenstein 1981). Here, a new allele is fixed in all populations that permits individuals to recognize and mate with those having similar phenotypes or genotypes. Unlike "no-gene" models, the preference for similar individuals thus depends on a specific allele whose alternative allele confers no preference.

3. "One-locus, two-allele" models (Felsenstein 1981; Johnson and Gullberg 1998). Carriers of allele A_1 mate preferentially with each other, as do carriers of allele A_2. Such models—which can involve more than two alleles—promote sympatric speciation only if each allele becomes nonrandomly associated with alleles for habitat preference or adaptation.

4. "Multilocus models" (P. A. Johnson et al. 1996). These more realistic models are based on quantitative genetics, involving a male trait and a female preference for that trait, with changes in each involving several to many genes. Again, speciation requires that traits and mating preferences be nonrandomly associated with niche adaptation and niche preferences.

Combining the traits. It turns out that full sympatric speciation involves the evolution of at least two of the three traits described above. There are models for all three pairwise combinations of these traits.

The earliest models combined the evolution of niche adaptation and assortative mating (Maynard Smith 1966; Felsenstein 1981). Speciation begins with the establishment of a balanced polymorphism involving one or more genes whose alleles have different fitnesses in two niches. This itself is not speciation, but intraspecific variation maintained by ecological diversity. After such a polymorphism is established, assortative mating can evolve, creating two reproductively isolated "habitat races." Full speciation, however, is improbable. For one thing, the conditions for maintaining niche polymorphism are restrictive. For another, unless one adopts no-gene or one-locus, one-allele models of assortative mating, the development of an association between habitat adaptation and assortative mating requires very strong habitat selection, very strong assortative mating, and very tight physical linkage between the two sets of genes. The concatenation of these conditions in nature seems unlikely.

The second class of models combines the joint evolution of niche adaptation and niche preference. An example is Diehl and Bush's (1989) simulation study. Their "HPS" (habitat-preference speciation) model assumes haploid organisms having one locus with alternative alleles conferring preference for one of two habitats, and two diallelic loci with alleles having habitat-specific fitnesses. Under some conditions, this model yields two sympatric forms showing substantial genetic association between alleles for preference and adaptation. (This association develops because genotypes having adaptations for one habitat but preferences for the other are selected against.) However, this model does not yield strong reproductive isolation except under extreme conditions (such as errorless habitat choice): what usually evolves are partially isolated "host races."

Two related models by Fry (2003) yield more complete reproductive isolation. Both involved multiple genes for habitat preference and adaptation in a haploid organism having recombination. The first model posited two loci for viability in a niche, with "+" alleles conferring adaptation to habitat 1 and "–"alleles to habitat 2. There were an additional four preference loci, with each + allele conferring a greater preference for habitat 1 and each – allele a greater preference for habitat 2. The population is initially fixed for + alleles at half of the relevant loci and – alleles at the other half (i.e., the ancestor is a generalist with no habitat preference), and new alleles are introduced at low frequency. As in the Diehl and Bush model, a genetic association develops between viability alleles and preference alleles, and under some circumstances one obtains sympatric host races showing complete reproductive isolation, with one race having all + alleles and the other all – alleles. This victory of selection over (free) recombination occurs only when selection on each viability locus is strong: around 0.5 per locus if natural selection acts between the zygote and adult stages (as occurs in many phytophagous insects), and around 0.26 under the less realistic assumption that selection occurs after adults choose hosts but before they reproduce. It is not clear whether this model can explain the evolution of one specialist from another (as in the host races of *Rhagoletis pomonella*), nor whether relaxing some assumptions—such as perfect habitat choice when all the right alleles are fixed—would reduce the probability of speciation.

By modifying this model to include varying numbers of loci, Fry (2003) showed that sympatric speciation occurs more easily when selection acts on early life stages, and, for a fixed fitness difference between generalists and specialists, when there are fewer genes affecting both preference and viability. Both this and the six-locus model assume that the habitats impose "soft selection," that is, each habitat produces the same number of individuals regardless of their genetic composition. It is not clear whether models of hard selection, in which the output of a habitat depends on the genotypes of its inhabitants, would yield sympatric speciation so readily.

Kawecki (1996, 1997) outlined a novel model of sympatric speciation involving adaptation and preference. To circumvent the difficulty of establishing a genetic polymorphism for niche use, Kawecki invoked the fixation of *the same* alleles in two habitats within one area. Such alleles could be fixed everywhere even though they have different fitnesses in different places. (For example, at locus A, carriers of allele A_1 might have the same fitness in both habitats, while those of allele A_2 might be fitter in habitat 2 but neutral in habitat 1. Allele A_2 will then replace A_1 in both habitats.) This habitat dependence of fitness can promote the evolution of alleles producing habitat *choice*. Kawecki made two different models of this choice, both involving alleles at a single locus. The first model posits two alleles, one of which (B) gives individuals a preference for the habitat in which they developed, while the alternative (b) gives no preference. In the second model, allelic variants confer habitat preference regardless of where the carrier grew up (i.e., carriers of allele B_0 have no preference, carriers of B_1 prefer habitat 1, and carriers of B_2 prefer habitat 2). Because the fittest individuals are those choosing habitats to which they are best adapted, the fixation of beneficial alleles imposes indirect selection for habitat-choice alleles. (For example, in the second case, the increase in frequency of allele A_2 will increase the frequency of B_2, for only carriers of B_2 prefer the habitat in which they have the highest fitness.) Yet because selection is "soft," the indirect selection for habitat preference is transitory, disappearing as soon as the "fitness allele" is fixed in both habitats. Both models thus yield two groups of individuals that have identical ecological adaptations (the fitness alleles are identical in both populations), but choose different habitats. In the first model, allele B becomes established; in the second, the frequency of preference allele B_1 increases in habitat 1 and that of B_2 increases in habitat 2.

As recognized by Kawecki, the problem with these models is that the populations are not reproductively isolated unless they mate almost exclusively with other individuals occupying the same habitat. Moreover, since selection on habitat preferences is ephemeral, these preferences will eventually disappear unless there is a constant influx of new beneficial mutations with habitat-specific fitnesses, as well as the development of an association (linkage disequilibrium) between the diverse alleles conferring preference to each habitat. It is not at all clear whether these assumptions are realistic. Kawecki (1998), however, suggested that coevolutionary interactions between hosts and parasites could yield continuing selection for new mutations.

The last class of pairwise models combines the evolution of niche preference and assortative mating. The key point here is that recombination becomes irrelevant if one assumes that assortative mating is a byproduct of niche use. The most common version assumes that individuals mate preferentially with others who choose the same niche, so that assortative mating automatically results from habitat choice. Rice (1984a, 1987) has been the most avid proponent of this scenario. In his simulation model (1984a), populations can rapidly fragment into races showing strong reproductive isolation. This result, however, requires a combination of strong selection against intermediate habitat preference and nearly perfect niche-based assortative mating. Rice noted that this analysis (and its subsequent experimental test by Rice and Salt 1990) was intended not to model sympatric speciation per se, but to explore how pleiotropy reduces the conflict between selection and recombination. Although the model does not yield "good" species, Rice (1984a) suggested that restricted gene flow between habitat races could foster the evolution of other isolating barriers—such as temporal isolation or physiological adaptation to new habitats—that could complete speciation. Rice's model is often invoked to support sympatric speciation in phytophagous insects that mate on their hosts.

Only the five-locus model of P. A. Johnson et al. (1996) explores the simultaneous evolution in sympatry of alleles affecting habitat preference, habitat adaptation, and non-habitat-related assortative mating. In this model, speciation occurs under a fairly broad range of conditions, but only if some assortative mating occurs as a byproduct of habitat preference. By removing individual genes from the model, Johnson et al. showed that sympatric speciation is unlikely unless there is genetic variation for all three traits. This is probably the most biologically realistic of all discrete-resource models.

CONTINUOUS-RESOURCE MODELS. Discrete-niche models do not seem appropriate for many organisms that might have speciated sympatrically. Sympatric species of African lake cichlids, for example, do not show dramatic differences in habitat; indeed, they are said to be almost ecologically identical (see below). The existence of species that do not use different hosts, nor have divergent niches, prompted the creation of continuous-resource models. In these models, the simultaneous coevolution of ecological traits and assortative mating splits a population into two sympatric groups, each using different parts of a resource distribution.

The most widely cited model of this type is that of Dieckmann and Doebeli (1999), which assumes that an important resource (such as seed size) is distributed unimodally. An ecological trait important in using that resource (such as beak size) is also distributed unimodally. As expected, the mean value of the trait is the one having the highest fitness on the key resource. As the population grows, density- and frequency-dependent selection favors individuals with more extreme values of the trait (e.g., larger and smaller beaks), because extreme individuals can use unexploited resources. However, random mating prevents the population from splintering into discrete groups.

Such splitting can occur, however, if individuals with extreme traits prefer to mate with others having similar traits. Selection favors such associations between trait genes and mating genes because like-with-like couplings avoid the production of intermediate offspring that have low fitness because they experience greater competition. Dieckmann and Doebeli model two types of these associations. First, assortative mating may be "nongenetic": a byproduct of the ecological trait (e.g., birds tend to mate with others having beaks of similar size). More realistically, assortative mating may involve two sets of genes, one producing ecologically "neutral" traits (e.g., red or black body color) and another promoting preferences for those traits. In the nongenetic model of mating, the population quickly fragments into two reproductively isolated groups. In the second model, genetic drift eventually produces a nonrandom genetic association between the marker traits and the extreme ecological traits. When this happens, mate preferences become part of the association, and nonrandom mating quickly causes the emergence of two sympatric, behaviorally isolated populations that use different resources. This form of speciation is unique because, in contrast to common wisdom, reproductive isolation is not the byproduct of ecological divergence. Instead, reproductive isolation (assortative mating) allows the population to *begin* diverging ecologically, and thereafter the two traits evolve in concert.

Despite the popularity of this model, there are questions about its plausibility and generality. Reanalysis of the asexual version of the model shows that it does not produce discrete clusters; instead, the long-term equilibrium is a continuous distribution of genotypes that matches the distribution of resources (Polechova and Barton 2004). In addition, the sexual version of the model works best when individuals have an innate preference for mating with others having similar values of the "resource" trait—a situation that, while occurring in some species, seems unlikely to be a general phenomenon. Moreover, Bolnick (2004) notes that the model assumes extremely strong trait-based assortative mating. If mate preference becomes weaker, the time required for speciation becomes so long (e.g., 10^6 generations) that one expects the initial conditions to change. Speciation under the alternative "neutral-trait" model of mating is even slower.

Another troubling issue afflicts the sexual model. As sympatric speciation occurs and the ends of the resource distribution become occupied, there is less competition for "central" resources, and selection might produce additional species using these resources. It is not clear whether the final result will be the formation of reproductively isolated clusters, or only a continuum of interbreeding forms. (However, other isolating barriers, such as postzygotic incompatibilities or enhanced sexual isolation due to reinforcement, might arise before the populations collapse into this continuum.) Gavrilets and Waxman (2002) describe other difficulties, such as Dieckmann and Doebeli's assumptions that all females have equal mating success regardless of the rarity of their preferred males, that the initial genetic variation for mating preference is maximized (all alleles have a frequency of 0.5), and that mutation rates are two orders of magnitude higher than observed in natural populations. Waxman and Gavrilets (2004) suggest that relaxing these assumptions might make speciation unlikely.

Kondrashov and Kondrashov (1999) propose a related model also based on disruptive selection for resource use. They posit not only a "neutral-trait" form of mating similar to that of Dieckmann and Doebeli (1999), but also a more realistic form in which females carry genes affecting their preference for males with different values of the neutral trait. Assume that color is such a trait. Random genetic drift may allow genes producing different colors to become weakly associated with genes for using different parts of a resource distribution. Females of a given color will now be selected to prefer males of a similar color, as this preference prevents the production of maladapted offspring having intermediate values of the ecological character. If disruptive selection is sufficiently strong, both models of mating readily yield two ecologically discrete populations that are behaviorally isolated.

But this model also rests on questionable assumptions (Turelli et al. 2001). The quantitative-genetic model of the ecological trait assumes that all loci have equal effects and that all alleles that increase (or decrease) the value of the trait have equal frequencies. More important, the model assumes that the strength of disruptive selection is *constant* during speciation. This seems unlikely, for competition surely becomes weaker as incipient species approach their different optima. Finally, while behavioral isolation is proportional to the difference between male and female neutral traits, such isolation is assumed to be complete only between individuals at the two ecological optima. It is possible that allowing some mating between individuals at different extremes would drastically reduce the probability of speciation.

Conclusions

What can we make of these models? We can obviously conclude that sympatric speciation is *theoretically* possible since some models produce reproductively isolated groups in the face of gene flow. Hence, Mayr's famous critique of sympatric speciation (1963) seems overly dismissive, though his analysis was limited to the verbal models then available. But we can also deduce that the conditions for sympatric speciation are more stringent than those for allopatric speciation. This justifies our use of allopatric speciation as a null hypothesis when we deal with data from nature. Finally, if sympatric speciation does occur, it is unlikely to be driven solely by sexual selection. Speciation requires coexistence, and coexistence often requires ecological difference. We thus expect that the earliest stages of sympatric speciation will involve selection for niche divergence.

The goal of theory, however, is to determine not just whether a phenomenon is theoretically possible, but whether it is *biologically reasonable*—that is, whether it occurs with significant frequency under conditions that are likely to occur in nature. It is here where prevailing theories run into trouble. Nearly every model suffers from unique problems involving either the realism of its assumptions or the robustness of its results. Too few models have relaxed or varied their initial conditions or parameter values, examining the effect of these changes on the probability of speciation.

Models involving sexual selection, for example, have assumed complete dominance of traits, or initial conditions that are genetically unstable. More important, they assume that sympatric speciation proceeds without the ecological differentiation that allows taxa to coexist. Models involving ecological selection have different problems. They may depend, for instance, on the recurrent fixation of mutations with habitat-specific fitnesses, on near-perfect habitat choice by individuals, or on unrealistically symmetrical gene frequencies and effects. Will the models work when these conditions are altered? We have few answers. How fast does speciation occur? Even if sympatric speciation is reasonably likely, the expected time to branching (Chapter 12) may depend critically on many assumptions, such as the degree of assortative mating, the

tion design can be seen as simple two-way directional selection without gene flow—an allopatric model.

Nevertheless, the success of the single-selection experiment shows that, even with gene flow, habitat isolation can evolve if selection is strong and assortative mating a byproduct of habitat choice. This result is not trivial. Although Rice and Salt's experiment involved extreme conditions, had it failed we would have almost no experimental support for ecologically based models of sympatric speciation.

Surveying all relevant experiments, then, we conclude that strong disruptive selection is necessary but not sufficient for the evolution of reproductive isolation in sympatry. The failure of nearly all such experiments to yield isolation clearly shows that assortative mating is not an inevitable byproduct of disruptive selection. However, because experiments often use small populations, they may be biased *against* the evolution of reproductive isolation. Reviewing 63 disruptive selection experiments exhibiting varying degrees of gene flow (i.e., experiments on both sympatric and parapatric speciation), Ödeen and Florin (2000) found that successful experiments had significantly higher effective population sizes than unsuccessful experiments. Nevertheless, the success of Rice and Salt's (1990) experiment supports the view that sympatric speciation becomes much more probable when assortative mating is an automatic byproduct of selection.

Rice and Salt (1990, p. 1151) advocate a largely experimental approach to sympatric speciation:

> It seems to us that the only way to make progress [in determining whether sympatric speciation is important in nature] is to use laboratory model systems to determine whether sympatric speciation is biologically plausible. If this can be shown, it would seem reasonable to conclude that sympatric speciation has probably played a nontrivial role as a speciation mechanism.

But this claim faces the same problem as does a purely theoretical approach: the value of models and theory is only as good as the biological realism of their assumptions. We simply do not know how often the conditions modeled by Rice and Salt occur in the wild. There are surely many species, for example, that do not show a complete correlation between habitat choice and mate choice.

Evidence from Nature

Until recently, many proposed cases of sympatric speciation rested on weak evidence—sometimes merely the observation that two related species are sympatric. But much of the data were equally compatible with allopatric speciation. Thus, the main difficulty in confirming sympatric speciation in nature is rejecting the alternatives.

Accordingly, we will consider allopatric speciation as the null hypothesis when evaluating examples in nature. That is, we will deem allopatric speciation as the most likely explanation unless sympatric speciation appears more plausible. Some may find this strategy unacceptable, but we feel it is reasonable on several grounds.

First, allopatric speciation is "easier" than sympatric speciation. All that is required for allopatric speciation is time: selection or drift acting on two geographically separated populations will eventually produce isolating barriers. Sympatric speciation, on the other hand, is not inevitable: its success depends on factors like gene number, the nature of epistasis, the strength of selection, the linkage relationships of genes producing isolating barriers, the distribution of ecological resources, the genetic basis of mate discrimination, the stability of the environment, and so on.

Second, as shown in the previous chapter, there is strong and pervasive evidence for allopatric speciation. Although the lack of similar evidence for sympatric speciation might reflect only the difficulty of getting good data, it seems reasonable to accept a controversial scenario only when it appears more likely than a well-established one. While most proposed cases of sympatric speciation are consistent with an alternative story involving allopatry, the converse is not true. It would be hard to argue, for example, that an island endemic whose closest relative lives on a nearby continent arose by sympatric speciation.

Finally, it is not obvious that opportunities for geographic isolation have been limited. In the past two million years, for example, there have been around 20 major glacial advances, producing changes in climate and geology that drastically altered the distribution of populations and species. There have also been more frequent but less extreme cycles of temperature change. All of these changes affected terrestrial, freshwater, and marine species living in temperate and tropical regions (Roy et al. 1996; Cronin 1999; Jablonski 2000). Considerable fragmentation of species' ranges must have occurred over evolutionary time.

What evidence could lead us to reject our null hypothesis? We conclude that two species probably arose sympatrically if they meet four criteria:

1. The species must be largely or completely sympatric. Although presently allopatric species may have originated in sympatry, this will usually be impossible to determine.
2. The species must have substantial reproductive isolation, preferably based on genetic differences.
3. The sympatric taxa must be sister groups. The genetic similarity used to establish this fact must not result from hybridization.
4. The biogeographic and evolutionary history of the groups must make the existence of an allopatric phase *very unlikely*.

Criterion 3 is hard to meet if reproductive isolation is incomplete. For example, if non-sister taxa originated allopatrically, secondary contact followed by some hybridization could make them appear to be sister species. Takahata and Slatkin (1984) show that, if interactions between autosomes make hybrids very unfit but mtDNA is neutral, a small amount of introgression will homogenize mtDNA, making species appear closely related. Unfortunately, many phylogenies used to support sympatric speciation are based on mtDNA or its analogue in plants, chloroplast DNA.

Criterion 4 is the most demanding: only rarely can one rule out a period of allopatry. Arguments about the likelihood of allopatry have dominated discussions of sympatric speciation, because what seems reasonable to one biologist can appear unreasonable to others. Whether criterion 4 is satisfied usually involves a somewhat subjective judgment about the likelihood of events in the unrecoverable past.

We must note at the outset that several unusual forms of speciation are almost certainly sympatric. These include autopolyploidy, allopolyploidy and other forms of speciation involving hybridization (Chapter 9). In birds, "cultural species" of nest parasites can arise sympatrically through the use of novel hosts (Chapter 6).

The evidence for sympatric speciation not involving polyploidy or nest parasitism falls into two categories. The most compelling is the presence of endemic, monophyletic groups of species occupying isolated patches of habitat. These patches include not only oceanic islands, but also continental "islands" such as isolated lakes, or hosts that can be considered islands for their parasites. The second line of evidence involves monophyletic groups on continents whose origin seems better explained by sympatric than by allopatric speciation. These include cases of host-related speciation in phytophagous insects, as well as allochronic speciation involving a divergence in life cycle or mating period. We limit ourselves largely to "classic" examples that have been widely cited as evidence for sympatric speciation, but also discuss a few lesser-known examples. If we have omitted a case, it is because we find it implausible, unconvincing, or lacking adequate documentation.

At the end of this chapter we consider comparative analyses of entire groups, which can give us an idea of the relative frequency of sympatric speciation.

Evidence from habitat "islands"

SPECIES ON OCEANIC AND CONTINENTAL ISLANDS. The most convincing evidence for sympatric speciation would be a monophyletic group of species confined to a small, isolated habitat. Oceanic islands are ideal places to gather such evidence. The islands must be small enough, however, to rule out "microallopatric speciation" caused by minor geographic barriers acting on organisms of limited mobility. Moreover, such islands should not have been part of an ancient archipelago that may have allowed allopatric speciation.

Although it is easy to collect the relevant data, which involves only species distributions and phylogenies, there have been few attempts to do so. White (1978, pp. 244–249) claimed that some radiations of insects on isolated oceanic islands give strong evidence for sympatric speciation. One example is the South Atlantic island of St. Helena, 1960 km from the nearest continent and only 122 km^2 in area. White notes that St. Helena harbors several endemic genera of Coleoptera and Orthoptera, some of which occur in only small areas of the island. He concludes (p. 247) that "the balance of probability certainly points to a sympatric interpretation," although he notes that there is not enough information to rule out the possibility of speciation caused by "very minor geographic barriers, on an extremely local scale." While the monophyly of some of the insect groups is fairly convincing, the main problem is that the species cited are flightless. This raises the possibility that their mobility is sufficiently limited to permit allopatric speciation on this rugged island.

White also considers sympatric speciation responsible for a radiation of small flightless weevils on the volcanic island of Rapa. Rapa is very small (40 km^2), lies 560 km from the nearest land (another island), and harbors 41 endemic species of phytophagous weevils in the genus *Microcryptorhynchus*, most restricted to single species of plants. But again one cannot exclude microallopatric speciation. The terrain of Rapa is rugged, having many valleys dissected by ridges. A resurvey of the genus by Paulay (1985) revealed an additional 26 species, but Paulay also cautioned that "although Rapa is a small island compared to many in the Pacific, it is quite large to a 2 mm-long flightless weevil" (p. 102). The situation becomes even more complicated when one considers the historical geology of Rapa. White concluded that the ridges do not constitute geographic barriers because they were connected around the coast and were originally forested. Paulay, however, notes that the island has several satellite islets that were connected to low mountain forest on the island during the Pleistocene. These islets may have been sites of allopatric speciation. Similarly, many species are restricted to high-elevation cloud forest, and many patches of mid-elevation forest lack weevils, again raising the specter of microallopatry. Finally, some weevils in this genus are found on other Pacific islands, and molecular phylogenies are required to establish the monophyly of species on Rapa.

These cases demonstrate just how hard it is to get evidence for sympatric speciation, even on islands. Indeed, Paulay considers allopatric speciation the most plausible explanation for the weevil radiation on Rapa. Ideally, one could judge the likelihood of sympatric speciation by using as a control a group more mobile than small flightless weevils. Fortunately, Rapa has just such a group: Lepidoptera. Clarke (1971) showed high levels of endemism of these species, including 15 endemic species within one genus of microlepidopterans (*Dichelopa*). The evidence for sympatric speciation would be compelling were the genus itself endemic to Rapa, but 21 additional species of *Dichelopa* occur in Australia and on other Pacific islands. Without a phylogeny of the genus, the possibility of speciation through multiple invasions cannot be ruled out. Moreover, while 60% of microlepidopterans on Rapa are endemic, this is true for only 25% of macrolepidopterans. One expects this negative correlation between mobility and endemism under allopatric, but not sympatric, speciation.

As noted in Chapter 3, surveys of endemic birds and *Anolis* lizards on oceanic and small continental islands fail to provide evidence for sympatric speciation (Diamond 1977; Coyne and Price 2000; Losos and Schluter 2000). Although birds have been dismissed as unlikely candidates for sympatric speciation owing to their mobility, mobility is in fact irrelevant to most theories of sympatric speciation. Moreover, some species of island birds show resource-use polymorphisms, often considered important for sympatric speciation (Grant and Grant 1989; Skúlason and Smith 1996; Coyne and Price 2000). Diamond (1977) concludes that speciation in birds occurs only on islands larger than New Zealand, while Losos and Schluter (2000) note that speciation in lizards occurs only on Caribbean islands larger than 3000 km^2. Such area requirements do not support the idea that sympatric speciation is common in these groups.

FISHES IN POSTGLACIAL LAKES. Many North American and Eurasian lakes were formed after the last glaciation, about 15,000 years ago, and endemic fish species in these lakes are likely to have evolved since that time. Some lakes contain two or more closely related fish differing in morphology, behavior, habitat, or life history (Schluter 1996a,b; Smith and Skúlason 1996; McKinnon and Rundle 2002). The distinct morphologies often include a slender pelagic form foraging on zooplankton and a heavier-bodied form eating larger invertebrates, molluscs, or algae. This diversity reaches its extreme in the Arctic charr (*Salvelinus alpinus*) of Lake Thingvallavatn, Iceland. This lake harbors four closely related morphs (Skúlason et al. 1999): two limnetics (planktivore and piscivore) and two benthic (dwarf and normal). Sympatric morphs of sockeye salmon (*Oncorhynchus nerka*) in Alaskan and Canadian lakes differ in both morphology and breeding behavior. An "anadromous" form migrates to sea in its second year, returning to the lake to spawn, while the smaller "kokanee" form lives permanently in the lakes (Wood and Foote 1996).

These fish are often considered cases of sympatric speciation, but before determining whether they meet our criteria, one must ascertain whether the morphs are genetically differentiated forms or merely developmental polymorphisms. Although there is some support for developmental plasticity of fish morphs (Hindar and Jonsson 1993), in other cases there is good evidence that morphs are genetic (Hatfield 1996; Gíslason et al. 1999).

But were sympatric fish morphs (whether or not one considers them species) *formed* in sympatry? Most glacial lakes are connected to the sea or river systems, allowing multiple invasions by allopatrically formed taxa. Moreover, gene exchange after secondary contact can yield the misleading conclusion that morphs are sister taxa. Reliable phylogenies can often settle both issues.

In only one case do we have reasonable evidence for the sympatric origin of fish morphs in a glacial lake (Gíslason et al. 1999). Lake Galtaból in Iceland, formed less than 10,000 years ago, harbors limnetic and benthic morphs of Arctic charr (*Salvelinus alpinus*). Genetic analysis indicates that, compared to morphs of *S. alpinus* in four nearby lakes, those in Lake Galtaból appear to be sister taxa. The genetic similarity of benthic and limnetic forms probably does not reflect current hybridization, because they share no alleles at one of six nuclear loci (suggesting complete reproductive isolation), and are strongly differentiated at other loci. Although we cannot rule out the possibility that the morphs arose in allopatry (fish in adjacent watersheds were not examined), sympatric speciation is certainly plausible.

Other sympatric morphs, however, seem more likely to have arisen in allopatry. Molecular phylogenies show that in some Canadian lakes, dwarf versus normal morphs of whitefish (*Coregonus clupeaformis*) are derived from double invasions of allopatric forms (Bernatchez et al. 1996, 1999). Similarly, DNA-based phylogenies of sockeye salmon (*Oncorhynchus nerka*) from Russia and North America also suggest an allopatric origin of sympatric anadromous and kokanee morphs (Taylor et al. 1996). Identical salmon morphs from different lakes are more closely related to each other than are different morphs within the same lake. This suggests that the now-sympatric morphs arose via allopatric differentiation in glacial refugia followed by double colonizations of lakes.

The most famous example of reproductively isolated fish morphs in glacial lakes is the threespine stickleback, *Gasterosteus aculeatus* (Bell and Foster 1994; McPhail 1994; Schluter 1998; McKinnon and Rundle 2002). Benthic and limnetic morphs occur together in six lakes in British Columbia. Within each lake, the morphs differ in not only morphological traits (which are genetic) but also in allozyme frequencies (McPhail 1984, 1992; Hatfield 1996). Although the morphs hybridize to a limited extent, their gene pools are kept distinct by disruptive selection favoring the different morphologies associated with habitat and feeding differences, and by assortative mating based on color and size (Schluter and Nagel 1995; Boughman 2001).

The presence of both morphs in several lakes has been seen as evidence for sympatric speciation (e.g., Jones et al. 2003)—evidenced buttressed by mtDNA analysis showing greater genetic similarity between benthic and limnetic morphs in the same lake than between identical morphs sampled from different lakes (Taylor et al. 1997). However, a more recent study using nuclear DNA suggests that sympatric morphs resulted from multiple invasions of each lake by groups that evolved allopatrically (Taylor and McPhail 2000). The similarity in mtDNA almost certainly reflects hybridization between sympatric forms.

The most likely scenario is that the lakes were originally colonized by a marine stickleback, which then evolved into the benthic form. A second wave of marine sticklebacks subsequently invaded the lakes and evolved into limnetics, perhaps with the aid of selection to avoid competition. The view that the two morphs originated from well-spaced double invasions is supported by the greater genetic similarity of present-day limnetics to marine forms than to sympatric benthic forms (McPhail 1992; Taylor and McPhail 2000) and by the observation that limnetics have greater physiological tolerance to seawater than benthics (Kassen et al. 1995). McPhail (1994) also noted that both morphs occur only in lakes less than 50 m above sea level; these low-lying lakes may have been submerged during a postglacial rise in sea level that allowed the second invasion of marine sticklebacks. He argued that if morphs formed in sympatry, there would be no reason why both should coexist only in lower-elevation lakes (but see James et al. 2002).

There is also little evidence for sympatrically formed morphs of other species in postglacial lakes. The Arctic charr of Lake Galtaból, then, appears to be the only feasible case of sympatric speciation in these lakes.

CICHLID FISH IN AFRICAN LAKES.   No radiation is more impressive than that of cichlid fish in East African freshwater lakes. The three major lakes alone (Victoria, Malawi, and Tanganyika) probably contain more than 1500 species, many still undescribed. While this radiation was once thought to reflect allopatric speciation (Mayr 1963; Fryer and Iles 1972), many now consider it the result of sympatric speciation.

The literature on the biology and biogeography of African cichlids is enormous (e.g., Fryer and Iles 1972; Echelle and Kornfield 1984; Meyer 1993; Martens et al. 1994; Seehausen 1996; Fryer 1999; Turner 1999; Barlow 2000; Kornfield and Smith 2000). Here we summarize the salient features of the story.

Table 4.1 lists six lakes and their endemic cichlid fauna. It also provides estimates of the net diversification interval in each lake (i.e., the expected wait-

Table 4.1 *Endemic cichlid diversity in African lakes*

Lake	Area (km^2)	Endemic cichlid species[a]	Age of lake	Age of flock	Net diversification interval (years/1000)[b]
Victoria	68,635	ca 500[c]	0.0146–0.75myr[d]	14,600–0.75 myr	2.6 (135)[d]
Malawi	29,604	659–1000	0.7–2 myr	0.7–1.8 myr	101–277
Tanganyika	32,893	170–250	5–12 myr	5–12 myr	764–1139[e]
Nabugabo	29	5	0.004 myr	<0.004 myr	<4[f]
Barombi Mbo	4	11	1 myr	0.9 myr	375
Bermin	0.6	9	<2.5 myr	0.5 myr	<227

Source: Data taken from Greenwood (1965), Stiassny et al. (1992), Schliewen et al. (1994), McCune and Lovejoy (1998), Turner (1999), Turner et al. (2001), Seehausen (2002), and D. Tautz (pers. comm.). Age of flock is estimated from either molecular data, or (as an upper bound) the age of the lake.

[a]When two figures are given, lower figure is current number of described (monophyletic) species, and higher number is estimated total species including those yet undescribed (Turner 1999; Turner et al. 2001). However, Turner et al. (2001) also note that species numbers in Lakes Victoria and Malawi may be substantially overestimated because of the questionable status of allopatric populations (see text).

[b]See Chapter 12 for calculation of net diversification intervals (average interval between successive branching events).

[c]Verheyen et al. (2003) give genetic evidence that the Lake Victoria flock is diphyletic instead of monophyletic. We thus approximate the speciation interval by assuming two monophyletic flocks of 250 species each.

[d]While Greenwood (1981) estimated the age of Lake Victoria at 750,000 years, Johnson et al. (1996, 2000) give evidence that the deepest portion of Lake Victoria (and hence, the entire lake) was completely dry 14,600 years ago. The recent-drying theory, however, is disputed by Fryer (2001) and Kornfield and Smith (2000), who suggest a maximum species-flock age of 0.5 to 1 million years. Assuming a flock age of either 14,600 or 0.75 million years, and an estimate of 250 species produced from each of the two colonizing lineages, one obtains speciation intervals between 2600 years and 135,000 years.

[e]Ages recalculated from McCune and Lovejoy (1998), who analyzed monophyletic flocks in Lake Tanganyika.

[f]Nabugabo (see text) contains ten species, five of which are endemic. Endemics in the lake are not monophyletic: each is more closely related to a sister species in adjoining Lake Victoria. Thus, we calculate speciation intervals as simply the age of Lake Nabugabo.

ing time until a single species divides into two; see Chapter 12). The three large lakes were formed when tectonic movement in Africa created depressions or valleys subsequently filled by rivers. Lake Nabugabo is a small satellite lake separated from Lake Victoria by a sand bar that formed about 4000 years ago, while Barombi Mbo and Bermin are tiny lakes inside volcanic craters. With the exception of Victoria, lake ages are fairly well established, while ages of the resident cichlid flocks are estimated from their genetic divergence and a molecular clock. Each lake has experienced large changes in water level, most recently during a period of dryness during Pleistocene glaciation. Changes in lake levels have led to the creation and disappearance of satellite lakes, to the appearance of intralacustrine islands and new areas of shoreline, and to the isolation of habitats along the lakeshore. These observations are important when considering the biogeography of speciation.

The cichlids fall into closely related groups called "species flocks," including the famous species-rich haplochromines of Lakes Victoria and Malawi, and its subset of *mbuna*, the group of roughly 300 species of colorful cichlids inhabiting the rocky shorelines of Lake Malawi. Closely related species of haplochromines differ mainly in breeding coloration of males. In contrast, Lake Tanganyika is dominated by cichlids of the Lamprologini tribe, containing more sexually monochromatic species.

Many cichlids are specialized feeders, including rock scrapers, predators on adults or young of other species, bottom feeders, pelagic forms that eat zooplankton, mollusk eaters, leaf choppers, and even fish that eat the scales or fins of other individuals. It has been suggested that the evolution of diverse feeding habits was facilitated by the cichlid's pharyngeal jaws: a second set of food-processing jaws in the throat (Liem 1974). Despite trophic differentiation of the group, closely related sympatric species in the largest lakes appear to differ little in diet, breeding site, and habitat use, which has led some workers to speculate that cichlids can coexist without niche partitioning (Fryer and Iles 1972; Genner et al. 1999a,b; Seehausen and van Alphen 1999; but see Kornfield and Smith 2000). However, there have been no studies of competitive exclusion in any African cichlid, and we badly need studies of the ecology of closely related sympatric species.

The reproductive behavior and sexual dimorphism of cichlids may also be important in understanding their speciation. Nearly all female cichlids in the large lakes give parental care, involving either mouthbrooding or guarding of nest sites and young. Males often form leks, defending territories or nests where females go to spawn. In some species many males do not find mates (Kornfield and Smith 2000), providing ample opportunity for sexual selection. This probably explains the bright male coloration (expressed either permanently or during the breeding season) that is the most obvious diagnostic trait of cichlids in Lakes Victoria and Malawi (Ribbink et al. 1983; Dominey 1984; Seehausen et al. 1998). Females, on the other hand, are often cryptic. This dichromatism has led to speculations that sexual selection was the main engine of speciation in this group (Dominey 1984; Turner 1994; Fryer 1999).

This idea may well be correct, at least for the most recently formed species. Many closely related cichlids produce viable and fertile hybrids in the laboratory, but do not hybridize in nature (Crapon de Caprona and Fritzsch 1984; Parker and Kornfield 1997). Observations in the field and laboratory often show strong sexual isolation between closely related, sympatric species (Marsh et al. 1981; Seehausen et al. 1998). Assortative mating in the laboratory can be lost when fish are exposed to light spectra that obscure differences in male color (Seehausen and van Alphen 1998). In Lake Victoria, species diversity and color morph diversity are positively correlated with water clarity, implying a breakdown in reproductive isolation in murkier waters (Seehausen et al. 1997). Nevertheless, at least in Lake Malawi, species within genera are distinguished largely by differences in male breeding color, but the genera themselves are distinguished mainly by differences in trophic morphology (Albertson et al. 1999, 2003; Allender et al. 2003). This implies that the radiation involved two mechanisms of speciation: an initial differentiation involving natural selection for feeding efficiency, followed by more recent differentiation involving sexual selection.

The fish of Lake Nabugabo also suggest that sexual isolation currently evolves before other reproductive barriers, at least in allopatry. As noted above, Nabugabo is a small satellite lake of Victoria, and harbors five endemic species of haplochromine cichlids (see Table 4.1). Four of the Nabugabo species are members of geminate pairs, each of whose closest relative lives in Lake Victoria. The members of each pair differ in male coloration but not ecology (Greenwood 1965). Similarly, the southern part of Lake Malawi contains endemic *mbuna* taxa currently regarded as species. The areas of endemism are believed to have been dry within the last thousand years (Owen et al. 1990). As with the Lake Nabugabo forms, these *mbuna* are classified as species largely by male breeding color.

The emphasis on color, however, may have inflated estimates of species numbers. Although there are certainly good species not yet described, many named "species"—especially in Lakes Victoria and Malawi—may be only allopatric populations differing slightly in color or trophic morphology. Such features sometimes appear as intraspecific polymorphisms (Turner et al. 2001). The one existing study found that allopatric populations of *Pseudotropheus zebra* from Lake Malawi did indeed show behavioral isolation in the laboratory (Knight and Turner 2004).

To evaluate the likelihood of sympatric speciation, one must first determine whether sympatric taxa are reproductively isolated sister species. As noted above, reproductive isolation appears common. Still, the degree of relatedness of similar forms in the large rift lakes is rarely known because good species-level phylogenies are not available (for an exception, see Allender et al. 2003). Most species are too genetically similar to resolve shallow branches, and phylogenies are further confused by hybridization or the incomplete sorting of mtDNA haplotypes from recent ancestors (e.g., Rüber et al. 2001). The idea of sympatric speciation in the largest lakes is drawn instead from three other lines of evidence: monophyly of groups within single lakes, high rates of speciation, and features of cichlid biology and biogeography that seem to rule out allopatric speciation.

Phylogenetic studies strongly support the view that large groups of cichlids in Lakes Victoria and Malawi descend from one or a few ancestral species. All endemic haplochromine species in Lake Malawi seem to be monophyletic (Moran et al. 1994; Kocher et al. 1995; Albertson et al. 1999), while 500-odd haplochromine species in Lake Victoria seem to have descended from only two ancestral species (Nagl et al. 2000; Verheyen et al. 2003). While the cichlids of Lake Tanganyika probably derive from multiple invasions, many species also arose within the lake (Meyer 1993). Species in the small crater lakes of Bermin and Barombi Mbo are monophyletic (Schliewen et al. 1994). Could such profuse speciation within single lakes have been sympatric? This depends on whether individual lakes had sufficient physical barriers to promote allopatric speciation, a question we address below.

In some lakes, cichlid speciation has been amazingly fast. While speciation intervals in Lakes Malawi, Tanganyika, Barombi Mbo, and Bermin are in line with those of many other animals—between 0.1 and 1 million years (Chapter 12)—McCune and Lovejoy (1998) note that African cichlids appear to speciate faster than fish in other freshwater lakes. But the most surprising result occurs in Lake Victoria, which appears to show amazingly rapid speciation. This conclusion is based on recent seismic surveys and core sampling suggesting that Lake Victoria was completely dry around 14,600 years ago, refilling thereafter from rivers (T.C. Johnson et al. 1996, 2000). Its endemic fauna of around 500 haplochromine species must therefore have evolved within the last 15,000 years!

This "complete desiccation" hypothesis, however, may be flawed. Fryer (1997, 2001) argues that any refugia for *recolonizing* cichlids must also have dried out, leaving the source of the colonists a mystery, and gives other geological evidence that contradicts a recent drying-out of the lake. Moreover, while the deepest part of the lake may indeed have dried out, it is possible that springs or other "lakelets" remained as refugia for fish (Seehausen 2000). By analogy, consider the Death Valley region of California, the remnant of a large lake that dried up about 15,000 years ago. This region contains a complex of endemic species and subspecies of pupfish living in isolated springs, ponds, and streams on the valley floor. If the lake refilled tomorrow, one would conclude that all of the newly sympatric taxa had formed within one day: geological studies would surely miss the small and scattered patches of water and would show instead that the deepest part of the lake was dry only 24 hours before.

Genetic evidence also casts doubt on the recent desiccation of Lake Victoria. In haplochromines, Verheyen et al. (2003) found endemic but old haplotypes of mtDNA that apparently originated within the lake. These haplotypes probably arose between 98,000 and 133,000 years ago, an observation that, according to the authors (p. 328), "strongly argues against the view that LV

[Lake Victoria] dried out completely." Seehausen et al. (2003) reach a similar conclusion based on nuclear DNA.

But it is the remarkable rate of speciation that casts the most doubt on the recent-refilling hypothesis. If the lake indeed dried up 14,600 years ago, and all present species derive from only two ancestors, then on average a lineage branched every 2600 years (see Table 4.1; these figures are actually overestimates as they ignore extinction). Such an interval is almost beyond belief given speciation intervals in other groups (Chapter 12). Although models of both sympatric and allopatric speciation can sometimes yield new species at this rate, it is hard to believe that such rates can be sustained over thousands of years. More important, it is hard to see how new mutations fuelling such a radiation could arise so quickly. Addressing this problem, Seehausen (2002) suggested that all the genetic variation for producing hundreds of species *pre-existed* in the colonizing ancestor, and was partitioned into new species during radiation. But the pre-existence of sufficient genetic variation to build every adaptation seems unlikely given the cichlids' immense diversity of morphology and behavior, which includes novel and complex traits. New mutations were almost surely involved. Yet even if Lake Victoria's radiation were as fast as claimed, this would say little about whether speciation was sympatric, parapatric, or allopatric. Theory shows that, under the right circumstances, all three modes can occur quickly (Chapter 12).

Finally, do the biological and biogeographic data in large lakes make sympatric speciation seem more plausible than allopatric speciation? Here the evidence is contradictory. The following observations have been used to support sympatric speciation in cichlids:

1. It is hard to envision sufficient geographic barriers within lakes to allow the formation of hundreds of species in less than 2 million years.

2. Speciation has also occurred in limnetic and deep-water forms that seem less likely to encounter geographic barriers than do fish living inshore.

3. Movement between habitats may be common enough to prevent allopatric speciation (McKaye and Gray 1984).

4. Some closely related species that differ greatly in color but only slightly in trophic requirements, anatomy, and habitat use are completely sympatric (Seehausen 1996).

5. Some pelagic species show no perceptible genetic structuring across the large Lake Malawi, suggesting that these fish are so mobile that "the possibilities for divergence in allopatry do not occur" (Shaw et al. 2000, p. 2279).

6. Monophyletic groups of pelagic species occur throughout these lakes. This shows that speciation occurred in pelagic ancestors, presumably mobile enough to preclude geographic isolation.

But there are also observations that support allopatric speciation:

1. Geological evidence shows that the levels of all three major lakes have risen and fallen repeatedly, creating isolated basins or small satellite lakes where

speciation could occur allopatrically. Lake drying might also expose new isolated stretches of shoreline that could promote allopatric speciation (Owen et al. 1990; Sturmbauer et al. 2001 and references therein).

2. Many species have highly localized ranges, even when pelagic. In all three large lakes, the shoreline alternates between stretches of rocky and sandy habitat, papyrus swamps, bays, and river estuaries. Each lake also contains offshore islands. Such habitats often harbor a unique cichlid fauna. Many littoral species, for example, are restricted to stretches of coastline less than 1 km long. Species are rarely found outside their normal habitats.

3. Habitat specificity can severely restrict gene flow. Molecular work has shown that significant genetic differentiation can develop over very short distances (thirty-five to several hundred meters) between conspecific populations separated by barriers such as stream inlets or unsuitable shoreline (van Oppen et al. 1997; Markert et al. 1999; Rico and Turner 2002; Rico et al. 2003).

4. Genetic patterns within or between species sometimes mirror the ancient fragmentation of lakes, implying differentiation during periods of geographic isolation (Sturmbauer and Meyer 1993).

5. Littoral fish have limited migration but occasionally colonize new habitats (McKaye and Gray 1984). This, along with their habits of mouthbrooding and nest guarding, suggest that new colonizing populations of cichlids will remain fairly isolated from others, promoting allopatric or parapatric speciation.

6. Some pelagic species return to the littoral zone for spawning, which could cause allopatric speciation through habitat segregation.

7. Some allopatric populations of what is considered a single species have evolved behavioral isolation during periods when they were probably isolated by barriers to dispersal (Knight and Turner 2004).

In view of the continuing debate, it seems premature to conclude that sympatric speciation is a major mode of cichlid speciation in rift lakes. We thus render the Scottish verdict of "not proven." A more systematic approach would use comparative analyses similar to those described at the end of this chapter. Sympatric speciation would be implied if the most recently formed sister species had sympatric ranges, while more distantly related sisters showed less overlap due to post-speciation range expansion. But these tests are impossible without species-level phylogenies. The only good phylogeny of this type, involving several dozen named species or color morphs in the *Maylandia zebra* species complex of Lake Malawi haplochromines, was unable to resolve the biogeography of speciation (Allender et al. 2003).

Despite the ambiguous situation in large lakes, the tiny caldera lakes of Barombi Mbo and Bermin in Cameroon (see Table 4.1) provide stronger evidence for sympatric speciation (Schliewen et al. 1994). In fact, we know of no more convincing example in any group. Each lake was formed by inflow from a river penetrating the crater wall. Barombi Mbo harbors 11 endemic species of mouthbrooding cichlids in 6 genera, while Bermin contains 9 substrate-spawning species (falling into two feeding types) in the genus *Tilapia*. All of the crater-lake fish are tilapiine cichlids, a group that has formed few species in the larger lakes.

Unlike cichlids in large lakes, those of Bermin and Barombi Mbo differ largely in feeding habits and morphological features related to diet (Trewavas et al. 1972; Dominey and Snyder 1988; Stiassny et al. 1992). There is little sexual dimorphism: all but one or two of the nine species in Lake Bermin are sexually monochromatic (Stiassny et al. 1992). The coexistence of such distinct forms without evidence for current hybridization shows that they are biological species. Moreover, analysis of mtDNA suggests that the fish in Lake Bermin are derived from a single ancestor. Some molecular and morphological data, however, imply that Barombi Mbo was invaded twice, yielding two smaller intralacustrine radiations (Trewavas et al. 1972; Kornfield and Smith 2000).

These fish thus meet the first three criteria for sympatric speciation. Moreover, the fourth—the unlikelihood of allopatry—is relatively easy to satisfy because of the unusual nature of the habitat. Unlike the large African lakes, these crater lakes have no obvious geographic barriers that could promote allopatric speciation. Each lake is in effect a cone, so that changes in water level could not create satellite lakelets; and the shorelines are homogeneous. Finally, given the mobility of the fish, the lakes seem too small to allow microallopatric speciation.

The only way to salvage the possibility of allopatric speciation would be to argue either that the species resulted from multiple invasions and that the riverine ancestors are extinct, or that the genetically based monophyly is spurious because hybridization has erased the genetic differences that could reveal multiple invasions. The first possibility seems remote. The second is somewhat more plausible given the ease with which mtDNA spreads across species boundaries in fish.

In sum, sympatric speciation seems the most parsimonious explanation for cichlids in these crater lakes. (See Schliewen et al. 2001 for a possible case of *incipient* sympatric speciation in a small non-crater lake in Cameroon.) But the sympatric scenario remains tentative because phylogenies were based solely on mtDNA, and because genotypes were obtained for only one individual of each species. In other taxa, mtDNA-based evidence for sympatric speciation has evaporated when nuclear DNA was studied as well (Shaw 2002). The crater-lake example would become even more convincing if the monophyly remained after analysis of nuclear DNA from several individuals of each species.

If sympatric speciation has occurred in these crater lakes, one might conclude that it must also have happened in the larger lakes, where it is simply harder to detect. This conclusion seems reasonable, but we are hesitant to infer that sympatric speciation has been *common* in Lakes Victoria, Malawi, or Tanganyika. All three possible cases of sympatric speciation in cichlids involve differentiation in ecology rather than male color. In contrast, the best-documented case of *allopatric* speciation (Nabugabo vs. Victoria) involves changes in male color but not ecology. This is in line with theory showing that sympatric speciation is more likely to occur through divergent ecological, rather than sexual, selection.

It is worth emphasizing again the lack of differentiation in resource use among sympatric, closely related species in the largest lakes. Seehausen et al. (1999, p. 376) suggest that ecological differences in Lakes Victoria and Malawi are "not of the kind that is expected in ecological speciation initiated by resource polymorphism." Genner et al. (1999b, p. 291) claim that in Lake Malawi "there is no published evidence to assign a competition-driven basis to the ecological and morphological differentiation that has occurred in the haplochromine cichlid flock," although there are niche differences between some closely related species (Kornfield and Smith 2000). There is thus a disparity between the traits involved in well-supported cases of sympatric speciation and the traits involved in recent speciation events within the major lakes.

Studies of African cichlids have been caught up in the revival of enthusiasm for sympatric speciation. Unfortunately, the largely inconclusive arguments about biogeography have diverted attention from a question that may be more interesting: why have cichlids, but no other group, speciated so profusely in these lakes? The solutions offered include pharyngeal jaws, sexual dimorphism, and the existence of maternal mouthbrooding and other unique reproductive habits (Galis and Metz 1998). All of these features are closely related to forms of selection that could cause speciation.

HOST-SPECIFIC PARASITES. We can think of a host-specific parasite as occupying an island that is geographically isolated from other species of hosts. The "host geography" of related parasites might thus give evidence for sympatric speciation. One line of evidence has been the observation of closely related parasite species that are each limited to a different but sympatric host. White (1978, p. 242) offers several examples, concluding that "in many groups such as this [nematode parasites in the genus *Procamallanus*] adaptation to new host species may have led to sympatric speciation, since the different host species would constitute so many sharply isolated ecological niches in the same region." But as Mayr (1963) noted, whether such speciation is considered sympatric depends on how often a parasite has the opportunity to colonize a new host. Without such information we cannot infer that speciation was sympatric.

Savolainen and Vespäläinen (2003) propose that sympatric speciation can promote *inquilism*, the phenomenon whereby one species of social insect becomes an obligate parasite (inquiline) on another. The parasite loses the worker caste and all females become reproductives (queens), producing only males and other queens that are tended by the host. In one pair of ants in the genus *Myrmica*, molecular evidence indicates that host and parasite are sister species. Given the permanent association of inquiline with host, Savolainen and Vespäläinen suggest that the inquiline evolved in sympatry. But a sympatric origin for inquilism is biologically difficult, requiring an improbable sequence of genetic preconditions and evolutionary events (Wilson 1971; Ward

1996). Wilson (1971, p. 360) suggests an allopatric alternative that he considers "far likelier": a free-living ancestor could speciate allopatrically and, after secondary contact, one species could become parasitic on the other. It is likely that parasitic ant queens could invade host colonies more easily if they are closely related species, so we might expect that inquilines and their hosts might often be sister species.

The more interesting cases involve related species of parasite that are confined to a *single* species of host. Host-specific species of bird lice, for instance, are often restricted to different parts of the body, such as the head, neck, back, or wings (Clay 1949). An extreme example is the sandhill crane (*Grus canadensis tabida*), which has four endemic species of mite on the primary feathers alone, each occupying a unique portion of the feather (Atyeo and Windingstad 1979). If parasite species endemic to one host species prove to be true sister taxa, this would suggest that the parasite speciated sympatrically on the host.

This situation, however, might also result from allopatric speciation. First, the geographic isolation of host populations might permit speciation of the parasite but not of the host, for the host may evolve more slowly. Secondary contact between host populations could lead to infestation of the host by both parasite species, which, although true sister taxa, were products of allopatric speciation. This possibility can be studied through geographic surveys of host and parasite. An alternative allopatric explanation is cospeciation (Chapter 5). A parasitized ancestral host could speciate in allopatry, as would its parasite. Upon secondary contact, sister species of parasites might recolonize one or both of the sister hosts. Long-term allopatry and cospeciation certainly occur, so it is arguable whether the presence of sister species of parasites on a single host is stronger evidence for sympatric than for allopatric speciation. But such cases are at least *potential* indicators of sympatric speciation.

However, these cases are rare. In fact, we know of only one: wasps in the genus *Apocryptophagus* that parasitize figs in the genus *Ficus*. The mutualism and cospeciation of figs and pollinators is a famous evolutionary story that we explore further in Chapter 5. However, *Ficus* is also parasitized by wasps that insert their eggs into the fruit, but whose offspring do not transfer pollen. Mitochondrial-DNA-based phylogenies of Melanesian fig species and their parasites show three cases in which two sister species of parasite are limited to only one species of fig (Figure 4.2; Weiblen and Bush 2002). These wasps are possible cases of sympatric speciation. Members of sister pairs have significantly larger differences in ovipositor length than do non-sister species attacking a single species of fig. Weiblen and Bush suggest that these sister taxa evolved sympatrically by temporally partitioning the fig: wasps with shorter ovipositors attack the fig at an earlier stage of development when it has a thinner wall. (Molbo et al. 2003 also report cryptic sister species of wasp pollinating single species of Panamanian figs, a conclusion also based on mtDNA-based phylogenies.)

The crucial question here is whether the wasp taxa are reproductively isolated sister species. Since mtDNA sequence was obtained from only one indi-

Figure 4.2 Potential sympatric speciation of parasitic wasps on figs. Phylogeny of figs (*Ficus*) is on the left, parasitic wasps (*Apocryptophagus*) on the right. Wasp species are unnamed but fall into eight sister pairs (a–h). Lines connect fig species with the wasp species that parasitize them. Bootstrap values are given below the nodes. In at least three cases (d, e, and f), two sister species of wasp parasitize (and are endemic to) a single species of fig. (From Weiblen and Bush 2002.)

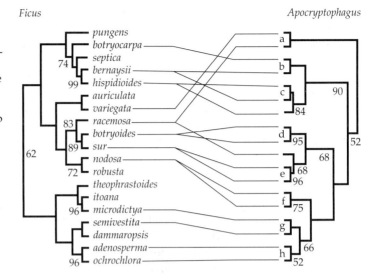

vidual per "species," observed genetic differences between sister taxa may not be fixed traits but only polymorphisms. The sister taxa are distinguished solely by differences in body size and ovipositor length, and are hence called "morphospecies." It is not clear whether these forms might actually represent morphological polymorphisms or host races that differ by one or a few genes. And even if there is reproductive isolation, hybridization between sympatric morphospecies may lead to erroneous conclusions about relatedness. Finally, Melanesia is an archipelago, raising the possibility that parasites might have speciated allopatrically on single species of fig. There have been no biogeographic studies of this possibility. We thus consider the three sympatric pairs of *Apocryptophagus* as potential but not airtight cases of sympatric speciation.

There has been an especially prolonged debate about whether human head and body lice formed sympatrically. These two taxa are usually classified as subspecies, *Pediculus humanus capitis* and *P. humanus corporis*, respectively (Busvine 1948; Levene and Dobzhansky 1959; Mayr 1963; White 1978). Their morphological similarity and difference from other louse species suggest that they are sister taxa. They are found in all human populations, feed on blood, and many people carry both forms. Head lice are generally restricted to the scalp and lay eggs on hair follicles. When not feeding, body lice live and lay eggs in clothing. When reared under constant laboratory conditions, the two forms differ in behavior, morphology, and physiology (Busvine 1948, 1978; Buxton 1948), implying genetic differentiation. This distinctness led Busvine (1978) to suggest elevating the two forms to species status. Forced hybridization yields the missing intermediates, which are fertile, indicating that head and body lice are isolated by habitat. Curiously, one form can evolve into the other in fewer than ten generations when forcibly reared on the "wrong" part of the human body (Levene and Dobzhansky 1959).

Because head and body lice appear to be sister taxa, and occur within each other's cruising range, White (1978) and others suggested that they evolved sympatrically. But Mayr (1963), with his characteristic skepticism toward such views, proposed that head lice evolved on scantily clad tropical populations, while body lice evolved in allopatric clothed populations, such as Eskimos.

Recent molecular data, however, suggest that head and body lice are not biological species, or even taxa with substantial reproductive isolation. Phylogenetic studies using both mitochondrial and nuclear DNA reveal neither fixed genetic differences between forms (Amevigbe et al. 2000; Leo et al. 2002; Kittler et al. 2003; Reed et al. 2004), nor any indication that worldwide samples of each form constitute a monophyletic group. These data, combined with Levene and Dobzhansky's (1959) observation of rapid evolutionary conversion between the forms, suggest that while head and body lice differ in key traits adapting them to their niches, they are probably weakly isolated ecotypes. Migration and hybridization presumably prevent them from becoming full species. Indeed, an ecotype may arise repeatedly when individuals accidentally enter the "wrong" habitat.

One more example will suffice to show the potential rarity of sympatric speciation in parasites. Cyamids ("whale lice") are amphipod crustaceans that live on various parts of cetaceans, including their callosities, blowholes, and eye folds (Rowntree 1996). Three species of whales, the northern right whale (*Eubalena glacilis*), the southern right whale (*E. australis*), and the gray whale (*Eschrichtius robustus*) each harbor three endemic species of cyamids occupying different parts of the body. But on no species of whale are the cyamids sister species. Instead, cyamid speciation has involved colonization from unrelated species of whales, or—in the case of right whales—allopatric speciation of the host (T. Haney, Z. Kaliszewska and J. Seger, pers. comm.).

In all other studies known to us, parasites endemic to a single host are never sister species, and often belong to different genera (e.g., Atyeo and Windingstad 1979; Page 1996; M. Hafner, pers. comm.). This seems to be almost a rule of parasitology. Moreover, the presence of related but non-sister species of parasites on the same host shows that the host offers distinct ecological niches. Thus, the absence of sympatric speciation cannot reflect the absence of ecological differences that could cause disruptive selection.

Evidence from host races and host-specific species

Host races, which usually involve two groups of phytophagous insects living on different host plants in the same area, are the most frequently cited evidence for sympatric speciation. Because they show moderate gene exchange, such taxa are not considered full species, but are often viewed as a stage in sympatric speciation. Host races are often used to motivate models of sympatric speciation since they show the resource differentiation that can promote reproductive isolation. Even Mayr (1963, p. 460) admitted that the formation of some host races "may constitute the only known case indicating the possible occurrence of incipient sympatric speciation." Because 25–40% of all animal species may be phytophagous specialists (Berlocher and Feder 2002), this possibility deserves careful examination.

Although "host race" has many definitions, Drès and Mallet (2002, p. 474) provide the most succinct: Host races are genetically differentiated, sympatric populations of parasites that use different hosts, and between which there is appreciable gene flow. These authors consider gene flow "appreciable" if it is greater than 1%. Jaenike (1981) and Drès and Mallet (2002) give similar criteria for recognizing host races. Modifying these criteria slightly, we suggest that two or more taxa should be considered sympatric host races if they meet all of the following criteria:

1. Their ranges overlap largely or completely.
2. They use different hosts for feeding and/or breeding.
3. They are sister taxa. This conclusion must reflect true phylogenetic status and not the genetic similarity of more distant species that can result from hybridization.
4. They show substantial but not complete reproductive isolation, preferably based on genetic differences. This isolation can be diagnosed indirectly by observing fixed or significantly different frequencies of alleles. (Such differentiation must involve more than one locus so that "host races" do not reflect polymorphism of a single gene.)
5. They must show assortative mating. This may involve either direct mate choice or simply the tendency of some species to mate on a preferred host.
6. They must show genetic adaptation to their respective hosts. This adaptation can include behavioral preference, oviposition preference, or adult or juvenile survival.
7. Each race must have higher fitness on its own host than on other hosts. That is, there must be tradeoffs in fitness for traits distinguishing the taxa.

In principle, it is easy to determine if six of these criteria are met, but criterion 3 is more problematic. Because host races exchange genes, they may appear to be sister taxa when actually they are not. However, races can sometimes be diagnosed as true sister taxa, as when they share complex, derived morphological traits.

Taxa can be considered *sympatrically formed* host races if they meet an additional criterion:

8. The biogeographic and evolutionary history of the taxa makes an allopatric phase *very unlikely*.

As expected, this is the hardest criterion to satisfy. In the famous case of host races in the fly *Rhagoletis pomonella*, for example, a large body of work convinced many evolutionists that allopatry was extremely unlikely—until recent genetic studies suggested otherwise (see below).

These eight criteria have been used in several ways. The least acceptable is to show that each criterion is satisfied by a different group. Some sympatric taxa, for example, show tradeoffs, while others show assortative mating. But this strategy fails because a convincing example of sympatric speciation must satisfy *all* criteria simultaneously.

A better, but still not ideal, way to infer sympatric host-race formation was used by Drès and Mallet (2002), who observed all degrees of reproductive isolation between host races in various groups, ranging from intraspecific polymorphism to a nearly complete lack of hybridization. They argue that this continuum is evidence for sympatric speciation via host-race formation, with different stages of the continuum representing varying degrees of progress toward full speciation. But as we noted in Chapter 3, such a continuum would arise *even if all speciation were completely allopatric*, or at least had begun in allopatry. Weakly isolated races might have evolved in allopatry and now be in the process of fusing. More strongly isolated races could represent cases of secondary contact in which the antagonism between hybridization and selection produces stable hybrid zones. Alternatively, strongly isolated sympatric races might have evolved some reproductive isolation in allopatry that was strengthened by reinforcement after secondary contact.

Invoking a diversity of sympatric host races as evidence for sympatric speciation thus raises two problems: we can almost never exclude the possibility that some reproductive isolation evolved in allopatry, and we can never be sure that host races will continue evolving into good biological species.

Because host races of phytophagous insects have been thoroughly reviewed by Berlocher and Feder (2002) and Drès and Mallet (2002), we examine only those cases that have attracted the most attention.

RHAGOLETIS POMONELLA. The most famous example of sympatric host races is the tephritid fly *Rhagoletis pomonella*, commonly known as the apple maggot fly. Two host races, inhabiting hawthorn (*Crataegus* spp.) and apple (*Malus pumila*), respectively, occur sympatrically in the northeastern and Midwestern United States, while the hawthorn race also lives in apple-free regions in the southern part of the country. There is also an isolated group of populations living on hawthorn in central Mexico (Bush 1993; Feder 1998; Berlocher and Feder 2002).

The ancestral hawthorn race inhabits at least 19 named "species" in the apomictic *Crataegus* complex (Bush et al. 1989). In the mid-nineteenth century, *R. pomonella* was found breeding on introduced apples in the Hudson Valley (Walsh 1867). Over the next 50 years, the apple race spread throughout the eastern United States, becoming a serious agricultural pest (Bush 1993). This well-documented spread, combined with genetic evidence for monophyly of the apple race, implies that it arose only once, probably in the last 150 years.

The life histories of the races are similar. Both have one generation per year. Adult flies congregate for mating on the plants, and females oviposit on ripening fruit. Larvae complete their development in the ripe fallen fruits, pupate in the soil, and undergo a facultative diapause until spring. Adults then spend

1–2 weeks moving about and feeding on diverse food before congregating again on the host plant for mating. Because apples and hawthorns are sympatric, Feder (1998) suggested that individuals with a genetic preference for apples gained a selective advantage, perhaps involving the use of an empty niche or escape from parasitism. This change in preference, coupled with a tendency to mate on the host plant, could have promoted the evolution of host races.

These groups satisfy most of our criteria for host-race status. First, they are broadly sympatric: in many areas of the northern United States, apple and hawthorn flies live in close proximity, with trees only yards apart. Individuals also move fairly large distances between eclosion and mating, so that flies of different races are within cruising distance of each other (Feder et al. 1994).

Although for decades the hawthorn race has been regarded as the ancestor of the apple race, no derived morphological or genetic characters group them as sister taxa. Most taxa related to *R. pomonella*, however, can be unambiguously excluded as ancestors of the apple race based on morphological and molecular traits. Although the hawthorn race of *R. pomonella* is more genetically similar to the apple race than to any other taxon of *Rhagoletis* that lives nearby, there are two or three species similar enough to the apple race to qualify as potential ancestors. Over the whole range of the species, however, some populations of hawthorn flies are more closely related to other taxa than to the apple race (Berlocher 2000). The problem is that there is geographic variation of the allozymes used to construct phylogenies, a lack of strong genetic differentiation between species, and a general lack of derived morphological traits (Feder 1998; Berlocher 2000). The close genetic similarity of apple and hawthorn races might thus reflect hybridization rather than sister-taxon status.

One can infer reproductive isolation between the host races from divergence of allozyme frequencies at six loci and of DNA markers linked to these loci (Feder et al. 1988; 2003a; McPheron et al. 1988). More direct studies show that this isolation involves several ecological barriers. Because apples and hawthorns fruit at different times, the life cycles of the two races are offset, with flies of each race appearing in synchrony with their host fruit. This seasonal difference reduces gene flow between the races by 20–30% (Feder et al. 1998), and is based on differences in larval development time and the sensitivity of diapause to temperature (Smith 1988; Feder et al. 1997a).

Ovipositing females also differ in their acceptance of fruits. While females of both races accept hawthorn fruits more readily than apples, apples are accepted more often by apple flies than by hawthorn flies (Prokopy et al. 1988). Flies derived from each host show a strong tendency to return to that host for mating and oviposition when released in the field. Finally, in both field and laboratory, each race is preferentially attracted to volatile chemicals produced by its host fruit (Linn et al. 2003). Because flies used in these homing studies were derived from field-collected larvae reared in the laboratory, we do not know if host fidelity is due to genetic differences in preference or merely to larval conditioning. But such conditioning is unknown in other host-specific insects (Futuyma and Mayer 1980), so the difference in host preference is probably genetic. Finally, the races show no differential survival in either fruit (Prokopy et al. 1988), no intrinsic assortative mating (that is, no assortative mating independent of host association), and no hybrid sterility or inviability (Feder et al. 1994).

These races thus show assortative mating by habitat preference, incomplete temporal isolation, and some adaptation to their respective hosts. Experimental evidence for tradeoffs rests on the apple and hawthorn races having pupal diapause regimes differentially adapted to the fruiting times of their respective hosts (Feder et al. 1997b; Filchak et al. 2000). In controlled rearing experiments, inter-race hybrids have eclosion times intermediate to those of parental races (Smith 1988). This intermediacy might reduce fitness in nature if hybrids appear between peak fruiting times of the two hosts. The racial differences persist despite gene flow of about 4–6% per generation (Feder et al. 1998).

With the possible exception of sister-taxon status, the apple and hawthorn forms of *R. pomonella* meet all criteria for sympatric host races. But was the apple race *formed* in sympatry? Carson (1989) suggested that this race might have descended not from the sympatric hawthorn race, but from a different allopatric hawthorn race that had previously evolved some reproductive isolation. The hawthorn genus *Crataegus* in North America is a classic apomictic complex comprising at least 20 core sexual species and many hybrids and apomictic forms (Chapter 1); over 1100 "species" have been named (Camp 1942; Muniyamma and Phipps 1984, 1985). Hawthorns are often early successional occupants of disturbed areas, and must have experienced many range shifts associated with glaciations and human activity. Different species and apomictic forms of *Crataegus* differ in habitat, fruiting time, fruit size, flower traits, and odor (Phipps and Muniyamma 1980; Berlocher and Enquist 1993). It is possible, then, that there may have been ample opportunities for the allopatric formation of host races on hawthorns.

Responding to this suggestion, Bush et al. (1989) cited a lack of evidence for host races of *R. pomonella* on various "species" of hawthorns. However, there had been no systematic search for such races, and there is now evidence for substantial genetic and temporal differentiation between neighboring *R. pomonella* populations that live on different hawthorns (Berlocher and Enquist 1993; Berlocher and McPheron 1996). Thus, we cannot reject Carson's (1989) suggestion that the apple and hawthorn races descend from two independent lineages that have become genetically homogenized through interbreeding.

At the same time, recent work has shown that evolution in allopatry may have promoted the formation of these races in another way. Feder et al. (2003b) found that genes causing the racial difference in diapause reside on three of the six chromosomes of *R. pomonella*. Genetic evidence suggests that each of these chromosomes harbors a polymorphic rearrangement—probably an inversion. It is likely that different diapause alleles reside in different rearrangements, and that this protects the alleles from being recombined between the races. (Reduction of recombination would work, of course, only if the inversions contain coadapted *sets* of genes. We do not know whether each inversion contains more than one gene causing the diapause difference.) Genetic surveys suggest that the rearrangement containing the "apple-race" diapause alleles arose within the last 1.6 million years in an isolated population of *R. pomonella* from the highlands of Mexico (Feder et al. 2003a). The Mexican arrangement presumably contained alleles better adapted to warmer climates and would be more advantageous for individuals living on the earlier-fruiting apples. ("Apple" pupae must endure more hot weather.) Introgression of the Mexican rearrangement into North American populations could thus have initiated the host shift onto apples. It is possible that an adaptive valley lay between the diapause regimes for hawthorn and apple, and that this valley could be bridged only by an allopatrically evolved gene complex of large effect.

As suggested by Noor et al. (2001b) and Rieseberg (2001), if genes causing reproductive isolation become associated with different chromosomal arrangements in allopatry, this isolation will tend to persist upon secondary contact, allowing reinforcement to strengthen the barriers to gene flow. This may be what is now happening in *R. pomonella*. We are not suggesting that all or even most isolating barriers between the host races arose in allopatry. The homing of apple flies to the appropriate host and their preference for its odor, for example, clearly must have evolved after the colonization of apples. But it is possible that the apple race could not have arisen without the infusion of genes from allopatric populations. Hence, the races of *Rhagoletis pomonella* may represent not sympatric host-race formation, but an early stage of "allo-sympatric speciation."

ACYRTHOSIPHON PISUM. The pea aphid, *A. pisum*, a pest on legumes, was introduced into the United States from Europe around 1860 (Thomas 1878; Davis 1915). It is a cyclical parthenogen, undergoing 10–12 asexual generations before forming a single sexually reproducing generation in the fall. Individuals feed by piercing plants and sucking sap from the phloem. Mating and oviposition occur only on the host plant, although individuals sometimes move between plants of different species. Eastop (1971) describes three subspecies of *A. pisum*, two of which are distinguishable only by the host plant they inhabit.

In the northeastern United States, two races of the subspecies *A. pisum pisum* live on alfalfa and clover, respectively. They are broadly sympatric, with alfalfa and clover fields often being adjacent. Via (2001) has proposed that these are host races that formed sympatrically. As there are no good phylogenies of *A. pisum*, it is not clear whether these closely related races are sister taxa. They do, however, show adaptive genetic differences. When offered a choice of plants in the laboratory, individuals of each race strongly prefer their native host. Reciprocal transplants show that each race suffers low fecundity and high mortality when forced to live on the wrong host (Via 1991; Via et al. 2000). These tradeoffs in fitness reflect the unwillingness of aphids to feed on the inappropriate host, leading to starvation, lower fecundity, and death (Caillaud and Via 2000). It is not known if the phloem of the two hosts is differentially toxic to the host races or if the aphids show other adaptations to their normal hosts.

Reproductive isolation is also implied by differences in allozyme allele frequencies between the races. Using this divergence, Via (1999) estimated that, in northern New York, around 10% of winged aphids settling on one crop originated on the other. The actual rate of genetic introgression between races, however, is certainly much lower, because aphids landing on the wrong host continue searching for the right one and, if unsuccessful, often die before reproducing. Finally, on both hosts interracial hybrids have lower fitness than individuals of the resident race (Via et al. 2000). With the exception of confirmed sister-taxon status, then, the alfalfa and clover forms qualify as sympatric host races.

The question remains as to whether the races arose in sympatry. Genetic studies of host acceptance and performance have been used to support this idea (Hawthorne and Via 2001). QTL (quantitative-trait locus) mapping of F_2 hybrids showed that the races differ by at least five genes affecting each of two traits: acceptance of the host and performance on the host. Within the limitations of gene mapping (necessarily crude given the small samples), loci affecting acceptance and performance seem to reside in similar regions of the genome. Moreover, performance and acceptance alleles were linked in the expected way: alleles characterizing each host race were fitter on the correct host than on the incorrect host. Hawthorne and Via conclude (2001, p. 904) that loci (QTL) "with antagonistic effects on performance on the two hosts are linked to QTL that produce assortative mating (through habitat choice)." Because such linkage is expected under sympatric speciation (it reduces the antagonistic effect of recombination), it has served as evidence for the sympatric origin of the host races.

An alternative explanation, however, is that host acceptance and host performance are not *linked* traits, but the *same* trait. Acceptance was measured by offering clonal individuals a choice between a clover and an alfalfa plant, and recording the proportion of individuals observed on each plant after 70 hours. Performance was measured as the fecundity of a clonal genotype constrained to live on a single clover or alfalfa plant. However, Caillaud and Via (2000) show that unrestrained aphids of a given race placed on a "wrong" plant will soon abandon it because it provides an inappropriate feeding stimulus (lack of acceptance). If *forced* to live on the wrong plant, aphids will not feed and thus produce few offspring (poor performance). Thus, in Hawthorne and Via's study, acceptance and performance are probably products of the *same* genes—those genes that determine whether a plant provides the correct feeding stimulus. This is neither linkage nor pleiotropy, but synonymy.

However, even if genes for acceptance and performance *were* different and genetically linked, this need not imply that host races formed in sympatry. As noted above, such linkage, whether caused by chromosomal rearrangements or not, may facilitate the *persistence* of allopatrically evolved races when they come into secondary contact, and this persistence can allow reinforcement to create other isolating barriers.

But the main problem with accepting a sympatric origin of these races is that we cannot rule out the allopatric alternative. We are still profoundly ignorant of the natural history, biogeography, and phylogeny of *A. pisum*, its three subspecies, and other relatives in the genus. *A. pisum pisum* lives on a variety of wild, domesticated, and feral legumes throughout the world. Eastop (1971) notes that at least 36 plant species serve as hosts for this subspecies. Moreover, both this subspecies and its alfalfa and clover hosts almost certainly came to North America from Europe. Unlike *Rhagoletis*, we have no data on the location and date of aphid host-race formation, which could have occurred outside of North America, and perhaps more than once.

Indeed, recent work shows that genetically differentiated alfalfa, clover, and pea host races of *A. pisum* exist in several locations in France, and that these races show host-related adaptations in feeding behavior and performance similar to those seen in America (Simon et al. 2003). As Simon et al. (2003, p. 1707) note, their study "suggests that host specialization in pea aphids may have preceded their introduction to North America. Alternatively, it may be that only one host-adapted race was introduced in North America, where the diversification between alfalfa and clover races took place secondarily." Some genetic evidence, however, suggests that the American races both descend from those found in Europe. At the single allozyme locus for which data are available in host races from both continents (*dipeptidase Gly-Leu*), host races show parallel genetic differences, with the slow allele absent or nearly so in the alfalfa race but common in the clover race. Although more genetic data are badly needed, this observation implies that the two American races did not evolve in situ, but descended from those living in Europe.

The situation in *A. pisum* thus resembles that of any pair of closely related sympatric taxa with a nebulous biogeographic history. It seems likely that the host races in America descended from those already existing in Europe, and we cannot exclude an allopatric origin of these races in either location. Further genetic work on many populations of *A. pisum* may settle the issue.

OTHER HOST RACES. Many of the other cases reviewed by Drès and Mallet (2002) and Berlocher and Feder (2002) fulfill our criteria for sympatric host races. Surveying 21 proposed cases, Drès and Mallet conclude that at least two are well-supported host races and another eight are strong candidates lacking only minor documentation. However, in none of these cases can one rule out an allopatric origin.

For example, the goldenrod ball gallformer, *Eurosta solidaginsi*, is another tephritid fly that forms two genetically differentiated, divergently adapted, and reproductively isolated races inhabiting two species of North American goldenrod, *Solidago* (Abrahamson and Weis 1997; Craig et al. 2001). The races have overlapping distributions and are often considered a case of sympatric host-race formation. But there is no strong evidence that the races formed sympatrically, and an allopatric origin is plausible. As Abrahamson and Weis (1997) note, *Solidago*, like hawthorn, is an early successional plant that grows in naturally disturbed areas such as forest canopy openings and river floodplains. These authors suggest that before human disturbance cleared large areas of

North America, *Solidago* lived in small, allopatric patches. Moreover, the two races of *E. solidaginsi* may have independent origins, deriving from pre-existing races that used a possible ancestral host, *S. altissima*.

The single proposed case of host races in plants, involving the parasitic desert mistletoe *Phoradendron californicum* (Glazner et al. 1998), suffers from the same problem. While genetically differentiated races parasitize two desert shrubs with overlapping distributions, an allopatric origin is plausible because, as Glazner et al. note (p. 19), "the two host species only occasionally occur together at a given site."

HOST-SPECIFIC SPECIES.   Good species living on different hosts and showing little or no gene flow have also served as examples of sympatric speciation. But in no case can we rule out allopatry with any confidence. We briefly discuss the two most frequently cited cases.

Wood and colleagues (Wood 1993, and references therein) have studied the *Enchenopa binotata* species complex, a group of nine species of treehoppers in eastern North America. Each of six species (two pairs of which appear to be sisters) inhabits a different species of tree, some of which grow sympatrically. The nine treehopper taxa appear to be strongly reproductively isolated based on differences in morphology, host attraction, and oviposition preference. They also show assortative mating in field and laboratory tests, and high mortality when transferred to the wrong hosts. Mating occurs on the host, and asynchrony of host phenology is another isolating barrier. But while some *Enchenopa* species are sympatric and host specific, there is little evidence that host shifts occurred in sympatry. The ranges of host trees are known to have changed with climate, and even today host species show only moderate range overlap. Claridge (1993) describes species of leafhoppers and planthoppers (related families of ecologically similar insects) that show geographic variation in host use, suggesting the possibility of allopatric speciation.

The two lacewings, *Chrysoperla plorabunda* (formerly *Chrysopa carnea*) and *C. downesi*, which differ in color, microhabitat, and seasonality, occur sympatrically in the northeastern United States (Tauber and Tauber 1977, 1989). *C. plorabunda* is a multivoltine meadow-dwelling species that is light green, changing to reddish brown when it moves to deciduous trees at the end of summer. *C. downesi*, on the other hand, is univoltine, dark green, and lives exclusively in conifers. Tauber and Tauber suggested that selection for color, microhabitat preference, and breeding season were the main forces propelling sympatric speciation.

The genetic differences in color appear to result from changes at a single locus, while the variation in life cycle is due to at least two unlinked loci (Tauber et al. 1977). Tauber and Tauber (1977, 1989) propose that these forms speciated sympatrically according to Maynard Smith's (1966) scenario, with natural selection initially causing a habitat-related polymorphism in color and later promoting asynchronous breeding that caused reproductive isolation between color morphs.

But this scenario is doubtful. Reproductive isolation between the two species is also caused by differences in courtship song (Wells and Henry 1998). More important, phylogenies based on song structure and molecular data indicate that these species are not sister taxa. Each is actually part of a geographically widespread complex of other species. There are two cryptic species of "*C. downesi*" and three of "*C. plorabunda*." At least one species of each complex is allopatric to one or more from the other complex (Henry et al. 1999). Henry (1985) argues that one cannot rule out an allopatric origin of *Chrysoperla* species that involved sexual isolation. Finally, the potentially large difference between conifers and deciduous trees in postglacial spread and distribution raises the possibility that *Chrysoperla* species may have formed in allopatry or parapatry.

CONCLUSIONS. In no case do we have compelling evidence that two host races or host-specific species experienced complete divergence in sympatry. Indeed, the best example of sympatric host races, *Rhagoletis pomonella*, almost certainly involved some evolution in allopatry.

Allochronic (temporal) isolation in sympatry

Although temporal isolation has played a role in the reproductive isolation of host races or species such as *Rhagoletis pomonella* and *Enchenopa*, other isolating barriers were also involved. But it is also possible that large and rapid changes in life cycle could be a *primary* cause of sympatric speciation. This might occur through the rapid fixation of an allele of large effect that alters life cycle, or through an environmental perturbation that divides a population into two sympatric entities breeding at different times. The former scenario seems nearly impossible in sympatry: Mayr (1963, pp. 474–476) emphasizes several difficulties, including the requirements that such a genetic change must occur in several individuals (so that a new population can arise) and must create a completely disjunct breeding season.

The second scenario begins with nongenetic change, usually a catastrophic environmental or developmental event that splits a population into two temporally isolated segments. (These changes are often assumed to become heritable through genetic assimilation.) Some might consider such an event equivalent to a geographic barrier, for populations isolated in this way are initially genetically identical and are outside each other's cruising range during the breeding season. Moreover, drastic allochronic changes immediately eliminate all possibility of interbreeding, so that geographic overlap poses no problem for the evolution of reproductive isolation. However, because a change in breeding time involves the biology of the organism, it could also be considered as a reproductive isolating barrier. Allochronic speciation thus differs from both sympatric and allopatric speciation. We consider the topic here because it involves speciation within a restricted geographic area, and creates the phylogenetic pattern of sympatric speciation: newly formed species having high range overlap.

Several proposed cases of allochronic speciation have not survived scrutiny. One involves the field crickets *Gryllus pennsylvanicus* and *G. veletis*. These species are largely isolated by differences in breeding season that result from differences in diapause: the former species overwinters as eggs and the latter as nymphs (Alexander 1968; Harrison 1985). This was once considered an example of sympatric speciation promoted by environmental cooling (through either climatic change or invasion of a northern habitat) that killed off all adults of the ancestral species, leaving only eggs and late juveniles (Alexander and Bigelow 1960). Recent work, however, shows that *G. pennsylvanicus* and *G. veletis* are not sister species but part of larger European and North American clades that differ in diapause regime (Harrison and Bogdanowicz 1995; Huang et al. 2000). Moreover, there is evidence for substantial postzygotic isolation (Alexander 1968). Thus, while temporal isolation may have been involved in the split between the larger clades, there is no reason to think that *G. pennsylvanicus* and *G. veletis* were formed by sympatric, allochronic speciation.

The most famous candidate for sympatric, allochronic speciation is the group of 13- and 17-year periodical cicadas in the genus *Magicicada*. The situation here is complex, involving three layers of genetic isolation (Alexander and Moore 1962; Lloyd and Dybas 1966a,b; Williams and Simon 1995; Cooley et al. 2003). First, three distinct species (*M. septendecim*, *M. cassini*, and *M. septendecula*) have a 17-year life cycle. Individuals spend nearly all of this period as underground nymphs feeding on root xylem. Adults emerge synchronously in huge numbers over a period of one to two weeks, and live for about three weeks. (This synchronicity is presumably important in understanding the evolution of the prolonged life cycle: sporadic off-year "stragglers" are quickly consumed by predators who might be satiated by a large simultaneous emergence.) These three species are differentiated morphologically and behaviorally, and can be distinguished using mtDNA. When different species emerge at the same time, few hybrids are formed. This appears to reflect sexual isolation and the inviability of hybrid eggs (White 1973).

Second, each 17-year species has a sister species with a 13-year life cycle (*M. tredecim*, *M. tredecassini*, and *M. tredecula*, respectively). Members of each sister pair are nearly identical in morphology. Where sister species are sympatric, the opportunity for hybridization occurs only once every 221 years, when 13- and 17-year cicadas co-emerge. Little is known about what forms of reproductive isolation, if any, prevent hybridization in this circumstance.

Finally, each species of 13- and 17-year cicada belongs to several different "broods": groups of cicada species having the same life cycle and emerging at the same time. Different broods emerge in different years. Most include all three species, are scattered throughout the eastern United States, and have largely parapatric distributions. Because "stragglers" that emerge off schedule are rare, populations of the same species that belong to different broods are temporally isolated.

A likely explanation for the existence of the six species is that each 17-year species originated independently from its 13-year sibling by adding a four-year developmental diapause in the second nymphal instar. This idea is supported by the greater genetic divergence among populations of *M. tredecim* (13-year) than among populations of *M. septendecim* (17-year), implying that *M. tredecim* is older (Martin and Simon 1990). The evolutionary and genetic bases of the change in life cycle remains a mystery, although there are theories involving an environmental resetting of the developmental clock that later became genetically encoded (Lloyd and Dybas 1966b; Cooley et al. 2001; Marshall et al. 2003, but see Cox and Carlton 2003).

It is also unclear whether the major allochronic transitions—between 13- and 17-year sister taxa and between conspecific populations belonging to different broods—occurred sympatrically. While all species sharing a single periodicity are broadly sympatric, 17- and 13-year sister taxa are parapatric, with ranges that overlap only slightly. New 17-year species could have originated through rapid climate-induced shifts in the periodicity of allopatric 13-year populations, with ranges that subsequently expanded. (This is the "long-term climate shock" model of Marshall and Cooley 2000.) On the other hand, the ecological similarity of nymphs may have forced taxa that arose sympatrically into parapatry. Because the divergence between the 13- and 17-year species occurred roughly 1–2 million years ago (Martin and Simon 1990), all traces of the species' original ranges have been obliterated by Pleistocene glaciation.

We still know too little about the ecology and historical biogeography of *Magicicada* to determine whether speciation was sympatric or parapatric. Nevertheless, this group is one of the few good examples of strong reproductive isolation produced by rapid changes in life cycle.

Comparative studies of the biogeography of speciation

Establishing that even a single speciation event was sympatric requires a great deal of work. But the preoccupation of evolutionists with potential cases of sympatric speciation might give a false impression of the frequency of this process. To judge the relative importance of sympatry versus allopatry in speciation, evolutionists have adopted the less laborious method of comparative analysis.

The idea of using *groups* of species to infer the biogeography of speciation is not new. In 1889, Wagner (quoted in Mayr, 1963, pp. 484–485) discussed the geographic ranges of what he considered closely related pairs of species. Observing that such species are usually allopatric and separated by geographic barriers, Wagner inferred that almost all speciation is allopatric. Jordan presented similar data, concluding that (1905, p. 548): "Always the species nearest alike in structure are not found together, nor yet far apart, and always a barrier lies between." While it was insightful to realize that the biogeography of close relatives might suggest how they formed, progress was stalled by the lack of good phylogenies.

The rise of molecular systematics, as well as the accumulation of better range maps, now permits fairly rigorous comparative studies. J. D. Lynch (1989) pioneered this work. Using range overlaps of close relatives, he tried to estimate the relative frequencies of vicariant, peripatric, and sympatric speciation. While Lynch concluded that most speciation was allopatric, his methods have been criticized on several counts (Chesser and Zink 1994; Losos and Glor 2003). First, Lynch used an arbitrary degree of range overlap to discern the mode of speciation (for example, a pair of sister taxa was said to have speciated sympatrically if their ranges overlap by 60% or more). In addition, in more than half of his cases, at least one member of a sister "pair" included more than one species. In such cases the meaning of "range overlap" is unclear, as it involves the overlap between the *sums* of areas of species that may have formed by different processes. Finally, taxa of diverse ages were used instead of only the youngest pairs. This is problematic because, as we have emphasized, species ranges have often changed drastically. One might expect, for example, that sympatrically formed species might expand their ranges as they age, reducing their range overlap. These problems became clear when Chesser and Zink (1994) applied Lynch's methods to birds, finding that between 21 and 35% of speciation events were "sympatric." This figure is clearly too high in view of other studies that show virtually no sympatric speciation in birds (Diamond 1977; Coyne and Price 2000).

To address range changes, Barraclough and Vogler (2000) modified the comparative approach to deal with taxa of different ages. Figure 4.3 shows their technique. One begins with a molecular phylogeny in which the genetic divergence between a pair of clades gives an idea of their relative age; Figure 4.3A gives a sample phylogeny of 12 species. For each independent pair of sister clades, one then calculates the degree of range overlap. Because sympatric spe-

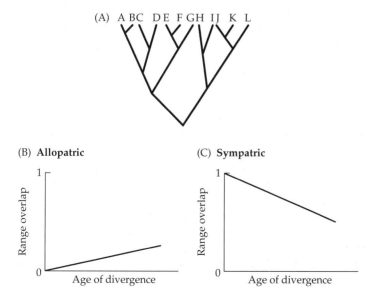

Figure 4.3 Barraclough and Vogler's (2000) method for inferring the biogeography of speciation. (A) A hypothetical molecular-based phylogeny of 12 species (A–L). (B) The expected relationship between age of divergence of sister taxa and degree of sympatry (range overlap) under pure allopatric speciation. Range overlap begins at zero and increases with time. (C) The expected relationship between age of divergence of sister taxa and degree of sympatry (range overlap) under pure sympatric speciation. Range overlap begins at one and decreases with time.

ciation can occur anywhere in a species' range, Barraclough and Vogler's estimate of overlap was the fraction of the smaller clade's range that lay within that of its sister clade. One thus obtains a list of sister clades, with each having a relative age ("node depth") calculated from the degree of molecular divergence, and a measure of range overlap varying between zero for complete allopatry and one for complete sympatry.

One can then infer modes of speciation by plotting range overlap against divergence time. If all speciation is allopatric, taxa begin with a range overlap of zero and, after speciation is complete, descendant groups invade each other's range, so that overlap increases between older taxa. In this case sister taxa should fall along a curve beginning at the origin and rising with time (Figure 4.3B; this relationship need not, of course, be linear). When all speciation is sympatric, taxa begin with range overlap of one. After speciation is completed, range movements will *reduce* the overlap between older taxa (Figure 4.3C). The goal, then, is to determine whether a group of taxa conforms to the pattern of sympatric or allopatric speciation. (A more precise test involves estimating the *intercept* of the linear regression on the "range overlap" axis [i.e., the amount of range overlap at a divergence time of zero]. Allopatric speciation should yield intercepts close to zero; sympatric speciation close to one.)

This method is far from perfect. If sympatric taxa exchange genes, they will become genetically similar and thus appear younger than they really are, even if they formed allopatrically. Hence one could erroneously infer that some "young" taxa showing high range overlap were cases of sympatric speciation. However, if no such cases exist, and the pattern conforms to that of increasing overlap with time, one can be confident that speciation is mostly allopatric. There is also a bias in the opposite direction. Subspecies and morphologically differentiated populations are always allopatric, and may be misclassified as good species. This would produce "young allopatric" taxa that falsely imply allopatric speciation.

Another problem is that, like Lynch (1989), Barraclough and Vogler use not just sister *species*, but sister *clades*, so that one must sometimes calculate range overlaps between *groups* of species (for example, see A–D versus E–G in Figure 4.3A). Barraclough and Vogler dealt with this problem by summing all the overlaps (e.g., considering groups A–D and E–G, the range overlap would simply be that between the summed ranges of each group). This method assumes that each daughter species inherits part of the ancestral range—an assumption unlikely to hold for older pairs of taxa.

But the most serious problem is that of range changes themselves. As we show in Chapter 12, the time scale for speciation is often on the order of 0.1–1 million years, and during that period environmental change can cause frequent and large shifts in ranges, effacing the biogeography of speciation. Recognizing this problem, Barraclough and Vogler made models of the effect of range shifts on their plots. Not surprisingly, they found that if range shifts are frequent, the biogeographic signature of speciation is lost, yielding a scatter of points that conform to neither the allopatric nor the sympatric pattern.

This biogeographic analysis was applied to ten diverse groups of insects, birds, and fish whose phylogenies were derived from allozymes or DNA sequence (Barraclough and Vogler 2000; Ribera et al. 2001). The results are mixed (Figure 4.4). Three taxa—fairy wrens, cranes, and swordtail fish—show patterns broadly consistent with allopatric speciation. But in other groups the patterns are less clear, with the points showing more scatter. All groups, however, show positive correlations between range overlap and time, although only five of these correlations are significant. The intercepts are small, with a mean of 0.11. Barraclough and Vogler (2000, p. 419) conclude that these groups show a "predominantly allopatric mode of speciation."

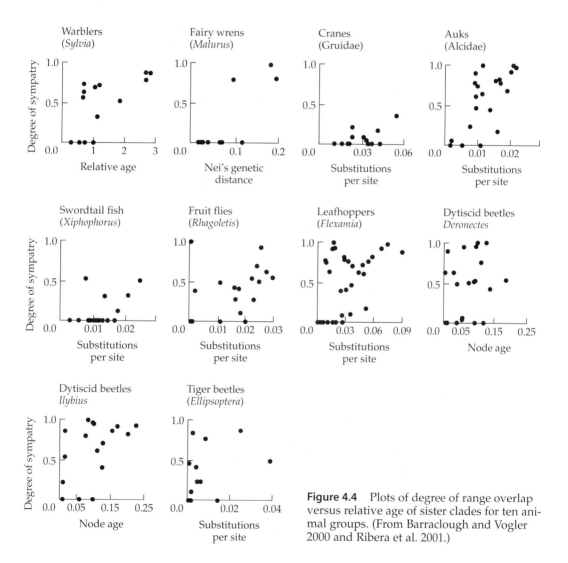

Figure 4.4 Plots of degree of range overlap versus relative age of sister clades for ten animal groups. (From Barraclough and Vogler 2000 and Ribera et al. 2001.)

Figure 4.5 Plot of range overlap versus relative divergence time for six mammal clades. Divergence times for mtDNA are taken relative to the deepest node (assigned a value of 100; Sanderson 2002), while divergence times for allozymes are estimated from Nei's *D*. Each point represents a phylogenetically independent pair of species or clades. Two values of range overlap are shown for each point: the average among all comparisons used (filled circles) and the maximum value observed (open circles). Arrow in plot (F) marks a putative case of sympatric speciation in kangaroo rats (*D. microps*/*D. ordii*). (After B. Fitzpatrick and M. Turelli.)

This interpretation, however, must be qualified in view of the number of young pairs of taxa having high range overlaps, which are especially common in the five insect groups. Such cases may represent sympatrically formed species (in which case one might infer that 10–25% of all speciation events are sympatric) or only recent range shifts (since all of these groups are in the northern hemisphere). The data thus show that only a few taxa conform to either a *pure* allopatric or sympatric pattern. This intermediacy, however, may not reflect a mixture of sympatric and allopatric speciation, but merely the movement of already-formed species. But two results are worth noting. First, the positive slopes for all taxa show that ranges do tend to expand over time, invalidating the idea that once a species is formed, its range remains static. Second, two of the three groups that show a clear allopatric pattern—fairy wrens and swordtail fish—occupy tropical habitats. It is possible that, if temporal variation in climate is smaller in the tropics than in the temperate zone, the current ranges of tropical species carry more information about speciation.

Fitzpatrick and Turelli (2004) performed a similar analysis of 12 mammal clades, with relative ages estimated from allozymes and mtDNA. The main difference between their method and that of Barraclough and Vogler (2000) is that when a sister-group comparison involved more than two species, the range overlaps between pairs of species were averaged instead of summed. In addition, they calculated the *maximum* range overlap between all species pairs separated by a given node in the phylogeny. These maxima can reveal outliers having high range overlap—potential cases of sympatric speciation.

Their data show that when the biogeographic signal is clear, speciation appears to be allopatric or parapatric. No group displays a consistent signal of sympatric speciation. Figure 4.5 shows a selection of six clades. One—pocket gophers in the genus *Orthogeomys*—shows not only allopatric speciation but also *perpetual* allopatry: even old taxa have no range overlap. Other taxa, including gazelles (*Gazella*), ground squirrels (*Spermophilus*), deer mice (*Peromyscus boylii* group), and shrews (*Sorex*), correspond to the "classic" pattern of allopatric speciation shown in Figure 4.3B: range overlap is virtually zero in recently diverged taxa, and increases with divergence time. This pattern is also seen in pikas (*Ochotona*; data not shown). (In some groups, like *Spermophilus*, taxa with no range overlap sometimes have abutting distributions, suggesting either parapatric speciation or competition that prevents sympatry.)

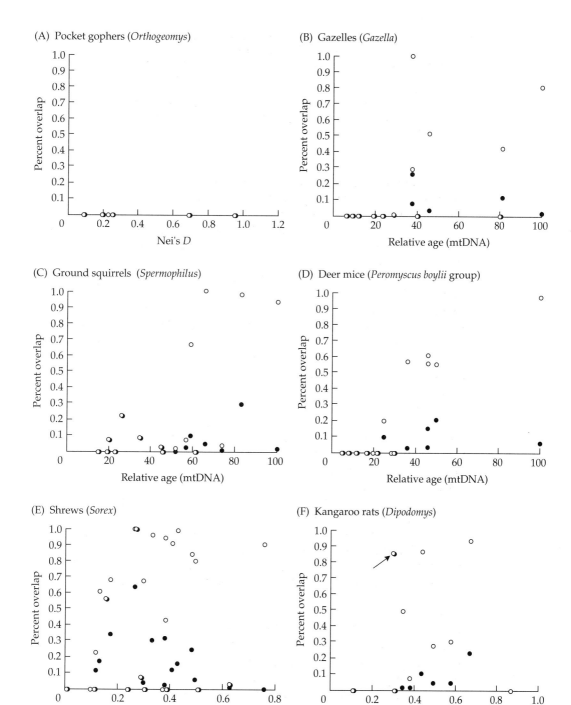

On the other hand, kangaroo rats in the genus *Dipodomys* (Figure 4.5F) form a scatter of points lacking a clear biogeographic pattern of speciation. This group, however, contains one possible case of sympatric speciation: the range of *D. microps* overlaps extensively with that of its presumed sister species *D. ordii*, and they are fairly closely related. Scattered data with no clear signal are also seen in bats of the genus *Myotis* and family Mormoopidae, and in the pocket mouse genera *Chaetodipus* and *Perognathus* (data not shown). Yet none of these groups evinces the telltale sign of sympatric speciation: very recently diverged taxa having high range overlap.

That signal, however, appears in one dataset involving mtDNA-based distances in North American chipmunks (Figure 4.6A). This pattern is based largely on one pair of species (*Tamias dorsalis* and *T. cinericollis*, indicated by the arrow) that show almost no mtDNA divergence but have a range overlap of about 80%. It is possible, however, that this outlier point does not reflect high sympatry of young taxa, but a misleadingly low estimate of divergence

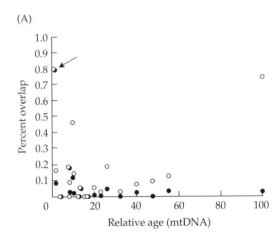

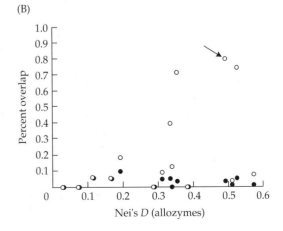

Figure 4.6 Plot of range overlap versus relative divergence time for phylogenetically independent comparisons between taxa of North American chipmunks (subgenus *Eutamias*). Filled circles are average range overlaps between taxa, and open circles are the maximum range overlap. (A) Relative age estimated from mtDNA divergence. (B) Relative age estimated from allozyme divergence. One comparison, indicated by the arrow (*Tamias dorsalis/T. cinericollis*), appears to show the signature of sympatric speciation using mtDNA. The allozyme plot, however, shows that these species are long diverged, having high genetic distance. The low mtDNA distance is therefore almost certainly due to introgression. (After B. Fitzpatrick and M. Turelli.)

time caused by introgression of mtDNA. This, in fact, appears to be the case: when analyzed with allozymes, *T. dorsalis* and *T. cinericollis* are not sister taxa, and the divergence between them is quite old (Figure 4.6B, arrow). The plot using allozymes shows that chipmunks fall squarely into an allopatric pattern, underscoring the problem of basing such plots on mtDNA data, particularly when taxa hybridize in nature. In general, then, the mammalian data provide strong evidence for allopatric speciation, with some observations consistent with parapatric speciation. There is virtually no evidence for sympatric speciation.

We suggest that, when possible, such comparisons be limited to tropical groups, or at least to those having fossil evidence that range shifts were small. Comparisons should involve pairs of species rather than pairs of clades, so that one need not calculate range overlaps between groups of species. Moreover, one should avoid using the intercept from regression analysis to infer the mode of speciation, as this statistic is heavily affected by older taxa that carry little information about the mode of speciation. It seems more reasonable to infer the mode of speciation from only the very youngest pairs of species. Firm conclusions about the biogeography of speciation can then be drawn only if nearly all of these pairs show either high or low range overlap. In mammals, all the groups that satisfy this criterion show the pattern of allopatric speciation.

Conclusions

Although the resurgence of interest in sympatric speciation has produced a deluge of new information about ecology, biogeography, and systematics, these data have not supported the view that sympatric speciation is frequent in nature, either overall or in specific groups.

Some simulations show that under certain conditions sympatric speciation can occur with high frequency, but virtually all theories depend on assumptions about nature whose reality cannot be adequately judged, or on parameters that are unknown or unknowable. A reasonable conclusion from existing theory, however, is that if sympatric speciation occurs, it must almost certainly involve resource differentiation and thus should often produce species with ecologically-based isolating barriers. Indeed, all three possible cases of sympatric speciation (see below) involve striking divergence in resource use. Moreover, it is clear that any biological feature of organisms that promotes assortative mating without requiring genetic change will also promote sympatric speciation. As has been noted many times, sympatric speciation is most likely in organisms that mate on or near their preferred resources, or have an innate tendency to mate with others of similar appearance. We should add that resource-based sympatric speciation could potentially occur in almost any group. Despite claims that such speciation is unlikely in mobile creatures such as large mammals and birds (e.g., Losos and Glor 2003), the essence of the sympatric-speciation hypothesis is that mobility is irrelevant.

Sexual selection alone seems unlikely to cause sympatric speciation because such selection does not allow the coexistence of newly formed species. We therefore doubt whether the explosive speciation of cichlids in large African rift lakes, which produce relatives differing primarily in breeding color, occurred sympatrically.

Experimental work shows that groups with substantial reproductive isolation can form sympatrically. The one really successful experiment, that of Rice and Salt (1990), involved divergent selection on resource use coupled with a design that permitted individuals to mate only on their chosen resource. While one can argue about the realism of this experiment, it does support the conclusions from theory about the importance of resource divergence and "no-gene" assortative mating.

In the end, conclusions about the occurrence and likelihood of sympatric speciation must rest on data from nature. Such work almost always encounters one problem: it is hard to rule out either allopatry or reinforcement. *Rhagoletis pomonella*, for example, was considered a paradigm of sympatric host-race formation until recent work showed that critical evolutionary changes might have occurred in allopatry. For most species, however, we may never be able to get this type of information.

However, there are at least three examples in which sympatric speciation seems plausible: tilapiine cichlids in the crater lakes of Cameroon (Schliewen et al. 1994), Arctic charr in Lake Galtaból in Iceland (Gíslason et al. 1999), and parasitic fig wasps in the genus *Apocryptophagus* (Weiblen and Bush 2002). (We regard the wasps as less convincing than the other examples because of the greater possibility that speciation was allopatric.) In all three cases, the species differ in resource use, supporting the importance of ecologically based selection in sympatric speciation. However, all three cases require further documentation. In cichlids and fig wasps, sister species were diagnosed using mtDNA-based phylogenies, and only the cichlids seem free from the objection that speciation was not really sympatric but appears to be so because the mtDNA of taxa was homogenized via hybridization. Moreover, in wasps and cichlids only one individual per species was examined, and firmer conclusions require examining several individuals per species and using phylogenies based on nuclear DNA. In the Arctic charr, there has been no systematic attempt to survey surrounding water systems, so there is some possibility that, like sticklebacks, the benthic and limnetic morphs are not sister taxa. Thus, although none of these examples is an airtight case of sympatric speciation, additional work could strengthen each.

Beyond these possible examples of "normal" sympatric speciation, there are several cases that are more unusual. These include sympatric "cultural" speciation in the African village indigobird (Chapter 6), allochronic isolation in periodical cicadas and pink salmon (Chapter 5), and, of course, polyploidy and diploid hybrid speciation (Chapter 9). All of these processes except polyploidy and diploid hybrid speciation depend on unusual features of the organism's biology, and so are unlikely to be common. Diploid hybrid speciation is probably also rare. But polyploidy is certainly a frequent mode of sympatric speciation in plants. Finally, new taxa of *asexual* organisms may arise sympatrically; in Chapter 1 we described how this can happen when a new mutation allows invasion of an unfilled niche.

If sympatric speciation has indeed occurred in crater-lake cichlids, Arctic charr, and fig wasps—which involve habitat islands most favorable for detecting the process—then it has certainly occurred in other places where it is hard or impossible to distinguish from allopatric speciation. This is why comparative studies, such as those of Barraclough and Vogler (2000) and Fitzpatrick and Turelli (2004) are important: in principle they allow us to estimate the frequency of sympatric speciation in entire groups of species. The problem with such studies is that unless the results are unambiguous, patterns could reflect range shifts and not the biogeography of speciation. Nevertheless, at least some taxa (especially in mammals) show a consistent pattern of allopatric speciation, while no group shows a consistent pattern of sympatric speciation. Future work should concentrate primarily on recently evolved taxa and on groups living in areas that have experienced minimal climatic or geological change.

One might be able to rule out sympatric speciation in cases where two sympatric or parapatric taxa show *intrinsic* postzygotic isolation. Such isolation cannot evolve in sympatry because the genes involved would be purged from interbreeding populations (Chapter 8), but can appear upon secondary contact between allopatric populations. (There is one caveat: the taxa studied must show some hybridization. If no gene flow is possible, then intrinsic postzygotic isolation could have evolved after reproductive isolation was complete.) Although we currently lack any theory telling us how much intrinsic postzygotic isolation can evolve under a given level of hybridization, the absence of intrinsic postzygotic isolation would seem to represent a good litmus test for sympatric speciation. We suggest, then, that sympatric taxa showing both gene flow and intrinsic postzygotic isolation must have experienced a period of allopatry or parapatry.

Although some have suggested that understanding the genetic basis of reproductive isolation can tell us whether speciation was sympatric or allopatric (e.g., Via 2001), we do not find this strategy promising. Different theoretical models, even of the same general type, predict different genetic bases of reproductive isolation. Even the observation of close linkage between genes causing reproductive isolation, once seen as a signature of sympatric speciation, can be produced by parapatric speciation or allopatric differentiation followed by reinforcement (Noor et al. 2001b; Rieseberg 2001; Navarro and Barton 2003).

Theoretical studies, which generally show that sympatric speciation occurs quite readily, stand in stark contrast to the paucity of evidence for sympatric speciation in nature. There are two explanations: either the theories are often incorrect or incomplete, or the theories are correct and there are many cases of sympatric speciation that cannot be unambiguously distinguished from other modes of speciation. However, the absence of sympatric speciation in situations favorable for its detection (e.g., oceanic islands and endemic parasites on hosts) leads us to suspect that theory has provided an overly optimistic view of the process.

In sum, the evidence for sympatric speciation is still scant. While we can point to a few promising cases, they do not add up to strong support for the idea that this process is common. While additional work may provide more compelling evidence, it is hard to see how the data at hand can justify the current wave of enthusiasm for sympatric speciation.

5
Ecological Isolation

Compared to more intellectually engaging areas of speciation, such as the role of biogeography, discussions of individual isolating barriers can seem dry. This is probably why major figures in speciation, like Dobzhansky and Mayr, often dealt with these barriers in a perfunctory way. With the exception of hybrid sterility and inviability—which engaged geneticists because these phenomena are easy to study in the laboratory—most discussions of reproductive isolation have been limited to producing a list of the various forms along with one or two examples of each. Yet isolating barriers must be a central object of work on speciation. While one can infer the presence of isolating barriers from genetic or phenotypic data, this gives no information about which barriers were the first to arise or what evolutionary forces created them.

In the next two chapters we discuss isolating barriers that act before fertilization—the "prezygotic barriers" described in Table 2.2. This chapter deals with the three "ecological" barriers: habitat, pollinator, and temporal isolation. These barriers arise directly from ecological differences between closely related species. By describing them as "ecological," we do not mean to imply that ecology played no role in the evolution of other barriers or is not involved in how they impede gene flow. On the contrary: virtually all barriers can be considered ecological in the sense that they may arise from environmentally imposed selection. Nevertheless, habitat, pollinator, and temporal isolation share a particularly intimate connection with the environment.

It is often observed that ecology has been neglected in studies of speciation (e.g., Morell 1999). This is partially true—at least until the last two decades. While both Darwin and Fisher proposed explicit ecological models of speciation, the role of ecology was largely ignored during the Modern Synthesis. To be sure, some founders of the Synthesis paid lip service to ecology:

The genotype of a species is an integrated system adapted to the ecological niche in which the species lives. Gene recombination in the offspring of species hybrids may lead to formation of discordant gene patterns (Dobzhansky 1951, p. 208; see also Dobzhansky 1946a, p. 210).

The shift into the new ecological niche and the need to become fully adapted to it sets up a considerable selection pressure. This leads to a genetic reconstruction of the population until it is ultimately genetically as well as ecologically so different from the parental population that the new species can return to its point of origin and coexist with the parental species without being seriously in competition with it. Although prevention of interbreeding with the parental species is dependent on the acquisition of isolating mechanisms, one should not minimize the importance of ecological factors in the process of geographic isolation (Mayr 1963, p. 575; see also Mayr 1947).

But beyond these general statements, one finds little consideration of exactly *how* adaptation to different niches leads to speciation. Dobzhansky concentrated less on the evolution of reproductive isolation than on its genetic and developmental basis. He viewed prezygotic isolation mainly as a product of reinforcement: such barriers arose as responses to maladaptive hybridization, not to environmental differences (viz. Dobzhansky 1951, p. 211). Although Mayr (1947, 1963) viewed adaptive differences between populations as signs of incipient speciation, he shed little light on how the former produce the latter. Instead, Mayr saw "founder effects" as the most important cause of speciation (Chapter 11). In his view, adaptation of genes to the external environment was far less important than their adaptation to the internal genetic environment, leading to the evolution of "harmonious epistatic interactions" largely divorced from ecology (Mayr 1963, p. 533). From about 1965 to 1985, many evolutionists were drawn to nonecological theories involving chromosome rearrangements, genetic drift, and various types of founder effects (Carson 1975; White 1978; Templeton 1980b). Stebbins's discussion of "the origin of isolating mechanisms" in *Variation and Evolution in Plants* (1950, pp. 236–250) says almost nothing about the role of natural selection.

Besides this preoccupation with genetics, there are other reasons for the early neglect of ecology. Most important were ecologists' lack of interest in speciation, and evolutionary ecologists' concern with demonstrating natural selection in the wild. Ecologists now realize, however, that the connection between adaptation and reproductive isolation is largely unexplored, and many are making a committed effort to forge this link. This has produced a welcome revival of interest in the ecology of speciation.

Before discussing specific isolating barriers, we must address some issues that have muddled discussions of ecological isolation. First, not all niche differences, including those that allow species to coexist, constitute isolating barriers. Consider two desert plants that partition the environment by having roots of different length, taking water from different depths in the soil. This niche difference will not cause reproductive isolation unless hybrids with roots of intermediate length suffer a fitness disadvantage.

Second, ecological differences can cause reproductive isolation either prezygotically or postzygotically. Consider two species of host-specific insects adapted to the fruits of their respective plants. If the hosts bear fruit at different times, preventing contact between adults of the two species, this constitutes temporal isolation—a prezygotic barrier. But if the fruits appear simultaneously, the insects may hybridize. If the hybrids perform poorly on either fruit, they suffer a form of *postzygotic* isolation: extrinsic hybrid inviability. This failure to distinguish between prezygotic and postzygotic ecological isolation has led to some confusion in the literature. Stebbins (1950), for example, applied the term "ecological isolation" to both the spatial restriction of individuals to different habitats (a prezygotic barrier) and the inability of hybrids to thrive in parental environments (a postzygotic barrier). But the two forms of isolation have different evolutionary consequences. Only postzygotic isolation, for instance, can promote the evolution of reinforcement (Chapter 10). Recognizing the importance of ecologically based postzygotic isolation has been an important advance in studying speciation.

Finally, just as ecological differences promoting coexistence need not cause reproductive isolation, so ecological differences causing reproductive isolation need not promote coexistence. Although *Mimulus cardinalis* and *M. lewisii* show strong pollinator isolation, their coexistence almost certainly depends on differential adaptation to climate and soil (Ramsey et al. 2003).

When dealing with isolating barriers, we focus on five areas. First, we define the barrier and discuss how to detect it and estimate its strength. Classifications of isolating barriers have often been ambiguous in ways that compromised our understanding of speciation. Measuring the strength of a barrier is important for comparative work that reconstructs the history of speciation. For example, the conclusion that behavioral isolation evolves faster in sympatric than in allopatric species of *Drosophila* depends entirely on literature estimates of the strength of these barriers (Coyne and Orr 1989a, 1997).

Second, we provide a few examples of each isolating barrier. Because other authors have supplied long lists of these (e.g., Dobzhansky 1937b; Stebbins 1950; Levin 1978), we concentrate on well-established cases or on barriers with interesting evolutionary features.

Third, we compare the importance of each barrier relative to others, in terms of both its current effect on gene flow and its likely importance in speciation. As we argued in Chapter 3, these effects need not be correlated.

Fourth, we discuss possible scenarios for the evolution of these barriers, trying to determine how they might have arisen as a byproduct of selection or drift. Until recently, this important endeavor—the real study of the origin of species—has been largely neglected. Understanding how barriers arise is usually easier for prezygotic than for postzygotic isolation, since some prezygotic barriers, such as pollinator and habitat isolation, are byproducts of well-understood forms of selection.

Finally, we describe what is known about the genetic basis of each barrier, which may help us understand the evolutionary path to speciation. For instance, if behavioral isolation is based on changes in many genes, it is likely to have evolved via gradual coevolution of male traits and female preferences.

Habitat Isolation

Many related, sympatric species occupy preferred habitats within a small area, even though individuals or their gametes can disperse more widely. Allopatric species, on the other hand, may have genetically based habitat preferences or tolerances that prevent them from becoming sympatric. If such spatial separation is based on biological differences, and reduces gene flow between taxa, then these species show *habitat isolation*. Although habitat isolation involves spatial separation of taxa, it is not identical to a geographic barrier. Habitat isolation is based on genetic differences between taxa, geographic isolation on historical accidents. Allopatric species ranges can result from either or both factors.

Habitat isolation is generally based on the inability of a species to use another species' environment. This may rest on genetically based differences in fitness associated with habitat use, on competition that forces species to use different niches where they are sympatric, and (in animals) on the ability to seek out the habitats to which they are best adapted.

In animals, habitat isolation limits reproductive encounters between heterospecific compared to conspecific individuals; in plants, it reduces the relative amount of heterospecific pollen reaching the stigma. In all cases, *hybridization is reduced solely because individuals or their gametes fail to encounter heterospecifics as often as conspecifics during the breeding season*. When dealing with such isolation, it is important to recognize the difference between "habitat" (an ecologically distinct area that is large compared to an individual's cruising range) and "niche" (an organism's way of life). Although habitat isolation is a form of niche differentiation, not all niche differentiation involves habitat isolation. We have already noted that the use of different resources need not reduce gene flow.

Habitat isolation comes in two forms. In "microspatial" isolation, members of two species occupy the same general area but their reproductive encounters are reduced by adaptations to or preferences for ecologically different parts of this area. Mayr (1947, p. 265) gives an example in two species of drongo birds, *Dicurus ludwigii* and *D. admilis*: "The former, in my experience, never ventures outside the edge of the evergreen forest, the latter seems never to perch even on the outside edge of a forest tree. Yet, both may occur within 50 yards of each other and the latter species is quick to occupy a clearing made in evergreen forest." Although plants cannot home to preferred habitats, they can be microspatially isolated by using different niches in a single habitat. Figure 5.1 shows a possible case: six species of the goldenrod *Solidago* distributed along a moisture gradient on a prairie slope (Werner and Platt 1976). Host-specific

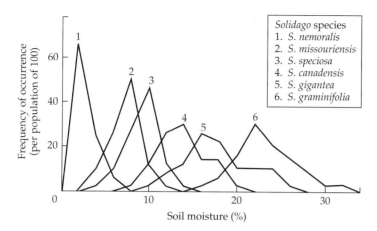

Figure 5.1 Microspatial segregation of six species of goldenrod (*Solidago*) along a soil moisture gradient. This gradient lies along a slope in an undisturbed tall-grass prairie in Iowa. Graph shows frequency of occurrence of individual plants of each species at various positions on the gradient. (From Werner and Platt 1976.)

insects, such as the alfalfa and clover races of the pea aphid *Acyrthosiphon pisum pisum*, cannot feed on each other's plants, limiting the possibility of interbreeding.

In "macrospatial" isolation, two taxa cannot interbreed because their habitats are allopatric. This form of isolation is important in parasites and host-specific insects dependent on allopatric hosts. For example, chrysomelid beetles in the genus *Ophraella* are fairly host specific, with larvae and adults of each species feeding on one or a few plant species. Laboratory studies show that some allopatric species of *Ophraella* cannot survive or oviposit on each other's hosts (Futuyma et al. 1994, 1995). This differential adaptation prevents all contact between these species.

Finally, gene flow can be restricted by both micro- and macrospatial isolation. For example, two species may be adapted to different habitats that are largely allopatric, but their ranges may overlap in an area containing patches of both habitats. The crickets *Gryllus pennsylvanicus* and *G. firmus* are largely allopatric in the eastern United States, with the former preferring loamy soils and the latter sandy soils. In places where both habitats exist in close proximity, each species is largely restricted to its respective soil type (Rand and Harrison 1989). The oaks *Quercus gambelii* and *Q. grisea* show a similar pattern (Howard et al. 1997).

Habitat isolation is surely an important barrier to gene flow, for many related species of plants and animals exhibit spatial separation that seems related to their adaptations. Moreover, in its macrospatial form, habitat isolation acts as a geographic barrier, preventing gene flow and promoting the evolution of other isolating barriers. Depending on the environment, only a slight difference in habitat use can create substantial allopatry. This is in contrast to other barriers: slight differences in flowering time, for example, can still allow appreciable hybridization.

Yet compared to other isolating barriers, habitat isolation has been sorely neglected (Schemske 2000). While its existence among sympatric species is easy to document, its strength is hard to measure, requiring technically demanding estimates of encounter rate or pollen movement. In addition, because allopatry can be caused by geographic barriers, macrospatial habitat isolation, or a combination of both, distinguishing among these factors requires laborious transplant experiments.

Detecting and measuring habitat isolation

Demonstrating *microspatial* isolation requires showing that

1. spatial separation between members of different species is, on average, larger than that between members of the same species;
2. this spatial separation impedes gene flow during the period when both breed simultaneously;
3. reduced gene flow occurs solely through decreasing the chance of cross-mating or cross-pollination;
4. the spatial separation is based on genetic differences between the species.

Satisfying these requirements may be difficult in plants, especially insect-pollinated ones in which habitat isolation may be confused with floral constancy, the tendency of individual pollinators to move between members of a single plant species. In such a case the dearth of hybridization might falsely be ascribed to spatial isolation of plant species. Nevertheless, habitat isolation can be demonstrated through transplant experiments: if cross-pollination and/or hybridization increases when heterospecific individuals are moved closer together, then they are habitat isolated. An alternative is removal experiments: if two related species are found in nearby habitats, and removing one causes the other to expand into the vacated area, this suggests that microspatial isolation is enforced by interspecific competition. We describe examples of such "niche expansion" below.

Like all forms of prezygotic isolation, habitat isolation can restrict gene flow only during the breeding season, and this must be taken into account when spatial overlap varies over time. The most dramatic fluctuations occur in migratory birds, which can have sympatric feeding ranges in winter and allopatric breeding ranges in spring and summer. The Golden-crowned sparrow (*Zonotrichia atricapilla*) and White-crowned sparrow (*Z. leucophrys*) are broadly sympatric in Alaska and western Canada, but differ in breeding habitat (Kessel, 1998, and pers. comm.). However, when overwintering farther south, these species associate much more closely, often forming mixed flocks (Rising 1996). Conversely, benthic and limnetic morphs of the threespine stickleback differ considerably in feeding habitat, but this spatial isolation largely disappears during the breeding season, when different morphs sometimes occupy adjacent territories (McPhail 1994). Gene flow is restricted not by habitat isolation, but by sexual isolation and ecological inferiority of hybrids (Schluter 1995; Schluter and Nagel 1995).

For a pair of animal species a and b, a measure of habitat isolation during the mating period is

$$1 - \frac{p_{ab}}{2p_a p_b}$$

where p_{ab} is the observed proportion of encounters between individuals of opposite sex that involve heterospecific partners, p_a is the proportion of total individuals that are members of species a, p_b is the proportion of individuals of species b, and $2 p_a p_b$ is the proportion of *expected* heterospecific encounters if the two species mix randomly. (These encounters must reflect proximity alone and not factors like differential responses to mating calls, which constitute behavioral isolation.) This statistic will usually vary between zero (individuals encounter heterospecifics randomly at mating time) and one (no heterospecific encounters). A similar statistic could be calculated for plants using the amount of conspecific versus heterospecific pollen landing on stigmas, but only if there is no pollinator constancy.

The problem of allopatry

It is hard to identify habitat isolation when species are allopatric. Although reproductive isolation is usually considered to include those genetic barriers that can act in sympatry, the evolution of habitat isolation may *prevent* such sympatry. Permanent allopatry can thus result from ecological isolation.

Merely observing that the habitats of allopatric taxa are different does not necessarily indicate habitat isolation. Such a difference may simply reflect geographic isolation and provide no indication whether the species would occupy different habitats in sympatry. Conversely, if the habitats of allopatric taxa seem similar, the species may nevertheless be divergently adapted to cryptic ecological factors that could cause microspatial isolation. Alternatively, habitat segregation could be enforced in sympatry by differences in competitive ability that are completely undetectable in allopatric taxa.

There are two ways to address these difficulties. The best is through transplant or laboratory experiments. If allopatry is strictly enforced through habitat isolation, then species transplanted to each other's habitats, or grown under conditions favorable to the other species, will not thrive. *Drosophila sechellia*, for example, lives only on the Seychelles Islands, while its close relative *D. mauritiana* inhabits the distant island of Mauritius. *D. sechellia* feeds and breeds only on the rotting fruits of *Morinda citrifolia*, which contain chemicals toxic to all related species (Louis and David 1986; Legal et al. 1992, 1994; D. Lachaise, pers. comm.). The difference between *D. sechellia* and its relatives in oviposition preference, larval survival, and adult survival on *Morinda* is genetic. It is reasonable to infer that were *D. mauritiana* and *D. sechellia* to become sympatric in an area harboring *M. citrifolia*, they would show complete habitat isolation. Similar studies in plants include the classic transplant experiments of Clausen et al. (1940; see also Schluter 2000).

Alternatively, habitat isolation might be inferred if habitat destruction leads to massive hybridization between taxa that formerly occupied distinct niches, as in *Iris fulva* and *I. hexagona*. The problem with this observation is understanding how much of the post-disturbance hybridization is due to the elimination of spatial separation and how much to the ability of previously inviable hybrids to occupy new and possibly intermediate habitats (implicating extrinsic hybrid inviability). Moreover, increased hybridization after habitat disturbance can be caused by a breakdown of sexual or temporal isolation (Seehausen et al. 1997; Lamont et al. 2003).

Measuring habitat isolation between taxa that are fully or partly allopatric is also hard unless each species is completely unable to survive in the other's habitat. Ramsey et al. (2003) calculated habitat isolation between the monkeyflowers *Mimulus cardinalis* and *M. lewisii*, which are largely allopatric but have overlapping ranges. Their statistic, which takes into account both macro- and microspatial isolation, involved determining whether the species co-occurred in large quadrats and then comparing this co-occurrence with the amount expected if individuals were located randomly in space. But this procedure is valid only if all allopatry is due to habitat isolation and not history. If one cannot use transplant experiments to distinguish these factors, it is better to estimate habitat isolation only in areas of sympatry, where the strengths of diverse isolating barriers can be compared directly.

Examples of habitat isolation

Unequivocal examples of habitat isolation are rare, mainly because there are few data showing that the spatial separation of species reduces gene flow. For example, the segregation of goldenrod species shown in Figure 5.1 may be ineffective as an isolating barrier: the species are pollinated by generalist insects that move easily between microhabitats. Levin (1978) cites many other cases of microhabitat segregation in plants as examples of "isolating mechanisms," but gives no evidence that this segregation restricts gene flow.

Evidence for habitat isolation (as opposed to simple geographic isolation) becomes stronger when observations in nature are coupled with laboratory studies showing differential adaptation. Chapter 7 discusses two parapatric subspecies of sagebrush, *Artemisia tridentata tridentata*, and *A. tridentata vaseyana*, segregated by altitude in northern Utah (Wang et al. 1997). A hybrid zone exists at intermediate elevations. When the two subspecies and their hybrids are transplanted to common gardens containing soils from all three locations, each of the three forms has the highest fitness on soil from its native habitat. Although extrinsic postzygotic isolation clearly plays a role in keeping these subspecies distinct, altitudinal separation caused by local adaptation is surely an important isolating barrier. Mooney (1966) gives a similar example in two species of composite (*Erigeron*) with overlapping ranges.

Transplant experiments have also been useful in revealing macrospatial isolation between allopatric species. Clausen et al. (1940) showed pronounced differences in habitat use between several related species having allopatric distributions. In California, for example, *Horkelia fusca* occupies meadows and gravelly slopes above 1400 m, while its two relatives *H. californica* and *H. cuneata* live along the coast below 1000 m. After one generation in a common garden, seedlings of these species were transplanted to various habitats. *H. fusca* showed poor viability and very little flowering at low elevations, while the low-elevation species did not survive in the alpine zone (Clausen et al. 1940). It seems unlikely that this habitat specificity would permit *H. fusca* to ever hybridize with its congeners.

Assessing habitat isolation in sympatry is easier in animals, as one need only observe a difference in mating and breeding sites. Chapter 4 describes several cases of host-specific insects whose mating and oviposition are restricted to a single plant, as well host-specific parasites whose hybridization is prevented by infrequent contact between hosts. While *Rhagoletis pomonella* inhabits apples and hawthorn, its close relative *R. mendax* mates and oviposits only on blueberry. Hybrids are readily produced in field cages but are never found where the species co-occur in Michigan (Feder and Bush 1989).

Lynch (1978) describes habitat isolation in the frogs *Rana blairi* and *R. pipiens*. Where their ranges overlap in Nebraska, the species occupy different habitats: *R. blairi* lives and breeds largely in turbid, silty streams while *R. pipiens* prefers clear streams. In a few areas having intermediate habitats, the species co-occur and occasionally hybridize. Habitat isolation is also seen in the fire-bellied toad *Bombina* in Eastern Europe and in *Epilachna* ladybird beetles in Japan (Katakura et al. 1989; Katakura and Hosogai 1994; MacCallum et al. 1998). Finally, some of the best examples of habitat isolation involve obligatory mutualisms between insects and plants (see below).

Even if two closely related species are able to use the same habitat—thus sharing part of their "fundamental niche"—interspecific competition in sympatry may restrict them to different habitats ("realized" niches). This also constitutes habitat isolation, although differences in competitive ability may reflect traits like size and aggression that might not be direct adaptations to the local habitat. In birds, there are several examples of "habitat displacement" in which related species occupy more divergent habitats when sympatric than when allopatric. Robinson and Terborgh (1995) showed size-related habitat segregation of birds along a successional area of Amazonian flood plain. Evidence that this segregation was maintained at least partially by competition came from observations of interspecific aggression. Similarly, Diamond (1973) found that congeneric bird species in New Guinea are often altitudinally segregated. This is apparently caused by competition and not innate adaptation to particular elevations, because species expand their altitudinal range when congeners are absent.

Artificial removal experiments can also show whether habitat expansion accompanies the absence of a competitor. This has been shown in species of bluegill sunfish (*Lepomis*), rockfish (*Sebastes*), and damselfish (*Stegastes*) (Werner and Hall 1977; Larson 1980; Robertson 1996). In rockfish, the role of genetically based homing could be distinguished from that of interspecific competition through observations of larval settling.

Relative importance of habitat isolation

The correlation between invasion of new habitats and the appearance of new species, as occurs during adaptive radiations on islands and archipelagos (Schluter 2000), has led some to conclude that habitat isolation is an important cause of speciation. This conclusion, however, does not follow, because natural selection accompanying the invasion of new environments can produce almost any form of reproductive isolation. Evaluating the role of habitat isolation in speciation must thus rest on comparative studies in which several isolating barriers are measured.

Only one study of this sort exists, and its results were negative: Barraclough et al. (1999a) failed to find a correlation between occupation of particular habitats and the age of speciation events among 13 species of tiger beetles (*Ellipsoptera*). This is not surprising, as the dearth of species allowed only a few phylogenetically independent comparisons. Studying the birds of New Guinea, Diamond (1986) concluded that the first stages of ecological differentiation during speciation entailed divergence of habitat, followed only later by divergence of food type and prey size. As Schluter (2000, pp. 50–51) notes, Diamond's study had several problems, including using range overlap as an estimate of divergence time. Nevertheless, Diamond's conclusions may well be correct. A more rigorous phylogenetic analysis of greenish warblers (*Phylloscopus*) by Richman and Price (1992) revealed that habitat isolation was the first form of niche differentiation to evolve, with differences in prey size and feeding method appearing later.

There are several famous cases of obligatory mutualism in which plants and their insect pollinators—which breed in or on the plants—have evolved and speciated in tandem ("cospeciation"), yielding not only a one-to-one relationship between plant and insect, but a congruence of their phylogenies. The two most famous cases are figs and fig wasps and yuccas and yucca moths (see below). Because in many mutualisms—including these two—adults mate only or primarily on their host plant, this cospeciation leads to habitat isolation of the pollinator species and pollinator isolation of the host species. In such cases, habitat isolation must have been an important factor in initially restricting gene flow between diverging species of insects. Other isolating barriers that are byproducts of adaptation to diverging plants, such as hybrid inviability, could evolve no faster than barriers involving host use.

The evolution of habitat isolation

There is little doubt that whether species form in allopatry, parapatry, or sympatry, habitat isolation is due to natural selection. As the nature of selection can depend on the biogeography of speciation, we divide our discussion along biogeographic lines.

ALLOPATRY AND PARAPATRY. The evolution of habitat isolation in allopatry or parapatry is conceptually simple: two populations with limited gene flow adapt to local habitats to the extent that they show microspatial isolation when they reattain sympatry—if they are able to attain sympatry. In animals, physiological adaptation to the habitat will often be reinforced by behavioral preferences.

The initial stages of habitat isolation can be seen in adaptive differences in habitat use among populations of single species. Classical studies involve plants that are locally adapted to temperature or soil chemistry. These include *Potentilla glandulosa* (Clausen et al. 1940) and *Hieracium umbellatum* (Turesson 1922). Other instances include adaptation to heavy metals on mine tailings and serpentine soils (Kruckeberg 1967; Macnair 1981, 1987; Macnair and Gardner 1998). Transplant or common-garden experiments (e.g., Macnair and Christie 1983) invariably demonstrate that habitat isolation is based on genetic differences between populations. Chapter 11 describes genetically based geographic variation in host use by the phytophagous beetle *Necoclamisus bebbianae* (Funk 1998).

Kruckeberg (1984, p. 44) gives several examples of plants endemic to serpentine soil whose relatives cannot live on serpentine. These could be cases of parapatrically evolved habitat isolation if the sister species live on adjacent, non-serpentine soils. Unfortunately, we lack the phylogenetic information that could confirm this.

Cospeciation that involves host-specific mating bears close study because it almost certainly occurred in allopatry, and because it implies that habitat isolation was important in speciation itself. Even if no other isolating barriers exist, parasites are prevented from interbreeding by their close ties to the host. There are several requirements for establishing cospeciation. First, there must be an obligatory one-to-one relationship between host and parasite (or between host and pollinator), and the parasite must mate primarily or exclusively on the host. Second, the phylogenies of host and parasite (or pollinator) should be largely congruent. Finally, sister species of hosts and of their pollinators should have formed at roughly the same time. This last prerequisite rules out the possibility that parasites or pollinators radiated much later than their hosts, and that phylogenies are congruent only because newly formed parasite species are more likely to use more phenotypically similar (and hence more closely related) host species.

We know of five cases that meet these criteria. We discuss two—the yucca/yucca moth and fig/fig wasp systems—in the section on pollinator isolation later in this chapter. The other three involve parasitic chewing lice. As wingless ectoparasites whose entire life cycle occurs on a single host, these lice are good candidates for allopatric speciation.

The best-studied example involves 15 species of gophers and 17 species of lice (Page 1994, 1996; Hafner and Page 1995; Huelsenbeck et al. 1997). There is a statistically significant (but not perfect) congruence between the host and parasite phylogenies (Figure 5.2). Because most species of these gophers are

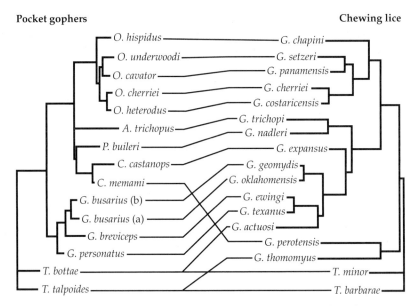

Figure 5.2 Parallel phylogenies of pocket gophers and chewing lice showing cospeciation. Thin lines connect parasite species that infest gopher species. The genera of pocket gophers are *Orthogeomys, Zygogeomys, Pappogeomys, Cratogeomys, Geomys* (two subspecies), and *Thomomys*. Chewing lice are in the genera *Geomydoecus* and *Thomomydoecus*. (From Hafner and Page 1995.)

allopatric, one cannot definitively test whether the parasites are habitat isolated, but surely contact between gopher species is limited in sympatry.

Similar congruence was found for 11 species of seabirds and 14 species of chewing lice (Paterson et al. 2000). In this case, some of the seabird species are sympatric. The specificity of parasites on sympatric hosts demonstrates the efficacy of habitat isolation. As sister species of lice never occur on the same host species, the lice clearly speciated allopatrically. Finally, Page et al. (1998) found phylogenetic congruence between 12 species of swiftlets (Collocalliinae) and 13 species of lice (*Dennyus*). Cospeciation has also involved asexual "species" and their sexually reproducing hosts, including mutualisms between bacteria and hydrothermal-vent clams or aphids (Peek et al. 1998; Clark et al. 2000).

Although cospeciation almost certainly involves habitat isolation, the process remains an evolutionary puzzle. What are the evolutionary forces that caused divergence? It seems reasonable that the speciation of parasites follows that of their hosts, but we do not know which changes in the hosts provoked evolution in the parasites. The situation is even murkier in cospeciation between plants and pollinators.

Beyond cospeciation, evidence for the allopatric origin of habitat isolation is seen in species whose overlapping ranges resulted from secondary contact between formerly isolated populations. These include the habitat-isolated toads *Bombina bombina* and *B. variegata* (MacCallum et al. 1998), and the columbines *Aquilegia formosa* and *A. pubescens* (Hodges and Arnold 1994).

SYMPATRY.   There are additional ways of evolving habitat isolation during a sympatric phase of speciation. If sympatry is primary—that is, the taxa undergo complete sympatric speciation—selection for niche divergence can be the initial force driving the process. If sympatry results from secondary contact, two factors can promote the evolution of habitat isolation: reinforcement and competition.

The most popular models of sympatric speciation involve either disruptive selection for resource use (Dieckmann and Doebeli 1999; Kondrashov and Kondrashov 1999), or disruptive selection on host use coupled with a tendency to mate on the host (Rice 1987). In both cases, habitat isolation can be an integral part of speciation. Of the three well-established cases of non-polyploid sympatric speciation discussed in Chapter 4, one (fig wasps) clearly involves habitat isolation, one (Arctic charr) is equivocal, and the other (crater-lake cichlids) shows no habitat isolation. Although allo- and autopolyploid speciation, which are clearly sympatric, must involve niche differentiation that permits coexistence of polyploids and ancestors, we do not know whether this differentiation involves adaptation to spatially distinct habitats.

Has reinforcement produced habitat isolation between sympatric species? One signature of reinforcement is the observation, in species with overlapping ranges, of greater premating isolation in areas of sympatry than of allopatry (Chapter 10). But although there are many cases of habitat displacement in sympatry, they can result from natural selection to reduce competition between either incipient or full species. This process would constitute not reinforcement but adaptation occurring after speciation.

The genetics of habitat isolation

Habitat isolation, like other ecological barriers, is often a direct byproduct of adaptation to the environment. Thus, one might expect the genetics of such isolation to reflect the genetics of adaptive differences between species (Orr 2001). For example, habitat isolation based on adaptation to a very different environment (e.g., following island colonization), might often be due to the substitution of alleles of large effect, since strong and novel selective pressures can favor alleles having large effects on phenotypes (Orr and Coyne 1992). On the other hand, some cases of habitat isolation might be gradual, as when parasites or pollinators cospeciate with their hosts. We suspect that such coevolution often involves adaptation to a slowly changing optimum, and will yield changes in many genes of small or moderate effect.

Table 5.1 lists all existing genetic studies of ecological isolation, including habitat isolation. The included studies meet two criteria. First, the analyzed trait must be known to cause ecological isolation between species in either

Table 5.1 *Genetic analyses of traits demonstrably or probably involved in ecological prezygotic isolation between closely related species*

Species pair	Trait	Number of genes	Reference
Drosophila simulans/ D. sechellia	Habitat isolation (use of novel food)		
	Adult tolerance	≥5	Jones 1998, 2001, 2003, 2004
	Larval tolerance	≥3	
	Adult oviposition preference/	≥2	
	Rate of egg production	≥4	
Acyrthosiphon pisum pisum alfalfa versus clover race	Fecundity on host plant	≥5	Hawthorne and Via 2001
	Acceptance of host plant	≥5	
Chrysoperla plorabunda/ C. downesi	Photoperiod response (temporal isolation)	≥2	Tauber and Tauber 1977; Tauber et al. 1977
Mimulus guttatus populations	Copper tolerance	1[a]	Macnair 1983; Christie 1983
Mimulus lewisii/ M. cardinalis	12 floral traits involved in pollinator isolation	—[b]	Bradshaw et al. 1995, 1998; Schemske and Bradshaw 1999
Aquilegia species	Presence of nectar spurs	1	Prazmo 1965
	Shape of spurs	1	
	Flower position	1	

[a] In crosses between tolerant and nontolerant populations, tolerance generally segregates as if it were due to a single dominant Mendelian allele, although there are anomalous ratios in some crosses.

[b] Trait genes, localized by QTL analysis, include flower color (anthocyanins and carotenoids), corolla width and projected area, upper and lower petal reflexing, nectar volume, stamen and pistil length, and height and width of the corolla aperture. All of these traits may affect whether a flower is bumblebee-pollinated (*M. lewisii*) or hummingbird-pollinated (*M. cardinalis*); species are largely isolated by this pollinator difference. Differences in all traits were controlled by between one and six loci, and for nine traits a major portion of the species difference (>25%) resided in one chromosomal region. (The difference in carotenoid concentration in petals was governed by only one gene, *yup*.)

nature or the laboratory, or be plausibly involved in such isolation. Second, the genetic analysis must have been fairly rigorous, using one of three methods:

1. Simple Mendelian analysis, in which segregation ratios in backcrosses or F_2s show that an isolating barrier is caused by changes at one or two genes.

2. Backcross/F_2 analysis, in which fertile F_1 hybrids are backcrossed to parental species or intercrossed to produce an F_2 generation yielding many combinations of chromosome regions from the two species. The genotype of each hybrid is determined using either morphological or molecular markers, and the phenotype scored for a trait causing reproductive isolation. This method allows one to identify chromosome regions affecting the trait, as well as to estimate the size of their effects.

3. Biometric analyses, in which measurement of character means and variances in F_2s or backcrosses yields a rough (and minimum) estimate of gene number.

Unfortunately, there are only three genetic studies of habitat isolation (see Table 5.1). Macnair (1983) found that copper tolerance in individuals of the monkeyflower *Mimulus guttatus* growing on a copper mine was based on a single dominant allele. Although there are no data on the reduction of gene flow caused by restriction to contaminated soil, it seems likely that this gene produces some habitat isolation.

Jones (1998, 2001, 2003, 2004) studied the genetics of the adaptation of *Drosophila sechellia* to the fruit *Morinda citrifolia*, which, as noted above, creates potential habitat isolation between the fly and its allopatric relatives. Separate analyses were done on the ability of adults and of larvae to survive exposure to octanoic acid (the toxic component of the fruit), and on the willingness of females to oviposit in the presence of octanoic acid. Although several genes were involved in the evolution of each of the three traits, the number of genes was probably not large because substantial portions of the genome had no effect.

Finally, Hawthorne and Via (2001) reported that genetic differences in host acceptance and fecundity between alfalfa and clover races of the pea aphid *Acyrthosiphon pisum pisum* involved at least five loci for each trait. As we noted in Chapter 4, however, acceptance and fecundity may actually be the same trait.

Thus, in the two cases where we can guess that reasonably strong selection occurred following invasion of a new niche (*M. guttatus* and *D. sechellia*), habitat isolation seems to be based on relatively few genes. While this provides weak support for the view that adapting to novel habitats often provokes a simple genetic response, we obviously need more studies.

Pollinator (Floral) Isolation

In this form of isolation, gene flow between angiosperm species is reduced because different species use different pollinators. These interactions can involve pollination by completely different insects, different frequencies of pollination by the same group of insects, or the use of diverse body parts of a single species of pollinator.

Grant (1949, 1994) distinguished between *ethological pollinator isolation* and *mechanical pollinator isolation*. In the former, isolation involves differential visitation by pollinators based on differences in flower color, odor, nectar, or morphology. Classic examples include differences between plants that are pollinated by hummingbirds (often having red flowers with narrow corolla tubes and high nectar volume), and their bee-pollinated relatives (often having pink or yellow flowers with visible nectar guides, and modified petals that provide landing platforms). Grant (1994) gives many examples.

Differences in pollinator behavior that enforce reproductive isolation can be either genetic or learned. Cases of extreme "monotropy," in which there is a one-to-one relationship between pollinator and angiosperm species, are almost certainly the result of hardwired preferences in the pollinator, including the mutualisms between fig wasps and figs (Wiebes 1979), and between euglossine bees and bee-mimicking orchids in the genus *Ophrys* (Paulus and Gack 1990). Pollinator isolation may also be enforced by pollinator constancy, the tendency of individual pollinators to visit flowers of a single species in sequence, even though different individuals concentrate on different species of plants. For example, although some bee species forage on both of the sympatric species *Aquilegia formosa* and *A. pubescens*, individual bees tend to show fidelity to one of the two species (Fulton and Hodges 1999).

Mechanical pollinator isolation, on the other hand, involves interspecific differences in flower or pollinator morphology that prevent cross-pollination even when a pollinator visits more than one species of plant. Mechanical isolation includes the physical inability of a pollinator to pollinate the "wrong" species of flower. Moths, for example, often fail to pollinate flowers adapted to bees, because the moths' long tongues prevent them from approaching the flower closely enough to contact the anthers. Alternatively, different species of flowers can deposit pollen on different parts of the body of a single species of pollinator, with the plants' stigmas positioned to receive only conspecific pollen (Nilsson 1983). Figure 5.3 shows an example involving a euglossine bee; Grant (1994) gives other examples.

Detecting and measuring pollinator isolation

In principle, demonstrating pollinator isolation is easy, but has rarely been done rigorously. One can infer its existence if closely related, sympatric species of animal-pollinated angiosperms bloom at the same time in nature and can produce viable hybrids in the greenhouse—but no hybrids are found in the wild.

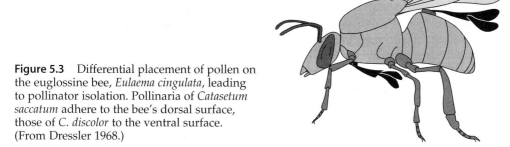

Figure 5.3 Differential placement of pollen on the euglossine bee, *Eulaema cingulata*, leading to pollinator isolation. Pollinaria of *Catasetum saccatum* adhere to the bee's dorsal surface, those of *C. discolor* to the ventral surface. (From Dressler 1968.)

Such inference becomes stronger if completely nonoverlapping sets of pollinators visit each species in sympatry, but such specificity is not the rule for closely related species of plants. Waser (2001) argues that the more common situation involves related plants being visited by a similar or identical *group* of pollinators. In this case, pollinator isolation depends on different pollinators having different average preferences.

When strict pollinator specificity is lacking, demonstrating pollinator isolation can be tricky, for different insect species can vary strongly in their ability to pollinate a single species of plant. Counting pollinator visits can thus yield misleading information about pollination frequency. For example, Lepidoptera make 29% of all insect visits to flowers of the neotropical herb *Calathea ovandensis*, but effect only 1% of successful pollinations (Schemske and Horvitz 1984).

Pollinator isolation is best measured by the transfer of differentially dyed pollen grains among species (Goulson and Jerrim 1997). The index of isolation between two species can be roughly estimated as $1 - P_{het}/P_{hom}$, where P_{het}/P_{hom} is the ratio of the proportions (or numbers) of heterospecific to homospecific pollen grains found on the stigmas of one species. This index ranges from $-\infty$ (complete disassortative mating) through 0 (no pollinator isolation) to 1 (complete pollinator isolation). It assumes that plant species are fully sympatric, so that pollen flow is not impeded by distance, and have equal amounts of pollen. Dyed pollen has been used to show pollinator isolation between red and white campions (*Silene dioica* and *S. latifolia*), which share some pollinators but not others. The strength of this barrier is about 0.45.

A strong dependence of a flower on a particular pollinator can have effects similar to those of habitat isolation, limiting flowers to the range of that pollinator. Pollinators can thus keep plant species allopatric, raising the problem of whether the allopatry of related angiosperms is due to pollinator specificity, habitat isolation, biogeographic history, or a combination of these factors.

Examples of pollinator isolation

The most striking cases of pollinator isolation involve obligatory mutualisms in which each species in a group of closely related plants relies on a single, unique species of pollinator. The classic examples are figs and fig wasps (Janzen 1979; Wiebes 1979; Weiblen 2002), and yuccas and yucca moths (Powell 1992; Pellmyr 2003). In these systems, there appears to be a largely one-to-one relationship between host and pollinator species, as well as mutualism: the plant benefits by pollination, while the insect gains a breeding site. (Both wasps and moths breed only on their host, with the plant's ovules serving as larval food.)

In the fig/fig wasp mutualism, nearly all of the 750 species of *Ficus* are pollinated by different wasp species in the family Agaoinidea. Although the conventional wisdom is that there is a one-to-one relationship between fig species and wasp species, the recent discovery of sibling species of wasps pollinating

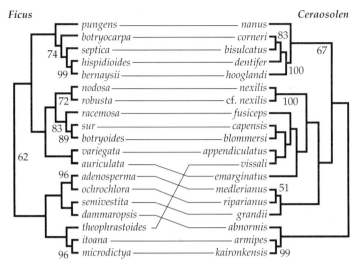

Figure 5.4 Parallel phylogenies of 19 species of Melanesian figs (genus *Ficus*) and 19 species of pollinating fig wasps (genus *Cerasolen*) showing a one-to-one relationship between species and parallel phylogenies, implying cospeciation. Thin lines connect figs with the wasps that pollinate them. Bootstrap values are given below the nodes. (From Weiblen and Bush 2002.)

single figs (Molbo et al. 2003) somewhat qualifies this conclusion. There have been few molecular studies of cospeciation in this system, but, as shown in Figure 5.4, there is a nearly perfect congruence between the phylogenies of 19 species of Melanesian figs and the phylogenies of their pollinators (Weiblen 2001; Weiblen and Bush 2002), as well as a general worldwide congruence between wasp genera and fig subgenera (Machado et al. 2001). We still lack extensive phylogenies for studying cospeciation in the yucca/yucca moth system. Despite a general one-to-one relationship between moth and yucca species, several cases of phylogenetic incongruence between a yucca and its resident moth rule out perfect cospeciation (Pellmyr 2003). Nevertheless, in both figs and yuccas, closely related sympatric species usually show complete pollinator isolation.

Mimulus lewisii and *M. cardinalis*, described in Chapter 2, also exhibit nearly complete pollinator isolation in sympatry. *M. cardinalis* (with red, tubular flowers, excerted anthers, and high nectar volume) is pollinated almost exclusively by hummingbirds, and *M. lewisii* (with broad, pink flowers, recessed anthers, and low nectar volume) almost exclusively by bees (see Plate 2). The strength of this isolation, estimated by direct observation of pollinator visits, is 0.98. This surely underestimates the true value because the difference in flower shapes also causes some mechanical isolation. Indeed, despite the fact that these species can be crossed easily in the greenhouse, only 2 out of 2000 seeds collected in the wild produced F_1 hybrids (Ramsey et al. 2003). Genetic and observational studies demonstrate that pollinator isolation is produced largely by differences in flower color and nectar volume (Schemske and Bradshaw 1999).

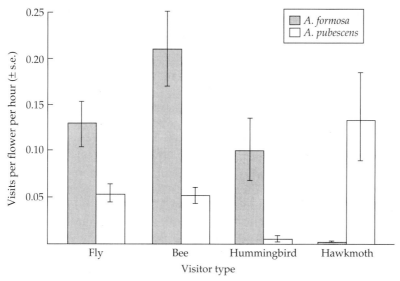

Figure 5.5 Pollinator isolation between *Aquilegia formosa* and *A. pubescens* in areas of artificial sympatry in transplant experiments (bars represent mean rate of visitation; lines show one standard error). *A. formosa* is adapted to hummingbird pollination, *A. pubescens* to hawkmoths. Judged by mean visits per flower per hour, each flower species is visited almost exclusively by its own class of pollinators. Flies and bees, on the other hand, show some differential visitation but cause little pollinator isolation. (From Fulton and Hodges 1999.)

Also isolated by habitat in the Sierra Nevada Mountains, *Aquilegia formosa* and *A. pubescens* nevertheless form some hybrids where their ranges overlap (Hodges and Arnold 1994). In this area there is moderate pollinator isolation, with *A. pubescens* pollinated largely by hawkmoths and *A. formosa* by hummingbirds. Transplant experiments, in which flowering individuals of one species were moved next to those of the other, showed strong differential visitation by hawkmoths and hummingbirds, but a smaller difference for flies and bees (Figure 5.5; Fulton and Hodges 1999). Although Hodges and Arnold characterize pollinator isolation as "strong," a fair number of hybrids are seen in nature. The observation that morphological differences between the species are maintained despite hybridization implies that there is some form of postzygotic isolation—perhaps the unattractiveness of hybrid flowers to pollinators. The existence of such extrinsic postzygotic isolation is supported by observations that genes affecting floral traits introgress less readily between the species than do random DNA markers (Hodges and Arnold 1994).

Relative importance of pollinator isolation

Compared to other barriers, how important is pollinator isolation as a cause of speciation? A common view holds that pollinator isolation is unimportant, even between distantly related species, because some pollinators have broad preferences (Ollerton 1996; Waser 1998). As we have seen, however, there are good examples of pollinator specificity. But generalizations are risky since we do not know what pollinates the vast majority of angiosperms (Johnson and Steiner 2000).

Studies of *Silene, Mimulus,* and *Aquilegia* all show that pollinator isolation is a strong barrier to gene flow, but these cases involve substantial habitat isolation as well. Indeed, if pollinator isolation evolves in allopatry, we can often expect it to be accompanied by habitat isolation. However, because many species of pollinator-isolated plants produce viable and fertile hybrids when pollinated artificially, we conclude that pollinator isolation often precedes the evolution of intrinsic hybrid inviability and sterility. We can sometimes infer that pollinator isolation is important without directly observing visitation or pollen transfer. For example, related species of orchids that use the same pollinator often do not live sympatrically unless they either place their pollinaria on different parts of the insect's body or if other isolating barriers act after pollination (Dressler 1968; Paulus and Gack 1990). Grant (1992) notes that related species within the genera *Ipomopsis* and *Aquilegia* are pollinated by either hawkmoths or hummingbirds, and that congeneric species are sympatric only if they use different pollinators. This implies that coexistence is difficult when pollinators are shared, and thus that pollinator isolation is currently important. But these observations are anecdotal and need more systematic analysis.

Along with studying taxa at different stages of divergence, the best way to demonstrate that pollinator isolation has been an important cause of speciation is comparative analysis of the species richness of sister clades. Two studies show a significant association of species diversity with either animal pollination itself (Dodd et al. 1999) or with traits associated with animal pollination (Hodges 1997). In Chapter 12 we show why this association implies that pollinator isolation was instrumental in angiosperm speciation.

The evolution of pollinator isolation

ALLOPATRY AND PARAPATRY. The most widely accepted scenario for the evolution of pollinator isolation involves geographic separation of plant populations—either parapatry or allopatry. Divergence in floral traits and pollinator preference could then occur through various forms of selection. For example, an isolated population of plants could experience a completely novel pollinator or group of pollinators, causing strong selection for traits that attract these new species. Alternatively, the species of pollinators might be identical, but their relative *proportions* could differ, thereby imposing weaker selection on floral traits that match the preferences of more important pollinators. Extinction of either a local plant or pollinator could also produce selection for new floral traits (Cruden 1972). Moreover, isolation could result from evolution of the pollinator. A change in the pollinator's preference due to changes of the visual or chemical environment, for example, could be followed by evolution of the flower to match the new preference.

The allopatric/parapatric scenario for the evolution of pollinator isolation predicts that one should find its incipient stages in different populations of some plant species. That is, one should find a correlation among populations between the array of local pollinators and the structure of the flowers they pollinate. Few studies have looked for such a correlation: the best is Johnson's (1997) work on the African orchid *Satyrium hallackii*. The subspecies *ocellatum*, living in the grasslands of northern and eastern South Africa, has long nectar spurs and is pollinated mainly by hawkmoths and long-tongued flies. In contrast, the coastal subspecies *hallackii* has much shorter spurs and is pollinated by carpenter bees. There is a cline of decreasing nectar spur length from north to south (Figure 5.6). Interpopulation divergence appears to have been driven by pollinator availability: the coastal region harbors woody shrubs that serve

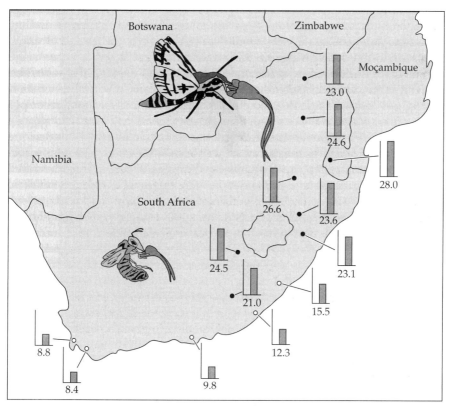

Figure 5.6 Geographic differentiation of flower spur length in *Satyrium hallackii* associated with the presence of different pollinators. Mean spur length (mm) is given for each sampled locality (indicated by a circle). The grassland-dwelling subspecies *ocellatum* (filled circles) has longer spur length and is pollinated by hawkmoths with a long proboscis. The southern coastal subspecies *hallackii* (open circles) has shorter spur lengths and is pollinated by carpenter bees with short tongues. The variation in plant morphology may represent an early phase in the evolution of pollinator isolation. (From Johnson 1997.)

as nest sites for bees, but lacks suitable host plants for hawkmoth larvae, while the opposite is true in the grasslands. (This study suffers somewhat from the lack of evidence that differences in spur length are genetic. Robertson and Wyatt 1990, for example, found that local environment causes substantial developmental differences in spur length of the orchid *Platanthera ciliaris*.)

Mazer and Meade (2000) reported heritable differentiation for flower size among California populations of the wild radish *Raphanus sativus*. Mean flower size was positively correlated with the local density of large pollinating honeybees, suggesting the initial stages of pollinator isolation. Similarly, Galen et al. (1987; pers. comm) found altitudinal differences in flower size, scent, and morphology between populations of the alpine skypilot, *Polemonium viscosum*, in Colorado. These changes correspond to a switch from pollination by flies at lower altitudes to bumblebees at alpine sites. However, neither study examined the crucial test of pollinator isolation: whether pollinators from different populations, when given a choice, prefer the local flower morph. Geographic variation of flower morphology and pollinator array are also seen in *Disa draconis* (Johnson and Steiner 1997), *Aquilegia caerulea* (Miller 1981), and *Saltugilia grinnellii* (formerly *Gilia splendens*; Grant and Grant 1965).

Pollinator isolation based on cospeciation, as in figs, seems unlikely to evolve in sympatry. Given the strict interdependence of plants and pollinators, it is hard to see how one mutualism could become two in the presence of gene flow. A possible scenario is the origin of a new fig by polyploidy, which could offer a new sympatric niche for a pollinator, but this must be ruled out because nearly all figs have the same chromosome number (Weiblen 2000). Most probably, cospeciation in *Ficus* involves allopatric fragmentation of fig populations, followed by coevolutionary changes in both fig and wasp. Although we know some of the morphological and physiological changes involved in the coevolution of figs and wasps (Ramirez 1970; Jousselin et al. 2003), we know nothing about the evolutionary forces that drove such changes.

SYMPATRY. The problem with sympatric speciation of plants via pollinator isolation is that it is hard to understand how reproductive isolation can evolve when at least one other species—the pollinator—is a key player. Suggestions that a new mutation changing flower color or odor could cause instantaneous sympatric speciation by recruiting a new pollinator (e.g., Dressler 1968) seem unrealistic: they depend on the fortuitous occurrence of a new mutation, a pollinator preadapted to that mutation, and pollinators sufficiently specific to effect complete reproductive isolation. Indeed, some data suggest that mutations of large effect are unlikely to trigger speciation. Using introgression experiments, Bradshaw and Schemske (2003) showed that the mutation with the largest effect on reproductive barriers between *Mimulus cardinalis* and *M. lewisii* (*yup*, which changes flower color from pink to orange-red) caused large changes in the attraction of hummingbird and bumblebee pollinators. However, new "mutant" flowers always had lower fitness than their ancestors, and could not be established without a significant shift in the existing proportions of pollinators. This strongly implies that the color change distinguishing these species evolved in allopatry, where different pollinator arrays might have preexisted. Moreover, it is not clear how a sympatric shift in pollinator use can allow a new plant to coexist with its ancestor in the absence of other ecological differences. In models of sympatric speciation via differential resource use (e.g., Dieckmann and Doebeli 1999), reproductive isolation is completed by the fortuitous linkage of alleles causing assortative mating to those alleles affecting resource use. Such linkage is impossible when another species is required for mating.

Nevertheless, both polyploidy and diploid hybrid speciation (Chapter 9) occur in sympatry, and can create new species differing in floral traits within a few generations. Although one reproductive barrier—hybrid sterility—is nearly complete at the outset, reproductive isolation could be enhanced by differential pollination. Husband and Schemske (2000) studied pollination by bumblebees (*Bombus*) of the diploid fireweed (*Chamerion angustifolium*) and its sympatric autopolyploid. While bees tended to preferentially visit plants of a given cytotype, producing substantial reproductive isolation, this was largely explained by the spatial clumping of diploids and tetraploids and by the short flight distances of bees. This is a likely case of microhabitat rather than pollinator isolation. Straw (1955) suggested that *Penstemon spectabilis* was a diploid hybrid derivative of *P. grinnellii* and *P. centranthifolius*, and that the hybrid's new floral traits allowed it to be pollinated by a novel species (a wasp) that did not use the parental species. Recent molecular work, however, shows that *P. spectabilis* is not a hybrid species (Wolfe et al. 1998a,b).

The genetics of pollinator isolation

Adaptive differences between species based on genes of large effect (relative to genes causing variation within a species) imply either that selection was very strong (that is, the new optimum was far from the old one), or that the evolution of a new trait involved crossing an adaptive valley in which intermediate phenotypes are less fit.

We would expect that the allopatric evolution of pollinator isolation might occasionally involve strong selection, especially if it is due to the presence of novel pollinators in a newly colonized habitat. Moreover, a shift in pollinators might also involve crossing an adaptive valley between an old and a new optimum: a shift from bee to hummingbird pollination, for example, might require genes of large effect simply because flowers of intermediate morphology are less attractive to both pollinators.

Although there are only two genetic studies of traits involved in pollinator isolation (see Table 5.1), both appear to involve genes of large effect. QTL analysis of 12 floral and nectar traits differentiating *Mimulus lewisii* and *M. cardinalis* showed that, while each trait difference was caused by between one and six QTL, nine of the differences involved a QTL with an effect larger than 25% of the phenotypic difference between the species (Bradshaw et al. 1995, 1998). This analysis was followed by the creation of F_2 interspecific hybrids that were transplanted into the area where the species are sympatric. (Plate 2 shows some of these hybrids.) Observing pollination of these hybrids, Schemske and Bradshaw (1999) found that four traits—anthocyanin concentration, carotenoid concentration, flower size, and nectar volume—affected the relative number of bee and pollinator visits, causing pollinator isolation in nature. Since each F_2 hybrid had a known genotype, Schemske and Bradshaw were able to show that genetic differences at two QTL (the *yup* locus affecting carotenoid concentration and a factor altering nectar volume) significantly influenced pollinator visitation in the expected direction. Clearly, genes of large phenotypic effect also had large effects on reproductive isolation (see also Bradshaw and Schemske 2003).

Prazmo (1965) crossed several *Aquilegia* species having floral differences adapting them to either long-tongued pollinators (moths and hummingbirds) or short-tongued pollinators (bees and flies; see Hodges and Arnold 1994). Each of the three traits given in Table 5.1 showed Mendelian inheritance, producing phenotypic ratios consistent with single loci having two alleles, one completely dominant. We suspect that, except for cases of cospeciation (which may involve the gradual coevolution of flower and insect), pollinator isolation may often rest on floral genes of large effect.

Temporal (Allochronic) Isolation

In temporal isolation, gene flow is impeded because members of different species breed at different times. In animals, this involves differences in mating period or season, spawning time, and so on, while in plants it involves differences in the periods of flowering, pollen shedding, or ovule receptivity. The magnitude of differences in breeding period that cause reproductive isolation differ greatly. In free-spawning marine animals, only a few hours can suffice (Knowlton et al. 1997; Clifton and Clifton 1999); isolation between closely related flowering plants can involve days or weeks (Opler et al. 1975); related fish taxa can breed in alternate years (Aspinwall 1974); and some species of periodical cicada breed simultaneously only every two centuries (Lloyd and Dybas 1966a).

The biological bases of temporal isolation can also be diverse. Such isolation can, for instance, rest on differential responses to identical environmental cues, such as tidal or lunar cycles, as is presumably the case for many related marine species having offset spawning periods. Alternatively, temporal isolation might be based on direct responses to *different* environmental cues. On the Galápagos islands of Daphne and Daphne Major, the first brood of the cactus finch *Geospiza scandans* often appears when cacti flower before the annual rains. Its relative *G. fortis*, however, breeds only after the rains begin (Grant and Grant 1996). Finally, temporal isolation can be a pleiotropic byproduct of habitat differences. In plants, for example, flowering time can depend on soil moisture, so that the same species growing in different microhabitats may flower at different times (Macnair and Gardner 1998). Vasek and Sauer (1971) found that four sympatric species of *Clarkia* in California flower at different times based on differences in soil temperature and moisture.

The timing of reproduction in phytophagous insects can be profoundly changed by their choice of host plants, with insect reproduction correlated with plant phenology. This has been shown by transferring eggs of the treehopper *Enchenopa binotata* among different species of *Viburnum* in a greenhouse. Surviving adults showed substantial differences in time of egg hatch, maturation, and mating in the first generation after transfer (Wood 1993). Here, temporal isolation was caused not by genetic changes affecting development time, but by developmental plasticity triggered by plant phenology. These changes produce habitat isolation as a byproduct (Wood et al. 1999). This plasticity could lead to nongenetic sympatric speciation, but only if the flowering times of host trees were completely overlapping or if juvenile conditioning led adults to oviposit on the same species in which they developed.

Temporal isolation has two unusual features. First, as shown in *Enchenopa*, it can be completely nongenetic. If a population has evolved to reproduce every other year, environmental or developmental accidents can produce rare "off-year" breeders. If such occurrences are rare, they could yield two populations breeding in alternate years and thus showing substantial temporal isolation. The even and odd year classes of the pink salmon, discussed below, may be such an example. Second, apart from polyploidy, temporal isolation is the only isolating barrier that can by itself cause sympatric speciation. This can lead to the production and coexistence of two sympatric species having identical ecology. This type of speciation may have occurred in 13- versus 17-year broods of periodical cicadas and in pink salmon.

Detecting and measuring temporal isolation

To show that temporal isolation *currently* impedes gene flow, one must demonstrate that reducing or eliminating the barrier would lead to the production of viable and fertile hybrids. This may be difficult, for breeding cues may depend on unknown environmental factors that cannot be manipulated in the laboratory. But even if other isolating barriers prevent gene flow when temporal barriers are overcome, this does not mean that temporal isolation was not an important component of speciation. The evolution of complete hybrid sterility, for example, may have occurred well after a complete disjunction of breeding seasons.

As always, the potential importance of temporal isolation in speciation is best shown by studying sister species. The importance of a correct phylogeny is demonstrated by the work of Mosseler and Papadopol (1989). These authors found genetically based differences in flowering times among six sympatric species of North American willows (*Salix*), with each species falling into either an early or late-flowering clade. However, taxonomic work (Dorn 1976; Argus 1986) suggests that every pair of sister species belongs to either the early or the late group, so that temporal isolation was potentially important in only the single speciation event that originally produced the two clades.

For a single species, the strength of temporal isolation from a related species is $1 - p$, where p is the proportion of pure-species offspring produced during the overlap in breeding seasons.

Examples of temporal isolation

Temporal isolation has not been studied as thoroughly as other isolating barriers. The technical difficulties of manipulating breeding periods are compounded by our lack of good species-level phylogenies in many groups (such as plants) where this barrier might be important. Nevertheless, we know of several examples in which temporal isolation seems an important *current* barrier to gene flow.

Knowlton et al. (1997) found temporal isolation between two sympatric species of tropical Atlantic corals (*Montastraea annularis* and *M. franksi*) that appear to be sister species. Individuals release their gametes synchronously into the water column over a 15–30 minute period. The peak spawning times of *M. annularis* and *M. franksi*, however, are 1.5–3 hours apart. This gap renders sperm from the earlier-spawning species too dilute to effectively fertilize eggs of the later-spawning species. Artificial hybridizations produce viable hybrid larvae, although the fitness of adults has not been studied (Oliver and Babcock 1992). Levitan et al. (2004) found much less temporal isolation beween *M. annularis* and its more distant relative *M. faveolata*. Because these two species have incompatible gametes and cannot produce hybrids, it is possible that the higher temporal displacement between *M. annularis* and *M. franksi* may have resulted from selection to prevent hybridization.

Convincing evidence for temporal isolation is rarer in plants than in animals. Levin (1978) provides several examples of seasonal offset in flowering time among congeneric species. The problem is that some of the species also show habitat isolation, so one cannot tell whether temporal barriers might significantly reduce gene flow in sympatry. Moreover, the evolutionary relationships of temporally isolated species are often unknown. Opler et al. (1975), for example, show some differences in flowering time—on the order of months—between eight species of *Cordia* in Costa Rica. While some species show no temporal overlap, others display extensive overlap, but we do not know whether overlapping species are the most closely related.

An intriguing example of temporal isolation is seen in green algae inhabiting coral reefs (Clifton 1997; Clifton and Clifton 1999). Like corals, each algal species shows mass spawning that is sporadic and brief (5–20 minutes). Gametes remain motile for about an hour after release. Although several species may discharge gametes on the same day, the most closely related species always spawn at different times of day. One explanation for this offset is direct selection for temporal isolation in sympatry, which could reduce the possibility of forming unfit hybrids. If there is some hybrid viability and fertility, this selection could contribute to speciation, but if postzygotic isolation were already complete, such selection would only create new breeding niches for already-existing species. Unfortunately, there is no information about the ability of these species to hybridize. Alternatively, the pattern could reflect "differential fusion": only those closely related species that had evolved temporal divergence of spawning time in allopatry could become sympatric without the possibility of fusing through hybridization. Similar explanations could account for Rabinowitz et al.'s (1981) observation that in a Missouri prairie community, wind-pollinated species had significantly shorter flowering periods than did insect-pollinated species.

Stebbins (1950) and Waser (1983) describe two other examples of temporal isolation in plants. In both cases, hybrids are viable and fertile.

Relative importance of temporal isolation

Reviews of temporal isolation in animals have concluded that this barrier is relatively unimportant. Mayr (1963, p. 94) argued that, in animals, "The actual contribution of seasonal isolation to the maintenance of reproductive isolation between species is largely unknown. In some cases in which the seasonal isolation broke down owing to unusual weather conditions, reproductive isolation was maintained by other factors." But this refers only to current barriers and does not rule out a pivotal role for temporal isolation in speciation. In groups such as birds and mammals, the reproductive period in one locality may be closely tied to both photoperiod and the timing of food availability for offspring, and many species breed synchronously, ruling out temporal isolation. However, when the diet or life cycle of an animal is tied to a specific host with distinct phenologies, as in the host races of *Rhagoletis pomonella*, temporal isolation may be more pervasive. Unfortunately, there have been few studies of temporal isolation between close relatives.

Reviewing the timing of flowering in plants, Rathcke and Lacey (1985) argue that temporal isolation is unimportant in preventing gene flow between related species, suggesting that habitat or pollinator isolation may be more significant. Waser (1983), on the other hand, concludes that temporal differences in flowering time are vital in preventing cross-pollination between plant species. The conflicting conclusions probably reflect Waser's inclusion of species that are either incapable of producing hybrids or whose hybrids are inviable or infertile. In such cases temporal isolation may be frequent, but may have evolved to prevent hybridization between already-isolated species. This sort of divergence can easily evolve when reproductive isolation is complete.

If the loss of temporal isolation after habitat disturbance causes species to fuse, it is reasonable to infer that this barrier was important in speciation. We know of just one case, involving two species of the Australian flowering plant *Banksia*. These species show some habitat isolation but also occur in sympatry, where hybrids are never formed. In the greenhouse, however, hybrids can be made readily and are highly fertile. The absence of hybrids in the wild reflects the nonoverlapping flowering seasons of the two species. In disturbed areas, however, this temporal isolation is lost, apparently through an increase in flower number leading to longer, overlapping flowering periods. In such areas the species cross readily, forming hybrid swarms that threaten to replace the parental species (Lamont et al. 2003). Clearly, temporal isolation was an important barrier in speciation, since the species cannot remain distinct in its absence.

Determining whether temporal isolation has been *generally* important as a historical cause of speciation must rest on either (1) finding many cases of closely related sympatric species in which reproductive isolation is incomplete without temporal isolation, or (2) showing through comparative studies that the youngest pairs of sister taxa typically exhibit temporal isolation. While some pairs of species satisfy the first criterion, no group of species has been studied using either method.

The evolution of temporal isolation

ALLOPATRY AND PARAPATRY. Many forms of divergent selection can produce temporal isolation between allopatric populations. In plants, abiotic cues for flowering time include photoperiod, temperature, and moisture; and geographic differences in these factors could produce differences in flowering time. In some cases, adaptive differences in flowering time are part of an entire syndrome of adaptations to a distinct habitat, adaptations that prevent plants from achieving sympatry (Turesson 1922, 1930; Clausen et al. 1940). In these cases habitat isolation is the most important current barrier to gene flow since evolved differences in flowering time do not restrict gene flow between taxa that are "genetically" allopatric.

However, if habitat adaptation is not so extreme as to prevent secondary contact, the environmental factors that select for different flowering times in allopatry may converge in sympatry, and temporal isolation may disappear. This fusion can be avoided in several ways. First, plants may have become adapted to specific niches or habitats, such as soil types, that provide different cues for flowering in sympatry or parapatry. Serpentine and other soils containing heavy metals, for example, are usually drier and warmer than adjacent "normal" soils, and usually select for earlier flowering time. Second, temporal isolation may be due to biotic factors, such as pollinators, that remain distinct in sympatry. Finally, if hybrids are formed but there is incomplete postzygotic isolation, natural selection can maintain or increase temporal isolation to prevent maladaptive hybridization. We discuss this form of reinforcement below.

In animals, the evolution of temporal isolation in allopatry must often reflect local differences in temperature or food availability. Several studies show the first step in this process: genetically based differences in breeding time among geographic populations. These include the Japanese cricket *Teleogryllus emma* (Masaki 1967), the American cricket *Gryllus firmus* (Harrison 1985), and the moth *Hyalophora cecropia* (Sternburg and Waldbauer 1978). As in plants, such barriers can diminish when sympatry is reattained. The crickets *Gryllus firmus* and *G. pennsylvanicus* are temporally isolated in Virginia, where they live at different altitudes and are adapted to different temperatures. This temporal isolation, however, largely vanishes in Pennsylvania, where the species are not altitudinally segregated (Harrison 1985).

REPRODUCTIVE CHARACTER DISPLACEMENT AND REINFORCEMENT. There is evidence in plants that good sympatric species can diverge in flowering time via reproductive character displacement. Such divergence can occur by selection to reduce competition for pollinators, to reduce the probability of forming sterile or inviable hybrids, or to prevent the clogging of stigmas with foreign pollen. Waser (1978) showed that in sympatric populations of *Delphinium nelsonii* and *Ipomopsis aggregata* having overlapping flowering times, individuals flowering during the period of overlap set fewer seeds than those flowering earlier or later. Cross-pollination is probably caused by hummingbirds. Waser (1983) also observed that when *D. nelsonii* is sympatric with a distant relative, *Penstemon barbatus*, the two species flower at different times, but in different orders in different localities. This suggests that geographic differences in flowering order result from reproductive character displacement and not selection for an optimal flowering time in each species.

While temporal displacement of good species may represent a post-speciation phenomenon—as it surely does in Waser's studies—one cannot always conclude that such displacement is irrelevant to speciation. Temporal divergence may be the "ghost of reinforcement," having evolved before speciation was completed with additional isolating barriers.

There are only a few well-established cases of reinforcement having increased temporal isolation. One is described in Chapter 4: the evolution of flowering-time differences between metal-tolerant and metal-susceptible populations of *Agrostis tenuis* and *Anthoxanthum odoratum* (McNeilly and Antonovics 1968). Displacement of flowering time at mine boundaries has presumably evolved to prevent the flow of deleterious genes between adjacent habitats.

In animals, reinforcement can increase temporal isolation not only to prevent maladaptive hybridization, but also to reduce the interference between mating calls (Noor 1999). Both factors may have operated in a possible case of reinforcement in the leopard frog, *Rana* (Hillis 1981). The breeding periods of three species overlap extensively when they are allopatric. However, in central Texas where the species are sympatric, breeding periods are displaced: two of them (*R. berlandieri* and *R. sphenocephala*) reduce their breeding schedules from twice to only once per year, while *R. sphenocephala* shortens its winter breeding period (Figure 5.7). These changes reduce the overlap of breeding seasons between species. In the laboratory, interspecific crosses show substantial but not complete postzygotic isolation (Hillis 1988). Some backcross hybrids are observed in nature, mostly in disturbed areas (Platz 1981; Kocher and Sage 1986). Given that hybrids are not completely sterile or inviable, this may well be a true case of reinforcement. Hillis (1981, p. 315) argues that "tem-

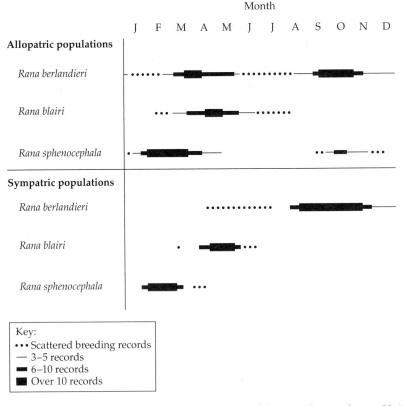

Figure 5.7 Temporal isolation in three species of frogs in the southwest United States. Allopatric populations of all three species show extensive overlap of breeding seasons. In sympatry, this overlap is eliminated by reducing the number of breeding seasons or shortening the winter breeding period. (From Hillis 1981.)

poral isolation is probably the single most important isolating mechanism among sympatric leopard frogs in central Texas." This may well be true at present, but enhanced temporal isolation in sympatry would probably not have evolved without some pre-existing postzygotic isolation. Moreover, these species also show habitat isolation and call differences that may cause behavioral isolation (Hillis 1981). Thus, it is not clear which isolating barriers were instrumental in speciation.

While we can point to only a few known examples of reinforcement and reproductive character displacement for temporal isolation, it is likely that many cases lie undiscovered, particularly in wind-pollinated plants and broadcast-spawning marine species.

SYMPATRY. Although temporal isolation alone can cause sympatric speciation, this is undoubtedly rare. Such an event requires a complete and virtually instantaneous shift in breeding season. Mayr (1963, pp. 474–477) points out the difficulties of sympatric speciation via "seasonal isolation." Nevertheless, temporal isolation is important as one of several barriers to gene flow between sympatric host races of insects, some of which may have speciated sympatrically (Drès and Mallet 2002). Moreover, models of sympatric speciation based on differential resource use have largely ignored the possibility that the resources (including pollinators) could appear at different times. In such cases, adaptation to the resource spectrum could lead to temporal isolation as a byproduct—a "no-gene" model of assortative mating evolving in sympatry. This scenario has been used to explain two sympatric forms of Madeiran storm petrel (*Oceanodroma castro*) that breed at different times of the year (Monteiro and Furness 1997), although allopatric speciation seems equally likely.

The pink salmon, *Oncorhynchus gorbuscha*, is an intriguing case of what seems to be incipient sympatric speciation involving allochrony. This species has a two-year life cycle, with spawning occurring in rivers of Asia and northwestern North America. Many streams harbor even- and odd-year "broodlines" that are reproductively isolated by their two-year life cycle and the low rate of gene flow occurring via very rare three-year breeders (Aspinwall 1974). These broodlines may have originated from a developmental or environmental event that delayed or accelerated breeding of several individuals. In both Asia and North America, alternate-year broodlines within a single stream show more genetic difference than do fish within a region that breed in the same year but in different streams (Aspinwall 1974; Beacham et al. 1988; Brykov et al. 1996). It is widely assumed that these genetic differences evolved in the last 10,000 years, after the last glaciation (Gharrett et al. 1988).

There is little morphological difference between alternate-year broodlines within a single stream (Beacham et al. 1988). F_2 individuals produced in artificial crosses between broodlines, however, show reduced survival (Gharrett and Smoker 1991; Gharrett et al. 1999). If reproductive isolation were to evolve between broodlines in the same streams, one would expect it to be exactly of this type—intrinsic postzygotic isolation. There can be no average difference between even- and odd-year environments, and so differentiation must occur through different mutations adapting broodlines to the same habitat. Sympatric broodlines thus offer a unique chance to see how rapidly reproductive isolation can evolve in populations inhabiting identical environments.

Unfortunately, the pink salmon story may not be so simple. It is possible that even- and odd-year broodlines have allopatric origins, each deriving from a different glacial refuge (Beacham et al. 1988). This could make them much older, and would imply that this is not a case of sympatric speciation. Indeed, two studies of divergence in mtDNA (Polyakova et al. 1996; Brykov et al. 1996) suggest that the broodlines separated between 10,000 and 1 million years ago. Clearly, we need more molecular analysis to resolve this issue.

Finally, during polyploidy or diploid hybrid speciation, temporal isolation may arise as a direct byproduct of chromosome doubling or the mixing of divergent genomes. Husband and Schemske (2000) found that the bee-pollinated diploid *Chamerion angustifolium* and its autotetraploid derivative showed considerable divergence in flowering time in sympatry, but it is not clear whether this is a byproduct of polyploidy or a result of later selection to reduce hybridization.

The genetics of temporal isolation

We know less about the genetics of temporal isolation than about the genetics of any other isolating barrier. The only existing study is that of Tauber et al (1977) on the diapause difference between two species of lacewing, *C. downesi* and *C. plorabunda*, which appears to rest on changes at only two loci (Table 5.1). But, as we noted in Chapter 4, these are not sister species and the difference in diapause may not be a barrier to gene flow in many locations.

Conclusions

Because they are closely connected to well-understood forms of adaptation, ecological barriers offer us the best chance to understand the evolutionary origin of reproductive isolation. Indeed, we already have examples of plant species in which habitat isolation seems to be a byproduct of adaptation to local soil types and climates. Similarly, the apple and hawthorn races of *Rhagoletis pomonella* are clearly habitat isolated by virtue of their genetic adaptations and preferences for different hosts. However, the study of habitat isolation is plagued by two technical problems. In allopatric populations, it is hard to detect such isolation without common-garden experiments. In sympatric or parapatric populations, on the other hand, ecological isolation may reflect either divergent adaptation to the habitat or competitive differences that have little to do with such adaptation. These alternatives can in principle be disentangled by removal or transplant experiments.

Adaptation to different pollinators has obviously created barriers to gene flow in plants. But we do not know whether such adaptations were promoted by the appearance of a novel, previously unused pollinator or by a less drastic change in the frequency of pollinators in an already-existing group. Some genetic evidence suggests that strong selection may be involved, supporting the first scenario.

The evolutionary forces creating temporal isolation are unclear. Temporal barriers may often evolve by reinforcement to prevent maladaptive hybridization. To determine this, we badly need information on the breeding seasons of related species from areas of both sympatry and allopatry, particularly in wind-pollinated plants that have no opportunity for pollinator isolation.

Finally, we have almost no measurements of the relative strengths of different forms of ecological isolation between recently-diverged pairs of species. This information is essential for understanding which barriers were important in speciation.

6

Behavioral and Nonecological Isolation

In classifying isolating barriers, we consider behavioral, mechanical, postmating prezygotic, and mating-system isolation as "nonecological." This is largely a term of convenience, for the evolution of many of these barriers surely involved ecologically based selection. Nevertheless, these barriers can be considered less ecological than, say, habitat isolation, as their evolution may have involved sexual selection, which—compared to natural selection—is related less directly to the environment. While many biologists regard breeding systems as isolating barriers, we take up this issue below and conclude that, although changes in the level of outcrossing may have important effects on speciation, they do not cause reproductive isolation.

Mating System "Isolation"

Lists of reproductive isolating barriers often include "mating system isolation," the reduction of gene flow between two taxa caused by a change in breeding system. The most frequently cited barriers involve changes from outbreeding to asexuality (apomixis) or selfing (autogamy). Because autogamy and apomixis reduce gene flow between individuals, it is sometimes assumed that they also reduce gene flow between populations and species. Levin (1978, p. 223), for example, notes that "by virtue of the fact that autogamy restricts interpopulation gene flow via pollen . . . it serves to isolate species as do other differences in reproductive biology, and thus should be included in a list of isolating mechanisms." Fishman and Stratton (2004, p. 297) observe that "the evolution of selfing clearly contributes to premating barriers between diverging populations, but may also accelerate the development of postzygotic isolation."

We argue, however, that selfing or apomixis differ profoundly from other isolating barriers, and in fact are *not* isolating barriers, for these mating systems impede gene flow between individuals of different taxa no more than they impede gene flow between individuals of the same taxon. For example, a set of individuals that are completely selfing or apomictic do not constitute a biological species, as they are not a group of interbreeding individuals. They are instead a collection of "microspecies," with each individual propagating its own genetically isolated lineage. Each selfer is isolated from other selfers just as much as it is from its outcrossing relatives. The same is true of partial selfing or asexuality: gene flow is reduced precisely as much within as among taxa. These changes in mating system are therefore neither isolating barriers nor components of speciation.

Although changes in mating system do not constitute isolating barriers, they can nevertheless promote speciation in diverse ways. Here, for example, are some possible effects of an increase in self-fertilization:

- Selfing can evolve through reinforcement to reduce gene flow from populations not adapted to the local habitat. This reduced gene flow can allow organisms to adapt more readily to the local habitat (as in grasses on mine tailings in Britain), with adaptation producing reproductive isolation as a byproduct.
- Selfing can allow a single individual to successfully colonize and produce a population in a new habitat, also leading to reproductive isolation as a byproduct of adaptation. This is impossible for dioecious or self-incompatible species, in which successful colonization requires at least two individuals (or one inseminated female). The "reproductive assurance" conferred by self-fertilization is surely the explanation for the high proportion of self-compatible species among plants that are weeds or endemics on oceanic islands (Baker 1955; Stebbins 1957a).
- Selfing can create new selective pressures that produce reproductive isolation. For example, predominant selfers often have smaller flowers than do outcrossers, and thus shorter pollen tubes. On the longer styles of outcrossers, pollen from selfers can be outcompeted by pollen from outcrossers, yielding gametic isolation through conspecific pollen precedence (Kiang and Hamrick 1978; Cruzan and Barrett 1993).
- Selfing reduces the effective population size of a species. This in turn increases the effects of genetic drift, which can fix chromosomal rearrangements that are deleterious as heterozygotes (see Gottlieb 1973, 1984a). Such rearrangements may be common causes of hybrid sterility in plants (Chapter 7). This chromosomal sterility also promotes speciation by polyploidy. Moreover, any selfing of a new polyploid will reduce gene exchange with its ancestors, facilitating its establishment as a species (Chapter 9).

There are five genetic studies of changes in mating system between species, all involving floral traits in pairs of species (one mainly selfing and the other mainly outcrossing) in the genus *Mimulus* (Macnair and Cumbes 1989; Fenster and Ritland Del 1994; Fenster et al. 1995; Lin and Ritland 1997; Fishman et al. 2002). Every floral difference was based on changes at several to many loci, so that the selfing/nonselfing transition was polygenic. In contrast, a polymorphism for selfing vs. outcrossing morphs of *Arenaria uniflora* may have involved changes in relatively few genes (Fishman and Stratton 2004). There is thus a striking difference between the polygenic basis of floral traits causing selfing and the oligogenic nature of floral traits causing pollinator isolation. This may reflect the possibility that pollinator isolation requires crossing adaptive valleys (thus necessitating allelic substitutions of large effect) while even a slight increase in selfing (involving alleles of small effect) can be advantageous (Fishman et al. 2002).

Behavioral Isolation

Behavioral isolation (sometimes called *ethological isolation*) includes all species differences that reduce the attraction—and therefore mating—between heterospecific individuals during the breeding period. Although this category is sometimes called "sexual isolation," this label is somewhat inaccurate because sexual interactions that prevent gene flow can also act *after* copulation but before fertilization. We consider post-copulatory sexual barriers as forms of gametic isolation.

Behavioral isolation is obviously limited to animals, and its key feature is that it usually involves the interaction between traits in different sexes. Typically, one sex (usually male) has a signal that stimulates a preference in conspecific but not heterospecific individuals of the opposite sex. Early in the Modern Synthesis, the failure to appreciate how sexual selection might affect these interactions hampered our understanding of the origin of behavioral isolation. As we discuss below, this isolation was invariably seen as either a byproduct of natural selection for local adaptation or as a product of selection for "species recognition."

Detecting and measuring behavioral isolation

One can infer the existence of behavioral isolation between two species if they live in sympatry, breed at the same time, encounter each other in the wild, but rarely or never hybridize. Measuring the *strength* of behavioral isolation, however, must nearly always be done in the laboratory. A variety of tests have been used, including those in which pairs of individuals of different sex are confined together ("no-choice" experiments) or in which either one or both sexes are given a choice between conspecific and heterospecific mates ("male-choice," "female-choice," or "multiple-choice" studies). Which test is most realistic depends on the conditions under which individuals encounter each other in the field, conditions that are often unknown. Studies of *Drosophila*, for example, have used all four tests, usually based on the preference of the investigator rather than conditions in the wild. Do natural encounters occur only between isolated single individuals (justifying no-choice tests), or between groups of individuals of different species (justifying choice tests)? In a few cases, such as cactophilic *Drosophila* in North America, related species can be found aggregating at single sites, implying that multiple-choice tests are more realistic (Heed and Mangan 1986). It is critical to employ realistic tests because the strength of behavioral isolation, at least in *Drosophila*, appears to depend strongly on the experimental design (Coyne et al., unpublished data). A further problem is that the artificial conditions under which these tests are carried out probably lead to underestimating the strength of behavioral isolation in nature. Confined individuals cannot always avoid unwanted matings. There have been many suggestions for how to measure behavioral isolation, but it is not clear which statistic is best (Gilbert and Starmer 1985; Rolan-Alvarez and Caballero 2000).

While demonstrating behavioral isolation is easy, it is appreciably harder to show which traits are involved. In the genus *Drosophila*, subject to hundreds of studies of behavioral isolation, we know which traits are involved in only a handful of cases (wing vibration and contact phermones; e.g., Coyne et al. 1994; Tomaru et al. 1995; Ritchie et al. 1999). Yet such traits must be identified if we are to understand the evolution of behavioral isolation. The best studies involve species in which one can manipulate traits like song or plumage and observe the effects on interspecific mating.

Much of the evidence for behavioral isolation in the earlier literature (e.g., Mayr 1963), rests on weak inference: closely related species differ in male-specific traits such as plumage color, and these trait differences were assumed to cause behavioral isolation. The underlying assumption is that females prefer the traits of conspecific males. This inference, however, may be wrong. First, even a signal that differs among species and acts as a cue for intraspecific mating cannot cause behavioral isolation unless the preferences of one sex coevolved with the signals of the other. Females of different species, for example, may not differ in preferences for male traits that diverged for ecological reasons (Price 1998). Sexually selected traits that vary among taxa but do not cause behavioral isolation include tail shape in swordtails versus platyfish (*Xiphophorus helleri* and *X. maculatus*; Basolo 1990, 1995), forehead crests in three species of the auklet *Aethia* (Jones and Hunter 1998), and breast color among populations of house finches (*Carpodacus mexicanus*; Hill 1994). Ryan (1998) and Ryan and Rand (1993a) supply further examples. In addition, many traits such as size, song, and weaponry, are deployed in male-male competition. In such cases, preferences for the traits of conspecific males may not have evolved: females might be constrained to mate with the winners of contests, thereby being unable to choose among males.

Examples of behavioral isolation

One of the best studies of behavioral isolation is that of Wiernasz and Kingsolver (1992) on the butterflies *Pieris occidentalis* and *P. protodice*, which are sympatric in the western United States. While females of these two species have nearly identical wing patterns, the males differ, with those of *P. occidentalis* having much darker forewings. Field observations reveal that newly eclosed *P. occidentalis* females readily mate with conspecific males, but reject nearly all heterospecific males. Experimentally darkening the forewings of *P. protodice* males, however, made them significantly more acceptable to *P. occidentalis* females. Similar manipulations have shown an isolating role for differences in ultraviolet patterning in the sulfur butterflies *Colias eurytheme* and *C. philodice* (Silberglied and Taylor 1978), as well as in body configuration in the Galápagos finches *Geospiza scandens* and *G. fortis* (Ratcliffe and Grant 1983). Although mating displays themselves—many of them highly ritualized—are important factors in intraspecific mating, we know of no studies showing that interspecific differences in these rituals cause behavioral isolation.

Auditory isolation is often detected by offering individuals (usually females) a choice between conspecific and heterospecific calls using loudspeakers. In this way Martinez Wells and Henry (1992) demonstrated isolation between three sympatric taxa of the lacewing *Chrysoperla plorabunda*. Examples of auditory isolation in other taxa include the frog *Physalaemus* (Ryan and Rand 1993b), the crickets *Teleogryllus commodus* and *T. oceanicus* (Hoy and Paul 1973), the indigo bunting and lazuli bunting, *Passerina cyanea* and *P. amoena* (Baker and Baker 1990), and subspecies of greenish warbler, *Phylloscopus trochiloides* (Irwin et al. 2001a).

Studies of isolation based on pheromones and other chemicals have concentrated on insects, although odor-based reproductive isolation is surely important in other animals as well as plants (odor differences can cause pollinator isolation). Chemical isolation can involve either volatile compounds detected at great distances or contact pheromones perceived at close range. The best example of the former involves pheromonal differences between two sympatric races of the European corn borer, *Ostrinia nubilalis* (Roelofs et al. 1985, 1987, 2002). Strong behavioral isolation is caused by an interspecific difference in female pheromones. This difference is subtle, based only on *cis* and *trans* isomers of a hydrocarbon pheromone. Females of one race produce the isomers in a 97:3 ratio; females of the other in a 1:99 ratio. Most males detect and respond only to the mixture released by females of their own race. Behavioral isolation caused by differences in contact pheromones is also seen in sulfur butterflies (*Colias*; Silberglied and Taylor 1978; Grula et al. 1980), sea snakes (*Laticauda*; Shine et al. 2002) and several species of *Drosophila* (Coyne et al. 1994; Coyne 1996a).

This leaves only the sense of touch as a cause of behavioral isolation. We know of no examples of "tactile isolation" during pre-mating courtship. However, females in cave-dwelling Atlantic mollies (*Poecilia mexicana*) appear to choose larger males using cues from their lateral line (Plath et al. 2004). Such tactile stimulation may be important in preventing mating between closely related species that live in caves or the deep sea. We discuss below other cases in which touch may cause reproductive isolation during the act of copulation.

Relative importance of behavioral isolation

Impressed by the lack of sexual attraction between sympatric animal species, and by the association of many adaptive radiations—such as the Hawaiian

Drosophila and New Guinea birds of paradise—with the evolution of striking sexual dimorphisms, zoologists have often considered behavioral isolation the main component of speciation. Mayr, for example (1963, pp. 106–107), claimed that "if we were to rank the various isolating mechanisms of animals according to their importance, we would have to place behavioral isolation far ahead of all others."

It is likely that, as a form of reproductive isolation acting early in the life cycle, behavioral isolation may often be an important *current* impediment to gene flow in animals (see Chapter 2). Three lines of evidence also imply that behavioral isolation can be instrumental in *initiating* speciation. The first comes from Coyne and Orr's studies in *Drosophila* (1989a, 1997) summarized in Chapter 2. In currently sympatric taxa, behavioral isolation is far stronger than postzygotic isolation. Although the elevated behavioral isolation in sympatry may have evolved through reinforcement (and thus have required the earlier evolution of some postzygotic isolation), behavioral isolation is nevertheless uniformly high among sympatric but not among allopatric species. This implies that *Drosophila* species cannot coexist in the absence of substantial mate discrimination.

Additional evidence for the primary importance of behavioral isolation comes from Seehausen et al.'s (1997) work on color differences in males of the Lake Victoria cichlids *Pundamilia nyerei* and *P. pundamilis*. In the laboratory, females strongly prefer intraspecific males when they can distinguish their colors, but this assortative mating breaks down under monochromatic light, yielding interspecific copulations and viable and fertile hybrids. (The importance of color rules out factors like mating displays as key causes of isolation.) Moreover, in Lake Victoria one finds fewer species of cichlids in turbid than in clear water, perhaps implying that species fuse when they cannot distinguish color. The inability of species to coexist in the absence of behavioral isolation further implies that this isolation was instrumental in speciation.

Finally, comparative studies in birds and insects show a consistently positive association between indices of sexual selection and rates of biological diversification (see Chapter 12). The increase in diversification probably reflects increased rates of speciation resulting from behavioral isolation that is itself the byproduct of sexual selection.

The evolution of behavioral isolation

Until recently, evolutionists were hard pressed to explain the evolution of behavioral isolation. If stabilizing selection acts on both signals and preferences, any alteration of either trait would reduce the ability to find a mate, and one must thus devise scenarios in which both signal and preference can coevolve. During the Modern Synthesis, this problem was "solved" by two suggestions. First, divergent natural selection might yield behavioral isolation as a pleiotropic byproduct. Behavioral isolation among geographical races was, as Mayr (1963, p. 495) put it, "nothing but a reflection of an over-all difference in their genetic constitution." It is not clear, however, how "general genetic divergence" could yield matching signals and preferences, although this seems to have happened in some artificial selection experiments (Chapter 3). Alternatively, behavioral isolation was explained as an evolutionary byproduct of "species recognition mechanisms." While this is more plausible, there was no attempt to explain how "species recognition" traits—which are presumably under stabilizing selection—could diverge.

It is curious that an important solution to this problem—the idea of sexual selection—was introduced by Darwin in 1871, but not applied to speciation until a century later. Although this idea was mentioned in 1949 by Haskins and Haskins, it is only in the last 25 years, beginning with the work of Ringo (1977), Lande (1981, 1982), and West-Eberhard (1983), that biologists have begun to examine the connection between sexual selection and behavioral isolation.

Table 6.1, modified from Kirkpatrick and Ryan (1991), summarizes the major forces that can produce concurrent divergence of traits and preferences among populations. We categorize these processes by the *initial* evolutionary force that triggers behavioral isolation, and, when this process is selection, whether the initial change occurs in the trait (signal) or preference.

Table 6.1 *Evolutionary forces that can cause behavioral isolation*

1. **Initial selection on preference**
   A. Direct selection on preferences: genetic change in mate preference increases immediate fitness of chooser
      1. Improves ability to acquire mates or resources
      2. Improves adaptation to local habitat (e.g., sensory bias)
      3. Improves ability to avoid deleterious features associated with mating (e.g., chase-away sexual selection)
   B. Indirect selection on preferences: genetic change in mate preference does not alter the chooser's immediate fitness (good-genes and runaway models)
2. **Initial selection on trait**
   A. Direct selection on trait improves the attractiveness of bearer to members of the opposite sex
   B. Direct selection on trait facilitates intrasexual competition: members of one sex compete directly with each other for resources or access to other sex
      1. Competition involving preference in other sex
      2. Competition not involving such preferences
   C. "Species recognition"
      1. Trait changes through natural selection, with preferences coevolving via natural selection.
      2. Reinforcement: selection improves discrimination against heterospecific individuals to prevent maladaptive hybridization
3. **Genetic drift**
   Nonselective changes in allele frequencies affect signal or preference
4. **Nongenetic mechanisms**
   A. Cultural drift: nongenetic change in a behavioral trait that produces behavioral isolation through learning or imprinting
   B. Host parasitism: young imprint on behavior and appearance of diverse host species, thus forming "cultural species"

The first two sets of forces shown in Table 6.1, with initial selection on either preference or trait, largely involve sexual selection. We follow Andersson's (1994, p. 7) definition of sexual selection on a trait "as a shorthand phrase for *differences in reproductive success, caused by competition over mates, and related to the expression of the trait.*" For convenience, we assume that males have the signal and females the preference, but the opposite can also occur, as can "mutual sexual selection" in which both sexes exhibit traits and preferences (Jones 1993). We will not describe the various forms of sexual selection in detail or give the evidence for each; these issues are fully discussed by Kirkpatrick and Ryan (1991), Andersson (1994), and Rice (1998). Rather, we describe how each form of selection can cause sexual isolation, providing both hypothetical scenarios and possible examples from nature. We emphasize that there are almost no cases in which we understand how sexual selection caused behavioral isolation in the wild. Our purpose is merely to describe various ways that this might happen.

DIRECT SELECTION ON PREFERENCES. In this scenario, selection initially changes female preference in a way that drives the evolution of the male trait. By "direct" selection, we mean that the alleles that alter preference also confer a direct fitness benefit to the bearer by increasing survival or fecundity. For example, selection can favor a female's ability to gain mates or resources by allowing her to choose males with the best territories or the brightest color (intense color might indicate the absence of transmissible parasites or diseases). Alternatively, female preference could be a byproduct of adaptive changes in other aspects of the sensory system. Female preference for male calls, for example, may change as a result of natural selection to detect and avoid the sounds of predators, with male calls evolving in response to this change. This process is called *sensory exploitation* or *sensory drive* (Ryan 1990, 1998). An example is Proctor's (1991, 1992) work on tactile stimulation in water mites. Finally, selection could change preferences if the reproductive interests of males and females do not coincide. In such cases males can be selected to reduce aspects of female fitness during mating, and females can regain fitness benefits by counteracting this evolution. This process is called *antagonistic sexual selection* (Rice 1998; Holland and Rice 1998).

Nearly all of these forms of selection, if acting in different directions in different populations, could yield behavioral isolation, since females will prefer the traits of males with which they have coevolved. Antagonistic sexual selection, however, seems less likely to promote behavioral isolation, as it is hard to see how externally visible traits or behaviors that enhance male fitness could be detrimental to females. (One possibility is that persistent males could force females to mate suboptimally [Holland and Rice 1998]). Nevertheless, antagonistic sexual selection operating *after* copulation but before fertilization is a potentially important cause of gametic isolation, and may have promoted speciation in some molluscs and insects (see below and Chapter 12). But sensory bias and antagonistic sexual selection can also *inhibit* speciation. Sensory bias may lead females to have "open ended" preferences, so that they always prefer more extreme versions of a trait, but conspecific males may be subject to evolutionary constraints (such as predation) that prevent evolution of that trait, making them less desirable than males from another species. This may have happened in the swordtail fish *Xiphophorus nigrensis* and *X. pygmaeus* (Ryan and Wagner 1987). Moreover, during antagonistic coevolution, females evolve resistance to conspecific males, which may accidentally lead them to prefer heterospecific males.

INDIRECT SELECTION ON PREFERENCES. In this form of selection, genes for female preference are not the direct object of natural selection, but evolve because they become genetically correlated with genes for male traits that are either more attractive to females (*runaway sexual selection*; Fisher 1930, Lande 1981) or indicate high fitness (*good genes*; Pomiankowski 1988). Explicit genetic models show that runaway sexual selection can cause divergent evolution—and hence behavioral isolation—under several circumstances: (1) genetic drift changes either the trait or the preference; (2) divergent, environmentally based selection alters the fitness of alleles affecting either the preference or the trait; (3) different mutations affecting traits or preferences occur in different populations (Lande 1981, Pomiankowski and Iwasa 1998). Similar processes could cause divergent evolution under good-genes models. In birds, for example, different populations could experience mutations affecting different plumage traits that signal a male's fitness (Schluter and Price 1993). Female preferences would then evolve to prefer the local plumage, yielding behavioral isolation between populations. Something of this sort may explain the spectacular divergence in plumage and display among groups of closely related species like the birds of paradise (Frith and Beehler 1998). Yet we know of no examples strongly implicating indirect selection on preferences as a cause of behavioral isolation. In the birds of paradise, for example, we do not know if exaggerated male plumage is correlated with genetic quality, nor whether species differences are involved in mate discrimination.

DIRECT SELECTION ON TRAITS. There are many types of selection that could cause divergent evolution of traits among populations, leading to selection on females to prefer males with locally adapted traits. Natural selection, for example, could make males more detectable in the local environment by changing their color, pheromones, or calls, setting off a bout of runaway sexual selection. Alternatively, relaxed or altered selection following invasion of a new habitat might allow a trait to evolve beyond its previous optimum. Irwin et al. (2001a) suggest that such a process may have altered song in the greenish warbler, yielding behavioral isolation between terminal populations of the "ring species" (Chapter 3).

Selection could also act on traits affecting intrasexual competition. Relatively little attention has been paid to this form of selection, in which members of one sex (usually males) compete directly for access to members of the opposite sex or for resources attractive to the opposite sex. Intrasexual competition can promote the evolution of armaments, aggressive behavior, territoriality, new mating songs, and so on. The question here is whether such selection can cause evolutionary changes in female preference. This depends on whether females are forced to mate only with the winners of such competitions, or whether they actively *choose* to mate with them. Only in the latter case can divergent evolution of traits cause divergence in preferences and hence behavioral isolation. Although Andersson (1994) finds little evidence that females actively choose the winners of male contests, and victors often appear to exclude other males from mating, it seems unlikely that females are invariably forced to mate with dominant males. The prevalence of extra-pair copulations in fish, birds, and other groups (Qvarnström and Forsgren 1998), and the observation that males often compete for resources and territories that are important for female fitness, imply that female choice can accompany male-male competition. Thus, although we lack concrete examples, we cannot exclude intrasexual competition as a cause of behavioral isolation.

SELECTION FOR SPECIES RECOGNITION. Discussions of premating isolation have sometimes been muddled by confusion between sexual selection and selection for *species recognition*. Both of these processes alter traits and preferences, yielding behavioral isolation as a byproduct. What distinguishes them is that selection for species recognition involves natural rather than sexual selection. Before biologists accepted the importance of sexual selection, the evolution of species recognition was seen as the primary cause of behavioral isolation.

There are several scenarios for the evolution of species recognition. One involves two successive bouts of natural selection, the first favoring a change in certain traits in both sexes. An increase in food availability, for example, can select for increased body size. Selection will then alter mating preferences of both sexes to favor individuals with the fitter trait, in this case larger males and females. Such preference evolves because it enables individuals to find appropriate mates more easily or to produce better offspring by mating with individuals having a more optimal phenotype. Divergent selection in different habitats could cause coordinated changes in trait and preference, leading to behavioral isolation between populations. In contrast to other forces yielding behavioral isolation, this type of selection need not produce sexual dimorphism. Moreover, although we call this process "selection for species recognition," this is somewhat of a misnomer. No related species need be present during the process: selection operates solely on individuals of a single species to find an appropriate mate.

Such selection is surely common—after all, natural selection must often alter phenotypes involved in mate recognition—but few cases have been described. In the Galápagos finches *Geospiza scandens* and *G. fortis*, behavioral isolation is based largely on differences in body size and beak shape, features that have diverged via natural selection for resource use (Ratcliffe and Grant 1983a,b; Grant 1986). Despite these interspecific differences, there is little sexual dimorphism within species. Variation in size, based on differential use of resources, also causes behavioral isolation between benthic and limnetic morphs of the threespine stickleback (Schluter 1993; Nagel and Schluter 1998; Rundle 2002). Finally, natural selection promoting behavioral isolation occurred in two sister species of Amazonian butterflies, *Heliconius melpomene* and *H. cydno*. Each species is a Müllerian mimic of a different sympatric (but more distantly related) species of *Heliconius*. Experiments with butterfly models show that the color differences involved in mimicry cause assortative mating between these species (Jiggins et al. 2001b).

The second form of selection for species recognition is reinforcement, which *does* depend on sympatry with a related species. Here there is a fitness disadvantage to mating with members of another taxon because hybrids are unfit. This disadvantage can cause selection to increase the preference for conspecific mates, perhaps by leading to a closer inspection of the phenotype. Since reinforcement can also trigger sexual selection (Liou and Price 1994; Kelly and Noor 1996), it is an especially powerful cause of behavioral isolation, and can produce sexual dimorphism. Chapter 10 summarizes the evidence for the role of reinforcement in speciation.

GENETIC DRIFT. It seems unlikely that random genetic drift could cause both a preference and a trait to evolve in tandem. But two simple genetic models (Nei et al. 1983; Wu 1985) show that a related phenomenon can occur in both sympatry and allopatry. However, both models assume that while changes in female *preference* are neutral, the male trait is selected to keep up with the drifting preference. Selection thus plays an integral role in behavioral isolation. However, as one might expect, in these models behavioral isolation evolves very slowly.

NONGENETIC MECHANISMS. Populations can become behaviorally isolated without *any* genetic change. One example involves the gradual divergence of calls or songs that attract mates. If there is no direct selection on calls, and female preference is broad, then isolated or semi-isolated populations may diverge through "cultural drift," with calls and preferences passed between generations by learning or imprinting. This is analogous to the evolution of human languages, in which dialects of different populations gradually diverge until they become mutually unintelligible. Behavioral isolation between "song races" of Nuttall's white-crowned sparrows (*Zonotrichia leucophrys nuttalli*) may have evolved this way (Baker 1983; Cunningham and Baker 1983). Female sparrows prefer songs of males from their own race, and different local songs are not genetically hardwired but learned from parents.

Imprinting can also produce a unique, nongenetic form of behavioral isolation in nest parasites—birds that lay eggs in nests of other species. A single species of parasite can use the nests of several different host species. This can cause the young male parasite to mimic the song of his foster father, and the young female parasite to imprint on both that song and the nest or phenotype of the host. Given this imprinting, sympatric, behaviorally isolated forms of the parasite—each using a different host and mating assortatively—can arise almost instantly. Although this process may be rare, it appears to have produced several "cultural species" of African indigobirds of the genus *Vidua*, in which the fidelity of parasites to hosts is greater than 99% (Payne et al. 1998, 2000; Sorenson et al. 2003). Although one might be reluctant to recognize species distinguished solely by differential imprinting, this nongenetic isolation subsequently allowed the parasites to perfect their mimicry through genetic evolution: young birds of each cultural species have evolved deceptive mouth markings resembling those of their host's offspring. This constitutes one of the few plausible cases of sympatric speciation.

TESTING THE ALTERNATIVES. Not surprisingly, it is usually impossible to decide which of the many mechanisms discussed above produced a given case of behavioral isolation. The situation is complicated because natural selection can trigger sexual selection, sexual selection can trigger *other* forms of sexual selection, and several processes can operate simultaneously. The difficulty of untangling these factors is seen in Endler and Houde's study (1995) of the guppy, *Poecilia reticulata*. Females from different populations show weak mating preference for the color and pattern of males from their own population, but the authors conclude that every conceivable type of sexual selection could have caused this isolation. In some species we have a reasonable understanding of the traits and preferences involved in behavioral isolation, and in others we have a good idea of how selection molded traits and preferences within species. But in almost no case have these studies involved the same species.

A notable exception is the work of Wiernasz and Kingsolver (1992) on the butterflies *Pieris occidentalis* and *P. protodice*. As noted above, differences in wing coloration produce behavioral isolation between these species. Wiernasz (1989) also found that females of *P. occidentalis* prefer darker wing patterns on males of their own species, thus connecting intraspecific sexual selection to behavioral isolation. However, we do not know what type of sexual selection was involved, nor whether the species divergence began with a change in color or preference.

There are several strategies for understanding the evolution of individual cases of behavioral isolation. The first is to concentrate on species in which natural selection probably drove behavioral isolation, as it does during the evolution of species recognition. It is much easier to understand isolation based on obvious adaptation to the environment than on sexual selection of nebulous origin. Another promising area includes cases of sexual selection in which the initial evolution of the trait or preference was probably caused by natural selection. This line of research will, of course, bias our view toward the importance of selection, which may be less involved in cases that are harder to understand.

Boughman's (2001) study of the threespine stickleback, *Gasterosteus aculeatus*, may be the best understood case of selection causing behavioral isolation. Nuptial male coloration is correlated with water clarity in six surveyed populations from British Columbia. Males living in clear water, in which red color is conspicuous, have large areas of intense red color. In contrast, males living in darker, organic-stained water—in which red is less visible but black more so—are blacker. Female color preferences differ in a correlated way: fish from clearer water have higher sensitivity to red light and show a stronger preference for spawning with red males. (The converse occurs in dark-water females.) These differences in color and preference cause substantial mate discrimination between "red" and "black" populations. Boughman argues that behavioral isolation began with divergent selection for female visual acuity based on water clarity. This in turn may have driven the evolution of divergent male coloration, producing behavioral isolation based on sensory bias. But there is (as usual) an alternative explanation: natural selection could have first caused divergence of male color, making them more visible to females in the local waters (Ryan 2001). This explanation involves species recognition instead of sensory bias.

Another route to understanding behavioral isolation involves good-genes models. One can sometimes determine whether the expression of a male's trait correlates with his genetic quality (e.g., Welch et al. 1998) and whether females prefer males with more highly expressed traits (e.g., Andersson 1982). If one can also show that such traits function as isolating barriers, then one has forged a reasonable evolutionary link between sexual selection and behavioral isolation. But there remains one complication with this scenario: genetically superior males may be more successful at mating even if female preferences evolved in other ways (Kirkpatrick and Ryan 1991). If, for example, genetically superior males produce longer calls, females may prefer these calls not because they are signs of high fitness, but simply because they are either more conspicuous or appeal to pre-existing sensory biases.

Since behavioral isolation is almost certainly a byproduct of the evolution of mate choice within species, understanding this isolation requires that we know how preferences and traits change within a lineage. This task is far more difficult than understanding the origin of ecological isolation.

The genetics of behavioral isolation

Table 6.2 summarizes genetic studies of nonecological barriers, all but one of which (Zeng et al. 2000) involve behavioral isolation. The criteria for inclusion are identical to those used for genetic studies of ecological isolation, but include another genetic method, introgression analysis. This technique requires repeatedly backcrossing fertile hybrids to a parental species, yielding lines in which marked chromosome regions are introduced (more or less singly) onto the genetic background of another species. These introgressed regions are typically made homozygous on this foreign background and genetically defined individuals are then scored for traits causing reproductive isolation. This yields a minimum estimate of the number of relevant genes.

The data are heavily biased toward *Drosophila*—the object of 21 of 25 studies—and the other four analyses also deal with insects. We do not include all the

Table 6.2 *Genetic analyses of traits demonstrably or probably involved in behavioral or mechanical isolation between closely related species*

Species pair	Trait
Drosophila heteroneura/D. silvestris	Head shape
D. melanogaster: two "races"	Male sexual isolation
	Female sexual isolation
Drosophila melanogaster/D. simulans	Female pheromones
Drosophila mauritiana/D. simulans	Male sexual isolation
	Female sexual isolation
	Shortened copulation
	Genital morphology
Drosophila mauritiana/D. sechellia	Female pheromones
Drosophila simulans/D. sechellia	Female sexual isolation
	Female pheromones
Drosophila mojavensis/D. arizonae	Male sexual isolation
	Female sexual isolation
Drosophila pseudoobscura/D. persimilis	Male sexual isolation
	Male courtship song
	Female sexual isolation
Drosophila virilis/D. littoralis	Male courtship song
Drosophila virilis/D. lummei	Male courtship song
Drosophila auraria/D. biauraria	Male courtship song
Drosophila ananassae/D. pallidosa	Female sexual isolation
	Male courtship song
Ostrinia nubilalis, Z and E races	Female pheromones
	Male response and attraction to pheromones
Laupala paranigra/L. kohalensis	Song pulse rate
Spodoptera latifascia/S. descoinsi	Pheromone blend

cases cited in the reviews of Ritchie and Phillips (1998) and Prowell (1998), as many of these fail to meet our criteria for reasonably rigorous genetic analysis.

Bearing in mind that these studies are heavily weighted by a single genus, we can nonetheless draw several conclusions. First, behavioral isolation requires changes at more than one gene. This is no surprise given that isolation is usually based on differences in at least one trait and one preference. But even single traits or preferences usually involve at least several genes. In the most closely related taxa, the Zimbabwe and "cosmopolitan" races of *Drosophila melanogaster*, behavioral isolation is based on a minimum of nine loci, and probably many more (Wu et al. 1995; Ting et al. 2001). Unfortunately, this genetic result tells us nothing about the evolution of behavioral isolation, as virtually all forms of sexual and natural selection can cause changes in few or many genes.

Number of genes	Reference
9	Templeton 1977; Val 1977
> 5	Wu et al 1995; Ting et al. 2001
> 4	Wu et al 1995; Ting et al. 2001
≥ 5	Coyne 1996a
≥ 4	Coyne 1992b, 1996b
≥ 7	Coyne 1992b, 1996b
≥ 6	Coyne 1992b, 1996b
≥ 14	Zeng et al. 2000
≥ 6	Coyne and Charlesworth 1997
≥ 2	Coyne 1992b
≥ 1	Coyne et al. 1994
≥ 2	Zouros 1973, 1981
≥ 2	Zouros 1973, 1981
≥ 3	Noor 1997a; Noor et al. 2001a
≥ 2–3	Williams et al. 2001
≥ 2	Noor et al. 2001b
≥ 3	Liimatainen and Hoikkala 1998; Hoikkala et al. 2000
≥ 4	Hoikkala and Lumme 1984; Liimatainen and Hoikkala 1998
≥ 2	Tomaru and Oguma 1994; Tomaru et al. 1995
≥ 2	Doi et al. 2001; Yamada et al. 2002
≥ 2	Doi et al. 2001; Yamada et al. 2002
1	Roelofs et al. 1987; Lofstedt et al. 1989
2	Roelofs et al. 1987; Lofstedt et al. 1989
≥ 8	Shaw 1996
1	Monti et al. 1997

However, single genes of large effect are involved in at least five traits causing sexual isolation. Differences in pheromone blends producing behavioral isolation of noctuid moth species (*Spodoptera*) and races of European corn borers (*Ostrinia nubilalis*) result from changes at single loci (Monti et al. 1997; Roelofs et al. 1987), although changes in pheromone structure in *Drosophila* are polygenic (Coyne 1996a; Coyne and Charlesworth 1997). Female preference for courting male wingbeat ("song") differences between *Drosophila ananassae* and *D. pallidosa* involves at least one gene of very large effect and one of smaller effect (Doi et al. 2001). Finally, the single gene causing pheromonal differences among females in corn borers is matched by a single gene responsible for differences in male antennal perception of those pheromones, and by yet another gene affecting how males respond to these perceptions.

These monogenic differences are puzzling, especially in corn borers. One would not expect key components of a coadapted signal-receiver system to undergo large changes, which might reduce mate attraction. A partial solution is offered by work on pheromonal differences that cause reproductive isolation between *Ostrinia nubilalis* and its Asian relative *Ostrinia furnacalis* (Roelofs et al. 2002; Roelofs and Rooney 2003). Molecular evidence reveals that this pheromonal difference—larger than that seen between the races of *O. nubilalis*—results from changes in two desaturase genes, with a different gene inactivated in each species. Roelofs et al. (2002) also observed that a few *O. nubilalis* individuals are attracted to pheromones of *O. furnacalis* females. They argue that if a few males were pre-adapted to respond to a novel pheromone, this could lead to the evolution of a new pheromone/response system. This scenario, however, does not explain how a new female pheromone can become fixed when only a few males respond to it.

A second conclusion is that behavioral isolation may involve genes on the X chromosome more often than expected based on it size (Reinhold 1998). This result is in line with several theoretical models that predict a disproportionate X-effect on traits causing behavioral isolation (Charlesworth et al. 1987; Rice 1984b), but not on other traits. This empirical pattern, however, rests largely on reciprocal F_1 hybrids in butterflies, which show strong differences in pheromones, calling time, mate selection, and other traits presumably involved in behavioral isolation (see Prowell 1998). Yet X-effects in Lepidoptera are not limited to traits yielding behavioral isolation: a disproportionate number of *all* species differences in this group, including those affecting morphology, physiology, and oviposition preference, show strong X-linkage (Prowell 1998). Because the X chromosome in butterflies is only a small fraction of the genome (3% on average), these pervasive X-effects remain a mystery. Because virtually all genetic analyses are based on Lepidoptera or *Drosophila*, we badly need similar studies in other groups.

Third, the genetic data do not support suggestions that the same genes may cause behavioral isolation in both sexes (i.e., the same genes affect both trait and preference). Genes causing behavioral isolation in males and females usually map to different chromosomes (as in *Ostrinia nubilalis*) or to different regions of the same chromosomes (Butlin and Ritchie 1989; Ritchie and Phillips 1998). This result is not unexpected: it would indeed be surprising if interspecific differences in traits such as morphology, song, behavior, or chemistry were based on the same genes used to perceive and evaluate those traits.

Fourth, behavioral isolation is often asymmetric, being much stronger in one direction of a species cross than in the other. A ubiquitous feature of studies in *Drosophila* (Watanabe and Kawanishi 1979; Kaneshiro 1980), this pattern is also seen in salamanders, parasitic wasps, and snakes (Arnold et al. 1996; Bordenstein et al. 2000; Shine et al. 2002). McCartney and Lessios (2002) note that such asymmetry is also common for sperm-egg incompatibility among free-spawning marine organisms. Arnold et al. (1996) explain this asymmetry as a result of runaway sexual selection, but in their model the asymmetry disappears after the earliest stages of speciation, failing to explain asymmetry among long-diverged species. If behavioral isolation has evolved through reinforcement, one would expect greater discrimination in females of the rarer species, who encounter heterospecifics more frequently, but this explanation fails because asymmetry is also common among allopatric species. Nevertheless, if asymmetry proves to be the rule in other taxa, it may give important clues about how selection causes behavioral isolation.

Finally, in virtually every case examined (all in *Drosophila*), F_1 hybrid females between behaviorally isolated species show no strong preference, mating readily with males of both parental species (Kyriacou and Hall 1986; Noor 2000; Coyne et al. 2002). This suggests that preference per se is based on alleles acting recessively in hybrids, so that hybrid females show *no* preference (Noor 2000 provides an evolutionary explanation). Regardless of its cause, the recessivity of female preferences may imply that speciation was allopatric. If speciation were sympatric, female preferences based on recessive alleles could not diverge: heterozygous females would mate readily with males of both taxa, preventing the evolution of assortative mating.

Deliberate hybridization of taxa whose behavioral isolation involves sex-limited traits, such as male plumage, might tell us something about the nature of genes producing such traits. Male-limited traits can evolve in two ways: the alleles producing them could be male-limited from the outset, or they could initially be expressed in both sexes and then later repressed in females (where they are maladaptive) by the accumulation of modifier alleles. In the second scenario, one might expect to see the expression of male-like traits in hybrid F_1 or backcross females, for the suppression could break down in a foreign genetic background. These crosses are relatively easy to make in some sexually dimorphic species—indeed, many must have been made already—but there are few reports about the phenotypes of hybrid females. Two intergeneric crosses in birds of paradise, however, show no expression in female hybrids of spectacular male plumage, implying either that the plumage differences are based on mutations with male-limited effects or that any modifier alleles are completely dominant in F_1 hybrids (Frith and Beehler 1998, Plate 14).

Mechanical Isolation

Mechanical isolation is the inhibition of fertilization between two species due to incompatibility between their reproductive structures that prevents normal copulation or pollination. Our view of "reproductive structures" is broad, including any morphological, behavioral, or neurological feature that promotes sperm or pollen transfer. In many animals, for example, sperm transfer occurs not through male genitalia, but through other appendages such as pedipalps or even spermatophores deposited on the ground (Eberhard 1986).

Mechanical isolation comes in two forms: structural and tactile. In structural isolation, gene flow is impeded by an incompatibility between morphological features of different species. This is the classic "lock and key" mechanism first suggested by Dufour (1844), who argued that isolation occurs because the genitalia of males and females of different species do not fit properly. As described in Chapter 5, structural isolation in plants occurs when pollinators cannot cross-fertilize flowers of different shape, or when different species of plants place pollinaria on different parts of a pollinator. Finally, structural isolation also includes what we call "aerodynamic isolation" in wind-pollinated plants: morphological changes in pollen and female reproductive structures that hinder the capture of pollen from other species.

Tactile isolation is limited to animals. In these cases, fertilization is impaired because one partner (usually female) detects aberrant morphology or inappropriate movement of the partner after copulation begins, prompting her to terminate copulation or expel sperm.

Examples of mechanical isolation

Although pollinator isolation involving structural incompatibility is common in plants, there are few convincing cases of mechanical isolation in animals. The best occurs in two species of Japanese carabid beetles with overlapping ranges, *Carabus maiyasanus* and *C. iwakianus* (Sota and Kubota 1998). Interspecific matings occur readily in the laboratory, but mated females suffer high mortality and/or low fertility. This occurs because males of the two species have differently shaped "copulatory pieces"—elaborations of the genitals that fit into special pouches off the vagina. Within each species, the size of the copulatory piece matches that of the vaginal pouch. In one of the two interspecific copulations, the poor fit causes the copulatory piece to break off, tearing the female's reproductive organs and sometimes killing her. The efficiency of sperm transfer is low in both reciprocal hybridizations, which produce few fertilized eggs. (Some tactile discrimination by females may also be involved here.) These factors reduce gene flow between the species by roughly 65%. It is likely that mechanical isolation contributes substantially to reproductive isolation in the hybrid zone, where one can find males with broken copulatory pieces and females whose vaginas contain fragments of these structures.

An oft-cited example of structural isolation is Paulson's (1974) study of ten damselfly species in Washington State. When mating begins, the male grasps the female's thorax with his abdominal appendages. This grasping does not occur properly with heterospecific females, preventing copulation. Grasping might, however, involve female cooperation, and thus could constitute tactile as well as structural isolation.

A little-known form of structural isolation is the preferential capture of conspecific over heterospecific pollen by wind-pollinated plants, based on differences in the structure of pollen and female parts. The latter differences can dramatically affect the shape of air currents around ovules or stigmas of closely related plants. If pollen from different species also has different aerodynamic properties, these phenomena can produce assortative pollen capture. This type of isolation occurs between two sympatric species of *Ephedra* (Niklas and Buchmann 1987) and between five species of pines (Niklas 1982), all studied in the laboratory, and between four species of pines analyzed in the wild (Linder and Midgley 1996). Given that closely related species of wind-pollinated plants often live sympatrically and overlap in flowering period, there may be a substantial fitness advantage to excluding foreign pollen from the stigma. Aerodynamic isolation might thus be fairly common, but is not often studied.

Tactile isolation is implied when interspecific matings fail to transfer sperm even though there is no structural misfit between male and female genitalia. In scarab beetles of the genus *Macrodactylus* (Eberhard 1992), the flies *Drosophila simulans* and *D. mauritiana* (Coyne 1993), and the butterflies *Erebia nivalis* and *E. cassiodes* (Lorkovic 1958), male but not female genital structures differ among closely related species, and females appear to prevent or prematurely terminate interspecific matings. This rejection may be based on the aberrant "feel" of foreign genitalia.

Finally, tactile isolation may involve both structural differences in genitalia and how they are used during copulation. The sepsid flies *Microsepsis eberhardi* and *M. armillata* differ not only in male genital shape, but also in the sequence in which these genitals are squeezed and flexed during copulation (Eberhard 2001). These differences could impede copulation between the two species, although this has not been proven. Eberhard (1996) describes other forms of "copulatory courtship" whose divergence among species might cause reproductive isolation.

Relative importance of mechanical isolation

Until recently, mechanical isolation was considered unimportant. This view is no longer warranted. Even if structural isolation is rare, tactile and aerodynamic isolation may well be common. Indeed, we have described several good examples of these latter barriers. Moreover, there is almost certainly an ascertainment bias against mechanical isolation: it is hard to detect and study, occurring as it does on a microscopic scale involving interactions between male and female reproductive parts. Finally, mechanical isolation can be studied only when behavioral or temporal isolation is incomplete.

There are several reasons for thinking that mechanical isolation may be of considerable importance in speciation. As described below, such isolation probably results from sexual selection acting on "cryptic" male traits and female preferences, and there is no reason why sexual selection should not operate as strongly on genital morphology as on traits like plumage and song. All that is required is that a female or her reproductive system is able to discriminate among the genitalia of different males. Thus one might expect mechanical isolation, at least of the tactile variety, to evolve about as rapidly as behavioral isolation, which appears to be a significant cause of speciation (Chapter 12). Moreover, in many groups male genital morphology is the most important trait for diagnosing species, differing even when all other traits are identical (Eberhard 1986). In these species male genitalia thus evolve faster than other features, including other sexually selected traits. If sexual selection is indeed the cause of this rapid genital evolution, it means that a female can discriminate among the genitals of different males from her species. This suggests in turn that females might discriminate strongly against *interspecific* differences in genitalia.

As Eberhard (1986) notes, although male genitalia and other structures involved in sperm transfer often evolve rapidly, the same is not true for female genitalia, which are often nearly identical among related species. This suggests that "lock and key" and related types of structural isolation are rare (see Shapiro and Porter 1989 for a review). In this sense, we agree with Mayr's (1963) conclusion that mechanical isolation is unlikely to play a major role in animal speciation. Yet Mayr did not consider *tactile* isolation, which is likely to be far more important.

Although the genitalia of males often evolve more rapidly than those of females, this does not imply that the sexes themselves have evolved at different rates. Evolution in males can produce easily observable changes in morphology, but these can elicit coevolutionary change of neurological features in females—features involved in recognizing male genitalia. Such neurological changes, which must surely be involved in most cases of behavioral isolation, are cryptic, detectable only through tests of mate discrimination.

The most important test of the role of mechanical isolation in speciation has not been performed: a comparative study determining whether the species richness of sister groups is correlated with variation in genital morphology among members of a group. Other sexually selected traits, however, have shown positive correlations in such tests (Chapter 12).

The evolution of mechanical isolation

Even in pre-Darwinian times, reproductive isolation was thought to result from differences in genitalia. As von Siebold and Stannius claimed (1854, p. 462), "They [insect genitalia] prevent allied species from producing bastards by adulterous connexions; for the hard parts of the male correspond so exactly with those of the female, that the organs of one species cannot fit those of another." For obvious reasons this view was called the "lock and key" hypothesis. After Darwin, its explanation involved either reinforcement or reproductive character displacement that prevented the formation of unfit hybrids. But apart from the lack of evidence for such structural misfit (Shapiro and Porter 1989), these theories cannot explain why species lacking sympatric relatives, such as oceanic-island endemics and host-specific parasites, also show elaborate, species-specific male genitalia (Eberhard 1986). Although reinforcement of mechanical isolation could evolve when interspecific copulation has maladaptive effects, as in *Carabus* beetles, studies of carabids and spiders show no enhanced divergence in genital pedipalp structure where the ranges of related species overlap (Ware and Opell 1989; Sota and Kubota 1998).

In *Sexual Selection and Animal Genitalia* (1986), Eberhard provides abundant evidence that the evolution of male genitalia is promoted by sexual selection involving cryptic female choice. As he explains (p. 183), "My hypothesis, sexual selection by female choice, proposes that male genitalia function as 'internal courtship' devices to increase the likelihood that females will actually use a given male's sperm to fertilize her eggs rather than those of another male." Other evidence for sexual selection on male genitalia includes direct studies of paternity in water striders (Arnqvist and Danielsson 1999); the observation that in some insects genital morphology varies far less with body size than do other morphological traits, suggesting selection for an optimal genital morphology that best stimulates females (Eberhard et al. 1998; Bernstein and Bernstein 2002); and a larger degree of divergence in genital morphology among species of insects whose females mate multiply than among those in which females mate only once (Arnqvist 1998).

Sexual selection on genitalia may involve many of the processes also causing behavioral isolation. Eberhard (1986, 1993) suggests that runaway sexual selection, perhaps connected to sensory bias, is most important: changes in male genitalia are driven by pre-existing female preferences, so that trait and preferences coevolve until the process is stopped by countervailing natural selection on males. Antagonistic coevolution could also operate if male genitalia either took advantage of a female's tactile "preferences," bypassing her behavioral preferences for a different mate, or gave males a fertilization advantage that is injurious to females (Hosken and Stockey 2004). Finally, male-male competition to fertilize might also promote the evolution of genital shape. It is hard to discriminate among these possibilities, although Eberhard (1993) rules out good-genes models on the reasonable grounds that genital structure is unlikely to reflect male fitness. (However, the vigor with which genitalia are wielded during copulation might reflect genetic quality.)

Aerodynamic isolation in plants may evolve through reinforcement or reproductive character displacement that prevents maladaptive hybridization, through natural selection to prevent clogging of the stigma by foreign pollen, or through natural selection to avoid wasted pollen and unfertilized ovules. The first scenario requires the sympatric presence of closely related species that produce unfit hybrids, the second the presence of sympatric species that release pollen at the same time, while the third can occur without any sympatric relatives. It is not hard to experimentally distinguish among these alternatives, but no such studies exist.

The genetics of mechanical isolation

We suggested above that mechanical isolation of pollinators may often be based on strong natural selection and thus involve floral changes based on genes of large effect. Mechanical isolation in animals, however, probably involves sexual selection yielding a gradual coevolution of male trait and female preference. We therefore suggest that genitalic differences between closely related species will often be polygenic, as will any behavioral isolation based on these differences. The only existing study (Zeng et al. 2000) supports this hypothesis. Genetic analysis of the difference in male genitalia between *D. simulans* and *D. mauritiana* revealed that at least 14 chromosome regions are involved, all having roughly equal effect. Moreover, all genomic segments from a given species affected genital morphology in the same direction, implying that the species difference involved selection and not drift (Orr 1998a). If, as seems likely, this divergence involved sexual selection based on female choice, the results suggest that females can detect very small changes in genital shape—at least 1/14th of the species difference. The short duration of interspecific copulations, which result in reduced sperm transfer, may be related to this genitalic difference (Coyne 1993).

Gametic (Postmating, Prezygotic) Isolation

Until recently, evolutionists have largely neglected isolating barriers that act between copulation and fertilization, a neglect evident in the common practice of dividing isolating barriers into the dichotomous categories of premating versus postmating, or prezygotic versus postzygotic. Nevertheless, the existence of gametic barriers was occasionally recognized. Stebbins (1950), for example, discussed "interspecific incongruity" between pollen and pistils, and *Drosophila* geneticists have long been aware of "insemination reactions," in which some interspecific matings produce a physiological reaction that sterilizes the female (Patterson 1946).

Gametic isolation was largely overlooked for the same reasons as was mechanical isolation: both act late in the life cycle and are hard to study. Recently, however, evolutionists have begun to appreciate the ubiquity of gametic isolation, to understand that it can evolve by forces identical to those causing behavioral and mechanical isolation, and to realize that it may have been important in speciation (Birkhead and Møller 1998; Simmons 2001).

Our definition of gametic isolation (also called *postmating prezygotic isolation*) includes all reproductive barriers acting between pollination and fertilization in plants and between spawning or copulation and fertilization in animals. We further divide gametic isolation into *noncompetitive* and *competitive* forms. The former operates during single interspecific pollinations or copulations, and thus does not require the simultaneous presence of conspecific and heterospecific gametes. Noncompetitive isolation includes, for example, the failure of pollen to germinate on a foreign stigma. In contrast, competitive isolation requires the simultaneous presence of conspecific and heterospecific male gametes, and causes isolation that may not be predictable from the results of single heterospecific matings. For example, pollen from one species may grow only slowly down the style of a different species, yet may eventually effect fertilization, producing many hybrid seeds. However, if conspecific pollen tubes grow more rapidly, they will win the race to fertilize when pollen from both species lands on a single style. Thus the reproductive barrier appears only when gametes compete.

In some cases there is no clear distinction between gametic and mechanical isolation, as when ejaculated heterospecific sperm are not stored because the female perceives the "wrong" male genitalia.

Examples of gametic isolation

There are many possible types of gametic isolation. Table 6.3 lists the known forms, both competitive and noncompetitive, and gives at least one example of each. For noncompetitive isolation, the forms are listed in temporal order beginning with pollination or copulation and ending with fertilization.

NONCOMPETITIVE ISOLATION. Nearly every step between copulation or pollination and fertilization can malfunction in heterospecific crosses. First, few sperm might be transferred. Matings between *Drosophila simulans* females and *D. sechellia* males result in transfer of either no or very few sperm, yielding fewer offspring than do conspecific matings (Price et al. 2001). If the female is also fertilized by a *D. simulans* male, almost no hybrid offspring are formed, so that these species show both noncompetitive and competitive gametic iso-

Table 6.3 *Forms of gametic (postmating prezygotic) isolation*[a]

Noncompetitive Isolation

1. **Poor transfer or storage of sperm**
   A. Reduced gamete transfer during copulation (Grimaldi et al. 1992; Price et al. 2001)
   B. Sperm lost or expelled from reproductive tract of heterospecific female (Vick 1973; Price et al. 2001)
2. **Inviability of gametes in foreign reproductive tract**
   A. Sperm not fully motile in heterospecific females (Gregory and Howard 1994)
   B. Sperm causes "insemination reaction" that sterilizes females (Patterson 1946; Patterson and Stone 1952)
   C. Pollen fails to germinate on foreign stigma (Levin 1978; Niklas 1997)
3. **Inability of gametes to effect fertilization due to poor movement or cross-attraction**
   A. Lack of cross-attraction of sperm and eggs (Miller 1997)
   B. Pollen tubes grow slowly down foreign style (Levin 1978; Williams and Rouse 1988)
4. **Failure of fertilization when gametes contact each other (intrinsic gametic incompatibility)** (Heslop-Harrison 1982; Williams and Rouse 1990; Palumbi and Metz 1991; Niklas 1997; Vacquier 1998; Levitan 2002)
5. **Foreign ejaculate fails to stimulate oviposition or reduces rate of oviposition** (Fuyama 1983; Gregory and Howard 1993; Price et al. 2001)

Competitive Isolation

1. **Conspecific sperm precedence:** females inseminated by both conspecific and heterospecific sperm produce few hybrid offspring (Wade et al. 1994b; Price 1997; Howard et al. 1998a)
2. **Conspecific pollen precedence:** flowers pollinated by both conspecific and heterospecific pollen produce few hybrid offspring (Rieseberg et al. 1995a; Carney et al. 1996; Diaz and Macnair 1999)

[a] References give examples of each barrier.

lation (Price 1997). Problems with interspecific sperm storage have also been reported in dermestid and ladybird beetles (Vick 1973; Katakura 1986). On the other hand, when a *D. simulans* female mates with a *D. mauritiana* male, copious sperm are transferred and stored, but disappear rapidly from the storage organs before they can fertilize eggs (Price et al. 2001).

In some instances, gametes are inviable or behave improperly in the reproductive tract of a heterospecific individual. In the crickets *Allonemobius fasciatus* and *A. socius*, stored sperm is less motile in the storage organs of heterospecific than conspecific females, perhaps explaining the few hybrids produced by interspecific crosses (Gregory and Howard 1994). The equivalent situation in plants occurs when pollen fails to germinate on the stigmas of closely related species (Levin 1978; Niklas 1997). In *Drosophila*, the insemination reaction involves the formation of a hard reaction mass in the vagina after heterospecific copulation, obstructing fertilization and even sterilizing females (Patterson and Stone 1952).

Transferred sperm or pollen can fail to reach female gametes because of a lack of cross-attraction or other problems. When tested in the laboratory, ovarian extracts from nine species of the free-spawning starfish *Macrophiothrix* attract heterospecific sperm much less strongly than conspecific sperm (Miller 1997). This is apparently based on differences in chemical attractants that promote normal fertilization in an environment that quickly dilutes expelled gametes. In crosses among species of *Rhododendron*, pollen tubes from species with short styles often fail to reach the ovules of long-styled species. Presumably, style length and pollen-tube length are coadapted within species (Williams and Rouse 1990).

The most thoroughly studied form of gametic isolation is *intrinsic gametic incompatibility*: the failure of heterospecific gametes to effect fertilization when encountering each other. These incompatibilities are most likely due to the failure of biochemical recognition mechanisms involved in normal fusion of sperm and egg. Palumbi (1998) and Vacquier (1998) review these mechanisms and their implications for speciation.

The best understood example of such isolation involves the interaction between sperm and eggs in abalones (reviewed by Kresge et al. 2001). This is in fact the only instance in which the proteins involved in conspecific recognition and heterospecific rejection have been identified. Gamete recognition has been studied most thoroughly in seven species in the genus *Haliotis* from California that diverged approximately 1–2 million years ago (Metz et al. 1998a). Abalones are broadcast spawners in which sporadic environmental signals induce mass spawning, with gametes fusing in the water column.

Successful fertilization in these abalones requires an interaction between the sperm protein lysin and the egg protein VERL ("vitelline envelope receptor for lysin"). Lysin is a dimeric, nonenzymatic molecule found in the sperm acrosome. When a sperm contacts the outer coat of the egg, lysin is released from the acrosome and reacts with VERL, a large glycoprotein constituting nearly a third of the egg envelope. The released lysin dissociates into monomers, causing separation of the VERL fibers and allowing the sperm to penetrate the egg. Kresge et al. (2001) suggest a biochemical mechanism for this interaction.

In vitro studies show that the interaction between lysin and VERL is highly species-specific, with purified lysins unable to dissolve the vitelline envelope of heterospecific eggs (Vacquier and Lee 1993; Swanson and Vacquier 1997). The VERL/lysin interaction is thus a potentially important component of gametic isolation, an idea supported by evidence that eggs are fertilized far more readily by conspecific than by heterospecific sperm.

Does this gametic isolation act in nature? Although some species live at different depths, implying habitat isolation, up to three species can coexist in the same small area (V. Vacquier, pers. comm.). Nothing is known, however, about whether sympatric species spawn at the same time. But even if temporal isolation exists, it may have evolved after gametic isolation (perhaps via reinforcement), so that the VERL/lysin system could still have been important in speciation.

Similar analysis of lysins in tegulid gastropods, a sister group of abalones, also shows species-specificity in the ability of lysin to dissolve vitelline envelopes, and, in some species comparisons, rapid evolution of the protein via selection (Hellberg and Vacquier 1999). The female receptor of lysins is unknown in these gastropods.

Gametic isolation has also been described in four species of the sea urchin *Echinometra*, a tropical Pacific group that diverged within the last 3 million years (Palumbi 1998). Like abalones, these urchins are broadcast spawners. They possess bindins: acrosomal proteins released to the outside of the sperm when it contacts an egg, presumably promoting adhesion of the two gametes. *Echinometra* eggs are fertilized far more easily by conspecific than by heterospecific sperm, which attach less efficiently to the egg's vitelline envelope. However, there has been no confirmation in vitro that the divergence in bindins plays a role in gametic isolation, nor has the cognate female protein or proteins been identified. Related species of these urchins often live within meters of each other around Pacific islands, and although it is not known whether they spawn at the same time, the presence of rare hybrids does imply some temporal overlap (Palumbi and Metz 1991).

As judged by the excess of replacement over silent substitutions in bindin DNA, selection has caused these proteins to diverge rapidly between urchin species (Metz and Palumbi 1996). Unlike lysin, however, bindins show significant intraspecific polymorphism in protein sequence. This is puzzling for a protein important in fertilization, especially because different variants have different probabilities of fertilization (Palumbi 1999).

In plants, the analog of sperm-egg incompatibility is the failure of pollen tubes to penetrate the ovule. This has been described in interspecific crosses in *Rhododendron* and grasses (Heslop-Harrison 1982; Williams and Rouse 1990).

Finally, even if gametes unite in interspecific matings, the rate of oviposition may be reduced. This barrier is known to occur in crickets and *Drosophila* (Fuyama 1983; Gregory and Howard 1993; Price et al. 2001). It might reflect gametic incompatibility, with only fertilized eggs being laid, or the malfunction in heterospecific females of seminal proteins that stimulate oviposition (Herndon and Wolfner 1995). Indeed, in matings between *Drosophila pulchrella* females and *D. suzukii* males, normal oviposition was restored by injecting females with extracts from the reproductive tract of *D. pulchrella* males (Fuyama 1983).

COMPETITIVE ISOLATION. Competitive gametic isolation occurs only in females simultaneously inseminated or pollinated by both conspecific and heterospecific individuals. The strength of such isolation can be gauged by the reduction in the number of hybrids below that predicted from single conspecific and heterospecific matings. If, for example, conspecific matings produce an average of 60 offspring and heterospecific matings an average of 20 (and there is no sperm precedence *within* species based on mating order), competitive isolation occurs only if fewer than 25% (20/80) of the offspring produced by doubly inseminated females are hybrids. This form of reproductive isolation, also known as *conspecific sperm precedence* (CSP) or *conspecific pollen precedence* (CPP), is reviewed by Howard (1999). Because fertilization of females by multiple males and pollination by multiple individuals is common in nature, competitive gametic isolation may be an important reproductive barrier.

Detecting CSP or CPP requires controlled crosses in which one can distinguish the origin of offspring and ensure that a deficit of hybrids is due at least partly to gamete competition and not simply to any reduced fitness of hybrid zygotes or offspring. Figure 6.1 shows a case of strong CSP involving *Drosophila simulans* and its sister species *D. mauritiana* (Price 1997). In double *conspecific* inseminations, the male who mates second produces most of the offspring, a common feature of many insects (Simmons and Siva-Jothy 1998). Single heterospecific inseminations of *D. simulans* females by *D. mauritiana* males yield many hybrids. However, a *D. simulans* female inseminated by both *D. simulans* and *D. mauritiana* males produces almost no hybrids, even when the heterospecific male is the second to mate. CSP in this case is caused by both the physical displacement and incapacitation of heterospecific sperm by conspecific sperm (Price et al. 2000).

Because these species are allopatric, CSP is only a potential barrier to gene flow. However, CSP has also been found in two sister species of *Drosophila* whose ranges overlap in nature (*D. yakuba* and *D. santomea*), and it may account for the paucity of hybrids in this location (Chang 2004). Likewise, CSP is an important isolating barrier in the zone of sympatry in the crickets *Allonemobius fasciatus* and *A. socius* (Howard et al. 1998a): despite the absence of temporal, habitat, and behavioral isolation, one finds few hybrids in nature. In mixed laboratory cages, many female crickets mate multiply, but females inseminated by males of both species produce almost no hybrids. This CSP occurs in addition to noncompetitive gametic isolation, for single heterospecific matings produce fewer offspring than single conspecific matings (Gregory and Howard 1993). Possible cases of CSP have also been described in flour beetles (Wade et al. 1994b), grasshoppers (Hewitt et al. 1989; Bella et al.

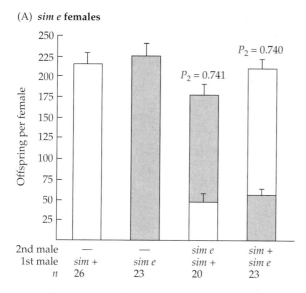
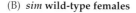
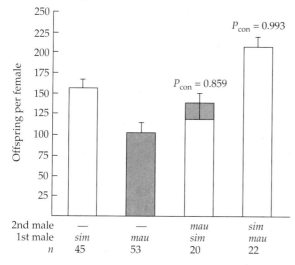

Figure 6.1 Conspecific sperm precedence, a form of competitive gametic isolation, in *Drosophila*. sim = *D. simulans*; mau = *D. mauritiana*; e = *ebony* males; + = wild-type males; n = sample size of females tested. All matings are to *Drosophila simulans* females. Where no genotype is given, individuals are wild type. For single matings, bars show the average number of offspring of each female produced *after* the first four days (lines show one standard error). For two successive matings (the second four days after the first), bars show the average numbers of offspring of each genotype produced after the second mating. (A) Second male precedence in conspecific matings of *D. simulans*. Females mated doubly to a wild-type or *ebony* male produce most of the offspring from the second male, regardless of his genotype (P_2, the proportion of total offspring sired by the second male, exceeds 0.7). (B) Conspecific sperm precedence in mixed matings. *D. simulans* females produce many progeny when mated singly to either *D. simulans* or *D. mauritiana* males. However, when mated twice in succession to *D. simulans* and *D. mauritiana* males, *D. simulans* females produce the vast majority of their offspring from the *conspecific* mating, regardless of which male mates first (P_{con}, the proportion of total offspring after the second mating sired by the *D. simulans* male, exceeds 0.85). (From Price 1997.)

1992), and mussels (Bierne et al. 2002). (As the grasshopper and mussel studies lacked singly mated heterospecific controls, gametic isolation might be noncompetitive.)

In plants, conspecific pollen preference has been seen in crosses between species of *Iris* (Carney et al. 1996), monkeyflowers (*Mimulus*; Diaz and Macnair 1999), and sunflowers (*Helianthus*; Rieseberg et al. 1995a). In all three cases, related species are partially sympatric and show overlap in flowering time, so that CPP probably causes isolation in the field. In *Iris* and *Mimulus*, CPP is due largely to the relatively slow growth of pollen tubes in foreign styles. (Differences in the swimming speed of animal sperm could yield the same result.) This shows that competitive gametic isolation need not involve direct interaction between gametes from different species.

Relative importance of gametic isolation

As noted above, gametic isolation is likely to be important in preventing current gene flow in sea urchins, abalones, and crickets. But we might also expect gametic isolation to be *historically* important in speciation. Such isolation probably evolves rapidly, at least in animals, because gametic barriers may be a byproduct of sexual selection at the molecular level—selection that might occur as rapidly as sexual selection affecting phenotypic characters like plumage, pheromones, or mating behavior. Like males themselves, male gametes can compete for access to eggs, and eggs can show biochemically based "preference." In many groups, including protozoa, fungi, plants, marine gastropods, and mammals, proteins involved in conjugation, gamete fusion, self-compatibility, and gametogenesis evolve rapidly (Swanson and Vacquier 2002a,b; Swanson et al. 2004). To demonstrate stronger selection on reproductive proteins, however, requires comparing their evolutionary rates with those of nonreproductive proteins.

To detect positive selection on a protein sequence, one calculates the ratio of the number of nonsynonymous DNA substitutions between species (those changing protein sequence) to the number of synonymous DNA substitutions between species (those not changing protein sequence). After correcting for the number of *potential* nonsynonymous and synonymous sites, one obtains the "d_N/d_S" ratio. High values imply a history of positive natural or sexual selection. Such high ratios have been seen in several sperm proteins, including abalone lysins, echinoderm bindins (see below), and self-incompatibility alleles in plants (Richman and Kohn 2000).

The evidence for selection is especially strong in *Drosophila*. Male reproductive proteins evolve rapidly: elevated d_N/d_S ratios are seen in 19 of 176 proteins found in the accessory gland, which produces seminal-fluid proteins. *Secreted* accessory gland proteins (*Acp*s) evolved even faster, showing more than twice the divergence rate of nonsecreted accessory-gland proteins, which in turn diverged twice as fast as nonreproductive proteins (Swanson et al. 2001a,b). *Acp*s in *D. melanogaster* and its relatives are numerous—there are more than 80—and are of special interest because they affect sperm storage, sperm competition, the rate of oogenesis, the release of oocytes, and the attractiveness and longevity of mated females (Chapman et al. 1995; Clark et al. 1995; Neubaum and Wolfner 1999; Chapman et al. 2000, 2001; Wolfner 2002). An important question is whether reproductive proteins evolve faster in males than in females, as this might imply specific forms of sexual selection (see below and Chapter 8). The answer appears to be yes, at least in *Drosophila*.

Only 14 of 461 genes in the female reproductive tract have elevated d_N/d_S ratios, a proportion significantly lower than that seen in reproductive proteins of males (Swanson et al. 2001a, 2004).

Faster evolution of reproductive proteins is not limited to invertebrates: three female reproductive proteins in mammals also show high d_N/d_S ratios. Two of these are egg coat glycoproteins that bind sperm, and for one of these, ZP3, in vitro assays suggest that evolutionary change is concentrated in regions that interact directly with sperm (Swanson et al. 2001b; Jansa et al. 2003). Likewise, a sample of 18 proteins from primates involved in male reproduction (including protamines) has a higher rate of amino acid substitution than a group of 30 nonreproductive "control" proteins (Wyckoff et al. 2000). Although Wyckoff et al. suggest that this pattern implies sexual selection acting on sperm, Clark and Civetta (2000) argue that the rapid evolution of protamines might reflect natural selection on their ability to bind DNA during spermatogenesis.

The rapid evolution of reproductive proteins is one of the most important findings in the last decade of work on speciation. It may imply that there is swift biochemical coevolution between male and female proteins—coevolution that could promote the evolution of gametic isolation. Isolating barriers like the insemination reaction, conspecific sperm precedence, and hybrid male sterility almost certainly rest largely on such evolution.

In a laboratory demonstration of this rapid evolution, Rice (1996) used a complex genetic design to prevent *Drosophila melanogaster* females from coevolving with males, while allowing males of the same strain to adapt to the "nonresponding" females. After only 30 generations, experimental males had become better than control males at both mating with previously mated females and at displacing the sperm of previously mating males. Females prevented from coevolving with males had a higher mortality rate after remating than did control females, perhaps because experimental females were prevented from adapting to the evolution of toxic compounds in male sperm. Rice argues that these results are evidence for antagonistic coevolution between males and females: males are selected to outcompete other males to fertilize eggs, even when the selected traits reduce female fitness. (The reciprocal antagonism must be less frequent because it is harder for females to reduce the fitness of males.)

In a related experiment in *D. melanogaster*, when male-male competition was prevented by enforcing monogamous mating for 47 generations, male ejaculates evolved a *reduced* ability to harm females, and females had a higher rate of mortality from the ejaculates of unselected males (Holland and Rice 1999). Monogamy appears to have obstructed antagonistic coevolution between the sexes. Taken together, the two experiments suggest that sexual coevolution occurs constantly, even in laboratory populations of *Drosophila*. Given that isolated populations in nature are likely to experience such evolution, but in diverse directions (i.e., different mutations may fuel the process in different locales), it is easy to see how genetic divergence could quickly promote gametic isolation.

Indeed, the best way to measure the evolutionary rate of gametic isolation is to determine whether it occurs in crosses between populations of a single species. There has been no systematic search for this phenomenon, but three examples suggest that it could be common. Brown and Eady (2001) observed both competitive and noncompetitive gametic isolation in crosses between two populations of the bruchid beetle *Callosobruchus maculatus*. In *Drosophila mojavensis* and *D. arizonae*, the reaction mass in the vagina produced after conspecific insemination was larger and lasted longer in interpopulation than in intrapopulation crosses (Knowles and Markow 2001). Because females do not lay eggs until the reaction mass subsides, random mortality of mated females would reduce the reproductive output of females mated to males from a different population, forming a weak reproductive barrier. Finally, Cordero Rivera et al. (2004) found that, in the damselfly *Calopteryx haemorrhoidalis*, males from two populations use two different mechanisms for displacing the sperm of previously-mating males, a difference correlated with interpopulation divergence in genital morphology. More studies like these will reveal whether gametic isolation is one of the fastest-evolving reproductive barriers. After all, laboratory experiments show that reproductive *systems* can evolve substantially in only a few dozen generations when coevolution is altered.

Comparative analysis can test whether gametic isolation is instrumental in speciation. Studying five orders of insects, Arnqvist et al. (2000) compared clades whose females mate only once per lifetime (monandrous) with clades of equal age containing females who mate multiply (polyandrous). On average, polyandrous groups contained four times more species. Assuming that extinction rates are equal in the two groups, this implies that polyandry facilitates speciation. Arnqvist et al. argue that females who mate multiply would experience higher levels of antagonistic sexual selection owing to competition between ejaculates of different males, and that this selection caused gametic isolation and speciation.

While suggestive, these results do not necessarily show that antagonistic sexual selection was the engine of speciation. This is because polyandry can also promote *non-antagonistic* sexual selection based on female choice or male-male competition. Moreover, these forms of selection can involve not only ejaculates, but also phenotypes and behaviors that could cause speciation via premating isolation. Finally, polyandrous species may also be more likely to be polygynous, so that the greater species richness of polyandrous taxa may actually be caused by multiply mating males. Arnqvist et al. (2000) do not give data that rule out these confounding effects, nor show that the species analyzed have gametic isolation. Thus, although this study implicates sexual selection as a cause of speciation, it does not tell us which form of sexual selection was important. A better way to judge the importance of gametic isolation would be to see whether it is a common barrier between recently diverged taxa.

The evolution of gametic isolation

Except when caused by reinforcement, gametic isolation must be a pleiotropic byproduct of traits that have evolved within species. In a few cases we have an idea of what those traits might be. The pattern of gametic incompatibilities in some groups of plants, for example, tends to follow the "SI × SC" rule (Lewis and Crowe 1958): crosses between self-compatible (SC) species and their self-incompatible (SI) relatives are much less likely to succeed when the female parent is SI than SC. This suggests that the genes controlling intraspecific incompatibilities may also be involved in interspecific gametic isolation (incompatibilities require SI genes acting in the stigma). This notion is supported by two further observations: (1) some mutations that convert an SI into an SC plant also restore its gametic compatibility with other species, and (2) genetic variability for rejection of heterospecific pollen sometimes maps near "S-loci" that control intraspecific incompatibilities (Martin 1967; McClure et al. 2000). However, self-incompatibility within species involves rejection of pollen that is genetically similar, while interspecific incompatibility involves rejection of pollen that is genetically different. Thus, interspecific and intraspecific incompatibilities may involve diverse evolutionary forces operating on the same biochemical pathways. Indeed, it has been suggested that genes producing self-incompatibility in stigmas (often producing enzymes that destroy RNA) evolved first as a defense against pathogens and only later were co-opted as self-incompatibility genes that reduce inbreeding (Pastuglia et al. 1997). Conversely, genes used in self-recognition may have evolved from genes used to reject pollen from other species.

Other forms of gametic isolation resemble phenomena occurring within species. Conspecific sperm precedence between *Drosophila* species, for example, appears to involve the same traits mediating sperm competition among males within species—physical displacement and inactivation of foreign sperm—but in exaggerated form (Price et al. 1999, 2000).

As with behavioral isolation, we categorize the causes of gametic isolation through the processes that trigger its evolution.

GENETIC DRIFT. Because the gametes and reproductive systems of males and females must be compatible, it would seem that natural selection would oppose their divergence by genetic drift. Nevertheless, at least one model shows that gametic isolation can result from genetic drift producing interpopulation divergence in the "preferences" of female gametes, followed by selection on male gametes to match these new preferences (Wu 1985). Although this process is glacially slow, Swanson and Vacquier (1998) suggested that it explains the divergence among abalone species of lysins and VERL receptor proteins, a divergence that, as noted above, probably causes gametic isolation.

VERL is a modular protein containing up to 28 tandem repeats, each of about 450 base pairs. Within each species, VERL sequences are homogeneous among repeats and among individuals, but the repeat sequences differ profoundly among species. This pattern suggests that VERLs have diverged by "concerted evolution," a process involving mutation, genetic drift, and a combination of gene conversion and unequal crossing-over that homogenizes repeats within a species (Li 1997). Low d_N/d_S ratios of VERL suggest that the interspecific divergence of this protein did not involve positive selection (Swanson and Vacquier 1998). Lysins, on the other hand, do evolve rapidly, with d_N/d_S ratios significantly higher than those expected under drift (Lee et al. 1995; Vacquier et al. 1997; Yang et al. 2000).

Swanson and Vacquier (1998) originally suggested that as VERL proteins change neutrally, the lysins are selected to keep up with their ever-changing need to dissolve the evolving female protein. But this scenario has problems. d_N/d_S ratios may have low power to detect selection (Biermann 1998), and any change in VERL, by reducing the likelihood of fertilization, could be maladaptive. Indeed, recent sequencing studies and more sensitive statistical tests suggest that, while most of the tandem repeats of VERL may show concerted evolution, repeats 1 and 2 appear to have experienced directional selection (Galindo et al. 2002, 2003). This raises the possibility that selection is more directly involved in gametic isolation of abalones.

NATURAL SELECTION AND REINFORCEMENT. DNA-sequence data reveal that the divergence of many reproductive proteins clearly involved positive selection, but was this natural selection, sexual selection, or both? The distinction between these types of selection becomes fuzzy at the level of gametic evolution and fertilization. Should we consider selection for faster-swimming sperm as natural selection for increased fertility, or sexual selection involving male-male (gamete) competition?

In some instances, it seems likely that gametic isolation resulted from natural selection. In the *Vireya* section of *Rhododendron*, cross-fertilization occurs only when style lengths of the two species are not too different (Williams and Rouse 1988, 1990; see Diaz and Macnair 1999 for a similar example in *Mimulus*). If style lengths are determined by ecological factors such as pollinators, natural selection on style length may be followed by selection on pollen tubes for compatible growth rates.

Unfertilized eggs or ovules—particularly in organisms with external fertilization—may be susceptible to pathogens or other harmful environmental factors. These factors may select for changes in surface proteins that reduce infection but also reduce sperm penetration, leading to compensatory natural selection on male gametes (Howard 1999). This constitutes a form of species recognition. Although this scenario seems plausible, there is little evidence that it has occurred in animals. But, as noted above, selection to avoid pathogens may have operated in plants, perhaps producing divergent self-incompatibility alleles that yielded gametic isolation as a byproduct (Pastuglia et al. 1997).

Reinforcement is often considered the most likely cause of gametic isolation, perhaps because it is easy to imagine selection on female gametes to resist fertilization by another species when hybrids are unfit. Yet reinforcement that increases gametic isolation may be less probable than reinforcement that increases *premating* barriers such as behavorial isolation. This is because selection will be stronger on traits that cause an individual to avoid mating with a heterospecific individual than on traits that increase gametic incompatibility when copulation or pollination has already occurred. By avoiding mating with members of another species when hybrids are unfit, individuals bypass the added fitness costs of copulation, gamete wastage, and so on. But when ejaculation or pollination has already occurred, selection becomes weaker. Imagine, for example, two sympatric, broadcast-spawning marine species whose hybrids are semisterile. Males and females in both taxa will experience selection to avoid cross-fertilization, which could increase habitat or temporal isolation. Eggs will also be selected to avoid penetration by foreign sperm. But selection will be weak or absent on *male* gametes, because once they have fused with an egg they have no second chance at reproduction. There would thus be little or no selective advantage to evolving increased incompatibility of male traits. (This is also true for pollen.)

Thus, while reinforcement for premating isolation can act on both sexes (but will still be stronger in females), reinforcement for *gametic* isolation is largely limited to eggs. It is surprising, then, that all four proposed cases of reinforcement of gametic isolation involve male traits, although the male gametes may have changed in response to the evolution of female receptors that have not been examined. Two studies are based on biogeographic patterns that are suggestive but not definitive. In abalones of the genus *Haliotis*, 19 species that occur sympatrically with congeners carry the species-specific terminal domain of lysin whose divergence causes gamete isolation. But this region of the protein is missing in the one species, *H. tuberculata*, that is allopatric to all others (Lee et al. 1995). Unfortunately, nothing is known about gametic isolation between *H. tuberculata* and other species. The second pattern occurs in sea urchins: species in the genus *Arbacia*, which are allopatric, show less differentiation of bindins and greater interspecific crossability than sympatric species of similar age in the related genera *Echinometra* and *Strongylocentrotus* (Metz et al. 1998b).

Stronger evidence for reinforcement comes from Geyer and Palumbi's (2003) work on two species of the sea urchin *Echinometra*. Compared to an undescribed but related species ("species C"), populations of *E. oblonga* have more divergent bindin alleles when sympatric than when allopatric. In fact, only allopatric populations of *E. oblonga* have bindin alleles identical to those found in species C. The authors propose that this is a case of reinforcement. As required for reinforcement, the species show postzygotic isolation: laboratory hybrids are viable and fertile, but very poor at fertilizing eggs of both parental species. Yet it is hard to see a selective advantage to reducing sperm binding in sympatry, for sperm containing more divergent bindin alleles would seem to suffer a selective *disadvantage*. (When encountering a heterospecific egg, sperm carrying such alleles would be less likely to form hybrids. But because *some* hybrids can pass on their genes, a reduced ability to hybridize entails a reduced fitness.) The sympatric divergence of bindins may result instead from reinforcement acting on eggs. Such reinforcement, by causing sympatric divergence of molecules on the egg envelope, could drive divergence of the bindins that interact with these molecules. A similar explanation could apply to the observation that a sperm lysin in the mussels *Mytilus galloprovincialis* and *M. trossulus* shows more difference in DNA sequence in sympatry than in allopatry (Springer and Crespi 2004). Sympatry has occurred very recently, almost certainly caused by the movement of mussels in ship ballast during the last 200 years.

SEXUAL SELECTION. Given the important role of sperm competition and selective fertilization within species (Simmons 2001; Birkhead and Pizzari 2002), sexual selection may be a major factor in the evolution of gametic isolation, especially in animals.

Coordinated changes in male and female gametes or reproductive structures can in principle be caused by any form of sexual selection. Good-genes models, however, seem unlikely. It is hard to see how females could recognize and reject genetically inferior sperm or pollen after a single mating or pollination. In some groups, such as *Drosophila*, the genes of a sperm itself are not expressed, and the external proteins among sperm in a single ejaculate are identical, reflecting the father's diploid genotype. Discrimination among gametes is easier when females are pollinated or inseminated by multiple males, but it would seem much harder to judge the fitness of a mate from the biochemistry of his gametes than from the vigor of his courtship or the color of his plumage.

The remarkable success of Rice's (1996) and Holland and Rice's (1999) experiments on *Drosophila melanogaster* has focused attention on antagonistic sexual selection—a result of male-male competition—as a likely cause of gametic isolation. Certainly, such selection must act on the gametic level: it is hard to believe that *Drosophila* seminal-fluid proteins that reduce female lifespan have evolved in any other way (Chapman et al. 1995). Another form of antagonism may be triggered by female avoidance of multiple fertilization. In some species, fertilization of an egg by several sperm from the same species ("polyspermy") produces a lethal zygote (Jaffe and Gould 1985; Howard et al. 1998b). In such cases selection may make eggs less easily penetrated by conspecific sperm. Sperm will then undergo selection to counter these defenses. This yields a continuing "arms race" that causes not only coordinated evolution of conspecific males and females, but also divergence in gametic proteins *among* species that could produce gametic isolation as a byproduct. Gavrilets's (2000a, 2002) models of antagonistic sexual selection show that such isolation can evolve rapidly in both sympatry and allopatry.

If one knows the proteins involved in gametic isolation, one can infer that this barrier evolved via antagonistic sexual selection when the following conditions are met:

- Sperm or pollen compete for access to female gametes.
- There is a known form of antagonism between gametes, such as polyspermy.

- One can demonstrate through statistical analysis of DNA sequences that selection has occurred on both the male and female proteins that interact during fertilization. There are now sensitive statistical methods for such analysis (Yang et al. 2000).
- One can exclude natural selection (e.g., pathogen resistance) that triggers the coevolution of male and female reproductive proteins. One might find, for example, that amino acid substitutions are concentrated at the sites involved in interaction between male and female gametes.

Lysins and VERL proteins in abalones fulfill most of these criteria, and thus were suggested by Galindo et al. (2003) as an example of antagonistic coevolution. Given the mass spawning and the excess of sperm over eggs, sperm competition surely occurs. Polyspermy is also a problem in abalone fertilization. Abalones have electrical blocks to polyspermy that operate after the first sperm penetrates the egg envelope, implying that polyspermy was a problem for females in the past (Jaffe and Gould 1985). These blocks, however, are not completely effective, so that polyspermy can still occur when sperm concentration is high, as can happen in nature (Breen and Adkins 1980; Stephano 1992). Molecular data show that lysins and VERLs have diverged among species by positive selection. What is not clear is whether the rapidly evolving parts of these proteins are those involved in fertilization. Galindo et al. (2003) suggest that antagonistic coevolution is based on polyspermy: new VERL alleles are selected when they slow the dissolution of the egg's vitelline envelope, preventing several sperm from simultaneously penetrating the egg before the polyspermy block is activated. Males could then undergo antagonistic sexual selection to produce lysins that counteract the increasingly refractory VERLs. The one remaining mystery is whether the part of the VERL receptor that has changed by natural selection—the first two repeats in the multi-repeat sequence—is the critical section involved in binding lysin.

Conclusions

Laboratory studies and theory both suggest that gametic isolation can evolve rapidly. Such isolation should be common among closely related taxa, especially geographically isolated populations of a single species. Several studies confirm this prediction, but more work is needed. One tactic is to look for gametic isolation in populations that are deemed conspecific, but have been geographically isolated for relatively long periods.

It is important to recognize that gametic isolation may entail more than evolutionary change of gametes, or even of genes directly involved in reproduction. In *Drosophila*, for example, reproductive proteins evolve faster in males than in females (Swanson et al. 2001a,b, 2004). One might conclude that this evolution resulted from male-male competition. Yet such a conclusion is premature. In animals with internal fertilization, changes in male gametes or ejaculates can act on female traits that are not only far removed from the reproductive tract, but also difficult to detect. Lung and Wolfner (1999), for example, show that some *Drosophila* seminal-fluid proteins that stimulate female egg-laying disappear from the female's reproductive tract within minutes after copulation, making their way into her circulatory system. It is likely that the target of these proteins resides somewhere in the female's nervous system. As with behavioral isolation, male/female coevolution may involve traits that are obvious in males but cryptic in females. Thus one cannot rule out such coevolution even if the evolution of reproductive proteins is faster in males than in females.

7

Postzygotic Isolation

Of all reproductive isolating barriers, postzygotic isolation most starkly reveals the problem posed by speciation to Darwinism: How could natural selection allow—much less favor—the evolution of unfit offspring?

Although Darwin did not strictly adhere to the biological species concept, at least in *The Origin of Species* (Sulloway 1979), he was well aware of this problem. Darwin realized that he asked his readers to believe both that most evolution is due to natural selection and that this evolution routinely yields species that produce unfit hybrids. He also saw that this problem is not trivial: "On the theory of natural selection the case [of hybrid sterility] is especially important, inasmuch as the sterility of hybrids could not possibly be of any advantage to them, and therefore could not have been acquired by the continued preservation of successive profitable degrees of sterility" (Darwin 1859, p. 245). (Although Darwin and his contemporaries were obsessed with hybrid sterility, many forms of postzygotic isolation pose the same problem.) Darwin spent an entire chapter of *The Origin* trying to explain away this dilemma. His solution—that postzygotic isolation, like the failure of grafts to take among plants, is not the direct result of selection but an incidental by-product of "unknown differences" between taxa—is only partly satisfactory. After all, differences causing one plant to reject a graft from another can evolve unopposed by natural selection because grafting does not occur in nature. But hybrids *do* occur in nature. What we want to know is how these kinds of differences, ones that might be selected against, evolve.

The problem of hybrid sterility also perplexed Darwin's contemporaries. Indeed, the problem amounted to whether Darwinism can explain species: if natural selection can account for the evolution of unfit hybrids, all is well; but if it cannot, Darwinism fails at one of its most critical tasks, for the species problem represented perhaps the most serious challenge confronting naturalists of the nineteenth century. T. H. Huxley, for instance, emphasized that Darwin's views left species unexplained, declaring that Darwinism's Achilles' heel lay in "the group of phenomena which I mentioned to you under the name of Hybridism, and which I explained to consist in the sterility of the offspring of certain species when crossed with one another" (Huxley 1896–1902, Vol. II, p. 463). Indeed:

> From the first time I wrote about Darwin's book in the Times . . . until now, it has been obvious to me that this is the weak point of Darwin's doctrine. He *has* shewn that selective breeding is a *vera causa* for morphological species; but he has not yet shewn it is a *vera causa* for physiological species" (Huxley 1901, letter of April 30, 1863).

(By "physiological species," Huxley meant those that produce sterile hybrids.)

The problem only intensified over the following decades. William Bateson, the great champion of Mendelism, repeatedly reminded biologists that Darwinism had provided no solution to the problem of hybrid sterility. Although Bateson, as we will see, played an important part in the problem's ultimate solution, he remained unconvinced by any Darwinian explanation. As late as 1922, he lamented:

> [T]hat particular and essential bit of the theory of evolution which is concerned with the origin and nature of *species* remains utterly mysterious. We no longer feel, as we used to do, that the process of variation now contemporaneously occurring is the beginning of a work which needs merely the element of time for its completion; for even time can not complete that which has not yet begun. The conclusion in which we were brought up, that species are a product of a summation of variations, ignored the chief attribute of species that the product of their crosses is frequently sterile in greater or less degree. . . .
>
> [O]ur knowledge of evolution is incomplete in a vital respect . . . though our faith in evolution stands unshaken, we have no acceptable account of the origin of "species."

He concluded, "When students of other sciences ask us what is now currently believed about the origin of species we have no clear answer to give. Faith has given place to agnosticism " (Bateson 1922).

Most theories of speciation offered over the last century have tried, in one way or another, to resolve Darwin's dilemma. As we will see, some of these attempts have been more successful than others. These attempted solutions fall into two general classes, which correspond to two distinct forms of postzygotic isolation (Table 7.1; see also Table 1.2).

In the first class, hybrids are unfit because they fall between parental niches: Hybrids, although suffering no inherent developmental defect, have an intermediate phenotype that fares poorly in the present environment (Chapter 2). This is *extrinsic* postzygotic isolation. In the second class, hybrids are unfit

Table 7.1 *Classification of postzygotic reproductive isolating barriers*

1. Extrinsic
    A. **Ecological inviability.** Hybrids enjoy normal development but suffer decreased viability, as they cannot find a suitable ecological niche.
    B. **Behavioral sterility.** Hybrids enjoy normal gametogenesis but suffer lowered effective fertility because they cannot obtain mates. Hybrids might present an intermediate courtship behavior or other phenotype that renders them unattractive to individuals of the opposite sex.
2. Intrinsic
    A. **Hybrid inviability.** Hybrids suffer developmental defects causing full or partial inviability.
    B. **Hybrid sterility.**
       1. **Physiological sterility.** Hybrids suffer developmental defects in their reproductive system causing full or partial sterility.
       2. **Behavioral sterility.** Hybrids suffer a neurological or physiological defect that renders them fully or partially incapable of courtship.

because they suffer inherent developmental defects that render them partially or completely inviable or sterile. This is *intrinsic* postzygotic isolation.

Before one can appreciate the solutions to Darwin's dilemma, one must distinguish among the many varieties of postzygotic isolation and understand why it is useful to divide them into extrinsic and intrinsic forms. We thus begin this chapter by describing the various postzygotic isolating barriers and providing examples of each. We then consider how often each barrier acts in nature. Finally, we turn to the genetic bases of postzygotic isolation, considering chromosomal, genic, and endosymbiont-based models. In the next chapter, we focus on Haldane's rule and on the results of recent genetic and molecular studies of postzygotic isolation.

Extrinsic Postzygotic Isolation

Extrinsic postzygotic isolation comes in several forms (see Table 7.1). In one, hybrids suffer low fitness for ecological reasons. This represents an extrinsic version of hybrid inviability: hybrids suffer in the viability component of fitness but for reasons having to do with ecological fit, not developmental defect.

Perhaps the simplest example involves the butterflies, *Heliconius melpomene* and *H. cydno*. These species are broadly sympatric throughout Central America and, although closely related, mimic model species that look profoundly different from each other (one model species is black, red, and yellow, while the other species is black and white). Hybrids, which occasionally form in nature, are non-mimetic intermediates and so presumably suffer low fitness (see Jiggins et al. 2001b; Naisbit et al. 2001).

A better-known example involves two closely related morphs of the stickleback *Gasterosteus aculeatus* that inhabit several lakes in British Columbia (Hatfield and Schluter 1999). These morphs, described briefly in Chapter 2, differ both ecologically and morphologically. The limnetic morph feeds on plankton in open water and has a tapered body with a narrow jaw gape. The benthic morph feeds on small invertebrates in the littoral zone and has a broader body with a wide jaw gape. Hybrids are found in the wild at low frequency (McPhail 1994). Crosses in the laboratory reveal that F_1 hybrids have an intermediate morphology, including jaw structure. Although hybrids are fit in the laboratory (most eggs are fertilized and virtually all hatch), hybrids perform poorly in the wild. Hatfield and Schluter (1999) placed F_1 hybrids versus pure parental individuals in enclosures that forced fish to feed in either of the two parental habitats. In both habitats, F_1 hybrids grew more slowly than did the relevant parental morph (limnetic in open water or benthic in the littoral zone).

Unfortunately, Hatfield and Schluter's design did not allow them to distinguish between two possibilities: (1) hybrids show low fitness because they fall between parental niches (extrinsic isolation), or (2) hybrids suffer inherent fitness problems that are exacerbated in the wild (intrinsic isolation). Though both hypotheses feature a role for the environment, the second has nothing to do with failure to fit parental niches. (This is why we prefer the "extrinsic/intrinsic" nomenclature over Rice and Hostert's 1993 "environment dependent/independent" nomenclature.)

More recently, Rundle (2002) distinguished between these possibilities experimentally. His test used backcross, not F_1, hybrids: if postzygotic isolation is extrinsic, the relative fitnesses of the two reciprocal backcross hybrids (those carrying mostly limnetic genes versus those carrying mostly benthic genes) should depend on the habitat in which they find themselves. In particular, the relative fitnesses of backcross hybrids should *switch* between the two parental habitats: backcross hybrids carrying mostly limnetic genes should perform best in open waters, while backcross hybrids carrying mostly benthic genes should perform best in the littoral zone. This is precisely what Rundle observed. These results thus provide strong evidence for a large extrinsic component to postzygotic isolation. In fact, we can imagine only one further test of this claim: if F_1 hybrids suffer in parental habitats because they possess intermediate morphology, hybrids should enjoy *higher* fitness than pure species individuals in intermediate habitats (if such habitats exist).

Although this prediction has not been tested in the stickleback, it has been tested—and confirmed—in a plant system, the big sagebrush *Artemisia tridentia*. Two subspecies, basin and mountain, form a narrow hybrid zone along an elevational gradient (Wang et al. 1997). The subspecies differ both morphologically and physiologically and hybrids are intermediate for many characters. Transplant/common-garden experiments along the elevational gradient showed that, for several fitness components, hybrids are *more* fit than parentals at the intermediate elevations characterizing hybrid zones. Indeed, a composite fitness index showed that basin plants are most fit in the basin garden, mountain plants are most fit in the mountain garden, and hybrids are most fit in the intermediate garden (Wang et al. 1997; Figure 7.1). Although

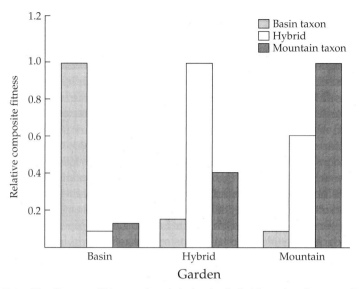

Figure 7.1 The fitness of big sagebrush in basin, hybrid garden (intermediate), and mountain environments. While each parental taxon (basin, mountain) performs best in its "normal" environment, hybrids perform best in the intermediate environment. (From Wang et al. 1997.)

total lifetime fitness was not measured, the results of this study appear inconsistent with intrinsic developmental problems and suggest instead that hybrids are selected against in parental habitats for ecological reasons. (We do not wish, however, to give the impression that hybrids usually perform better in hybrid zones. They do not; e.g., see Bronson et al. 2003 in the chickadee *Poecile* and Barton 1980 in the grasshopper *Podisma*.)

Extrinsic hybrid inviability might also reflect intermediate *behavior*. Helbig (1991), for example, showed that F_1 hybrids between two populations of blackcap birds (*Sylvia atricapilla*) display intermediate migratory behavior. Blackcap populations from Central Europe migrate either in a southeastern (SE) or in a southwestern (SW) direction. Direction of migration appears heritable. Helbig produced F_1 hybrids between SE and SW migrators and showed that they usually try to migrate in an intermediate direction. Such behavior would certainly be deleterious in the wild, sending hybrids into unsuitable wintering grounds.

Hybrids can also suffer another form of extrinsic postzygotic isolation (see Table 7.1): behavioral sterility. Hybrids, in other words, might suffer in the reproductive component of fitness, not because they are physiologically sterile, but because they are behaviorally (or pheromonally, etc.) intermediate between the parental species and, so, are often rejected by potential partners.

A good example occurs between two species of wolf spider, *Schizocosa ocreata* and *S. rovneri*. Though morphogically similar, males of the two species have different courtship behaviors. These differences cause strong prezygotic isolation, although hybrids are occasionally formed in nature. Stratton and Uetz (1986) produced F_1 hybrids that appear generally, though not entirely, fit. Most important, F_1 hybrid male courtship behavior is intermediate between that of the parental species. In fact, hybrids often switch between behaviors typical of each species. Even the sounds produced by hybrid males are intermediate: males sometimes begin singing like one species but finish singing like the other. Although F_1 males court as often as pure-species males, all females—*S. ocreata*, *S. rovneri*, and hybrid F_1—routinely reject them. While we cannot be sure that this rejection reflects intermediate behavior and song, this seems likely. Similar cases are known in birds. Blue-winged and Golden-winged warblers of the genus *Vermivora*, for instance, sometimes hybridize in North America. The parental species differ in color and pattern and F_1 hybrids of both sexes are usually phenotypically intermediate. While F_1 hybrids suffer little or no intrinsic sterility, they leave fewer offspring than do pure-species individuals (Ficken and Ficken 1968a,b). This probably reflects intermediate courtship behavior of hybrids. Finally, two species of green tree frog, *Hyla cinerea* and *H. gratiosa*, differ in male advertisement call. Hybrid males produce calls that differ from both parental species and that are unattractive to females of both species (Hobel and Gerhardt 2003).

Phenomena analogous to extrinsic behavioral sterility occur even in plants, where we loosely consider improper presentation of pollen as behavior. Nilsson (1983), for instance, showed that hybrids between the orchids *Platanthera bifolia* and *P. chlorantha* are effectively male sterile. Sterility does not result, however, from failed pollen production but from improper presentation of pollen to pollinators due to intermediate flower morphology (see Chapter 2).

Finally, we note two potential problems with our classification of postzygotic isolation. The first is that extrinsic isolation is not a fixed phenomenon, but one that might appear and disappear as ecological conditions change. In certain years, the environment may be such that hybrids perform poorly because they fall between available niches, while in other years the environment may be such that hybrids perform well because they fit an available niche. Grant and Grant (1993), for example, showed that hybrids between two species of Darwin's finches that co-occur in the Galápagos, *Geospiza fortis* and *G. fuliginosa*, almost never survived before the El Nino event in 1982–1983. After the event, however, hybrids survived well; indeed, hybrids appeared to briefly enjoy higher fitness than the parental species. This temporal variation in hybrid fitness seemed to reflect their varying fit to the environment: because hybrids are of intermediate size, they consume seeds of different size than do the parental species. The spectrum of seed sizes available on the Galápagos shifted dramatically following the El Nino event.

Second, it is sometimes hard to distinguish extrinsic from intrinsic postzygotic isolation. A striking example involves two subspecies of house mouse, *Mus musculus musculus* and *M. m. domesticus* that form a hybrid zone in Europe.

Mice from at least two sites in the hybrid zone suffer much greater loads of intestinal parasites than do nearby pure subspecific individuals. Several lines of evidence suggest that these increased parasite loads reflect genetically based reduced resistance to parasitism in hybrids (Sage et al. 1986; Moulia et al. 1991; see Whitham 1989, for an analogous case in plants). Similarly, Bert et al. (1993) describe a hybrid zone between two species of clam of the genus *Mercenaria*. Tumors caused by a microbial infection occur at high frequency in the hybrid zone, with hybrids afflicted at much higher rates than pure-species individuals. Postzygotic isolation in both of these cases seems simultaneously extrinsic and intrinsic: extrinsic because it depends on parasites or infectious agents in the environment, and intrinsic because hybrids are inherently more vulnerable to these agents. We do not believe, however, that such cases invalidate the usual—and typically useful—distinction between extrinsic and intrinsic isolation.

Intrinsic Postzygotic Isolation

Intrinsic postzygotic isolation has been more thoroughly studied than extrinsic. Intrinsic isolation is also usually divided into two classes: hybrid inviability and hybrid sterility (see Table 7.1).

Examples of hybrid inviability abound. One of the best known involves the cross of *Drosophila melanogaster* with its sister species *D. simulans*, first studied by Sturtevant in 1920. When *D. melanogaster* females are crossed with *D. simulans* males, only hybrid daughters appear; hybrid males die at the larval to pupal transition. The reciprocal cross produces only hybrid sons and hybrid females die as embryos (Sturtevant 1920). As this case suggests, hybrid inviability can appear at various stages of development and affect one sex only or both. Intrinsic hybrid inviability might also be complete or partial.

Intrinsic hybrid sterility itself takes two forms (see Table 7.1). The better known is physiological, in which hybrids fail to produce functional gametes. The classic example—discussed even by Aristotle—is the mule, a hybrid between a female donkey and a male horse. This example has, however, misled biologists in one respect. Because mules have high viability (they are an equally classic example of heterosis for vigor), the impression has arisen that hybrid sterility might reflect the greater inherent sensitivity of gonadal tissue to hybridization (i.e., the gonads may represent the first tissue to collapse under the stresses of a hybrid genome). As Dobzhansky (1937b, p. 259) emphasized, this impression is false. Lack of hybrid vigor is not invariably accompanied by sterility. Instead, many cases are known in which hybrids are partially inviable but fully fertile (see Bock 1984 for examples from *Drosophila*). Any contrary impression likely reflects an observational bias: we can observe the sterility of live hybrids, but we cannot observe the fertility of dead hybrids. Physiological hybrid sterility has been, for both historical and practical reasons, the focus of genetic analyses of reproductive isolation, especially in *Drosophila*.

The second form of intrinsic hybrid sterility is behavioral, i.e., hybrids fail to reproduce because they suffer a neurological (or pheromonal, etc.) defect that renders them incapable of effective courtship, although they produce functional gametes. (The constrast with extrinsic behavioral sterility is subtle: in the extrinsic form, hybrids present *intermediate* behaviors or pheromones, etc.; in the intrinsic form, they present ablated or disrupted behaviors or pheromones.) In *Drosophila pseudoobscura* and *D. persimilis*, males of each species court females of either species with equal intensity. But F_1 males having a *D. persimilis* mother do not court vigorously (Noor 1997a). This phenomenon cannot be fully explained by an overall decrease in robustness, but appears partly specific to courtship. (F_1 males show some reduction in overall locomotor activity, but a much larger decrease in courtship intensity.) Behavioral sterility also appears among F_2 hybrid males of the parasitic wasps *Nasonia vitripennis* and *N. giraulti* (P. O'Hara and J. H. Werren, pers. comm.). Some hybrid F_2 males fail to orient during courtship and so fail to find females; among those that do orient, other disruptions occur in subsequent steps of courtship. Similar examples occur in Neotropical butterflies of the genus *Anartia* (Davies et al. 1997), and in the fall armyworm *Spodoptera* (Pashley and Martin 1987).

A more dramatic example is seen in hybrids between two parrot species, *Agapornis roseicollis* and *A. personata fischeri* (Buckley 1969). Hybrid females respond very slowly to male courtship displays. They are also largely incapable of building a nest: while *Agapornis* builds nests in the laboratory by cutting strips of paper and either tucking them into their rump feathers or carrying them with their bills to the new nest, hybrids cut strips of the wrong size and tuck them improperly, almost always dropping them before reaching the nest. Worse, on the rare occasions when females lay eggs, they fail to incubate them.

Given that intrinsic behavioral sterility involves a physiological or neurological defect, it stands to reason that this isolating barrier might share certain features with more narrowly defined "classical" postzygotic isolation. One such pattern is Haldane's rule, the preferential sterility or inviability of hybrids of the heterogametic (XY or ZW) sex (Chapter 8). It is interesting that all of the above examples of intrinsic behavioral sterility involve hybrids of the heterogametic sex (or in the haplodiploid *Nasonia*, the haploid sex), suggesting that intrinsic behavioral sterility might well conform to Haldane's rule.

Many early writers identified another form of intrinsic postzygotic isolation: "hybrid breakdown," the loss of fitness among backcross, F_2, or other later generation hybrids given high F_1 hybrid fitness. We find this category confusing. Hybrid breakdown necessarily involves either hybrid inviability or hybrid sterility and so does not represent an additional isolating barrier. By "hybrid inviability" and "hybrid sterility," we include cases in which hybrids of *any* generation are affected; indeed, we will see that later-generation hybrids often suffer more severe fitness problems than F_1 hybrids (Chapter 8).

The Frequency of Various Forms of Postzygotic Isolation

Authors often emphasize a particular isolating barrier when discussing a hybridization although other barriers are also present. Thus, while the wolf spider hybrids discussed above show extrinsic behavioral sterility, they also show intrinsic fitness problems (Stratton and Uetz 1986). Similarly, while *Drosophila pseudoobscura–D. persimilis* hybrid males suffer behavioral sterility, they also suffer physiological sterility, and in fact produce no motile sperm (Dobzhansky 1936). And while the parrot hybrids discussed above show behavioral sterility, they also suffer nearly complete physiological sterility, as well as severe viability problems (Buckley 1969).

It has become fashionable to suggest that extrinsic, and especially ecological, postzygotic isolation is more common or more important than intrinsic in nature (see Chapter 3). This might well be true. But at present, such assertions rest more on intuition than data. When one considers the kind of detailed biological information—especially ecological—required to accurately ascertain the relative effects of extrinsic and intrinsic postzygotic isolation in a large group of species, it becomes clear that no strong conclusions are currently possible. Assertions to the contrary are usually based on one or a few favorite hybridizations, not on large and careful comparative contrasts.

The Evolution of Extrinsic versus Intrinsic Postzygotic Isolation

As Darwin's confusion revealed, it is not obvious a priori how postzygotic isolation can evolve. But as there are two classes of postzygotic isolating barriers, extrinsic and intrinsic, it should not be surprising that there are two general solutions to the dilemma. Indeed, the fact that extrinsic and intrinsic isolation involve distinct explanations provides one of the best reasons for carving up postzygotic isolation in this way.

As Hatfield and Schluter (1999) and others have emphasized, extrinsic postzygotic isolation may be a direct result of adaptive evolution. As an ancestral population radiates into different ecological niches, we have no guarantee that hybrids, which are often phenotypically intermediate between pure species, will fare well in either parental habitat. Note that the intermediate hybrid phenotype may well have existed in the history of these lineages; indeed it may have represented a step in the evolution of the present species. But this does not ensure that the intermediate phenotype functions well in the present environments: the present environments may not, after all, have existed in the distant past. With extrinsic isolation, then, adaptation itself may act as a direct engine of speciation (Schluter 2000; see also our extended discussions of sympatric speciation and ecological reproductive isolation in Chapters 4 and 5).

Extrinsic isolation can also evolve under strictly additive gene action. One need not invoke epistatic interactions between loci to explain either the phenotypic divergence of parental species or the production of phenotypically intermediate hybrids. Indeed, extrinsic postzygotic isolation might even be underlain by a single locus in which one species is *AA*, the other *aa*, and the unfit hybrid *Aa*. As we will see, such additivity is generally impossible with intrinsic postzygotic isolation.

The evolution of intrinsic postzygotic isolation poses a harder problem—and it was this problem that Darwin and his contemporaries labored over. With intrinsic isolation, hybrids suffer low fitness more or less independently of the environment: it is not that hybrids perform worse than the parental species in the present environment; they perform worse in *every* environment. The problem, then, is to explain how two populations separated by a "fitness valley" can evolve from a common ancestor without either lineage passing through the valley. There have been many attempts to solve this problem.

These attempts differ according to the presumed genetic basis of intrinsic isolation: one kind of solution has been offered if isolation involves chromosomal rearrangements, another if it involves genic incompatibilities, and so on. We must therefore consider these different genetic modes of speciation to determine how successfully each overcomes Darwin's dilemma. As we will see, the most plausible path to the evolution of intrinsic isolation involves epistasis among the genes causing reproductive isolation. We emphasize that our focus on intrinsic postzygotic isolation does not reflect its greater importance in nature, but the fact that it has been the object of a tremendous amount of experimental and theoretical work.

Genetic Modes of Intrinsic Postzygotic Isolation

In intrinsic postzygotic isolation, something goes awry in hybrids. Moreover, this something has a genetic basis—it consistently occurs in hybrids but not in pure species individuals. Evolutionists have identified four kinds of genetic problems as likely causes of these hybrid difficulties: different ploidy levels (i.e., different numbers of chromosome sets); different chromosomal rearrangements; different alleles that do not function together in hybrids; or infection by different endosymbionts. While each of these mechanisms has enjoyed its advocates, some play a more important part in speciation than others. Polyploidy, for instance, plays a major role in plant, but not animal, speciation. Because of its many unique features, including this unusual phylogenetic distribution, we defer discussion of polyploidy until Chapter 9. In this chapter, we review three genetic modes of speciation that have been proposed to act in a wide variety of organisms—chromosomal, genic, and endosymbiont-based.

Chromosomal speciation: theory

Chromosomal rearrangements could play a role in speciation in different ways. Structural changes in chromosomes might directly cause reproductive isolation. Alternatively, structural changes in chromosomes might merely tie up sets of *genes* that cause reproductive isolation (Noor et al. 2001b; Rieseberg 2001; Navarro and Barton 2003). Only the first scenario represents true chromosomal speciation. We consider the second scenario in the next chapter on the genetics of postzygotic isolation.

Many species, of course, do differ by chromosomal rearrangements. Moreover, some of these rearrangements, particularly certain inversions and translocations, can cause semisterility *within* species. For instance, individuals heterozygous for a pericentric inversion (one that includes the centromere) are often effectively semisterile because recombination within the inverted region yields gametes that suffer duplications or deficiencies. Because rearrangements only decrease the effective fertility of chromosomal *heterozygotes*, it is tempting to think that such rearrangements might cause hybrid sterility: once a population becomes fixed for a new chromosome arrangement, it is partially reproductively isolated from other populations. Indeed, plant workers have long known that one can produce reproductively isolated lines from a common ancestor by X-raying the ancestor and then selecting lines homozygous for the required structural rearrangements (Gerassimova 1939). Several workers have offered different varieties of chromosomal speciation theory, and the older literature is filled with discussion of the differences between, say, Wallace's (1959) triad model, Lewis's (1962, 1966) saltatory model, and White's (1969, 1978) stasipatric models (for reviews, see Spirito 2000 and Rieseberg 2001). Each of these verbal models was inspired by a particular biological system: the fly *Drosophila*, the angiosperm *Clarkia*, and the grasshopper *Vandiemenella*, respectively. Because White's model received the most attention and because similar ideas have recently resurfaced (King 1993), we describe it here. But the differences between the above models are of little significance for present purposes: most of the problems afflicting chromosomal speciation affect all of these models.

White's stasipatric model posits that chromosomal evolution causes speciation among taxa that remain in geographic contact. The model considers a scenario in which new underdominant rearrangements (those that are unfit as heterozygotes) arise and become fixed *within* the range of a widely distributed species. According to White, daughter species (and there may be several) typically appear in the central parts of an ancestral species' range and expand their ranges at the expense of the ancestor. Hybridization among daughter species along often-narrow hybrid zones leads to partially or fully sterile F_1 hybrids, reflecting a disruption of meiosis among rearrangement heterozygotes.

The stasipatric model was inspired by White's studies of the *viatica* group of Australian morabine grasshoppers *Vandiemenella*, which includes at least seven species (White 1978, pp. 178–186). These species are wingless, have partially overlapping geographic distributions, and differ by a number of rearrangements, including a pericentric inversion and several centric fusions (in which nonhomologous chromosomes are attached to a common centromere) (White 1969, 1978). White's biogeographic and karyotypic studies led him to conclude that these species arose within the range of a hypothetical *proto-viatica* ancestor in southern Australia and that the appearance of underdominant chromosomal rearrangements within small local populations played a causal role in speciation. Each of these nascent species then extended its range in a wave of expansion along "an advancing frontier which is the narrow zone of overlap" (White 1969, p. 89) *despite* the underdominance that, by hypothesis, appears in this zone of overlap. Parts of this story have been contested. Key (1968), for instance, questioned whether these taxa represent true species, as many produce hybrids showing few meiotic irregularities. Mayr (1969) and Key (1968) further argued that divergence in *viatica* was likely allopatric. Here, however, we focus on the larger theoretical and empirical problems besetting the attempt to explain speciation by underdominant rearrangements in any group.

The most serious problem is that chromosomal speciation runs squarely into Darwin's dilemma. Because a new rearrangement arises as a heterozygote, which is presumably semisterile, how can it increase in frequency? Indeed, as Sturtevant (1938) and Muller (1940) emphasized long ago, the greater the reproductive isolation ultimately caused by a rearrangement, the more it is selected against when it appears. This problem is particularly serious because new arrangements are selected against until they reach intermediate frequencies (Futuyma and Mayer 1980).

The traditional response to this objection has been that new rearrangements might be fixed within small populations, where genetic drift can overwhelm natural selection. The solution to Darwin's dilemma is thus stochastic, featuring a rapid excursion through an adaptive valley. Unfortunately, population genetic theory shows that the effective population size required to yield strong isolation in a reasonable time under this model is exceedingly small ($N_e < 50$) (Walsh 1982). This reflects the fact that the probability of fixation of an underdominant rearrangement falls off very rapidly with population size. Indeed, Lande (1979) showed that this probability is

$$\Pi \approx \frac{1}{N_a} e^{-N_e s} \sqrt{N_e s / \pi}$$

where N_a is the actual population size, N_e is the effective population size, and s is the selection coefficient against rearrangement heterozygotes. Thus the probability of fixation falls nearly exponentially with population size. It therefore seems likely that populations of the requisite size would go extinct or fuse with others long before they fixed an underdominant arrangement. Theory also shows that isolation is unlikely to result from a few rearrangements of large effect as suggested by White. Rather, a given amount of reproductive isolation is more likely to involve many rearrangements, each having a small effect on heterozygote fitness (Walsh 1982).

There are at least two ways to overcome these objections, both suggested by White (1978, pp. 174, 183). First, heterozygote disadvantage might be offset by an advantage among *homozygotes* for the new rearrangement. But theory shows that this makes little difference unless the homozygous advantage is very large (Hedrick 1981; Walsh 1982). The reason is that rearrangements are most vulnerable to loss when rare; but when rare, rearrangements occur almost exclusively in heterozygotes, and homozygous advantage is irrelevant. Alternatively, underdominant rearrangements might be swept to fixation by meiotic drive (see also King 1993). Calculations show that such driven arrangements must arise frequently to yield strong isolation in reasonable time periods (Futuyma and Mayer 1980; Walsh 1982). Although this once seemed unlikely, there is now good evidence for (normally cryptic) meiotic drive factors separating several closely related species (Chapter 8). Although we do not know if these meiotic drive factors were expressed in the history of any pure species (or merely represent a hybrid pathology), we cannot rule out the possibility that meiotic drive factors do commonly appear within species. Indeed, there is now good evidence that many chromosomal rearrangements cause meiotic drive. Centric fusions, for instance—which arise at very high rates in human meiosis (Hamerton et al. 1975)—often cause segregation distortion in females, apparently reflecting the presence of homologous chromosomes having unequal numbers of centromeres at meiosis (de Villena and Sapienza 2001; Henikoff et al. 2001; see Chapter 8).

In summary, although theory suggests that chromosomal speciation is difficult, we cannot rule it out—especially if effective population sizes are small or meiotic drive is common. In the end, the importance of chromosomal speciation must be determined empirically.

Chromosomal speciation: data

There is no question that chromosomal rearrangements sometimes contribute to reproductive isolation. This is essentially proved by recent work in yeast. Several species of the genus *Saccharomyces* differ by reciprocal translocations, while the chromosomes of other species do not (they are "colinear"). Delneri et al. (2003) engineered strains of *S. cerevisiae* that were either colinear or rearranged with respect to two other species. They could thus compare the fertility of F_1 species hybrids that were heterozygous for naturally occurring rearrangements with the fertility of hybrids that were colinear in an otherwise identical genetic background. They found that fertility was generally higher in colinear hybrids. These results leave little doubt that naturally occurring rearrangements can, and sometimes do, lower hybrid fertility.

It does not follow, however, that chromosomal speciation is common. Indeed, even in yeast, colinear species sometimes produce sterile hybrids (Delneri et al. 2003). To assess the importance of chromosomal speciation in nature, we find it useful to discuss animals and plants separately.

ANIMALS. Several facts pose immediate problems for the idea that chromosomal speciation is common in animals. The first is so obvious that it is rarely noted. Chromosomal speciation cannot explain the *inviability* of F_1 hybrids; it is solely a theory of F_1 hybrid *sterility*. This also means that, to the extent that F_1 hybrid sterility and inviability share certain patterns, these similarities are not easily explained by chromosomes. Chapter 8 will show that hybrid inviability and sterility in animals *do* show such similarities; most strikingly, both obey Haldane's rule. Second, even if hybrid sterility is caused by chromosomal differences, why are male hybrids (at least in taxa in which males are the XY sex) sterile so much more often than female hybrids? Why do chromosomal rearrangements only disrupt hybrid male meiosis? This is not, after all, how heterozygous rearrangements act *within* many species. Indeed, rearrangements typically show the opposite behavior in *Drosophila*, "sterilizing" only females, as male flies do not undergo recombination and so do not suffer deficiencies and duplications as a result of exchange within rearrangement heterozygotes. (There is an exception: X-autosomal translocations are often male sterile in both *Drosophila* and mammals, for reasons that remain poorly understood [Ashburner 1989, p. 566; Forejt 1996].)

Third, chromosomal speciation cannot explain one of the most striking patterns characterizing the *genetics* of postzygotic isolation in animals: the X chromosome has a disproportionately large effect on hybrid male sterility (Chapter 8). The problem is that males in *Drosophila*—the genus from which almost all the relevant data derive—are *hemizygous* for the X chromosome, making heterozygosity for chromosome rearrangements impossible. (A similar argument applies to haplodiploid species, like the wasp *Nasonia*. F_2 hybrid males are sometimes less fit than F_2 hybrid females [Breeuwer and Werren 1995]; but these males are haploid and so cannot be heterozygous for chromosomal rearrangements.) Fourth, hybrid male sterility often involves *postmeiotic* problems. At least in flies, hybrid sterility often results from problems during spermiogenesis that leave hybrids with immotile sperm (Chapter 8). This is also true of induced genic (but not chromosomal) sterility mutations within *Drosophila* species (Lindsley and Tokuyasu 1980; Perez et al. 1993). Finally, several factors having large effects on hybrid fitness have recently been mapped, and some have been molecularly characterized (Chapter 8). These factors have no obvious association with chromosomal rearrangements. In fact, some cause hybrid problems only when *homozygous*. Indeed, as we will see, there is growing evidence from a variety of taxa that introgressions of foreign genetic material typically have greater fitness effects when homozygous than heterozygous, an observation that is completely at odds with the theory of chromosomal speciation. Taken together, these facts suggest that chromosomal speciation is not common in animals.

White, however, argued that two observations support the opposite conclusion. The first is that most species differ by chromosomal rearrangements (White 1969, p. 77). The problem is that this numerology may be misleading. As Futuyma and Mayer (1980) emphasized, chromosomal differences can accumulate *after* speciation (see also Carson 1981). The point is that many, if not most, of the taxa that White considered are old; they have been full species for a long time. It is remarkable that no one has systematically studied the number of rearrangement differences between taxa as a function of divergence time (measured molecularly). While White's theory requires that rearrangements appear before or at speciation, it seems entirely possible that they instead accumulate in a clock-like manner, with many arising long after strong (non-chromosomal) isolation has evolved. Indeed, a casual glance at the *Drosophila* literature suggests that young species pairs are less likely to differ by chromosomal rearrangements (though reproductively isolated) than are older species pairs. A rigorous comparative analysis of this problem is needed.

Second, White emphasized that species hybrids often suffer meiotic problems in which chromosomes fail to pair properly. This is undoubtedly true. But this does not mean that chromosomal speciation occurred. As Dobzhansky (1937b) emphasized, pairing problems in hybrids can have two causes: structural *or* genic incompatibilities between species. Meiosis, after all, requires the coordinated action of many genes, and incompatibilities between these genes in hybrids might derail chromosome pairing. Fortunately, this issue can be resolved by experiment.

The most direct test was suggested and performed by Dobzhansky (1933). Consider the fate of a *tetraploid* hybrid between two species whose normally diploid hybrids suffer chromosome pairing problems and sterility. In such $4x$ hybrid individuals (or cells), all chromosomes have a conspecific pairing partner. Thus, if hybrid pairing problems reflect structural chromosomal differences, pairing should be restored in tetraploid hybrids. But if hybrid pairing problems reflect genic incompatibilities, pairing will not necessarily improve in tetraploid hybrids. This test has been performed only once in animals: Dobzhansky (1933) analyzed hybrids of *Drosophila pseudobscura* and *D. persimilis*, which differ by at least four inversions and which produce sterile male hybrids. He found that meiotic pairing in diploid hybrids is abnormal and that univalents (unpaired chromosomes) are common. But he also found that islands of *tetraploid* spermatocytes occur in hybrid testes and that univalents are just as common in these tetraploid cells as in diploid cells. Therefore, the meiotic problems of *D. pseudobscura–D. persimilis* hybrids do not have a chromosomal basis.

Recent work also suggests *why* some natural chromosome rearrangements are unlikely to cause hybrid sterility in animals. Although there is no doubt that many new pericentric inversions are underdominant, there is good reason to think that those that become fixed are not a random sample of newly arising inversions. Recent studies show that, contrary to expectation, many (ca. 40%) pericentric inversions have no detectable effect on fitness when heterozygous. In both mammals and *Drosophila*, females that are heterozygous for even large pericentric inversions sometimes have as many offspring as do homokaryotypic controls (Sites and Moritz 1987; Nachman and Myers 1989; Coyne et al. 1991a, 1993). This high fitness appears to reflect a lack of crossing over in the inverted region. The absence of crossing over does not reflect genes within the inversions that suppress recombination as *homozygotes* for the inversions show normal levels of recombination (Coyne et al. 1991a). Rather, inverted chromosomes apparently do not form loops when heterozygous; instead they pair "straight" (with nonhomologous regions aligned), precluding recombination. Consequently, no aneuploid gametes are produced and no effective sterility results (see also Hale 1986 and Reed et al. 1992). These surprising findings are of considerable import for the theory of chromosomal speciation: it can no longer be assumed that pericentric inversions will cause sterility when heterozygous in species hybrids. Indeed, John (1981) has argued that there are *no* cases—including those highlighted by White—in which fixed pericentric inversions have been shown to cause reproductive isolation in animals.

Although the above data provide little support for chromosomal speciation in animals, there are several reasons for thinking that it may occur occasionally. The first is that chromosomal speciation seems more likely in mammals than in, say, invertebrates. The reason is that exchange between homologous chromosomes is required in eutherian mammals for the completion of meiosis. Although this was suggested as long ago as 1937 by Darlington, it is now clear that exchange—even between the pseudoautosomal regions of the X and Y chromosomes—is obligate (Burgoyne 1986; Rouyer et al. 1986). Without proper pairing and subsequent recombination, meiosis in mammals is typically disrupted, giving rise to inviable germ cells (Searle 1993). Sterility does not result, therefore, only from production of aneuploid gametes but also from germ cell death. In most invertebrates, on the other hand, chiasmata formation is not required for the completion of meiosis; indeed, the heterogametic sex is often achiasmate (White 1973; Gethmann 1988). Thus, all else being equal, chromosomal rearrangements might cause greater hybrid sterility in mammals than in invertebrates.

Also, certain types of chromosomal rearrangements might cause meiotic problems in hybrids without causing severe problems *within* species. Centric fusions, also known as Robertsonian fusions or translocations, occur when two acrocentric chromosomes (those with their centromeres near the tip of the chromosome) fuse to form a single metacentric chromosome (one with its centromere near the center of the chromosome). Centric fusions appear to have a small effect on heterozygous fitness, since most chromosomes segregate properly at meiosis (Barton 1980; Capanna 1982; Baker and Bickham 1986; Searle 1988; Hauffe and Searle 1998). Such fusions might therefore accumulate in populations with little opposition from natural selection. This accumulation seems more likely in taxa, like mammals, with small effective population sizes, since selection is weaker in small populations. Centric fusions could, then, cause hybrid sterility in two ways. First, although heterozygotes for single fusions may enjoy reasonably high fertility, heterozygotes for *many* fusions suffer considerable sterility (see below). Thus, two allopatric populations could gradually accumulate different centric fusions and, upon secondary contact, produce hybrids that are nearly, or completely, sterile. Second, populations might come to differ by "monobrachial fusions," in which different centric fusions involve some, but not all, of the same chromosomes (Figure 7.2). One allopatric population, for instance, might experience a fusion of chromosomes 1 and 2, while

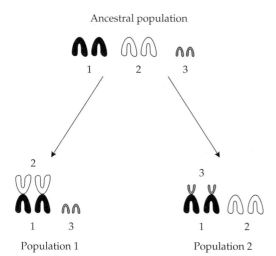

Figure 7.2 The evolution of a monobrachial fusion. One allopatric population experiences a centric fusion between chromosomes 1 and 2, and the other allopatric population experiences a centric fusion between chromosomes 1 and 3. Because chromosome 1 is involved in both fusions, hybrids between the populations suffer complex multivalents at meiosis and thus low fertility. (From Baker and Bickham 1986.)

another population might experience a fusion of chromosomes 1 and 3. Because these fusions share a chromosome, any hybrid between the populations would form quadrivalents or more complex multivalents at meiosis, causing improper chromosome segregation and low fertility (Capanna 1982; Baker and Bickham 1986; King 1993, pp. 234–238). (As we will see, speciation by monobrachial fusions represents a chromosomal form of the "Dobzhansky–Muller" model.)

There are several reasons for thinking that centric fusions might play a role in animal, and especially mammal, speciation. For one thing, they are common, arising spontaneously at a rate of 10^{-4} per gamete per generation in humans (Lande 1979), and at even higher rates in some house mice (Nachman and Searle 1995). Not surprisingly, centric fusions are often polymorphic in mammalian populations (Nachman and Searle 1995; Baker and Bickham 1986), and also appear to be the most common type of rearrangement fixed between mammalian species: monobrachial fusions, for example, are known between species of bats, mice, shrews, and rats, among others (Baker and Bickham 1986; Moritz 1986; Searle 1988).

The idea that centric fusions cause reproductive isolation in mammals has considerable support (reviewed in King 1993, and Searle 1993). Although good data are available in a number of systems (e.g., in the shrew *Sorex araneus* [Searle 1988]), to keep our treatment manageable we will only describe the best-studied case, that of the European house mouse, *Mus musculus domesticus*. The standard (and predominant) race of this human commensal has a diploid number of $2n = 40$, with all chromosomes acrocentric. But at least 40 "races" of house mouse in Europe and North Africa have a lower chromosome number; indeed, some races have numbers as low as $2n = 22$. This reduced chromosome number reflects the accumulation of centric fusions (Searle 1993; Nachman and Searle 1995; Piálek et al. 2001). Molecular evidence also suggests that many of these chromosomal races—which are highly localized geographically, often being restricted to single human villages—are young, having arisen within the last 10,000 years (Nachman and Searle 1995).

M. m. domesticus forms a number of hybrid zones throughout Europe and North Africa, with contact occurring either between the all-acrocentric race and a centric fusion race, or between different centric fusion races (Searle 1993). Although early laboratory work showed that hybrids heterozygous for centric fusions suffer frequent meiotic problems and sterility (Gropp and Winking 1981), this finding proved hard to interpret. The difficulty is that hybrids were typically produced by introgressing wild chromosomes into laboratory stocks or by crossing geographically distant populations. As Sites and Moritz (1987), Searle (1993), Everett et al. (1996), and others argued, hybrid sterility might therefore reflect genic, not structural, incompatibilities. More recent work has focused on *naturally* occurring hybrids (either taken from the field or produced in laboratory crosses between geographically contiguous races). Although this work has yielded conflicting findings about the fitness effects of one or a few centric fusions (Capanna et al. 1976; Britton-Davidian et al. 1990; King 1993, pp. 135–142; Searle 1993; Hauffe and Searle 1998; Castiglia and Capanna 2000), one result is clear: hybrids heterozygous for many centric fusions—and especially monobrachial ones—suffer significantly reduced fertility (King 1993, pp. 141–144; Searle 1993; Hauffe and Searle 1998; Castiglia and Capanna 2000; Piálek et al. 2001). We believe that these results provide strong support for the role of chromosomal rearrangements in reproductive isolation between mouse races.

This is not to say, however, that chromosomal rearrangements are the sole cause of sterility in mouse hybrids, much less in mammalian hybrids generally. Said et al. (1993), for instance, showed that F_1 hybrid males between a Tunisian race of *M. m. domesticus* that carries many centric fusions and the neighboring all-acrocentric race produce fewer offspring than do pure species individuals. Although this sterility appears to partly reflect aneuploidy among hybrid sperm, a spermatogenic failure phenotype that appears in hybrids does *not* correlate with structural heterozygosity. Instead, the latter defect seems to have a genic basis. Similarly, Borodin et al. (1998) studied two races of house musk shrews (*Suncus murinus*) that differ by many centric fusions and Y chromosome rearrangements, and that produce sterile male hybrids. Several lines of evidence suggest that hybrid sterility is not caused by structural problems: among later-generation hybrids, heterozygotes for all chromosome rearrangements appear in both fertile and sterile hybrid males; there is no correlation between number of heterozygous regions in a hybrid individual and its fertility; and pairing is often disrupted in both heterozygous *and* homozygous chromosome regions. Finally, fine-scale genetic analyses of the *Mus musculus musculus*–*M. m. domesticus* and of the *M. m. domesticus*–*M. spretus* mouse hybridizations have localized what appear to be ordinary *genes* that cause the sterility of male hybrids: *Hst-1* in the former species pair (Forejt and Iványi 1975) and *Hst-3*, *Hst-4*, *Hst-5*, and *Hst-6* in the latter (reviewed in Forejt 1996).

In summary, reproductive isolation in mammals may sometimes, though certainly not always, have a chromosomal basis. There is particularly good evidence for the role of centric fusions in hybrid sterility between races of house mice. Given that newly arising centric fusions may be slightly underdominant, hybrid sterility in mammals may evolve with the aid of genetic drift (see Chapter 11). It is far from clear, however, if this situation is common in animals generally; indeed, we know of no compelling evidence for chromosomal speciation in animals other than mammals.

PLANTS.   Although one of the key tests of chromosomal speciation—the chromosome doubling test—has been performed only once in animals, it has been performed many times in plants. Surprisingly, the results differ from that in animals. Although artificial doubling of chromosome number fails to restore chromosome pairing in some tetraploid plant hybrids (e.g., in *Lycopersicum*, *Aegilops*, and *Triticum* [Stebbins 1950, pp. 218–227]; Stebbins [1958, p. 171]), it *does* restore proper chromosome pairing in other species hybrids (Stebbins 1950, pp. 218–227; Stebbins 1958, pp. 174–175).

One need not, however, perform the chromosome-doubling test in the greenhouse. Instead, it has been performed many times in nature. Many fertile allopolyploid species (tetraploid, etc. species of hybrid origin) exist in cases in which the relevant *diploid* hybrids are sterile. Indeed, Clausen, Keck and Hiesey coined the term "amphiploid" to describe such cases. These cases appear reasonably common (e.g., in *Tragopogon* and *Gilia* [Stebbins 1950, pp. 218–227; see especially, Stebbins 1958, pp. 174–175]). In a recent literature review of many species crosses, Ramsey and Schemske (2002) found that, while diploid species hybrids have a mean pollen fertility of only 17%, newly formed tetraploid hybrids between the same species enjoy a mean pollen fertility of 71%, a highly significant difference. There can be little doubt, then, that tetraploidy often restores hybrid fertility in plants.

Although these findings are consistent with chromosomal sterility, there are two complications. Both involve *genic* problems that might also give rise to a pattern of sterile diploid, but fertile tetraploid, hybrids. The first complication involves haploid gene expression in plant gametes (Stebbins 1958, p. 167). (There is little haploid expression in animal gametes.) Under certain simplifying assumptions, one can show that genic incompatibilities in gametes will cause greater gametic inviability when gametes come from diploid than tetraploid hybrids (Stebbins 1958). The fact that chromosome doubling rescues fertility in taxa having gametic expression (plants) but not in taxa lacking gametic expression (animals) might therefore reflect the action of (genic) gametic incompatibilities in plants. Indeed, some genetic evidence consistent with this hypothesis has appeared in yeast, although alternative interpretations are possible (Greig et al. 2002a,b).

But more careful tests of Stebbins's hypothesis can be performed in plants. The haploid expression idea, after all, can explain the rescue of hybrid *fertility* but not the rescue of hybrid chromosome *pairing* per se (since pairing occurs in the diploid, not haploid, phase). And it is clear that in many cases in which natural allotetraploids enjoy both fertility and regular pairing, diploid hybrids suffer both sterility *and* pairing problems. Indeed, Stebbins (1950, p. 327) concluded that the typical allopolyploid derives from hybridization between taxa whose chromosomes fail to pair properly in diploid hybrids.

The second complication is that sequence divergence might disrupt fertility in diploid, but not tetraploid, hybrids. Indeed there is some evidence from yeast (though not from plants) that the mismatch repair (MMR) system may contribute to hybrid sterility (Hunter et al. 1996). (This system corrects mismatched bases after DNA replication. When a mismatch is detected, the MMR system blocks recombination, causing chromosome loss and aneuploidy in gametes.) Although *Saccharomyces cerevisiae* and *S. paradoxus* produce mostly sterile F_1 hybrids, fertility increases slightly if the MMR system is knocked out (Hunter et al. 1996). But MMR-based sterility would presumably *not* appear in tetraploid hybrids, as all chromosomes have perfect pairing partners. While it is hard to exclude the possibility that MMR effects contribute to the pattern of diploid sterility/tetraploid fertility in plants, this seems doubtful to us. As Sniegowski (1998) emphasized, *S. cerevisiae* and *S. paradoxus* show 30% sequence divergence; most crossable plant species, on the other hand, show far less sequence divergence and MMR effects would probably be small. (Greig et al. 2003 suggest that MMR knockouts also increase the fertility of *population* hybrids, but their data are not entirely convincing.) In any case, we know of no evidence that the MMR system causes postzygotic isolation in metazoans.

For now, then, we interpret findings of diploid sterility/tetraploid fertility in plants in a traditional way: as evidence for the role of chromosomal rearrangements in speciation (see also Rieseberg 2001; Levin 2002, Chapter 4.)

In summary, it is odd that an entire generation of animal evolutionists (including the present authors) took polyploid—but not chromosomal—speciation seriously. This view seems untenable. In plants, the best line of evidence for chromosomal speciation *is* polyploid speciation. The restoration of normal pairing and fertility in allotetraploid hybrids is not readily explained by any other hypothesis.

In the end, we are left with a puzzle. Why is chromosomal speciation more common in plants than in animals? Several explanations have been suggested. First, the disparity might reflect different degrees of pleiotropy in plants versus animals (Stebbins 1958, p. 197): genic changes might have more profound pleiotropic effects in animals than in plants, presumably reflecting the fact (or at least the claim) that animal development is "closed" (i.e., rigidly determined) while plant development is "open" (i.e., resilient to perturbation; see also Gottlieb 1984b). Genic changes would therefore be more likely to cause hybrid problems in animals than in plants. We know of no evidence for this hypothesis. Second, certain plant groups may be subject to large fluctuations in population size, which would increase the chance of fixing underdominant rearrangements. There is some evidence to support this idea: both Grant (1981) and Stebbins (1958, p. 198) note that chromosomal evolution and hybrid sterility seem more common in herbs, particularly annuals, than in trees and shrubs (see also Darlington 1956, pp. 103–104). Stebbins speculates that these differences reflect the fact that woody plants, which dominate their habitats, typically have stable populations while short-lived plants suffer intense fluctuations in population size. While this seems plausible, we suspect that there is another and more important explanation for this pattern. As Grant (1981, pp. 145–146) noted, "most of the trees, shrubs, and perennial herbs . . . have an outcrossing breeding system, whereas many of the groups of annuals considered contain some partially or predominantly self-fertilizing species." He suggests that there may be "a more fundamental relationship between type of breeding system and the formation of sterility barriers." Grant comes close here to what we suspect is the main reason for more frequent chromosomal speciation in plants than animals: many plants can self while most animals cannot. Self-fertilization, even at modest rates, greatly increases the probability of fixation of underdominant rearrangements, especially if the new homokaryotype enjoys an advantage (see also Dobzhansky 1951, p. 203).

Rieseberg (2001) has identified another factor that might explain why chromosomal speciation is more common in plants than animals. He speculates that the disparity reflects the ubiquity of degenerate sex chromosomes (Y or W) in animals and their rarity in plants. Degenerate sex chromosomes may speed the evolution of genic incompatibilities for reasons we will consider below. If so, animals may evolve genic isolation so fast that there is little opportunity for chromosomal speciation. The above two hypotheses are not, of course, mutually exclusive. Selfing *and* the absence of sex chromosomes may allow more frequent chromosomal speciation in plants.

Despite the finding that chromosomal speciation is common in plants, we have not yet explained hybrid sterility in most animals (and some plants) or F_1 hybrid inviability in *any* group.

Genic incompatibilities

We now turn to genic speciation. When considering how alleles from different species might cause intrinsic hybrid sterility or inviability, we have to weigh two possibilities. The first is *within-locus* incompatibilities, in which allele *A* from one species is incompatible with allele *a* from another species. The cross of species *AA* with species *aa* thus produces unfit *Aa* hybrids. Unfortunately, this model also runs squarely into Darwin's dilemma: how can we evolve two species that are separated by an unfit heterozygote without one of them passing through this genotype (Muller 1942, p. 84)? If, for instance, we try to evolve the *aa* taxon from an *AA* ancestor we must pass through the sterile or inviable *Aa* genotype. There are at least two ways of trying to save the within-locus hypothesis, but neither is attractive: (1) species, including many animal ones, often *do* pass through adaptive valleys; or (2) multiple substitutions occur at a single locus often enough to yield single-gene speciation. (For example, *A* may give rise to *a* in one lineage and to *A'* in the other. Although both substitutions are fit, the new aA' heterozygote might be sterile or inviable; see also Bordenstein and Drapeau 2001.)

Alternatively, hybrid sterility or inviability might result from *between-locus* incompatibilities. An allele at the *A* locus from one species may not interact properly with an allele at the *B* locus from another species. Although this model may at first sound contrived, recent work shows that it is almost certainly correct in many cases. Indeed this may well represent one of the most important results from the last two decades of work in the genetics of speciation. Despite frequent disagreement over details, a broad consensus has developed on this point: *hybrid sterility and inviability often result from between-locus incompatibilities*.

Several of the founders of the Modern Synthesis, especially Dobzhansky (1937) and Muller (1940), guessed as much. Indeed each presented experimental evidence showing that sterility or inviability in certain species crosses is caused by incompatibilities between different chromosome regions and hence between different loci. The accumulated evidence, much of which comes from *Drosophila*, is now overwhelming. (This evidence was largely ignored by King 1993 [p. 279] who oddly claimed that genic incompatibilities are a "red herring" and a ploy "more suited to politics than the scientific approach"; for more, see Coyne 1994b.) Pantazidis et al. (1993) and Pantazidis and Zouros (1988), for instance, have mapped an autosomal factor in *D. mojavensis* that causes hybrid male sterility when brought together with the Y chromosome of *D. arizonae* in hybrids. Orr (1987) showed that X-linked genes from *D. persimilis* interact with Y-linked genes from *D. pseudoobscura* to cause hybrid male sterility. Orr and Coyne (1989) found that the X chromosome of *D. virilis* is incompatible with the Y chromosome of *D. texana*, again causing hybrid male sterility. Zeng and Singh (1993) and Johnson et al. (1992) showed that hybrid sterility in the *D. simulans* subgroup involves incompatibilities between X-linked and autosomal genes from different species. Similarly, Presgraves (2003) mapped many autosomal factors from *D. simulans* that cause hybrid inviability when brought together with an X chromosome from *D. melanogaster*. Additional examples from the *Drosophila* literature could be multiplied almost indefinitely (e.g., Dobzhansky 1936; Crow 1942; Pontecorvo 1943a,b).

Between-locus incompatibilities are not limited to flies. They have also been found in hybrids between the platyfish (*Xiphophorus maculatus*) and swordtail (*X. helleri*) (Wittbrodt et al. 1989), and between species of the wasp *Nasonia* (Breeuwer and Werren 1995; Gadau et al. 1999). Nor are they limited to animals. Indeed, the first genic incompatibility was found in the composite *Crepis* (Hollingshead 1930). Between-locus interactions also cause the inviability of cotton hybrids (Gerstel 1954), of certain backcross hybrids between wild and cultivated rice (Matsubara et al. 2003), and of interpopulation hybrids of the monkeyflower *Mimulus guttatus* (Christie and Macnair 1984). Perhaps the most thorough analysis of between-locus incompatibilities in plants is that of Fishman and Willis (2001), who studied male and female sterility in hybrids of *M. guttatus* and *M. nasutus*.

In summary, hybrid incompatibilities between loci appear common. Indeed, we know of no case in which intrinsic postzygotic isolation in any outbreeding species is caused by heterozygote disadvantage. The key question facing us is therefore: How can evolution within populations yield between-locus incompatibilities between populations?

The evolution of genic incompatibilities: the Dobzhansky–Muller model

This question was essentially answered when Dobzhansky (1934) sketched a genetic model showing how hybrid sterility could evolve *without selection opposing any intermediate step*. Because Muller (1939, 1940, 1942) later elaborated the idea, it is usually called the "Dobzhansky–Muller" model. This is, unfortunately, a misnomer. Given his role in decrying the sad state of our understanding of speciation, it is ironic that William Bateson first arrived at the "Dobzhansky–Muller" model in 1909, a quarter century before Dobzhansky or Muller and only nine years after the rediscovery of Mendelism. In a remarkably prescient essay, Bateson (1909) labeled the evolution of hybrid sterility *the* problem that he would most like to see solved, as it remained the sole evolutionary phenomenon that had not been duplicated in artificially selected varieties. He then laid out his solution to the "secret of interracial sterility," a solution that proved essentially identical to that later offered by Dobzhansky and Muller. Bateson's argument was apparently unknown to both Dobzhansky and Muller and was only recently rediscovered (Orr 1996). Although the name "Bateson's model" would be more historically accurate, the term "Dobzhansky–Muller model" is too imbedded in the literature to remove. (Poulton 1908 also glimpsed the essence of the Dobzhansky–Muller model, though his treatment was less explicit than Bateson's.)

In its simplest form, the Dobzhansky–Muller model is little more than common sense. Consider two allopatric populations that evolve independently. Each experiences many substitutions over long periods of time until the populations become distinct genetically. If we could take a diverged gene from one population and place it in the genome of the other, would it work? It is easy to imagine that the gene might be reasonably effective on this related genetic background. It is also easy to imagine that it would not work well. But it is hard to imagine that it would often work better than on its own genetic background. This simple asymmetry forms the basis of Dobzhansky and Muller's model. Genes within a population are selected to work well together, while genes from different populations have not been tested together. On average, then, we expect a mixture of genes from two species to be less well adjusted than those from a single species. Hybrid sterility or inviability might therefore be a simple byproduct of the divergence of genomes that are geographically isolated. Nowhere do we have to suggest that natural selection opposed any step in this process.

Dobzhansky and Muller formalized this idea with a simple genetic model. Suppose a species is divided into two allopatric populations (Figure 7.3). The

Figure 7.3 The Dobzhansky–Muller model. The model shows how a genic incompatibility between two loci (*A* and *B*) can evolve unopposed by natural selection.

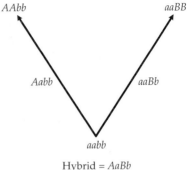

populations start out genetically identical, with an *aa* genotype at one locus and *bb* genotype at another. An *A* allele appears and becomes fixed in one population; the *Aabb* and *AAbb* genotypes are perfectly viable and fertile. In the other population, a *B* mutation appears and becomes fixed; the *aaBb* and *aaBB* genotypes are also perfectly viable and fertile. The key point is that, although the *B* allele is compatible with *a*, it has not been tested with *A*, as these alleles have never "seen" each other. It is entirely possible that *B* has a deleterious effect that appears only when *A* is present. If the two populations meet and hybridize, the resulting *AaBb* hybrid might well be inviable or sterile. In the imagery of Wright's adaptive landscape, we have evolved two taxa separated by an adaptive valley, although no genotype ever passed through the valley. Darwin's dilemma is resolved.

Despite its simplicity, several aspects of the Dobzhansky–Muller model have been misunderstood and are worth clarifying at the outset. First, the model does not require that the *A* and *B* alleles have a devastating effect on hybrids. Instead, a genic incompatibility might cause only slight hybrid inviability or sterility, with complete postzygotic isolation resulting from the cumulative effects of many such incompatibilities. Second, there is no reason to think that any particular genic incompatibility must involve only two genes, as in the simplest version of the model. As Muller (1942) emphasized, it is entirely possible that a hybrid must carry the appropriate alleles at three or more loci for any hybrid sterility or inviability to occur. Third, many variations on the simple scenario sketched above are possible. Both substitutions might occur, for example, in one population and none in the other. (Consider the sequence: *aabb* → *AAbb* → *AABB*, while the other population remains *aabb*. *B* has never been tested with *a*.) The Dobzhansky–Muller model does not, therefore, require equal rates of evolution in the two lineages or indeed any evolution at all in one lineage (Orr 1995).

Next, the genes causing hybrid sterility or inviability need not be fixed within species. It has long been known that there is standing genetic variation within some species for the severity of intrinsic postzygotic isolation when crossed to another species. In fact, the first example of a Dobzhansky–Muller incompatibility discovered—that in *Crepis*—involved polymorphism within species. Stebbins (1958, p. 189) noted several other examples in plants, and Patterson and Stone (1952, Chapter 10) discussed many cases in *Drosophila* (see also Crow 1942). Standing variation for the strength of reproductive isolation has also been found in the flour beetle hybridization *Tribolium castaneum* × *T. freemani* (Wade et al. 1994a). Many of these examples, however, involve variants found in different populations, not variants segregating at appreciable frequencies within single populations. Indeed population genetic theory gives some reason for thinking that the alleles involved in Dobzhansky–Muller incompatibilities will not usually segregate together within populations (Nei 1976; Phillips and Johnson 1998). In any case, because the genes causing hybrid sterility or inviability often *do* appear fixed within species—mapping experiments performed in different decades and involving different strains often locate the same hybrid sterility genes (Dobzhansky 1974; Orr 1989b; Muller and Pontecorvo 1942; J. P. Masly, pers. comm.)—we will usually assume that the genes causing hybrid problems are fixed. Last, belief in the Dobzhansky–Muller model does not require belief in segregating "coadapted gene complexes" within species. It is one thing for alleles at two or more loci to function well together within but not between species and quite another for blocks of coadapted alleles to segregate in linkage disequilibrium within species. In the first case, selection need not maintain linkage disequilibrium, as sets of alleles are separated by geography, while in the second case selection must maintain disequilibrium in the face of recombination.

Despite wide acceptance in the 1930s and 1940s, the Dobzhansky–Muller model faded from prominence in the following decades and, by 1983, Futuyma lamented that speciation workers had succumbed to a fondness for vague, holistic theories despite the availability of simple genetic models of hybrid problems. It is not clear why these simple models were ignored for nearly forty years (though see Futuyma 1983; Howard et al. 2002). But whatever the reason, the status of the Dobzhansky–Muller model dramatically changed course over the last two decades. Indeed, a consensus has grown that not only are hybrid sterility and inviability often caused by incompatibilities between loci, but that these incompatibilities evolve in the way described by Dobzhansky and Muller's model.

Several lines of evidence support this conclusion. The first is indirect. We know that the genetic ingredients required by the model exist. This is not as trivial as it might seem. It may be worth asking when a peak shift or "genetic revolution" was last observed; these may occur, but there is little direct evidence on either point (see Chapter 11). But alleles that are fit on one genetic background and unfit on another are well known (see Thompson 1986 for a list of such so-called complementary lethals in *D. melanogaster*). Although many of these interactions involve artificially induced mutations, complementary genes have also been found in natural populations (Dobzhansky 1946b; Dobzhansky et al. 1959; Krimbas 1960; Thompson 1986). The genetic material required by the theory is therefore real.

In addition, the sequence of events posited by Dobzhanksy and Muller has been observed at least once in nature. Two populations of the monkeyflower, *Mimulus guttatus*, produce hybrids that die as embryos (Christie and Macnair 1984, 1987). One population, in response to human disturbance (copper contamination of soil at a mine site), has fixed a new copper resistance allele within the last 200 years. This substitution causes no serious fitness problems in the disturbed population. But when introduced onto the genetic background of other populations, the copper-tolerance allele, or something linked to it, causes hybrid inviability.

Mathematical models of genic speciation

The Dobzhansky–Muller model is also important because it reduces the complex problem of the evolution of reproductive isolation to a far simpler one—the accumulation of genic incompatibilities. Oddly, however, evolutionary genetics fell down on the job here. Much of the older literature on the population genetics of speciation can be viewed as an attempt to construct theories of postzygotic isolation while ignoring the underlying genetic *mechanism* of isolation. Although one can obviously do interesting work on the ecology or biogeography of speciation while ignoring the underlying genetics, it seems odd to think that one can build relevant evolutionary theories of intrinsic isolation that neglect its underlying mechanism. Instead, it seems clear that we should *begin* with the known genetic mechanism and build up our theories from there, asking if any interesting patterns emerge. Until recently, little of this was attempted. Indeed as late as 1974, Lewontin noted that it is "an irony of evolutionary genetics that, although it is a fusion of Mendelism and Darwinism, it has made no direct contribution to what Darwin obviously saw as the fundamental problem: the origin of species" (Lewontin 1974, p. 159). And earlier, Lewontin (1961, p. 383) complained that "[d]espite the great amount written about speciation, there is as yet no mathematical theory of species formation." In the case of the Dobzhansky–Muller model, this lacuna was especially odd as the idea is easily modeled mathematically. Despite this, the first mathematical treatment of the Dobzhansky–Muller model did not appear until Nei's (1976) little-known paper, which considered the fate of a pair of incompatible alleles in a single population.

More recent theoretical work takes a different approach. This theory follows the long-term fate of diverging populations, tracking numbers of substitutions and genic incompatibilities that arise over long stretches of evolutionary time (Orr 1995). These models resemble those in molecular evolution (Kimura 1983; Gillespie 1991) in that the fixation process is considered instantaneous and populations monomorphic at any point in time. It is usually assumed that the diverging populations are strictly allopatric, with no gene flow during the time studied. To visualize this process, it helps to consider Figure 7.4. Each of the two heavy lines represents a population diverging from a common ancestor. The two allopatric populations begin with identical ancestral lowercase alle-

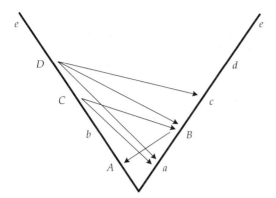

Figure 7.4 The evolution of genic incompatibilities between two allopatric populations. Time runs upward and the first substitution occurs at the *A* locus, the second at the *B* locus, and so on. Possible genic incompatibilities in hybrids are shown as arrows. (From Orr 1995.)

les at all loci (*a, b, c,* . . .) and time runs upward. The first substitution occurs at the *a* locus, the second at the *b* locus, and so on. For now, we only allow incompatibilities between pairs of loci.

The first substitution involves the replacement of the *a* allele by the *A* allele in population 1 (uppercase letters denote that an allele is derived; no dominance is implied). If the two populations were to come into contact and hybridize, this new *A* allele *cannot* cause any hybrid incompatibility: because *A* is compatible with the genetic background of population 1, it must be compatible with the identical genetic background of population 2. Now imagine that the second substitution occurs at the *b* locus in population 2: *b* is replaced by *B*. If the two populations were to hybridize now, there is one possible Dobzhansky–Muller incompatibility: as indicated by the arrow in the figure, allele *B* in population 2 has not been "tested" with allele *A* in population 1. Similarly, substitution *C* might cause two hybrid incompatibilities, and so on.

As this process continues, several patterns emerge. Some are obvious, others are not. Because discussion of these patterns is scattered across a technical literature, we summarize them here. We present no mathematical derivations, which can be found in the original papers. The Dobzhansky–Muller model predicts the following patterns:

SUBSTITUTIONS MUST OCCUR AT BOTH LOCI. Hybrid incompatibilities only occur between loci that have *both* experienced substitutions (Orr 1995). Arrows in Figure 7.4 never run up to loci that have not diverged in either population (e.g., *e*). This follows from the fact that the two populations carry identical alleles at undiverged loci. Any substitution that is compatible with these loci on its normal within-species genetic background (as it must be) will, therefore, be compatible with these loci in the other species' background.

DERIVED ALLELES CAUSE INCOMPATIBILITIES MORE OFTEN THAN ANCESTRAL ALLELES. Although a locus must diverge to be involved in a hybrid incompatibility, the allele causing the incompatibility need not be the new derived one.

Instead, ancestral alleles can also cause incompatibilities. Nonetheless, a glance at Figure 7.4 shows that derived alleles cause incompatibilities more often than ancestral. This is because the substitution of a new derived allele (say, C) might be incompatible with either a derived allele (B) or an ancestral one (a): both derived-derived and derived-ancestral incompatibilities occur. The only incompatibility that cannot occur is ancestral-ancestral: all ancestral alleles must be compatible with each other as they represent the (fit) ancestral genotype. With equal rates of evolution in the two lineages, one can show that derived alleles are three times more likely to be involved in incompatibilities than are ancestral alleles (Orr 1995).

INCOMPATIBILITIES ARE ASYMMETRIC. As Muller (1942) noted, hybrid incompatibilities are asymmetric: although B in Figure 7.4 might be incompatible with A, a cannot be incompatible with b. The reason is that *aabb* represents an ancestral step in the divergence of these two populations. More specifically, *aabb* represents either the common ancestor of the populations or, in more complex cases, an intermediate step in their divergence. There is now good evidence that hybrid incompatibilities show the predicted asymmetry. Wu and Beckenbach (1983), for instance, analyzed X-linked male sterility in *Drosophila pseudoobscura–D. persimilis* hybrids. They showed that if a region from the *D. pseudoobscura* X chromosome causes sterility when introgressed into *D. persimilis*, the same region from *D. persimilis* causes little or no sterility when introgressed into *D. pseudoobscura*. (For results from other species, see Vigneault and Zouros 1986; Orr 1989b; Wittbrodt et al. 1989; Johnson et al. 1992; Wu and Palopoli 1994; Gadau et al. 1999.) This observed asymmetry provides strong support for the Dobzhansky–Muller model. While other theories might explain the pattern post hoc, none other predicted it.

THE NUMBER OF HYBRID STERILITY AND INVIABILITY GENES "SNOWBALLS" THROUGH TIME. Figure 7.4 shows that later substitutions are more likely to cause hybrid problems than earlier ones. The first substitution at A cannot cause any hybrid incompatibility, while the second substitution at B might cause one, the third substitution at C might cause two, and, in general, the Kth substitution might cause K-1. The total number of incompatibilities, I, separating two taxa thus increases faster than linearly with the number of substitutions separating them. Indeed, the expected number of hybrid incompatibilities is

$$E[I|K] \cong K^2 p/2$$

where p is the probability that two diverged loci are incompatible (Orr 1995). While we have expressed this result as a function of K, it is easy to express it as a function of time. Assuming a standard Poisson molecular clock, it can be shown that

$$E[I] = (kt)^2/2$$

where k is the rate of substitution and t is time (Orr and Turelli 2001). *Thus, the expected number of hybrid incompatibilities rises exactly with the square of time.* We have of course assumed that hybrid incompatibilities involve pairs of loci; if triplets, etc. are allowed, the number of incompatibilities rises even faster (Orr 1995; Orr and Turelli 2001).

This "snowball effect" can in principle be tested by a simple contrast. Consider a species pair that is, say, four times older (more diverged molecularly) than another species pair. Without a snowball effect, the older species pair should be separated by about four times more genes causing hybrid problems than the younger pair. But with a snowball effect, the older species pair should be separated by many more than four times as many genes causing hybrid problems than the younger pair. The snowball effect can therefore be tested by a simple chi-square test, contrasting the numbers of substitutions separating two species pairs versus numbers of hybrid sterility/inviability genes separating the same species pairs. This simple test, which does not require knowledge of the time (in years) since species split, has not been recognized in the literature. Though present data on numbers of hybrid sterility or inviability genes are not of high enough resolution to allow rigorous testing of this prediction, the work of True et al. (1996), Presgraves (2003), and Tao and Hartl (2003) suggests that such tests are not far off. (Lijtmaer et al. 2003 recently claimed that there is no evidence for the snowball effect since, in several groups, the strength of postzygotic isolation grows linearly with time. This is incorrect. The snowball effect concerns the number of *genes* causing postzygotic isolation, not the resulting strength of that isolation. The latter quantity depends on an unknown fitness function and must asymptote at complete isolation—it *cannot* increase without limit.) Finally, the snowball effect characterizes evolution between allopatric taxa; if gene flow occurs between populations during divergence, the number of incompatibilities can grow linearly with time (Kondrashov 2003).

COMPLEX INCOMPATIBILITIES SHOULD BE COMMON. Hybrid incompatibilities need not involve pairs of genes. Hybrids may have to carry the appropriate combination of alleles at three or more loci to suffer any fitness loss. Complex incompatibilities might arise for several reasons (Orr 1995). One is that such incompatibilities could reflect the redundancy of biochemical pathways: hybrid sterility or inviability might require that both loci A and B and redundant loci C and D be "knocked out" in hybrids. This might be especially likely if duplicate genes are common in key developmental pathways (Orr 1995; Lynch and Force 2000). Complex incompatibilities might also arise for more abstract reasons. Most important, the evolution of complex incompatibilities is "easier" than the evolution of simpler ones. To see this, reconsider single-gene versus two-gene speciation. In the single-gene case, essentially all imaginable paths to speciation require passing through the unfit hybrid genotype. But in the two-gene case, many imaginable ways of getting two species from a common ancestor do *not* pass through the unfit hybrid genotype. Though not intuitively obvious, the same calculus holds for more complex incompatibilities. Of all imaginable ways of getting two species from a common ancestor, a greater fraction do not require passing through the sterile or inviable hybrid genotype in the three-gene than in the two-gene case (Cabot et al. 1994; Orr 1995). Thus, for the same reason that two-gene speciation is easier than one-gene, so three-gene is easier than two-gene, and so on. The evidence for complex incompatibilities is now strong (Chapter 8).

POPULATION SUBDIVISION MAY SLOW THE EVOLUTION OF INTRINSIC ISOLATION. The Dobzhansky–Muller model yields counterintuitive predictions about the effects of population subdivision on postzygotic isolation. While we have focused on divergence of two taxa from a common ancestor, we can also consider divergence of many initially identical taxa. Orr and Orr (1996) modeled a species subdivided either into a few large populations or into many small populations. They found that, contrary to many popular models, postzygotic isolation does not evolve fastest in species subdivided into small populations (see Chapter 11). Instead, if substitutions are driven by natural selection, the mean time to speciation increases as a species is split into more small populations and is shortest when a species is divided into two large populations. If, on the other hand, substitutions are neutral and fixed by genetic drift, the mean time to speciation is either independent of population subdivision or declines modestly as a species is split into many small populations. Church and Taylor (2002) suggest that some of these results may depend on the assumption of no gene flow between diverging populations.

CONCLUSIONS. We have spent a good deal of time on the Dobzhansky–Muller model for two reasons. First, it has not been widely appreciated that the model provides a basis for a mathematical theory of speciation by postzygotic isolation. Second, the lion's share of genetic studies of postzygotic isolation over the last two decades has focused on genic incompatibilities. An understanding of this literature is impossible without an understanding of the Dobzhansky–Muller model. We consider these studies in Chapter 8.

Wolbachia *and cytoplasmic incompatibility*

Finally, we consider a somewhat surprising cause of intrinsic isolation—cytoplasmically inherited endosymbionts. The isolation resulting from such endosymbionts is called "cytoplasmic incompatibility" or CI. Although CI may be caused by different kinds of organisms—a recent report shows that *Bacteroidetes* are sometimes involved (Hunter et al. 2003)—almost all work has focused on *Wolbachia*. *Wolbachia* are Rickettsia-like bacteria that inhabit the cells of many eukaryotes (reviewed in O'Neill et al. 1992, 1997; Werren 1997a; Stouthmaer et al. 1999; Bordenstein 2003). While best known in insects, *Wolbachia* have also been found in isopods, mites, and nematodes (Sironi et al. 1995;

Werren et al. 1995a; Breeuwer and Jacobs 1996; Werren and Windsor 2000). *Wolbachia* have many effects on reproduction, depending on the strain and host species involved. These include parthenogenesis, feminization, and male-killing (Werren 1997a; Stouthmaer et al. 1999; Weeks et al. 2002). But the effect most relevant to speciation is CI: when uninfected females are crossed to infected males, many if not all progeny die as embryos. The reciprocal cross of infected females to uninfected males shows no lethality. (As we will see, however, bidirectional incompatibilities are possible in certain circumstances.) Antibiotic treatment of infected strains both cures *Wolbachia* infection and rescues normally dead progeny (O'Neill et al. 1997; Werren 1997b). Formally, *Wolbachia*'s effects on host reproduction can be viewed as manipulations of the host by the endosymbiont that encourage the endosymbiont's spread. Darwin's dilemma is thus resolved in a particularly subtle way: hybrid inviability at the level of hosts reflects natural selection at the level of parasites.

BIOLOGY. Several pieces of *Wolbachia* biology are clear. First, *Wolbachia* are maternally transmitted: infected mothers pass the bacteria to most of their progeny and paternal transmission is rare (Hoffmann and Turelli 1988). Second, although the sperm of infected males evokes CI, *Wolbachia* are shed during spermatogenesis and do not physically travel with mature sperm (Bressac and Rousset 1993); *Wolbachia* must therefore modify sperm. Third, the cellular target of this modification appears to be evolutionarily conserved: *Wolbachia* from the mosquito *Aedes*, for instance, induce CI when injected into *Drosophila* (Braig et al. 1994). (*Wolbachia*, however, do sometimes have different phenotypic effects when introduced into different species [Sasaki et al. 2002].) Fourth, the target of modification involves the male pronucleus, not any essential extranuclear component of the sperm (Presgraves 2000). Finally, CI involves a disruption of paternal chromosome processing and/or mitosis in early embryogenesis: paternal chromosomes often show condensation defects and are frequently lost in early divisions (Reed and Werren 1995; Lassy and Karr 1996; Callaini et al. 1997). The resulting haploidy appears to cause inviability.

Ironically, *Wolbachia*'s ability to kill offspring explains its ability to invade populations. Because infected females can produce progeny with *any* male, infected or not, while uninfected females can produce progeny only with uninfected males, females who carry *Wolbachia* enjoy an advantage. The infection thus increases in frequency (at least when starting above a usually low threshold; Caspari and Watson 1959). The rapid spread of *Wolbachia* is not mere theoretical expectation: Turelli and Hoffmann (1991, 1995) witnessed an astonishingly rapid sweep of *Wolbachia* through California populations of *Drosophila simulans*. The combination of ability to rapidly invade and to transfer horizontally among taxa—*Wolbachia* appear to have leapt taxonomic boundaries (Werren et al. 1995b; O'Neill et al. 1997)—surely explains *Wolbachia*'s broad taxonomic distribution. Indeed, *Wolbachia* are present in about 15%–20% of insect species (Werren et al. 1995a; Werren and Windsor 2000) and, not suprisingly, CI also appears fairly widespread, having been documented in at least five orders of insects (O'Neill et al. 1997). The prevalence of *Wolbachia* does not mean, however, that it plays a role in reproductive isolation. Infected species are known in which *Wolbachia* plays no apparent role (e.g., field crickets of the genus *Gryllus* [Mandel et al. 2001], and the fruit flies *Drosophila santomea* and *D. yakuba* [J. A. Coyne, pers. comm.]).

Our understanding of *Wolbachia* and CI derives almost entirely from three systems, the mosquito *Culex*, the wasp *Nasonia*, and the fly *Drosophila* (reviewed in Werren 1997b; Hoffmann and Turelli 1997). CI was first discovered by Laven (1951) in crosses between strains of the species *Culex pipiens*. Laven's later work revealed an enormously complex network of crossing relations among mosquitoes from different locations. Although he ultimately concluded that the incompatibility factor was cytoplasmically inherited (Laven 1957), the causative agent, *Wolbachia*, was not identified until the early 1970s (Yen and Bar 1971, 1973).

Wolbachia's possible role in speciation has been studied most thoroughly in the wasp *Nasonia* (see Werren 1997b, and references therein). This genus includes three species, the outgroup *N. vitripennis* (cosmopolitan), and the sister species *N. giraulti* (eastern North America) and *N. longicornis* (western North America). These species are morphologically distinct but crossable. Populations of all three species are infected with *Wolbachia*. The crosses *N. vitripennis* × *N. giraulti*, *N. vitripennis* × *N. longicornis*, and *N. giraulti* × *N. longicornis* all show partial to complete CI. Because *Nasonia* is haplodiploid and CI causes loss of paternal chromosomes, incompatible crosses can involve either conversion of diploid females into haploid males (Reed and Werren 1995), or embryonic lethality (Bordenstein et al. 2001). CI is, as expected, reversible by antibiotic treatment, although later generation hybrids often suffer further breakdown due to genic incompatibilities. Bordenstein et al. (2001), however, showed that CI sometimes appears before other, genic-based, forms of postzygotic isolation in *Nasonia*. (As the authors emphasize, however, the species studied are allopatric and show some prezygotic isolation.)

The population biology of *Wolbachia* has been studied most thoroughly in the fly *Drosophila simulans* (Hoffmann et al. 1986; O'Neill and Karr 1990; Montchamp-Moreau et al. 1991; Turelli and Hoffman 1995; Hoffmann and Turelli 1997). Crosses between populations of *D. simulans*, a widespread human commensal, often show CI. As noted above, Turelli and Hoffmann (1991) documented the rapid invasion of *Wolbachia* in previously uninfected California populations. Turelli and Hoffmann (1995) estimated a number of parameter values affecting *Wolbachia* population dynamics (e.g., the severity of CI, the effect of age on this severity, the fidelity of maternal transmission, and female fecundity). They also compared many of these parameter values between laboratory and wild flies (maternal transmission, for instance, appears essentially perfect in the laboratory but imperfect in nature).

ROLE IN SPECIATION. Despite their frequency in nature and their undoubted ability to kill certain progeny, endosymbionts might not seem to provide a particularly effective isolating barrier for two reasons. First, cytoplasmically inherited factors would appear, at first glance, to only halve gene flow: infected and uninfected populations are incompatible in only one direction of the hybridization. Second, any reproductive isolation between populations may be ephemeral: *Wolbachia* rapidly invade uninfected populations and what today are one infected and one uninfected population may tomorrow be two infected (and hence, compatible) populations.

The first objection is not as serious as it seems. Surprisingly, *Wolbachia* can cause *bidirectional* incompatibility. Indeed such reciprocal incompatibilities appear common, occurring in *Culex* (Yen and Bar 1973), *Nasonia* (Breeuwer and Werren 1990; Bordenstein et al. 2001), and *Drosophila* (O'Neill and Karr 1990). Bidirectional incompatibilities reflect the fact that *Wolbachia* are not all the same. Instead, females harboring one variant of *Wolbachia* are usually "immune" to the effects of the same variant but not to the effects of a different one. Indeed, *Wolbachia* fall into several to many incompatibility types. Females bearing *Wolbachia* type "A," for instance, are usually incompatible with sperm from males bearing *Wolbachia* type "B" and, similarly, B-bearing females are usually incompatible with A sperm. Doubly infected females are compatible with A or B sperm (Hoffmann and Turelli 1997; Werren 1997b).

The second objection is more serious. Early theory showed that coexistence of infected and uninfected populations is unstable (Caspari and Watson 1959). But the most important issue concerns the maintenance of two or more variants of *Wolbachia* (e.g., those causing bidirectional incompatibilities). There are two scenarios to consider. In the first, rare *Wolbachia* variants that are incompatible with a "wild-type" variant might appear within an already infected host population. Turelli (1994) showed that such new variants are selected against when rare (assuming infection is costly to the host). In the second scenario, different microbes might become fixed in different allopatric populations. The question is whether the resulting bidirectional incompatibility can be maintained upon secondary geographic contact. Early theory suggested that the answer is no: upon contact, all but one infection type is usually eliminated, implying that *Wolbachia* alone are unlikely to provide a stable basis for speciation (Rousset et al. 1991). The same property that allows rapid invasion (infectiousness) thus seemed to ensure that a single *Wolbachia* type ultimately sweeps through all populations, erasing CI (Shoemaker et al. 1999; Weeks et al. 2002). More recent work, however, suggests that bidirectional incompatibilities can be maintained in the face of gene flow between populations, at least under certain conditions involving spatial structure (Hoffmann and Turelli 1997, p. 71; Keeling et al. 2003), or simultaneous selection at nuclear loci (Telschow et al. 2002).

But even if stable coexistence of two *Wolbachia* types proves rare, it does not follow that *Wolbachia* plays no role in reproductive isolation. Instead, it seems they can and do. For whether or not *Wolbachia* yield complete, stable reproductive isolation by themselves, they can contribute to total isolation between taxa. Indeed Shoemaker et al. (1999) have described a likely case between the sister species *Drosophila recens* and *D. subquiniaria*. *D. recens* is infected with *Wolbachia* while *D. subquinaria* is not. As expected, the *D. subquinaria* female × *D. recens* male cross produces few viable hybrids. But while the reciprocal cross does yield viable hybrids, the cross is difficult, reflecting strong prezygotic isolation. Moreover, hybrid males are completely sterile in both directions of the cross. The combined action of these barriers—CI, prezygotic isolation, and hybrid sterility—provides strong, species-level reproductive isolation. It would be absurd to deny a role for endosymbionts in reproductive isolation in this case.

The real question, then, is whether *Wolbachia* play a *frequent* part in speciation. Our best guess is that, while *Wolbachia*'s role may be non-trivial, it is far from ubiquitous. Although the *D. subquinaria–D. recens* case shows that CI can cause reproductive isolation between species in nature, several broader patterns suggest that CI does not play a particularly common role in speciation, even in insects.

For one thing, the observation that 15%–20% of insect species carry *Wolbachia*, means that 80%–85% do not. More seriously, *Wolbachia* cannot explain several striking patterns that characterize postzygotic isolation. The first is the commonness of hybrid sterility: endosymbionts do not appear to often cause sterility (although there are exceptions; Somerson et al. 1984) and so cannot make a substantial contribution to what appears to be the most common form of intrinsic postzygotic isolation. The second is the striking pattern of sex-limited hybrid inviability. As we will see, hybrid inviability in virtually all animal taxa shows a sex-limited pattern, such that, if only one sex is afflicted, it is heterogametic (XY or ZW) (Haldane's rule; Chapter 8). Although male-killing endosymbionts are known (Hurst and Jiggins 2000)—and not unexpected, given the maternal inheritance of endosymbionts—female-specific killers are not known and would be strongly selected against under maternal inheritance. Endosymbionts cannot therefore explain Haldane's rule for inviability in female heterogametic taxa (Coyne and Orr 1999; Bordenstein 2003; Keeling et al. 2003). This fact is not easily reconciled with the idea that endosymbionts represent a common path to animal speciation, and is particularly important since Haldane's rule represents an *early* stage in the evolution of postzygotic isolation (Coyne and Orr 1999; Bordenstein 2003). Finally, surveys in several animal groups (e.g., *Drosophila*, Lepidoptera) show that intrinsic postzygotic isolation evolves gradually: there is a "speciation clock," with young taxa pairs showing little or no postzygotic isolation, intermediate-aged taxa showing intermediate levels of isolation, and older taxa showing complete isolation (Chapter 2). This clock would presumably not arise if endosymbiont-based speciation were the norm, as CI represents an essentially instantaneous form of speciation.

Conclusions

We have considered the contributions of extrinsic and intrinsic postzygotic isolation to speciation, and have described examples of each. We have spent most of our time on intrinsic isolation, not because it is the more important barrier, but because it is the best studied. We have seen that intrinsic isolation might involve at least three genetic mechanisms: chromosomal, genic, and endosymbiotic. Each mechanism features a different solution to Darwin's dilemma—the evolution of an apparently maladaptive phenotype, production of unfit hybrids. Chromosomal speciation usually involves a stochastic solution: hybrid sterility might evolve by genetic drift in opposition to natural selection. Genic speciation involves epistatic interactions that can evolve unopposed by natural selection and that might even be driven by natural selection. Endosymbiont-based speciation involves natural selection at the level of parasites that causes hybrid inviability at the level of hosts.

Some of these genetic mechanisms play a larger role in postzygotic isolation than others. Chromosomal changes clearly play an important role in plants but a smaller one in animals. Centric (and especially monobrachial) fusions may, however, cause hybrid sterility in some mammals. Endosymbionts like *Wolbachia* sometimes cause reproductive isolation between taxa in nature. Several comparative patterns suggest, however, that they do not play a prominent part in speciation. Finally, genic incompatibilities play a common role in both hybrid sterility and inviability, in both animals and plants. It seems then, that genic incompatibilities may be the most important cause of intrinsic postzygotic isolation, at least between taxa of the same ploidy level. In the next chapter we turn to genetic studies of these incompatibilities.

8

The Genetics of Postzygotic Isolation

There are many things that we would like to know about the genetics of postzygotic isolation: How many genes are involved in these isolating barriers? How do these genes interact with each other? And *which* genes cause postzygotic isolation? The last question opens the door to a large set of questions that, until recently, was almost completely neglected: Do specific types of genes (say, transcription factors) typically cause hybrid sterility and inviability? Do these genes usually diverge at regulatory or coding regions? Does natural selection or genetic drift drive the divergence of these genes? This chapter considers these questions. We also consider Haldane's rule—the preferential sterility or inviability of hybrids of the heterogametic (XY or ZW) sex—a pattern that characterizes postzygotic isolation in almost all animals.

We focus again on *intrinsic* postzygotic isolation. As before, we do so not because intrinsic is more important than extrinsic isolation but because almost all genetic analyses of postzygotic isolation have focused on this type of barrier. Because these studies have mostly involved *Drosophila*, this chapter is inevitably more fly-heavy than our others.

Work on the genetics of postzygotic isolation can be divided into two periods: that from the 1920s through the 1940s, and that from the 1980s through the present. The literature from each of these periods poses its own problems to a reviewer. Material from the earlier period is almost entirely unknown to most readers. Material from the later period, on the other hand, has grown very sophisticated, reflecting intense work over the last two decades. This rapid progress has left many evolutionists out of date and, sometimes, confused. Here we try to remedy both of these problems. We try, that is, to revive a nearly forgotten body of early work and to demystify a complex body of recent work.

A disproportionately large share of studies of the genetics of postzygotic isolation, both early and recent, has been inspired by a single problem—Haldane's rule. Indeed, it might be argued that geneticists studying speciation have been unduly obsessed with this problem. In any case, no understanding of the genetical literature seems possible without an understanding of its usual historical context, the attempt to unravel Haldane's rule. We thus begin this chapter with a discussion of this rule; our treatment largely follows Orr (1997). In the latter half of the chapter, we turn to other genetical problems. We conclude with a discussion of the recent molecular identification of several genes that cause intrinsic isolation. This work may represent the start of an important new phase in the study of speciation.

Haldane's Rule

In 1922, J. B. S. Haldane observed that

> When in the offspring of two different animal races one sex is absent, rare, or sterile, that sex is the heterozygous [heterogametic, i.e., XY or ZW] sex.

Haldane showed that this rule is obeyed in several animal groups, and more recent surveys suggest that it holds in all animals possessing sex chromosomes.

Although "Haldane's rule" has been the focus of tremendous work, the reason has not always been understood. It is not that the pattern itself is considered inherently fascinating, or that evolutionary geneticists see their most important task as the explanation of sex-limited fitness effects. Rather, Haldane's rule has been the focus of much work because it implies a surprising uniformity in the genetic changes that cause hybrid problems in very different kinds of animals—crickets, birds, butterflies, and mice, among others. The important question, then, is: Why *do* hybrid sterility and inviability evolve in the same way in such different organisms?

Although textbooks invariably offered pat explanations of Haldane's rule, work in the mid-1980s surprisingly revealed that we did not, in fact, understand why the rule is obeyed. A wave of experiment and theory followed, a wave that was largely responsible for the modern renaissance of interest in the genetics of speciation.

We begin by summarizing the phenomenology of Haldane's rule: How strong is the pattern, where is it seen, and how many times has it evolved independently? We then turn to the causes of the rule. We emphasize those explanations that now appear correct and, to keep our treatment manageable, allude only briefly to hypotheses that have been rejected. As usual, we emphasize places where more decisive experiments are needed.

The phenomenon

No theorist predicted Haldane's rule. The pattern was simply noticed by Haldane (1922), who presented data from Lepidoptera, birds, flies, mammals, and

Table 8.1 *The frequency of Haldane's Rule*

Group	Phenotype	Asymmetric hybridizations[a]	Number obeying Haldane's Rule
Heterogametic males			
Drosophila	Sterility	114	112
	Inviability	17	13
Mammals	Sterility	25	25
	Inviability	1	1
Heterogametic females			
Lepidoptera	Sterility	11	11
	Inviability	34	29
Birds	Sterility	23	21
	Inviability	30	30

Source: Modified from Coyne (1992a).
[a] Hybridizations in which only one hybrid sex is sterile or inviable.

other groups. Since then, several reviewers have tabulated the frequency with which the rule is obeyed in various taxa (Craft 1938; Coyne and Orr 1989a; Coyne 1992a; Wu and Davis 1993; Laurie 1997; Presgraves 2002; Price and Bouvier 2002; Tubaro and Lijtmaer 2002; Lijtmaer et al. 2003). Table 8.1 is adapted from Coyne (1992a), but similar results have been obtained in all reviews. Despite claims to the contrary (White 1978), Haldane's rule is well obeyed in all taxa surveyed.

The most important fact emerging from Table 8.1 is that Haldane's rule is obeyed in taxa in which males are the heterogametic sex (e.g., *Drosophila*, mammals) *and* in taxa in which females are the heterogametic sex (e.g., Lepidoptera, birds). The cause of the rule cannot therefore be connected to sex per se; rather it must be connected to the sex chromosomes.

Another important fact is not apparent from Table 8.1: Haldane's rule represents an early stage in the evolution of reproductive isolation. This does not follow automatically from the rule itself. As Coyne and Orr (1989a) emphasized, there might be two paths to postzygotic isolation. In one, the heterogametic sex would become sterile or inviable early, with homogametic hybrids affected later. In the other, both hybrid sexes would become sterile or inviable simultaneously. Even if both of these paths were common—and even if the second were *more* common—Haldane's rule would still be obeyed: when only one sex is afflicted, it is the heterogametic sex. To determine if Haldane's rule appears before both-sex problems, Coyne and Orr (1989a, 1997) collected data on the age and strength of reproductive isolation between many pairs of *Drosophila* species, as described in Chapter 2. The results showed that, among young taxa, almost all species pairs showing any postzygotic isolation conform to Haldane's rule: of 24 phylogenetically independent taxa pairs showing weak postzygotic isolation, 22 are cases of Haldane's rule. Hybrid problems affecting both sexes usually appear only later. Similar analyses in Lepidoptera (Presgraves 2002), ducks

(Tubaro and Lijtmaer 2002), pigeons and doves (Lijtmaer et al. 2003), and birds in general (Price and Bouvier 2002) have arrived at a similar conclusion: Haldane's rule represents a nearly obligatory first step in the evolution of intrinsic postzygotic isolation.

Because so many species pairs obey Haldane's rule—242 out of 255 in Table 8.1—it may seem odd to ask if the pattern is significant. However, Read and Nee (1991) argued that the association between postzygotic isolation and heterogamety might be illusory. Their argument was that the relevant sample size is not number of hybridizations but number of cases in which the sex that is heterogametic evolved *independently*. The reason is that birds and *Drosophila* differ in many ways, not just heterogamety. The fact, then, that all *Drosophila* (male heterogamety) and all birds (female heterogamety) obey Haldane's rule shows only that *something* shared by all flies makes them conform to the rule, and that something shared by all birds makes them conform to the rule. This "something" may or may not be heterogamety. Read and Nee argued that there were only four independent associations between heterogamety and the hybrid sex afflicted. As Coyne et al. (1991b) pointed out, however, Read and Nee overlooked cases of Haldane's rule in at least one other group with independently evolved male heterogamety: salamanders of the genus *Triturus* (Spurway 1953). Haldane's rule has since been described in other taxa, including lizards of the genus *Lacerta* (female heterogamety; Rykena 1991), and nematodes of the genus *Caenorhabditis* (male heterogamety; Baird et al. 1992). These cases bring the number of independent associations between heterogamety and the hybrid sex afflicted to seven.

But we need not rely on comparative data to prove an association between postzygotic isolation and heterogamety. The idea that Haldane's rule is connected to heterogamety can be tested by direct genetic analysis. If, for instance, such analyses showed that the factors causing hybrid problems were always X-linked or always recessive (and thus expressed only in the heterogametic sex), we would understand *causally* why heterogametic hybrids suffer low fitness. As we will see, there is now a large body of genetic data supporting a causal connection between heterogamety and Haldane's rule. We assume, then, that Haldane's rule is "significant." The important question is *why* the heterogametic sex suffers disproportionately when species are crossed.

The causes of Haldane's rule

Many explanations of Haldane's rule have been offered. Most have now been falsified. It is now clear that Haldane's rule is not invariably caused by incompatibilities between X- and Y-linked genes (Orr 1989b; Orr and Coyne 1989; Johnson et al. 1992, 1993; Zeng and Singh 1993; indeed, the rule is obeyed in XO taxa; Mantovani and Scali 1992). Neither is the rule caused by a disruption of dosage compensation or sex determination (Orr 1989a; though, see Sturtevant 1946 and Baird 2002); or by species-specific chromosomal rearrangements, including translocations between the X chromosome and autosomes (indeed, the rule is obeyed between species that are homosequential; Coyne and Orr

1989b). Each of these mechanisms may act now and then, but none accounts for the ubiquity of Haldane's rule.

Only four ideas—the dominance theory, the faster-male theory, the faster-X theory, and the meiotic drive theory—remain viable general explanations of Haldane's rule. There is little doubt that both dominance and faster-male effects play an important role in Haldane's rule. Although evidence for the faster-X and meiotic drive ideas is weaker, they may also contribute to the pattern.

THE DOMINANCE THEORY.   Muller (1940, 1942) was the first to consider how the Dobzhansky-Muller model might help explain Haldane's rule. Although his insight was simple, the history of his theory and its modern descendant—the dominance theory—is complex (Zeng 1996; Orr 1997). The theory was first sketched verbally by Muller; became the leading textbook theory of Haldane's rule; was rejected 50 years later; was then experimentally resurrected; and was finally formalized mathematically. (To make matters worse, the mathematics revealed that the original verbal version was slightly wrong.)

To see Muller's idea, consider two genes that interact to cause hybrid inviability. For concreteness, assume that males are heterogametic and write one species' genotype as $A_1A_1B_1B_1$ and the other as $A_2A_2B_2B_2$. Allele A_1 from the first species is incompatible with B_2 from the second species. First, consider the case in which both loci are autosomal. If females of species 1 mate with males of species 2, the resulting hybrids are $A_1A_2B_1B_2$. Because A_1 and B_2 occur as heterozygotes, hybrids will be inviable only if both factors are fairly dominant. (Dominance here refers to an allele's effect on fitness on a hybrid genetic background. Nothing is assumed about the dominance of alleles on their normal within-species genetic background.) Note that we have no reason to expect a difference in the fitness of hybrid males versus females since both sexes have the same genotype.

Now consider the case where one locus (A) is X-linked and the other (B) is autosomal. If females of species 1 mate with males of species 2, male hybrids have genotype $A_1B_1B_2$ (where we ignore the Y chromosome), and female hybrids have genotype $A_1A_2B_1B_2$. As before, females will be inviable only if A_1 and B_2 are both fairly dominant. If, for instance, B_2 is dominant, but A_1 is recessive, hybrid females remain fit. But under the same conditions, hybrid males are dead: because the X chromosome is hemizygous in males, hybrid males are inviable *regardless* of the dominance of A_1.

Muller's point is clear: hybrid males are affected by all X-linked genes involved in genic incompatibilities, dominant *and* recessive, whereas hybrid females are affected only by that subset of genes that are fairly dominant. So long as some fraction of the genes causing hybrid problems are recessive, hybrid males (heterogametic) should fare worse than hybrid females (homogametic), giving rise to Haldane's rule. Despite the confusing terminology, note that the "dominance" theory hinges on alleles that act *recessively* in hybrids.

Perhaps the simplest prediction of the dominance theory was noted by Coyne (1985): if hybrid males are sterile or inviable because they express reces-

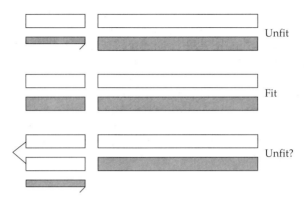

Figure 8.1 The "unbalanced female" test in a case of Haldane's rule. The hybrid F_1 males shown at the top are unfit, while the F_1 females shown in the middle are fit. If hybrid male problems are due to expression of recessive X-linked genes, then the "unbalanced" hybrid females (bottom) who carry both X chromosomes from the same species should also be unfit. The short bars represent sex chromosomes (the Y is shown with a hook), while the long bars represent haploid sets of autosomes. (From Coyne 1985.)

sive X-linked genes causing hybrid problems, then hybrid *females* that are homozygous for the same X should also be sterile or inviable (Figure 8.1). These "unbalanced" hybrid females, after all, express the same X-linked recessives as F_1 males. Coyne tested this prediction in two hybridizations obeying Haldane's rule for hybrid sterility: *Drosophila simulans–D. mauritiana* and *D. simulans-D. sechellia*. Using an attached-X chromosome, he produced unbalanced hybrid females that carry two *D. simulans* X chromosomes on an otherwise hybrid genetic background. Surprisingly, these females remained fertile in both hybridizations, contradicting Muller's hypothesis.

Although this result was unexpected—indeed so unexpected that it single-handedly reignited interest in the genetics of speciation—it was quickly confirmed by analogous tests in other *Drosophila* species. The unbalanced female test (or close variations on it) was performed in at least six evolutionarily independent hybridizations, and in all cases unbalanced females remain fit (Orr 1987, 1989b; Coyne and Orr 1989b; Orr and Coyne 1989; Turelli and Orr 1995; J. A. Coyne et al., unpublished results). These findings appeared to falsify Muller's theory.

As Wu and colleagues (Wu 1992; Wu and Davis 1993) emphasized, however, the above tests suffer one limitation: all involve hybrid sterility. It remained possible, then, that Muller's theory could explain Haldane's rule for inviability. Unbalanced female tests for inviability had not been performed for largely technical reasons: Haldane's rule for inviability is rarer than for sterility in *Drosophila* and attached-X stocks were not available in the relevant species.

Orr (1993b), however, found two *Drosophila* hybridizations in which the test could be performed: *D. simulans–D. teissieri* and *D. melanogaster–D. simulans*. Surprisingly, in both crosses, females who are homozygous for an X on an otherwise hybrid background are *inviable*. Moreover, these females die at the same developmental stage as F_1 males, suggesting that hybrid males and unbalanced hybrid females die from the same problem. Taken together, these results provide strong support for Muller's theory of Haldane's rule for inviability.

In retrospect, it is clear why the hybrid inviability versus sterility experiments yield different results. As we discuss below, studies in *D. melanogaster* show that lethal mutations almost always affect both sexes, whereas sterile mutations typically affect only one sex. If the same pattern holds in species hybrids—and there is now good evidence that it does (see below)—an X that carries a recessive allele killing hybrid males will, when made homozygous, kill hybrid females. But an X that carries a recessive allele *sterilizing* hybrid males need not sterilize hybrid females when made homozygous: a female is none the worse for being homozygous for a male sterile.

Muller's theory of Haldane's rule has been formalized mathematically (Orr 1993a; Turelli and Orr 1995; Orr and Turelli 1996). Although confirming the essence of Muller's argument, this work revealed an error in his intuition. While Muller correctly noted that the genome of hybrid females, unlike that of males, partly masks the effects of X-linked genes causing hybrid problems, he overlooked the fact that females, by carrying twice as many X chromosomes, suffer twice as many possible hybrid incompatibilities involving the X (Orr 1993a). If the alleles causing hybrid problems act additively in hybrids (no dominance), these two tendencies balance and hybrid males and females have equal fitness. But if the alleles causing hybrid problems act partially recessively, hybrid males suffer lower fitness than females. Muller's theory thus requires more than the existence of some recessive alleles affecting hybrid fitness—it requires that the genes causing hybrid problems are *on average* partially recessive. In particular, Haldane's rule occurs when $d < 1/2$, where the parameter, d, incorporates both simple dominance and any covariance between deleterious effects in hybrids and dominance (Turelli and Orr 1995).

These calculations assume that genes afflicting hybrid males versus females accumulate at the same evolutionary rate. Although this is presumably true for hybrid inviability (as the same loci affect both sexes), it is not necessarily true for sterility (as different loci affect each sex). Indeed, there is good evidence that male and female steriles accumulate at different rates (see below). Taking this into account, Turelli and Orr (1995) showed that Haldane's rule occurs when

$$d < \frac{\tau p_x}{2[1-\tau(1-p_x)]}$$

where p_x is the proportion of Dobzhansky–Muller incompatibilities that involve X-linked loci, and τ is the ratio of male to female hybrid steriles (if hybrid male steriles accumulate faster than female steriles, $\tau > 1$). If male and female steriles accumulate at the same rate ($\tau = 1$), the above equation reduces to $d < \frac{1}{2}$, as expected.

The more interesting cases occur when $\tau \neq 1$ (i.e., when there is both dominance *and* differential accumulation of sex-specific factors). The qualitative effect is straightforward: as the rate of male evolution exceeds female evolution, it is easier to obtain Haldane's rule (i.e., the mean dominance consistent with the rule rises above $d = ½$). Similarly, as the alleles causing postzygotic isolation become more recessive, Haldane's rule holds even if male steriles accumulate more slowly than female steriles ($\tau < 1$). Thus, either force alone can yield Haldane's rule; their joint action merely makes it easier to obtain Haldane's rule in male heterogametic taxa.

Additional support for the dominance theory comes from tests of one of its simplest predictions: if the genes causing postzygotic isolation are recessive, it should be possible to find autosomal alleles whose effects, while masked in F_1 hybrids, cause postzygotic isolation when made homozygous or hemizygous on a hybrid background. Such normally masked factors have now been found in backcross experiments involving nearly intact chromosomes (Muller and Pontecorvo 1942; Orr 1992; Davis et al. 1994), in deletion mapping experiments that make chromosome regions from one species hemizygous in the genetic background of another species (Coyne et al. 1998; Presgraves 2003), and in introgression experiments involving small chromosome regions (Breeuwer and Werren 1995; Hollocher and Wu 1996; True et al. 1996; Sawamura et al. 2000; Tao et al. 2003a).

Support for the dominance theory also comes from comparative contrasts. Turelli and Begun (1997) pointed out that, all else being equal, taxa with large X chromosomes should, under the dominance theory, evolve Haldane's rule faster than taxa with small X chromosomes. This prediction was confirmed by comparing the time needed to evolve Haldane's rule in pairs of *Drosophila* species having 40% versus 20% of their genome on the X chromosome.

Further evidence for the dominance theory appears to come from the so-called large X-effect. Genes having a large effect on postzygotic isolation often map to the X chromosome, e.g., *Hst-3*, which causes male sterility in hybrids between the mouse species *Mus musculus domesticus* and *M. spretus* (Guenet et al. 1990). Moreover, backcross analyses in *Drosophila* typically show that substitution of an X chromosome from one species versus the other has a larger effect on postzygotic isolation than substitution of an autosome (Coyne and Orr 1989b). Figure 8.2 shows an example. Similarly, studies of hybrid zones in mice, butterflies, and birds have repeatedly revealed less introgression of X-linked genes than autosomal genes across the zones (Tucker et al. 1992; Dod 1993; Hagen and Scriber 1989; Saetre et al. 2003; Payseur et al. 2004). However, interpretation of this large X-effect has proved difficult. Initially, it was thought to reflect rapid evolution of X-linked genes causing postzygotic isolation (Charlesworth et al. 1987; Coyne and Orr 1989b). While possible, this idea is controversial (see below). Wu and colleagues instead argued that the large X-effect reflects an observational bias: in backcross analyses, one can only compare hemizygous-X with *heterozygous*-autosome substitutions, not with *homozygous*-autosome substitutions (Wu and Davis 1993; Hollocher and Wu 1996; Wu et al. 1996).

But it is important to recognize a distinction. If the genes causing hybrid problems act additively (no dominance), one might expect hemizygous-X sub-

Figure 8.2 Genetic evidence from *Drosophila pseudoobscura–D. persimilis* hybrids for a "large X-effect." Hybrid male genotypes produced in a backcross are shown along the Y-axis, and fertility (measured as percentage of males with motile sperm) is shown along the X-axis. Males from the top half of the plot carry a *D. pseudoobscura* X chromosome and are often fertile, whereas males from the bottom half of the plot carry a *D. persimilis* X chromosome and are almost always sterile. *D. pseudoobscura* chromosomes are shown in white, *D. persimilis* in gray. (From Orr 1987.)

stitutions to have twice the effect of same-sized heterozygous-autosome substitutions; such a difference might legitimately be called an artifact of hemizygosity. However, if the genes causing hybrid problems are recessive, hemizygous-X substitutions might have (and do have) *much* larger effects than autosomal substitutions. In this case, the large X-effect is not an artifact of dominance but *evidence for* dominance (Orr 1997; Turelli and Orr 2000). An observational bias also cannot explain one of the best-known facts about the X-effect:

it is seen only with hybrid sterility and inviability, not with morphological differences between species (Charlesworth et al. 1987; Coyne and Orr 1989b; but see Reinhold 1998 for possible exceptions involving sexually selected traits). If the X-effect is an artifact, why isn't it seen with all characters? The likely answer is that morphological species differences often involve alleles that act roughly additively in hybrids, while postzygotic isolation involves alleles that act fairly recessively in hybrids.

While providing qualitative support for the dominance theory, the above work does not tell us if the quantitative criterion for conformity to Haldane's rule—$d < 1/2$—is met. Recent analyses provide the first quantitative estimates of d. Tao and Hartl (2003) used a large battery of mapped P element inserts to study the effects of hundreds of introgressed chromosome regions on sterility between *D. simulans* and *D. mauritiana*. They showed that hybrid male steriles have a dominance coefficient (d) of about 0.25, and that hybrid lethals have a dominance coefficient of about 0.35. Both values are well within the range required by the dominance theory.

Recent work has thus produced a strong consensus that dominance alone can explain Haldane's rule for inviability. The data further suggest that hybrid steriles are on average partially recessive.

THE FASTER-MALE THEORY. Haldane's rule would obviously arise if incompatibilities afflicting heterogametic hybrids were more common than those afflicting homogametic hybrids. Wu and colleagues (Wu and Davis 1993; Wu et al. 1996) have championed a version of this idea. Hybrid male steriles, they suggest, are more common than hybrid female steriles. They argue that two processes might give rise to such "faster-male" effects:

1. Spermatogenesis might be an inherently sensitive process that is easily perturbed in hybrids (Wu and Davis 1993). If so, male- and female-expressed genes may diverge at the same rate, but a larger fraction of the former would cause hybrid problems.

2. Sexual selection might cause faster evolution of male- than female-expressed genes (since males are involved in male-female, as well as male-male interactions). These more diverged male genes would be more likely to cause problems in hybrids (Wu and Davis 1993).

For the time being we ignore the distinction between these possible causes and focus on the evidence for or against the predicted pattern: Are there more genic incompatibilities afflicting male than female hybrids?

The faster-male theory makes a simple prediction: chromosome regions moved from one species into another should include hybrid male steriles more often than hybrid female steriles. This prediction was verified in True et al.'s (1996) large introgression experiment between *D. mauritiana* and *D. simulans*, in which they found a nine-fold excess of hybrid male steriles over female steriles. True et al.'s results were confirmed by Hollocher and Wu (1996), who introgressed material from *D. mauritiana* and *D. sechellia* into *D. simulans* and, more recently still, by Tao et al. (2003a) and Tao and Hartl's (2003) analysis of hundreds of introgressions between *D. mauritiana* and *D. simulans* (see also Sawamura et al. 2000).

The above data all derive from *Drosophila*. To test the generality of faster male evolution, Presgraves and Orr (1998) performed a comparative test outside *Drosophila*. They compared the incidence of Haldane's rule in two types of mosquitoes: those having "normal" degenerate Y chromosomes in which males are hemizygous for the X (genus *Anopheles*), and those having genetically active Y chromosomes in which males are not hemizygous for the X (genus *Aedes*, in which sex is determined by a small chromosome region). In *Anopheles*, both the dominance and faster-male theories can act and these mosquitoes should conform to Haldane's rule for both hybrid inviability and sterility. They do. In *Aedes*, on the other hand, the dominance theory cannot act— since no sex is hemizygous—and there is no reason to expect *Aedes* to conform to Haldane's rule for inviability. They don't. There is, however, reason to expect *Aedes* to conform to Haldane's rule for *sterility*: the faster-male theory can act whether the X chromosome is hemizygous or not. As predicted, *Aedes* does conform to Haldane's rule for sterility. These results provide strong support for both the dominance and faster-male theories. A similar pattern appears to hold in a second group. Frogs of the genus *Xenopus* also have non-degenerate Y chromosomes, although *females* are heterogametic (Kelley 1996). As predicted by the faster-male theory, hybrid *males* appear to be sterile more often than females (Kobel 1996; Kobel et al. 1996; Orr and Presgraves 2000), although quantitative data have apparently not been published.

Finally, recent work on the misexpression of genes in hybrids has provided support for the faster-male theory. Using *Drosophila* microarrays, Michalak and Noor (2003) compared gene expression patterns in *Drosophila simulans*, *D. mauritiana*, and their hybrids. Interestingly, genes that are expressed in males only are the most likely to be misexpressed in hybrids, as expected under the faster-male theory. Indeed this finding appears fairly general, characterizing expression patterns in both *D. pseudoobscura–D. persimilis* (Reiland and Noor 2002) and *D. melanogaster–D. simulans* (Ranz et al. 2004) hybrids.

Despite the above support for the faster-male theory, there are several caveats. The first is that the theory has limited applicability. Faster-male evolution cannot explain Haldane's rule for hybrid inviability— it is a theory of hybrid sterility. More important, the theory cannot explain Haldane's rule for sterility in taxa in which *females* are heterogametic. In Lepidoptera and birds, for example, faster-male evolution would work *against* Haldane's rule, as it is hybrid females that are preferentially sterile. (And this preference is strong: Presgraves 2002 showed that 29 out of 30 Lepidopteran hybridizations obey Haldane's rule for sterility; see Price and Bouvier 2002 for birds.) This fact does more than show that faster-male evolution cannot offer a general explanation of Haldane's rule; it shows that faster-male evolution is weaker than whatever force—presumably dominance—causes female sterility in Lepidoptera and birds.

Second, we still do not know what force drives faster-male evolution. Although sexual selection has received more attention than the spermatogenesis-is-special idea, the evidence discussed above is consistent with either possibility. Recent molecular data do provide, however, some support for the sexual selection hypothesis. Although early studies of DNA and protein divergence between species revealed that reproductive tract genes and proteins evolve faster than non-reproductive ones, male and female reproductive genes often evolved at *similar* rates, contrary to faster-male expectations (Coulthart and Singh 1988; Civetta and Singh 1995; Swanson et al. 2001b). More recent data, however, yield a different picture. Zhang et al. (2004) compared patterns of sequence evolution among *Drosophila* genes that show male-biased, female-biased, or no sex-bias expression patterns, as determined from cDNA microarray experiments. They found that male-biased genes evolve significantly faster between *Drosophila* species than either female-biased or nonsex-biased genes; this faster evolution mainly reflects more rapid change at replacement sites. Similarly, Ranz et al. (2003) found that male-biased genes have diverged in *expression* pattern between *D. melanogaster* and *D. simulans* more than have female-biased or nonsex-biased genes. Both of these results are obviously consistent with the idea that sexual selection drives rapid evolution of "male genes."

THE FASTER-X THEORY. Charlesworth et al. (1987) argued that Haldane's rule may be a consequence of the large X-effect: if X-linked genes have a disproportionately large effect on hybrid fitness, it is hardly surprising that heterogametic hybrids suffer more than homogametic hybrids. The problem, then, is to understand why the X plays such a large role in postzygotic isolation. Charlesworth et al. speculated that X-linked genes simply evolve faster than autosomal ones. They showed that, when evolution is driven by natural selection of new favorable mutations, X-linked substitutions occur at a higher rate than autosomal ones when favorable mutations are on average partially recessive ($\bar{h} < 1/2$; where dominance now refers to alleles on their normal genetic background and nothing is assumed about dominance in hybrids).

The faster-X theory differs in an important way from the dominance and faster-male theories: by itself, it cannot explain Haldane's rule. The problem is that, if genes affecting males and females evolve at the same rate and act additively in *hybrids*, male and female hybrids are equally fit regardless of the rate of evolution on the X (Orr 1997). The faster-X theory can, however, be modified in two ways to yield Haldane's rule. First, the genes causing hybrid problems might also act recessively in hybrids. While the faster-X theory now depends on dominance, faster-X evolution can exaggerate the effect of dominance (Orr 1997). Alternatively, many of the genes that afflict hybrids might be expressed in one sex only. If so, X-linked genes expressed in the heterogametic sex will evolve faster than those expressed in the homogametic sex when new favorable mutations are partially recessive (Coyne and Orr 1989b).

Introgression experiments provide a potentially powerful test of faster-X evolution. By comparing hemizygous-X with homozygous-autosomal introgressions, one can assess the density of X versus autosomal genes causing hybrid problems. Unfortunately, these experiments have yielded mixed results. Hollocher and Wu (1996) argued that hybrid sterility genes are no more common on the X than autosomes in the *D. simulans* clade. True et al. (1996), on the other hand, found that hybrid male sterility genes are 50% more common on the X than autosomes in *D. mauritiana–D. simulans* hybrids. More recently, Tao et al. (2003a), in their study of the same hybridization, found that hybrid male steriles are 2.5-fold more dense on the X than on an autosome.

The faster-X theory does, though, make an additional prediction that can be tested without studying postzygotic isolation: if new favorable mutations are partially recessive, we should see more rapid molecular evolution at X-linked than autosomal genes. Unfortunately, tests of this prediction have also yielded mixed results. Betancourt et al. (2002) studied divergence of over 200 genes between *D. melanogaster* and *D. simulans* and found no evidence of faster replacement (amino acid changing) evolution at X-linked than autosomal loci. Counterman et al. (2004), on the other hand, found some evidence for faster-X evolution, although the effect was small.

In summary, support for the faster-X theory is mixed. Although faster-X effects may well contribute to Haldane's rule, the evidence is weaker than that for dominance and faster-male effects.

MEIOTIC DRIVE. The idea that selfish genetic elements play a role in speciation has deep roots in evolutionary biology (Grun 1976, pp. 352–354; Hurst and Pomiankowski 1991; Cosmides and Tooby 1981; reviewed in Hurst and Werren 2001). Although this idea takes many forms, involving infectious endosymbionts, mitochondrial lesions causing cytoplasmic male sterility, or transposable elements, one version of the idea—that meiotic drive factors cause postzygotic isolation generally, and Haldane's rule specifically—has received a great deal of attention (Frank 1991; Hurst and Pomiankowski 1991; Henikoff et al. 2001; Tao et al. 2001; Henikoff and Malik 2002; Tao and Hartl 2003). Meiotic drive alleles distort Mendelian ratios to their own advantage, often by inactivating sperm that carry a homologous chromosome (e.g., X chromosome-bearing sperm might inactivate Y chromosome-bearing sperm). While advantageous for the driving gene, meiotic drive imposes a fertility cost on its bearers and on most other genes in the genome (Jaenike 2001). When on the sex chromosomes, meiotic drive factors also distort sex ratios away from the Fisherian 50:50 sex ratio. There will thus often be strong selection to suppress drive (Sandler and Novitski 1957).

It is easy to imagine that two allopatric populations might evolve different meiotic drive factors, each of which then becomes suppressed, leaving both populations with normal segregation ratios. However, if these populations were to later hybridize, normally masked meiotic drive might become unmasked (if the suppressors are less than fully dominant). One can even imagine that X-linked drive factors would inactivate Y-bearing sperm while Y-linked factors would inactivate X-bearing sperm, rendering hybrid males sterile and giving rise to Haldane's rule (Frank 1991; Hurst and Pomiankowski 1991).

Although this idea is attractive—especially as meiotic drive occurs in many organisms—it fell out of favor in the early 1990s. Johnson and Wu (1992), and Coyne and Orr (1993) showed that, in several evolutionarily independent *Drosophila* hybridizations, meiotic drive does not occur in hybrid males who are partially fertile. These results were widely taken as fatal to the meiotic drive hypothesis.

Recent findings, however, suggest that Johnson and Wu (1992), and Coyne and Orr (1993) were unfortunate in their sampling of species pairs or hybrid genotypes. Indeed, normally cryptic meiotic drive systems have now been found in several *Drosophila* hybridizations. In *D. simulans*, flies sampled from within populations show little or no drive, while those produced by crossing individuals from Tunisia with individuals from Seychelles or New Caledonia show drive (Mercot et al. 1995; Cazemajor et al. 1997; Montchamp-Moreau and Joly 1997). Similarly, while drive is rare in *D. simulans* and *D. sechellia*, some hybrid introgression lines between the species show strong meiotic drive (Dermitzakis et al. 2000).

Neither of these cases, however, implicates meiotic drive as the *cause* of hybrid sterility. But three others do. Hauschteck-Jungen (1990) showed that introduction of a known meiotic drive X chromosome rearrangement from Tunisian populations of *D. subobscura* into European populations causes hybrid male sterility. Similarly, Tao et al. (2001) found that a small (< 80 kb) region of the third chromosome from *D. mauritiana* causes meiotic drive when homozygous in *D. simulans*: hybrid males carrying this region produce mostly daughters (the region presumably includes a suppressor of sex chromosome drive). Interestingly, this same small region also causes hybrid male sterility. Because Tao et al. were unable to separate hybrid meiotic drive and hybrid male sterility, they conclude that the same gene—*tmy* (*too much yin*)— probably causes both. Last, Orr and Irving (2004) recently found that F_1 hybrid males between the Bogota and USA subspecies of *D. pseudoobscura* show X chromosome meiotic drive. The genes causing hybrid meiotic drive map to the same X chromosome regions—and show the same pattern of epistasis—as the genes causing hybrid male sterility.

The fact that meiotic drive might sometimes cause hybrid sterility does not, however, mean that meiotic drive drove the *evolution* of postzygotic isolation. The problem is that the above findings are consistent with two scenarios. One is that meiotic drive factors arose within populations and were suppressed, as described above. The other is that meiotic drive never occurred in the history of either species and instead represents a hybrid pathology (Dermitzakis et al. 2000). In this case, hybrid meiotic drive is a special case of a Dobzhansky–Muller incompatibility and neither lineage passed through the adaptive valley represented by the driving genotype. Unfortunately, we know of no way to distinguish between these scenarios. Thus, while meiotic drive could turn out to be common in hybrids—and might even cause the sterility of some hybrids—we do not yet know if meiotic drive was the evolutionary cause of the *substitutions* that ultimately cause postzygotic isolation.

Two variations on the traditional meiotic drive theory of Haldane's rule have recently appeared. In the first, Tao and Hartl (2003) argue that, because different regions of the genome "prefer" different sex ratios (the Y chromosome in mammals, for instance, prefers more sons), the resulting genomic conflicts might drive rapid evolutionary change. Such conflicts, they suggest, are especially likely in the heterogametic sex. Tao and Hartl argue that this "faster-heterogametic" theory—which could explain Haldane's rule for sterility, not inviability—makes several unique predictions. Perhaps most important, in taxa having heterogametic females, hybrid *female* steriles should accumulate faster than hybrid male steriles. Unfortunately, though, this prediction is not unique to the meiotic drive idea: it also follows from the faster-X theory, as extended by Coyne and Orr (1989b; see also Orr 1997). In any case, there is some evidence against the above prediction, at least in one group. In the genus *Xenopus*, females are heterogametic, the Y is not degenerate, and dominance cannot cause Haldane's rule. The faster-heterogametic theory would thus seem to predict an excess of hybrid female steriles, while the faster-male theory predicts an excess of hybrid male steriles. As noted, hybrid male sterility appears more common than female sterility (Kobel 1996; Kobel et al. 1996), although more quantitative data are badly needed.

In the second recent theory, Henikoff et al. (2001) and Henikoff and Malik (2002) argue that centromeric DNA sequences might play a special role in Haldane's rule. Their argument starts from the observation that the repetitive DNA that makes up centromeres evolves rapidly, as does the histone (Cid) that binds to these sequences (Malik and Henikoff 2001; Malik et al. 2002). Henikoff and colleagues suggest that this rapid evolution reflects an arms race. Homologous chromosomes compete with each other during female meiosis: because only one of four meiotic products arrives in an egg in most species (the others are degraded), centromeric sequences compete for spindle attachment and so evolve in an attempt to be the "lucky" product that enters the egg. While this type of meiotic drive is well known in mammals (de Villena and Sapienza 2001; Henikoff et al. 2001), it comes at a cost: misalignment of centromeres can derail male meiosis, causing sterility. There will thus be selection to suppress "selfish" centromeres—suppression that might well involve evolution at Cid. The argument now takes a familiar form: bouts of coevolution between centromeric sequences and Cid maintain fair meiosis within populations; but there is no guarantee that Cid from one population can suppress the meiotic-drive of centromeric sequences from another. Because heterogametic hybrids suffer a greater imbalance in centromere strength, such hybrids will often be rendered sterile (or produce biased sex ratios), yielding Haldane's rule.

Although we would not be surprised if centromeric sequences sometimes cause postzygotic isolation, there are some reasons for doubting that this idea provides a general explanation of Haldane's rule. First, the model cannot easily explain the *inviability* of F_1 hybrids. Second, it is not clear that the model can operate in XO species, where competition between X and Y centromeres cannot occur; nonetheless, such species obey Haldane's rule (Coyne and Orr 1989b). Third, the model would not seem to apply to haplodiploid species where, again, centromeric competition cannot occur in haploid males; nonetheless, such males often suffer genic-based postzygotic isolation (Gadau et al. 1999). Fourth, it is not clear that the model works in taxa having single-locus sex determination; nonetheless, such taxa obey Haldane's rule for hybrid sterility (Presgraves and Orr 1998). Finally, recent work shows that *Drosophila* stocks that carry hybrid rescue mutations (see below) do not differ from wild-type stocks in Cid sequence; moreover, species that have different Cid sequences can produce fit hybrids (Sainz et al. 2003). Although none of these problems is fatal, some special pleading seems required to rescue the centromeric hypothesis in each case. In the end, the question of how often centromeric arms races cause postzygotic isolation will be settled by molecular identification of the genes causing hybrid problems. So far, such work provides no strong evidence either for or against the hypothesis (see below).

Indeed this seems a fair assessment of all the meiotic drive theories of Haldane's rule, old and new. We have no compelling reason for rejecting them nor particularly good reasons for embracing them, at least as general explanations. Meiotic drive theories represent one of the few areas of Haldane's rule that demands further work.

Conclusions

After two decades of intensive study, a consensus has emerged that two forces, dominance and faster-male evolution, cause Haldane's rule. Two other forces, faster-X evolution and meiotic drive, may also play a role, although the evidence is more ambiguous.

One of these forces—dominance—differs from the others in several ways. First, only dominance can explain Haldane's rule for hybrid sterility *and* inviability and in taxa with heterogametic males or females. Second, several of the other theories rely, in one way or another, on dominance. (The faster-male theory, for instance, requires dominance to explain Haldane's rule in taxa having heterogametic females.) One could argue, therefore, that dominance plays a more fundamental role in Haldane's rule than the other theories.

There can be no doubt, however, that faster-male evolution is also involved in Haldane's rule for sterility. Thus, at least in some taxa, Haldane's rule must reflect the simultaneous action of both dominance and faster-male affects. It is reassuring to note that this conclusion—reached by sometimes arcane genetic studies—makes sense of one of the simplest patterns associated with Haldane's rule. For if both forces play a role in the pattern, one might expect more cases of Haldane's rule for sterility than inviability (Wu et al. 1996): with sterility, after all, both dominance and faster-male evolution can act; while, with inviability, dominance alone can act. Moreover, this excess should be limited to taxa having heterogametic males, the only groups in which both forces act in the same direction. This is precisely the pattern seen (see Table 8.1): there are many more cases of hybrid sterility than inviability in mammals and *Drosophila*, but not in birds and Lepidoptera (Wu et al. 1996).

In summary, we suspect that the main causes of Haldane's rule have been identified. Future work should focus on controversial theories like meiotic drive (not on well supported ones like dominance), and on poorly studied female-heterogametic taxa like Lepidoptera (not on well studied male-heterogametic ones like *Drosophila*).

The Genetic Basis of Postzygotic Isolation

Although geneticists studying postzygotic isolation have perhaps been obsessed with Haldane's rule, the consequences have not been altogether bad. The burst of work inspired by this rule led to the collection of enormous quantities of data that bear on many other problems in the genetics of speciation. Because the resulting literature is both large and technical, we limit our discussion to the most important of these topics. These are: the number of genes causing postzygotic isolation, the complexity of hybrid incompatibilities, the probability that genes are incompatible in hybrids, the location of the genes causing postzygotic isolation, the developmental basis of hybrid problems, the role of duplicate genes, and the molecular identity of the genes causing postzygotic isolation. Our discussion generally moves from more traditional to more modern topics.

How many genes cause postzygotic isolation?

Although this might seem one of the simplest questions that one could ask about hybrid sterility and inviability, it is not. The problem is that the question means different things to different people. To introduce this topic, we first survey various views about gene number that have appeared since the Modern Synthesis. We then turn to several important distinctions that have often been obscured in these discussions. Finally, we review experimental attempts to count the genes causing hybrid problems.

THE VIEWS. The Dobzhansky–Muller model shows that intrinsic isolation must involve at least two genes, but it does not tell us if it typically involves two or two thousand. Nonetheless, the founders of the Synthesis had strong views on the matter. Mayr (1963, p. 543) summarized the consensus: "species differences ... seem to be controlled by a large number of genetic factors with small individual effects. The genetic basis of the isolating mechanisms, in particular, seems to consist largely of such genes." Given that the first genetic dissection of a genic incompatibility—in the composite *Crepis* (Hollingshead 1930)—involved a single incompatibility between two Mendelizing genes, this polygenic view may seem puzzling. But the founders' views on the genetics of speciation were closely allied with their views on the genetics of adaptation. As is well known, the founders championed an extreme brand of micromutationism, positing that adaptive evolution is underlain by a very large number of genes, each of small effect (Muller 1923, p. 543; Dobzhansky 1937, p. 26; Wright 1948; for reviews, see Turner 1985; Orr and Coyne 1992; Orr 1998b). It seems clear that this micromutationism colored the founders' views of *all* species differences, including reproductive isolation. But there is no necessary connection between the genetic basis of adaptation and that of intrinsic isolation. Species might diverge adaptively via many genes of small effect and yet hybrids might suffer low fitness due to only two or three of these genes. Conversely, species might diverge adaptively via a few genes of large effect and yet hybrids might suffer low fitness due to the effects of many genes that diverged by genetic drift. The founders' views were nonetheless influential and, by the 1940s, widely accepted (e.g., Mather 1943, 1949).

Recent views fall into two camps. One camp remains polygenic and itself includes two variants. The "strong" variant is exemplified by Naveira and colleagues (Carvajal et al. 1996; Naveira and Maside 1998). They suggest that postzygotic isolation is not only polygenic but that the factors involved are "interchangeable" (i.e., hybrid sterility involves many genes and the identity of these genes is of little significance; it is their *number* that matters). Naveira and Maside (1998) go so far as to suggest that introduction of arbitrary non-coding DNA from a foreign species can cause postzygotic isolation, so long as there is enough of it. The "weak" polygenic view is less radical. While maintaining that the total number of genes involved is large, it argues that the identity of these genes matters: some genes cause hybrid problems, while others do not (Wu et al. 1996). The second camp argues that postzygotic isolation sometimes has a simple genetic basis and sometimes involves genes of major effect (Orr 1992).

DISTINCTIONS.   Before we can answer questions about gene number, we have to clarify several distinctions. In the past, these distinctions were often obscured, causing considerable confusion.

The most important distinction is: Are we interested in the number of genes that *originally* caused complete postzygotic isolation, or the number that *now* cause postzygotic isolation? To see the difference, consider two diverging taxa. Imagine that after 100,000 years these taxa evolve complete hybrid male sterility due to a single incompatibility between two genes. These species will not stop diverging at additional loci that can also cause hybrid male sterility. Indeed, if we genetically analyze this species pair after, say, 1 million years of divergence, they may well now differ by many genes capable of sterilizing hybrid males. We must therefore be clear about the quantity of interest. And surely the quantity of most evolutionary interest is the number of genes that *initially* caused complete postzygotic isolation. There are also good theoretical grounds for thinking that this "overcounting problem" is serious. The main reason is the snowball effect discussed in Chapter 7: under the Dobzhansky–Muller model,

Reference
Pontecorvo 1943b
Presgraves 2003
Coyne 1984
True et al. 1996; Wu et al. 1996
Tao and Hartl 2003; Tao et al. 2003a, b
True et al. 1996
True et al. 1996
Coyne and Kreitman 1986; Coyne and Charlesworth 1989; Cabot et al. 1994; Hollocher and Wu 1996
Hollocher and Wu 1996
J. A. Coyne et al., unpubl. results
Vigneault and Zouros 1986; Pantazidis et al. 1993
Orr 1987; Orr 1989c; Noor et al. 2001b
Orr 1987
Noor et al. 2001b
Orr 1989b; Orr and Irving 2001
Kaufmann 1940
Sturtevant 1946
Crow 1942
Carvajal et al. 1996
Naveira and Fontdevila 1986, 1991
Khadem and Krimbas 1991
Mitrofanov and Sidorova 1981
Heikkinen and Lumme 1991
Schäfer 1978
Schäfer 1978
Schäfer 1979
Patterson and Griffen 1944; Patterson and Stone 1952
Lamnissou et al. 1996
Dobzhansky 1975
Gadau et al. 1999
Wittbrodt et al. 1989
Macnair and Christie 1983; Christie and Macnair 1984, 1987
Macnair and Christie 1983; Christie and Macnair 1984, 1987
Rieseberg 1998; Rieseberg et al. 1999
Barton and Hewitt 1981
Szymura and Barton 1991

2. Introgression analysis, in which partially fertile hybrids are repeatedly backcrossed to a parental species, yielding lines in which chromosome regions are introduced (more or less singly) onto the genetic background of another species. These introgressed regions are typically made homozygous on this foreign background and then scored for hybrid fitness.

3. Deficiency analysis, in which chromosome deletions having known breakpoints are used to find hybrid sterility or inviability genes whose effects are normally masked in F_1 hybrids. This approach allows precise mapping but requires sophisticated genetic tools, like those available in *Drosophila melanogaster*.

All three of these approaches can detect factors that might not cause hybrid F_1 problems; the reason is that all three produce genotypes that do not exist in the F_1 generation. It should be understood, then, that few of the studies listed in Table 8.2 count the number of genes causing F_1 problems. With this in mind, the first conclusion to be drawn from Table 8.2 is clear: postzygotic isolation involves a fair number of genes. Rarely do experiments reveal only two genes causing hybrid problems, the fewest possible under the Dobzhansky–Muller mechanism.

Table 8.2 also shows that, as expected, the number of factors detected usually increases with experimental resolution. This can be seen in two pairs of studies. The first involves hybrid sterility between *D. simulans* and *D. mauritiana*. In early work, Coyne (1984) used five markers to study the genetic basis of hybrid male sterility. He found that all markers are linked to genes causing hybrid sterility and, thus, at least five genes are involved. In recent work, Tao et al. (2003a,b), and Tao and Hartl (2003) used a large collection of *P* element inserts to study the effects of hundreds of introgressed chromosome regions on sterility between the same species. Remarkably, about half of the introgression lines suffered substantial hybrid male sterility, leading Tao and Hartl to estimate that about 60 genes are involved. Similarly, early work by Pontecorvo (1943b) showed that at least nine factors play a role in the inviability of *D. melanogaster*–*D. simulans* hybrids. Nevertheless, higher resolution work by Presgraves (2003) arrived at a very different figure. Using a large set of chromosome deletions, he found 20 chromosome regions that cause hybrid inviability (Figure 8.3). Extrapolating from the fraction of the genome studied, he concluded that nearly 200 genes cause hybrid lethality.

Although these findings might seem to suggest that the number of genes causing postzygotic isolation is limited only by experimental resolution, some caution is called for. The problem is that the cases just described involve species that diverged long ago. The presence of many hybrid sterility or inviability genes, therefore, could reflect the overcounting problem noted earlier. Indeed, such overcounting is essentially guaranteed by the design of introgression and deficiency experiments: in both, regions are said to include hybrid steriles or lethals only if they lower hybrid fitness *independently* of other tested regions. (Though introgression of small chromosome segments will sometimes fail to detect alleles that are involved in complex incompatibilities.) Indeed, Tao et al.

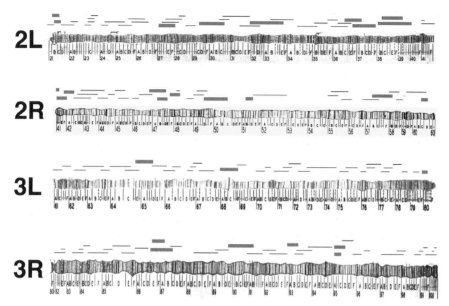

Figure 8.3 Regions of the *Drosophila simulans* genome that cause hybrid inviability in a partly *D. melanogaster* genetic background. These regions were identified by introducing chromosomal deletions into species hybrids. Deletions that uncover gene(s) causing hybrid inviability genes are shown as bold bars above chromosomes, while deletions that do not uncover hybrid inviability are shown as light bars. (From Presgraves 2003.)

(2003a) find that *D. simulans* and *D. mauritiana* are separated by 15 hybrid male sterile "equivalents" (i.e., hybrid males are sterile 15 times over). The real question, then, is what happens in younger taxa. Are evolutionary younger taxa—but ones that still show strong F_1 sterility or inviability—*also* separated by many genes causing postzygotic isolation?

To answer this question, Orr and Irving studied two subspecies of *Drosophila pseudoobscura*. The Bogota and USA subspecies diverged recently (Schaeffer and Miller 1991; Wang et al. 1997; Machado et al. 2002; Machado and Hey 2003) and are considerably younger than the taxa considered above. Despite this, these subspecies form sterile F_1 males in one direction of the hybridization and so have been viewed as a paradigmatic case of incipient speciation (Lewontin 1974). Orr and Irving (2001) concluded that postzygotic isolation between these taxa involves a modest number of genes. Although at least five chromosome regions cause hybrid male sterility, large regions of the genome—including entire chromosomes—have *no* discernible effect on hybrid fertility. This result qualitatively differs from that seen in older species pairs. Also, the factors found in this study do not include ones that accumulated after the evolution of complete hybrid male sterility. Instead, four of the five regions found are required for *any* hybrid sterility and the fifth is required for strong hybrid sterility. Sterility between young taxa thus appears to involve fewer genes than between older taxa.

HYBRID STERILITY VERSUS HYBRID INVIABILITY. Early studies probably overestimated the number of genes needed to cause postzygotic isolation in another way: they focused on hybrid male sterility. It now appears that, at any point in time, hybrid male sterility involves more genes than hybrid female sterility or hybrid inviability. The best evidence comes from *Drosophila simulans* and *D. mauritiana*. In their large introgression analysis, Tao and Hartl (2003) and Tao et al. (2003a) found at least ten-fold more genes causing hybrid male sterility than hybrid female sterility or hybrid inviability (see also True et al. 1996; Wu and Ting 2004). Similar results have been obtained in two other hybridizations: *D. melanogaster–D. simulans* (Sawamura et al. 2000) and *D. sechellia–D. mauritiana* (J. P. Masly and D. Presgraves, pers. comm.).

RESULTS FROM HYBRID ZONES. Although early studies of hybrid male sterility probably overestimated the number of genes required to cause reproductive isolation, a different kind of study—statistical analysis of hybrid zones—suggests that reproductive isolation *is* often fairly polygenic. Barton (reviewed in Barton and Gale 1993) showed how information on dispersal rate (inferred from the strength of linkage disequilibrium) and the distance over which hybrids suffer low fitness (inferred from cline width and shape) can be used to estimate the number of genes lowering hybrid fitness. Barton and colleagues used this approach to estimate the number of genes decreasing hybrid fitness between two races of the Alpine grasshopper, *Podisma pedetris*, and between two species of the fire-bellied toad, *Bombina bombina* and *B. variegata*, both pairs of which meet in hybrid zones. This work yielded high estimates of gene number (Table 8.2): 150 in *Podisma* (Barton and Hewitt 1981b) and 55 in *Bombina* (Szymura and Barton 1991). Although this work is far less direct than the mapping experiments considered above—and the results surely involve a larger margin of error—it does appear that many genes are involved.

Given that hybrid zones by definition involve taxa that are incompletely isolated (negating concerns about overcounting), these high estimates may seem surprising. But the explanation is probably simple. While genetic mapping studies focus on a single isolating barrier (e.g., hybrid male sterility), the hybrid zone approach considers *all* components of hybrid fitness, whether intrinsic or extrinsic, and involving inviability or sterility.

SUMMARY. The main reason we have had a hard time answering "How many genes cause postzygotic isolation?" should now be clear: It is not a single question. Instead, this query masks a large number of questions that may have very different answers. If the question is, "Do direct genetic analyses of postzygotic isolation usually implicate many genes?", the answer is "yes." But if the question is, "How many genes *initially* caused complete postzygotic isolation?", the answer is less certain, although several studies suggest it may be "only a few." Despite this, there is good reason to think that *total* postzygotic isolation often involves many genes.

Below we consider two related questions: (1) Does the identity of individual genes matter?, and (2) Do major genes cause postzygotic isolation?

Complexity of hybrid incompatibilities

We now turn to the number of genes involved in a *single* genic incompatibility, not the total number of genes causing hybrid problems. Do particular incompatibilities involve two loci—as in the simplest version of the Dobzhansky–Muller model—or more than two loci? As noted in Chapter 7, theory suggests that incompatibilities might often be complex (i.e., hybrids must carry the appropriate alleles at three or more loci for sterility or inviability to appear). Because in such cases at least two genes must come from the same species, this pattern is sometimes referred to as "complex conspecific epistasis."

Muller (1942) understood that incompatibilities might involve three or more genes and discussed two cases in detail. Both involved *Drosophila pseudoobscura* and *D. persimilis* and, in both, at least three genes are required for the appearance of hybrid lethality (see also Gottschewski 1940; Mampell 1941). Subsequent work has uncovered many similar cases. Dobzhansky (1975), for instance, showed that the third chromosome from *D. willistoni* has no effect on male fertility when moved alone onto a *D. quechua* genetic background; but it has a considerable effect if the *D. willistoni* second chromosome is also present. These *D. willistoni* chromosomes obviously interact with some third (or additional) factors from *D. quechua*.

Wu and colleagues have uncovered several complex interactions in the *D. simulans* clade. One involves a region of the X chromosome that includes the *Odysseus-H* (*OdsH*) locus. As we will see, this locus probably corresponds to a hybrid sterility gene: males carrying *D. mauritiana* material from this region on an otherwise *D. simulans* background are often sterile. High-resolution analysis shows, however, that males who carry very small introgressions that include *OdsH* are sterile only 50% of the time, while males carrying *OdsH and* a more distal region are completely sterile (Perez and Wu 1995). Thus, at least two genes from the *D. mauritiana* X chromosome are required for complete sterility. For other examples of complex conspecific epistasis in *Drosophila*, see Cabot et al. (1994); Palopoli and Wu (1994); Carvajal et al. (1996); Davis and Wu (1996); Orr and Coyne (1989); Orr and Irving (2001); Tao et al. (2001, 2003a); Sawamura et al. 2004; for a non-fly example, see Gadau et al. (1999) in the wasp *Nasonia*.

Wu and colleagues have argued that epistatic interactions among tightly linked genes are so strong that fine mapping of the genes causing intrinsic isolation is compromised (Davis and Wu 1996): the effect of a chromosome region on hybrid fitness tends to dissolve among interacting factors. While such cases may occur—and might even turn out to be common for hybrid male sterility (Sawamura et al. 2004)—they are not the rule: several recent studies reviewed below show that the effects of small chromosome regions on postzygotic isolation map to single loci (Wittbrodt et al. 1989; Barbash et al. 2003; Presgraves 2003; Presgraves et al. 2003).

Incidentally, this work also proves that major genes—those having very large effects on hybrid fitness—exist. Although previous work suggested as much (e.g., Forejt and Iványi 1974; Guenet et al. 1990; Orr 1992), the recent molecular work is definitive. Finally, this recent work also proves that gene *identity* matters: certain genes—but not others—cause postzygotic isolation.

Probability of hybrid incompatibilities

What is the probability that two diverged sites will, when brought together in hybrids, be incompatible? (We emphasize sites since, formally, it is substitutions, not genes, that are incompatible.) We can estimate this quantity from genetic and molecular data from the same species pair. In particular, given estimates of the numbers of substitutions and genic incompatibilities separating two species, we can calculate the probability, p, that two randomly chosen diverged sites will cause hybrid problems. We do this by rearranging a result from Chapter 7: $p \approx 2I/K^2$, where I is the number of incompatibilities, K is the number of substitutions separating species, and we assume the simplest scenario in which incompatibilities involve pairs of substitutions (Orr and Turelli 2001).

Perhaps the best estimates of the numbers of substitutions and incompatibilities separating two species come from *Drosophila simulans–D. mauritiana*. Orr and Turelli (2001) made several conservative assumptions that allowed them to place a crude upper bound on p of $\sim 10^{-7}$ for hybrid male sterility between these species. Values of p might of course differ for other forms of postzygotic isolation, such as hybrid inviability. Indeed, there is some evidence (albeit weak) that they do. Presgraves (2003) studied hybrid inviability between the older species pair, *D. melanogaster* and *D. simulans*. He found that two diverged nonsynonymous sites cause complete or nearly complete hybrid inviability with a probability of $\sim 10^{-8}$.

But the important point is that—whether looking at hybrid sterility or inviability—p is small. Indeed, Orr and Turelli (2001) concluded that p ranges between 10^{-6} and 10^{-10}. If hybrid interactions involve triplets (or more) of sites (not pairs), p is yet smaller. The overwhelming majority of diverged sites are therefore compatible in species hybrids. This explains, of course, why hybrid problems evolve slowly.

Where are the genes causing postzygotic isolation?

Are the genes causing hybrid problems scattered randomly throughout the genome, or are they concentrated in certain chromosomal regions? There has been much speculation about two types of nonrandom distribution. The first is that the genes causing hybrid problems might be concentrated on the sex chromosomes. The second is that such genes might be concentrated in or near chromosomal rearrangements. We have already considered the evidence for the first idea (the large X-effect) in our discussion of Haldane's rule.

There is some reason to think that chromosomal rearrangements might play a special role in speciation. Some of the relevant evidence comes from plants. Rieseberg and colleagues, for instance, tested the role of rearrangements in slowing gene flow between the sunflower species *Helianthus annuus* and *H. petiolaris*. These taxa have 17 linkage groups, seven of which are colinear (i.e., show no rearrangements) and ten of which are rearranged (usually by a mixture of translocations and inversions). Rieseberg et al. (1995b) showed that, in hybrid backcrosses in the greenhouse, introgression is prevented more in rearranged than in colinear chromosomes. Using many molecular markers, they found that 40% of the genome from colinear chromosomes introgressed, while only 2% of the genome from rearranged chromosomes introgressed. Similar results were obtained from nature: gene flow across three natural hybrid zones between *H. annuus* and *H. petiolaris* was lower on rearranged than on colinear linkage groups (Rieseberg et al. 1999).

Analogous patterns have been seen in animals. *Drosophila pseudoobscura* and *D. persimilis*, for instance, are sympatric in parts of the American West, where they hybridize at a low rate. Interestingly, genes residing in chromosome regions having fixed inversion differences between the species show less interspecific gene flow than do genes in colinear regions (Wang and Hey 1996; Machado et al. 2002). The reason seems clear: several forms of reproductive isolation between these species *also* map to these inversions (Noor et al. 2001a,b). More suggestive still, a survey of the *Drosophila* literature reveals that young species pairs that are sympatric differ by more chromosomal rearrangements than do young species pairs that are allopatric (Noor et al. 2001b).

While these findings suggest that rearrangements slow gene flow, it is not clear why. The traditional guess would be that structural chromosome differences cause postzygotic isolation, representing true chromosomal speciation. As emphasized in Chapter 7, however, the theory of chromosomal speciation is beset by many problems and there is little evidence that it occurs in non-mammalian animals. Rieseberg (2000), Noor et al. (2001b), and Navarro and Barton (2003), however, recently proposed a set of related hypotheses that, while explaining the apparently large effect of rearrangements on isolation, escapes the problems plaguing traditional theories of chromosomal speciation. Although these models differ somewhat, all suggest that rearrangements might be especially likely to harbor *genic* incompatibilities. If so, chromosomal rearrangements—because they effectively suppress recombination in hybrid heterozygotes—would impede gene flow between taxa over a larger region than if the same incompatibilities resided in non-rearranged regions. Because we described the Rieseberg, Noor, and Barton and Navarro hypotheses in Chapters 2 and 3, we do not repeat our discussion here.

Developmental basis of postzygotic isolation

Are certain developmental processes especially likely to be disrupted in hybrids? This question has been surprisingly neglected given that hybrid defects provide a rare window on those developmental processes and pathways that diverge rapidly between taxa. It is entirely possible that these processes differ from those that have been intensively studied by developmental geneticists: the latter are, after all, often early acting, functionally important, and so, at least potentially, slowly evolving (though see Castillo-Davis and Hartl 2002).

HYBRID INVIABILITY. Two patterns characterize the developmental biology of hybrid inviability. The first is that the genes involved are expressed in both sexes, as shown by studies of unbalanced hybrid females (see Figure 8.1): if the genes causing F_1 hybrid male inviability are also expressed in females, these unbalanced hybrid females should also be inviable. They are (Orr 1993b). Similarly, introgression experiments show that small chromosome regions that are lethal when made homozygous in another species almost always kill hybrids of both sexes (Wu and Davis 1993; True et al. 1996). Finally, genetic manipulations show that hybrids that have an appropriate genotype die regardless of whether they are somatic males or females (Orr 1999). Interestingly, the expression of hybrid lethals between species mirrors that of lethals *within* species: the overwhelming majority of lethal mutations in *D. melanogaster* also kill both sexes (Ashburner 1989, Chapter 10).

Second, maternally expressed genes are often involved in hybrid inviability (Hutter 1997). Virtually every exception to Haldane's rule for inviability in *Drosophila*, for instance, involves hybrid females that are dead in only one direction of a cross (Wu and Davis 1993; Turelli and Orr 2000, Table 1). Because females from reciprocal crosses have identical nuclear genotypes, this asymmetry provides prima facie evidence for the role of the cytoplasm in postzygotic isolation. Although this asymmetry might involve maternally acting nuclear genes (a true maternal effect), cytoplasmic organelles, or endosymbionts, true maternal effects are common.

A good example involves *Drosophila montana* and *D. texana*. When *D. montana* females are crossed to *D. texana* males, hybrid females die as embryos; the reciprocal cross yields hybrids of both sexes (Patterson and Griffen 1944). Hybrid inviability here depends on the mother's nuclear genotype: *D. montana* maternal factor(s) are incompatible with X-linked gene(s) from *D. texana* (Patterson and Griffen 1944; Patterson and Stone 1952, pp. 469–470). Time-lapse photography reveals abnormal movement of the yolk in the blastoderm of hybrids, which gives rise to unusual positioning of the cortical cytoplasm, cellular proliferation without differentiation, and ultimately death (Kinsey 1967). For other examples of maternal effects on hybrid viability in *Drosophila*, see Sturtevant (1920, 1946); Orr (1991a); Crow (1942); Kaufmann (1940); Mitrofanov and Sidorova (1981). Work in other taxa, such as amphibians, also suggests frequent maternal effects on hybrid fitness (see Stebbins 1958, pp. 150–151).

It would be misleading, though, to suggest that maternal effects account for all cytoplasmic effects in species hybrids. In addition to endosymbionts (Chapter 7), cytoplasmic organelles sometimes cause hybrid problems. Burton and colleagues have studied intrinsic isolation between populations of an intertidal copepod, *Tigriopus californicus* (Burton 1990; Edmands and Burton 1999; Rawson and Burton 2002). Even geographically contiguous populations often produce F_2 hybrids that suffer many problems, including delayed development (Burton 1997). Rawson and Burton (2002) showed that mitochondrially derived cytochrome *c* oxidase and nuclear-derived cytochrome *c* suffer low enzyme activity when the two proteins come from different populations (although it is not clear that this is the cause of the observed hybrid fitness problems). Cytoplasmic organelles also play a role in postzygotic isolation in plants. Reciprocal hybrids sometimes suffer very different fates (e.g., in *Gilia* [Grant 1964] and *Epilobium* [Michaelis 1954]). Because most plants lack heteromorphic sex chromosomes, these differences almost certainly involve the cytoplasm. In some cases, mitochondrial-nuclear incompatibilities appear involved (e.g., Schmitz and Michaelis 1988). Moreover, cytoplasmic male sterility in plants—whether within or between species—appears to *always* involve the mitochondrion (Saumitou-Laprade et al. 1994; Schnable and Wise 1998; Hurst and Werren 2001).

Although we know of no developmental pathway that plays a consistent role in hybrid inviability—and there may be none—we do know something about the developmental basis of hybrid problems. Perhaps the clearest picture comes from *D. melanogaster* and *D. simulans*. The larval/pupal lethality seen in these hybrids probably involves a mitotic defect. This defect prevents the formation of imaginal discs, the structures that give rise to the adult fly (Orr et al. 1997).

HYBRID STERILITY. The genes causing hybrid sterility usually affect one sex only. This is supported by many genetic manipulations. The simplest involves the unbalanced female test: when F_1 males are sterile but F_1 females fertile, unbalanced F_1 females that carry both of their X chromosomes from the same species are almost always fertile (Coyne and Orr 1989b). Similarly, introgression experiments show that chromosome regions that cause male sterility when moved into a foreign species typically do *not* cause female sterility, and vice versa (Hollocher and Wu 1996; True et al. 1996; Tao et al. 2003a). This pattern of sex-limited expression again mirrors that seen within species: the overwhelming majority of mutations causing sterility in *D. melanogaster* affect either males or females but not both (Ashburner 1989, Chapter 10). It is also clear that hybrid steriles, unlike lethals, do not often show maternal effects. This is not surprising given that fertility is a late (adult) phenotype and maternal effects on fertility are rare among *D. melanogaster* mutations.

We have some information on the developmental basis of hybrid sterility. In *Drosophila pseudoobscura–D. persimilis* hybrid males, all reproductive structures appear normal except the testes. In one direction of the cross, hybrid testes are of nearly normal size, while in the other direction they are small. In both cases, hybrid males are sterile and the proximate cause of sterility involves spindle malfunction at the first meiotic division (Dobzhansky and Boche 1933; Dobzhansky 1934; Dobzhansky and Powell 1975), which leaves spermatids with an incorrect number of nuclei (Dobzhansky and Boche 1933; Dobzhansky 1934). Hybrids thus suffer both meiotic and post-meiotic problems. In a largely forgotten experiment, Dobzhansky and Beadle (1936) showed that these defects are autonomous; that is, testes transplanted from hybrid larvae into pure species individuals remain sterile, while those transplanted from pure species individuals into hybrids remain fertile. Sterility is thus a property of the genotype of the testes, not of the surrounding tissues.

In one respect, the above results are unusual: genic hybrid sterility in animals does not usually involve meiotic problems. Rather, sterility typically involves pre- and postmeiotic defects. In *D. melanogaster–D. simulans* hybrids, for instance, both spermatogenesis and oogenesis are arrested before meiosis (Kerkis 1933). Indeed, female sterility appears to be caused by mitotic defects during early oogenesis (Hollocher et al. 2000). (Transplant experiments show that this female sterility is also autonomous [Ephrussi and Beadle 1935; Dobzhansky and Beadle 1936].) Pontecorvo (1943a) similarly showed that many "pseudo-backcross" hybrids between these species fail to proceed past the oogonial or spermatogonial stage. (See Johnson et al. 1992 and Cabot et al. 1994 for similar data from *D. simulans* group hybrids.)

More often, though, hybrid males suffer *post*meiotic problems. Perez et al. (1993) showed that F_1 hybrid males from both *D. simulans–D. mauritiana* and *D. simulans–D. sechellia* usually complete meiosis normally, but suffer problems in sperm bundling and motility. Cabot et al. (1994) and Palopoli and Wu (1994) described similar phenotypes in several *D. mauritiana–D. simulans* hybrid introgressions. Similar postmeiotic problems were seen in Y- or fourth-chromosome introgressions from *D. simulans* into *D. melanogaster* (Muller and Pontecorvo 1942; Pontecorvo 1943a; J. P. Masly, pers. comm.). Interestingly, this pattern of postmeiotic problems *also* mirrors that seen within species: male-sterile mutations in *D. melanogaster* usually act postmeiotically (Lindsley and Tokuyasu 1980; Lifschytz 1987).

In summary, the developmental basis of intrinsic postzygotic isolation shows at least three parallels with patterns seen *within* species. First, genes causing inviability typically affect both sexes, while genes causing sterility typically affect one sex only. Second, maternal effects on viability are common, while those on fertility are rare. Third, male sterility typically involves postmeiotic problems. These parallels strongly suggest that intrinsic postzygotic isolation involves "ordinary" genes that have normal functions within species. As we will see, this inference has been confirmed by recent molecular work.

Are duplicate genes important?

Several evolutionists have suggested that duplicate genes might often cause intrinsic postzygotic isolation (Werth and Windham 1991; Lynch 2002; Lynch and Conery 2000; Lynch and Force 2000), an idea that has received considerable attention (Taylor et al. 2001). While the duplicate gene hypothesis is a special case of the Dobzhansky–Muller scenario, it offers a unique explanation of *how* hybrid incompatibilities arise. In particular, duplicate genes may experience "divergent resolution." If genes *A* and *B* are functionally redundant duplicates that are present in two species, locus *A* might lose part or all of its function in one species, while locus *B* might lose part or all of its function in the other species. Loss of particular functions could occur by accumulation of neutral mutations (Lynch and Force 2000), or by purifying selection to maintain proper dose balance across loci (Lynch and Conery 2000). The important point is that if each species silences one copy at random, it will be a *different* copy half the time. The result is that, although F_1 hybrids enjoy high fitness, some F_2 hybrids will carry silenced copies at *both* loci, causing sterility or inviability. Indeed, with independent assortment, one-sixteenth of F_2 hybrids will suffer such problems. Lynch and Conery (2000, p. 1154) suggest that divergent resolution of duplicate genes may therefore "provide a common substrate for the passive origin of isolating barriers."

Although the divergent resolution hypothesis is intriguing, and we would be surprised if it does not explain some cases of postzygotic isolation, there are some reasons for doubting that it represents a particularly common cause of hybrid problems. For one thing, the hypothesis does not easily explain postzygotic isolation in F_1 hybrids. Although Lynch and Force (2000) argued that F_1 sterility might result from the fact that one-fourth of the gametes from F_1 hybrids lack functional copies of either *A* or *B*, this argument cannot apply to animals, where there is almost no gene expression in sperm. Also, the claim that one-sixteenth of F_2 hybrids lack function is somewhat charitable: many duplicate genes will not assort independently since the majority may be physically near each other (Thornton 2002); but when tightly linked, *A* and *B* from the same species almost always arrive together in F_2 hybrids, ensuring high hybrid fitness. Finally, as discussed below, molecular characterization of several genes causing postzygotic isolation provides little clear support for the divergent resolution hypothesis (though see our discussion of *OdsH*).

It does not of course follow that duplicate genes play no special role in speciation. They may well. But the evolutionary changes underlying such a role may have little or nothing to do with the silencing of genes or gene functions by divergent resolution. Instead, duplicate genes may be likely to evolve novel (neomorphic) functions, as suggested long ago by Ohno (1970). This rapid *adaptive* evolution might make duplicate genes particularly important in speciation.

Which genes cause postzygotic isolation?

One of the oddest things about our discussion of the genetics of speciation so far is that it does not look like a discussion of developmental genetics, molecular genetics, or, in fact, any other kind of genetics. Those discussions are almost always held in terms of actual genes, that is, actual DNA sequences of known identity and function. Until recently, the genetics of speciation remained a form of classical genetics: "speciation genes" were black boxes of unknown function that resided in poorly defined chromosome regions and that interacted with other poorly defined chromosome regions. Consequently, we have been unable to say anything of substance about the identities, functions, and evolutionary histories of the genes causing postzygotic isolation.

The reason is easy to understand. Study of the genetics of speciation is, almost by definition, an attempt to do genetics where it cannot be done—

between species. The very phenomenon under study, reproductive isolation, precludes most genetic analysis, a problem that Lewontin (1974, p. 163) called the "methodological contradiction" sitting at the heart of speciation research. Although this problem is eased somewhat because certain taxa produce partially viable, fertile hybrids, it is worsened again by a particularly unfortunate fact: one of our most sophisticated genetic model systems, *Drosophila melanogaster*, forms only sterile or inviable hybrids when crossed to any other species, barring almost all genetic analysis. Historically, then, *D. melanogaster* has proved nearly useless in the genetic study of speciation. There can be no doubt that our genetical understanding of reproductive isolation would have progressed far faster if, for instance, *D. simulans*—which *does* form viable, fertile hybrids with other species—had been developed as a model system.

These difficulties have left us unable to answer a large set of fundamental questions about the factors causing intrinsic postzygotic isolation: Do these factors correspond to "ordinary" genes that have normal functions within species, or to unusual factors like repetitive DNA sequences or transposable elements (Engels and Preston 1979; Kidwell 1983; Rose and Doolittle 1983; Mayr 1988, p. 374)? If ordinary genes *are* involved, do they belong to certain functional classes? Are they, for instance, always transcription factors? Is divergence at the genes causing postzygotic isolation concentrated in regulatory or structural gene regions? Do the genes that cause postzygotic isolation fail to function in hybrids or do they produce a "poison" in hybrids? Do these genes correspond to single-copy genes or to duplicate genes? Are the genes that cause postzygotic isolation rapidly evolving or do they show normal rates of evolution? Finally, and perhaps most important, do the genes that cause postzygotic isolation diverge by natural selection or genetic drift?

Our inability to answer these questions has represented an embarrassing shortcoming in our understanding of speciation. Not surprisingly, it has also allowed nearly unchecked speculation about their answers. Over the last decade, however, geneticists have finally identified several genes that cause intrinsic isolation, allowing preliminary answers to these questions. Four genes have been identified, one in fish and three in *Drosophila*. Although we have no guarantee that these genes played a role in the earliest stages of speciation, they at least represent a sample of loci involved in Dobzhansky–Muller incompatibilities between crossable species. We briefly review each of these genes and then draw some tentative conclusions.

XIPHOPHORUS: XMRK-2. Hybrid inviability in *Xiphophorus* fish—which was widely discussed in the older literature (Dobzhansky 1937b; Stebbins 1958; Mayr 1963)—provided our first glimpse at a gene that causes postzygotic isolation.

Some species of *Xiphophorus*, like the platyfish, *X. maculatus*, are polymorphic for spots composed of black-pigmented cells called macromelanophores. Other species of *Xiphophorus*, like the swordtail, *X. helleri*, lack macromelanophores. When spotted *X. maculatus* and *X. helleri* are crossed, all F_1 hybrids show an increased number of spots. When F_1 hybrids are backcrossed to *X. hel-*

leri, half of the resulting progeny lack macromelanophores, while the other half develop phenotypes ranging from F_1-like to extreme invasive malignant melanomas that are often lethal.

The genetic basis of hybrid lethality is well understood (Anders 1991; Malitschek et al. 1995; Schartl 1995). Classical work suggested a simple genetic basis and it is now clear that spotted *X. maculatus* fish carry a sex-linked complex, the *Tumor* (*Tu*) locus, that specifies macromelanophores and that is regulated by an autosomal suppressor locus, *R* (genotype *RR*). *X. helleri*, on the other hand, lacks macromelanophores, lacks the *Tu* locus, and lacks dominant suppressor alleles at *R* (genotype *rr*). Certain backcross hybrids thus inherit *Tu* but lack *R* suppressors (genotype *Tu*; *rr*); they consequently develop melanomas. Hybrid inviability in *Xiphophorus* hybrids thus behaves as a two-locus Dobzhansky–Muller incompatibility.

Wittbrodt et al. (1989) isolated a candidate gene in the *Tu* region by positional cloning. They showed that the *Tu* complex is composed of several tightly linked genes (Weis and Schartl 1998; Gutbrodt and Schartl 1999): the pigment-encoding *Macromelanophore determining locus (Mdl)* and two duplicate copies of a novel receptor tyrosine kinase. These duplicate genes are called *Xmrk-1* and *Xmrk-2* (Wittbrodt et al. 1989). *Xmrk-1* homologues reside on the sex chromosomes of all *Xiphophorus* species; *Xmrk-2*, however, only resides on the sex chromosomes of some species (Weis and Schartl 1998). The two *Xmrk*s are nearly identical except that *Xmrk-2* has come under the control of a new promotor (Adam et al. 1993). It is therefore a chimeric gene.

Several lines of evidence show that *Xmrk-2* causes tumors in species hybrids. First, mutant *X. maculatus* lines that fail to induce melanomas when hybridized to *X. helleri* carry mutations at the *Tu* locus (Schartl et al. 1999); second, overexpression of *Xmrk-2* causes tumors (Malitschek et al. 1995); finally, transcription levels of *Xmrk-2* correlate with the severity of melanomas across genotypes (Malitschek et al. 1995).

DROSOPHILA: ODSH. Coyne and Charlesworth (1986) performed an early genetic analysis of male sterility in *D. simulans*–*D. mauritiana* hybrids, mapping a putative hybrid male sterility gene to an approximately 2 cM segment of the *D. mauritiana* X chromosome. Perez et al. (1993) confirmed and refined these results, showing that the effect of this region is due to a putative gene, which they called *Odysseus* (*Ods*). Males carrying *Ods* from *D. mauritiana* on a largely *D. simulans* genetic background appeared completely sterile. Later, Perez and Wu (1995) showed that matters are more complex. In finer-scale experiments, they localized *Ods* to a small region that included only 250 kb of DNA. Surprisingly, males who carry very small introgressions that include *Ods* are sterile only 50% of the time, while males who carry *Ods* and a more distal region are completely sterile. Thus, at least two genes in the region are required for complete hybrid sterility.

Ting et al. (1998) took the study of *Ods* to the molecular level. They showed that the small region harboring *Ods* includes only two exons, which belong to a homeobox gene (i.e., this gene encodes a homeodomain, a long stretch of amino acids that binds DNA; genes carrying this motif are often transcription factors). They named this gene *OdsH* (for *Ods*-site Homeobox gene). *OdsH*, which is a duplicate gene, is expressed in the fly testis (and other tissues), as expected if it plays a role in hybrid male sterility. There is some evidence that *OdsH* is misexpressed in hybrids (Wu and Ting 2004).

OdsH has evolved at a remarkably fast rate between *D. simulans* and *D. mauritiana*: in the roughly half million years separating these species, 15 amino acid replacements have occurred in the homeodomain alone—more than between distantly related taxa like worms and mice. Moreover, replacement substitutions greatly outnumber silent ones in the branch leading to *D. mauritiana*, a telltale sign of positive Darwinian selection (see Chapter 6).

As Ting et al. note, however, they have not yet proved that *OdsH* is the hybrid sterility gene *Ods*. *OdsH* sits in the right place and several of its features—particularly its rapid evolution—are very suggestive. However, as Wu and Palopoli (1994) emphasize, functional data implicating *OdsH* are currently lacking and "final proof by transgenic study will be necessary" (i.e., one must show that a normally sterile hybrid genotype is rendered at least partially fertile when introducing a compatible [*D. simulans*] copy of *OdsH*). This has not yet been reported.

DROSOPHILA: NUP96. In his deficiency screen of *Drosophila melanogaster*–*D. simulans* hybrids, Presgraves (2003) identified 20 regions from *D. simulans* that cause hybrid inviability on a *D. melanogaster* genetic background (see Figure 8.3). One of these regions is on the right arm of the third chromosome. Further fine-scale deletion mapping and complementation tests in this region showed that its effect is due to a single gene (Presgraves et al. 2003); this gene obviously interacts with another gene(s) that resides elsewhere in the genome.

Presgraves et al. (2003) showed that this third chromosome gene encodes the nucleoporin *Nup96*, a component of the nuclear pore complex (NPC). The NPC, one of the largest structures in the eukaryotic cell, regulates the passage of proteins and RNAs in and out of the cell nucleus. *Nup96* is one of about 30 different proteins making up the NPC; Nup96 protein is therefore almost certainly produced in all *Drosophila* cells and is present throughout the life cycle. Genetic analysis revealed that *Nup96* from *D. simulans* is incompatible with an unknown gene or genes on the X chromosome of *D. melanogaster*: hybrids that carry *Nup96* from *D. simulans* and an X chromosome from *D. melanogaster* are lethal, while otherwise genetically identical hybrids that carry an X chromosome from *D. simulans* are viable. Because both of the partner genes involved in this incompatibility act recessively in hybrids, this incompatibility can cause the inviability only of F_2 or backcross, not of F_1, hybrids. Further analysis suggested that this hybrid incompatibility involves the amino terminus of the Nup96 protein.

A molecular population-genetic study of *Nup96* revealed that it has evolved rapidly since the split of *D. melanogaster* and *D. simulans*—which is surprising given that the NPC is generally conserved in eukaryotes. More important, this rapid divergence reflects adaptive evolution: statistical tests of DNA sequence data show that *Nup96* evolved by positive natural selection in both the *D. melanogaster* and the *D. simulans* lineages (Presgraves et al. 2003).

In recent work, D. Presgraves (pers. comm.) has identified a gene on the *D. melanogaster* X chromosome that represents a candidate for the partner gene with which *Nup96* interacts. Analysis of this candidate gene shows that it has *also* evolved rapidly under positive Darwinian selection.

DROSOPHILA: HYBRID RESCUE MUTATIONS AND *HMR*. While the genes discussed above were identified because they cause postzygotic isolation, another set of genes does the opposite—they suppress postzygotic isolation. More precisely, certain mutations are known that restore the viability or fertility of normally dead or sterile hybrids. All of the "hybrid rescue mutations" recovered so far occur in *Drosophila* (Sawamura et al. 1993b; Hutter 1997). This surely reflects the intense genetic scrutiny of this group and there is every reason to think that such mutations occur in other taxa. We briefly review what is known about rescue mutations and then describe the molecular identification of one.

The best-studied rescue mutations restore hybrid viability. At least four such mutations have been found, all in the *D. melanogaster* complex (Sawamura et al. 1993b; Hutter 1997). This complex includes four species, the outgroup *D. melanogaster*, and three more closely related species, *D. simulans*, *D. mauritiana*, and *D. sechellia*. We refer to the latter three species as "sib" (for sibling species) when we do not need to distinguish them. Crosses within the sib species produce viable offspring, but crosses between sib and *D. melanogaster* produce some inviable offspring.

When *D. melanogaster* females are crossed to sib males, only hybrid females are produced; males die at the larval-pupal transition (Sturtevant 1920). The reciprocal cross of sib females to *D. melanogaster* males also involves hybrid inviability; while hybrid males survive, hybrid females die as embryos (Sturtevant 1920).

Four mutations that rescue the viability of *D. melanogaster*-sib hybrids are known:

1. *Lethal hybrid rescue* (*Lhr*): A *D. simulans* mutation that rescues hybrids that die at the larval-pupal transition. *Lhr* is autosomal and behaves as a single gene (Watanabe 1979).

2. *Hybrid male rescue* (*Hmr*): A *D. melanogaster* mutation that rescues hybrids that die at the larval-pupal transition. *Hmr* is on the X chromosome and behaves as a single gene (Hutter and Ashburner 1987; Hutter et al. 1990; Orr and Irving 2000; Barbash and Ashburner 2003; and see below).

3. *maternal hybrid rescue* (*mhr*): A *D. simulans* mutation that rescues hybrids that die as embryos. *mhr* is autosomal and maternally acting: homozygous females produce viable daughters when crossed to *D. melanogaster* males (Sawamura et al. 1993a,b).

4. *Zygotic hybrid rescue* (*Zhr*): A *D. melanogaster* mutation that rescues hybrids that die as embryos. *Zhr* is X-linked and zygotically acting (Sawamura et al. 1993b,c; Sawamura and Yamamoto 1993, 1997).

Hybrid *fertility* can also be rescued with single mutations:

1. *167.7*: A *D. simulans* mutation that rescues the fertility of *D. simulans*–*D. melanogaster* hybrid females (Davis et al. 1996). *167.7* is maternally acting and rescues fertility in the *D. simulans* female × *D. melanogaster* male direction of the species cross. Rescue is weak.
2. *Hybrid male rescue* (*Hmr*): This *D. melanogaster* hybrid viability rescue mutation can also rescue the fertility of hybrid females produced in the cross of *D. melanogaster* females to sib males (Barbash and Ashburner 2003).

Molecular analysis of hybrid rescue mutations has been slow. One of the reasons is that it was unclear if these mutations represent alleles of true "speciation genes." The problem is that rescue mutations might act in either of two ways: they might represent mutations at the actual loci that cause the death or sterility of hybrids, or they might represent second-site suppressors (i.e., alleles that suppress the effects of the loci causing hybrid problems). Recent work, however, strongly suggests that the first interpretation is correct: rescue mutations are almost certainly alleles of the genes that cause postzygotic isolation (Sawamura et al. 1993c; Sawamura and Yamamoto 1997; Barbash et al. 2000; Orr and Irving 2000). This finding renders the molecular isolation of these genes important.

Recognizing this, Barbash et al. (2003) recently isolated the best studied of rescue mutations, *Hmr*. The gene identified includes two regions homologous to a family of transcriptional regulators (the so-called MYB-related regulators) and is expressed at all stages of the life cycle. Barbash et al. (2003) speculate that the wild-type allele of *Hmr* plays a normal role in gene regulation and that this regulation is disrupted in species hybrids.

Barbash et al. (2003) also show that *Hmr* is rapidly evolving: *D. melanogaster* and its sister species in the *D. simulans* clade differ by many replacement substitutions, as well as by many insertions/deletions (all of which are in frame, suggesting that *Hmr* remains functional). More remarkable still, *Hmr* has evolved rapidly even at one of its putative DNA-binding domains. Looking over the entire locus, *Hmr* appears to be one of the fastest evolving genes known in *Drosophila*. Recent polymorphism data show that this rapid evolution reflects positive natural selection in both the *D. melanogaster* and *D. simulans* lineages (D. Barbash and P. Awadalla, pers. comm.). Finally, excluding its DNA-binding domains, *Hmr* shows no homology with any other gene outside *Drosophila*. This implies that *Hmr* is either a young gene or is so rapidly evolving that similarity with distant homologues has been obscured.

CONCLUSIONS. *Xmrk-2*, *OdsH*, *Nup96*, and *Hmr* provide our first glimpses at the factors causing postzygotic isolation. They also let us answer, at least provisionally, some of the questions posed at the start of this section. Less provisional answers will obviously emerge only after analysis of many additional genes—particularly those that cause postzygotic isolation between very closely related taxa.

But it is already clear that the factors causing intrinsic postzygotic isolation are ordinary genes having normal functions within species. There is no evidence so far for the once-popular idea that postzygotic isolation involves novel genetic factors like repetitive DNA sequences or the mass mobilization of transposable elements in hybrids (Engels and Preston 1979; Kidwell 1983; Rose and Doolittle 1983; Mayr 1988, p. 374).

It is also clear that the genes causing intrinsic postzygotic isolation do not belong to a single functional class. Although several of the above genes appear to play a role in transcriptional regulation (*OdsH*, *Hmr*)—and it remains entirely possible that disruption of gene regulation is a common cause of hybrid problems—not all do. *Nup96*, for example, is a structural gene with no predicted regulatory function. Similarly, it is already clear that divergence of non-coding regulatory sequences does not provide a universal path to postzygotic isolation. As *Nup96* shows, structural divergence of proteins is sometimes involved (see also our earlier discussion of Rawson and Burton's 2002 analysis of enzyme activity in copepod hybrids). Finally, the genes causing intrinsic postzygotic isolation are not invariably duplicates. Although *OdsH* is a duplicate gene, *Nup96* and *Hmr* are both single copy genes, and *Xmrk-2* is a chimeric gene whose evolution cannot be explained by divergent resolution.

Although it is obviously hard to generalize from a sample of four, two patterns do appear to characterize the genes causing intrinsic postzygotic isolation. The first is that these genes are rapidly evolving. This is perhaps not surprising. It seems likely that the genes that disrupt hybrid development would be those that have diverged most between species. (It is also worth noting that both hybrid steriles (*OdsH*) *and* hybrid lethals (*Nup96*, *Hmr*) are fast evolving. Though Wu and colleagues speculated that *OdsH*'s rapid evolution might reflect sexual selection on reproduction-related genes, it appears, so far at least, that *any* gene causing postzygotic isolation is likely to be rapidly evolving.)

Second and more important, the genes causing postzygotic isolation diverge by positive Darwinian selection. Thus, as both Presgraves et al. (2003) and Barbash et al. (2003) emphasize, direct molecular evolutionary data now support one of the central tenets of the neoDarwinian view of speciation—that reproductive isolation results from natural selection within species. This may well represent the most important finding to emerge from the last decade of work on the genetics of speciation.

9

Polyploidy and Hybrid Speciation

Although we have devoted almost all of our attention to the origin of new species by the splitting of old ones, there is another possibility. New species might arise by the *hybridization* of older species. While both of these paths can lead to an increased number of species, they differ profoundly. One of the most surprising facts about speciation is that this second path—which at first might seem a mere formal possibility—is taken in nature. Indeed, hybrid speciation is common in certain groups.

Here we discuss various kinds of hybrid or "reticulate" speciation. As we will see, there are two main types, polyploid speciation and recombinational speciation. Polyploid speciation is common, well known, and has received a great deal of attention. Recombinational speciation is of unknown frequency, is less well-understood, and has received less attention, at least until recently. We also discuss non-hybrid forms of polyploidy.

Polyploidy

Polyploidy is the condition in which individuals carry three or more complete sets of chromosomes. Polyploid speciation is the origination of taxa having an increased number of chromosome sets, an increase that might yield nearly instantaneous reproductive isolation. To see why, consider a new tetraploid individual ($4x$) that has just arisen from diploid ($2x$) ancestors (x is the "base" chromosome number in a lineage; this differs somewhat from n, which is best used to indicate the haploid number of a species, whether or not this corresponds to the base number of the lineage; Ramsey and Schemske 1998). The cross between a new tetraploid individual and an ancestral diploid individual produces triploid (3x) offspring. These progeny are often sterile because meiosis cannot proceed properly due to problems in chromosome pairing/segregation in a genome having an odd number of chromosome sets; moreover, many of the resulting gametes are aneuploid and thus inviable (at least in plants wher... ploid is therefore reproducti... ming it can give rise to an in... s.

Th... polyploid speciation is instar... tion bottlenecks. While evolu... context of polyploidy, they w... y other context. Polyploidy a... een plant and animal speciatio... non in plants but rare in animal... came as a surprise. Indeed, Ha... neous form of speciation would... most important correction which... of species."

Polyploidy... volutionary genetics. Indeed, the disco... ed the first major triumph in the genetic... creases in ploidy level provide a nearly ins... first offered by Winge (1917), who noted the occurrence of polyploid series (i.e., arithmetic series of chromosome number) in two plant genera, an early example of the power of comparative contrasts. Winge's hypothesis was verified experimentally by Clausen and Goodspeed (1925), who formed artificial tetraploids in the tobacco *Nicotiana*. Shortly thereafter, Muntzing (1930b) artificially reconstructed a known natural tetraploid species by hybridizing two wild diploid species. From then on, there could be little doubt that polyploid speciation occurs in nature.

Here we review what is known about polyploid speciation. We consider the types of polyploids, the genetic paths to polyploidy, the incidence of polyploid speciation, the ecology of polyploidy, and why polyploidy is common in plants but rare in animals.

Classification

There are two classes of polyploids: those whose origin involves species hybridization and those that do not. While only the former represents hybrid speciation, it would be awkward to split our discussion, ignoring non-hybrid polyploids in this chapter. We thus review both forms, comparing and contrasting hybrid polyploid speciation with its non-hybrid cousin.

Non-hybrid polyploids—*autopolyploids*—result from an increase in ploidy level within a species and are the easiest to understand. In the simplest case, a diploid species might spontaneously give rise to a tetraploid by mechanisms described below. Because such tetraploids now carry four copies of each chromosome, they may often suffer fertility problems: all four homologous chromosomes try to pair at meiosis (forming "quadrivents"), and chromosome segregation is compromised; aneuploid gametes result. The effect on fertility varies across autopolyploid species. In some, seed set is reduced to near 0%, while in others it remains near 100% (Stebbins 1950; Ramsey and Schemske 2002). Examples of autopolyploids include the wild species *Galax aphylla* and the cultivar alfalfa (Stebbins 1950; Grant 1981). Autopolyploids need not stop at the $4x$ level. Instead, $4x$ species might give rise to other species that carry an even higher multiple of chromosome complement.

Allopolyploids are polyploids that arise by hybridization between species. Stebbins (1950) further divided such polyploids into several types and his classification remains popular. *Genomic allopolyploids* carry entire (and at least diploid) chromosome sets from two or more species. In the simplest case, all of these chromosomes are sufficiently diverged structurally that chromosomes from different species do not pair. Consequently, only bivalents form at meiosis, gametogenesis proceeds fairly normally, and fertility should be high (we temporarily ignore genic hybrid sterility). Divergent haploid chromosome sets from each species are typically symbolized by different upper case letters (e.g., A and B). Genomic allopolyploids include the wild species *Iris versicolor* and the cultivars *Nicotiana tabacum* (tobacco) and *Gossypium hirsutum* (cotton) (Stebbins 1950).

In *segmental allopolyploids*, some chromosomes are sufficiently diverged structurally to prevent pairing between chromosomes from different species, while other chromosomes are not. Thus, some bivalents and some multivalents form at meiosis. The set of chromosomes that forms multivalents is typically symbolized by different subscripts (e.g., A_s and A_t). New segmental allopolyploids are often unstable and continue to evolve until meiosis involves most, or all, bivalents. Segmental allopolyploids include the wild species *Delphinium gypsophilum* (Stebbins 1950). Allopolyploids, whether genomic or segmental, are permanently heterozygous for those diverged genes in the diploidized part of the genome (i.e., that part of the genome that forms bivalents).

Finally, *autoallopolyploids* are polyploids that involve both autopolyploidy and allopolyploidy. One (but not the only) route to their formation involves doubling the chromosome complement of a "regular" allotetraploid. Examples of autoallopolyploids include the wild sunflower *Helianthus tuberosus* and (probably) most of the ornamental *Chrysanthemum* varieties (Stebbins 1950). Once again, allopolyploids need not stop at the $4x$ level and polyploid series reaching high ploidy levels are well known.

Following Grant (1981), we can summarize the types of polyploids as follows:

I. Autopolyploids A A A A

II. Allopolyploids
    Genomic A A B B
    Segmental $A_s\ A_s\ A_t\ A_t$
    Autoallopolyploid A A A A B B B B

This classification scheme is not entirely uncontroversial: some botanists prefer to define auto- versus allopolyploidy not on the basis of whether intra- or interspecific crosses were involved but on the basis of chromosome behavior at meiosis (Jackson 1982; Levin 2002, p. 102). We obviously prefer, and will adhere to, the more evolutionary approach of Clausen et al. (1945) and Ramsey and Schemske (2002).

Finally, we note that nature is less exact than mathematics. Although repeated rounds of polyploidy are expected to yield polyploid series, these series are often approximate (e.g., a genus might include species having haploid chromosome counts of 7, 14, *and* 13). The main reason appears to be "polyploid drop": although diploids typically cannot tolerate chromosome loss (aneuploidy), polyploids, which have duplicated genomes, sometimes can (Darlington 1973). The result is imperfect arithmetic series.

Pathways to polyploidy

There are three genetic paths to doubling chromosome number, whether polyploidy involves hybrids or not: somatic doubling, nonreduction during meiosis, and polyspermy (Grant 1981; Ramsey and Schemske 1998). Somatic doubling occurs when mitotic products in a diploid cell fail to segregate to opposite poles. If doubling occurs in an early zygote, a tetraploid individual results. If doubling occurs in somatic tissue that gives rise to reproductive tissue, tetraploid tissue (producing $2x$ gametes) results. Meiotic nonreduction occurs when a cell wall fails to form late in meiosis, yielding diploid gametes. Tetraploids could then form via two routes, involving either one or two steps (Figure 9.1; Grant 1981; Levin 2002). In the one-step route, two unreduced gametes fuse, forming a tetraploid. In the two-step route, an unreduced gamete and a normal gamete fuse, to form a triploid individual. This triploid might in turn produce an unreduced $3x$ gamete which, when fertilized by a normal $1x$ gamete, forms a tetraploid individual. (There are other possibilities: a tetraploid might for instance produce two $2x$ gametes which, upon fusion, form a

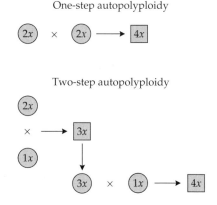

Figure 9.1 Formation of an autopolyploid by meiotic nonreduction (the scenario for allopolyploids is similar). Both one-step and two-step pathways are shown. The two-step pathway, which appears to be common, involves a "triploid bridge." (The triploid is shown producing a $3x$ gamete, although other pathways are possible.) Circles represent gametes; squares represent zygotes.

tetraploid individual.) Polyspermy occurs when two sperm fertilize a single egg, yielding a triploid.

Although polyspermy appears to be a rare path to polyploidy, it is harder to distinguish the frequency of somatic doubling from that of meiotic nonreduction. Thompson and Lumaret (1992) and Ramsey and Schemske (1998), however, argue that meiotic nonreduction is probably the most common path to polyploidy. This view is supported by the observation that individual parent plants of spontaneous polyploids sometimes produce a substantial number of unreduced gametes. The fact that certain individuals are prone to producing unreduced gametes suggests that genetic variation for nonreduction resides in natural populations, variation that plays a critical role in the origins of polyploids. There is now good evidence for such genetic variation (reviewed in Ramsey and Schemske 1998, p. 476, and Levin 2002, pp. 103–104). Indeed, this variation may often involve a few major genes, suggesting that deleterious mutations that segregate at low frequency might cause nonreduction. This idea is supported by the fact that $2x$ gamete formation appears more common in perennials than annuals: the former often reproduce vegetatively and so may experience relaxed selection on meiosis (Ramsey and Schemske 1998).

But polyploids need not form in a single step. As noted above, there are several paths to polyploidy that involve passing through an intermediate triploid state (see Figure 9.1). The role of these "triploid bridges" might seem paradoxical since reproductive isolation between a new tetraploid species and its diploid progenitors depends on the presumed *low* fitness of triploids. The question, then, is whether triploids are fertile enough to act as a plausible bridge to polyploidy. Ramsey and Schemske (1998) argue that the answer is yes. Their review of the literature shows that, although triploids suffer meiotic problems, both auto- and allotriploids enjoy a mean pollen fertility of about 30%. Ramsey and Schemske (1998) further show that about 6% of the progeny of triploids are tetraploids. Triploid bridges therefore play a plausible, and probably important, role in the origin of polyploids.

Although our interest in this section is in the genetic *origins* of polyploids, we would be remiss if we did not mention recent work on the fate of the duplicated genes that arise by polyploidy. An interesting pattern has been seen. Singly duplicated genes (those that arise on a gene-by-gene basis) have short half-lives. In a study of nine species (including three that have been completely sequenced: the worm *C. elegans*, the fly *D. melanogaster*, and the yeast *S. cerevisiae*), Lynch and Conery (2000) found that the average duplicated gene survives only 3–7 million years. But duplicated genes that arise on a genome-wide basis by polyploidy are longer-lived. Roughly half of all duplicated genes in tetraploid fish, for instance, remain functional after 50 million years (Ferris and Whitt 1977). (See also Amores et al. 1998, Wagner 2001, and references in Lynch and Conery 2000.) As Lynch and Conery note, this difference likely reflects the fact that polyploidy maintains proper dosage balance across loci (i.e., maintains the proper stoichiometric relationships among the products of interacting genes). Duplications of single genes do not.

Incidence

Polyploidy occurs in ferns, mosses, algae, and virtually all groups of vascular plants (Stebbins 1950; Grant 1981; Levin 2002). Among the latter, polyploidy is best known in the angiosperms, including many ornamental plants, e.g., roses, bearded iris, *Chrysanthemum*, and primrose (Stebbins 1950; Darlington 1956; Grant 1981). As suggested earlier, polyploidy is also common in crop plants, including maize, sorghum, wheat, sugar cane, coffee, tobacco, cotton, soybean, and potatoes (Hilu 1993; Otto and Whitton 2000; Soltis and Soltis 2000). Polyploidy also occurs occasionally outside plants. Indeed, it now appears that a polyploidization event occurred in the yeast *Saccharomyces cerevisiae* roughly 100 million years ago (Wolfe 1997; Seoighe and Wolfe 1998). It also appears that two rounds of ancient polyploidization occurred in vertebrates (Sidow 1996). These events, which occurred before the split of the common ancestor of bony fishes, birds, and mammals, probably explain why vertebrates contain about four times as many genes as do well-studied invertebrates, such as *Drosophila*. Fish appear to have experienced another, and later, polyploidy event after the split of lobe-finned and ray-finned fish but before the radiation of the teleosts (Amores et al. 1998).

The frequency of polyploidy varies greatly from group to group. Polyploidy is, for example, the norm in ferns, common in angiosperms, but rare in gymnosperms. But even *within* angiosperms, the frequency of polyploidy varies greatly from family to family and even from genus to genus. Overall, polyploidy seems most common in perennial herbs, then annuals, and then woody plants (Muntzing 1936; Stebbins 1938, 1950; Levin and Wilson 1976). This pattern was first interpreted to mean that polyploidy more often leads to perennial (and thus often vegetative) lifestyles. As we will see, however, the causal relation likely runs the other way around: perennial lifestyles and vegetative reproduction probably promote polyploidy (Stebbins 1950, p. 355).

Beyond these qualitative observations, there have been several attempts to estimate the frequency of polyploidy in particular groups. Darlington (1937), for example, estimated that about half of all angiosperm species are polyploid, while Stebbins (1950) put the figure at 30%–35%, Masterson (1994) at 70%, Goldblatt (1980) at 70%–80% (restricted to the monocots), and Grant (1981) at 43%–60%. In ferns, on the other hand, about 95% of all species are polyploid (Grant 1981).

Otto and Whitton (2000) discuss problems with all of these estimates, including the fact that most surveys used different criteria. More important, however, the above numbers do *not* estimate the percentage of speciation events involving polyploidy. Stebbins's method, for example, involves tallying the percentage of species within a genus that show higher multiples of a basal chromosome number, which then gives the percentage of species in a genus that experienced a polyploidy event anywhere in its evolutionary history. This method suffers from two problems. First, it cannot recognize a group as showing a history of polyploidy if the progenitor diploid species is extinct. (We know this problem is real. Using guard cell size as a proxy for DNA content, Masterson 1994 provided strong evidence for extinct diploid progenitors: fossil angiosperms in three families had lower ploidy levels than all extant species.) Second, consider how Stebbins's metric behaves in two genera, one in which one species has 12 chromosomes and four species have 24 chromosomes, and another in which one species has 12 chromosomes and the others have 24, 48, 96, and 192 chromosomes. In both cases, Stebbins's metric yields 80% polyploidy, though the latter genus may well have experienced more bouts of polyploid speciation than the former (Otto and Whitton 2000).

What we would like is a method that estimates the percentage of speciation events in a group that involve polyploidy. Otto and Whitton (2000) proposed such a method. The essence of their approach is that the fraction of speciation events involving polyploidy can be estimated as the product of two quantities: the fraction of speciation events that involve *any* change in chromosome number and the fraction of changes in chromosome number that involve polyploidy. The first quantity can be roughly estimated by dividing the minimum number of chromosome changes that occurred in a group by the total number of speciation events required to account for the present species. The second quantity demands a more subtle approach. It requires realizing that polyploid speciation leaves a signal: an even number of haploid chromosomes. Other changes in chromosome number (not involving polyploidy) need not result in an even haploid chromosome number. Using this approach, Otto and Whitton estimated that 16% of all speciation events in ferns involve a change in chromosome number, and 41.7% of these changes in number reflect polyploidy. Thus, approximately 7% (= 16% × 41.7%) of speciation events in ferns appear due to polyploidy. A similar analysis suggests that 2%–4% of speciation events in angiosperms involve polyploidy (Otto and Whitton 2000).

This approach is not, however, without problems. It relies, for example, on literature reports of the number of species per genus. But plant systematists rarely use the BSC. In particular, they almost never name autopolyploids as distinct species and instead often speak of "ploidal races" within species. Moreover, the Otto-Whitton method cannot detect independent origins of the same ploidy level. Both of these problems probably cause underestimation of the frequency of polyploid speciation (D. Schemske, L. Rieseberg, pers. comm.). Despite this, we believe the Otto-Whitton approach represents a step in the right direction: it attempts to find the evolutionarily relevant quantity and provides a reasonable minimum estimate of it.

A good deal of attention has been devoted to explaining *why* the incidence of polyploidy varies among plant groups. What biological characteristics explain why polyploidy is common in certain plants but rare in others? Although this literature is large and contentious—and focuses almost entirely on "incidence" in the Stebbins, and not in the Otto-Whitton, sense—at least two reasonably clear factors have been identified. The first is long life, including the tendency to propagate vegetatively via rhizomes or bulbs. Grant (1981), in particular, argued that extended life span increases the opportunity for somatic doubling. Long life and vegetative propagation might help explain why perennials show more polyploidy than annuals (Grant 1981). The second factor is self-fertilization. Selfing, as we will see, greatly increases a new polyploid's chances of successful establishment (see also Stebbins 1950 and Grant 1981).

Frequency of auto- versus allopolyploidy

Does polyploid speciation usually involve species hybridization? Although earlier authors consistently claimed that the answer is yes (Stebbins 1950; Grant 1981), this view has recently been challenged.

The relative rates at which allo- versus autopolyploids form depend on two quantities (Ramsey and Schemske 1998): (1) the frequency of species hybridization; and (2) the rate at which polyploids emerge from hybrid versus nonhybrid crosses. The second quantity clearly favors allo- over autopolyploidy. Although spontaneous autopolyploids arise at a surprisingly high rate—Ramsey and Schemske put the figure at 10^{-5} per plant per generation—the frequency of spontaneous tetraploids in *hybrid* $2x \times 2x$ crosses is very high (see also Grant 2002 in *Gilia*). Indeed, tetraploids are often the *only* offspring that appear in such crosses (Ramsey and Schemske 1998). But the former quantity—the rate at which species hybridize in nature—is probably too low to make allopolyploidy more common than autopolyploidy. To show this, Ramsey and Schemske (1998) perform a comprehensive analysis in which they estimate parameters characterizing key steps in the formation of allo- versus autopolyploids. For outcrossing species, they conclude that 2.7% of all crosses would have to be interspecific to bring the rate of allopolyploid formation up to that of autopolyploid formation. For selfers, the figure drops to 0.2%. Ramsey and Schemske argue that natural rates of hybridization are unlikely to be this high and conclude that autopolyploids are probably formed more often than allopolyploids (see also Thompson and Lumaret 1992).

Ramsey and Schemske's analysis highlights a surprising fact and raises a difficult question. The surprising fact is that polyploids constantly appear in natural populations, arising, at least as autopolyploids, at rates equivalent to that of genic mutation. (This idea has recently received support from an unexpected source: molecular data suggest that polyploid species may sometimes have multiple origins [Soltis and Soltis 1993, 1999, 2000]. A polyploid species, in other words, may not derive from a single event but may be polyphyletic.)

The difficult question raised by Ramsey and Schemske's (1998) analysis is: Where are all the autopolyploids? If autopolyploids arise at such high rates, why did almost all plant evolutionists conclude that *allo*polyploidy is the norm? There are two possible answers.

One is that autopolyploids *are* more common in nature than traditionally thought but have not been identified. There is surely some truth to this, as even early authors admitted (e.g., Stebbins 1950, p. 328). Autopolyploids might be missed for several reasons. Allopolyploids often differ morphologically from their diploid progenitors, while autopolyploids may not; indeed, many autopolyploids are morphologically indistinguishable from their diploid progenitors (see Husband and Schemske 1998). To make matters worse, when morphologically cryptic autopolyploids are identified, most botanists do not consider them new species, reflecting their frequent adherence to a morphological species concept. As noted above, almost no autopolyploid plants are named as separate species. (This is not as true in animals. There are a fair number of named autopolyploid frog and toad species [Keller and Gerhardt 2001].) Moreover, cytological attempts to identify autopolyploids often rely on detecting multivalents at meiosis. But autopolyploids may ultimately diploidize, causing misclassification. Given all this, it is hardly surprising that recent authors conclude that autopolyploids are more common than once thought (Thompson and Lumaret 1992; Ramsey and Schemske 1998; Otto and Whitton 2000; Soltis and Soltis 2000).

Second, Ramsey and Schemske's (1998) findings may not suggest what they first seem to. Ramsey and Schemske suggest that autopolyploids *form* at a faster rate than allopolyploids, but formation is not equivalent to establishment or *persistence*. Autopolyploid individuals might, in other words, appear at higher rates than allopolyploid individuals, but also go extinct more often. (By analogy, deleterious mutations appear at higher rates than favorable ones but go extinct more often.) There are good reasons for thinking that new autopolyploids may disappear more often than new allopolyploids:

1. Autopolyploids might often suffer multivalents at meiosis (at least early on), causing sterility; allopolyploids, on the other hand, should enjoy more bivalent formation and thus higher fertility.
2. As noted, autopolyploids may differ less from their diploid progenitors both morphologically and ecologically than do allopolyploids (Levin 2000, p. 19); autopolyploids may thus be less likely to persist in the face of competition from related diploids or to colonize novel habitats.

The first reason has been questioned. In a recent literature review, Ramsey and Schemske (2002) found that multivalents occur at a frequency of 29% in autopolyploids and only 8% in allopolyploids. Although large and in the right direction, this cytological difference did *not* translate into fertility differences: neither mean pollen viability nor mean seed fertility differs between new auto- and allopolyploids (Ramsey and Schemske 2002). (One possible reason, not discussed by these authors, is that allopolyploids may also suffer *genic* hybrid sterility. Autopolyploids cannot suffer these incompatibilities.)

The idea that autopolyploids are established less often than allopolyploids is supported by a little-noted biogeographic pattern: polyploid series are absent from, or at least rare on, oceanic islands (see Schluter 2000, p. 79, and below). This finding is easily explained if most successful polyploid speciation involves allopolyploidy: early in the history of oceanic islands, hybrid speciation is essentially impossible. Lone colonizing species invade but the absence of sister taxa precludes allopolyploidy. Allopolyploid speciation requires a long wait until one of two rare events occurs: either the arrival of a sister species by colonization or the splitting of an early colonizing species into two sister taxa having structurally different chromosomes. Allopolyploid speciation will eventually become possible on oceanic islands or archipelagos, but its rate must be greatly reduced. It is important to note that the claim is not that there are no polyploids on oceanic islands; there are. The claim is also not that there are no endemic polyploids; again, there are. The point is that *endemic polyploid series* are rare or absent on islands or archipelagoes. The famed silversword alliance of Hawaii, for instance—a complex of at least 30 species showing remarkable morphological diversity—all share the same ploidy level, although the group as a whole may have an allopolyploid origin (Carr et al. 1996). But the presumed diploid progenitor species occurs not in Hawaii, but in western North America; Carr et al. (1996) argue that the polyploid first appeared on the continent and then dispersed to Hawaii. The allopolyploid cotton *Gossypium tomemtosum*, also endemic to Hawaii, appears to have arisen in a similar way (DeJoode and Wendel 1992; Levin 2000, p. 107). (See also Wendel and Percy 1990 on the cotton *G. darwinii*, endemic to the Galápagos.) Such cases incidentally prove that the absence of polyploid series on islands has nothing to do with their lack of ecological suitability for island life: when introduced, polyploids do well. The absence of polyploid series on oceanic islands is most easily explained by the idea that most *successful* polyploids are allopolyploids.

In the end, the possibilities discussed above are not incompatible: new autopolyploid individuals may appear more often than once believed *and* fail to establish populations more often than new allopolyploids. Similarly, autopolyploid *species* might be more common than traditionally thought but still less common than allopolyploid species. Indeed this seems likely.

Ecology and persistence

Because they arise in sympatry with their diploid progenitors, all polyploids—both auto- and allopolyploid—face a serious challenge to survival. Polyploids do not benefit from the (perhaps extended) breathing period enjoyed by the products of allopatric speciation. Instead, they are thrown into immediate competition with their diploid ancestors. Most surely do not survive the test. (This is a necessary consequence of Ramsey and Schemske's origination rate of 10^{-5}; polyploid species clearly do not establish themselves at anything like the rate implied by this value.) But two factors increase the odds of survival of a new polyploid: prezygotic isolation and ecological divergence from diploid progenitors.

Some prezygotic isolation seems almost necessary for survival. The difficulty is that the postzygotic isolation that automatically follows from polyploidy (triploid hybrid sterility) bars gene flow between the new species and its diploid ancestors, but paradoxically also puts the polyploid at risk of extinction. The reason is that a population harboring both tetraploids and diploids is subject to dynamics similar to those characterizing heterozygote disadvantage (underdominance). Such a system is unstable, with selection driving the initially rare type (e.g., the new tetraploid) out of the population: the new tetraploid is "mated to death." The result is "minority cytotype exclusion," the selective elimination of rare polyploids (Levin 1975). (Note, however, that if the tetraploid *does* reach high frequency, minority cytotype exclusion will then act to exclude the diploid parental species.)

Although a small body of theory considers the problem of overcoming minority cytotype exclusion (Levin 1975; Fowler and Levin 1984; Felber 1991; Rodriguez 1996; Husband and Sabara, 2003), the general solution seems clear: polyploids must abandon random mating. There are several ways to do this. The most effective are vegetative reproduction and selfing. Not surprisingly, both are common in polyploids (Stebbins 1957a; Ramsey and Schemske 2002). Taxa that can self or reproduce vegetatively are preadapted to escaping minority cytotype exclusion and thus to successful polyploid speciation.

Minority cytotype exclusion might also be avoided by prezygotic isolation. Such isolation between tetraploids and diploids could arise in a number of ways. One is flowering asynchrony: taxa that flower at different times will not randomly interbreed. Such asynchrony is well known in natural polyploids (Segraves and Thompson 1999; Levin 2002). Husband and Schemske (2000) showed that the autotetraploid fireweed *Chamerion* (formerly *Epilobium*) *angustifolium* begins flowering one week later than diploids in mixed populations. Tetraploids and diploids consequently show an overlap in flowering time of only about 50%, decreasing hybridization; indeed, triploid hybrids are found less often in mixed natural populations than expected with random mating (Husband and Schemske 1998). Interestingly, flowering asynchrony is common even in new artificial polyploids. As in *Chamerion* and certain other natural polyploids, artificial polyploids often flower later than their diploid ancestors (Stebbins 1950, p. 304; Levin 1983; Husband and Schemske 2000). This suggests that flowering asynchrony might be a physiological consequence of increase in ploidy—perhaps reflecting a general slowing of development (Stebbins 1950; Otto and Whitton 2000; Ramsey and Schemske 2002)—and not the result of subsequent selection on a polyploid lineage. (Note, however, that differences in flowering time between ploidy levels might come at a cost: if triploids and diploids *also* flower at different times, the rate of formation of polyploids via a triploid bridge slows [D. Schemske, pers. comm.].)

In plants, prezygotic isolation may also result from pollinator specificity, which might also reflect an immediate consequence of change in ploidy: polyploids often show an increase in overall size (see below) or flower morphology that could affect pollinator visitation. Segraves and Thompson (1999) showed that autotetraploids of the perennial *Heuchera grossularifolia* have flowers of different size, shape, and color from their diploid progenitors. Certain lepidopteran, dipteran, and hymenopteran pollinators consequently visit plants of one ploidy level more often than those of other ploidy levels in sympatry. We hasten to add, however, that because these autotetraploids were natural, not spontaneous, the observed floral differences may have evolved *after* polyploid formation, a problem that plagues all studies of non-spontaneous polyploids.

Pollinator isolation might also arise as an automatic consequence of vegetative reproduction. Taxa that can reproduce vegetatively are often clumped spatially. Because insect pollinators frequently visit local clusters of plants, insects may move pollen between clones of the same ploidy level more often than expected when ignoring spatial structure. Husband and Schemske (2000) show that such spatial clumping contributes to prezygotic isolation between tetraploid and diploid *Chamerion* in mixed populations. Indeed, only 15% of bee flights that involved visits between different inflorescences included both tetraploids and diploids. More recent analyses suggest that pollinator fidelity plays an even more important role in prezygotic isolation between diploid and tetraploid *Chamerion* than does flowering asynchrony (Husband and Sabara 2003). Moreover, prezygotic isolation generally appears more important between these species than postzygotic isolation, at least currently (Husband and Sabara 2003; Whitton 2004).

Interestingly, there is even evidence that assortative mating in *animals* sometimes reflects an automatic consequence of polyploidy. Males of the gray tree frog, *Hyla versicolor*, which is tetraploid, produce calls that differ from those of its diploid progenitor, *Hyla chrysoscelis*. It appears that this difference partly reflects a difference in cell size between the two species, which is probably a direct consequence of ploidy level (Keller and Gerhardt 2001, and see below). (Of course females would have to evolve a preference for the new male call, presumably by selection for "species recognition," as described in Chapter 6.)

Even if new polyploids are completely isolated prezygotically from their progenitors, their establishment is far from assured. Establishment of a stable population also requires *ecological* divergence. Moreover, this divergence must occur on a much faster timescale than with allopatric taxa. Thus, while one can debate whether ecological establishment and persistence are legitimate parts of speciation—and we have argued that they are not—the debate is essentially moot when considering polyploidy. Without ecological differentiation, we are unlikely to get a *population* of polyploids and so, under the BSC, we do not have two species.

But rapid ecological divergence could be common. The reason is that some ecological differentiation may be an immediate consequence of polyploidy. It is well known that polyploidy increases cell size (the "gigas effect"). This increase has an automatic effect on surface-to-volume ratios and, consequently, on certain metabolic rates. There is also growing evidence that newly formed polyploids show unusual patterns of gene expression, including frequent gene silencing and organ-specific changes in transcription levels (e.g., Adams et al., 2003; Osborn et al. 2003). Not surprisingly, polyploidy also appears to change a number of phenotypic traits. Polyploids are often larger and more robust than their diploid progenitors, with bigger flowers, stems, seeds, and more luxurious foliage (reviewed in Levin 2002). Even adult polyploid invertebrates are larger than their diploid progenitors (Levin 1983; Otto and Whitton 2000; Ramsey and Schemske 2002). These phenotypes may be connected to an overall slowing of development often seen in polyploids, a slowing that, in plants, frequently causes delayed flowering and fruiting. As Levin (1983) emphasizes, a simple one-week delay in these times might have a host of ecological consequences, including changing the identity of pollinators, seed dispersers, and herbivores. Increases in ploidy also often affect many features of plant physiology, including hormone concentrations, water regulation and rates of CO_2 exchange, transpiration, and photosynthesis (Tal 1980; Levin 1983, 2002).

The physiological consequences of polyploidy are probably, then, common enough and dramatic enough that new polyploids differ ecologically from their progenitors. Indeed, plants of different ploidy level clearly sometimes have different ecological preferences or tolerances, though few or none of the relevant data come from newly formed polyploids (Whitton 2004). Polyploids, for instance, may differ from their progenitors in substrate preferences, temperature tolerances, and so on (Levin 1983, 2002). Even when such differences are too small to ensure ecological persistence, they may slow competitive exclusion, providing an opportunity for the evolution of further ecological differences. Indeed, multigenerational experiments on populations containing mixtures of diploids and tetraploids often show that tetraploids persist better than one would expect under random mating in the absence of ecological differences (Levin 2002, pp. 111–113). It is also worth noting that, because polyploid species may have multiple origins—and each such event may yield polyploids that differ slightly genetically and ecologically from the others—each origination provides a new chance for establishment (Segraves and Thompson 1999). Polyploidy may be nearly unique as a form of speciation, then, in that the very genetic changes that give rise to reproductive isolation may simultaneously increase the odds of ecological establishment and persistence. (The other case is ecologically mediated sympatric speciation.)

Finally, we note a possibility that seems to have been neglected in the literature. New polyploids might form in parapatry, not sympatry. Long-distance movement of a pollen grain onto a parapatric sister species could, for example, ultimately give rise to a new allopolyploid. The important point is that this new polyploid might only have to compete ecologically with a single (or perhaps with no) parental species, increasing its odds of establishment. The tradeoff, of course, is that the rate of origination of polyploids is probably lower in parapatry than in sympatry. It is unclear if this would more than compensate for the easier establishment of polyploids in parapatry. This problem deserves further study.

Why is polyploidy rarer in animals than in plants?

The rarity of polyploidy in animals compared to plants, which was recognized early on (Gates 1924; Muller 1925), probably represents "the greatest known difference between the evolutionary patterns in the two kingdoms" (Dobzhansky 1937b, p. 219). Not surprisingly, evolutionists have speculated at length about the cause of this difference. Here we review these speculations. We emphasize, however, that the correct answer remains unknown. This is not to say that no progress has been made; on the contrary, several hypotheses have been rejected. But it remains unclear which of the remaining hypotheses, if any, is correct. In the end, we will argue that there may be no single reason why animals speciate by polyploidy less often than plants. Nonetheless, certain factors seem more important than others; we try to identify these. We also introduce a new hypothesis.

We first consider three ideas that, while probably wrong, still appear in the literature. One of the oldest involves development. Wettstein (1927) and Stebbins (1950) argued that plant development is simple, "open," and indeterminate, while animal development is complex, closed, and determinate. Consequently, plants should be better at tolerating the developmental shock of the diploidy-to-tetraploidy transition than animals. Though enjoying the virtue of simplicity, this explanation succumbed to the later finding that parthenogenetic and hermaphroditic animals *do* show polyploidy (see Otto and Whitton's 2000 review; their list of insect and vertebrate cases shows that most polyploid animals are parthenogenetic). The barrier to animal polyploidy, then, does not seem to reflect a general property of animal development.

A second hypothesis maintains that animal species do not hybridize in nature as often as plant species (Dufresne and Hebert 1994), slowing the formation of allopolyploids. While there may be some truth to this, it does not seem a major reason for the rarity of animal polyploidy. For one thing, some animal groups *do* hybridize at appreciable rates (e.g., fish, birds, toads, snails, and Lepidoptera [Stebbins 1950, p. 254; Dowling and Secor 1997; Presgraves 2002]). For another, this hypothesis deals only with *allo*polyploid speciation. If many polyploids are of autopolyploid origin, the disparity between animals versus plants is left unexplained: where are all the animal autopolyploids?

A third hypothesis is that plants speciate by polyploidy because they can self, whereas animals do not because they cannot usually self. While there is again surely some truth to this idea—polyploidy is more common in animals that can self or that are asexual (White 1978; Dowling and Secor 1997)—it partly sidesteps the critical question. Selfing may increase the odds of polyploid speciation, but what we really want to know is: What is it about outcrossing that *blocks* polyploid speciation?

H. J. Muller (1925) was the first to seriously address this question and his answer represents the most popular explanation of the rarity of animal polyploids. Muller's hypothesis hinges on the fact that animals are usually outcrossing—and in particular dioecious—while plants are not. More specifically, Muller argued that the diploid sex-determining mechanism present in dioecious animals blocks polyploid speciation. Using *Drosophila* as a model, he showed that the cross of a new tetraploid individual to a diploid produces sterile intersexes, supermales, superfemales, etc.—everything but fertile tetraploids. Genotypic sex determination thus seems to present an insurmountable barrier to the establishment of polyploids.

Unfortunately, Muller's theory suffers several problems. For one thing, polyploid series turn out to be fairly common in dioecious plants with genotypic sex determination: although dioecy itself is rare in plants, ten or more independent polyploid series have been identified in dioecious plants (Stebbins 1950, p. 367; Westergaard 1958, table 21; Ehrendorfer 1961; Dempster and Ehrendorfer 1965). If polyploidy is rare in animals because they are dioecious and have genotypic sex determination, why isn't polyploidy just as rare in dioecious plants that have genotypic sex determination? Second, Westergaard (1940, 1958) and Stebbins (1950) pointed out that Muller's theory holds only for organisms having *Drosophila*-like sex determination. In *Drosophila*, sex is determined by the ratio of X chromosomes to autosomes (females = 1:1, and males = 1:2) and the diploid-to-tetraploid transition disrupts these ratios. In particular, the progeny of a $4x \times 2x$ cross includes 2X:1Y:3A individuals who are sterile intersexes. But if sex is instead determined by the presence or absence of a Y chromosome, polyploidy would *not* disrupt sex determination. Unfortunately for Muller's theory, it is now clear that such "Y-dominant" sex determination systems are far more common than the *Drosophila*-like system. Thus, genotypic sex determination does not present an insuperable barrier to polyploid speciation.

Orr (1990) suggested an alternative hypothesis. According to this idea, the rarity of animal polyploids is, as Muller thought, connected to dioecy and sex chromosomes but not to a disruption of sex determination. Instead, the transition from diploidy to tetraploidy disrupts *dosage compensation*, the mechanism by which the dose of X chromosomal gene products is equalized between the sexes, one having a single X and the other two X's. Although the details of dosage compensation vary across taxa—in *Drosophila*, the single X in males is transcribed at twice the rate of each X in females, while in mammals, one X is inactivated in females—polyploid speciation seems to always disrupt the balance of X to autosomal gene product. Orr's argument differs from Muller's in that it can be shown that the establishment of a new polyploid species disrupts dosage balance *whether or not* it disrupts sex determination. The theory also makes several predictions that distinguish it from Muller's theory. Most important, the theory holds only for taxa having degenerate Y chromosomes—only these taxa have dosage compensation—and not for all taxa with genotypic sex determination. (The theory would predict that polyploidy is possible in taxa having single-locus sex determination, in which the Y chromosome is genetically active.) This theory thus claims that polyploidy might be common in plants because degenerate sex chromosomes are rare (both because dioecy is rare and because extreme Y degeneracy is rare in dioecious plants). Polyploidy might be rare in animals, on the other hand, because degenerate sex chromosomes are common.

The hypothesis's predictions appear at first to have a mixed record. Perhaps the clearest prediction is that polyploidy should be possible in animals that lack dosage compensation. This prediction is borne out in Lepidoptera, which possess small sex chromosomes and show no dosage compensation (Johnson and Turner 1979): a surprising number of polyploid Lepidoptera are known (Otto and Whitton 2000), and artificial polyploids have been produced (Bundenberg 1957). But as Otto and Whitton (2000) point out, the prediction seems to fail in birds: birds also possess small sex chromosomes and have been thought to lack dosage compensation. Yet, polyploidy is unknown in birds. Otto and Whitton (2000) thus conclude that dosage compensation may not represent a critical barrier to polyploidy. Recent evidence, however, invalidates this objection. The earlier conclusion that birds lack dosage compensation was based on a single sex-linked gene in several species (Baverstock et al. 1982). Recent work shows that at least six of nine sex-linked genes in the chicken *are* dosage compensated (McQueen et al. 2001). If this finding is confirmed in other birds, the absence of polyploidy in birds cannot be taken as evidence against the dosage compensation idea.

At least one other factor might help explain the rarity of animal polyploidy, one that has not been discussed in the literature. Allopolyploid speciation is possible only when two species differ by structural rearrangements but *not* by factors causing genic sterility (Stebbins 1950, p. 330). If species produce genically sterile diploid hybrids, they may also typically produce genically sterile allopolyploid hybrids, thus barring successful polyploid speciation. There is, therefore, a race between the evolution of genic incompatibilities on the one hand, and polyploid speciation on the other. Any factor that accelerates the evolution of genic incompatibilities will thus close the window of opportunity on allopolyploid speciation. It seems clear that one such factor is sex chromosomes: degenerate Y (or W) chromosomes allow expression of normally masked hybrid incompatibilities involving partially recessive alleles on the X (or Z) chromosomes—an idea that forms the basis of the dominance theory of Haldane's rule (Chapter 8). As there is good evidence that the genes causing hybrid problems act fairly recessively in hybrids, degenerate sex chromosomes will speed the rate at which taxa produce genically sterile hybrids. Allopolyploid speciation should thus be rarer in taxa possessing degenerate sex chromosomes.

Given that this hypothesis, like the dosage compensation one, hinges on possessing degenerate Y (or W) chromosomes, the two hypotheses make many of the same predictions. In particular: polyploidy should be rare in animals having degenerate sex chromosomes; polyploidy should be rare in plants having degenerate Y chromosomes; polyploidy might occur in parthenogenetic animal taxa (either because they require no dosage compensation or because they cannot produce heterogametic hybrids, at least when females are the homogametic sex); and so on. One difference, however, distinguishes the dosage compensation and "genic incompatibility" hypotheses. The former applies to both allo- and autopolyploidy, whereas the latter applies only to allopolyploidy. Because the cases needed to test these hypotheses are so rare—plants with sex chromosomes, or parthenogenetic animals—we currently have too little data to empirically distinguish these hypotheses.

In summary, it seems likely that the rarity of polyploidy in animals reflects the consequences of a commitment to chromosomal sex determination. Dioecious organisms having chromosomal sex determination (1) are usually incapable of selfing; (2) typically evolve dosage compensation; and (3) probably rapidly evolve hybrid genic incompatibilities. These factors may prevent polyploidy.

It remains worrisome, however, that so little effort has been devoted to testing these hypotheses. Indeed, we possess so few critical data that no one should be surprised if all combinations of the above hypotheses turn out to be wrong. Although there has been a good deal of sensible speculation, and although we suspect that we have at least identified some of the factors blocking polyploidy in animals, it is sobering to realize that we can conclude so little about this most profound difference between animal and plant speciation.

Recombinational Speciation

"Recombinational" speciation—the diploid form of hybrid speciation—is less well known than polyploidy. There are two possible reasons: recombinational speciation might be rare in nature or it might be hard to detect (Rieseberg 1991). The latter hypothesis is certainly plausible: while one can easily detect polyploid speciation with cytological techniques, detecting recombinational speciation typically requires molecular techniques, which became available only later.

What is recombinational speciation?

In recombinational speciation (a term coined by Grant 1981), hybridization between two species gives rise to a new lineage that is both fertile and true breeding, but that is reproductively isolated from both parental species. This new species has the same ploidy level as the parental species. (For this reason recombinational speciation is sometimes called "homoploid hybrid speciation.") The process occurs in sexual taxa and, as traditionally understood, involves forming a novel genotype that is homozygous for several chromosomal rearrangements that differentiate the parental species (Grant 1981, p. 250). Because recombinational speciation requires hybridization between species, it occurs in sympatry or parapatry. As we will see, this fact poses the single greatest hurdle to recombinational speciation: although sympatry increases the frequency of hybridization at the start of the process, it threatens the survival of any new hybrid lineage, as this lineage is constantly assailed by gene flow from the parental species.

The notion that hybridization between species might give rise to a new species has a long history (Grant 1981, pp. 244–247; Rieseberg 1997). The earliest mention of the idea seems to be that of Linnaeus in 1744. A century later, Mendel speculated that the preservation of certain hybrids between plant taxa might represent the origin of new species. But as Rieseberg (1997) emphasizes, it was only with the rediscovery of Mendelism that the problem was stated in its modern form: How can a true-breeding hybrid lineage having a "normal" ploidy level arise that is both fit *and* reproductively isolated from its progenitor species?

We first consider theoretical models, both verbal and mathematical, that try to answer this question. We then turn to the evidence from the greenhouse and nature.

Theory

EARLY EFFORTS. Muntzing (1930a) noted that species whose F_1 hybrids are not completely sterile might produce a "multitude of new forms," by which he meant recombinant homozygous types that breed true. But the first systematic explanation of how recombinational speciation might occur appeared much later, in Stebbins (1957a) and Grant (1958). Their model involves three steps (Rieseberg 1997):

1. Two parental taxa differ by some number of chromosomal rearrangements (e.g., reciprocal translocations).

2. Although these rearrangements lower the fertility of hybrids between the parental species, segregation and assortment in F_1s give rise to novel, "genetically balanced," and thus viable gametes. Selfing of these gametes produce new homozygous genotypes.

3. These novel diploid genotypes are fit within the lineage, but reproductively isolated from both parental species.

To see how this process works, it is best to first look at a simple case where recombinational speciation does *not* occur (our treatment here follows Grant 1958). Consider two species of the same ploidy level that are sympatric. Imagine further that these species differ by a single reciprocal translocation: one has genotype *A A B B*, and the other has genotype *Ab Ab Ba Ba* (where "*Ab*" and "*Ba*" represent translocated chromosomes). If these species hybridize, the resulting F_1 hybrid is *A Ab B Ba*. This hybrid can form four kinds of gametes: two parental (*A B* and *Ab Ba*) and two recombinant (*A Ba* and *Ab B*). Note that the latter two kinds of gametes suffer duplications and deficiencies and so are inviable (where we assume gene expression in gametes, as in plants). The F_1 hybrid is, consequently, only half as fertile as normal and—given selfing—the only fit F_2 lineages that appear are those that reconstitute the parental genotype. We have *not* produced a fit hybrid lineage.

All of this changes if we consider a slightly more complex scenario, in which sympatric species differ by *two* reciprocal translocations (Figure 9.2). One species has genotype *A A B B C C D D*, and the other has genotype *Ab Ab Ba Ba Cd Cd Dc Dc*. This hybrid can form 16 kinds of gametes. Of these, only four are genetically balanced and viable: two parental (*A B C D* and *Ab Ba Cd Dc*), and two recombinant (*A B Cd Dc* and *Ab Ba C D*). The F_1 hybrid is, consequently, only one-fourth as fertile as normal. But if this F_1 selfs, it will give rise to four fit homozygous F_2 lines. As shown in Figure 9.2, two of these—*A A B B Cd Cd Dc Dc* and *Ab Ab Ba Ba C C D D*—represent novel true-breeding lines. Most

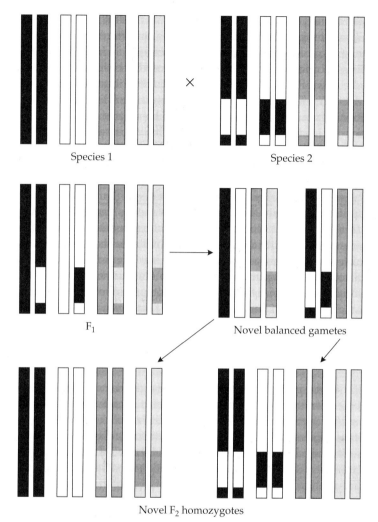

Figure 9.2 Recombinational speciation. One parental species has genotype $A\,A\,B\,B\,C\,C\,D\,D$ and the other $Ab\,Ab\,Ba\,Ba\,Cd\,Cd\,Dc\,Dc$ (i.e., each chromosome is rearranged). The resulting F_1 hybrid can form four genetically balanced gametes: two parental and two recombinant. By selfing, this F_1 hybrid can give rise to four fit homozygous F_2 lines. Two of these, shown at the bottom, represent novel true-breeding lines. (From Rieseberg 1997.)

important, these new lines are at least partially reproductively isolated from both parental species since hybridization with the parentals again yields hybrids suffering lowered fertility due to production of unbalanced gametes.

This scenario is similar to the Dobzhansky–Muller model of speciation. In both cases, speciation is due to intrinsic postzygotic isolation and is essentially impossible when taxa differ by a single genetic factor, but possible when they differ by two. This point was understood by both Stebbins (1957a) and Grant (1958), who emphasized that the above argument still stands if the reciprocal translocations are replaced by "complementary gametic lethals" (but see below).

The above scenario can be generalized in several ways. We could include, for instance, other types of chromosomal rearrangement (e.g., transpositions) (Grant 1981). We could also consider any number of independent reciprocal translocations. This last extension is of some significance. It turns out that, as the number of reciprocal translocations differentiating two taxa grows, the proportion of fit F_2 types that represent novel genotypes increases rapidly. Recombinational speciation might therefore become the rule, not the exception. This argument was first quantified by Stebbins (1957a), who showed that the proportion of fit gamete types that represent novel genotypes is $p = (2^n - 2)/2^n$, where n is the number of reciprocal translocations distinguishing the parental species. This proportion rapidly approaches one as the number of reciprocal translocations grows.

The above scenario, however, rests on several assumptions. It assumes, for example, that F_1 hybrids between taxa showing many genetic differences remain partially fertile. If reciprocal translocations are involved, this should be true, as $1/2^n$ of all gamete types remain viable. This proportion decreases rapidly with n, but the inviability of most gamete types is irrelevant as these types represent genotypes that, upon inbreeding, would not yield balanced F_2 genotypes anyway. It is worth noting, however, that the algebra leading to Stebbins's result above predicts an inverse relation between the number of novel, fertile species that can be "extracted" from two parental species and the fertility of their F_1 hybrids (Grant, 1958). It is also assumed that F_1 individuals are partially or fully selfing. This is important as selfing increases the odds that a largely sterile F_1 hybrid will produce balanced true-breeding F_2 offspring. If the F_1 hybrid instead outcrosses to a parental species, the fraction of balanced stabilized lines falls (Grant 1981, p. 260), and stabilized types often emerge only in later generations. We thus expect a strong association between selfing and recombinational speciation.

RECENT THEORY. After Stebbins (1957a) and Grant (1958), theoretical study of recombinational speciation ceased for almost 40 years. But McCarthy et al.'s work in 1995 provided some long needed rigor to a field previously dominated by verbal theories. McCarthy et al.'s (1995) goals were to determine (1) the conditions under which recombinational speciation is likely; and (2) if gene flow from the parental species swamps formation and establishment of a new hybrid lineage.

They addressed these questions with computer simulations that considered a spatial model in which two parental species come into contact along a geographic interface. Species were monoecious and diploid, and were allowed different degrees of selfing. As in Stebbins's (1957a) and Grant's (1958) verbal models, these species differed at a variable number of chromosome rearrangements that cause fertility problems in heterozygotes. Balanced homozygous recombinants between these parental species enjoy high fitness—indeed higher than the parentals. Successful hybrid speciation thus resulted in fixation of the new type and extinction of the parental species.

Some of McCarthy et al.'s findings were expected. They found, for instance, that low F_1 fertility slows hybrid speciation but does not prevent it; recombinational speciation is precluded only when F_1 hybrids are *completely* sterile. Also, the time required for establishment of a successful new recombinant decreases as the fitness of novel balanced recombinants rises. McCarthy et al. also confirmed Grant's intuition about the role of selfing: all else being equal, greater selfing rates speed the time to hybrid speciation. But if the novel balanced type enjoys a large enough advantage, recombinational speciation can occur even in an outcrosser, although many generations are often required. Similarly, the longer the geographic contact zone between the parental species, the faster hybrid speciation occurs, as longer contact zones allow more hybridization events.

Other findings of McCarthy et al., however, were surprising. For instance, speciation was generally faster with *fewer* chromosomal differences between parental species. The explanation seems to be that—if the recombinant type enjoys a selective advantage—it will sooner or later sweep to fixation once it successfully establishes a small local population. But the time required to form this local population grows with the number of rearrangement differences. Second, McCarthy et al. found that the time course of recombinational speciation is punctuated: a new balanced recombinant type spends a long time at low frequency and then suddenly speeds to fixation. This suggests that recombinant speciation involves a critical threshold or positive feedback loop. When rare, a novel type is essentially a hopeful monster: it is surrounded by a sea of individuals of different genotype and any cross between rare and common types yields mostly unfit progeny. But when common, novel-type individuals can breed with each other, yielding high fitness progeny that also carry the novel genotype.

Although McCarthy et al.'s work considered many genetic aspects of recombinational speciation, it ignored most ecological aspects. This is unfortunate since, as Grant (1981, pp. 251–253) emphasized, a new hybrid type might be partially or fully isolated from its parental species by "external" or ecological factors. If a new hybrid genotype can, for instance, occupy a habitat in which the parental species are fairly unfit, its chances of survival increase. Buerkle et al. (2000) considered such ecological factors. Building on McCarthy et al.'s computer simulations, they introduced one major variation: an ecologically novel (and initially empty) habitat. In particular, Buerkle et al. considered a spatial model in which the parental species enjoy high fitness at each "end" of the environment while hybrids enjoy high fitness in an intermediate region. The ecological performance of a plant thus depends both on its genotype and on the habitat in which it finds itself.

Buerkle et al.'s simulations showed that the greater the intensity of habitat-specific ecological selection, the greater the chance of hybrid speciation.

Indeed, with weak ecological selection, hybrid speciation almost never occurred, while with strong ecological selection it occurred in approximately 20% of simulation runs. Most important, in all cases the novel hybrid genotype was successfully established—and persisted for many generations—*without* extinction of either parental species.

Before leaving the theory of recombinational speciation, we must consider one other issue: Can genic factors (i.e., Dobzhansky–Muller incompatibilities) underlie recombinational speciation? The answer is a qualified yes. Although genic recombinational speciation is theoretically possible, the genic and chromosomal cases are not identical. For one thing, genic incompatibilities show an asymmetry that reciprocal translocations do not. Given, for instance, two species—one of genotype *A A b b*, and the other *a a B B*—where the *A* and *B* alleles are incompatible—the reciprocal *a b* combination of alleles must be compatible (Chapter 7). However, given two species separated by a reciprocal translocation, *both* hybrid combinations are inviable. Similarly, Buerkle et al. (2000) noted that more complex combinatoric scenarios (involving more factors) are generally required to yield both high F_2 fitness and reproductive isolation from parentals under genic than chromosomal sterility. Nonetheless, there is no insurmountable barrier to recombinational speciation involving genic incompatibilities and, as we saw in Chapter 8, *Helianthus* (sunflower) data that were originally interpreted as evidence for chromosomal hybrid speciation may instead reflect genic incompatibilities. Indeed, there is some evidence that new recombinant lines can arise and persist that show *no* postzygotic isolation, whether chromosomal or genic. Such lines may instead persist solely by prezygotic isolation from their parental species. There is, however, no theory addressing these unorthodox modes of recombinational speciation.

In summary, although little theoretical attention has been paid to recombinational speciation, the key point seems clear: recombinational speciation *is* possible and should be more common in situations involving selfing and greater ecological heterogeneity.

The data: frequency and artificial hybrids

FREQUENCY AND FITNESS OF HYBRIDS. The theory of recombinational speciation would be irrelevant if species do not hybridize in nature. But as we saw in Chapters 1 and 2, many do. The theory of recombinational speciation would also be irrelevant if hybrid genotypes of reasonably high fitness do not appear. Although the founders of the Modern Synthesis generally viewed hybrids as unfit, there was some recognition that certain genotypes might be exceptions. Dobzhansky (1970), for instance, noted, "a minority of the gene combinations formed by the hybridization of species might be fit, perhaps fit enough to spread onto as yet unoccupied adaptive peaks." Unfortunately, attempts to directly measure the fitness of hybrids have been controversial. Perhaps the best known is that of Arnold and Hodges (1995) (see also Table 5.1 in Arnold 1997), who surveyed the fitness of natural plant and animal hybrids. Arnold and Hodges concluded that hybrids were clearly worse off than parentals in only 13 of 44 cases. Surprisingly, hybrids were as fit or *fitter* than parentals in 24 cases. Arnold and Hodges concluded that their survey did not support the notion that hybrids are generally unfit. However, this conclusion, which follows a long botanical tradition of thinking that hybridization might be beneficial (Anderson 1949), is compromised in two ways. First, the taxa surveyed were not a random sample (Day and Schluter 1995). Instead, Arnold and Hodges focused on hybrid zones (i.e., cases in which hybrids are common in nature and so, a priori, may be less likely to suffer catastrophic fitness effects). If 99% of all species crosses yielded completely inviable hybrids, these cases would not show up in studies of hybrid zones (and we know from hundreds of *Drosophila* hybridizations that such severe fitness effects are common; see Bock 1984). More important, 17 out of 19 of the studies considered by Arnold and Hodges assayed only a *single* fitness component. It seems likely that, if total lifetime fitness had been measured, the results would have been different (mules, after all, are vigorous, but sterile).

But these problems do not affect Arnold and Hodge's second conclusion: various hybrid genotypes formed between two species show a *range* of fitnesses. A good example is seen in natural hybrids between *Iris fulva* and *I. brevicaulis*. Cruzan and Arnold (1994) molecularly genotyped individuals from a hybrid population and found that fitness varies with marker genotype: "intermediate" individuals suffer a higher rate of seed abortion than those having more "extreme" genotypes (those that approximate one of the pure species' genotypes). A similar pattern characterizes many genetic analyses of backcross hybrids in *Drosophila* (Chapter 8).

Arnold and Hodges's two conclusions can therefore be decoupled. Hybrids are almost surely generally less fit than parentals, but certain rare hybrid genotypes may enjoy higher fitness than the parentals. This view is supported both by recent theory (Barton 2001) and by direct evidence in the sunflower, in which artificial hybrid lines between *Helianthus annuus* and *H. petiolaris* revealed that 5% of hybrid gene combinations appear more fit than parental species (see below). The basic requirements of recombinational speciation—hybridization and the rare production of fit hybrid genotypes—thus seem satisfied.

We now turn to a search for cases of recombinational speciation. We first look at the attempt to artificially produce hybrid species in the laboratory or greenhouse. We then look in nature.

ARTIFICIAL RECOMBINATIONAL SPECIATION. The earliest experimental work on recombinational speciation appeared before the idea of non-polyploid hybrid speciation was clearly articulated. Gerassimova (1939) (see Grant 1958, p. 356) began with two mutant lines of the cereal *Crepis tectorum*, each homozygous for a different reciprocal translocation. As expected, these lines were fertile, but produced partially sterile F_1 hybrids. The F_2 generation yielded one plant that was homozygous for both reciprocal translocations, and was fertile. When this line was crossed to wild-type *C. tectorum*, however, the resulting progeny were partially sterile. This work proved that one could pass through a partially sterile F_1 genotype on the way to an F_2 of higher fitness.

While more work on recombinational speciation followed (e.g., in *Nicotiana* and *Collinsia*), we consider only the most important of these experiments, involving *Elymus*, *Gilia*, and the yeast *Saccharomyces*. All involve strong—and thus biologically realistic—levels of sterility between parental taxa. The first experiment was performed by Stebbins (1957b), who crossed the grasses *Elymus glaucus* and *Sitanium jubatum*, recovering a small number of F_1 hybrids. Although these hybrids showed low pollen fertility and nearly all F_1 florets failed to produce seed, a few such plants successfully backcrossed to *E. glaucus*. While most of the resulting progeny appeared to be revertants to pure *E. glaucus*, one backcross individual was both morphologically intermediate between *E. glaucus* and F_1 hybrids and was reasonably fertile. Two generations of selfing yielded a line showing both high vigor and fertility. Most important, when this line was crossed to *E. glaucus*, the resulting progeny were unfit, suffering 0%–3% pollen fertility.

While this work suggested that hybrid speciation is possible, Grant (1966b, 1981) was concerned that it did not fully mimic models of recombinational speciation since the experiment featured backcrossing, not production of a hybrid F_2 generation by selfing. Grant (1966b) redressed this concern in a ten-year experiment involving two desert annuals, *Gilia modocensis* and *G. malior*. These plants are largely selfing. Upon hybridization, Grant (1966a) obtained vigorous but mostly sterile F_1 hybrids. A few F_2 plants were obtained, however. These plants, which were not polyploids, were used to establish lines that were inbred and selected for vigor and fertility for additional generations. Several of these lines became reasonably fit. One line sometimes produced individuals having nearly normal pollen fertility. Morphological work suggested that this line was not a revertant to a pure parental genotype but a novel hybrid genotype. Most important, crosses between this line and each of the parental species produced mostly sterile progeny.

More recently, Greig et al. (2002) produced artificial hybrid "species" of yeast. Although *Saccharomyces cerevisiae* and *S. paradoxus* are postzygotically isolated (F_1 hybrids have about 1% fertility), Greig et al. were able to recover a number of viable F_1 gametes. By taking advantage of "autofertilization" in yeast (unfertilized gametes can switch mating type after mitotic division, allowing a gamete to fertilize a clone of itself), they were able to produce diploid, homozygous F_2 hybrids. Many of these hybrid lineages were highly fertile but reproductively isolated from both parental species.

The data: natural recombinational speciation

While the above work shows that recombinational speciation is possible, it suffers one shortcoming: it is unclear how readily the results, like those from any greenhouse or laboratory experiment, can be extrapolated to the wild. Grant's experiment, for instance, required many generations of intense artificial selection and constant nursing of sick genotypes. The important question, then, is if recombinational speciation is not only possible but occurs in nature. The answer is yes. Two cases of natural recombinational speciation—one involving the sunflower *Helianthus* and the other irises—have been sufficiently well-analyzed to merit special attention. We review these cases and then briefly discuss several others that are less well documented.

HELIANTHUS. *H. annuus* and *H. petiolaris* are diploid ($2n = 34$), annual, hermaphroditic, and (usually) self-incompatible species that are widespread throughout North America. Both are often found in disturbed habitat. While *H. annuus* usually occurs at elevations lower than *H. petiolaris*, the species often overlap geographically. As expected given these different habitats, the species have different ecological preferences: *H. annuus* prefers heavier soils while *H. petiolaris* prefers drier, sandier soils (Schwarzbach et al. 2001). DNA and morphological evidence show that *H. annuus* and *H. petiolaris* are not sister species (Rieseberg 1991; Rieseberg et al. 1991). Molecular data further show that the two species differ by at least ten chromosome rearrangements (Rieseberg et al. 1995c).

A third species, *H. anomalus*, also an outcrossing annual, is rarer than *H. annuus* and *H. petiolaris*. It occurs in Arizona and Utah, well within the range of both *H. annuus* and *H. petiolaris*, and is morphologically uniform throughout its range. Although it differs morphologically and ecologically from the above two species, molecular data show that *H. anomalus* is a hybrid between *H. annuus* and *H. petiolaris* (Rieseberg 1991; Rieseberg et al. 1991; Schwarzbach and Rieseberg 2002). But *H. anomalus* is diploid ($2n = 34$), so its origin does not involve polyploidy.

Molecular data suggest that *H. anomalus* is young. The best estimate is that *H. anomalus* appeared 100,000–160,000 years ago (Rieseberg et al. 1991; Ungerer et al. 1998; Schwarzbach and Rieseberg 2002); indeed, there is some evidence that *H. anomalus* arose multiple times (Schwarzbach and Rieseberg 2002). *H. anomalus* also differs from its parental species at the chromosomal level: while *H. anomalus* carries some rearrangements derived from *H. annuus* and some from *H. petiolaris*, it also carries several unique rearrangements (Rieseberg et al. 1995b). (Schwarzbach and Rieseberg [2002] suggest, based on the above molecular and chromosomal data, that the hybrid species formed before humans appeared in North America, precluding a role for human disturbance in the species' origin. While plausible, it is hard to exclude the possibility of human disturbance entirely, especially given the imprecise calibration of the molecular clock in plants.)

How did this hybrid species arise? Hybrid speciation would obviously be impossible if reproductive isolation between the parental species were complete. But it is not. Although *H. annuus* and *H. petiolaris* differ somewhat in flowering time (Heiser 1947), they do not show strong prezygotic isolation. In fact, pollinators barely discriminate between the two (Rieseberg et al. 1998), and hybrid swarms are common. Studies of natural hybrid zones show that hybridization rates between parental-like individuals are about 4%–15%

(Rieseberg et al. 1998). And while F$_1$ hybrids are semisterile (Heiser 1947; Rieseberg et al. 1995c), F$_2$ and other later generation hybrids are often fit and found in nature (Rieseberg et al. 1995b,c). It thus seems possible that *H. annuus* and *H. petiolaris* came into contact, allowing certain chromosomal regions to flow fairly freely from one species into the other while other regions did not, giving rise to a subset of fit recombinant genotypes. Indeed, something like this can be seen today: using many markers distributed over all chromosomes, Rieseberg et al. (1999) characterized gene flow across three *H. annuus–H. petiolaris* hybrid zones, showing that while many chromosome regions flowed readily between species, many did not.

The production of a fit recombinant genotype, however, is not enough to yield a new species. The new hybrid species must be immediately reproductively isolated from its parental species. In fact it is. F$_1$ hybrids between *H. anomalus* and *H. annuus* are almost completely sterile, while F$_1$ hybrids between *H. anomalus* and *H. petiolaris* are semisterile (Rieseberg et al. 1995c). There is also good evidence for ecological differentiation from the parental species: *H. anomalus* is xerically adapted and is found in sand dune and swale habitats, unlike its parents (Heiser et al. 1969; Rieseberg 1991). It seems, then, that the final stabilized hybrid genotype can maintain its genetic distinctness from its parental species and that extrinsic, as well as intrinsic, isolation play some role.

Recombinational speciation is more likely if it is fast: if stabilization of a recombinant genotype that is also reproductively isolated from both parentals required, say, a thousand generations, speciation would probably not occur. But two lines of evidence suggest that *H. anomalus* hybrid speciation was fast. The first is genetic. Ungerer et al. (1998) showed that the *H. anomalus* genome is composed of long blocks of contiguous markers from one species or the other. This pattern is unexpected if stabilization of the *H. anomalus* genotype took many generations: if chromosomal regions from *H. annuus* and *H. petiolaris* remained heterozygous for many generations, recombination would result in fine-scale scrambling of material from the two species (except in rearranged regions). Comparison of marker data from *H. anomalus* with computer simulations suggests that the *H. anomalus* genome was stabizilized in fewer than 60 generations and, probably, in about 25 generations. Those results are, however, subject to one concern: while the "large blocks" finding is consistent with rapid stabilization of the *H. anomalus* genotype in the distant past, it is also consistent with recent introgression from the parental species into *H. anomalus* (Ungerer et al. 1998). We would like, therefore, more direct evidence of rapid hybrid speciation.

Fortunately, there is such evidence. Rieseberg et al. (1996) synthesized three hybrid lineages between *H. annuus* and *H. petiolaris* in the greenhouse by crossing the parental species and then allowing further generations of intercrossing and backcrossing. These artificial hybrid lineages were molecularly genotyped (see also Ungerer et al. 1998). The recombinant genotypes arrived at were nearly identical across the hybrid lineages, showing that natural selection—not chance—governs the fate of introgressed regions. Moreover, the recombinant genotypes were similar to that of the natural hybrid species, *H. anomalus*. This essentially proves that the hybrid species' genotype can be stabilized in a few generations. The new hybrid genotype was also fit: all three lines enjoyed pollen viability of greater than 90% by the fifth generation. The artificial hybrid lines were also reproductively isolated from the parental species, though not as strongly as are the natural hybrid species. Finally, the artificial hybrid lines were cross-compatible with the natural hybrid *H. anomalous* (Rieseberg 2000). In sum, there are good reasons for thinking that the hybrid speciation event that produced *H. anomalous* was fast and that the genotype of the resulting hybrid species was shaped by selection.

There is also evidence for more than one hybrid speciation event in the genus *Helianthus*. Rieseberg's (1991) early work revealed three unambiguous cases of hybrid speciation—yielding *H. anomalus*, *H. deserticola*, and *H. paradoxus*, all of which result from hybridization between the same two species, *H. annuus* and *H. petiolaris*—as well as several other possible, but less well-supported, cases. In total, 3 to 5 of the 12 species in the section may have hybrid origins. As one might also expect, both *H. deserticola* and *H. paradoxus* reside in novel habitats (the former in desert and the latter in brackish marshes) and are reproductively isolated from their parental species (Rieseberg 1997). Molecular data also suggest, but do not prove, that both *H. deserticola* and *H. paradoxus* formed before humans arrived in the Americas (Welch and Rieseberg 2002; Gross et al. 2003).

IRISES.   Several species of the genus *Iris* reside in Louisiana. Two species, *Iris fulva* ($2n = 42$) and *I. hexagona* ($2n = 44$), come into contact in southern regions of Louisiana where bayous deriving from the Mississippi River touch freshwater marshes and swamps. *I. fulva* is restricted to shaded margins of the bayou above sea level, while *I. hexagona* is found in marshier freshwater habitats (Riley 1938; Arnold et al. 1990; Arnold and Bennett 1993; see also Chapter 5). A third species, *I. brevicaulis* ($2n = 44$), is found nearby but in drier habitats. From morphological and fitness evidence, Riley (1938), Anderson (1949), and others suggested that these three species hybridize in nature. Although a large and somewhat controversial literature has grown up around this hybridization (reviewed in Arnold 1992, 1997; Arnold and Bennett 1993), we are concerned here only with whether hybridization between these taxa has yielded a new, stabilized recombinant species. The answer seems to be yes.

There is now good evidence that a fourth species of Louisiana iris, *I. nelsonii*, is of hybrid origin, deriving from some combination of genes from the above three species. *I. nelsonii* has the same ploidy level as its congeners ($2n = 42$; Randolph 1966), so its origin does not involve polyploidy. *I. nelsonii* co-occurs in southern Louisiana with the above species (Arnold and Bennett 1993), but occupies a novel habitat: areas of deep water, showing considerable fluctuations in water level and heavy shade (Randolph 1966; Arnold et al. 1990). The first evidence that *I. nelsonii* is hybrid was cytological: *I. nelsonii* shows a unique karyotype, with one chromosome apparently deriving from *I. fulva* and another from *I. hexagona* (Randolph et al. 1961). Further evidence for hybrid status came from morphological studies. Randolph (1966) showed that natural populations of *I. nelsonii* are intermediate between *I. fulva* and *I. hexagona* at several characters, while one character seems most similar to *I. brevicaulis*. More convincing evidence of *I. nelsonii*'s hybridity did not appear for another 30 years when Arnold et al. (1990) studied allozyme variation at many loci in the above species. Although few loci proved diagnostic for species identity, *I. nelsonii* carries alleles typical of both *I. fulva* and *I. hexagona*; overall, *I. nelsonii* is most similar to *I. fulva* (Arnold et al. 1990). Later nuclear DNA data confirmed that *I. nelsonii* is a hybrid between *I. fulva* and *I. hexagona* (Arnold et al. 1991). Interestingly, though, chloroplast DNA data showed that *I. nelsonii*'s cytoplasm derives entirely from *I. fulva* (Arnold et al. 1991). It thus appears that *I. nelsonii* derived from *I. fulva* seed and *I. hexagona* pollen, and that early generation hybrids backcrossed mostly to *I. fulva*. Later data revealed that at least some nuclear alleles were derived from *I. brevicaulis* (Arnold 1993).

The finding that *I. nelsonii* is a hybrid between *I. fulva* and *I. hexagona*, and to some extent *I. brevicaulis*, makes good sense given what is known about present hybridization between these species in nature: although *I. fulva* and *I. hexagona* are partially reproductively isolated, they share pollinators and flower nearly synchronously. Similarly, although *I. fulva* and *I. brevicaulis* flower largely asynchronously, they hybridize at some rate (Arnold 1997).

While this case represents a fairly compelling example of natural hybridization, it suffers some shortcomings. First, other than the habitat isolation noted above, we know little about reproductive isolation between *I. nelsonii* and its progenitor species. Second, an *I. nelsonii* shows a distinctive morphology and habitat use, suggesting at least partial reproductive isolation. In addition, an *I. nelsonii*-like genotype has not been artificially reconstructed from the parental species in the laboratory, although contemporary hybrids found segregating in the wild are similar to *I. nelsonii* (Arnold 1993). Finally, it is unclear if contemporary rates and patterns of hybridization between the parental species are representative of those that held when *I. nelsonii* formed: contemporary hybridization at least partly reflects human disturbance (Riley 1938, p. 733; Randolph 1966, p. 145), raising the possibility that this case of hybrid speciation may not be natural.

OTHER CASES. There are many other alleged cases of hybrid speciation. But as Rieseberg (1997) emphasizes, many are unconvincing. Most rely solely on morphological intermediacy of the presumed hybrid species—an intermediacy that might have causes other than hybridity. Better evidence of hybrid status requires multilocus molecular data. Unfortunately, few putative cases have been studied with molecular methods and, among those that have, species often turn out to be non-hybrid. In fact, Rieseberg (1997) identified only eight well-supported cases of recombinational speciation, although several others have been identified since. We discussed four cases above—*Helianthus anomalus*, *H. deserticola*, *H. paradoxus*, and *Iris nelsonii*. We briefly describe several other cases, which have been less well studied.

Perhaps the clearest involves Asian pines, which are wind-pollinated and largely outcrossing. Several sympatric species of pine hybridize in China and Tibet, yielding largely fertile hybrids (Wang and Szmidt 1994). Several workers have suggested that certain pine species are products of hybridization (Mirov 1967). This claim was originally based on morphology, but in one case—*Pinus densata*—good molecular evidence has now appeared. *P. densata* is restricted to high mountains (Wang et al. 1990), elevations at which its putative parental species, *P. tabulaeformis* and *P. yunnanensis*, do not grow. Both parental species are, however, fairly common in Asia and each overlaps geographically with *P. densata* (Wang and Szmidt 1994). All three species are diploid. The first molecular evidence of *P. densata*'s hybrid status appeared in Wang et al.'s (1990) analysis of allozyme variation, which showed that *P. densata* carries a mixture of alleles from *P. tabulaeformis* and *P. yunnanensis*. Wang and Szmidt (1994) later showed that *P. densata* carries chloroplast DNA haplotypes characteristic of both *P. tabulaeformis* and *P. yunnanensis*. Recent allozyme and mitochondrial DNA work confirms *P. densata*'s hybrid status (Wang et al. 2001; Song et al. 2002). This work further shows that there is little present gene flow among the three species. *P. densata* appears, therefore, to be a fairly ancient and stable recombinant genotype.

Another reasonably well-established case of recombinational speciation involves *Stephanomeria exigua* and *S. virgata*, and their hybrid *S. diegensis*. The parental species differ in morphology, as well as in geographic distribution, although they overlap in Southern California, where they hybridize. The resulting F_1 hybrids are largely sterile (Gottlieb 1971). The third species, *S. diegensis*, is common along the southern California coast, often occurring in disturbed habitat (e.g., road embankments; once again raising the possibility that these hybrids may be less than natural). All three species are diploid, so polyploidy is not involved. But two lines of evidence suggest that *S. diegensis* is hybrid. First, *S. diegensis* is a morphological amalgam of characters from *S. exigua* and *S. virgata* (Gottlieb 1971). Second, allozyme work shows that, while all of the species are closely related, *S. diegensis* represents a mixture of alleles from the other two species (Gallez and Gottlieb 1982). Indeed, all alleles but one in *S. diegensis* are found in *S. exigua* and *S. virgata*. *S. diegensis* is also strongly reproductively isolated from its parents: F_1 hybrids between *S. diegensis* and either parent suffer low pollen viability (Gottlieb 1971).

The peony species, *Paeonia emodi*, also appears to be a hybrid between two congeneric peonies (*P. lactiflora* and either *P. veitchu* or *P. xinjiangensis*, or their common ancestor). All of these species are outcrossing perennial herbs and all are diploid. But DNA data show that *P. emodi* is a mixture of sequence from the above taxa (Sang et al. 1995). Interestingly, *P. emodi* is now allopatric with its putative parents; it appears, however, that the parental species may have been sympatric in the Himalayas during the Pleistocene. Sang et al. also provide some sequence evidence suggesting that an entire clade of ten species of peonies (also outcrossing perennial herbs) is of hybrid origin. A single recombinational speciation event may therefore have given rise to a lineage that continued to speciate, presumably by more ordinary means. At least one other peony species may have arisen by an independent hybrid speciation event: there is some molecular evidence for a recombinational origin of *Paeonia officinalis*, a widespread Mediterranean species (Ferguson and Sang 2001). *Paeonia officinalis* can reproduce vegetatively by rhizomes.

There is also some molecular evidence that the angiosperm *Penstemon clevelandii* is of hybrid origin; the presumed progenitor species are known to hybridize in the wild and *P. clevelandii* is ecologically isolated from both of these species (Wolfe et al. 1998b). *P. clevelandii* appears to be outcrossing and is primarily bee-pollinated. Similarly, the composite *Argyranthemum sundingii*, endemic to the Canary Islands, may be a diploid hybrid between *A. brousonetii* and *A. frutescens* (Brochmann et al. 2000; this case may, however, reflect human disturbance in the form of deforestation).

Recent evidence also suggests that several *animal* species may be products of recombinational speciation. A combination of nuclear and mtDNA data, for instance, suggest that both the cyprinid fish *Gila seminuda* and the crustacean *Daphnia galeata mendotae* may be diploid hybrids (DeMarais et al. 1992; Taylor et al. 1996). Molecular, morphological, and biogeographic data also suggest that the parasite *Schistosoma sinensium* may be a diploid hybrid (Hirai et al. 2000). These animal examples are not, however, as well supported as several of the plant examples discussed above.

The data meet the theory

Looking across the above cases of recombinational speciation, several theoretical predictions fare well, while others do not. The first class of predictions includes the simple claim that recombinational speciation *does* occur in nature. Less trivially, it is clear that recombinational speciation can be fast. Finally, habitat differentiation appears to ease recombinational speciation, as predicted by Buerkle et al. (2000). Indeed, new stabilized species seem typically ecologically differentiated from their parents, as shown by *Helianthus*, *Iris*, *Pinus*, *Stephanomeria*, and *Penstemon*. This differentiation itself represents an isolating barrier (i.e., habitat isolation).

Among those predictions that fare poorly, one stands out as particularly surprising: inbreeding is *not* clearly associated with recombinational speciation, despite the expectations of Stebbins (1957a), Grant (1958, 1981), and McCarthy et al. (1995). Instead, as Rieseberg (1997, see his Table 2) emphasized, many cases of natural recombinational speciation involve *outbreeding* species (although irises and the peony *Paeonia officinalis* can reproduce asexually by rhizomes; Arnold and Bennett 1993; Ferguson and Sang 2001). It is unclear why theory fares so badly here. Rieseberg (1997) speculates that an outcrossing lifestyle greatly increases the initial rate of species hybridization—an effect that might more than compensate for the lowered rate at which recombinant genotypes are later made homozygous. Though possible, it is unclear why this effect did not show up in McCarthy et al.'s (1995) computer simulations.

Finally, it is hard to judge the frequency of recombinational speciation in nature. Though we know of only a handful of good cases, this may mean only that hybrid speciation is difficult to detect and document. There are, after all, many suggested cases of recombinational speciation, particularly in plants, and few have been carefully studied molecularly.

On the other hand, there are some reasons for believing that recombinational speciation is rare in nature. For one, phylogenetic reconstruction would often break down if hybrid speciation were common. Although such breakdowns do occur, they are obviously not the rule. For another, there are sound reasons for thinking that recombinational speciation *should* be fairly rare, or at least far rarer than allopolyploid hybrid speciation. Grant (1966b, p. 1198)—who once thought recombinational speciation common—in the end concluded that it was a "real and interesting but relatively rare process in nature." The reason for his change of heart is that recombinational speciation differs in an important way from allopolyploidy. Though both forms of hybrid speciation offer a way out of the chromosomal sterility of F_1 hybrids, allopolyploidy also avoids later generation hybrid breakdown (a new tetraploid line is often immediately stabilized and fit), while recombinational speciation does not: only a small proportion of later generation hybrids are typically fit and these may often cross with individuals having different genotypes, giving rise to yet more hybrid breakdown. This point was, of course, evident in Grant's own work with recombinant *Gilia* lines, lines that suffered low fitness for many generations. (Grant 1966b estimated that only 2% of his *Gilia* zygotes were viable and fertile in the F_1 to F_6 generations.) If these results are at all representative, recombinational speciation should be difficult and, thus, rare.

These lines of argument, however, are just that—lines of argument, not evidence. We simply do not know if recombinational speciation is moderately common in certain groups or exceedingly rare in all taxa, little more than an evolutionary curiosity. It is remarkable that we cannot yet answer this question. The matter of how often new species arise by the fusion of old ones is not, after all, some arcane issue, but one that is central to our view of the origin of species—and one that can be readily resolved by experiment.

10
Reinforcement

The theory of reinforcement—the enhancement of prezygotic isolation in sympatry by natural selection—has had an extraordinarily tortuous history. Even the origin of the idea is more complex than generally appreciated. Although A. R. Wallace, the codiscoverer of natural selection, is typically given credit for the idea—reinforcement is still sometimes called the "Wallace effect"—Wallace's hypothesis differs from the modern one. Wallace argued for the enhancement of *postzygotic* isolation by *group* selection (Littlejohn 1981; Howard 1993), a scenario that has little to do with the modern theory of reinforcement. The modern theory instead dates to Dobzhansky (1937b), who described the following scenario: two taxa diverge in allopatry; upon secondary geographic contact, hybridization occurs at some rate, yielding unfit hybrids; because production of hybrids is maladaptive, individuals who mate only with their own taxon enjoy a fitness advantage; natural selection thus favors the evolution of enhanced prezygotic isolation.

Enthusiasm for this new idea quickly outstripped all evidence for it and, by 1940, Dobzhansky went so far as to suggest that even *allopatric* divergence in mate preference may sometimes reflect the evolutionary consequences of rare migrants from another taxon. Dobzhansky also came to believe that reinforcement (a term coined by Blair 1955) represents a nearly obligate last step in speciation, a view that remained popular into the 1970s (e.g., Lewontin 1974). Indeed, Dobzhansky ultimately concluded not only that prezygotic isolation evolves as a direct product of natural selection, but that it is an ad hoc contrivance designed to protect the integrity of species, a remarkably teleological view (e.g., Dobzhansky 1951, p. 208).

In the 1980s, however, the popularity of reinforcement plummeted. As Noor (1999) and Marshall et al. (2002) note, this change had little to do with new data and everything to do with new theoretical objections and verbal arguments.

While we review many of these objections and arguments below, it is worth clarifying one point at the outset. Butlin (1987a) emphasized that cases in which taxa produce completely unfit hybrids must be distinguished from those in which taxa produce partially unfit hybrids. In the first case, no gene flow is possible and the taxa are already good species; in the second case, gene flow is possible and the taxa are not yet good species. Butlin argued that enhancement of prezygotic isolation in the first case has nothing to do with speciation and should be kept distinct from reinforcement proper. He suggested the term "reproductive character displacement" for an increase in isolation between taxa that are already good species. True reinforcement is restricted to cases in which isolation is enhanced between taxa that can still exchange genes; reinforcement might therefore complete speciation. We will abide by Butlin's distinction: we are interested in speciation, not in changes that occur afterwards. This does not mean, however, that we can learn nothing about reinforcement from reproductive character displacement. On the contrary, we will see that both the experimental and theoretical study of reinforcement can benefit from analysis of cases in which hybrids are completely unfit.

By about 1990, reinforcement once again became popular, reflecting the emergence of new data revealing that *something* interesting was happening in sympatry and that, at the least, this something resembled reinforcement. On the heels of these data, many theorists found that reinforcement was possible after all. Indeed, several found that it was easy.

Although we have played some part in the revival of enthusiasm for reinforcement, we are not uncritical champions of its importance. We believe the present data and theory show that reinforcement is possible—and must be taken seriously—but they do not show that reinforcement is common, much less ubiquitous (Marshall et al. 2002; Servedio and Noor 2003). Here, we critically review these data and theory.

Our approach is somewhat unusual. Inverting our usual order, we consider the data before the theory. There are several reasons for this. First, this order better reflects the history of research on reinforcement. Second, it is important to establish if patterns *consistent* with reinforcement are common in nature before worrying about whether these patterns are best explained by reinforcement or by many alternative hypotheses. Our presentation thus proceeds in several steps: (1) we ask if there is evidence consistent with an increase of prezygotic isolation in sympatry; (2) finding that there is, we ask if reinforcement provides a plausible explanation; (3) we review alternative hypotheses that might also explain the data; and (4) finally, we suggest a test to empirically distinguish reinforcement from these alternatives.

The Data

Data testing reinforcement come in two forms: selection experiments and observations from nature (Rice and Hostert 1993).

Selection experiments

Before considering the frequency of reinforcement in nature, it is worth knowing if reinforcement can happen under *any* circumstances. The most direct way to find the answer is clear: one must attempt to reproduce the phenomenon in the laboratory. One should be able to show that enhanced prezygotic isolation evolves in response to *artificial* selection against hybrids in the laboratory, just as it allegedly does to natural selection in the wild. As Rice and Hostert (1993) emphasize in their review of such experiments, these efforts fall into two classes. The first involves "destroy-the-hybrids" designs, and the other involves disruptive selection on arbitrary traits. For the moment, we include experiments in which hybrids are completely unfit (reproductive character displacement, not reinforcement). If no response to selection occurs in this extreme case, we need not concern ourselves with less extreme, but more biologically relevant, cases.

The first destroy-the-hybrid experiment was that of Koopman (1950), who simulated complete hybrid inviability between two species of *Drosophila*, asking if production of unfit hybrids leads to the evolution of enhanced prezygotic isolation. Koopman studied replicate population cages containing both *D. pseudoobscura* and *D. persimilis*, each of which carried a different recessive eye color mutation. This allowed him to distinguish the progeny of conspecific versus heterospecific matings by eye color. Each generation, Koopman discarded all offspring of interspecific couplings, breeding only from the progeny of conspecific couplings. Enhanced prezygotic isolation evolved remarkably quickly: in the first generation between 22% and 50% of all progeny were hybrid, but within six generations the figure dropped to less than 5%.

Most destroy-the-hybrid experiments show at least some enhancement of prezygotic isolation, albeit often weaker and more slowly evolving than in Koopman's study. A good example is that of Knight et al. (1956) who successfully selected for sexual isolation within *D. melanogaster* over nearly 40 generations. Similarly, Crossly (1974) successfully selected for increased sexual isolation between two mutant strains of *D. melanogaster*. Such experiments are not limited to *Drosophila*. Indeed, one of the most impressive involved selection for prezygotic isolation between two strains of maize, *Zea mays* (Paterniani 1969). (Plants provide excellent material for such experiments. Unlike many animals where a female mates with one or a few males, plants may receive pollen from many individuals. By scoring the frequency of hybrid kernels in maize, one obtains a quantitative measure of prezygotic isolation per *individual*.) After sowing non-hybrid kernels from individuals showing low rates of intercrossing, Paterniani saw a remarkable 30%–40% decrease in the frequency of intercrossing in six generations. Rice and Hostert (1993) review similar experiments.

Despite this record, destroy-the-hybrids experiments have been criticized on several grounds. First, there is a suspicion that their success rate reflects a publication bias. Responses to selection for greater isolation are exciting and are published, while failed responses are uninteresting and may go unpublished. Although there are some exceptions to this pattern (see Robertson 1966, and Fukatami and Moriwaki 1970 for negative, or at least non-significant, results), it does seem likely that failed responses are underreported. (Indeed one of our own early experiments involved a failed—and unpublished—attempt to replicate Koopman's results in the wasp *Nasonia*.) Second, as noted, these experiments test reproductive character displacement, not reinforcement. One cannot, then, necessarily extrapolate from these findings to true reinforcement.

The obvious experimental modification—a destroy-the-hybrids design in which a few hybrids are allowed to breed—has been attempted, so far as we know, only once. Harper and Lambert (1983) selected against hybridization between two lines of *D. melanogaster*, allowing variable degrees of gene flow. Though no response to selection was seen, this may have reflected a lack of genetic variation in their marker stocks, which are usually highly inbred.

The closest analogue to modified destroy-the-hybrid experiments involves allopatric populations that are first adapted to different environments (e.g., food with EDTA versus food without EDTA) and then brought into contact (e.g., Robertson 1966a; Wallace 1982; Ehrman et al. 1991). If population hybrids suffer intermediate phenotypes and so are poorly adapted to available environments (extrinsic postzygotic isolation), increased prezygotic isolation should be favored. Although such experiments allow gene flow, they probably allow far too much of it: it seems unlikely that, after a brief period of adaptation in the laboratory, most hybrids would die. In any case, these experiments are more complex than needed to address the above concern. What we want to know is if response to selection disappears when going from no gene flow to low gene flow in Koopman-like experiments.

In the absence of such work, attention shifted to a second class of selection experiment—that in which an arbitrary character experiences disruptive selection in a single population. Although such experiments are usually considered tests of sympatric speciation, they bear on reinforcement: they ask if assortative mating evolves when intermediate "hybrid" phenotypes are selected against. The first and most celebrated of these experiments was that of Thoday and Gibson (1962). Because we described this experiment in Chapter 4, we only sketch it here. Thoday and Gibson performed disruptive selection on bristle number in *D. melanogaster*. Virgin flies with the highest and lowest bristle number were selected from a base population; they were then placed together and allowed to mate freely; high and low bristle females were separated and allowed to produce progeny. This was continued for 12 generations. Although this design allows random mating among flies, Thoday and Gibson saw a rapid response to selection: by the end of the study, high bristle females produced almost all high bristle progeny and low bristle females almost all low bristle progeny. Natural selection appeared to promote the evolution of assortative mating.

Given their dramatic results, Thoday and Gibson's work spawned a minor industry. Unfortunately, this work arrived at a consistently depressing result:

every attempt to replicate Thoday and Gibson's findings with new stocks failed (reviewed in Thoday and Gibson 1970; Scharloo 1971; for discussion of a possible exception, see Rice and Hostert 1993). In retrospect, the reason seems clear. As Felsenstein (1981) explained, the evolution of prezygotic isolation by disruptive selection requires that a population leap an especially high population-genetic hurdle. Reinforcement requires the establishment and maintenance of strong non-random associations (linkage disequilibria) between the alleles underlying the disruptively selected character and those underlying assortative mating. Although disruptive selection automatically generates such disequilibrium, recombination destroys it and—unless the two kinds of loci are tightly linked—recombination wins. We discuss Felsenstein's insight in more detail below. For now, it suffices to realize that there are good reasons why disruptive selection experiments *should* fail.

Another variety of disruptive selection experiment has been more successful. The key insight was provided by Slatkin (1982), who noted that disruptive selection on a single character might *automatically* yield prezygotic isolation. Slatkin gave the example of flowering time. If selection favors early or late flowering, reproductive isolation between subpopulations arises as an automatic consequence of response to selection, as early breeders will not hybridize with late breeders. Formally, this idea sidesteps Felsenstein's objection by allowing the rate of recombination between the genes underlying the character under selection and those underlying assortative mating to go to zero—they are the *same* genes. Rice and colleagues have been vocal champions of this idea. Most important, they have demonstrated the efficacy of the process in large disruptive selection experiments on habitat preference in *Drosophila* (reviewed in Chapter 4).

While such findings prove the efficacy of the single-character scenario—and have clear implications for sympatric speciation—the connection between disruptive selection and reinforcement has grown tenuous as we moved from two-character (Thoday-and-Gibson-like) to one-character (Rice-like) experiments. In the first case, one asks for the evolution of prezygotic isolation *in order* to prevent production of unfit hybrids. One asks, that is, for reinforcement. In the second case, prezygotic isolation is fortuitous—one disruptively selects on a character that just *happens* to cause prezygotic isolation. This is not, we would argue, reinforcement.

In summary, while selection experiments provide some support for the idea that prezygotic isolation can be enhanced by natural selection, those experiments that work best are only loosely connected to reinforcement, while those that work worst are more closely connected to reinforcement. More encouraging results emerge from the second kind of evidence, that from nature.

Evidence from nature: case studies

Almost all work on reinforcement in nature takes the same form: a contrast between the strength of prezygotic isolation in sympatry versus allopatry. In some cases the taxa compared are species while in others they are populations.

Such data can of course only detect a pattern, not identify a process. Several processes other than reinforcement may produce a pattern of increased prezygotic isolation in sympatry. To be clear, then, we will refer to a *pattern* of stronger prezygotic isolation in sympatry as "enhanced isolation," implying nothing about its evolutionary cause.

The earliest data bearing on enhanced isolation come from Dobzhansky and Koller's (1938) study of sexual isolation between *D. miranda* and *D. pseudoobscura*. (Two "races" of *D. pseudoobscura*, A and B, were used; race B was later renamed *D. persimilis*.) Early data suggested that *D. miranda* was geographically concentrated in the northwestern United States, near Puget Sound, while *D. pseudoobscura* had a broader distribution in the western United States, overlapping with that of *D. miranda*. Dobzhansky and Koller found that, as one moved from populations of *D. pseudoobscura* (race A or B) far from *D. miranda*'s range, to those closer to (and ultimately sympatric with) *D. miranda*, sexual isolation between the species increased. This discovery had a profound effect on Dobzhansky's view of speciation. Indeed, Howard (1993) dates Dobzhansky's conversion to reinforcement to this finding. Not surprisingly, the *D. pseudoobscura*–*D. miranda* story played a prominent part in Dobzhansky's (1937b) discussion of reinforcement in the first edition of *Genetics and the Origin of Species*, where many evolutionists first learned of reinforcement. (As the dates indicate, Dobzhansky and Koller's 1938 paper was not yet published at the time of Dobzhansky's book, but much of the data were in hand.)

A closer reading of the 1938 paper, however, yields a more complex story. The complication springs from the fact that—after the publication of the first edition of *Genetics and the Origin of Species*—Dobzhansky and Koller found that *D. miranda* was *not* restricted to Puget Sound. Instead, they report the discovery in 1937 of populations of *D. miranda* in the Sierra Nevada Mountains of California. As they note, this finding casts doubt on Dobzhansky's earlier interpretation. Dobzhansky and Koller tested the strength of sexual isolation between Sierra Nevada populations of *D. miranda* and *D. pseudoobscura*. Surprisingly, these sympatric flies crossed fairly freely—as freely as any allopatric combination—a result, they conclude, that is "the reverse of what one might have expected by analogy with the behavior of the different strains of *D. pseudoobscura* toward the [Puget Sound] race of *D. miranda*." Curiously, this finding seems to have had little effect on Dobzhansky's enthusiasm for reinforcement. While the second (1941) edition of his book includes a description of the troubling Sierra Nevada findings (p. 266), Dobzhansky still concludes that selection plays a likely role in the origin of isolation. Remarkably, the Sierra Nevada finding disappears from the third (1951) edition, with Dobzhansky merely reporting that *D. pseudoobscura* flies from regions close to *D. miranda* "show, in general, greater sexual isolation than do strains from distant regions" (p. 210). It is hard to see how this claim can be reconciled with the data.

Over the years, other reports of enhanced isolation appeared sporadically in the *Drosophila* literature. Ehrman (1965), for example, compared the strength of sexual isolation between sympatric and allopatric populations of several semispecies of *D. paulistorum*, finding greater isolation in sympatry in most cross combinations. Similarly, Wasserman and Koepfer (1977) found enhanced isolation between *D. mojavensis* and *D. arizonensis* (later renamed *D. arizonae*), two species that live in the American Southwest and that have partly overlapping distributions. One of the most recent reports of enhanced isolation involves *D. pseudoobscura* and *D. persimilis*. *D. pseudoobscura* is widely distributed throughout western North American while *D. persimilis* is found only in Pacific coastal states; the latter species is thus completely embedded within the distribution of the former. Because hybrids sometimes form in nature (albeit rarely) and hybrid males are sterile, selection surely punishes hybridization. One would thus predict that *D. pseudoobscura* females from regions sympatric with *D. persimilis* would discriminate against *D. persimilis* males more than do *D. pseudoobscura* females from regions of allopatry. This was precisely the pattern seen by Noor (1995): in five of six contrasts, sympatric females are choosier than allopatric females.

Reports of enhanced isolation have not been limited to *Drosophila*. In the damselfly *Calopteryx*, for instance, Waage (1975, 1979) showed that males who derive from regions of geographic overlap between two species distinguish females of the two species better than do males from non-overlapping regions; interestingly, females of these species differ in wing coloration in sympatry but not in allopatry. Similarly, songs of Hawaiian crickets of the genus *Laupala* differ more among sympatric populations than expected by chance (i.e., than if songs were randomly assigned without regard to geography) (Otte 1989). Also, the walking stick *Timema cristinae* shows greater prezygotic isolation between populations that exchange migrants at intermediate rates. Nosil et al. (2003) argue that this pattern reflects reinforcement, which should occur neither under negligible migration rates (as there is insufficient selection against hybridization), nor under high migration rates (as response to selection is swamped by gene flow; see also Cooley et al. 2001 on the periodical cicadas *Magicicada*.)

Beyond insects, some of the best-known cases of enhanced isolation involve gametic isolation in marine organisms. We have already described these cases in Chapter 6—they include the abalone *Haliotis* (Lee et al. 1995), the sea urchin *Arbacia* (Metz et al. 1998b), the sea urchin *Echinometra* (Geyer and Palumbi 2003), and the mussel *Mytilus* (Springer and Crespi 2004; see also Wullschleger et al. 2002 for results from snails).

Frogs have provided particularly good material for studies of enhanced isolation as the primary mechanism of prezygotic isolation is both known and easily studied: male calls attract females and, in a number of frog families, females favor conspecific over heterospecific calls (Blair 1974). A pattern of enhanced isolation is fairly common. In *Gastrophryne olivacea* and *G. carolinensis*, for example, calls of males from allopatric populations are similar, while those from sympatric populations differ in both duration and frequency. One of the best-studied cases involves the Australian tree frog *Hyla*. Littlejohn (1965) showed that calls of distant allopatric populations of *H. ewingi* and *H. verreauxi* are similar, while those of sympatric populations are different. Subsequent work revealed that, when sympatric females of each species are given a choice of sympatric males from both species, they invariably "choose" conspecific males (Littlejohn and Loftus-Hills 1968). Two other species of tree frog, *Hyla cinerea* and *H. gratiosa*, also show a pattern of enhanced isolation: female preference for conspecific male calls in *H. cinerea* is greater in sympatric than allopatric populations (Hobel and Gerhardt 2003; see also Hillis 1981 and Chapter 5 for a possible case of reinforcement in the leopard frog *Rana* involving temporal isolation; and see Pfenning 2003 for a possible case of reinforcement in the spadefoot toad *Spea*). In a review of the older literature, Blair (1974) emphasized that no sympatric species are known that have the same call. He concluded that reinforcement plays a probable role in the origin of frog species.

Similar cases occur in fish. Rundle and Schluter (1998), for example, found enhanced isolation in the threespine stickleback *Gasterosteus aculeatus*. Benthic and limnetic morphs co-occur in several coastal lakes in British Columbia; in other lakes, only single forms are found. Rundle and Schluter found that benthic females from lakes that include limnetics discriminate against limnetic males, while females from lakes that do not include limnetics, do not. This result is particularly interesting as the sympatric morphs cross at a low rate in nature and mtDNA data reveal past gene flow as well. Moreover, the resulting hybrids are unfit under natural conditions (Chapter 7).

In birds, males from two species of Darwin's finches (*Geospiza fuliginosa* and *G. difficilis*) that live in sympatry on the Galápagos archipelago prefer conspecific female models, whereas males of the same species from regions of allopatry show little preference (Ratcliffe and Grant 1983b). Unfortunately, though, this difference may not reflect response to selection: if it is genetically based, the findings are obviously consistent with reinforcement, but there is no hard evidence for such heritability. Instead, as Ratcliffe and Grant note, the difference might reflect learning, as sympatric (but not allopatric) males could learn to distinguish between conspecific and heterospecific females. (Indeed, such learning is known to occur in guppies of the genus *Poecilia*, yielding a false appearance of reinforcement; Magurran and Ramnarine 2004.) This problem afflicts all studies of reinforcement in organisms that can learn; unambiguous tests of enhanced isolation in such cases must involve naïve individuals.

In another bird study, Saetre et al. (1997) found evidence of enhanced isolation between two species of *Ficedula*, the collared and pied flycatchers. While females of both species are brown, males of the collared flycatcher are black and white, and males of the pied flycatcher are either black and white, or brown. Interestingly, brown males occur only where the species overlap. Saetre et al. argue that brown color evolved to prevent hybridization, which yields mostly sterile hybrids (Alatalo et al. 1990; Saetre et al. 1997). Saetre et al. show that, when given a choice of sympatric males, females of both species choose conspecific males. When given a choice of allopatric males, many females choose heterospecific males. While suggestive, this pattern might reflect a kind of mimicry: collared males are dominant to pied males in competition for nesting sites. But collared males tend to ignore brown pied males, possibly allowing pied males to acquire better nesting sites without interference (Alatalo et al. 1994). Brown color, then, might represent female mimicry that acts to reduce interspecific competition.

Finally, there are several reports of enhanced isolation in plants and fungi. Grant (1966c), for instance, studied nine species of leafy-stemmed *Gilia*, all of which produce sterile hybrids. Five species occur in the foothills and valleys of California and are sympatric; the remaining four species are maritime and allopatric to each other. Grant showed that the sympatric foothill-and-valley species are strongly isolated from each other while the allopatric maritime species are not. While this isolation probably involves a mixture of pre- and postzygotic barriers, at least part reflects prezygotic incompatibilities between pollen and pistils. Similarly, two species of tropical herbs, *Costus allenii* and *C. laevis*, occur sympatrically, flower at the same time, and share the same bee pollinator; despite considerable interspecific pollen flow in the wild, artificial hybridization is difficult, reflecting failed pollen-pistil interactions (Schemske 1981; Kay and Schemske 2003). Another species, *C. guanaiensis*, which is ecologically segregated from both *C. allenii* and *C. laevis*—and so shows little interspecific pollen flow with either species in nature—easily produces hybrids in the greenhouse with both *C. allenii* and *C. laevis* (Schemske 1981). Thus, the species that experience high pollen exchange in nature show strong prezygotic isolation, while those that experience low pollen exchange in nature do not. For other possible plant examples, see Levin's (1985) work in *Phlox*, and McNeilly and Antonovics's (1968) work in *Agrostis* and *Anthoxanthum* (Chapters 3 and 4). Recent work has also revealed a pattern of enhanced isolation in fungi. In particular, crosses between species belonging to the genus *Neurospora* show greater reproductive isolation when populations are sympatric than allopatric, although it is unclear how much of this enhanced isolation involves prezygotic barriers (see Dettman et al. 2003 and references therein).

While the above reports tend to support reinforcement, others do not. The literature includes a fair number of studies in which enhanced isolation was looked for but not found (Marshall et al. 2002). Patterson and Stone (1952, pp. 357–358 and 549–550) listed several cases in *Drosophila* (e.g., *D. microspina* × *D. limpiensis*) in which prezygotic isolation *increases* with geographic distance between two species, contrary to the pattern emphasized by Dobzhansky. Coyne et al. (2002) recently found that *Drosophila santomea* and *D. yakuba*, which have partially overlapping ranges on the volcanic island of Sao Tome, show no enhanced isolation in sympatry. Similarly, Walker (1974) found no evidence of song displacement among acoustic insects (although he studied fewer cases than often reported [Howard 1993]). Similarly, Loftus-Hills (1975) argued that evidence for enhanced isolation between the toads *Bufo americanus* and *B. woodhousii* is weaker than previously thought, while Sanderson et al. (1992) found no evidence for enhanced isolation in a hybrid zone between the toads *Bombina bombina* and *B. variegata*. In addition, Doherty and Howard (1996) found that, although male song in the cricket *Allonemobius fasciata* appears displaced in some populations that are sympatric with its sister species *A. socius*, females are mostly oblivious to these differences. Butlin and colleagues (Butlin 1989; Ritchie et al. 1989; Butlin and Ritchie 1991) further argued that a hybrid zone between two subspecies of the grasshopper *Chorthippus parallelus* provides little evidence of enhanced isolation. But most important of all, Butlin (1987a,b, 1989) argued that several alleged—and classic—cases of reinforcement are flawed in several ways, most seriously in that they often involve taxa that produce *completely* sterile or inviable hybrids. Such cases, he argued, represent reproductive character displacement, not reinforcement.

The case study data, then, do not present a particularly simple picture. Part of the problem is inherent in the approach itself. A case or two of enhanced isolation is seen in one small group, but not in another, quite different, group. Given that all else is surely not equal across these groups, it is hard to know if such inconsistencies reflect the rarity of reinforcement or are consequences of looking at taxa of different evolutionary ages, or taxa that have been in contact for different periods of time, or taxa that show different degrees of postzygotic isolation, and so on.

Evidence from nature: comparative studies

The solution to this problem is clear: we require larger, more systematic studies, ones that allow comparison of the strength of prezygotic isolation in sympatry versus allopatry over large groups, *controlling* for the age of taxa and the strength of postzygotic isolation between them, etc.

This was one of the motivations for our survey of patterns of speciation in *Drosophila*, in which we collected data on geography (allopatry versus sympatry), genetic distance (a molecular proxy for evolutionary age), and the strength of pre- and postzygotic isolation between many pairs of species. Our original survey included 119 species pairs (Coyne and Orr 1989a), while our subsequent update included 171 (Coyne and Orr 1997; see Chapter 3). Some of our results are shown in Figure 10.1. Prezygotic isolation is obviously much stronger between sympatric than allopatric species. Indeed, young sympatric taxa (Nei's genetic distance less than 0.5) show an average prezygotic isolation index of 0.83 (where 0 means no prezygotic isolation, and 1 means complete isolation), while young allopatric taxa show an average of 0.29. This difference cannot be explained by the age of species (i.e., sympatric taxa are not simply older than allopatric). Most important, *postzygotic* isolation does not differ between sympatry and allopatry. Just as expected under the theory of reinforcement, then, pre- but not postzygotic isolation is greater among sympatric taxa.

These data had some effect on the status of reinforcement among evolutionary biologists, an effect that had much to do with the theoretical climate in which these results appeared. As we will see, theorists had all but abandoned reinforcement by the late 1980s. Although the pattern shown in Figure 10.1 does not prove reinforcement—we will consider a long list of possible alternatives—it was, we suggested, sufficiently striking to require taking reinforcement seriously.

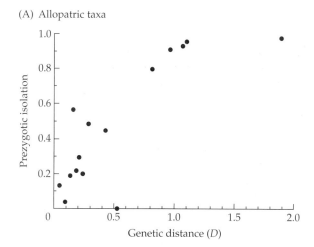
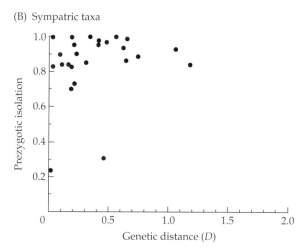

Figure 10.1 Enhanced prezygotic isolation in *Drosophila*. The top plot shows prezygotic isolation between allopatric pairs of species, while the bottom plot shows prezygotic isolation between sympatric pairs of species. (From Coyne and Orr 1997.)

Several other comparative tests of reinforcement have since appeared. The most exhaustive is that of Howard (1993), who searched the literature for cases that, by meeting certain criteria, allow one to test for enhanced isolation in sympatry with some confidence. (Howard required, for instance, that the character studied be known to at least sometimes play a role in reproductive isolation.) Howard identified 48 cases that satisfied his criteria. Remarkably, 33 of these cases (69%) showed a pattern of enhanced isolation. Because these data derive from a wide variety of organisms—insects, birds, mammals, amphibians, reptiles, fish, and plants—enhanced isolation is clearly not hard to find. One could argue, however, that Howard's findings reflect a publication bias: cases in which preliminary studies suggest greater isolation in sympatry might be more likely to be subjected to fuller analysis, resulting in publication. To guard against this possibility, Howard further scoured the literature for cases in which patterns of assortative mating had been assessed in hybrid zones: random *or* assortative mating may occur here and the results would presumably be published regardless. Howard found that, in a sample of 37 hybrid zones in which patterns of assortative mating had been characterized by genetic, morphological, or behavioral means, 19 showed positive assortative mating and 16 showed random mating, findings that are at least consistent with reinforcement.

Oddly, we know of only one comparative survey of isolation in sympatry vs. allopatry that focuses on plants. Although this study (Moyle et al. 2004) uncovered no evidence of enhanced isolation, only two genera were surveyed (*Silene* and *Glycine*) and aspects of prezygotic isolation that act before pollination (like pollinator behavior and flowering time) were not studied. It is hard, therefore, to know how representative these findings are. The relative neglect of plants by reinforcement workers is unfortunate. Indeed a number of novel predictions could be tested here (e.g., because wind-pollinated plants cannot achieve reinforcement by changing pollen vectors or flower color, reinforcement should often take the form of changed flowering time and/or "aerodynamic isolation" in which heterospecific pollen is less likely than homospecific pollen to land on the stigma; Chapter 6).

Although the above surveys reveal that sympatry can enhance prezygotic isolation, they do not tell us *how often* sympatry matters. The problem is that biases causing either over- or underestimation of the frequency of enhanced isolation probably exist. In Coyne and Orr's (1989a, 1997) work, for instance, single species were sometimes used in multiple comparisons with other taxa. A single bout of reinforcement between two species could, therefore, give rise to a pattern of enhanced isolation between several species: a species that experienced selection for greater prezygotic isolation might *generally* show greater choosiness. On the other hand, case studies might underestimate the frequency of enhanced isolation: if increased isolation evolves in sympatry by reinforcement but then spreads throughout a species range, contrasts between sympatric and allopatric populations will reveal no reinforcement although it occurred (Walker 1974).

Noting these problems, Noor (1997b) introduced a method that allows estimation of the frequency with which sympatry yields enhanced isolation. The idea is simple. Consider three species: an outgroup species A, and two ingroup species, B and C (Figure 10.2). Species A and B are sympatric while species A and C are allopatric. If reinforcement occurs, A-B will show more prezygotic isolation than A-C. In searching for such contrasts, Noor imposed two restrictions. First, sympatric species can be used in one comparison only and, second, the allopatric species cannot be sympatric with any other close relative (as such sympatry might fortuitously heighten isolation between species A and C). Surveying data from *Drosophila*, Noor concluded that sympatry yields enhanced isolation in 21% of phylogenies like those in Figure 10.2. Because this value is a lower bound on a confidence interval, and because his data incorporated only one component of prezygotic isolation (sexual isolation), Noor argues that it probably underestimates the true frequency of enhanced isolation.

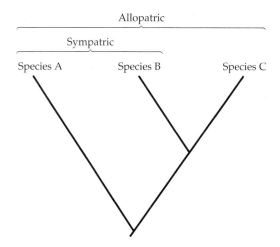

Figure 10.2 Noor's (1997b) test of the frequency of enhanced isolation. Species A and B are sympatric, while species A and C are allopatric.

Overall, then, the comparative data are consistent with reinforcement. While we have not yet considered possible alternative explanations, work over the last two decades strongly suggests that a *pattern* of increased prezygotic isolation in sympatry is reasonably common in nature. Before turning to theoretical attempts to explain this pattern, we briefly consider whether *postzygotic* isolation shows a similar pattern.

Reinforcement of postzygotic isolation

While the logic behind selection for increased prezygotic isolation is clear, the same cannot be said for postzygotic isolation. But Grant (1966c, p. 104) and Coyne (1974) argued that reinforcement of postzygotic isolation is possible in species having substantial parental investment. If hybrids are sterile or inviable (where lethality occurs late in development) and if parents sacrifice further reproduction during a period of pregnancy, child rearing, or seed/fruit production, it may pay to abort hybrid development as early as possible (i.e., to evolve hybrid lethality). (See Johnson and Wade 1995 for theoretical treatment of this idea.) Such reinforcement seems most likely in organisms like mammals or plants, where maternal investment is large.

Coyne described several possible examples. One involves the "corky" syndrome in cotton. Hybrids between two species of *Gossypium* suffer a syndrome of morphological and physiological problems (Stephens 1946, 1950). The alleles causing these problems are mostly restricted to regions of geographic overlap between the species. In another case, discussed earlier, Grant (1966c) showed that sympatric species of *Gilia* suffer greater barriers to hybridization than allopatric species. While some of these barriers are prezygotic, others appear to be postzygotic.

There are therefore hints that postzygotic isolation might sometimes be reinforced. Once again, however, the problem is that the approach taken so far has been entirely anecdotal. While we can point to isolated examples of heightened hybrid inviability in sympatry, we have no idea how common these cases are relative to those in which postzygotic isolation is greater in *allopatry*, cases that may often go unpublished. Again, the obvious solution is a more systematic approach. (Systematic data have been collected in *Drosophila* [Coyne and Orr 1989a, 1997], but flies do not provide promising material given that they do not necessarily sacrifice future reproduction by producing unfit progeny.) Fortunately, the Grant-Coyne idea makes two predictions that could be readily tested in plants. First, crosses between sympatric taxa should produce fewer viable seeds than crosses between allopatric taxa. Second, no such difference should appear in the strength of hybrid *sterility*, which cannot be reinforced (at least not in a straightforward way). These predictions could be systematically tested across plants having large fruit or seed investment.

The Theory

The data reviewed above pose a considerable challenge to theory. A pattern of stronger prezygotic isolation in sympatry seems reasonably common. Does this reflect reinforcement? Is reinforcement theoretically possible and, if so, under what conditions?

The history of theoretical work on reinforcement falls into three phases. In the first phase, attention focused on the case in which hybrids are completely unfit and "reinforcement" was found plausible. In the second phase, starting in the mid-1980s, attention focused on the case in which hybrids have non-zero fitness and reinforcement was found unlikely. In the last phase, beginning in the 1990s, attention focused on ever more realistic (or at least more complex) models, and reinforcement was generally found possible.

Early enthusiasm

Though he presented no formal theory, Fisher (1930) offered the first population-genetic model bearing on reinforcement. He considered a scenario closely allied to, if not identical with, reinforcement. In particular, he considered a widely distributed species that adapts over a continuously varying ecological gradient. A certain set of alleles is favored and so increases in frequency at one end of the species' range, while another set is favored and increases in frequency at the other end of the range. Fisher argued that under such conditions selection will favor the evolution of restricted dispersal (i.e., modifier alleles that reduce mobility will increase in frequency as such alleles reduce maladaptive hybridization; see also Balkau and Feldman 1973). As this process continues, the population might rupture into two distinct species. But, Fisher argued, the evolution of restricted dispersal is not the only imaginable outcome. One can also imagine direct selection against acceptance of a mate of the "wrong" type:

> The grossest blunder in sexual isolation, which we can conceive of an animal making, would be to mate with a different species from its own

and with which the hybrids are either infertile or, through the mixture of instincts and other attributes appropriate to different courses of life, at so serious a disadvantage as to leave no descendants (Fisher 1930, p. 130).

Fisher left little doubt about his belief in the efficacy of direct selection for prezygotic isolation.

No further theoretical work on reinforcement appeared until Wilson's (1965) presentation of Bossert's unpublished computer simulations. Wilson and Bossert emphasized what would become an important theme: reinforcement is a race between the enhancement of prezygotic isolation and the fusion of populations. One must do more, in other words, than show that there is a selective advantage for alleles that increase mate discrimination. One must show that these alleles reach high frequency *before* populations fuse through hybridization. (No such race occurs if hybrids have zero fitness, since fusion is impossible.) Wilson and Bossert simulated the case in which two populations come into contact and hybridize at an initial "error rate." Hybrids suffer low fitness due to heterozygote disadvantage at a single locus, while prezygotic barriers involve many genes. Wilson and Bossert found that low hybrid fitness and low error rates often lead to reinforcement, while high hybrid fitness and high error rates often lead to fusion (Figure 10.3). Crosby (1970) performed another early computer simulation of reinforcement. He considered two scenarios: one in which two taxa are sympatric, and another in which they meet in a hybrid zone. Modeling the evolution of flowering time in an annual plant and allowing for partial hybrid fertility, he showed that, in both scenarios, prezygotic isolation could evolve within several hundred generations.

These studies represent some of the earliest computer simulations in population genetics and, not surprisingly, they suffer the shortcomings that characterized such work: unrealistically small population sizes and little replication. More rigorous, and preferably analytic, work was needed.

Such analytic work initially focused on the case in which hybrids are completely unfit. Although this scenario is irrelevant to speciation, it at least enjoys the merit of being mathematically tractable. More important, it is clear that—if enhanced prezygotic isolation does not evolve under these extreme circumstances—it will not evolve at all. One of these early studies was population genetic and the other quantitative genetic.

In the population genetic study, Sawyer and Hartl (1981) analyzed change in allele frequency for a mutation that lowers the probability of mating with a second species, assuming a hybrid fitness of zero and initial random mating. They showed that, for a rare "choosiness" allele having frequency p, $\Delta p \approx p(1-f_1)m_1$ where f_1 is the fraction of the mating pool made up by the population in which the mutation arises, and m_1 is its probability of rejecting a mate from the "wrong" population. Because change in p is always positive, choosiness alleles always invade. Not surprisingly, then, prezygotic isolation readily increases when hybrids are completely sterile or inviable.

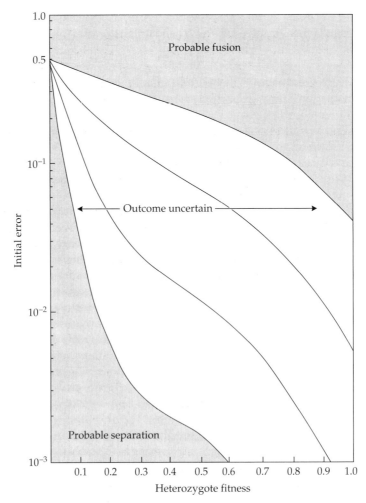

Figure 10.3 The probability of reinforcement versus fusion. (From Wilson 1965, based on computer simulations by Wilson and Bossert.)

Sved (1981) considered a quantitative-genetic model in which the characters affecting mate choice are underlain by polygenes. He allowed separate genes for male versus female mating behavior. Assuming that hybrids suffer zero fitness, he calculated the response to selection on female mating behavior due to hybridization. He found that the change in female mating behavior is proportional to the initial difference in male means between populations. As expected intuitively, populations that begin with more divergent mating behavior enjoy faster change. Once again, therefore, we arrive at an unsurprising answer: increase in prezygotic isolation is unproblematic when hybrids suffer zero fitness.

Objections to reinforcement

This optimistic assessment was vigorously challenged in the 1980's. Evolutionists raised many objections to reinforcement, some substantive, and some less so. By the end of the decade, reinforcement was considered unlikely at best, and an exercise in wishful thinking at worst (Noor 1999; Marshall et al. 2002). We discuss the most important objections below.

SELECTION-RECOMBINATION ANTAGONISM. While the above theory shows that prezygotic isolation can be exaggerated if there is no gene flow between populations, the effect of even minimal gene flow is unclear. Although Sved (1981) suggested that his results might be robust to low gene flow, two papers in the early 1980s suggested otherwise.

The first study was by Felsenstein (1981). Although Felsenstein did not explicitly consider reinforcement, his main finding played an important role in all subsequent thinking about the process. Because we have described this work in Chapter 4, we only sketch it here. Felsenstein considered two interbreeding subpopulations. The model featured three loci: one affecting prezygotic isolation, and two affecting postzygotic isolation. At the first gene, individuals carrying the *A* allele tend to mate amongst themselves, while those carrying the *a* allele tend to mate amongst themselves. At the other two loci, *BC* individuals enjoy high fitness in subpopulation 1 while *bc* individuals enjoy high fitness in subpopulation 2. Felsenstein's question was: Will alleles at the prezygotic isolation locus become non-randomly associated with those at the postzygotic isolation loci? That is, will the species split in two, with *ABC* predominating in subpopulation 1 and *abc* predominating in subpopulation 2? Felsenstein showed that, over part of the parameter space, non-random associations (linkage disequilibrium) between the pre- and postzygotic isolation loci do arise. But over much of the parameter space, they do not. More important, he explained the reason for these different outcomes: building an association between the genes causing pre- and postzygotic isolation involves a conflict between two forces, selection and recombination. Selection builds up such associations but recombination tears them down. As Felsenstein further showed, very strong selection is required to overcome recombination.

Felsenstein identified *the* key barrier to the evolution of prezygotic isolation in sympatry: recombination. This barrier cannot act, of course, when hybrids suffer complete hybrid sterility or inviability. In that case, the only breeding individuals are *BC* and *bc*, and—in each of these now distinct gene pools—alleles favoring assortative mating can increase in frequency. As Felsenstein also emphasized, recombination cannot block the evolution of increased prezygotic isolation if the *same* allele causes homogametic matings in both subpopulations (the "one allele" model).

Nearly simultaneously, Barton and Hewitt (1981a) noted a related problem. When postzygotic isolation is incomplete *and* based on many genes, the selection-recombination problem grows more acute: any single modifier of prezygotic isolation can realistically remain in linkage disequilibrium with only a few of the genes causing postzygotic isolation.

In summary, going from zero to non-zero hybrid fitness may fundamentally change the dynamics—and chances—of reinforcement.

THE RACE BETWEEN EXTINCTION AND REINFORCEMENT. While Wilson and Bossert recognized that reinforcement is a race against fusion, they did not discuss another kind of race: reinforcement must occur before one or both of the populations go extinct (Templeton 1981). The problem is that secondary contact between taxa producing unfit hybrids is analogous to heterozygote disadvantage (Paterson 1978, 1982; Harper and Lambert 1983). When heterozygotes at a locus are unfit, there is an unstable equilibrium: any departure from this equilibrium results in loss of one allele. As Paterson emphasized, taxa coming into secondary contact are susceptible to the same unstable equilibrium, with the rarer population prone to extinction. This is because, under random mating, the rarer population meets the "wrong" type more often and so produces unfit hybrids more frequently than the common population. This causes the rarer population to become rarer yet, triggering a positive feedback loop that ultimately dooms the rare population to extinction. Thus, if reinforcement is to evolve, it must do so quickly.

This argument assumes that the two populations are ecologically identical. But as Sved (1981) noted, if this is true, extinction has little or nothing to do with a race with reinforcement: extinction of one taxon is likely by competitive exclusion alone whether or not the populations evolve prezygotic isolation.

There can, however, still be a race between reinforcement and extinction. For it remains possible that extinction could occur before reinforcement *even when populations differ ecologically*. Indeed, extinction might still occur even when populations use completely separate resources and have independent density regulation: populations will still produce fewer progeny when females waste their gametes on males of the wrong type. Thus, populations might be sufficiently distinct ecologically to coexist in the *absence* of hybridization but not in the *presence* of hybridization. In this case, populations can coexist only if prezygotic isolation evolves quickly and we can sensibly speak of a race between reinforcement and extinction.

This point was highlighted in one of the most influential theoretical studies of reinforcement, that of Spencer et al. (1986). Focusing on the case in which hybrids have zero fitness, Spencer et al. performed computer simulations to determine if increased prezygotic isolation can evolve before hybridizing populations go extinct. They considered the case in which populations have independent logistic growth and thus suffer little chance of extinction in the absence of hybridization. With hybridization, however, each population's growth was reduced by the fraction of females who mated heterotypically. Male and female mating characters were polygenic.

Spencer et al.'s key result was simple. Even with independent population regulation, extinction often occurs before reinforcement. Indeed, reinforcement was likely only when populations already differed significantly in mating behavior and so already showed considerable prezygotic isolation. Spencer et al. thus concluded that coexistence without hybridization provides no guarantee against extinction *with* hybridization. After gene flow, the threat of extinction poses the most serious challenge to reinforcement.

THE SWAMPING EFFECT. As Moore (1957, p. 335) pointed out, the alleles causing reinforcement have a selective advantage only within the zone of contact between two populations. If these alleles were advantageous outside this zone they would presumably increase in frequency everywhere—and it would be absurd to contend that they were selected to avoid hybridization. The alleles underlying reinforcement are thus probably either neutral or deleterious outside the zone of contact. This poses two problems.

First, selection in the zone of overlap might be swamped by gene flow from allopatric regions (Bigelow 1965). This would seem particularly serious immediately after secondary contact, as the zone of overlap would presumably be narrow and the area of non-overlap large. Gene flow would thus be overwhelmingly *into* the zone of overlap. While Caisse and Antonovics (1978) showed that prezygotic isolation could evolve in sympatry despite gene flow if the alleles involved are neutral elsewhere, Sanderson (1989) showed that, as one might guess, the chances of reinforcement fall if these alleles are deleterious elsewhere. (The same problem arises if two populations do not physically overlap but exchange migrants at some rate. Reinforcement requires *some* gene flow, but not too much [Sanderson 1989; Servedio and Kirkpatrick 1997; Cain et al. 1999; Servedio 2000].)

Second, it is hard to see how the alleles underlying reinforcement could spread outside the region of overlap to the rest of the incipient species. One possibility, of course, is that they do not. Instead, the alleles conferring reinforcement could remain permanently trapped in the zone of contact. Alternatively, the populations might slowly migrate into a region of complete geographic overlap in which all individuals experience selection for prezygotic isolation. Finally, Howard (1993) suggested that *species*, not alleles, might spread from zones of contact: long selection in a zone of contact might yield reproductive isolation between individuals from the overlap zone and those from the remaining parental populations. Thus, a newly evolved species might ultimately migrate outwards, becoming sympatric with its ancestors. But it is not clear how this escapes the key problem: if swamping impedes reinforcement, it would probably also impede the evolution of reproductive isolation between individuals in the contact zone and those outside it.

REINFORCEMENT IS SELF-DEFEATING. Moore (1957, p. 336) and Spencer et al. (1986) emphasized another problem: reinforcement is self-defeating. Because the strength of selection for increased prezygotic isolation is proportional to the frequency of hybridization, any increase in prezygotic isolation automatically reduces the strength of selection for further reinforcement. Reinforcement thus pulls the rug out from under itself.

Spencer et al. saw this effect in their simulations. Among populations that survived extinction, mating behavior changed quickly early on, but then slowed as populations diverged. Spencer et al. (1986, p. 257) thus concluded that it is "extremely doubtful whether selection alone would lead to complete speciation."

PRE- VERSUS POSTZYGOTIC ISOLATION.   Finally, while there is a selective advantage to increasing prezygotic isolation, there is also an advantage to *decreasing* postzygotic isolation. Why not, then, eliminate the cost of hybridization by improving the fitness of hybrids, precluding reinforcement?

This idea has received little attention (but see Sanderson 1989). This may reflect the common intuition that it is hard to "undo" postzygotic isolation, at least in its intrinsic form (Muller 1939). But when postzygotic isolation is incomplete, it is obviously easy to purge incompatible alleles, leaving only compatible ones. Moreover, even when postzygotic isolation is complete (the case of reproductive character displacement), modifier alleles might accumulate that lessen the effects of genic incompatibilities. In addition, there is no reason for thinking that it is hard to reverse *extrinsic* postzygotic isolation, no matter how strong.

SUMMARY.   The above objections posed an enormous challenge to the theory of reinforcement, a challenge that was not immediately met. Consequently, by the late 1980s the weight of opinion swung against reinforcement. The theory appeared dying, if not dead.

The revival of reinforcement

The third and most recent phase in the history of reinforcement featured a sudden reversal of fortunes. Over a span of a few years, reinforcement went from seeming nearly impossible to seeming likely. This reversal reflected a burst of theoretical work, a burst that was triggered not by perceived flaws in the above objections but by the appearance of new data sets suggesting that reinforcement—or at least something *looking* like reinforcement—acts in nature (Coyne and Orr 1989a; Howard 1993). The patterns revealed by these studies stood in stark contrast to those expected under existing theory. As we will see, the theory that followed identified and incorporated several factors ignored by previous theory.

The first of these new studies was that of Liou and Price (1994). Their computer simulations roughly followed those of Spencer et al. (1986). Populations were sympatric and ecologically distinct (i.e., a population could not go extinct by competitive exclusion, though it could by hybridization). Mate choice featured a male trait (z) and a female trait (y), each underlain by multiple loci. Females had an absolute preference function such that females of phenotype y preferred males of phenotype $z = y$. In a departure from Spencer et al., Liou and Price studied the full range of hybrid fitness.

They showed that reinforcement occurs reasonably often if hybrid fitness is low and populations already differ in female preference and male character. Although reinforcement is most likely when hybrids have zero fitness, this is not required. Reinforcement *does* occur in the face of gene flow. Liou and Price also found that any factor that decreases the chances of extinction increases the odds of reinforcement. Such factors include those boosting the carrying capacity or the intrinsic rate of increase. As expected from Felsenstein (1981), greater recombination hinders reinforcement when hybrids have non-zero fitness; reinforcement, however, does remain possible with recombination. Liou and Price (p. 1451) concluded, "There is a wide range of genetic and ecological conditions under which reinforcement rather easily occurs." Indeed, they found that reinforcement occurs far more readily than in Spencer et al.'s study, even when using the same parameter values (Figure 10.4).

Given that this work signaled a reversal in opinion about reinforcement, it is worth understanding why Liou and Price's results differed from those of Spencer et al. The reason seems clear. Liou and Price's simulations allowed for sexual selection, while Spencer et al.'s did not. Recall that assortative mating gives rise to genetic correlations between male and female trait values (since females with large y values tend to mate with males with large z values). Because these associations arise automatically, it may not be obvious why they appear in Liou and Price but not in Spencer et al. The reason is that Liou and Price's simulations explicitly followed the loci encoding male and female traits, while Spencer et al.'s simulations did not. By recording the genotypic values of males and females who mated, and assuming a constant heritability, Spencer et al. calculated *separate* responses to selection for male and female traits. But this approach does not allow a correlation between male and female genotypic values, precluding runaway sexual selection.

The important point is that sexual selection boosts the odds of reinforcement because *two* forces now drive the evolution of female preference: direct

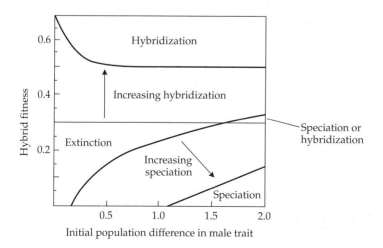

Figure 10.4 The parameter space over which reinforcement is likely to occur. (From Liou and Price's 1994 computer simulations.)

selection on females (reflecting the price paid for choosing a male of the wrong type), *and* indirect selection on females (reflecting a correlated response to sexual selection on males). Sexual selection also helps overcome the self-defeating nature of reinforcement. Although the intensity of natural selection for reinforcement decreases as reinforcement progresses, the intensity of sexual selection need not.

Later work by Kelly and Noor (1996) also emphasized the ease of reinforcement. Kelly and Noor focused on the evolution of mate discrimination, not preference. Maladaptive hybridization might, in other words, favor a narrowing of female taste—such that males that were previously acceptable become unacceptable—not a shift in mean female preference. Although Kelly and Noor's simulations were complex, their question was simple: Can a rare allele that increases female discrimination increase in frequency when hybrids suffer low fitness? They found that the answer is often yes. More surprisingly, they found that this remains true even when selection against hybrids is weak. The question is why.

In retrospect, the answer seems straightforward. Because reinforcement here reflects the evolution of greater female discrimination, not divergence in female preference, reinforcement involves the fixation of the *same* discrimination allele in both populations. And, as Felsenstein (1981) emphasized, this reduces the challenge posed by recombination: in a "one-allele" model of reinforcement, linkage disequilibrium need not be maintained in the face of recombination. Thus, while Liou and Price showed how the self-defeating nature of reinforcement could be overcome, Kelly and Noor showed how selection-recombination antagonism could be overcome. In each case, the mechanism involved—sexual selection, or female discrimination—seems biologically plausible.

These pro-reinforcement conclusions have been confirmed in the latest and most systematic theoretical studies of reinforcement, those of Servedio and Kirkpatrick. These authors have considered the effect of many factors on the probability of reinforcement: one-way versus two-way gene flow (Servedio and Kirkpatrick 1997); assortative versus preference-based mate choice (Servedio 2000); many versus few loci causing hybrid incompatibilities (Kirkpatrick and Servedio 1999); extrinsic versus intrinsic postzygotic isolation (Kirkpatrick 2001); and postmating, prezygotic versus postzygotic fitness problems (Servedio 2001). Their qualitative conclusion is simple: reinforcement can occur under a broad range of conditions. Indeed, reinforcement can occur under both two-way and one-way gene flow (though it is easier in the former); whether mating is assortative or preference-based (though it is often easier in the former); or whether many versus few loci cause hybrid incompatibilities (though it is slightly easier in the former). Moreover, reinforcement can occur regardless of the nature of selection against hybridization (extrinsic isolation will serve as well as intrinsic; indeed, hybrids might merely be unattractive to potential mates from the parental species [Coyne and Orr 1989a]).

In summary, recent theory leads to two broad conclusions, one satisfying and the other troubling. The satisfying conclusion is that reinforcement seems both possible and plausible. Indeed, while the details of the models differ wildly—complicating systematic assessment of the precise conditions under which reinforcement is likely—virtually all recent models suggest that reinforcement is easier than imagined a decade ago (Turelli et al. 2001; Kirkpatrick and Ravigné 2002). The troubling conclusion concerns the extraordinarily labile nature of this body of theory. While early theory mostly denied reinforcement and later theory mostly affirmed it, what occurred in the interim was the appearance of large empirical surveys suggesting that something that looks like reinforcement occurs. Put bluntly, theory said it could not happen until the data said it probably did. While a charitable interpretation of this history would emphasize the healthy dialectic between theory and data, we suspect that a more sober one would emphasize the often-misunderstood role of theory in evolutionary biology. Evolutionary theory is not, at least usually, in the business of telling us what is and is not biologically possible, a matter that depends on the particular suite of assumptions made. Rather, evolutionary theory tells us what does and does not follow from a given set of verbal assumptions. These roles were often confused in the history of reinforcement theory.

Alternative Explanations

The data reviewed earlier show that prezygotic isolation is often stronger in sympatry than allopatry. And the theory just reviewed shows that reinforcement is possible. But it does not follow that reinforcement is the correct explanation of the enhanced isolation seen in nature. Instead, several alternative explanations are possible. The primary task now confronting students of reinforcement is thus clear: to distinguish among these competing hypotheses.

As a step to this goal, we critically review several alternative hypotheses. Our discussion partly follows that of Noor (1999), although we consider several alternatives that he did not. In the end, we argue that reinforcement may yield unique (or nearly unique) predictions that allow us to distinguish it from the alternatives.

Publication bias

One of the most obvious but least discussed alternative explanations of the pattern of enhanced isolation is that it reflects a publication bias (though see Howard 1993). Findings that are consistent with reinforcement are exciting and so seem more likely to see the light of day than the opposite findings. Fortunately, this bias cannot explain all of the data. As Noor (1999) pointed out, a pattern of enhanced isolation appeared in Coyne and Orr's (1989a) literature survey of *Drosophila* long *after* the data were originally collected. Indeed, this is one of the advantages of such broad surveys: they are both less vulnerable to publication bias in the primary literature (as most data were collected for other reasons) and, as large and time-consuming enterprises, are likely to be published whatever the conclusion. We agree with Noor (1999) that the pattern of enhanced isolation is probably real, whatever its cause.

Differential fusion

Templeton (1981) suggested that enhanced isolation might reflect differential fusion, not reinforcement. According to this idea, different taxa coming into secondary contact might show a range of prezygotic isolation. Taxa that are strongly isolated can persist in sympatry while those that are weakly isolated cannot. Instead, the latter fuse, leaving us with nothing to study. In the end, a pattern of greater prezygotic isolation in sympatry emerges although there has been *no* response to individual selection for increased isolation.

While this seems a powerful explanation of enhanced isolation, several arguments have been made against it. Coyne and Orr (1989a) argued that the hypothesis cannot explain two patterns seen in *Drosophila*. First, differential fusion predicts that levels of prezygotic isolation in sympatry are a *subset* of those in allopatry: strongly isolated taxa that can persist in sympatry must, by hypothesis, preexist in allopatry. But in *Drosophila*, the high levels of prezygotic isolation seen in sympatry are not seen in allopatry, weighing against differential fusion. As Noor (1999) points out, however, this argument is not conclusive. Strongly isolated allopatric taxa might, after all, be rare. If, for instance, only 1 in 100 cases of secondary contact involves taxa that are isolated enough to persist, our sampling of allopatric taxa would probably not detect them.

Second, differential fusion would seem to predict that both pre- *and* postzygotic isolation would be greater in sympatry than allopatry: either barrier lowers the odds that taxa fuse. But the *Drosophila* data show a striking enhancement of pre-, but not postzygotic, isolation in sympatry. Again, this pattern seems more consistent with reinforcement. Although this argument has been weakened somewhat by Gavrilets and Boake's (1998) demonstration that prezygotic isolation prevents fusion more effectively than postzygotic isolation, postzygotic isolation should still have *some* effect under differential fusion. It does not appear to.

Noor (1999) offered a novel objection to differential fusion. While it might, in principle, explain an *among*-species pattern of enhanced isolation (as in Coyne and Orr 1989a), it has a hard time explaining a *within*-species pattern (as in Noor 1995) unless gene flow among populations is very low. The reason is that if populations of a species all enjoy reasonable levels of gene flow before secondary contact, they would probably not differ substantially in levels of prezygotic isolation when coming into secondary contact with another species—and thus differential fusion cannot occur. This is an important point and, at least in one case of enhanced isolation—that of *Drosophila pseudoobscura* and *D. persimilis*—we know that levels of gene flow between populations are high (Schaeffer and Miller 1992).

Thus, while we cannot rule out differential fusion, and while it surely occurs at some rate in nature, there is some reason to believe that it is not the leading explanation of enhanced isolation.

Direct ecological effects

If an ecological variable both allows coexistence of two species and fortuitously increases mate discrimination in one or both of them, a pattern of enhanced isolation might result.

As Noor (1995, 1999) notes, however, this hypothesis suffers an awkward problem: why should ecological factors always *increase* mate discrimination? Why don't the ecological conditions found in sympatry just as often decrease discrimination? One possible answer is that sympatric conditions *do* decrease versus increase mate discrimination equally often, but in the former case, taxa hybridize and so are lost to fusion. But in this case, the ecological effects hypothesis collapses into a special case of the differential fusion hypothesis.

The ecological hypothesis suffers another problem. It is unclear how well ecological factors can explain the actual observations. In *Drosophila*, at least, comparisons of the strength of prezygotic isolation in sympatric versus allopatric taxa are made under uniform laboratory conditions (i.e., vials) in which ecology is almost certainly beside the point.

Ecological character displacement

There is, though, a special sense in which ecological effects might systematically give rise to enhanced isolation. Under this "ecological character displacement" hypothesis, natural selection at the level of individuals *does* act in sympatry, yielding heightened isolation, but selection acts not to reduce maladaptive hybridization but to reduce niche overlap between species. Response to such selection might *incidentally* change mate signals, yielding greater prezygotic isolation.

We know of only one attempt to experimentally distinguish reinforcement from ecological character displacement. Rundle and Schluter (1998) compared the mate discrimination of sympatric versus allopatric stickleback females. Importantly, allopatric females were taken from populations that resemble sympatric females both morphologically and ecologically. Rundle and Schluter found that sympatric females remain choosier than allopatric females despite this attempted leveling of morphological and ecological differences.

A special case of the ecological character displacement hypothesis has received a good deal of attention. According to this "noisy neighbors" hypothesis, sympatric species that produce similar courtship songs might interfere with each other acoustically, leading females to waste time and resources tracking males of the wrong species. Thus—even if species do not hybridize—natural selection might favor song divergence in sympatry. Such interference could yield enhanced isolation, at least when examining proxies for prezygotic isolation like song (Otte 1989; Noor 1999). As noted earlier, there is in fact evidence that courtship song sometimes differs between species in sympatry more than expected by chance: Otte (1989) examined divergence in male song in Hawaiian crickets of the genus *Laupala*. Gathering data on 46 cricket taxa distributed over 68 sites where at least two species coexist, he showed that the "spacings" between sympatric songs are greater than expected by chance. *Some* force has displaced song in sympatry. The question is whether this force is selection against acoustic interference or selection against maladaptive hybridization.

In principle, the noisy neighbor and the reinforcement hypotheses can be distinguished. If taxa hybridize at some rate, both hypotheses remain viable. But if taxa do *not* hybridize, reinforcement cannot occur, though noisy neighbor displacement can. Thus, enhanced isolation among non-hybridizing taxa is more plausibly explained by noisy neighbor effects. Unfortunately, this test is imperfect, as we can usually only ask if taxa hybridize *now* (Noor 1999). But the fact that taxa do not presently hybridize does not mean they did not do so in the past. Later, we suggest a test that may allow one to better distinguish noisy neighbor effects from reinforcement.

Runaway sexual selection

Day (2000) showed that a type of runaway sexual selection could yield results similar to those expected under reinforcement. He considered a model in which female choice is costly and a population is distributed over an ecological gradient. This gradient causes different male trait values to be favored in different locations. Day showed that, although runaway selection does not usually occur when female choice is costly, it *does* occur in spatial models, even when female choice is costly. More important, if the male trait shows a sharp transition along the ecological gradient, a pattern resembling enhanced isolation can arise: females mate randomly near the ends of the gradient but assortatively near the transition zone.

As Day notes, sexual selection and reinforcement can, in principle, be distinguished. The sexual selection model holds only with a single population arrayed along an ecological gradient—not with two once-allopatric populations that come into contact. Thus, the runaway sexual selection model cannot account for enhanced isolation between taxa showing *postzygotic* isolation. More precisely, the model could explain such cases only if one believes that postzygotic isolation arose within a single continuously distributed population, not in allopatry. This seems to us unlikely, especially for intrinsic postzygotic isolation.

Sympatric speciation

Finally, we note a radical alternative: sympatric speciation. A pattern of strong prezygotic (but not postzygotic) isolation among young sympatric (but not allopatric) species might arise by the splitting of taxa in sympatry. Sympatric speciation, after all, is likely characterized by the rapid evolution of mate discrimination, not postzygotic isolation (at least of the intrinsic type).

The attractiveness of this hypothesis depends on the plausibility of sympatric speciation as determined by independent lines of evidence (Chapter 4). For present purposes, we assume that sympatric speciation is plausible, and merely ask if its signature can be distinguished from that of reinforcement. When dealing with data at the species level (as in the Coyne and Orr data sets), one approach is to take advantage of Noor's (1997b) biogeographic contrast as shown in Figure 10.2. If the outgroup species A speciated sympatrically from the common ancestor of B and C, one would predict strong prezygotic isolation between A and B *and* between A and C. If, however, all speciation were allopatric but reinforcement occurred upon geographic contact, one would predict stronger prezygotic isolation between species A and B than between species A and C. Noor's (1997b) results are, of course, consistent with this second prediction.

Matters are different at the *population level* (i.e., when explaining a pattern of enhanced isolation among sympatric populations of two species). The clearest scenario involves a cline. Here, there are at least two ways to distinguish reinforcement from sympatric speciation. The first concerns the key difference between the models: reinforcement posits that a cline reflects secondary contact, while sympatric speciation does not. One might therefore test for telltale signs of secondary contact: as Barton and Hewitt (1989) emphasized, secondary contact is supported if two taxa show congruent clines over many characters. Second, reinforcement and sympatric speciation make different assumptions about postzygotic isolation (Kirkpatrick and Ravigné 2002): the former requires its pre-existence, while the latter does not. Sympatric speciation is therefore unlikely if two young taxa show intrinsic postzygotic isolation in both allopatric and sympatric populations.

Distinguishing the Alternatives

Although we have described tests that allow us to distinguish reinforcement from a particular alternative, what we would most like to have is a way to distinguish reinforcement from *all* the alternatives. We would like, in other words, to have a pattern that could serve as a signature of reinforcement. We emphasize that the search for such a signature does not require us to believe that enhanced isolation is *always* caused by reinforcement. Some of the above alternatives can, and surely do, act. For instance, even if reinforcement is real and reasonably common, differential fusion surely acts as well. What we are looking for, then, are predictions that allow us to infer the action of reinforcement, *not* predictions that allow us to conclude that the alternatives never act.

Noor (1999) suggested such a prediction. If heightened prezygotic isolation in sympatry is due to natural selection for greater choosiness, the alleles underlying enhanced isolation should, he argued, be mostly dominant. This prediction follows from "Haldane's sieve," the fact that dominant favorable mutations enjoy greater probabilities of fixation than recessive favorable mutations. Noor reviewed the evidence bearing on this prediction from *Drosophila*, finding it mixed.

Unfortunately, this prediction suffers several problems. First, the degree of dominance of an allele may differ between pure species and hybrids, as Noor noted. Second, Haldane's sieve itself is not straightforward. Although dominant mutations do enjoy higher probabilities of fixation when mutations are unique, probabilities of fixation are nearly independent of dominance when mutations come from the standing genetic variation and were previously deleterious (Orr and Betancourt 2001). (This seems likely in the case of reinforcement: "choosiness" alleles are probably kept rare in allopatry because they are costly in the absence of the other species.) Finally, Haldane's sieve does not allow us to distinguish between reinforcement and ecological character displacement, including the noisy neighbors scenario. In *both* cases, enhanced isolation is due to natural selection at the individual level and so, in both cases, the alleles involved might be dominant.

Are there any other predictions that might distinguish reinforcement from the above alternatives? There seems to be at least one. In most species, females will pay a larger fitness cost for mating with the wrong species than will males. Thus, as Partridge and Parker (1999) emphasized, reinforcement should typically result in larger changes in female than male behavior, physiology, or morphology. In the simplest scenario, males of two species may have diverged in allopatry, but do not change at all upon secondary contact. Females, on the other hand, become choosier upon contact, as they are now confronted with two types of males. This sex asymmetry was already appreciated in some of the earliest papers on reinforcement (e.g., Sved 1981, p. 210), and on sexual selection in good genes models (e.g., Trivers 1972). But while Sved (1981), Trivers (1972), and Partridge and Parker (1999) all saw the point, none seem to have emphasized that it might provide a systematic way to distinguish reinforcement from the above alternative hypotheses. But it seems unlikely that any of these alternatives would predict female-specific changes, and some would seem to predict the opposite.

There is no reason, for instance, why a publication bias should result in more cases of female than male changes. Similarly, differential fusion would not seem to predict sex-differential effects. Fusion occurs because of gene flow and it is irrelevant whether that flow occurs because of female or male behavior. Similarly, there is no obvious reason why direct ecological effects would preferentially change female behavior or preference in sympatry. Similarly, fortuitous changes in mate choice as a result of ecological character displacement might affect either sex; indeed, with noisy neighbor effects, we would, if anything, predict more change in *male* not female traits in sympatry, as males are the sex that typically calls. Although we cannot rule out the possibility that some of the above hypotheses might, under some circumstances, lead to preferential change in female behavior or structure, the prediction does not seem to follow naturally from them. (The sole possible exception is sympatric speciation. We have argued on independent grounds, however, that sympatric speciation offers a fairly implausible explanation of heightened prezygotic isolation in sympatry when considering either species or population level data.) We suspect, therefore, that systematically greater change in female behaviors or structures in sympatry represents a reasonably robust signature of reinforcement. (Marshall et al. 2002 argue that conspecific gamete precedence might eliminate much of the cost of hybridization in females and increase it in males. They conclude, however, that this will probably prevent reinforcement. Here, we are interested only in those species pairs that *do* show enhanced isolation.)

In fact, we can recognize three sub-predictions. The first is the obvious one that, among taxa pairs showing greater sympatric isolation, enhanced isolation should be due to changes in female components of mate choice more often than to male components of mate choice. (Note that we are not arguing that males of the two taxa do not differ; females are obviously choosing on the basis of something. Our prediction is that enhanced isolation reflects greater female than male *changes* in sympatry. We are also not arguing that male characters do not change in sympatry. We are arguing that enhanced isolation is due *more* to female changes than to male changes.) Second, in cases in which hybrids have low fitness in one direction of a species cross (A female × B male) and high fitness in the other (B female × A male), A females will be more discriminating than B females. Again, this follows from the fact that males pay little or no price for hybridization. Finally, reinforcement due to greater changes in *male* behaviors or structures might occur in those rare taxa in which males mate only once, or invest significantly in offspring or copulation (e.g., by nuptial gifts). There should be, in other words, exceptions that prove the rule. The finding that sympatric males of the butterfly genus *Heliconius* reject or ignore females having mimetic patterns from the wrong species might well represent such an exception (Jiggins et al. 2001b) as *Heliconius* males transfer costly spermatophores (Naisbit et al. 2001).

As far as we know, none of the above predictions has been tested systematically. Until this is done, or until other unique predictions are offered and tested, we will not be able to conclude with confidence if the pattern of enhanced isolation seen in sympatry is usually—or only rarely—caused by reinforcement.

11

Selection versus Drift

One of the oldest and most hotly contested questions in speciation is whether natural selection or genetic drift plays a larger role in the origin of species. Unlike many of the questions that we have considered up to now, we believe that this one has been answered.

To arrive at this answer, we first catalogue the ways in which selection or drift might play a role in speciation. We then review the theory, both verbal and mathematical, on the plausibility of these roles and, finally, we turn to the evidence. Because the role of selection in speciation is, as we will see, clear and uncontroversial, we will spend most of our time trying to determine if speciation by drift is theoretically plausible and empirically supported. The amount of space we devote to selection versus drift should not therefore be taken as an indication of their biological significance.

Speciation by Selection

We must first clarify the ways in which natural or sexual selection might be involved in speciation. To do this, we make several distinctions. While most have been made before, the literature describing these distinctions is filled with confusing or conflicting nomenclature. One of our simpler goals is to standardize this nomenclature (Table 11.1).

There are two ways that selection might play a role in speciation. The first is direct: individuals showing greater reproductive isolation enjoy a fitness advantage and selection acts directly to increase this isolation. Direct selection obviously plays a role in both sympatric speciation and reinforcement. As we have already discussed these processes at length, we will not repeat our discussions here.

Table 11.1 *Classification of the possible roles of selection versus drift in speciation*

Evolutionary force	Definition
Selection	
Direct	Direct natural selection *for* reproductive isolation. Direct selection characterizes models of sympatric speciation and reinforcement.
Indirect	Reproductive isolation arises as a pleiotropic side effect of natural or sexual selection.
Primary	Selection acts on a character and the same character ultimately causes reproductive isolation.
Secondary	Selection acts on a character and the genes involved pleiotropically affect another character that ultimately causes reproductive isolation.
Tertiary	Selection acts on a character and linked genes that hitchhike along with those under selection ultimately cause reproductive isolation.
Genetic Drift	
Neutral	Reproductive isolation results from genes whose divergence was strictly or nearly neutral.
Peak shift	Reproductive isolation involves a period of maladaptive evolution, during which genetic drift must overcome selection. This class of model often involves founder-effects (e.g., Mayr's "genetic revolution," Carson's "founder-flush-crash," and Templeton's "genetic transilience" models).

Although direct selection might well occur, it is not ubiquitous. The clearest evidence is biogeographic: there can be no direct selection for reproductive isolation in allopatry. Nonetheless, allopatric populations of many taxa evolve both pre- and postzygotic isolation with the passage of time (Chapters 2 and 3). While one might argue that reproductive isolation between presently allopatric taxa evolved in sympatry, such special pleading is, in many cases, awkward, if not absurd (e.g., endemic insular taxa or those separated by the Isthmus of Panama). There is also strong genetic evidence against ubiquitous direct selection. This is clearest in the case of intrinsic postzygotic isolation. As we emphasized earlier, it is hard to directly select for postzygotic isolation. But as Muller (1942, pp. 101–103) further emphasized, the first substitution involved in any genic incompatibility simply *cannot* cause reproductive isolation. Thus, at least half of all substitutions underlying intrinsic isolation cannot have experienced direct selection for reproductive isolation. As Muller (1942) also emphasized, species are sometimes separated by many hybrid incompatibilities, each of which causes *complete* hybrid inviability or sterility. Such overkill is impossible under direct selection.

The second way that selection might play a role in speciation is indirect: isolating barriers could arise as a *byproduct* of selection for something other than reproductive isolation. Indirect selection is probably more common than direct. Indeed, the founders of the Modern Synthesis, including Dobzhansky, Mayr, and Muller, routinely invoked indirect selection in the origin of species. Oddly, however, they provided little evidence for it. We try to fill this gap later in the chapter.

At the risk of being splitters, we can distinguish between several forms of indirect selection. In the primary form, selection acts on a character and that character ultimately causes reproductive isolation. For example, natural or sexual selection on plumage color in birds might give rise to different color patterns between two allopatric populations; this difference might then cause sexual isolation between the taxa upon geographic contact. In the secondary form, selection acts on a character and the genes involved pleiotropically affect another character that ultimately causes reproductive isolation. Selection on nuclear pore function might, for example, pleiotropically give rise to an intrinsic incompatibility, as may have occurred at the *Nup96* gene in *Drosophila* (Chapter 8). In the tertiary form, selection acts on a character, and linked genes that hitchhike along with those under selection ultimately cause reproductive isolation. This case differs from the secondary form in that speciation does not involve a subset of the genes that were the immediate target of selection. Tertiary selection might explain, for instance, hybrid inviability between certain *Mimulus* populations (Macnair and Christie 1983; and see below).

There will surely be cases in which we will have a hard time deciding if reproductive isolation reflects direct or indirect selection and, if the latter, which variety of indirect selection. Moreover, a single bout of selection might give rise to multiple barriers belonging to different "classes" (e.g., selection on plumage color could give rise to sexual isolation [primary] and to postzygotic isolation due to linked genes [tertiary]). Despite such complications, we believe that the classification given in Table 11.1 is generally useful.

As we have already considered direct selection, we focus in this chapter on indirect selection. Our goal is to answer the following question: Is reproductive isolation usually a byproduct of selection or is it usually due to genetic drift? Throughout, we assume that evolution occurs in allopatry, unless stated otherwise.

Natural selection

It is easy to see how reproductive isolation could evolve as a by-product of response to natural selection. With prezygotic isolation, mate preferences might evolve as correlated responses to selection on morphology or physiology (Schluter 2000; see also Chapter 6). Similarly, with intrinsic postzygotic isolation, the substitutions underlying a hybrid incompatibility might well have been driven by selection. But it is *extrinsic* postzygotic isolation that most clearly reveals the hand of natural selection. It is uncontroversial that most phenotypic divergence in ecologically important traits is driven by natural selection. Consequently, cases in which species hybrids fall between parental phenotypic optima often provide good evidence for the role of selection in speciation (Schluter 2000). Because we reviewed extrinsic postzygotic isolation before (Chapter 7), we do not repeat our discussion here.

It is important to note, however, that even if speciation results from natural selection, it need not involve "ecological" selection (i.e., adaptation to different environments). Selection can drive the evolution of reproductive isolation between two populations even if they reside in *identical* environments (Muller 1942). The reason is that adaptation is not deterministic. This is most easily seen if adaptation involves new mutations. Replicate bouts of adaptation will often involve the substitution of different mutations, as favorable mutations appear in different orders in different populations, and each new mutation suffers a large probability of accidental loss (Fisher 1922; Haldane 1927). Replicate populations will consequently diverge even under identical conditions, possibly giving rise to reproductive isolation. Indeed, given enough time, this process is guaranteed to cause speciation. (See our discussion of even- and odd-year salmon that live in the same environment but breed in different years; Chapter 5.) It should also be noted that some adaptation, and perhaps much of it, might be in response to *internal* changes in organisms: if an initial substitution occurs in response to an environmental change, subsequent substitutions might occur that merely ameliorate the deleterious physiological or biochemical side effects of the first substitution. These compensatory substitutions might be the ones that ultimately cause reproductive isolation.

But far more attention has focused on the case in which allopatric populations experience different environments (Schluter 2000). Even in this case, however, reproductive isolation might reflect adaptation to features of the environment that are *shared* between populations. Nevertheless, there are good experimental, as well as theoretical, reasons for thinking that adaptation to different features of the environment will give rise to reproductive isolation more often than adaptation to shared features. While we will discuss the experimental reasons later, the theoretical reason is simple and can be stated now: the fact that a pair of populations adapts to different optima means that at least one of them is adapting to a *new* optimum. And adaptation to a new optimum probably involves a greater number of evolutionary substitutions than does continued residence near the same optimum. Under an indirect model of speciation, this greater number of evolutionary changes typically translates into a greater chance of reproductive isolation (see also Barton 2001).

Sexual selection

Sexual selection acting in allopatric populations could also indirectly give rise to reproductive isolation. While this is obvious in the case of prezygotic isolation (Chapter 6), there is also good evidence that sexual isolation sometimes drives the indirect evolution of postzygotic isolation, especially hybrid sterility (Chapter 8).

As Schluter (2000) emphasized, models of sexual selection can be divided into two classes: those that involve natural selection and those that do not. Natural selection is not involved in Fisherian runaway or chaseaway sexual selection, but it is involved in models that feature good genes, spatial variation in the optimal male trait, selection on the sensory system, or selection on the most efficient signaling system. We have little information on the relative roles of these two classes of models. But, if runaway and chaseaway varieties of sexual selection prove common causes of speciation, the popular intuition that ecological speciation is the norm would be partly undermined. It might well turn out that sexual selection plays at least as important a role in the evolution of reproductive isolation as does natural selection. We consider the evidence for each process later.

Mathematical theories of selection-based speciation

We have little mathematical theory describing how indirect selection in allopatry drives speciation. The only substantive exceptions involve the quantitative genetics of sexual selection (Chapter 6) and the population genetics of intrinsic postzygotic isolation (Chapter 7). At first sight, this paucity of theory might seem odd. But the explanation is simple: speciation by selection in allopatry is conceptually straightforward and little mathematics is required. The shortage of theory should not be confused, therefore, with doubts about the plausibility of the process.

Speciation by Drift

Just as there are several ways in which selection might cause speciation, so there are at least two ways in which genetic drift might create new species. In the first, divergence is simply neutral. In the second, genetic drift must *overcome* selection, that is, at least some evolutionary change is deleterious. These latter models involve "peak shifts" (see below).

Neutral models of speciation have received less attention than peak shift models. Indeed, the only neutral models we know of are those of Nei and colleagues (Nei 1976; Nei et al. 1983), who consider the evolution of both post- and prezygotic isolation. In their models of postzygotic isolation, evolution is characterized by a lack of transitivity: although alleles that are separated by one mutational step are compatible, those separated by *two* steps are not. (Nei et al. 1983 also consider similar two-locus models.) In their models of prezygotic isolation, evolution is again stepwise (allele i can evolve only to allele $i + 1$ or $i - 1$), but one locus is now expressed only in males and another only in females. Mate preference is absolute such that male and female genotypes that differ by zero or one step can mate, while those that differ by two or more steps, cannot. Evolution is thus nearly, though not completely, neutral: all alleles are equally fit and selection acts only in those rare instances when alleles that differ by two mutational steps segregate together in a population.

These models show that neutral evolution *can* cause reproductive isolation between allopatric populations. This is not surprising. As we emphasized in

Chapter 7, genic incompatibilities can evolve whether the substitutions involved are advantageous and fixed by selection, or neutral and fixed by drift. But Nei et al.'s findings do not encourage the idea that neutral speciation is common. There are several problems. The first is that reproductive isolation evolves slowly in these models. Moreover, because evolution is nearly (but not strictly) neutral, reproductive isolation arises ever more slowly in larger populations. (The rate of evolution of prezygotic isolation can be increased when runaway sexual selection is allowed [Wu 1985], but then the model is no longer neutral.) Second, the fitness schemes considered in these models are artificial (*all* alleles separated by two mutational steps are incompatible), and more realistic schemes would almost surely yield less reproductive isolation.

Peak shift models

In the mid-1950s, several evolutionary biologists began to question the Darwinian orthodoxy that natural selection—especially that based on ecology—is the engine of speciation. The result was the elaboration of peak shift models, in which a population must travel downhill on an adaptive landscape—movement that is *opposed* by natural selection—before traveling up a new fitness peak. (An adaptive landscape is a plot of mean fitness as a function of allele frequencies at many loci; selection will usually cause a population to climb a fitness peak; Figure 11.1.)

These new models of speciation had a number of radical consequences. First, because the pace of speciation is no longer set by the pace of adaptation (which was deemed slow), speciation might be very fast. Indeed, Carson (1971, p. 67), a champion of peak shift models, argued that "there is no incontrovertable [sic] evidence that speciation, like adaptation, is accomplished gradually in a microevolutionary fashion." Second, when natural selection does act during cladogenesis, it might be entirely non-ecological: selection in peak shift models is primarily "internal," involving the adjustment of one set of genes to earlier changes at another set of genes. Finally, genetic drift, not selection, triggers speciation in these models. Although the random genetic changes that underlie speciation might occur in small populations of stable size, most peak models emphasized a "founder event," in which a new population is established by one or a few individuals. These founder events in turn sometimes trigger founder *effects*—the rapid evolution of reproductive isolation as a result of a sudden shift to a new fitness peak.

Peak shift models grew in popularity over the decades that followed. Indeed, as Provine (1989, p. 72) noted, "some version of ... founder effect and genetic revolution has been the favored explanation for at least island speciation since 1954."

Here we review these peak shift models. Unfortunately, this task is complicated by the profusion of slightly different versions of the peak shift idea: the literature is filled with baroque discussion of such models as genetic revolution, founder-flush, flush-founder, and transilience to name a few. To keep

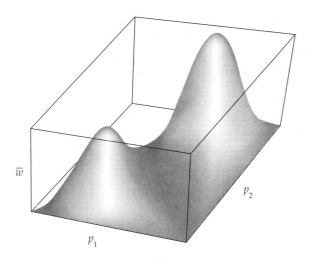

Figure 11.1 An adaptive landscape showing two fitness peaks separated by an adaptive valley. The x and y axes give allele frequencies at each of two loci (p_1 and p_2), while the z axis gives the mean fitness of a population having these allele frequencies, $\bar{w}$. (In reality, mean fitness is a function of allele frequency at many loci, but this is hard to depict.) Peak shift speciation requires that a population first move down from its current peak, cross an adaptive valley, and then ascend a new fitness peak.

our treatment manageable, we describe only the three most popular varieties of peak shift model, those offered by Mayr, Carson, and Templeton.

MAYR: GENETIC REVOLUTIONS. Mayr was struck by the fact that, although species are often undifferentiated over vast geographical regions, populations that are physically isolated from the rest of a species often appear distinct. Island populations, for example, often differ morphologically from mainland populations despite the fact that island and mainland appear similar ecologically. In one of his best known examples, Mayr argued that populations of the New Guinea kingfisher *Tanysiptera galatea* are morphologically uniform throughout the main island of New Guinea, but are morphologically distinct on nearby smaller islands (Mayr 1954a, p. 157; Mayr 1963, pp. 503, 522). Mayr labeled such geographically peripheral but morphologically distinct populations "peripatric."

Mayr believed that the phenotypic uniformity usually seen over large geographic regions reflects the action of several conservative evolutionary forces that maintain the "cohesion" of the gene pool. The most important of these forces is gene flow. Mayr believed that gene flow within a species prevents precise local adaptation: the genetic composition of a population represents a compromise between alleles favored in a habitat and those introduced from elsewhere by migration. When looked at over wide geographic areas, species will thus show little more than gradual, clinal variation. Mayr thus emphasized that substantial phenotypic divergence requires a complete or nearly complete break in gene flow: "The genuinely sharp break is not between the panmictic and the partially isolated system, but between the partially and the virtually fully isolated system" (Mayr 1963, p. 521). Another of the conservative forces that Mayr emphasized is epistasis: alleles are constrained to function on a particular constellation of background loci, a problem of co-adaptation that Mayr believed had been neglected by traditional "beanbag" population genetics

(Mayr 1959). To Mayr, the requirement of coadaptation across a network of loci severely constrains the possibilities for further evolution. Organisms are characterized by a "limited number of highly successful epigenetic systems and homeostatic devices which place a severe restraint on genetic and phenotypic change" and impede, if not block, evolutionary change (Mayr 1963, p. 523). The last conservative factor that Mayr emphasized was the slowness with which adaptive substitutions occur in large species (Mayr 1982, p. 604), a result he sometimes attributed to Haldane's (1924, 1932) calculations on the rate of spread of favorable alleles, and sometimes to Haldane's (1957) calculations on the cost of selection.

Given Mayr's belief that large populations suffer from a kind of evolutionary inertia, the problem of speciation is to explain how populations can overcome this inertia. Mayr's solution was the population bottleneck—a sudden and severe reduction in population size. He believed that such bottlenecks often coincide with the founding of a new population (e.g., on an island or on the periphery of a species' normal range). Although Mayr briefly sketched this idea in his 1942 book, *Systematics and the Origin of Species*, he did not develop it until his 1954a paper, and more fully still in his 1963 book *Animal Species and Evolution*. To Mayr, the break in gene flow associated with a bottleneck allows a founder population to be "emancipated" from the parental population. The key idea is that a small founder population suffers "a great loss of genetic variability" reflecting the fact that founders carry "only a very small proportion of the variability of the parent population" (Mayr 1942, p. 237; see also Figure 11.2). This loss of variation, in turn, changes the selection pressures experienced by alleles still segregating in the founder population. While alleles in the parental population were previously selected to function on a heterozygous genetic background, they are now selected to function on a more homozygous

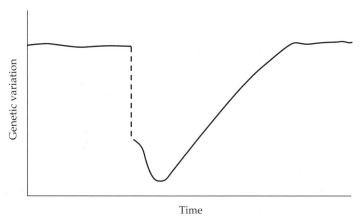

Figure 11.2 Mayr's plot showing the alleged loss of genetic variation following a founder event. The figure also shows the gradual increase in genetic variation as the founder population recovers from the population bottleneck. (After Mayr 1954a.)

background. Allele frequencies might therefore shift. But these evolutionary changes might then trigger changes at *other* loci, and so on, producing an evolutionary chain reaction. In Mayr's terminology, population bottlenecks can initiate a "genetic revolution"—a cascade of genetic changes that ultimately involves a wholesale reorganization of the genome. Although Mayr (1963, p. 554) emphasized that most founder populations would not survive such an evolutionary upheaval, those that do may have a new "balancing system," that is, a new system of coadapted genes (Mayr 1963, p. 539). Because the new and old populations reside on different fitness peaks, they are at least partially reproductively isolated: any hybrid is likely to fall into the fitness valley that separates the two peaks.

CARSON: FOUNDER-FLUSH-CRASH. Carson also believed that large populations suffer evolutionary inertia. Indeed, he spoke of older populations as suffering a kind of evolutionary senescence: "In the face of an ever-changing environment, a mature population may become saturated and be unable to respond to new conditions. Thus, it may decline to a state of genetic rigidity and risk extinction" (Carson and Templeton 1984, pp. 98–99). But Carson's solution to this problem differed from Mayr's.

Carson's views were deeply influenced by his classic work on the Hawaiian drosophilids (Carson 1992). The family Drosophilidae in Hawaii includes 500 named species and at least 350 known but undescribed species; the number of species may well exceed 1000 (Kaneshiro et al. 1995). Most of these species are endemic to single islands. Moreover, the closest relative (and presumed ancestor) of any given species usually resides on an adjacent island—typically one to the northwest, the direction in which the islands grow older. Faced with this pattern of single-island endemism, and with the vast stretches of water that separate islands, Carson concluded that speciation in Hawaii often involves founder events. Indeed, Carson and Templeton (1984, p. 127) concluded that "founder-induced speciation accounts for at least a quarter of the entire genus" on Hawaii.

Though Carson proposed several variations on his model, the most popular involves three phases. The first is the founder event. Carson often emphasized the extreme case of a single inseminated foundress. Indeed, he argued that the entire clade of Hawaiian *Drosophila* may have derived from a single female that arrived on the archipelago 5 million years ago (Carson 1971, p. 58). (We now know that the Hawaiian *Drosophila* predate the existing islands: molecular data suggest the group is roughly 20 million years old and presumably first lived on now submerged islands to the northwest [Hardy and Kaneshiro 1981; DeSalle 1995].) Carson also argued that the founding of new species *within* the archipelago—between islands and sometimes within islands—also often involves single foundress events. Such founders carry little of the genetic variation present in the ancestral species. But to Carson, the fact that new species are established by a few founders was only part of the story. An equally important part was that these populations expand rapidly.

Although many founder populations will go extinct, reduced intraspecific competition within those that survive allows dramatic increases in population size. These increases characterize the second, or "flush," phase of Carson's model. During this phase, many new mutations appear. Carson (1968, 1971) argued that the mildness of intraspecific competition within small but growing populations causes a relaxation of selection—a scenario that allows the persistence of mutations that would be purged in larger, more stable populations.

Following the flush phase, the population enters the third and final phase, the population crash, which Carson sometimes called "inevitable" (Carson 1971, p. 60). Such crashes reflect the reinstatement of strong intraspecific competition when founder populations exceed the local carrying capacity. The crash may leave the founder population fractured into several demes. Some of these demes might then go through subsequent flush phases, triggering another, or perhaps several, flush-crash cycles.

In different versions of his model, Carson sometimes changed the order of events. Thus, in his well-known 1975 paper, the founder-flush-crash model became the flush-crash-founder model (Carson 1975; see also Carson 1968). Carson also began to emphasize a distinction between "open" and "closed" systems of genetic variability. The open system, he claimed, comprises regions of the genome harboring genes of small effect that show little epistasis and that act late in development. This system provides the material for adaptive geographic variation. The closed system involves tightly linked factors showing strong epistasis. These associations are old and essentially form segregating clusters of "supergenes": recombination between these clusters usually yields unfit individuals. Carson emphasized that the essence of speciation involves the evolution of a *new* closed system, one that—when mixed with the old system in species hybrids—yields individuals who fall into adaptive valleys. This new closed system is thought to emerge during the flush phase, when relaxed selection allows the survival of recombinants that would ordinarily perish. To Carson (1975, p. 88), "this cycle of disorganization and reorganization should be viewed as the essence of the speciation process."

TEMPLETON: TRANSILIENCE. Templeton (1981) divides speciation events into two classes, "divergence," which is driven by selection, and "transilience," which occurs despite selection. Although Templeton describes several forms of transilience (genic, chromosomal, etc.), we sketch only the genic form, which he considered the most important.

The genetic transilience model is harder to understand than Mayr or Carson's models. Templeton (1980b, p. 1013) defines it as "a rapid shift in a multilocus complex influencing fitness in response to a sudden perturbation in genetic environment." Similarly: "Genetic transilience occurs when a founder event triggers a rapid adaptive shift in a previously stable genetic system or systems owing to fixation at some loci and a rapid increase in inbreeding levels relative to the ancestral population" (Templeton 1981, p. 36).

Founder events, according to Templeton, have two effects. One is random change in allele frequency, which leads to a net loss of genetic variation. The other is departures from Hardy-Weinberg equilibria in the direction of an excess of homozygotes. Templeton argues that this "accumulation of inbreeding" in turn has two consequences: (1) alleles are selected more for their homozygous than heterozygous effects; and (2) alleles function on a narrower range of genetic backgrounds. The latter effect means that

> ...epistatic terms that could not be effectively selected for in the ancestral population can now respond to selection and play a major role in restructuring the fitness properties of the genome. For example, fixation at one critical locus could have cascading effects in a strongly epistatic genetic system (Templeton 1980b, p. 1015).

Templeton thus imagines a scenario in which an allele—because it now resides on a more homozygous background than before—experiences novel selection pressures. The resulting evolutionary changes at this locus might then trigger changes at other loci, with effects cascading throughout an epistatic network. In the end, such changes might give rise to either pre- or postzygotic isolating barriers between founder and ancestral populations (Templeton 1981, p. 36).

At first sight, it is hard to distinguish Templeton's scenario from Mayr's. But there are differences. One is that while Mayr believed that genetic revolutions affect a broad cross-section of the genome, Templeton (1980a) does not. To clarify this difference, Templeton (1981) draws a distinction between several genetic architectures. Type I involves many genes (each of small effect); type II involves a few major factors and many epistatically interacting modifiers; type III involves two locus complementary systems. Templeton (1980b, p. 1013) concludes that transilience is likely to involve type II and type III genetic architectures, not genetic revolutions. Second, unlike Mayr, Templeton tries to couch his model in the language of mathematical population genetics. This, he argues, yields a model that can make clearer predictions.

Templeton's (1980a,b) mathematical argument hinges on the difference between two kinds of effective population size. The variance size, N_{ev}, measures random change in allele frequency; the inbreeding size, N_{ef}, measures increase in homozygosity. To Templeton, one can best gauge the change in genetic environment induced by a founder event by the drop in N_{ef}. But such a drop does not ensure a founder effect. The reason is that evolutionary *response* to a new homozygous background requires genetic variation—and that means sizable N_{ev}. Templeton thus concludes that speciation is most likely when N_{ef} falls dramatically but N_{ev} remains as large as possible given this drop.

Templeton (1980b) argues that this conclusion allows prediction of the conditions under which speciation by transilience is likely. These include many offspring per parent, overlapping generations, assortative mating, many chromosomes, and few crossover suppressors. He also concludes that genetic transilience will be rare. Indeed, "[i]f speciation follows a founder event, it is more likely to be through adaptive divergence" than through transilience (Templeton 1981, p. 36). However, in organisms having certain life histories and population structures, transilience might be more common. Not surprisingly, such organisms are said to include the Hawaiian *Drosophila* (Templeton 1980a,b; 1981).

Theoretical Criticisms

Founder effect theories have provoked a great deal of criticism. We sketch these criticisms here. Because this literature is large and technical, we do not attempt a complete treatment. Instead, we focus on what seem to us the most important objections.

Perhaps the simplest objection to founder effect theories is that they are not necessary: adaptive radiation can explain most or all of the patterns attributed by founder-effect theorists to genetic drift. The profusion of species on archipelagos, for instance, may simply reflect a felicitous combination of allopatry and novel habitat. Rapid speciation might then result from strong natural selection in the absence of gene flow. Mayr's (1954a) argument that explanations involving selection are unlikely because environments are so similar between island and mainland is less than compelling. Even in his classic case of *Tanysiptera* in New Guinea, Mayr admits that the island "flora is somewhat different and the fauna is somewhat impoverished" relative to the mainland (Mayr 1954a, p. 168). Note, too, that the idea that island speciation is adaptive does not deny the occurrence of founder *events*—new islands might well be founded by one or a few individuals. It denies only a role for founder *effects*—reproductive isolation does not evolve by genetic drift but by natural selection.

It is also far from clear that speciation requires small populations. Mayr (1954a, 1963) and Carson and Templeton's (1984) claim that large populations are unable or unlikely to evolve or speciate seems to have little theoretical or empirical justification. If anything, larger populations should respond more readily to selection, as genetic drift is weaker and favorable mutations more abundant (Barton and Charlesworth 1984); indeed, there is good experimental evidence for greater responses to selection in larger populations (Weber 1990; Weber and Diggins 1990). Contrary to Mayr, epistasis does not change this. As Barton and Charlesworth (1984) emphasize, epistasis merely means that selection "sees" the fitness of alleles averaged over genetic backgrounds, where we assume some recombination. Moreover, even in largely asexual organisms, selection can routinely drive evolution (Lenski and Travisano 1994; Holder and Bull 2001), even in the face of strong epistasis (Barton and Charlesworth 1984). In any case, an enormous literature on adaptive clines and response to artificial selection (Yoo 1980; Hill and Caballero 1992) contradicts the notion that large populations suffer poor evolutionary potential.

Mayr's claims about the population-genetic effects of founder events have also been severely criticized. While there is no doubt that founder events can and sometimes do reduce additive genetic variance (Goodnight 1988; Whitlock and Fowler 1999; Whitlock et al. 2002), Mayr's intuitions about drastic genome-wide reductions in genetic variability following founder events seem incorrect. This point was first grasped by Lewontin (1965, p. 481), who argued that "[i]f there is a colonization by a single fertilized female, there will be a loss of genes and a radical change in the gene frequencies. But the one thing that will not happen is a profound change in the total amount of genetic variation available." Mayr's theory, in other words, ignores a distinction. It is true that founder events cause a loss of alleles, especially rare ones. But it is not true that brief bottlenecks cause a precipitous decline in measures of population variability like heterozygosity, additive genetic variance, and heritability. The reason is that, while many alleles do not make it into the founder population, those that *do* begin at intermediate frequencies. In the extreme case of a single inseminated foundress, each of four alleles at any locus starts at a frequency of 0.25. Because heterozygosity, additive genetic variance, and heritability all depend on allele frequency—and are maximized at intermediate frequencies—they remain much higher following a founder event than one might guess. And because short-term response to selection is proportional to heritability, a founder population's ability to respond to selection should not differ radically from the ancestral population's ability to respond. Lewontin's objection is clearly correct and Mayr's famous plot (see Figure 11.1) cannot be taken literally (see also Nei et al. 1975; Lande 1980).

Perhaps the most important objection to peak shift models is that the chances of such shifts are very small and, even if they do occur, they yield only trivial reproductive isolation. To see why, first consider a small population of constant size (e.g., a peripheral isolate), and imagine an adaptive landscape with two peaks. The peripheral isolate and the mainland population both currently sit at the same adaptive peak having mean fitness W_p, while the adaptive valley has lower fitness W_s. The probability of a peak shift is a function of the chance that the small population is found in the valley separating the peaks: a population cannot be drawn to a new peak unless it first drifts into a valley. It can be shown that the probability of a peak shift is proportional to $(W_s/W_p)^{2N_e}$ (Barton and Charlesworth 1984; Barton 1996). Thus, the deeper the valley, the smaller the chances of a peak shift. However, if we measure the strength of reproductive isolation between populations on the two peaks (i.e., the rate of gene flow between them), we find that it is *also* proportional to W_s/W_p. Thus, the deeper the adaptive valley, the less gene flow there is. The lesson is clear: while deeper valleys yield greater reproductive isolation, they are less likely to be crossed. In fact, the situation is even worse than this implies. The number of peak shifts required to yield a given amount of reproductive isolation varies inversely with the size of each step. But the chance of a step happening falls *exponentially* with its size. Taking both factors into account, isolation is more likely to result from many small peak shifts than from a few large ones (Barton and Charlesworth 1984; Barton 1996). Single genetic "revolutions" are unlikely.

Turning to population bottlenecks, the mathematics change but the lesson stays the same. The probability of a peak shift now depends mostly on the distance between peaks (Barton 1989, 1996). But the chance of a peak shift is constrained by two factors. First, populations are likely to experience a shift only if they are very genetically variable, spilling into the valley and so into the domain of attraction of the other peak. But such populations suffer large genetic loads (a decrease in fitness below the optimal value possible). Second, as populations become more variable, the chance of a peak shift does not keep increasing. Instead, maxima in the surface of mean fitness begin to merge into a *single* peak, preventing peak shifts altogether. The genetic variation required for peak shifts is thus constrained to a narrow range: too small and shifts are rare; too large and peaks fuse.

Carson, of course, argued that the chances of a peak shift increase because founder populations experience relaxed selection during bouts of explosive growth: genotypes that are usually deleterious can consequently persist. It is hard to know what to make of this argument. On the one hand, while the nature of selection might change during periods of rapid population growth, there is no reason to think it will *decrease*: one allele can still outcompete another allele during population growth (Barton and Charlesworth 1984). On the other hand, genotypes that would ordinarily perish in a population of constant size *can* persist in a growing population (Ewens 1979, p. 26; Otto and Whitlock 1997). But persistence is not equivalent to an increase in frequency and, in the end, there is no obvious reason why peak shifts should be more likely to occur in growing populations. At the very least, the burden of theoretical proof seems to rest on those making such claims.

There are also good reasons for questioning another assumption of Carson's model—that cycles of flushing and crashing occur *repeatedly*. Although Carson refers to such crashes as inevitable, they do not occur in most ecological models of approach to carrying capacity. Instead, it seems more likely that founder populations would stabilize at an equilibrium size, as in the logistic model.

Templeton's (1980b; Carson and Templeton 1984) claim that the chance of a peak shift increases given a small inbreeding effective size (N_{ef}) but large variance size (N_{ev}) has suffered even harsher criticism. Barton (1996, p. 788) charges that "[t]here is a fundamental mistake here" and that, while differences between N_{ef} and N_{ev} following a founder event are formally possible, they are short-lived. More fundamentally, Templeton's attempt to separate the two effects of genetic drift—decreased variation and increased homozygosity—cannot be sustained. Both follow from the same phenomenon: random change in allele frequency. The effects, consequently, are necessarily entangled (Barton and Charlesworth 1984; Barton 1989, 1996).

Recent Peak Shift Models

Problems with "classical" peak shift theories inspired a new generation of models. These models—due mainly to Gavrilets and Hastings (1996) and Wagner et al. (1994)—are more successful than their predecessors. For brevity, we consider only one, that of Gavrilets and Hastings (1996).

Gavrilets and Hastings identify a shortcoming in previous theories: most models focus on one of two ways of mapping fitness onto genotype—heterozygote disadvantage at a single locus (or chromosome), or disruptive selection acting on a trait underlaid by additively acting alleles. These scenarios share an important property: both require passage through the *worst* possible genotype when moving from one peak to another. This is not, however, a necessary feature of peak shifts. Instead, evolution might move along a ridge of intermediate genotypes of fairly high fitness. But following such an excursion, the derived population might find itself separated from the ancestral population by a deep valley.

Although this idea is related to the Dobzhansky–Muller model, there are differences. First, while Dobzhansky and Muller suggested that high fitness intermediate genotypes *might* exist, Gavrilets (1997, 1999) emphasizes that they are nearly inevitable. (This follows from Gavrilets's analysis of "holey adaptive landscapes," a network of related genotypes.) Second, while the Dobzhansky-Muller model emphasizes neutral or adaptive evolution, the Gavrilets and Hastings model emphasizes crossing shallow adaptive valleys. In both cases, populations end up separated by deep valleys, but the path there was unopposed by selection in the first case and weakly opposed in the second case.

Gavrilets (1997) claims that founder events substantially increase the rate of evolution among genotypes of similar fitness. This is easily seen in a three-locus model (Gavrilets and Hastings 1996). Variation at the A locus is maintained by heterozygote advantage. On this segregating A/a background, the B and C alleles are favored at the other two loci. But there is epistasis: in an entirely a population, the b and c alleles would be favored. The landscape thus has two peaks: $A/a\ B\ C$ and $a\ b\ c$. Hybrids between these peaks are unfit; indeed, all $B/b\ C/c$ individuals are lethal. Gavrilets and Hastings show that transitions from the $A/a\ B\ C$ peak to the $a\ b\ c$ peak occur at surprisingly high rates in founder populations. The reason is that there is a good chance the A allele will be lost from a small founder population, causing b and c to go to fixation. Any hybridization between ancestral and derived populations now yields inviable hybrids.

Gavrilets and Hastings claim that their mathematics captures Mayr's (1954a) verbal model in which alleles, once favored because they were "good mixers" on a heterozygous background, are replaced by alleles that function best on a homozygous background. This claim seems correct. Their work thus shows that founder-effect speciation is both possible and, under some conditions, reasonably likely.

Nonetheless, the model and its relatives have been challenged. The first objection concerns Gavrilets's scheme for penalizing hybrids: he typically assumes that fitness declines with greater heterozygosity (Gavrilets 1999; Turelli et al. 2001). But hybrid sterility and inviability usually result from *between*-locus incompatibilities, not from heterozygote disadvantage at single loci (Chapter 7). This problem is probably not fatal, however. Indeed, founder-effect speciation remains possible when postzygotic isolation is caused by Dobzhansky–Muller incompatibilities, not from heterozygote disadvantage (H. A. Orr, unpublished).

Barton (1996) leveled other criticisms at the new peak shift models. One is that reproductive isolation resulting from the evolution along ridges of high fitness tends to collapse under gene flow. The problem is that, unless reproductive isolation is complete, interspecific gene flow allows reconstruction of high fitness intermediate genotypes, promoting fusion of the two "species." This objection is not, however, specific to founder effect models. Instead, it applies to *any* model in which populations traverse genotypes of high fitness. Barton (1996) also emphasized that Gavrilets and Hastings require the ancestral population to segregate for variation affecting fitness and thus to suffer a genetic load. Once again, however, this would not seem fatal. Founder-effect speciation would not seem to make any greater demand on a population's ability to tolerate load than does, say, balancing selection, an idea that has obviously been taken seriously.

Finally, Barton (1996) argued that it is unclear why stochastic loss of variation would be more likely to trigger adaptive evolution than direct changes in the environment or epistatic responses to adaptive evolution at other loci. In other words, there is nothing special about the stochastic mechanism championed by Gavrilets and Hastings: *any* change in the physical or genetic environment can trigger a bout of adaptive evolution that yields populations separated by an adaptive valley. This is an important point. Thus, although one of the standard objections to founder effect models—that they do not work when translated into mathematics—now appears incorrect, the *other* standard objection—that the models seem unnecessary when compared to adaptive ones—still carries some weight.

To conclude, we believe that the question of founder-effect speciation has become primarily empirical. While the last twenty years of theory has shown that founder-effect speciation is usually unlikely (Turelli et al. 2001), certain varieties of the model *do* work on paper. The key question now is whether these models operate in nature. As we will see, the data provide little evidence that they do.

The Data

Evidence on the roles of selection versus drift in speciation comes from both the laboratory and nature. We first review laboratory experiments and then turn to data from the field.

Evidence from the laboratory

SELECTION. Experimental tests of the role of natural selection in speciation are straightforward. Most involve subjecting replicate lines to divergent selection (e.g., positive and negative geotaxis) and measuring the strength of reproductive isolation between the resulting lines. As noted in Chapter 3, these experiments are often successful. Prezygotic isolation in particular evolves remarkably often.

In perhaps the best of these experiments, Dodd (1989) studied eight lines of *Drosophila pseudoobscura* that derived from a single natural population. Half of the lines were reared on stressful starch medium and the other half on stressful maltose medium. After about one year of adaptation, Dodd (1989) showed that sexual isolation had evolved between lines reared on different, but not on the same, media. The results were remarkably clear, with all starch × maltose combinations showing positive assortative mating (many significantly) and none of the starch × starch or maltose × maltose combinations showing significant assortative mating. Kilias et al. (1980) obtained similar results in a long-term study of *D. melanogaster* (Chapter 3).

Such experiments show two things. First, reproductive isolation evolved as a byproduct of selection, not by genetic drift. Second, adaptation to different environments yields more reproductive isolation than does adaptation to a common environment (see also Florin and Ödeen 2002).

DRIFT. The theory of founder-effect speciation enjoys at least one great strength: it is eminently testable. The theory depends neither on many ecological details nor on enormous spans of evolutionary time. Instead, it is strictly demographic and allegedly rapid. As Carson (1971) emphasized, "If single founders are involved in some way in the formation of species it should be possible, through laboratory manipulations, to bring the process under close observational and experimental scrutiny." And as Templeton (1980b, p. 1026) added: "Genetic transilience is not an equilibrium phenomenon. If it occurs at all, it will probably occur during the initial few generations immediately after the founder event." Carson (1971) laid out the obvious test: "Competent single propagules can be obtained, an artificial flush and crash cycle induced and the results examined."

Powell (1978) performed the first such test by mixing many lines from several populations of *Drosophila pseudoobscura*, and then extracting founder lines. These founder lines were allowed to "flush," breeding in large numbers for over a year. Many lines were taken through three additional founder-flush cycles. Prezygotic isolation was tested among many possible combinations of founder-flush lines. Eleven of 36 combinations showed significant, though weak, prezygotic isolation. This isolation appeared, however, entirely due to the behavior of a few founder lines. Large outbred control lines showed no isolation; nor did lines that were inbred but not allowed to flush. No postzygotic isolation was observed. Dodd and Powell (1985) later re-tested many combinations of these same lines. They found that some showed prezygotic isolation and, more important, that these sometimes involved the same line combinations that had shown isolation several years before.

Ringo et al. (1985) performed a similar experiment in *D. simulans*. They extracted "drift lines" established by single-pair matings from a large outbred base population; these lines were then grown to several thousand individuals. Ringo et al. simultaneously established large selection lines that were artificially selected for several traits. They found that only one drift line evolved prezygotic isolation from the base population, while none of the selection lines did. Similarly, the drift but not the selection lines showed some postzygotic isolation from the base population. In general, however, reproductive isolation was weak and inconsistent through time.

Similar weak effects were seen in Meffert and Bryant's (1991) study of houseflies subjected to serial bottlenecks. Of 15 combinations of bottleneck × bottleneck and bottleneck × outbred control crosses, one showed significant positive assortative mating and one showed negative assortative mating. But neither effect was strong; indeed, neither remains significant after correcting for multiple contrasts. Weak and erratic results were also seen in Galiana et al.'s (1993) large experiment. Following many populations of *D. pseudoobscura* through seven founder-flush-crash cycles, they found that positive assortative mating arose somewhat more often than negative. While suggestive, prezygotic isolation was rare and unstable through time, with isolation persisting in one case only. Galiana et al. conclude that their results "do not support the claim that the founder-flush-crash model identifies conditions very likely to result in speciation events." Even less encouraging results emerged from Moya et al.'s (1995) work on *D. pseudoobscura* lines that were followed through 13 founder-flush-crash cycles over ten years. Performing many hundreds of prezygotic isolation tests, Moya et al. found only weak and temporally unstable reproductive isolation.

As if this were not enough, Rundle et al. (1998) argued that the above experiments had features predisposing them to yield reproductive isolation. For instance, individuals from different geographic populations were often pooled to form a base population, yielding artificially high initial variation and linkage disequilibrium (nonrandom association of alleles across loci). Experiments might therefore simply involve the sorting of sexual isolation already present between different geographic populations, a process not relevant to founder events (Barton and Charlesworth 1984). Second, populations were often new to the laboratory, so that any effect of a bottleneck might be confounded by adaptation to the laboratory, which itself might yield reproductive isolation. Third, experiments often involved a small number of bottlenecked lines. Finally, experiments typically involved repeated bottlenecks, a scenario that seems unlikely in nature.

To obviate these problems, Rundle et al. (1998) performed a more realistic experiment involving single bottlenecks of *D. melanogaster* populations that had a common geographic origin and that were long adapted to the laboratory. They also tested many founder-flush lines. At the end of their study, Rundle et al. found that prezygotic isolation values between the founder-flush lines and the outbred control population were roughly normally distributed about zero (Figure 11.3). Only one line showed significant prezygotic isolation (before multiple contrast corrections) and this effect disappeared upon re-testing. Variation in isolation values does not, therefore, appear biologically meaningful and instead represents sampling variation about zero. Given Templeton's (1999) complaint that *D. melanogaster* provides unpromising material for founder-effect speciation, Rundle (2003) essentially repeated the above experiment in *D. pseudoobscura* (although with two bottlenecks per line). Once again, bottlenecks had no effect on prezygotic isolation. This new experiment also tested for possible interactions between bottlenecks and novel selection pressures (involving new environmental conditions); there were none (see also Mooers et al. 1999).

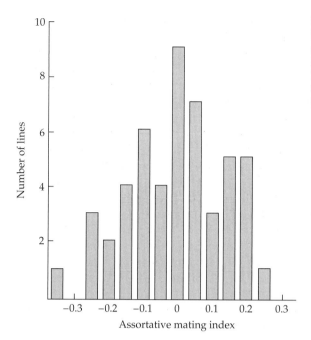

Figure 11.3 The distribution of prezygotic isolation between founder-flush lines and an outbred control population in *Drosophila melanogaster*. Variation is centered about zero and appears to reflect sampling error, not reproductive isolation. (After Rundle et al. 1998.)

In summary, bottleneck experiments provide little evidence for founder-effect speciation. Although reproductive isolation may arise sporadically in such work, it is almost always weak and transient. It is also worth noting that these experiments may have been unnecessary. *Drosophila* geneticists have, after all, inadvertently performed the requisite experiment thousands of times during the routine establishment of stocks: the *D. melanogaster* stock center at Bloomington, Indiana, for instance, maintains nearly 12,000 stocks and has sent out over 100,000 subcultures (though these lines are often inbred and related to each other). Needless to say, no new species are known to have emerged from this process.

Evidence from nature

Several lines of evidence from nature bear on the roles of selection versus drift in speciation. These include chromosomal and molecular analyses, comparative contrasts, and several case studies. We review each line of evidence below.

CHROMOSOMAL DATA. Chromosomal rearrangements once seemed to provide the best evidence for the role of genetic drift in speciation. Many rearrangements were thought to be strongly underdominant (as they render heterozygotes semisterile) and thus would be selected against when they arose. Nonetheless, many closely related species show fixed differences for such rearrangements, suggesting that genetic drift played a role in their fixation.

Indeed even such generally skeptical critics as Futuyma and Mayer (1980, pp. 262–263) concluded that "[i]t is not unreasonable to consider much of the corpus of cytogenetic data *prima facie* evidence that speciation occurs by the geographic isolation of small populations." Unfortunately, these data now appear more mixed.

On the one hand, there *is* good evidence for the role of genetic drift in the establishment of chromosome rearrangements in plants. As explained in Chapter 7, if the sterility of diploid hybrids involves rearrangements that are underdominant, *tetraploid* hybrids between the same species should be fertile, as all chromosomes have perfect pairing partners. This prediction is often upheld in plants. As also explained in Chapter 7, there is good evidence that centric fusions (which appear slightly underdominant) cause hybrid sterility between races of the European house mouse *Mus musculus domesticus*; these fusions may have accumulated by genetic drift.

On the other hand, there is little compelling evidence for the accumulation of underdominant rearrangements in animals other than mammals (Chapter 7). In particular, there is no compelling evidence that pericentric inversions (those that include the centromere) accumulate by genetic drift. As noted in Chapter 7, pericentric inversions were once thought to be universally underdominant as recombination within rearrangement heterozygotes would yield (aneuploid) gametes suffering duplications or deficiencies. Recent work in mammals and flies, however, reveals that many pericentric inversions are *not* detectably unfit as heterozygotes: surprisingly, chromosomes appear to align "straight" in heterozygotes, precluding recombination and thus the production of aneuploid gametes (Sites and Moritz 1987; Nachman and Myers 1989; Coyne et al. 1991a, 1993). Those pericentric inversions that differentiate species are unlikely, therefore, to be a random sample of new inversions; instead, they are almost surely those that had little or no effect on the fitness of heterozygotes.

MOLECULAR DATA. Molecular data allow several tests of the roles of selection versus drift. One of the oldest involves measuring levels of polymorphism in species that allegedly arose via founder effects.

It has often been claimed that, because the Hawaiian *Drosophila* speciated by founder effects, they have small effective population sizes (DeSalle and Templeton 1982; Ohta 1993) and so should show depressed levels of genetic variation. Despite this, we know of no strong evidence for reduced molecular variation in the Hawaiian *Drosophila*. On the contrary, these flies appear to show substantial variation. Rockwood et al. (1971) measured allozyme variation in more than 40 species of Hawaiian *Drosophila*. Although two species appeared monomorphic, mean heterozygosity across species was $H = 0.17$, as high as or higher than that seen in many continental species of *Drosophila*. Similarly, Sene and Carson (1977) found normal levels of allozyme heterozygosity and polymorphism in *D. silvestris* and *D. heteroneura*, two species from the Big Island of Hawaii.

DNA data also provide little evidence for decreased variation in the Hawaiian *Drosophila*. Sequence analysis of the *Adh* locus in *D. mimica*, for instance, shows that, while polymorphism levels are somewhat lower than in most *Drosophila*, the effective population size appears to be 1.1–1.7 million. Similarly, DeSalle et al. (1986) report considerable variability in mtDNA in both *D. silvestris* and *D. heteroneura*. And Hunt et al. (1989) report so much sequence and restriction site variation within five species of Hawaiian *Drosophila* that they lament that "finding fixed differences between close species is going to be very difficult" (see also Bishop and Hunt 1988).

Despite these findings, firm conclusions about founder-effect speciation are hard to come by. The problem is that neither positive nor negative results yield unambiguous interpretations. A decrease in molecular variation is consistent with either a founder effect or a founder *event* followed by adaptive radiation. Normal levels of variation, on the other hand, are consistent with either no bottleneck or a *single* bottleneck: theory shows that, while repeated bottlenecks substantially decrease heterozygosity, single bottlenecks often do not (Barton and Charlesworth 1984). Thus, while observed levels of molecular variation in Hawaiian flies provide no compelling evidence for founder-effect speciation, no stronger conclusion seems possible.

Somewhat firmer conclusions emerge from analysis of molecular variation in Darwin's finches (*Geospiza*) on the Galapagos archipelago. Vincek et al. (1997) document extensive variation at the major histocompatibility complex: at least 21 alleles segregate at the group 1 locus of *Mhc*. These alleles appear to coalesce to a most recent common ancestor about 15 million years ago—far too long ago to be consistent with a single founder on the islands, as the archipelago is only about 5 million years old. Computer simulations instead suggest that at least 30 founders settled the islands. (Errant flocks of birds—those that have presumably been blown out to sea—have often been observed [Vincek et al. 1997].)

Other molecular data have also been used to assess the roles of selection versus drift in speciation. Fitzpatrick (2002), for instance, showed that levels of both pre- and postzygotic isolation between allopatric species of *Drosophila* correlate more closely with allozyme than with silent DNA divergence. Because amino acid changes are almost certainly subject to stronger selection than are silent DNA changes (Gillespie 1991), Fitzpatrick concluded that reproductive isolation is a function not only of time, but also of selection in allopatry.

In a more direct molecular approach, one can ask whether divergence of the actual genes that cause reproductive isolation was driven by selection or drift. Answering this question requires use of statistical tests of DNA sequence polymorphism and divergence (e.g., the McDonald–Kreitman [1991] test). This approach has now been taken in a number of cases. As noted in Chapter 8, *OdsH*, *Nup96*, and *Hmr*—all of which appear to cause intrinsic postzygotic isolation between species of *Drosophila*—show strong evidence of rapid evolution driven by positive selection. And as discussed in Chapter 6, lysins and bindins in certain marine species show spectacularly elevated rates of replacement substitution, a pattern that clearly reflects positive selection. Such molecular population-genetic analyses could (and should) be replicated over many genes causing reproductive isolation as they are discovered. Though one should not over-interpret such findings—as we have seen, peak-shift models of speciation *also* involve a phase of positive selection—results that consistently point to adaptive evolution would at least close the window somewhat on the role of neutral evolution in speciation (Kirkpatrick and Ravigné 2002).

COMPARATIVE DATA: NATURAL SELECTION. The roles of selection versus drift in speciation have also been tested by comparative contrasts. Although these contrasts provide some evidence for the role of natural selection in speciation, that evidence is less impressive than one might expect. The comparative evidence for *sexual* selection, on the other hand, is stronger. We offer a possible reason for this difference below.

One of the problems facing comparative analyses is that many contrasts that first appear plausible are seen, on closer inspection, to be flawed. The classic example involves the comparison of species richness on islands versus continents. While founder effect theories predict explosive speciation on archipelagos so, of course, do adaptive theories: any sensible theory of ecological speciation predicts that invasion of open niches will increase speciation rates. To take a subtler example, Schluter (2000) argued that, if adaptation to novel habitats drives island speciation, the rate of ecological speciation on islands should increase while the background rate of nonecological speciation should not. He further suggested that the rate of polyploid speciation could serve as a proxy for the rate of nonecological speciation. He then noted that, despite explosive speciation, there are no known cases of polyploid speciation on the Hawaiian archipelago. But this analysis is flawed. The presumably constant "background" rate of polyploid speciation almost surely plummets on islands due to a paucity of closely related species with which to hybridize (where we assume that many polyploid events are allopolyploid).

Unfortunately, even more careful contrasts are sometimes hard to interpret, as we will see. Perhaps the most careful of comparative tests involves sister group contrasts. These contrasts allow us to ask if groups possessing a feature that, by hypothesis, leads to greater speciation rates are in fact more species rich than sister groups lacking the feature. Sister group comparisons automatically control for the age of clades since sister taxa are, by definition, equally old. Although we systematically review sister group analyses in Chapter 12, it is worth describing several studies here.

One of the best-known studies asks whether species richness is associated with plant-feeding in insects. Strong et al. (1984) first noted that, while phytophagy is largely limited to 9 of 30 orders of insects, roughly half of all insects are phytophagous. (Over 400,000 phytophagous insect species and 300,000 green land plant species have been described [Mitter et al. 1991].) Though suggestive, this numerology relies on fairly arbitrary taxonomic divisions. Mitter et al. (1988), however, performed a sister group analysis, showing that, in 11 of 13 contrasts, phytophagous clades include more species than do non-phytophagous clades; these differences are also often large, typically two-fold or more. Although this analysis was somewhat controversial, later studies (e.g., Farrell and Mitter 1993; Farrell 1998) largely confirmed the association between phytophagy and species richness.

While this finding is straightforward, its interpretation is not. One possible interpretation is that a phytophagous lifestyle involves confronting, and adapting to, new hosts over evolutionary time, a process that encourages indirect speciation. Another interpretation is that herbivores might have fragmented populations, reflecting the patchy distribution of host plants (Mitter et al. 1988; Funk 1998). The association between phytophagy and species richness might, in other words, reflect the role of *genetic drift* in speciation.

As Funk (1998) emphasized, demonstrating a role for selection requires a more careful contrast. One must compare insects that are adapted to *different* host species with those that are adapted to the *same* host species. Because the two groups presumably have similarly fragmented populations, insects adapted to different hosts should show greater species richness than those adapted to the same host only if adaptation to new hosts (not genetic drift) drives speciation. To test this prediction, Funk (1998) compared the strength of reproductive isolation between populations of the leaf beetle *Neochlamisus bebbianae* adapted to the same host plant (*Acer*) versus different host plants (*Betula* or *Salix*). He showed that populations adapted to different hosts show greater sexual isolation in the laboratory than those adapted to the same host.

The pattern found by Funk is not, however, universal. Populations of pine beetles that use different hosts, for example, show no pre- or postzygotic isolation (Langor et al. 1990; for other examples, see Funk 1998). Although there is no reason to think that adaptation to different hosts should *always* yield reproductive isolation, contrasts between different-host groups and same-host groups do not always reveal even a *statistical* association with species richness. Markow (1991), for instance, showed that populations of cactus-feeding *Drosophila mojavensis* adapted to different cactus species exhibit *less* sexual isolation than those adapted to the same cactus species. Sexual isolation instead seems connected to geography. The problem is that such a finding does *not* rule out a role for selection in speciation. Instead, as Markow (1991) notes, greater reproductive isolation between certain populations of cactus-feeding flies might well reflect a history of selection for desiccation resistance in some populations. The point is that all else may not be equal across different-host versus same-host contrasts.

Studies of between- versus within-host speciation represent a special case of a broader comparative test, the "parallel speciation" test. According to Rundle et al. (2000, p. 306), parallel speciation

> is a special form of parallel evolution in which traits that determine reproductive isolation evolve repeatedly in independent, closely related populations as a by-product of adaptation to different environments... The outcomes are reproductive compatibility between populations that inhabit similar environments and reproductive isolation between populations that inhabit different environments.

An intriguing case involves the freshwater amphipod *Hyalella azteca* (McPeek and Wellborn 1998). This species (or species complex) consists of two ecotypes: large- and small-bodied. Each lake usually harbors a single ecotype. The sorting of ecotypes into lakes appears adaptive: the large ecotype is found in lakes lacking fish while the small ecotype is found in lakes with fish. Phylogenetic analysis shows that large and small ecotypes do not form two separate clades, and McPeek and Wellborn (1998) thus speculate that large and small ecotypes may have evolved multiple times. Laboratory hybridizations show that crosses within ecotypes generally proceed well—whether individuals come from the same or different lakes—while crosses between ecotypes never proceed, even when involving genetically closely related or geographically nearby populations (McPeek and Wellborn 1998). Although more work is needed, the fact that reproductive isolation neatly maps onto ecotype, not onto genetic or geographic distance, suggests that natural selection had a hand in the evolution of this reproductive isolation.

A similar pattern appears among the limnetic and benthic ecomorphs of the threespine stickleback found in small lakes in coastal British Columbia. As discussed in Chapter 5, these fish are derived from the marine stickleback, which founded freshwater lakes as recently as 12,000 years ago (Schluter and McPhail 1992; McPhail 1994). Within lakes, benthic and limnetic forms show strong reproductive isolation. The morphs mate assortatively and F_1 hybrids suffer (probably extrinsic) postzygotic isolation and a likely mating disadvantage. Because phylogenetic analysis of mtDNA and microsatellite loci suggests that benthics and limnetics do not form distinct monophyletic clades (Taylor and McPhail 1999, 2000), it appears that each ecomorph evolved multiple times. Most important, mating tests reveal no significant prezygotic isolation *within* ecomorphs among lakes but significant prezygotic isolation *between* ecomorphs from different lakes (Rundle et al. 2000).

Finally, Nosil et al. (2002) studied populations of the American walking-stick *Timema cristinae* that use two different host plants. These populations differ in the frequency of two genetically based color morphs: the morph that is most common on each host plant is the one that is most cryptic on that plant. Walking-sticks that use different host plants also differ in host preference, body size, and certain behaviors. Both mtDNA and nuclear DNA indicate that populations using the same host plant do not form a monophyletic group; instead, host specialization has apparently evolved multiple times. Nosil et al. show that individuals from different host plants display significantly greater prezygotic isolation than those from the same host plant, across many combinations of populations. They conclude that parallel speciation probably occurred.

While studies of parallel speciation are suggestive, they are not always conclusive. The main worry is that gene flow might yield incorrect phylogenies that falsely imply a history of parallel speciation. To see this, consider the case in which each ecomorph had a *single* origin, precluding the possibility of parallel speciation. The problem is that gene flow between different ecomorphs within a locality might obscure the signal of monophyly, giving a false impression that each ecomorph had multiple origins. This problem disappears only if no gene flow occurred between ecomorphs, but we will rarely know if this is true: even if reproductive isolation is currently complete it may not have been in the past. As we have emphasized, mtDNA data are particularly vulnerable to the effects of gene flow. Although data from nuclear DNA in sticklebacks show that phylogenies having multiple origins of ecomorphs are *significantly better* than phylogenies having single origins (Taylor and McPhail 2000), we know of no other study of parallel speciation that rises to this high—and, in our opinion, necessary—evidentiary standard. (Indeed, in the walking-stick case, there is no evidence that any of the "adaptive" characters distinguishing host plant populations is genetic, except color; and prezygotic isolation does not involve color [Nosil et al. 2002, 2003].)

Several other comparative tests of the role of natural selection have been performed and, like some of the above, they often yield frustratingly weak results. Barraclough et al. (1999a), for instance, tested whether ecological differences played a part in the speciation of North American tiger beetles of the subgenus *Ellipsoptera*. In particular, they asked if changes in habitat mapped more often to the tips of the *Ellipsoptera* phylogeny than expected under a null model, as one would predict if such changes played a causal role in speciation. The answer is no. Barraclough et al. also failed to find any association between ecologically important characters like body size, mandible length, and the origin of new species.

COMPARATIVE DATA: SEXUAL SELECTION. Comparative evidence for the role of sexual selection in speciation is far more compelling. A number of studies have found a correlation between indicators of sexual selection and species richness (reviewed by Barraclough et al. 1999b). Because we describe these studies in Chapter 12, we merely list them here: using sister taxa comparisons, Barraclough et al. (1995) showed that sexual dichromatism (a presumed proxy for sexual selection) is correlated with species richness in passerine birds. Similarly, Mitra et al. (1996) showed that bird groups with promiscuous mating systems (also a presumed proxy for sexual selection) are more speciose than sister groups with non-promiscuous mating systems. Moreover, Arnqvist et al. (2000) showed that polyandrous insect clades (in which females mate multiply, allowing postmating sexual selection, e.g., sperm competition) are more speciose than monandrous insect clades (in which females mate once).

Although these studies measure species richness—which reflects both speciation and extinction rates—it seems unlikely that taxa experiencing more intense sexual selection experience lower extinction rates. If anything, sexual selection might *increase* extinction rates, making these contrasts conservative, as Mitra et al. 1996 emphasized. These studies thus suggest a causal connection between sexual selection and speciation. Often enough, then, sexual selection alone, unaided by ecology, appears to drive speciation. We believe that this represents one of the most important findings in the last decade of work on speciation (Chapter 12).

It is worth asking why comparative tests of sexual selection have been more successful than those of natural selection. Although we do not know the answer, we suggest that it might reflect a difference in the "dimensionality" of natural versus sexual selection. Comparative tests of either form of selection require the construction of "more-selection" and "less-selection" classes. But while it is fairly easy to find sensible proxies for the intensity of sexual selection (e.g., the promiscuity of mating systems), it is harder to find good proxies for the intensity of natural selection. The reason is that organisms are simultaneously adapting to many factors, both abiotic and biotic. Thus, in any particular contrast (e.g., different-host versus same-host), individuals belonging to what appears to be the less-selection class may well experience intense selection for other, orthogonal features of the environment, features that might *also* drive speciation. Given the multifarious ways in which natural selection can act, such contrasts may be less decisive, in general, than those that test for the signature of sexual selection.

CASES STUDIES: NATURAL SELECTION. Although we have repeatedly argued that comparative analyses are more compelling than case studies, the opposite admittedly appears true in the case of selection versus drift: some of the best evidence for the role of natural selection in speciation derives not from large comparative contrasts but from careful studies of particular systems in the wild. While Carson (1971, p. 68) once claimed that "[d]irect observation provides no clue as to whether the two sets of genetic differences [those leading to adaptation vs. those leading to reproductive isolation] are in fact correlated," in some cases this is clearly wrong. In some cases it is simply obvious that reproductive isolation reflects a history of adaptation. As we emphasized at the start of this chapter, this is often especially clear with extrinsic postzygotic isolation. Because we reviewed extrinsic isolation in Chapter 5, here we describe several case studies that involve other reproductive isolating barriers. We begin with prezygotic isolation.

In plants, the clearest evidence for the role of selection in prezygotic isolation involves pollinator isolation. *Mimulus lewisii* (primarily bee-pollinated) and *M. cardinalis* (primarily hummingbird-pollinated), for instance, have radically different floral morphologies. There is no doubt that many of the floral features of each species are adaptations to its preferred pollinator: changing such features profoundly affects pollinator visitation in the wild (Schemske and Bradshaw 1999). There is also no doubt that these pollinator adaptations cause strong prezygotic isolation in sympatry. Schemske and Bradshaw (1999) showed that where *M. lewisii* and *M. cardinalis* species co-occur in the Sierra Nevada mountains, bees are the sole visitor to *M. lewisii* and hummingbirds are the primary (97%) visitor to *M. cardinalis*. It is hard to believe that this situation is rare. Indeed, there is good evidence that adaptation to animal pollinators often increases the species richness of plants (Chapter 12).

In animals, some of the best evidence for the role of selection in prezygotic isolation comes from the threespine stickleback. The limnetic and benthic morphs differ in body size, which plays a key role in reproductive isolation: mating tends to occur only between the largest individuals of the small morph (limnetic) and the smallest individuals of the large morph (benthic). Indeed, species hybridization is roughly restricted to males and females of the same length (Nagel and Schluter 1998). There is strong evidence that these size differences reflect adaptation to foraging in different habitats (Nagel and Schluter 1998).

The role of natural selection in speciation also seems clear in *Heliconius melpomene* and *H. cydno*, two closely related and often sympatric butterfly species that mimic different lepidopteran taxa. The first species is often black, red, and yellow, while the second species is black and white. Interestingly, these mimetic differences play a large part in assortative mating between sympatric *H. melpomene* and *H. cydno* (Jiggins et al. 2001b; Naisbit et al. 2003). Unless one is willing to believe that mimicry evolved without the aid of natural selection, this case provides strong evidence for the adaptive origin of reproductive isolation.

Though not as compelling as the above, some case studies also suggest that *postzygotic* isolation evolves by indirect selection. Macnair and Christie (1983), for instance, studied hybrid inviability between two populations of the yellow monkeyflower, *Mimulus guttatus*. The Copperopolis population is resistant to copper contaminated soil (the population resides on the tailings of a California copper mine); conversely, the Cerig population (in Wales) is susceptible. F_1 and certain backcross hybrids between these populations are inviable. Macnair and Christie concluded that this inviability involves a simple Dobzhansky-Muller incompatibility and argued that (1) a single allele from Copperopolis is involved; and (2) the relevant locus cannot be separated from the copper tolerance locus. Given that the Copperopolis population is probably genetically similar to its conspecifics—the mine is only 100 years old—and that *M. guttatus* has 14 linkage groups, it seems unlikely that the association between copper tolerance and hybrid inviability is fortuitous. Instead, Macnair and Christie suggested that natural selection drove fixation of the copper tolerance allele and that this allele or something tightly linked to it (which perhaps hitchhiked to fixation) causes reproductive isolation when placed on a Cerig genetic background. Although this system could benefit from QTL analysis, postzygotic isolation here probably reflects adaptation.

Similarly, as we noted in Chapter 9, there is good evidence that the origin of the sunflower *Helianthus anomalus*, a diploid recombinational hybrid, involved natural selection. Crosses between the parental species, *H. annuus* and *H. petiolaris*, yield artificial hybrid lineages that are nearly identical in genotype to each other as well as to the natural hybrid *H. anomalus* (Rieseberg et al. 1996). The fact that hybrid lineages repeatedly arrive at such similar genotypes clearly reflects selection, not chance. Moreover, there is good evidence that this selection involves pollen viability (Chapter 9).

CASE STUDIES: GENETIC DRIFT. Ignoring chromosomal speciation in plants and mammals (which we have already discussed), we know of only one kind of case study that provides reasonable evidence for speciation by genetic drift.

Small snail populations are sometimes found that show a reversal of the usual direction of shell coiling (Gittenberger 1988; Orr 1991b; VanBatenburg and Gittenberger 1996; Ueshima and Asami 2003; our Chapter 3). Because snails of opposite coil have difficulty copulating with each other, these populations are reproductively isolated from nearby ones and so appear to represent new species. Since reverse-coiling individuals are selected against when rare, genetic drift likely played a role in the spread of new coiling types. Theory shows that such peak shifts are facilitated by a combination of unusual circumstances: single-locus control of coiling, maternal effects on coiling, and very small local population sizes (Orr 1991). This case does not provide good reason, therefore, for thinking that peak shift speciation is common.

Conclusions

There is now considerable evidence that natural and sexual selection play a role in speciation. This evidence derives from a wide variety of studies: laboratory experiments on divergent selection, molecular analyses, comparative contrasts, and, perhaps most important, case studies. In contrast, firm evidence for the role of genetic drift in speciation is rare. Indeed, it is hard to point to any examples other than chromosomal speciation in plants and mice, and unusual cases of shell coil-reversal in snails. Most important, direct laboratory tests provide little or no support for founder-effect speciation. It appears, then, that at least one important debate has been settled: selection plays a much larger role in speciation than does drift. It is also worth noting that genetic drift appears to play little part in morphological evolution (Coyne et al. 1997).

Finally, we note that Eldredge and Gould's (1972) controversial theory of punctuated equilibrium relied upon founder-effect speciation. While a pattern of punctuated change might well characterize the fossil record, the evidence summarized above suggests that any founder-effect explanation of this pattern is untenable. This point was recently conceded by Gould (2002).

12

Speciation and Macroevolution

So far we have concentrated on the evolutionary forces and genetic changes that produce new species, and on the search for patterns that might reveal how speciation happens. Such patterns might also be derived from macroevolutionary data on *groups* of species diversifying over time. In this final chapter we consider several related topics: (1) rates of species formation within clades; (2) attempts to correlate these rates with biological features of organisms that might promote speciation (key characters); and (3) species selection—the controversial idea that macroevolutionary patterns can result from selection operating not on individuals, but on species.

Rates of Speciation

Estimating rates of speciation can answer questions beyond simply determining how long it takes for a new species to arise. Different evolutionary forces or biogeographic scenarios, for example, might produce species at different rates. Sexual selection, sympatric speciation, and founder-effect speciation are often presumed to produce new taxa quickly. Adaptive radiation, involving natural selection acting on groups that invade a new habitat, is said to yield initially high rates of speciation that decrease as niches become filled (Schluter 2000). The evolution of novel traits or behaviors, such as flight, resembles adaptive radiations in allowing the invasion of empty "adaptive zones," yielding initially high rates of speciation. These scenarios predict that speciation rates will change over time, which can in principle be tested with data from the fossil record.

We first discuss various definitions of "speciation rate," describing the three rates used most often. We concentrate on one: the "biological speciation rate." We then summarize predictions that different theories make about rates of speciation. Finally, we give the data: estimates of speciation rates from taxonomic, molecular, and paleontological studies. We conclude that, while one can estimate biological speciation rates in a few groups, there are currently too few data to do this in most taxa.

What is a speciation rate?

Since we consider species to be taxa showing substantial reproductive isolation from other taxa, the ideal speciation rate is the average rate at which one species branches to produce two groups that are reproductively isolated. We call this the *biological speciation rate* (BSR), as it rests explicitly on the biological species concept. This rate is expressed as the number of new species produced per existing species per time period (typically a million years). By taking the inverse of this fraction, we obtain the mean waiting time for biological speciation: the average period elapsing between the origin of a new lineage and the next branching event in that lineage. We prefer to use this waiting time, which we call the *biological speciation interval* (BSI), as the metric for speciation rates: BSIs are usually expressed in years and are easy to understand.

It is important to distinguish the BSI from a different speciation interval: the *transition time for biological speciation*. In contrast to the BSI, which measures the interval between two branching events, the transition time measures the time required to evolve strong reproductive isolation *once the evolution of that isolation has begun*. The speciation interval begins when a new lineage is formed, but the transition interval starts when the first mutation arises that ultimately yields some reproductive isolation. (For allopatric speciation, this might be the time when gene flow is interrupted by geographic isolation: in many cases geographic isolation must be closely followed by the onset of reproductive isolation.) Transition times must be calculated post facto—only after a new species is formed—because the evolution of reproductive isolation may not go to completion. Such times can be estimated as the *minimum* divergence times seen between good sister species. The distinction between transition times and speciation intervals is equivalent to molecular evolutionists' distinction between the transit time for fixing a new mutation and the interval between successive fixations (1/substitution rate).

Transition times will usually be shorter than speciation intervals because lineages do not begin to branch immediately after they arise. (There are a few exceptions. A widespread species, for example, may produce many peripheral populations, each of which attains reproductive isolation very slowly but nearly simultaneously. Here, the average transition time may be longer than the speciation interval because several species arise at once.) The most extreme differences between speciation intervals and transition times occur in forms of speciation that involve crossing adaptive valleys, such as founder-effect or chromosomal speciation (Barton and Charlesworth 1984; Lande 1985b). In populations of moderate size, for example, one must wait millions of generations for an underdominant chromosomal rearrangement to arise and become fixed.

Yet those rare arrangements that become fixed and cause reproductive isolation do so very quickly (Lande 1985b). Similarly, the transition time for speciation by allopolyploidy may last only a few generations, but the interval for polyploid speciation may be much longer: the latter also includes the waiting time until the parental species hybridize.

Because we cannot usually observe when speciation begins, it is easier to measure BSIs than transition times. Data on speciation taken from the fossil record or from a combination of the fossil record and the number of extant species (see below) yield estimates of speciation intervals, not transition times.

The most easily measured—and most common—alternative indices of speciation rates are the *net diversification rate* and the *net diversification interval* (NDR and NDI, respectively). NDR is simply the change in the number of surviving lineages per unit time, and NDI is its reciprocal. These indices are usually calculated by estimating the age of the common ancestor of a group using either a molecular clock or fossil data, and then calculating the NDR from the number of extant species derived from that common ancestor. Net diversification rates thus measure the change in biodiversity, and are almost always positive since one usually deals with a single ancestor and its multiple descendants.

The ease of estimating NDIs and NDRs is outweighed by their inaccuracy. The most serious problem is that diversification results from two processes: speciation and extinction. NDRs estimate the difference between speciation rates and extinction rates. Since extinction rates may be impossible to obtain, the resulting metrics are usually poor estimates of speciation rates.

This problem is dealt with in the third pair of indices, the *apparent speciation rate* and its inverse, the *apparent speciation interval* (ASR and ASI, respectively). These involve estimating speciation rates by subtracting an estimate of the extinction rate from the easily measured net diversification rate. Estimates of extinction rates can be based on data from the fossil record or, less reliably, on assumptions about the possible range of values. Ideally, ASRs should be identical to BSRs. In practice, though, they will differ because ASRs involve estimating extinction not of biological species, but of morphologically distinguishable taxa. The problem is compounded if one estimates terminal biodiversity from fossil data rather than extant species. The number of fossil species at a given time may differ markedly from the number of biological species because of the incomplete fossil record, the likelihood that some reproductively isolated taxa cannot be distinguished as fossils, and the possibility that morphologically distinguishable populations of fossils may not be biological species.

Theory and speciation rates

Speciation rates produced by evolutionary models are usually transition times. In this respect they underestimate biological speciation intervals. On the other hand, almost all theories deal with a single isolating barrier. Since speciation usually involves a combination of isolating barriers, such theories can overestimate speciation intervals. Finally, different models of even the same form of speciation may make different assumptions, yielding speciation rates ranging from nearly instantaneous to extremely slow.

Our objective in this discussion is not to judge the biological reality of models of speciation, but to determine whether they yield BSRs that are relatively fast or slow. We would like to know these rates because they have been used to deduce the biogeography of speciation (e.g., McCune and Lovejoy 1998). For convenience, then, we group the models by biogeography. Unfortunately, because most of these models yield transition times and not BSIs, they are hard to compare with real data.

ALLOPATRIC SPECIATION. As one might expect from verbal models, allopatric speciation resulting from divergent natural or sexual selection can be either rapid (as when populations invade a new habitat) or slow (as when a geographic barrier divides a uniform habitat). Unfortunately, at present verbal theory must suffice. There are almost no genetically based models of how fast natural selection causes allopatric speciation, and existing models not only consider a single isolating barrier, but also depend heavily on unknown parameters (e.g., Orr and Turelli 2001).

Sexual selection in allopatry can theoretically yield very short transition times—between several hundred and several thousand generations, depending on initial assumptions about the costs of female preference and the mechanism of sexual selection (Pomiankowski and Iwasa 1998; Gavrilets 2000b). But BSIs may be much longer owing to prolonged waiting times until speciation begins.

Allopatric models involving genetic drift in large populations are, as expected, extremely slow (Nei et al. 1983). In smaller populations, however, founder-effect models involving a combination of drift and selection can yield rapid transition times. Indeed, in most of these models transition times *must* be short, for intermediate states are unstable and reproductive isolation must be achieved quickly if it is achieved at all. Such models usually include shifts between adaptive peaks involving genes or chromosome arrangements (Walsh 1982; Lande 1985a,b; Gavrilets and Hastings 1996). As with all peak-shift models, waiting times can be very long. In Lande's (1985a) model of speciation by movement between adaptive peaks, transition times can be as short as 100 generations, but BSIs can be as long as a million generations. Gavrilets and Hastings's (1996) two- and three-locus models of speciation by drift and selection yield probabilities of speciation *and* estimates of transition times, which in principle allow us to predict BSIs. These can be as short as a few hundred generations.

Except for models involving only genetic drift, then, allopatric speciation can occur quickly, yielding short transition times and BSIs. There is thus no theoretical support for the notion that allopatric speciation must be slow. Moreover, in some cases transition times may be close to speciation intervals, for genetic divergence by natural or sexual selection may begin as soon as populations are geographically isolated, and reproductive isolation can evolve hand in hand with this divergence. This congruence will be especially close when populations invade a novel habitat.

REINFORCEMENT. We expect that when speciation involves reinforcement, it will usually occur faster than pure allopatric speciation, as reproductive isolation during reinforcement is accelerated by both natural selection against hybrids and sexual selection that increases mate discrimination. Indeed, recent models show that reinforcement can complete speciation within a few hundred to a few thousand generations after populations attain secondary contact (Liou and Price 1994; Kelly and Noor 1996). These are transition times, calculated only for cases in which speciation is completed. Coyne and Orr (1997) show that the effect of sympatry on accelerating speciation may be considerable, at least in *Drosophila*. As suggested above, their data are best regarded as minimum estimates of BSIs. For *Drosophila* species whose ranges overlap, speciation is completed in about 200,000 years, while speciation in allopatric taxa requires roughly 2.7 million years. Both figures are surely overestimates as they are based on only two isolating barriers.

SYMPATRIC AND PARAPATRIC SPECIATION. Nearly all models show that sympatric speciation is fast (but see Bolnick 2004). Models involving sympatric divergence by sexual selection resemble peak-shift models in that the intermediate state of speciation is unstable and so transition times must be short. However, in many cases the *probabilities* of speciation might be low because the required initial conditions are unlikely or unknown. Thus, although some rates of sympatric speciation are characterized as "waiting times" (e.g., Dieckmann and Doebeli 1999), these are actually waiting times only when the proper ecological conditions and mutations are in place.

As noted above, the transition time for polyploid speciation, which is sympatric, lasts only a few generations, but the BSI for polyploidy can be much longer. Using known genetic pathways for creating polyploids, estimates of the frequency with which unreduced gametes are formed, and observations of the fertility of individuals involved in polyploid speciation, Ramsey and Schemske (2002) calculated that the proportion of autopolyploid individuals formed each generation is roughly 10^{-5}. In theory, *allopolyploid* individuals appear less often than this unless hybridization is fairly frequent—on the order of a few percent. However, it is impossible to convert these rates to BSIs without knowing the probability that a newly formed polyploid will establish itself as an interbreeding population. As discussed in Chapter 9, this probability may be small.

Recombinational hybrid speciation may also be rapid. In a simulation model by Buerkle et al. (2000), diploid hybrid speciation not only occurred frequently, but also began soon after initial contact between the parental species (within 10–500 generations). Moreover, once the first hybrids were formed, speciation was completed within about 50 generations. In this case, the BSI for recombinational speciation would be long only if there were extended intervals with no contact between potentially hybridizing species.

Many other models of sympatric or parapatric speciation yield longer—but still brief—transition times. These models include parapatric speciation by genetic drift or natural selection (Gavrilets 2000b; Gavrilets et al. 2000), parapatric and sympatric speciation via sexual selection (Lande 1985a; Turner and Burrows 1995; Lande et al. 2001; Gavrilets 2002), sympatric speciation via habitat selection with "no-gene" assortative mating (Rice 1984a), and sympatric speciation via disruptive selection (Dieckmann and Doebeli 1999; Kondrashov and Kondrashov 1999). Transition times in these models are on the order of several hundred to several thousand generations. In all of these models, however, speciation intervals again depend on the likelihood of the initial conditions and on the chance that speciation will occur if these conditions are met.

CONCLUSION. In theory, speciation can occur rapidly whether it is sympatric, parapatric, or allopatric, and whether it is caused by genetic drift, natural selection, or sexual selection. However, the transition times for speciation are nearly always short when the process involves either a sympatric phase (i.e., pure sympatric speciation, allo-sympatric speciation, or reinforcement) or crossing an adaptive valley. This is because both processes include an unstable intermediate stage that must be traversed rapidly. In contrast, transition times for allopatric speciation can be long: there is usually no unstable intermediate phase, and the environments of geographically isolated populations may be similar.

Despite these observations, we cannot conclude, at least from theory, that cases of rapid speciation must have occurred sympatrically. This is because we do not know how often the assumptions of sympatric-speciation theories hold in nature—information that is required to estimate BSIs. Moreover, speciation rates estimated in nature are more likely to approximate BSIs than the transition times predicted by most models. Nor can one conclude that rapid speciation on islands implies founder-effect processes, because rapid speciation can also occur by strong natural or sexual selection.

The main conclusion is thus rather depressing: it is unsafe to infer biogeographic or genetic modes of speciation from rates of speciation. Beyond this, we can make only one prediction: for a given divergence time, the amount of reproductive isolation between taxa with overlapping ranges should be greater than that of completely allopatric taxa. At least some of the former cases should involve reinforcement, which is rapid and—if there is genetic variation for increasing prezygotic isolation—begins to evolve as soon as populations achieve secondary contact.

Calculating speciation intervals

Here we explain how to calculate the three main forms of speciation rates and provide a sample of rates estimated from data on fossil and living species. Our goal is to get an idea of biological speciation intervals in different groups, and to determine if there are any broad patterns (e.g., do plants and animals speciate at different rates?). We begin with the rates that are easiest to measure.

NET DIVERSIFICATION RATES AND INTERVALS. NDRs and NDIs describe the rates at which surviving lineages are produced over time. These estimates depend on the assumed pattern of speciation. In a "comb phylogeny" (Figure 12.1A), all species are formed as branches off a single persisting lineage. This might occur, for example, when a mainland species produces a number of descendant species through repeated island colonization, while these descendants do not themselves speciate. To calculate the NDR from a comb phylogeny, one assumes that n extant species are formed over time period t from a single common ancestor. Thus NDR equals $(n-1)/t$, and the NDI is its reciprocal, $t/(n-1)$. This model seems unrealistic, and biologists have concentrated instead on more "balanced" phylogenies (Figure 12.1B), in which all lineages have roughly equal probabilities of branching in a given interval. If branching occurs at the same rate in comb and balanced phylogenies, the latter will produce species much faster. Real phylogenies are undoubtedly a mixture of these two extremes, but our intuition suggests that more balanced phylogenies are closer to the situation in nature.

NDRs and NDIs are calculated using the "pure birth model" introduced by Yule (1925) to compute speciation rates, later adopted by ecologists to model population growth. This model assumes that in each small time interval, each species has the same probability of producing a new species: D (the diversifi-

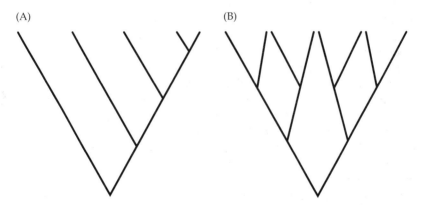

Figure 12.1 Two extreme phylogenies. (A) "Comb" phylogeny, in which new species are produced by repeated branching from a single lineage. (B) "Balanced" phylogeny in which each lineage itself divides at the same rate. If branching rates for each lineage in a balanced phylogeny are identical to the branching rate of the single lineage in a comb phylogeny, the balanced phylogeny will produce species much faster, yielding a shorter average speciation interval. In this example, branching of the single lineage in (A) occurs at the same rate as branching of each lineage in (B), yielding five species in the comb phylogeny and eight species in the balanced phylogeny.

cation parameter). The expected number of species thus increases according to $dN/dt = DN$, where N is the existing number of species. Starting at time zero with N_0 species, the number of species at time t, N_t, is $N_0 e^{Dt}$. The number of species thus grows exponentially.

In reality, the diversification parameter D equals $S - E$, the instantaneous rate of biological speciation (S, which is equal to the BSR) minus the instantaneous rate of true extinction (E). In this model, these values are also assumed to be constant among lineages. Thus, if we knew t, N_t, N_0, D, and E, we could estimate the biological speciation rate S. E, however, is usually unknown. For convenience, biologists often assume that there has been no extinction. Rearranging the above equation then gives D (the NDR) as $(\ln N_t - \ln N_0)/t$. Assuming $N_0 = 1$, (i.e., the phylogeny involves a monophyletic group beginning with a single ancestor), the maximum-likelihood estimate of D is $(\ln N_t)/t$. The net diversification interval ($1/D$) thus becomes NDI = $t/\ln N_t$.

NDIs are customarily expressed in millions of years (Table 12.1) and are easy to calculate. N_t is the recorded or estimated number of living species in a group, and t the time since the common ancestor, usually based on the fossil record or a calibrated molecular clock. Since we use this formula to estimate diversification rates, it is important to remember that the NDI is not the average time between appearances of new species in an entire clade, but *the average time elapsing between the origin of a new lineage and when that same lineage branches again.*

If based on living species, NDIs are always positive, which implies that species multiply without limit. This is of course absurd—there is surely extinction, so that net diversification intervals exceed biological speciation intervals. Diversification intervals are sometimes expressed as "doubling times" (t_d), the expected waiting time until the number of species in a phylogeny doubles. Doubling times can be obtained from the diversification interval simply by calculating the time until $N_t/N_0 = 2$, yielding $t_d = (\ln 2)/D$. Thus $t_d = \ln(2)t/\ln(N_t)$, or 0.693(NDI).

Doubling times are often viewed as NDIs because it is assumed that the average time for a single lineage to double is identical to the average time for a *group* of lineages to double. But these two rates are not equivalent. Doubling times are shorter than NDIs because the pure-birth process is exponential. Thus, during the average time it takes for any given lineage to divide, some members of an ensemble of lineages will have divided several times, and this more than compensates for those lineages that remain undivided. We provide NDIs rather than doubling times in Table 12.1 because NDIs seem easier to grasp.

A large literature exists on other ways to calculate NDIs, including using the distances between nodes in a reconstructed phylogenetic tree (Hey 1992; Barraclough and Nee 2001; Nee 2001), the waiting times between bifurcation events in these trees (Nee et al. 1994a,b) and imagining "reasonable bounds" for E/S, the ratio of extinction rates to speciation rates (Magallón and Sanderson 2001). Such methods may yield diversification rates more accurate than those calculated simply from the number of extant species and the time since common ancestry (Nee 2001).

Table 12.1 *Mean times to speciation or diversification*[a]

Taxon	Speciation or diversification interval (millions of years)	Reference
NET DIVERSIFICATION INTERVALS		
Protoctista		
Planktonic foraminifera	10.0	Stanley 1998
Nannoplankton	0.1	Van Valen 1985
Animalia		
Reef corals	6.7–11.1[b]	
Cheilostome bryozoans	6.7–8.3[c]	
Bivalves	16.4	
Veneridae	19.6	
Teredinidae	14.5	Stanley 1998
Mactridae	21.7	
Marine gastropods	14.9	
Echinoids	10.5	
Trilobites	0.8–1.6[d]	
Tetragnatha (Hawaiian spiders)	1.2	McCune 1997
Lake Baikal amphipods	5–7[c, d]	Turner 1999
Hadenoecini (cave crickets)	10.0	Hey 1992
Coleoptera (beetles)	13.6 (4.9–39.8)[b]	Wilson 1983
Cicindela (beetles)	4.5	Barraclough and Vogler 2002
Agrodiaetus (Lepidoptera)	0.3–0.7[c]	Kandul et al. 2004
Hymenoptera (mean of 7 families)	7.2 (4.0–9.6)[b]	Wilson 1983
Formicidae (ants)	9.4	P. Ward, pers. comm.
Diptera (mean of 22 families)	11.3 (4.3–27.1)[b]	Wilson 1983
Hawaiian *Drosophila*	0.8–4.4[c]	McCune 1997, and our estimate (see text)
Drosophila nasuta subgroup	8.3	Hey 1992
Drosophila melanogaster subgroup	6.2	Hey 1992
Ictalurus (N. American catfish)	13.2	Stanley 1998
Cyprinidae (N. American minnows)	5.6	Stanley 1998
Sculpins, Lake Baikal	0.6–0.9[c, d]	Turner 1999
Cichlids, Lake Tanganyika	0.7–1.1[c]	Turner 1999; McCune and Lovejoy 1998
Cichlids, Lake Malawi	0.1–0.3[c]	Turner 1999
Cichlids, Lake Nabugabo	0.004	Fryer and Iles 1972
Cichlids, Lake Barombi Mbo	0.4	Schliewen et al. 1994
Gambusia (mosquitofish)	1.6	
Rivulus	2.2	McCune and Lovejoy 1998
Xiphophorus (swordtails & platyfish)	1.4	
Colubridae (snakes)	1.8–3.4[c]	Stanley 1998
Family Gruidae (cranes)	2.6	Hey 1992

(continued)

Table 12.1 *Mean times to speciation or diversification[a] (continued)*

Taxon	Speciation or diversification interval (millions of years)	Reference
Galápagos finches	0.8–1.1[c]	Grant and Grant 2002
Dendroica warblers (New World)	0.8	Lovette and Bermingham 1999
Passerine birds	3.3	Stanley 1998
Mammals (spp. within families)	4.5	Stanley 1998
Bovidae	6.7	Stanley 1998
Muridae	2.8	Stanley 1998
Condylarths (fossil)	0.3	Van Valen 1985
Equinae (horses)	7.7	Hulbert 1993
Hipparionini (horses)	–71.5	Hulbert 1993
Equus (horses)	3.8	Hey 1992
Cebidae	5.3	Stanley 1998
Primates	2.9–14.2[b]	Purvis et al. 1995
Plantae		
Pteridophytes (ferns)	8.5–12.5[b]	Niklas et al. 1983
Gymnosperms	5.0–11.1[b]	Niklas et al. 1983
Angiosperms total	11.2–13.0[e]	Magallón and Sanderson 2001
Hawaiian angiosperms	0.5–5.0[f]	Levin 2000
Hawaiian silverswords	1.8	Baldwin and Sanderson 1998
Eudicots	10.5–12.3[e]	
Core Eudicots	8.4–9.8[e]	
Core Rosids	9.6–11.5[e]	Magallón and Sanderson 2001
Core Asterids	4.3–5.0[e]	
"Higher" monocots	8.7–10.6[e]	
Clarkia, sect. *Peripetasma*	1.6	Hey 1992
APPARENT SPECIATION INTERVALS		
Using assumptions about extinction rates		
D. melanogaster subgroup	3.2	
Drosophila nasuta subgroup	5.6	
Hadenoecini (cave crickets)	6.2	Hey 1992
Family Gruidae (cranes)	1.2	
Equus (horses; fossil)	5.0	
Clarkia, sect. *Peripetasma*	0.9	
Using data on extinction and speciation from the fossil record		
All taxa		
"Canonical speciation interval"	3.3	Sepkoski 1999
(Calculated net diversification interval)	125	See text
Mammals (spp. within families)	1.1–2.3[g]	Stanley 1998

Table 12.1 *Mean times to speciation or diversification[a] (continued)*

Taxon	Speciation or diversification interval (millions of years)	Reference
Bivalves (mean over 18 periods)	6.7–11.1[g]	Stanley 1998
Marine gastropods	2.3–4.3[h]	Jablonski 1986
Equinae	3.1	Hulbert 1993
Hipparionini	3.3	Hulbert 1993
Cycads	33.3	Levin and Wilson 1976
Conifers	25.0	Levin and Wilson 1976
Herbs	0.9	Levin and Wilson 1976
Shrubs	3.6	Levin and Wilson 1976
Hardwoods	8.3	Levin and Wilson 1976
TRANSITION TIMES FOR BIOLOGICAL SPECIATION[i]		
Drosophila (sympatric spp.)	0.08–0.2 [j,k]	Coyne and Orr 1997
Drosophila (allopatric spp.)	1.1–2.7 [j,k]	Coyne and Orr 1997
Frogs	1.5 [j,l]	Sasa et al. 1998; Lijtmaer et al. 2003
Lepidoptera	3.5–4.5 [j,l,m]	Presgraves 2002
Birds	5.0–5.5	Price and Bouvier 2002; Lijtmaer et al. 2003
North American songbirds	2.0–2.6 (0.4 – 5.4)[b]	Klicka and Zink 1997; Avise 2000
Mammals	2.2	Avise et al. 1998; Avise 2000

[a] Net diversification and most apparent speciation intervals are calculated as the reciprocal of either the net diversification rate or apparent speciation rate (both represented by "s"), where the intervals are calculated as 1/s. This is equivalent to $t/\ln(N_t)$ for net diversification rates. A single figure accompanied by a range is the mean interval among all taxa considered.

[b] Range among different taxa in group (if a single figure is given, it is the mean)

[c] Range depends on different age estimates of stem group

[d] Range calculated assuming different numbers of ancestors

[e] Range calculated from models using different ratios of extinction rates to speciation rates (0.1 and 0.9)

[f] Range of diversification rates among four islands

[g] Range depends on estimates of pseudoextinction rates versus real extinction rates

[h] Lower figure is for nonplanktotrophic gastropods, higher figure for planktotrophic gastropods

[i] As noted in the text, these are probably overestimates of transition times. Data from African lake cichlids, which approximate biological speciation intervals, are given in Table 4.2.

[j] Rates are for completion of reproductive isolation based on only one or two isolating mechanisms, and therefore overestimate the true transition time for speciation. Intervals for frogs are calculated from genetic distances corresponding to complete reproductive isolation.

[k] Rates depend on whether one uses Nei's (1987) or Carson's (1976) calibration of the allozyme clock; Nei's clock, which gives longer intervals, is based on mammalian data; Carson's clock is based on Hawaiian *Drosophila*.

[l] Estimates for postzygotic isolation only, therefore an overestimate of the true speciation interval

[m] Range reflects two different methods for calculating time to complete postzygotic isolation

Unfortunately, the fact of extinction renders NDIs virtually useless as estimates of biological speciation intervals. Extinction rates are often high, yielding a wide disparity between diversification and speciation intervals. For lineages that have formed in the very recent past, however, diversification rates may approximate speciation rates because recently formed species have not yet become extinct.

Calculating NDIs from living species also assumes that we know every species within the clade of interest, an assumption that is unrealistic for many groups such as tropical insects. This leads to overestimating NDIs, but because the number of species enters the NDI as a logarithm, this error is less serious than using inaccurate times since the common ancestor.

NDIs can also *underestimate* speciation intervals if there are problems with the fossil record. For example, the first appearance of an ancestor in the fossil record always underestimates the age of that lineage. This can be a serious problem because time enters the diversification equation linearly rather than logarithmically. Most diversification intervals in Hawaiian *Drosophila*, for example, are calculated using the age of the oldest extant island, Kauai (5.6 million years), giving a diversification interval of about 0.8 million years for the group. This interval is fairly short compared to biological speciation intervals in *Drosophila* presented below. But molecular evidence shows that Hawaiian Drosophilidae are much older, probably evolving on islands that eroded below sea level millions of years ago (DeSalle 1992). Using the older estimates gives an NDI of roughly 4.4 million years, closer to estimates of biological speciation intervals in allopatry (Coyne and Orr 1997).

Finally, unmeasured variables can bias NDIs in either direction. Most serious is the pure birth model's assumption that speciation and extinction rates are constant over time and among all lineages. This assumption is not only unrealistic, but is directly contradicted by considerable evidence (see, for example, Hulbert 1993 on horses and Sepkoski 1999 on marine invertebrates). In many cases, such as adaptive radiations, speciation may be rapid when a clade originates and decrease when niches become filled. NDIs calculated at the end of this process can seriously underestimate speciation intervals at the beginning of the radiation. Variation of speciation and extinction rates among taxa or lineages within a clade make estimates of NDIs even less accurate. In general, then, NDIs should be used to estimate speciation intervals only if there is evidence for no extinction or if the group in question arose very recently.

With these caveats in mind, we give in Table 12.1 a representative sample of NDIs calculated from plants, animals, and protists. Our impression is that, with the exception of African cichlids from Lakes Nabugabo and Victoria, the variation among groups seems surprisingly small. Neglecting the large negative value reflecting the extinction of horses in the Hipparionini, the mean diversification interval is 6.5 million years. Of 84 NDIs given, 72 fall between 0.5 and 20 million years. Plants in particular seem to vary little in diversification rates, even among major groups of angiosperms.

The shortest well-established diversification interval is 4000 years for each of the five cichlid species in Lake Nabugabo (Chapter 4). We have omitted the calculations for Lake Victoria cichlids because of the controversy surrounding the age of both the common ancestor and the lake, but if one accepts the most recent date of 14,600 years, the NDI is about 2600 years (see Table 4.1). The longest NDIs, averaging around 20 million years, occur in fossil bivalves. Marine invertebrates diversify more slowly than do terrestrial plants, which in turn diversify more slowly than terrestrial animals, who have NDIs between 1 and 8 million years.

When diversification rates vary little, as in Magallón and Sanderson's (2001) study of angiosperms, one can identify taxa having rates much higher or much lower than the average. Rosaceae, for example, diversify much faster than average, while Platanaceae diversify very slowly. Once one has identified such outliers, one may be able to determine whether they are due to changes in the rates of speciation, extinction, or both.

APPARENT SPECIATION INTERVALS. The problems with diversification rates have led researchers to derive *apparent* speciation intervals by estimating extinction rates. In the absence of direct fossil data, one can either assume these extinction rates, or try to calculate them from existing phylogenies. Table 12.1 gives a representative sample of data. Hey's (1992) model, for example, relies on the rather unrealistic assumption that each speciation event is accompanied by an extinction event, so that S and E are equal. As expected, this method yields ASIs much lower than NDIs: The 6.2 million year NDI of the *Drosophila melanogaster* subgroup, for example, becomes an ASI of 3.2 million years.

Mathematical analysis of existing phylogenies may also yield estimates of extinction rates. Unfortunately, all of the methods are subject to considerable uncertainty. The approach of Nee et al. (1994a,b) allows one to estimate both $S-E$ and E/S from the distribution of branch points in a reconstructed phylogeny. From these approximations one can calculate S and E separately. But, as the authors note, the estimates of both $S-E$ and E/S have such large variances that they should not be used for such calculations. Kubo and Iwasa (1995), Magallón and Sanderson (2001), and Barraclough and Nee (2001) suggest other ways to estimate extinction rates, but these appear either impractical or imprecise.

Owing to these problems, nearly all estimates of ASIs are based on extinction rates derived from fossils. Some groups appear to have a fairly complete fossil record, so that one can directly observe speciation and extinction. Less satisfactory but still acceptable estimates can be obtained by observing the average *duration* of species in the fossil record; the reciprocal of this duration estimates extinction rate. When added to diversification rates, these extinction rates yields ASRs. But many species that arise and disappear quickly are unlikely to be found in the fossil record, potentially leading to serious underestimates of speciation rates.

One of the best direct estimates of apparent speciation intervals, which can be compared to diversification intervals derived from the same data, is that of Hulbert (1993) on fossil horses over an 18 million year period. This group, which includes the monophyletic subfamilies Equinae and Hipparionini, has a remarkably complete fossil record in North America. In both clades, an initial burst of speciation was followed by a period of relatively constant species numbers and then by a period of extinction. In these taxa, the speciation interval among 17 time periods ranged between 0.57 million years and infinity (no speciation), with a mean of 3.1 million years for Equinae and 3.3 million years for Hipparionini (see Table 12.1). As this table shows, the NDIs for these groups are 7.7 and –71.5 million years, respectively, with the negative value reflecting a net loss of species over the period. This disparity exemplifies the problem of equating diversification and speciation intervals.

This difficulty becomes especially striking when one calculates the "canonical speciation interval": the average speciation interval across *all* organisms. Assuming an extinction rate of 0.25 species/species/million years (i.e., an average waiting time of 4 million years until a lineage becomes extinct), and a current number of ten million species, Sepkoski (1999) estimated that roughly three new species are formed each year. This yields a canonical speciation interval of about 3 million years (see Table 12.1). On the other hand, we can calculate the canonical diversification interval from the pure birth model by assuming that these ten million species derive from a single eukaryotic ancestor that lived about 2 billion years ago. This gives an NDI of roughly 125 million years, a clear overestimate of the waiting time to speciation. This 40-fold difference results from neglecting extinction.

There is thus no substitute for a good fossil record in accurately estimating speciation intervals. Unfortunately, data as good as that for fossil horses are rare. Nevertheless, with the exception of the remarkably slowly speciating cycads and conifers, the data point to apparent speciation intervals of between 1 and 10 million years.

BIOLOGICAL SPECIATION INTERVALS AND TRANSITION TIMES. Even when based on a good fossil record, ASIs are expected to exceed BSIs, for reproductive isolation may have evolved well before sympatric fossil taxa are recognizable as morphologically distinct species. BSIs are thus better derived from extant than from extinct species.

BSIs incorporate waiting times, which are hard to estimate. But one can occasionally get a rough idea of these times. In at least two instances, allopolyploids formed within 100 years after a new diploid species invaded the range of a relative (see Thompson and Lumaret 1992). And certainly the NDIs of cichlids in Lake Tanganyika and Malawi—ranging between 100,000 and 1,000,000 years—must be upper bounds of the BSIs. (This assumes that allopatric taxa described as species are in fact reproductively isolated.)

Minimum estimates of transition times for biological speciation in frogs, birds, *Drosophila*, and Lepidoptera can be calculated by extrapolating the relationship between divergence time and reproductive isolation to the time when reproductive isolation is nearly complete. In Chapter 2, we gave estimates of BSI for four groups. But all of these estimates neglect many potential isolating barriers, thus overestimating true BSIs. Nevertheless, in cases where reinforcement could not have promoted speciation because the species are allopatric or because the measure of reproductive isolation was postzygotic, BSIs for the four taxa are fairly consistent: between 1.1 and 5.5 million years (see Table 12.1; Lijtmaer et al. 2003).

We can estimate transition times for biological speciation (TTBS) from divergence times between pairs of sister species in a group. The best data are those of Klicka and Zink (1997) on North American songbirds (see Table 12.1). Molecular estimates of divergence times between 33 pairs of closely related species range from 0.35–5 million years, with a mean of 2.6 million years. These must overestimate the true TTBSs because these taxa are already good species. But the overestimate cannot be huge, for the mean divergence time between *subspecies* is 0.7 million years. For songbirds, then, complete speciation probably requires about 1–2 million years. This is supported by the diversification rate of Galápagos finches (0.8–1.1 million years), which is probably a TTBS because this group shows no evidence of extinction. The estimate of 5–5.5 million years from Price and Bouvier (2002) and Lijtmaer et al. (2003), however, is based solely on intrinsic postzygotic isolation. The disparity between these dates dates implies that this form of postzygotic isolation may not be important in avian speciation.

Extreme rates of speciation

There are many cases of extremely slow speciation, in which a clade persists for a long period without diversifying. Stanley (1998) provides a list, noting that members of such clades also show little morphological evolution within lineages. (This correlation between morphological stasis and low speciation rates, if it does not result from the presence of unrecognized sibling species, deserves further study.) According to Stanley, the record for failure to speciate is held by small crustaceans of the order Notostraca, whose two living genera extend back at least 300 million years, with no other fossil genera known. The order has always had only a few species (currently nine) and its extreme morphological conservatism is revealed by the presence of only trivial phenotypic differences between two Triassic and two living species.

Ginkgos are a similar example in plants. The only extant species, *Ginkgo biloba*, is a member of the family Ginkgoaceae, which appeared in the early Jurassic. Plants nearly identical to *G. biloba* existed about 120 million years ago. Over its history, the family included fewer than 20 known species, of which no more than 11 existed simultaneously. Although the group was widespread (fossils have been found in Europe, Asia, North and South America, Africa, India, and Australia), it showed a remarkable failure to diversify (Stewart and Rothwell 1993; Zhou and Zeng 2003).

In principle, slow speciation should be of considerable interest to biologists. After all, factors that *prevent* speciation can tell us what is required for speciation. Nevertheless, examples of rapid speciation have dominated the literature because they offer many chances to study the process itself.

The most rapidly evolving isolating barriers, as opposed to complete speciation itself, are seen in the two host races of *Rhagoletis pomonella*. Some factors preventing gene flow, such as the preference of the apple race for volatile compounds of apples, must have arisen within the last 200 years. Other reproductive barriers between these races, however, might be much older (Chapter 4).

Infectious speciation that could produce substantial hybrid inviability occurs very quickly. The bidirectional incompatibility between Hawaiian and non-Hawaiian populations of *Drosophila simulans*, caused by their infection with different strains of *Wolbachia* (O'Neill and Karr 1990), must have occurred rapidly. It is likely that, as a human commensal, *D. simulans* arrived in Hawaii within the last thousand years. Thus, the waiting time must have been short, and transition times even shorter. Models show that a new *Wolbachia* infection can become fixed in its host in about 10–100 host generations—roughly one to ten years in *D. simulans* (Turelli and Hoffmann 1995).

The record for rapid speciation is held by cases in which speciation follows hybridization. The two examples of allopolyploidy described above had BSIs of less than a century, although both involved human disturbance. Diploid hybrid speciation may have occurred even faster. Ungerer et al. (1998) suggest that the diploid hybrid sunflower *H. anomalus* arose only 25–60 years after its ancestors hybridized (Chapter 11). This, however, represents a transition time and not a BSI.

Until recently, the fastest example of "normal" speciation—that is, substantial reproductive isolation of diploid organisms caused by genetic change—was thought to involve five endemic species of Hawaiian moths in the genus *Omiodes* (formerly *Hedylepta*). Zimmerman (1960) claimed that these species fed and bred only on the banana plant, which was introduced by humans into Hawaii within the last 1000 years. This would suggest at least four speciation events during the last millennium. While this example is cited widely (e.g., Mayr 1963; Gillespie and Roderick 2002), it appears to be incorrect. Recent studies have shown that these species are not in fact confined to banana. A more parsimonious explanation is that they diversified originally on native palms and moved onto banana when palms became rarer (Foote 2004).

This record then devolves to the five cichlid species in Lake Nabugabo, which presumably evolved in less than 4000 years (Chapter 4). However, we know nothing about whether they are reproductively isolated from their presumed sister species in the adjacent large lake. This problem also afflicts another potential case of rapid speciation in cichlids, that of rock-dwelling species (mbuna) in Lake Malawi. Some of these species are endemic to islands and outcrops far from shore (Owen et al. 1990). Evidence from sediment cores suggests that these outcrops were completely dry within the last 200–300 years, implying that several allopatric speciation events occurred within that period. This represents an almost unbelievable rate of speciation, and must be viewed with skepticism since we do not know whether these species (defined largely by differences in color) are reproductively isolated from their allopatric relatives. These localized endemics may also be relicts of taxa that were once distributed more widely.

What is the effect of biogeography?

SYMPATRY VERSUS ALLOPATRY. Theory predicts that transition times for speciation events involving a sympatric phase will typically be shorter than for those involving purely allopatric speciation. The *Drosophila* data summarized by Coyne and Orr (1989a, 1997) support this prediction: taxa with overlapping ranges showed far higher reproductive isolation than allopatric taxa of equivalent age. McCune and Lovejoy (1998) found a similar result in fish, estimating divergence times of sister species from mtDNA sequences. Sequence divergence ranged from 2% to 6% in allopatric pairs, but from only 0% to 1.2% in sympatric pairs. The authors estimate that a 1% difference in sequence represents about 400,000 years of divergence. Because estimates of divergence time between sympatric species could be artificially low due to introgression of mtDNA, McCune and Lovejoy improved the test by comparing the maximum genetic divergence and average speciation interval (calculated as doubling time) for *clades* of fish that supposedly speciated sympatrically versus allopatrically. Doubling times averaged 0.37 million years for the 13 sympatric clades and 1.47 million years for the 6 allopatric clades.

It is important to realize that these results were not meant to test whether speciation was sympatric or allopatric—these modes of speciation were assumed—but to estimate the effects of biogeography on divergence times. The results do support a faster origin of species found in sympatry, but there are several caveats. First, this might reflect reinforcement instead of, or in addition to, sympatric speciation. In *Drosophila*, species with overlapping ranges appear to originate in less than a tenth of the time of allopatric species, but this effect is probably due to reinforcement (Chapter 10). Since reinforcement seems to have operated in some lake fish, such as the threespine stickleback, this is a viable alternative explanation to sympatric speciation. Second, the "sympatrically speciating" group was heavily weighted with African lake cichlids, which composed 12 of the 13 clades. Speciation in cichlids almost certainly involved sexual selection, which seems less likely in the allopatric group, whose members show less sexual dimorphism. (Indeed, within the cichlids themselves, speciation was faster in sexually dimorphic than in sexually monomorphic groups.) Since sexual selection can also speed speciation, the faster speciation of cichlids may reflect the form of selection rather than biogeography. Finally, speciation may have been rapid within African lakes because it involved adaptive radiation. The invasion of a new and largely unoccupied habitat may cause rapid speciation regardless of whether it is sympatric, parapatric, or allopatric. Thus, although McCune and Lovejoy's analysis suggests that speciation having a sympatric phase may be faster than allopatric speciation, this conclusion is compromised by uncontrolled factors in the comparison.

The acceleration of speciation in sympatry is supported by cases of closely related but reproductively isolated fish taxa in postglacial lakes, many of which must have formed within the last 15,000 years (Chapter 4).

IS SPECIATION FASTER WHEN ORGANISMS INVADE A NEW HABITAT? Schluter (2000, p. 2) defines adaptive radiation as "the evolution of ecological and phenotypic diversity within a rapidly multiplying lineage." Besides listing the criteria for whether a given radiation is adaptive, Schluter also discusses how to determine when speciation rates are high. One might expect that such rates would be elevated early in a radiation, and decrease as niches are filled. But tests of this prediction are difficult unless extinction rates are constant. The same pattern can also occur in "non-adaptive radiations," in which reproductive isolation following invasion of a new habitat results not from adaptation but from sexual selection (in this case niches still become filled, but the primary cause of speciation is behavioral isolation). The New Guinea birds of paradise might represent such a radiation (Frith and Beehler 1998).

Statistical tests for rapid speciation in new habitats yield mixed results. Schluter (2000) determined the relative numbers of species in 12 pairs of related clades, each pair including one clade on an archipelago and the other on the mainland. Because the clades were of different age, mainland clades were converted to "pseudo sister groups" by calculating how many species they would contain if they were as old as the archipelago clades. With this correction, 9 of the 12 comparisons showed more species on the archipelago, a suggestive but nonsignificant result. But even a significant outcome would be problematic. In effect, this test compares diversification rates. Higher diversification rates on islands might reflect not faster speciation, but slower extinction, perhaps resulting from fewer competitors. Alternatively, faster speciation on archipelagos might be due only to a greater opportunity for allopatry.

A better analysis involves determining whether reproductive isolation itself evolves faster during radiations. The only existing test has compared mainland with Hawaiian *Drosophila*, the latter a classic case of island radiation. Schluter (2000) found no difference in the evolutionary rates of behavioral or intrinsic postzygotic isolation between these two groups, though the data were scanty.

Given these inconclusive tests, we have surprisingly little support for the prevailing opinion that speciation rates are exceptionally high during adaptive or non-adaptive radiations.

Conclusions

Although we can date the origin of sister species fairly accurately, there are few good ways to calculate biological speciation intervals. At present, the best appears to involve estimating the time course for speciation by correlating the divergence time between sister taxa with their degree of reproductive isolation. A lower bound on BSIs in a group can then be estimated as the average time at which reproductive isolation is nearly complete. An upper bound can be estimated from the average divergence time between good sister species.

The best data now come from birds and *Drosophila*. In these groups, we have diversification intervals and either apparent or biological speciation intervals, and thus can bracket the BSIs. For both groups, the BSI for allopatric species appears to be about 1–2 million years, and about 0.2 million years (in *Drosophila*) when species are sympatric. Other animal taxa show BSIs on the order of 0.5 to 5 million years. The plant data are spottier, but on average show BSIs greater than 5 million years. One can provisionally conclude that the waiting time for speciation is on the order of 100,000 to a few million years in animals, and possibly longer in plants. All of these data, of course, are crude, and speciation rates vary considerably within a group. We now turn to research that may explain such variation in rates.

Factors Affecting Speciation Rates

Which characteristics of individuals, populations, or species increase the rate at which new species arise? Such factors, especially when they permit invasion of a new adaptive zone, are often called "key innovations." But since some of these are preexisting traits and not innovations, we prefer to call them "key factors." These factors are important because if they raise the rate of speciation without increasing the rate of extinction, clades containing them will become relatively larger over time, yielding an evolutionary trend. Key factors can also suggest which reproductive barriers were important in speciation.

Many traits have been suggested as major influences on speciation rates, including body size, reproductive mode, degree of specialization, intensity of mate choice, type of pollination, and migration rate. Such suggestions result from observing that one or more clades possessing the trait contain more species than do related clades lacking the trait. However, drastically unequal sizes of sister clades can be caused by many factors, including biogeographic history and differential rates of extinction.

In fact, at least four factors can cause an association between a trait and the species richness of a group possessing it:

1. *Properties of organisms that facilitate speciation.* These organismal traits speed up the evolution of reproductive isolation in populations. The most obvious candidates are traits that promote sexual selection in animals, for such selection can produce sexual isolation as an immediate byproduct. There are many indices of the potential for sexual selection, including the degree of multiple mating by males or females, and behavioral or morphological traits that indicate past or ongoing sexual selection. Similarly, pollinator isolation may be a frequent cause of speciation in flowering plants. We might thus expect to see higher rates of speciation among animal pollinated than among abiotically pollinated angiosperms.

Speciation might also be promoted when organisms are involved in biological interactions with other species. An organism adapted to the abiotic environment may reach an evolutionary plateau if the environment remains constant. But adapting to other species, which themselves evolve, can involve ever-changing selection pressures and create coevolutionary races, both leading to increased reproductive isolation. Schemske (2002) suggested that the higher number of biotic interactions in tropical groups—which may reflect greater climatic stability that allows more species to coexist—has made tropical groups more species rich than related temperate groups.

Some forms of sympatric speciation are promoted by a tendency of parasitic organisms to mate on their hosts. If such speciation is common, one might expect more species in clades of host-specific insects and parasites than in sister clades less tied to particular hosts.

Many cases of speciation may depend on environmental features such as the opportunity for allopatry. Thus, biotic traits like low vagility could promote speciation by reducing gene flow, allowing local populations to adapt more readily to the local habitat. (Low vagility, however, may also *reduce* speciation by leading to a lower rate of colonization.)

Key factors need not themselves be adaptations—all that is required is that they increase the rate of speciation. The susceptibility to infection by the endoparasitic bacterium *Wolbachia*, for example, is probably not an adaptation, but could facilitate speciation by postzygotic isolation (Chapter 7). Finally, key factors associated with diversification may be generalized behaviors or features involving *combinations* of different adaptations. Phytophagy in insects, a trait associated with rapid diversification, is based on several traits associated with chewing or sucking by adults and larvae.

2. *Properties of organisms that prevent extinction.* Any organismal trait that reduces the probability of extinction will increase in frequency among existing groups, and will be associated with increased species richness even without any effect on speciation rate. Such traits might include those increasing population size or geographic range, such as higher migration rates, greater chemical defenses against herbivores, small body size, self-fertilization, and the ability to find habitats to which the organism is adapted. Such properties, while of intrinsic interest to paleobiologists, concern us mainly as complications in calculating rates of speciation.

3. *Properties of organisms that open up new "adaptive zones."* Some features, like terrestriality in vertebrates, and wings in birds and insects, may be associated with increased rates of speciation because they allow a rapid invasion of new habitats. These features can trigger adaptive radiations without having a direct effect on reproductive isolation, simply because such radiations involve many open niches and a reduced likelihood of extinction.

Schluter (2000) discusses traits that may promote adaptive radiation. These include "preadaptations": traits evolved for a specific function that fortuitously assume another function when the environment changes.

Hunter (1998), for example, shows that in several groups of Cenozoic mammals the independent evolution of the hypocone (a cusp on the molar tooth) was consistently associated with increased species richness, but that hypocones evolved well before increased diversification. He suggests that hypocones evolved for more efficient chewing, which permitted adaptive radiation when fibrous plants subsequently became available with the rise of Eocene forests.

Finally, some traits can promote speciation by simultaneously allowing the evolution of reproductive isolation *and* opening new adaptive zones. Nectar spurs in flowering plants, for example, may allow the plants to enter the new "pollination zone" of long-tongued insects, while differences between the lengths of the spurs can cause speciation by pollinator isolation (Hodges 1997; von Hagen and Kadereit 2003).

Entry into a new adaptive zone is often a unique event. There is general agreement, for instance, about which "preadaptations" allowed freshwater organisms to invade land (sturdy limbs and lungs). But in most cases of adaptive radiation, the key traits—if any—are impossible to determine without the replication necessary for sister-group analysis (see below). Because some adaptive radiations have occurred repeatedly, such as the invasion of angiosperms by insects, one can test whether any traits are consistently associated with this diversification.

4. *"Species-level traits" that affect speciation or extinction rates.* Some macroevolutionists (e.g., Stanley 1998, Gould 2002) argue that speciation and extinction rates can be affected by "emergent" properties of entire species—traits that cannot be extrapolated from the properties of individuals. These include features like range size and the degree of range fragmentation. If these traits are also heritable among species (i.e., daughter species inherit the properties of ancestors), the proportion of species with these emergent traits will increase over time. This process is "species selection," which we discuss at the end of the chapter.

Tests for the effects of key factors

New forms of phylogenetic analysis—especially comparative methods—have taken theories about the effects of key factors from the realm of speculation to that of testable hypothesis. The most important tests are those comparing the species richness of sister clades whose ancestors differed in either the presence or the state of a postulated key trait. Such clades are of equal age because they descend from a common ancestor. These methods obviously demand accurate phylogenies and correct resolution of sister groups. Moreover, since the effect of the key factor is determined by comparing the number of *extant* species in two lineages, the test actually determines the effect of the factor on biodiversity, not on speciation rate. Without other information, we cannot determine whether a key factor affected speciation, extinction, or both.

SINGLE COMPARISONS. Biologists are often impressed by monophyletic groups that contain many more species than their sister groups. The classic example is angiosperms. This clade comprises about 250,000 species, while its sister group—either Gnetales (around 80 species in the genera *Ephedra*, *Gnetum* and *Welwitschia*) or, more likely, the group including conifers, Gnetales, ginkgos, and cycads (770 species)—is depauperate (Sanderson and Donoghue 1994; Donoghue and Doyle 2000). This 300- to 3600-fold difference in species richness between equal-aged clades has been attributed to various angiosperm traits that could promote speciation. Insect-pollinated flowers, for example, could produce pollinator isolation. Can one test such hypotheses?

Such tests usually begin by choosing two drastically unbalanced sister groups that differ in important biological features. However, imbalance itself gives no clue whether any interesting biology is involved. Surprisingly, under the null model of constant instantaneous speciation and extinction rates, *all possible degrees of inequality among the numbers of species descending from the initial bifurcation are equally probable*. So, for example, in a tree comprising n extant species, the comblike pattern of Figure 12.1A, in which the initial two branches produce 1 and $(n-1)$ extant species, is just as probable as any more balanced phylogeny (Slowinski and Guyer 1989).

To deal with this problem, Slowinski and Guyer (1989) suggested a statistical test for differences in diversification rate based on an observed phylogeny containing n species divided into two sister groups, one with r species (the larger number), and the other with $n-r$ species. One calculates the proportion of possible n-species phylogenies whose initial bifurcation produces two groups of terminal species whose sizes are as unequal as, or more unequal, than those seen in the observed phylogeny. If this proportion is smaller than 5% (a one-tailed test) or 2.5% (a two-tailed test), the phylogeny is considered significantly asymmetrical.

The proportion of phylogenies at least as unbalanced as that seen in the phylogeny of interest, is easily calculated as $(n-r)/(n-1)$. Thus, one needs at least 21 species (20 on one branch, 1 on the other) to see significant asymmetry under a one-tailed test, and 41 species under a two-tailed test. Slowinski and Guyer (1989) used this method to compare two sister groups of Caribbean lizards, the Cuban genus *Chamaeleolis* (3 species) and its sister group of Neotropical anoline lizards (250 species). The one-tailed test gives a probability of 0.012 of observing a more extreme phylogeny.

But there are problems with drawing biological conclusions from the relative species richness of a single pair of sister clades. Most important, although the test may indicate real differences in rates of diversification, it gives no evidence that these differences were caused by any particular feature. After all, angiosperms and their sister group, like all sister taxa, differ in several traits, any of which—or none of which—might be responsible for a difference in species diversity. Flowers, for example, are the most obvious feature that could accelerate angiosperm speciation, for their existence allows the evolution of pollinator isolation. But Doyle and Donoghue (1986) suggest that the key factor promoting angiosperm radiation may have been closed carpels. This feature could promote speciation by allowing the evolution of new ways to disperse seeds, leading in turn to greater migration and more chances for allopatric speciation. The point is that when one has but a single comparison, one can conclude only that the results are *consistent* with the effect of a proposed key factor.

Second, clades chosen for sister-group analysis may be nonrandom. After all, biologists pay more attention to species-rich clades because of their size. If such clades are analyzed merely because they are dramatically larger than their sister groups, then the Slowinski-Guyer test is biased in favor of asymmetry. Ideally, one would devise an a priori hypothesis about what factors influence species richness without knowing the relevant phylogenies. One could then search for appropriate groups to test this hypothesis.

Third, even if one interprets an extreme asymmetry between sister groups as the result of a key factor, it is not clear whether the relevant test should involve probabilities derived from "null clades" generated by a model of random branching and extinction. As de Queiroz (1998) notes, the proper test should not compare the imbalance of a real clade with that of a model clade, but rather the imbalance between *two* real clades (i.e., two pairs of sister groups), only one pair of which differs in the key trait. After all, a phylogeny can be unbalanced for reasons other than the presence of a key trait. One clade, for example, may have experienced more frequent allopatry, and thus produced more species. Real clades, then, may be more asymmetrical than those generated by null models, even if there is no key factor affecting diversification. This assertion is supported by the work of Stam (2002), who showed that a high proportion of real clades are significantly more unbalanced than those predicted by null models. Thus, it is not clear how to evaluate biologically the imbalance between two sister clades.

Sanderson and Donoghue (1994) developed likelihood models that provide other tests of the effects of key traits in single phylogenies. Such models can detect, for example, whether the origin of the key traits that distinguish sister groups coincides with the difference in their rates of diversification. These models show that, if one assumes that the sister group of angiosperms was Gnetales, the faster diversification of angiosperms was apparently not associated with the origin of the shared derived traits that distinguish the angiosperm lineage. Sanderson and Donoghue reached this conclusion because the rate increase occurred in only one of the first two branches that formed *within* the clade of angiosperms.

But the most serious problem with identifying key traits by comparing a single pair of sister clades is the lack of replication. To deal with this problem, one requires *multiple* sister groups. This method has the advantage of not relying on any model of diversification.

MULTIPLE COMPARISONS. Multiple-sister-group comparisons use data from several to many evolutionarily independent pairs of sister taxa, with the taxa in each pair differing in whether or not they carry the trait of interest. Statistical analysis is then used to determine whether the trait is consistently associated with a difference in species richness. This method eliminates the problem of non-independence of traits, because, as the number of sister clades increases, the chance of a consistent association between a key trait and other traits decreases, and thus the effects of other traits on diversification average out.

This method was first used by Mitter et al. (1988) to determine whether phytophagy in insects promoted diversification (see Chapter 10). Since then, it has been most extensively used and promoted by Barraclough and his colleagues (Barraclough et al. 1995, 1996, 1998). Such analyses usually take the form of a table of n comparisons involving two sister taxa that differ in the presence of a key trait proposed to affect diversification. Each comparison involves the number of extant species in the group containing the trait with the number of species in the sister group lacking the trait:

Sister pair	Number of species with trait	Number of species without trait
1	x_1	y_1
2	x_2	y_2
3	x_3	y_3
.	.	.
n	x_n	y_n
Species sum	$\sum_{i=1}^{n} x_i$	$\sum_{i=1}^{n} y_i$

The compared clades can be at any taxonomic level (families, genera, etc.), so long as each group is monophyletic and the true sister group of the other. The numerical comparisons need not involve species (one could, for example, compare the number of genera in pairs of sister families), but since the number of higher-level groups is likely to be more arbitrary than the number of species, almost all published comparisons involve species.

There are many ways to analyze such data. These include testing for a consistent direction of effects, using nonparametric sign tests, Wilcoxon tests, or parametric t tests comparing the distribution of $\ln(x_i/y_i)$ to its expected value of zero under the null hypothesis (Mitter et al. 1988; Gittleman and Purvis 1998). Other analyses incorporate both direction and size of the effect, including regression analysis and randomization tests (Nee et al. 1996; Gittleman and Purvis 1998). Still other methods are discussed by Slowinski and Guyer (1993), Nee et al. (1996), Barraclough et al. (1998), and Isaac et al. (2003).

Two points should be emphasized before we summarize existing studies. First, if a trait is associated with increased species richness, this gives no *historical* information about whether the appearance of the trait increased diversification or the loss of the trait decreased diversification. Such information must be derived from inspecting outgroups. In this way, Dodd et al. (1999) demonstrated that the positive association between biotic pollination and angiosperm diversification repeatedly involved reduced diversification associated with the loss of biotic pollination. (This, of course, does not affect the conclusion that biotically pollinated taxa diversify faster than non-biotically pollinated taxa.) Second, we emphasize again that diversification does not equal speciation. Increased diversity can result from increased speciation, decreased extinction, or both.

Distinguishing speciation from extinction

How can comparative methods tell us which key factors are the ones most interesting to students of speciation: those traits that promote the evolution of reproductive isolation? The solution is to examine those traits *whose effects on speciation and extinction rates are likely to be in the same direction*. After all, if one has good reasons for suspecting that a trait increases the rate of extinction, then any greater species richness of groups possessing that trait must mean that it also increases the rate of speciation. (In other words, since $D = S - E$, if both D and E are higher, then S must also be higher.) Such traits may thus have some connection with isolating barriers that were the primary causes of speciation. This conclusion becomes more plausible if there is an obvious connection between the diversification-promoting trait and reproductive isolation.

One might guess, for example, that mating systems or traits that allow stronger sexual selection might promote speciation by accelerating the rate of behavioral isolation. It also seems plausible that sexual selection could increase the rate of *extinction* by reducing the survival of males having extreme traits or visible displays, increasing the chance of population extinction if females are choosy, and so on. If sister-group analysis shows a positive association between traits indicating sexual selection and species diversity, one might thus conclude that those traits increased the rate of speciation. Likewise, increased specialization on habitats or resources may promote speciation, but may also enhance the likelihood of extinction if resource abundances change. Decreased migration may increase speciation by allowing populations to become more isolated, but might also increase extinction by creating smaller ranges that are more easily affected by local environmental perturbations. Those interested in comparative studies of speciation should concentrate on traits with such parallel effects.

Of course, any claim about the effect of a trait on extinction rates is only an informed guess, which may be hard to substantiate. Some data, however, are available from the fossil record. Jablonski (1986, 1995), for example, shows that groups of marine snails that either brood their larvae or have restricted larval dispersal show higher rates of speciation *and* extinction than do groups with free-swimming larvae.

Many traits, however, may have opposite effects on the rates of speciation and extinction. Small body size in animals might increase speciation by allowing larger population sizes that simultaneously enlarge the opportunity for allopatry and reduce the chance of extinction. Traits with such opposite effects cannot be used in comparative studies to identify features that promote speciation.

The data

Table 12.2 summarizes all published sister-group comparisons that reveal statistically significant effects of traits on species richness. (We mention below other studies that failed to find such associations.) We alluded to a few of these comparisons in earlier chapters. Along with the type of tests used and the num-

Table 12.2 *Multiple sister-group comparisons that show traits significantly associated with species diversity*[a]

Trait increasing species diversification and group	Statistical test used[b]	Number of sister-group comparisons	Relative species richness (N)	Reference
Resin canals, plants	S	16	4.9 (43,239)	Farrell et al. 1991
Nectar spurs, angiosperms	S	8	3.1 (2503)	Hodges 1997
Nondioecy, angiosperms	S[c]	28	7.6 (40,781)	Heilbuth 2000
Biotic (animal) pollination, angiosperms	F	11	1.2, 5.6[d]	Dodd et al. 1999
Herbaceous growth form, angiosperms	F	14	4.2	Dodd et al. 1999
Abiotic (vs. biotic) dispersal, herbs	F	3	12.5	Dodd et al. 1999
Two forms of dispersal versus biotic dispersal only, angiosperms[e]	F	20	2.6	Dodd et al. 1999
Increased rate of neutral evolution, angiosperms	R, S	89	4.1 (107,497)	Barraclough and Savolainen 2001
Phytophagy, insects	S	13	3.1 (436,376)	Mitter et al. 1988
Feeding on angiosperms, beetles (group Phytophaga)	S[f]	10	140.9 (115,549)	Farrell 1998; Farrell et al. 2001
Polyandry, insects	S, T	25	4.0 (35,479)	Arnqvist et al. 2000
Longer testes, hoverflies	T[g]	40	—	Katzourakis et al. 2001
Viviparity, actinopterygian fishes	F	10	0.7[h,i] (1336)	Slowinski and Guyer 1993
Promiscuous mating, birds	S	14	2.5 (550)	Mitra et al. 1996[j]
Sexual dichromatism, passerine birds	Ra, S, F	9–31	1.7[i] (4392)	Barraclough et al. 1995; (see also Owens et al. 1999)
Feather ornamentation, birds	T, W	70	2.4 (1045)	Møller and Cuervo 1998
Low body mass, insectivores, bats, rodents	S, F	96	—	Gardezi and da Silva 1999
Low body mass, carnivores	R, T	171	—	Gittleman and Purvis 1998
Lower-latitude ranges, Papilionidae (swallowtail butterflies)	S, W	13	1.7 (192)	Cardillo 1999

ber of sister groups compared, we give the "relative species richness" associated with the trait: the ratio of the number of species possessing the trait to the number of species lacking the trait, added across all sister groups, i.e. $\dfrac{\sum_{i=1}^{n} x_i}{\sum_{i=1}^{n} y_i}$ in the example above.

Table 12.2 *Multiple sister-group comparisons that show traits significantly associated with species diversity[a] (continued)*

Trait increasing species diversification and group	Statistical test used[b]	Number of sister-group comparisons	Relative species richness (N)	Reference
Lower-latitude ranges, passerine birds	S, W	11	7.2 (903)	Cardillo 1999
Increased annual dispersal, birds	S, W	28	—	Owens et al. 1999[k]
Larger range size, birds	S	28	—	
Greater range fragmentation, birds	S	28	—	

[a]The traits listed are those associated with an *increase* in species richness compared to the alternative state. Relative species richness is the ratio of total number of species possessing the trait to those lacking the trait, summed over all clades (the total number of species in the sample, N, is given in parentheses). Some studies did not provide species numbers or relative species richness.

[b]Key to tests used: "S" is sign test; "T" is t test or other two-group comparison of species richness; "W" is the Wilcoxon signed-ranks test; "R" is a regression of taxon-pair values or species richness on differences in a continuously varying trait; "Ra" is a randomization test; and "F" is Fisher's test on combined probabilities from test of Slowinski and Guyer (1993). (See text for references to these tests.)

[c]Dioecious taxa are significantly *less* species rich than their outbreeding sister taxa; this holds for comparisons including sister families or sister genera (we show only the family-level data, involving 28 comparisons).

[d]Lower species richness ratio is for total data from all families, higher ratio is from data omitting one comparison (Joinvilleaceae vs. Poaceae) that has a difference in the "wrong" direction. In a separate sister-group analysis of angiosperms, Verdú (2002) found that biotically pollinated genera had a significantly higher number species than abiotically pollinated genera (a ratio of 1.22 to 1).

[e]This association is probably a sampling artifact rather than a real effect of dispersal mode on species richness (see text).

[f]Test of association in Farrell (1998) is significant using a one-tailed sign test ($p = 0.03$), but not a two-tailed sign test ($p = 0.062$). Association in Farrell et al. (2001) is significant using a two-tailed sign test. Data presented are from Farrell et al. 2001, which includes results from Farrell (1998).

[g]Comparison not significant if Bonferroni correction used (seven other traits were examined).

[h]Sister-group comparison is significant only if Fluviphylacinae and not Apocheilichthyinae is used as the sister group, and only if the Slowinski-Guyer test is used ($p < 0.05$). Note that this significance occurs despite the observation that, overall, viviparous species are less numerous than non-viviparous species.

[i]Although Fisher's test of cumulative probability yields a significant difference between sister taxa, the randomization test of Goudet (1999) shows no significant effect of viviparity on fish diversity or of dichromatism on bird diversity.

[j]Owens et al. (1999) found no effect of the degree of social polygamy (or of sexual size dimorphism) on species richness of sister families of birds.

[k]Data presented are from all families of birds; significance levels are derived using sequential Bonferroni corrections (Rice 1989) of the probability values given by Owens et al. (1999).

A significant association with diversification rate is seen for 23 features, ranging from characters present in individuals (e.g., nectar spurs) to emergent traits characterizing entire species (range size and degree of fragmentation). The quality of the data among studies is uneven. In several cases (e.g., Mitra et al. 1996; Farrell 1998; Katzourakis et al. 2001), the probability of association is not far below the 0.05 level of significance, and becomes nonsignificant if one uses two-tailed tests or corrects for multiple comparisons. In addition, seven of the studies use Slowinski and Guyer's (1993) multiple-group analysis, a procedure that, unless combined with a randomization test, yields artificially low probabilities of association (Nee et al. 1996). Finally, at least one of the associations—the higher species richness of angiosperm families having two forms of dispersal instead of biotic dispersal alone (Dodd et al. 1999)—appears to be an artifact. Families characterized as having "more than one form of dispersal" are not those in which *all* species have more than one form of dispersal, but rather those in which some species have abiotic dispersal, some have biotic dispersal, and some have both. This automatically leads to a greater species richness of such families than those in which all constituent species have biotic dispersal.

Nevertheless, many studies yield results that are consistent with biological intuition. Perhaps the strongest conclusion is that taxa with traits indicating stronger sexual selection are consistently more species rich than sister taxa lacking such traits. All five relevant sister-group tests show that indices of sexual selection increase diversification: promiscuous mating systems in birds and insects, sexual dichromatism and feather ornamentation in birds, and testis length in hoverflies (considered an index of sperm competition by Katzourakis et al. 2001).

It seems reasonable to suppose that sexual selection is likely to either raise or have little effect on extinction rates. Extinction rates may increase because sexual selection can reduce effective population size by extreme polygamy or polyandry, or reduces census size by allowing the evolution of male traits that decrease viability or increase susceptibility to predators or parasites. Several studies support this hypothesis. Two surveys of birds introduced onto islands by humans showed that sexually dichromatic species become established less often than monochromatic species (McLain et al. 1995; Sorci et al. 1998). A 21-year analysis of North American bird communities found that sexually dichromatic species had a significantly higher rate of local extinction than did monochromatic species (Doherty et al. 2003). Finally, the mortality of males relative to females in North American ducks increases with the degree of sexual dichromatism (Promislow et al. 1994). If sexual selection increases the rate of extinction, then the comparative analyses suggest that sexual selection also increases the rate of speciation. This in turn implies that behavioral isolation, a byproduct of sexual selection, is an important cause of speciation in birds.

In a comparative study of mammals, spiders, and butterflies, however, Gage et al. (2002) claim that sexual selection is *not* involved in speciation. But their methods violate key assumptions of comparative analysis. First, they used genera not known to be monophyletic, and undoubtedly of different ages. Moreover, they do not support their conclusions by using sister clades, a method that reduces the effect of other biological factors on species richness.

There are only two other comparative studies of traits expected to increase both speciation and extinction rates: the presence of nectar spurs on flowers, and biotic (vs. abiotic) pollination in angiosperms (Hodges 1997; Dodd et al. 1999). In both cases, reliance on insect pollinators may facilitate speciation through pollinator isolation, but also increase extinction by creating extreme dependence on particular pollinators. These significant associations thus show the potential importance of pollinator isolation in plant speciation, an association supported by individual studies described in Chapter 5. These results are tentative, however, for comparative studies in other taxa did not find significant effects of nectar spurs or biotic pollination (Bolmgren et al. 2003; von Hagen and Kadereit 2003).

For most of the other studies, one can at least concoct a plausible story that the trait increases speciation rates but also decreases extinction rates, allowing no firm conclusion about which rate (or both) explains the difference in diversification. Resin canals in plants, for example, may reduce extinction by repelling herbivores, but might also accelerate speciation by increasing range size, thus allowing more opportunity for allopatry. In plants, nondioecious clades are consistently more species rich than their dioecious relatives, possibly because hermaphroditism allows successful colonization involving fewer individuals. But nondioecy may also reduce extinction rates. (Note, however, that these considerations predict that *selfing* plants should be more species rich than nonselfing relatives, an idea that Heilbuth 2000 failed to support in comparative studies of angiosperms.)

As we have seen, increased dispersal may have unpredictable effects on both speciation and extinction rates, either raising or lowering them. Thus, we cannot be sure about the cause of association between species richness and (1) increased dispersal in birds (Owens et al. 1999), or (2) abiotic (vs. biotic) dispersal in herbs (Dodd et al. 1999). Similarly, we do not know whether the association between species richness and greater range fragmentation in birds (Owens et al. 1999) has to do with increased speciation caused by greater allopatry. Range fragmentation could either increase extinction by making populations more susceptible to stochastic fluctuations in size, or decrease extinction by allowing isolated populations to escape predation, parasitism, or disease.

Surprisingly, given the huge and contentious literature on why the tropics are so diverse (Schemske 2002), there is only one study showing the expected effect of biogeographic region on diversification rate. Comparing members of sister taxa in passerine birds and in swallowtail butterflies, Cardillo (1999) found that temperate-zone clades are significantly less species rich than are clades living closer to the tropics. However, Farrell and Mitter (1993) failed to show an association between tropical habitat and species richness in five sister-group comparisons of phytophagous insects, although this test had low power. Higher species richness in the tropics, however, might not reflect higher rates of speciation. Tropical species might have lower extinction rates than their temperate counterparts because their populations may be larger (Wright 1983; Gaston 2000).

Two studies found an association between species richness and behavioral traits that probably promoted entry into a new adaptive zone: phytophagy in insects, and angiosperm (vs. conifer) feeding by phytophagous beetles (Mitter et al. 1988; Farrell 1998; Farrell et al. 2001). The latter comparison shows an astounding 141-fold difference in species richness—by far the greatest effect of any factor on diversification. It is likely that both associations represent both a direct effect of plant feeding on the opportunity for speciation (via adaptation to allopatric food resources or disruptive selection for divergent resource use in sympatry), and an indirect effect reflecting an initial radiation into a huge area of habitable space lacking competitors, predators, or parasites (there are vastly more species of angiosperms than of conifers). However, radiations into empty niches are also likely to reduce the rate of extinction. In this case, as with most of the adaptive radiations discussed by Schluter (2000), it is impossible to gauge the relative roles of speciation and extinction.

Two associations given in Table 12.1 are mysterious. There is no obvious reason for the correlation of species richness with herbaceous growth form in angiosperms. Dodd et al. (1999) propose that herbaceous growth may be associated with lower generation times and more ephemeral populations, both of which may increase speciation rate, but it would be easy to concoct an ad hoc explanation were the association in the opposite direction.

More intriguing, Barraclough and Savolainen (2001) found a positive association between "branch length" (the amount of noncoding genetic change occurring during the divergence between sister groups) and species richness. Webster et al. (2003) also report a correlation between species richness and the rate of molecular evolution in 28 diverse plant and animal groups. This association appears to be a general phenomenon and requires explanation.

It is not obvious why increased species diversity would be associated with genetic changes that have no effect on the phenotype. Barraclough and Savolainen offer three explanations:

1. Clades having shorter generation times might show faster neutral evolution (which is proportional to generation time, not clock time), and may also experience stronger natural selection. Natural selection could create species as a byproduct.

2. Genome-wide increases in mutation rate could increase the rate of neutral evolution as well as accelerate speciation by providing more raw material for natural selection.

3. Speciation involves splitting of populations and may lower population sizes. In small populations, alleles that are usually deleterious will behave as if they were neutral, so the rate of neutral evolution will increase. Thus, lineages experiencing more splitting events might show higher rates of "silent" substitutions.

The first hypothesis is supported by Verdú's (2002) observation that angiosperm diversification rates are negatively correlated with generation time. The second hypothesis is hard to test, but is corroborated by comparative data of Davies et al. (2004) showing that the species richness of angiosperm clades is significantly correlated with the energy input (UV light and temperature) in their habitat and the observation of Wright et al. (2003) in the plant genus *Mearnsia* that taxa living in warmer areas have a higher evolutionary rate of ribosomal DNA. The third hypothesis assumes that population size stays low for long periods after speciation. This seems unrealistic, but may be plausible if speciation usually occurs by colonization of new habitats by a few individuals.

Table 12.2 shows only comparative studies yielding significant associations between traits and diversity, but nearly as many studies showed no association. Surprisingly, some of the latter studies involve traits identical to those described in Table 12.2, but examined in other groups. For example, although small body size is associated with species richness in four groups of rodents (Gittleman and Purvis 1998; Gardezi and da Silva 1999), there is no similar association in birds (Owens et al. 1999). Results can differ even for the same trait analyzed in the same group: the significant association between polygamy and species richness in birds found by Mitra et al. (1996) was not seen by Owens et al. (1999) in a larger sample. Other traits showing no association with species richness include measures of ecological diversity in tiger beetles (Barraclough et al. 1999a), parasitism in carnivorous insects (Wiegmann et al. 1993), body size in several groups of mammals (Gardezi and da Silva 1999), range size in molluscs (Jablonski and Roy 2003), the presence of image-forming eyes (de Queiroz 1999), host specificity in fish ectoparasites (Desdevises et al. 2001), and self-compatibility versus self-incompatibility in plants (Heilbuth 2000).

Conclusions

Sister-group comparisons are a potentially powerful method of determining which biological factors increase speciation rates. Moreover, they may allow us to infer which isolating barriers are *primary* causes of speciation, because any factor that creates isolating barriers after reproductive isolation is complete will not increase the rate of speciation. However, such factors can still increase species richness if they reduce the rate of extinction, and thus can be associated with higher diversity in sister-group comparisons. Thus, to infer which factors promote speciation, one must concentrate on those factors that probably promote extinction as well.

Our major conclusion is that two sets of key factors—traits increasing sexual selection in animals and traits promoting animal pollination in plants—appear to increase the rate of speciation. However, this conclusion is tentative because it depends on assuming that these factors either reduce or have no effect on extinction rates. Strengthening this conclusion will require other tests. For traits of individuals that are potentially associated with reproductive isolation, these tests can involve time-course analyses like those described in

Chapter 2. One would expect, for example, that the earliest-appearing forms of reproductive isolation would include behavioral isolation in birds and pollinator isolation in nonpolyploid angiosperms.

Finally, conclusions derived from some associations between traits and diversity can be tested using other traits expected to have similar effects. As noted above, the presumed diversity-enhancing effect of nondioecy in plants was not supported by a similar analysis of the effect of selfing; yet in both cases the same forces should increase diversity. Given the predominance of sexually reproducing species—and the long-standing view that outcrossing taxa speciate more rapidly and resist extinction more strongly than their asexual or selfing relatives (e.g., Maynard Smith 1978)—it seems surprising that almost no sister-group studies have been performed on the relative diversity of predominantly sexual versus predominantly asexual or selfing taxa. Heilbuth (2000), in fact, observed no difference in plant species richness between clades of obligate outbreeders and their sister clades that can self-fertilize.

Species Selection

The existence of biological traits that facilitate speciation or diversification means that a higher-order form of selection—"species selection"—might produce evolutionary trends not predictable from selection acting *within* species. Consider the positive association between diversification rate in birds and their degree of sexual dichromatism or ornamentation (Barraclough et al. 1995; Møller and Cuervo 1998). These morphological traits presumably indicate stronger sexual selection, which could promote diversity by increasing the rate of speciation involving behavioral isolation. Within a species, sexual selection replaces alleles for sexually monomorphic traits with alleles producing dimorphic traits. Yet, dimorphic species might also speciate faster than monomorphic species. Over time, then, the proportion of bird species that are sexually dimorphic will increase, as will the proportion of all *individual* birds that have sexually dimorphic traits. Both of these changes could be considered evolutionary trends.

Such trends represent a higher-level form of sorting that can be viewed as species selection. In this process, entire species are selected as units having their own group fitness: D, the relative diversification rate. Differential diversification is not reducible to population-genetic phenomena occurring within species. The enhanced diversification of sexually selected bird species, for example, cannot be derived from the relative fitness of sexually selected genotypes *within* species. Nor can the increase in the total number of dimorphic individuals be understood without knowing the effect of sexual selection on speciation rate. The idea that species selection causes evolutionary trends was a major pillar of the theory of punctuated equilibrium (Eldredge and Gould 1972; Gould and Eldredge 1977), although Gould (2002) later acknowledged that he knew of no completely convincing cases of such selection.

Species selection requires the same three preconditions as individual selection: variation, heritability, and differential reproduction. Variation exists among species for both individual traits that may affect diversification (e.g., biotic vs. abiotic pollination) and probably for some "species-level" traits as well (e.g., the degree of range fragmentation). There will also be heritability for many species-level traits, so that daughter species acquire the traits of their ancestors. Finally, Table 12.2 gives ample evidence for differential reproduction (diversification) of species having different traits. This differential diversification might be seen as species selection acting on those traits. Table 12.2 thus gives strong evidence for this controversial process.

The literature, however, is filled with diverse and conflicting definitions of species selection (Vrba 1983, 1989; Damuth 1985; Lloyd and Gould 1993; Grantham 1995, 2001; Stanley 1998; Gould and Lloyd 1999; Jablonski 2000). Many of these arguments seem more semantic and philosophical than scientific. Nevertheless, there are two possible problems with defining species selection on a trait as the differential diversification of groups containing that trait. First, traits associated with diversification may not themselves cause speciation or extinction, but may merely be associated with other traits that are the real cause of diversification. Second, some biologists (e.g., Vrba 1983) make a distinction between what they consider to be "true" species selection and a different process, the *effect hypothesis* (sometimes called *species sorting*). The effect hypothesis refers to evolutionary trends caused by properties of *individuals* within a species, which in aggregate have a net effect on diversification. Most of the traits listed in Table 12.2, like sexual dimorphism, mode of pollination, and degree of specialization, are of this type. "True" species selection, on the other hand, is said to occur only when "the differential reproduction or extinction of species is caused by heritable fitness of *species-level traits*" (Grantham 1995, p. 305, our italics). Species-level traits are emergent properties of the species that cannot be reduced to the traits of its members. Such traits are said to include geographic range (Jablonski 1987), genetic variability of the species (Lloyd and Gould 1993), and population structure (Jablonski 1986). Emergent traits associated with diversification in Table 12.2 include range size, range fragmentation, and mean latitude.

The problem with identifying "true" species selection is thus deciding whether a trait is indeed emergent, that is, whether the trait interacts with the environment in a way that cannot be derived from the properties of individuals. This is both a conceptual and an empirical question. Its difficulties are best demonstrated by the trait most often cited as an emergent property affecting diversification—the geographic range of fossil marine molluscs, Here, range size is positively correlated with reduced extinction rate (Jablonski 1987). The problem is that one can view range size either as a species-level trait or as a product of the properties of individuals. Larger ranges, for example, might reflect individuals' greater ability to migrate, or their adaptation to a wider range of food or environments.

An empirical demonstration of the emergent versus nonemergent difficulty comes from marine gastropods of the late Cretaceous. In this group, range size is highly correlated with the mode of larval dispersal. Species with highly planktonic larvae have ranges nearly five times larger than those whose larvae are only temporarily planktonic or are brooded (Jablonski 1986). The important question is whether range size itself, rather than the traits of individuals, is "seen" by the environment as the unit of selection. In other words, is range size the *interactor*—the trait directly acted on by selection—while individual traits contributing to range size are only *replicators* (Hull 1980)? This distinction can be clarified by analogy with individual selection. In the evolution of cryptic coloration, for example, the phenotype detected by predators—the object of selection—is the interactor, while the genes producing crypsis, which increase in frequency as a result of selection, are the replicators.

Finding examples of "true" species selection is problematic because it is hard to separate the features of individuals from those of species. In the case of marine gastropods, we know of no way to determine whether individual or emergent traits were the real cause of enhanced diversification. As Gould and Lloyd (1999, p. 11908) note, "emergent characters are difficult to define because their elucidation usually requires considerable historical knowledge of details in the ancestral history of particular lineages to distinguish truly emergent traits from cascading effects of properties built at other levels. Such density of historical information can rarely be obtained from imperfect fossil records."

Lloyd and Gould's (1993) solution, which seems reasonable, is to consider species selection as resulting not from emergent *characters*, but from emergent *fitnesses*. Emergent fitnesses arise when species differing in a trait have different rates of diversification as a result of that trait. Under this characterization, formalized mathematically by Damuth and Heisler (1988), aggregate, nonemergent traits can still undergo species selection because they affect diversification rates in ways that cannot be extrapolated from the relative fitnesses of genotypes *within* species. To predict how many species will show sexually selected traits in the future requires knowing not only the relative fitnesses of individuals carrying different traits within species, but also the relative diversification rates of sexually dimorphic versus monomorphic species.

Thus, if we define species selection on a trait as *repeatable effects of that trait on the rate of diversification of species possessing it*, we avoid unproductive arguments about whether or not selection acts on emergent properties. Measuring emergent fitnesses is in principle straightforward: it is simply the relative species richness across taxa given in Table 12.2. Because various factors can affect diversification, differential species "fitness" can be calculated post facto even when there is no real effect of the character on diversification. Gould (2002) calls such evolution "species drift." What distinguishes species selection from species drift (and organismal selection from genetic drift) is the *repeatable* effect of the trait on diversification—an effect that can be shown by comparative analysis of sister groups.

In any study of species selection, however, one must consider the possibility that differential diversification might be caused not by differences in rates of speciation or extinction, but by the different evolutionary *lability* of traits. That is, even if alternative forms of a trait have equal effects on speciation and extinction rates, the trait that is less likely to revert to the alternative state will appear in more species as time goes on, forming an evolutionary trend. This process—called "directional speciation" by Gould (2002)—is the macroevolutionary analogue of microevolution resulting from unequal mutation rates between characters. The evolutionary lability of a trait can be seen in the fossil record by determining the relative number of switches to alternative states, regardless of their association with species richness. Duda and Palumbi (1998) for example, show that in marine gastropods of the genus *Conus*, nonplanktonic larvae originated far more often than planktonic larvae. Nosil (2002) found that the change from generalist to specialist habits in phytophagous insects occurred significantly more often than the reverse transition. Statistical analysis of phylogenies can be used to determine whether differential diversification rates are caused by differential probabilities of character change (Nosil 2002).

Is it possible for such differential character shifts to overcome the relative effects of the characters on diversification rate, so that a trait causing higher diversification rates can actually become less frequent over time because it evolves only rarely? We know of no cases of this phenomenon. There is, however, one example of the opposite type: a trait that evolves more easily than its alternative but is associated with *lower* diversification. That trait is asexual reproduction. Nearly all asexual eukaryotes evolved from species that reproduced sexually, and nearly all predominantly selfing plants evolved from outcrossing plants. In both situations, evolution has almost never gone in the opposite direction, probably because of the difficulty of re-evolving sexual reproduction and self-incompatibility once they are lost. This asymmetry, however, has not overcome the much higher diversification rate of sexually reproducing species compared to either asexuals or selfers.

In conclusion, we regard most of the examples given in Table 12.2 as true cases of species selection. However, we are most interested in those cases in which the character of interest is likely to have promoted the evolution of reproductive isolation. Most of our examples do not provide this information, but the best candidates for traits facilitating speciation are phytophagy and angiosperm feeding in insects, biotic pollination in angiosperms, and those features that increase the strength of sexual selection in birds and insects.

Those who continue to debate the possibility of species selection fail to realize that comparative studies have already settled the issue. What remains is to determine how often this type of selection has shaped evolutionary trends.

APPENDIX

A Catalogue and Critique of Species Concepts

Here we describe and evaluate eight species concepts that are considered serious competitors to the biological species concept (BSC) (Table 1.1). We describe the reasons why each concept was proposed (i.e., the "species problem" it was designed to solve), explain why its proponents see it as superior to the BSC, and assess its advantages and disadvantages. We also show how each concept deals with issues that are problematic for the BSC: allopatric populations, gene exchange between taxa that remain distinct, and uniparental organisms. We note how closely each concept coincides with the BSC—that is, whether it identifies the same species in sympatry. Finally, we note what process constitutes "speciation" under each concept. Throughout the discussion, we adhere to our version of the BSC, which allows limited gene exchange, rather than to the strict version that demands complete reproductive isolation between taxa.

We will not deal with strictly typological species concepts—those that define species by specifying an arbitrary degree of morphological or genetic difference. Mayr (1942, 1963) has explained the problems with such concepts. We do, however, discuss two somewhat typological concepts: the "genotypic cluster" species concept and several versions of the phylogenetic species concept.

Genotypic Cluster Species Concept (GCSC)

> A species is a [morphologically or genetically] distinguishable group of individuals that has few or no intermediates when in contact with other such clusters (Mallet 1995).

This concept was proposed in response to the observation that, while the BSC *defines* species by the presence or absence of interbreeding, it *recognizes* them as distinguishable clusters in sympatry. (These clusters can be seen in phenotypic data as a bimodal distribution of traits, and in genetic data as a deficit of heterozygotes or the presence of linkage disequilibrium among genes.) For advocates of both the BSC and the GCSC, the species problem is identical: understanding the origin of discrete entities in sympatry. Unlike the BSC, however, the GCSC defines species solely by the features used to recognize them. The GCSC does not specify how many traits and/or genes are required to diagnose sympatric clusters as species.

Advocates of the GCSC claim that it has several advantages over the BSC. First, the GCSC is supposedly independent of theories about speciation: it is presented as a way to recognize species rather than understand how they evolved. Defining clusters on the basis of interbreeding is said to lead the BSC into circularity: "Since theories of speciation involve a reduction in ability or tendency to interbreed, species cannot themselves be defined by interbreeding without confusing cause and effect" (Mallet 1995, p. 295; all quotations and page numbers refer to this paper). Mallet feels that the GCSC allows one to consider other causes of clustering besides reproductive isolation: "Gene flow is not the only factor maintaining a cluster; stabilizing selection will also be involved, as well as the historical inertia of the set of populations belonging to the cluster.... Clusters can remain distinct under relatively high levels of gene flow provided that there is strong selection against intermediates" (p. 296).

Proponents of the GCSC view the BSC's emphasis on isolating barriers as not only intellectually vacuous, but misleading: "To include such a number of different effects under a single label must be one of the most extraordinary pieces of philosophical trickery ever foisted successfully on a community of intelligent human beings" (pp. 297–298). The BSC is also considered unscientific: "Mayr has repeatedly stressed that the biological concept cannot be refuted by practical difficulties in its application; this means it is untestable" (p. 296). Moreover, the notion of species as reproductively isolated entities is said to impede our understanding of speciation: "Because no gene flow between species is conceptually possible under interbreeding concepts, it is extremely hard to imagine how speciation, which must often involve a gradual cessation of gene flow, can occur" (p. 295). Mallet notes that allopatric speciation is one such mode of speciation, but also argues that the BSC is biased against other modes of cluster formation—such as sympatric and parapatric speciation—that involve gene flow between incipient species. Finally, Mallet echoes the criticism of Sokal and Crovello (1970) that it is impossible to apply the BSC in practice, as this requires making or observing the crosses needed to test the reproductive compatibility of every pair of individuals.

It is important to recognize that, despite its emphasis on species recognition rather than reproductive isolation, the GCSC and our version of the BSC identify nearly the same set of species in sympatry. The real disparity between these concepts is in the amount of genetic difference required for species status. In principle, the GCSC could diagnose sympatric clusters that differ in only one or two genes or traits while exchanging alleles freely throughout the rest of the genome. Such clusters could be maintained by habitat-related selection. For example, Wilson and Turelli (1986) show that density- and frequency-dependent selection acting on alleles at one or two loci can lead to the stable coexistence of distinct genotypes, even though heterozygotes have the lowest fitness and appear in less-than-expected frequencies. This will produce statistically distinguishable clusters that might be considered species under the GCSC. If one claims, however, that such polymorphisms do not diagnose species because they represent only intraspecific variation, then one is reverting to the BSC. The failure to specify how many loci (or what degree of heterozygote deficit and linkage disequilibrium) are required to diagnose species is a problem for the GCSC. Setting such a threshold would involve an arbitrary decision.

In contrast, the BSC diagnoses species only if there is evidence that gene flow between them is strongly limited. This involves either observing many genetic differences between sympatric taxa—a degree of difference too large to be explained by disruptive selection alone—or observing isolating barriers so strong that gene flow is almost zero. This can also involve an arbitrary decision if there is any introgression between sympatric groups, but clearly genetic differentiation must be higher for recognizing biological than for recognizing genotypic-cluster species.

Another problem for the GCSC involves the *level* of clustering. Because of the hierarchical nature of evolution, genotypic clusters occur at many levels. These clusters can involve intrapopulation polymorphisms, local host races, species, or higher-level groups such as genera. Trying to apply the GCSC to the *Rhagoletis pomonella* complex of tephritid flies, Berlocher (1999, p. 661) observed a "continuum of decreasing degree of cluster overlap as level of genetic divergence increases from host race to distinct species. . . . No species threshold is apparent." Since the GCSC sees no fundamental distinction between species and higher-level groups (p. 296: "Whether species do have a greater 'objective' reality than lower or higher taxa is either wrong or at least debatable; the idea that taxa are qualitatively different from other taxa is therefore best not included within their definition"), the definition of species as "genotypic clusters" must be recast as "clusters that do not include other subclusters." But this would lead one to diagnose as species polymorphic forms such as beak morphs in *Pyrenestes* finches (Smith 1987) or Batesian mimicry phenotypes in the butterfly *Papilio memnon* (Clarke and Sheppard 1969). To get around this problem, one must then include a reproductive criterion: such polymorphisms do not diagnose species because their carriers readily interbreed. This, however, defeats the GCSC's goal of avoiding criteria based on reproductive compatibility.

As noted in Chapter 1, the presence of sympatric clusters involving several genes implies the existence of isolating barriers. Thus, disruptive selection that creates and maintains distinct clusters involves a form of reproductive isolation—extrinsic postzygotic isolation. This is recognized in statements such as, "The maintenance of sympatric species is not just due to reproductive traits, but also due to ordinary within-species, stabilizing ecological adaptations that select disruptively against intermediates or hybrids" (Mallet 1995, p. 296). There is no real difference between positive selection for alternative phenotypes and negative selection against intermediates between those phenotypes.

As for other processes that can create and maintain species, we do not understand how "historical inertia" can play such a role. The maintenance of distinct clusters in sexually reproducing taxa must always involve selection against intermediates (i.e., reproductive isolation). We also fail to see why the BSC is circular. If species are regarded simply as an advanced stage in the evolution of reproductive isolation, then no circularity ensues. We are baffled by the claim that lumping diverse phenomena under the category "isolating mechanisms" involves "philosophical trickery." As Harrison (1998, p. 24) argues: "[It] is the common effect of all these differences (limiting or preventing gene exchange) that provides the rationale for grouping them. I see no reason not to adopt a single term (e.g., 'barriers to gene exchange') to refer to the set of differences that have this very important effect."

Moreover, the fact that the BSC is not theory-free—that it immediately suggests a *process* of speciation—seems to us an advantage, not a problem. A theory-free definition of identical twins might be given as "two individuals, born of one mother at the same time, who are exceedingly similar morphologically." But this definition is surely less useful than one that incorporates process, such as "identical twins are the products of splitting of a single fertilized egg." The claim that the BSC is untestable holds, as Brookfield (2002) notes, for *all* species concepts: none can be falsified by experiment or observation. A species concept is a tool for research, not a hypothesis subject to refutation.

The claim that under the BSC "it is extremely hard to imagine how speciation, which must often involve a gradual cessation of gene flow, can occur," seems unfounded. Over the last 60 years, biologists have had no problem imagining how biological speciation can occur. There are well-established genetic and ecological models—both verbal and mathematical—for the origin of isolating barriers in sympatry, allopatry, and parapatry.

Finally, the view that the BSC is not useful because one cannot do breeding tests seems misguided. While breeding tests can be useful for identifying species (e.g., Dobzhansky and Epling 1944), biological species can also be identified by the concordance of many characters and genes that show the existence of isolating barriers, or by the consistent correlation between a group of traits on one hand and reproductive compatibility on the other. Once one has described such a group, even a single trait can then be used to diagnose species. Traditionally, this has been done with great success: genitalia are reliable indicators of biological species status in many organisms (Eberhard 1986), and chromosomal and molecular characters have served equally well.

How does the GCSC differ from the BSC? The most important aspect is how these concepts deal with sympatric taxa showing moderate to substantial gene exchange. Such taxa include sympatric host races of insects such as the apple and hawthorn races of *Rhagoletis pomonella* (Feder et al. 1998), which form GCSC species but would probably not be accorded species status by the BSC. Hybrid zones are another example. If one considers a wide area including the zone of hybridization, one can recognize two genetic clusters with various intermediate individuals (hybrids) falling between the clusters. The two clusters would be considered species under the GCSC but not the BSC unless there was little gene flow beyond the hybrid zone. However, introgression outside hybrid zones is often limited (Barton and Hewitt 1985), allowing one to diagnose two biological species that are identical to the two genotypic-cluster species.

Both the BSC and the GCSC have difficulty diagnosing *allopatric* taxa that are morphologically distinguishable. (Genotypic clusters are recognizable only in sympatry.) Here, however, the BSC has something of an edge: if allopatric populations form either inviable or sterile hybrids when artificially hybridized, one can say with assurance that they are biological species.

There are arguably two advantages of the GCSC over the BSC. First, the GCSC is less ambiguous at diagnosing species in problematic situations such as taxa that hybridize with limited gene flow. Such cases constitute a gray area for the BSC, but offer no problem to the GCSC if one can observe distinct clusters. But because we are more concerned with process than with diagnosis, we do not consider this a particularly meaningful advantage, especially because GCSC clusters may involve only one or two genetic differences. Second, the GCSC can also be applied to largely or completely asexual taxa, groups where the BSC is impotent. But strict use of the GCSC would diagnose each asexual clone as a different species. For such groups it seems preferable to adopt neither the BSC nor the GCSC, but an ecological species concept (see Chapter 1).

The GCSC is one of the few species concepts that come with an explicit definition of speciation: "Speciation is the formation of a genotypic cluster that can overlap without fusing with its sibling" (Mallet 1995, p. 298). This differs from our own notion of speciation only in that we define "clusters" as "groups between which reproductive barriers are very strong." However, Mallet adds (p. 298), "To understand speciation, we need to understand when disruptive selection can outweigh gene flow between populations." This applies only to parapatric and sympatric speciation, because the conflict between selection and gene flow does not exist during allopatric speciation.

The GCSC is the most serious competitor to the BSC because the two concepts share many features. But by concentrating on the identification rather than the origin of species, the GCSC does not yield a particularly fruitful program of research.

Recognition Species Concept (RSC)

> Species are the most inclusive population of individual biparental organisms, which share a common fertilization system (Paterson 1985; see also Lambert and Paterson 1984, Lambert et al. 1987, and Masters et al. 1987).

This concept is also motivated by the problem of organic discontinuity. The RSC resembles the GCSC and the cohesion species concept (see below) in that species are defined by those factors that hold populations together rather than by those that isolate them.

Under the RSC, these cohesive factors constitute their shared "fertilization system": the set of biological features that "contribute to the ultimate function of bringing about fertilization while the organism occupies its normal habitat" (Paterson 1985, p. 24). Important aspects of the fertilization system are included in the Specific-Mate-Recognition-System (SMRS), which includes all features by which organisms "recognize" each other as mates. This recognition can be either active (as in courtship signals and responses), or passive (as in biochemical processes of gamete fusion). Paterson's working definition of a species is a "field for gene recombination" (1985, p. 21). The RSC thus explicitly excludes ecological and temporal isolating barriers, as well as all postzygotic barriers.

Paterson and co-authors claim that the RSC remedies many of the weaknesses they find in the BSC (which they call the "isolation concept") because of the BSC's presumed concentration on species distinctness rather than cohesion. Detailed analysis and criticisms of this theory have been published elsewhere by ourselves and others (Butlin 1987a; Raubenheimer and Crowe 1987; Coyne et al. 1988; Templeton 1989). We refer the reader to these papers and to the counterarguments of Spencer et al. (1987) and Masters and Spencer (1989).

As noted by Coyne et al. (1988), the RSC can be considered a subset of the BSC that involves a limited set of isolating barriers (behavioral, pollinator, and gametic). The RSC excludes other barriers that can create and maintain discrete clusters in sympatry. In Chapter 3, we show that these excluded barriers have clearly played a major role in speciation. In polyploidy, for example, new taxa are created by a combination of intrinsic postzygotic and ecological isolation. We see no advantage, and considerable disadvantage, in concentrating on only the subset of isolating barriers that involve mating and fertilization. Situations that are problematic for the BSC are equally problematic for the RSC. Moreover, the RSC faces additional problems, such as how to deal with cases in which there is some hybridization but no introgression because hybrids are sterile or inviable.

Cohesion Species Concept (CSC)

> A species is the most inclusive population of individuals having the potential for phenotypic cohesion through intrinsic cohesion mechanisms (Templeton 1989).

The CSC gives a mechanistic underpinning to the GCSC by attempting to include all the factors that *preserve* morphological and genetic clusters in sexual and asexual organisms. It thus takes as its species problem the existence of discrete clusters, but, like the recognition concept, the CSC emphasizes factors keeping members of a cluster together more than those keeping members of different clusters apart.

The CSC is sometimes considered superior to the BSC for two reasons. First, it sees reproductive isolation as a misleading way to think about speciation (Templeton 1989, p. 5; all quotations and page numbers refer to this paper):

> For example, under the classic allopatric model of speciation, speciation occurs when populations are totally separated from each other by geographic barriers. The intrinsic isolating mechanisms given in Table 1 are obviously irrelevant as isolating barriers during speciation because they cannot function as isolating mechanisms during allopatry. Hence the evolutionary forces responsible for this allopatric speciation process have nothing to do with 'isolation.'

This is said to cause confusion for adherents to the BSC (p. 6):

> This is not to say that [reproductive] isolation is not a product of the speciation process in some cases, but the product (i.e., isolation) should not be confused with the process (i.e., speciation). The isolation concept has been detrimental to studies of speciation precisely because it has fostered that confusion (Paterson 1985).

We do not understand the rationale for separating the process of speciation (the evolution of barriers to gene exchange) from its product (species themselves). Under the BSC, speciation cannot be equated with simple differentiation of populations, because without the evolution of barriers to gene exchange, distinct taxa cannot coexist in sympatry. It is not difficult for us to see species as simply an advanced stage in the evolution of such barriers. And we cannot point to a single case in which research on speciation has been hindered by confusion between process and product.

Second, the CSC is deemed superior to the BSC because it can diagnose species in two difficult cases: asexuality, and hybridization between sympatric groups that nevertheless maintain their distinctness as clusters.

The difference between the GCSC and the CSC is that the latter incorporates explanations for why individuals within clusters remain genetically and phenotypically *similar*. Templeton describes a number of factors, called "cohesion mechanisms," that enforce this similarity. These mechanisms fall into two classes. "Genetic exchangeability" mechanisms include all factors "that define the limits of the spread of new genetic variants through *gene flow*"(p. 13). These include not only the complete list of reproductive isolating barriers characterizing the BSC, but also mechanisms facilitating gene flow *within* clusters, such as a common fertilization system ("the organisms are capable of successfully exchanging gametes") and a common developmental system ("the products of fertilization are capable of giving rise to viable and fertile adults"). But all these mechanisms are simply a different way of describing isolating barriers. A species can be seen as a group of interbreeding populations (i.e., conspecific individuals share "fertilization and developmental systems"), while *different* species can be seen as groups of populations whose fertilization and developmental systems are sufficiently diverged to prevent gene exchange.

The novel aspect of the CSC is the emphasis on cohesion mechanisms that enforce demographic exchangeability. This includes all factors "that define the fundamental niche and the limits of spread of new genetic variants through genetic drift and natural selection" (p. 13). (The fundamental niche is considered "the intrinsic [i.e., genetic] tolerances of the individuals to various environmental factors that determine the range of environments in which the individuals are potentially capable of surviving and reproducing" [pp. 14–15].) Within sexually reproducing populations, of course, selection and genetic drift promote genetic homogeneity of a species. But understanding how these forces operate within a cluster does not explain how *distinct* sympatric clusters arise.

Demographic exchangeability becomes more important when dealing with asexual or uniparental populations, because this factor—and not reproductive isolation—may limit the spread of alleles by natural selection and genetic drift. As noted in Chapter 1, the origin of a new adaptive mutation in a population of bacteria produces a periodic selection event, during which the mutant clone replaces all other clones having similar ecological properties. Such replacement can also occur through the asexual equivalent of genetic drift: random differences in reproductive rates among demographically exchangeable clones. As Templeton notes (p. 15), "Every individual in a demographically exchangeable population is a potential common ancestor to the entire population at some point in the future."

One might therefore use the CSC to demarcate species or taxa in asexually reproducing groups. It has been so used by Cohan (2001), who connects this species concept to an explicit mechanism for speciation in bacteria. (As we note in Chapter 1, we do not object to applying the words "species" and "speciation" to asexual groups, so long as one recognizes that these words have a different meaning in sexual groups.) We are less convinced that the CSC works better than the BSC in sexually reproducing organisms. Nor do we feel that a species concept is better when it applies to both sexual and asexual groups rather than to sexual groups alone. If the processes of cluster formation differ between these two groups, as we suspect they do, then adopting a single species concept for both groups may impede rather than promote progress.

In many cases, especially those involving sympatric sexual taxa, both the BSC and the CSC identify the same clusters, for the CSC considers isolating barriers to be "cohesion mechanisms." In other situations, however, the CSC encounters the same difficulties as does the BSC. When forced crosses show that allopatric populations have complete intrinsic postmating isolation, they would presumably be regarded as good species by both the CSC and BSC. But if isolation is not complete, there is no way to diagnose these populations under either concept, for it is impossible to determine the "fundamental niche" of allopatric taxa. Thus, the CSC also faces problems with allopatric populations. Both concepts also have difficulties when dealing with groups, such as host races, that show some gene exchange but that nevertheless remain distinct.

Such hybridizing entities show genetic exchangeability (an adaptive allele can spread between races), but not demographic exchangeability (one group cannot ecologically displace the other).

There are other situations in which the BSC can diagnose species but where the CSC fails because the criteria of genetic and demographic exchangeability conflict. Consider, for example, two sympatric, reproductively isolated species that compete for resources. A new mutation may arise in one species that allows it to outcompete the other, driving it to extinction. In such cases—and in any case in which an invader outcompetes a local species—the two groups are genetically nonexchangeable but demographically exchangeable. Under the BSC they are good species, but under the CSC their status is unclear.

The main problem with the CSC, however, is that it causes confusion, especially through its emphasis on "cohesion mechanisms." As Harrison (1998 pp. 24–25) notes:

> Many (perhaps most) biological properties of organisms that confer "cohesion" did not arise for that purpose. They are also effects not functions! Thus, life cycles that result in adults appearing at the same season, or habitat/resource associations which lead to aggregation of individuals in particular places, facilitate fertilization or lead to genetic and/or demographic cohesion. But in most cases, life cycles and habitat associations have not been molded by selection for the purpose of "cohesion."

It is hard to regard forms of natural selection that can create isolating barriers as "cohesion mechanisms." The fixation of adaptive alleles that cause reproductive isolation as a byproduct do not involve selection for cohesion. Rather, it is the reproductive cohesion of the group that allows such alleles to spread. In such cases the CSC reverses cause and effect. Moreover, not all aspects of sexual reproduction can be regarded as "cohesion mechanisms." Antagonistic sexual selection, produced by differing reproductive interests of males and females, may be important in speciation (see Chapter 8). But such selection is a manifestly *non-cohesive* evolutionary force.

Under the BSC, it is fairly clear when speciation has occurred—substantial barriers to gene flow exist. Under the CSC, however, speciation is seen as "the process by which new genetic systems of cohesion mechanisms evolve within a population," or as "the genetic assimilation of altered patterns of genetic and demographic exchangeability into intrinsic cohesion mechanisms" (p. 24). But how can one know whether these processes have caused speciation unless one observes isolating barriers between a population and its relatives?

Finally, the CSC does not seem to lead naturally to a research program that reveals the causes of clustering in sexually reproducing groups. The concept of demographic exchangeability, however, may give insight into the origin of clusters in asexual organisms.

Evolutionary Species Concept (EvSC)

A species is a single lineage of ancestral descendant populations or organisms, which maintains its identity from other such lineages and which has its own evolutionary tendencies and historical fate (Wiley 1978, modified from Simpson 1961).

The EvSC differs from the BSC and other concepts discussed above by endowing species with broader evolutionary significance. Under the EvSC, the species problem becomes the recognition of evolutionarily independent entities, and the species is the unit that evolves independently of other species. Wiley asserts that the EvSC is more universally applicable than the BSC because the EvSC deals with both sexual and asexual taxa. Moreover, the EvSC, unlike the BSC, is said to be "capable of dealing with species as spatial, temporal, genetic, epigenetic, ecological, physiological, phenetic, and behavioral entities" (Wiley 1978, p. 18; all quotations and page numbers refer to this paper).

The EvSC differs from the BSC by including no explicit mention of genetic interchange or reproductive isolation. Nevertheless, the arguments in favor of the EvSC show that in most cases it is equivalent to the BSC, at least for diagnosing species in sympatry.

Indeed, this equivalence is recognized by Wiley: "Separate evolutionary lineages (species) must be reproductively isolated from one another to the extent that this is required for maintaining their separate identities, tendencies, and historical fates"(p. 20). But the notion of "separate identities," which implies recognizable genotypic or phenotypic clusters, can conflict with the notion of separate "tendencies" and "historical fates." Two species that hybridize, for instance, may maintain separate identities, but even a small amount of hybridization can allow a generally advantageous mutation to spread from one group to the other, so that their evolutionary fates are connected. The grass *Agrostis tenuis*, for example, has developed local races that can survive high concentrations of lead on mine tailings, while adjacent populations lack the genes for tolerance. Tolerant and non-tolerant populations are often adjacent, and, being wind-pollinated, freely exchange most genes (McNeilly and Antonovics 1968). The populations have diverged in a few traits, but still hybridize pervasively. Are they different evolutionary species? The EvSC gives no clue. With gene flow, taxa may be evolutionarily independent at some loci and not others.

Allopatric populations that are genetically differentiated pose as many problems for the EvSC as for the BSC. While geographic isolation may seem to confer separate evolutionary fates, Wiley (1978, p. 23) notes that "we have no corroboration that this particular geographic event will lead to separate evolutionary paths and thus we have no reason to recognize two evolutionary species." Such recognition becomes possible only when "significant evolutionary divergence" occurs between allopatric populations (p. 23). Wiley, however, gives no idea of what constitutes significant divergence. If "significant" means "divergence that prevents the populations from exchanging genes were they to become sympatric," then the EvSC becomes the BSC.

Asexually reproducing taxa may be resolvable by the EvSC if significant divergence occurs between them. But the meaning of "significant" is again unclear. Are distinct clusters of many loci necessary to diagnose asexual species, or can they differ at only one or two loci?

The unique aspect of the EvSC is that it can deal with a single lineage evolving through time. According to Wiley, such a lineage is considered to be a single species so long as it does not branch, no matter how much evolutionary change it undergoes. This, of course, may result in some taxonomic confusion, as the same species name will often be used for very different organisms (consider the lineage leading to modern humans). But applying names to stages of a single evolving lineage is always an exercise in subjectivity.

The major problem with the EvSC, then, is that it cannot deal with gene flow between populations of sexually reproducing organisms. Unless one is precise about the meaning of "separate evolutionary tendencies and historical fates," decisions about species status become arbitrary. Clearly, greater evolutionary independence is conferred by stronger barriers to gene flow. In this sense, the EvSC approximates the BSC. Given the choice, we prefer the BSC because it is more useful: in sexually reproducing organisms, this concept explains the evolutionary independence of taxa (whose origin is a mystery under the EvSC) as a byproduct of isolating barriers.

Ecological Species Concept (EcSC)

> A species is a lineage (or a closely related set of lineages), which occupies an adaptive zone minimally different from that of any other lineage in its range and which evolves separately from all lineages outside its range (Van Valen 1976).

Van Valen proposed the EcSC to remedy the problem of ecologically differentiated entities that still exchange genes. He specifically mentions hybridizing oaks (see Chapter 1) as "cutting across the frame of reference of the now usual concept of species" (Van Valen 1976, p. 233; all quotations and page numbers refer to this paper). The EcSC resembles the EvSC except that the independently-evolving lineages are also characterized as occupying "minimally different adaptive zones." ("Minimal" is used so that higher taxa are not considered ecological species.) The species problem again seems to be that of explaining discontinuities among sympatric groups.

Requiring that different species occupy different adaptive zones imposes a severe burden on the EcSC. According to Van Valen, adaptive zones are defined a priori, independent of the organisms that inhabit them: "An adaptive zone is some part of the resource space together with whatever predation and parasitism occurs on the group considered. It is a part of the environment, as distinct from the way of life of a taxon that may occupy it, and exists independently of any inhabitants it might have" (p. 234). Van Valen suggests that occupants of different adaptive zones can be recognized by observing "a difference in the ultimately regulating factor, or factors, of population density" (p. 234).

This definition conflicts with the view that niches cannot be defined independently of their occupants, since many organisms, such as beavers and moles, change the environment to suit their needs (Lewontin 1983). Moreover, some groups can coexist as distinct entities in sympatry without gene flow, even though their adaptive zones are identical or nearly so. This is true, for example, of some temporally isolated species, such as periodical cicadas or the even- and odd-year races of pink salmon described in Chapter 5.

More important, it is often hard to determine whether two sympatric relatives occupy different adaptive zones, much less "minimally different" ones. The haplochromine cichlids of Lake Victoria, for instance, are often considered almost ecologically identical (see Chapter 4). In such cases, Van Valen suggests using a surrogate criterion: the coexistence of species in sympatry *proves* that they occupy minimally different adaptive zones. But this notion makes the EcSC operationally identical to either the GCSC or BSC, depending on the amount of hybridization. It is questionable, however, whether sympatric coexistence always constitutes evidence for "minimally different adaptive zones." As noted in Chapter 1, ecologists have suggested several ways that ecologically identical species can coexist. While we believe that differential resource use is widespread among closely related sympatric species, it may not be necessary.

Further, very different taxa may nonetheless occupy the same adaptive zone, as shown by competitive exclusion. Criticizing ecological species concepts, Wiley (1978, p. 24) notes,

> In the case where resources are limiting, one of the species could replace the other through interspecific competition from that portion of the range where they are sympatric, or entirely via extinction. Indeed, if interspecific competition causes at least some extinctions, it can work only where the niches of the competing species are similar enough for competition to occur or where one species' niche completely overlaps the other's . . . one might argue that a species forced to extinction through interspecific competition was not a species at all.

Such situations are common in nature, especially with introduced species. Should the Argentine ant, *Linepithema humile*, be considered conspecific with the unrelated *Pheidole megacephala* because the former outcompeted the latter in Bermuda (Crowell 1968)? It seems better to regard ecological difference as a criterion for species *persistence* than for species status.

Van Valen suggests that it is a matter of taste whether differentiated allopatric populations are considered different species, although it is, in principle, possible to determine whether such populations are regulated by different ecological factors. But he further argues that "reproductive isolation of allopatric populations is of minor evolutionary importance and hence needs little consideration" (p. 234). Thus, allopatric taxa that yield completely inviable or sterile hybrids in forced crosses might not be considered different ecological species. However, sympatric taxa that exchange genes but maintain phenotypic distinctness, such as oaks, *would* be considered separate species because their different traits imply different niches. Here the EcSC resembles the GSC if there is substantial introgression, but resembles the BSC if introgression is limited.

Finally, Van Valen suggests that the EcSC could be useful for distinguishing species in asexual groups (p. 235): "Species are maintained for the most part ecologically, not reproductively. Completely asexual communities would perhaps be as diverse as sexual ones, with numerous subcontinuities and even discontinuities. This suggests but does not require that the main criterion of species be ecological." We agree that the EcSC (and the CSC) might be more useful than the BSC in dealing with asexual groups, although, as noted by Van Valen, the EcSC encounters difficulties in agamic complexes.

Phylogenetic Species Concepts (PSCs)

PSCs differ markedly from the BSC and the five concepts discussed above, which take as their species problem the origin of discrete groups in nature. In contrast, PSCs are concerned with identifying historically related groups, and their species problem is reconstructing the history of life. Systematists are thus the main proponents of PSCs and the most severe critics of the BSC.

Systematists can be quite caustic when comparing the PSC to the BSC (e.g., Nelson 1989). This acrimony does not derive from their view that reproductive isolation is unimportant—for most of them admit that it is—but from the belief that it is largely irrelevant to reconstructing history. As Baum (1992, p. 1) notes, "The potential for gene exchange is only loosely coupled to historical relatedness—the central consideration of systematics." Wheeler and Nixon (1990, p. 79) state this position forcefully:

> The militant view that systematists need to embrace is that the responsibility for species concepts lies *solely* with systematists. If we continue to bow to the study of process over pattern, then our endeavors to elucidate pattern become irrelevant.

While most advocates of the BSC recognize that reproductive isolation may sometimes be inconsistent with evolutionary history (see below), they consider historical relationships as largely irrelevant to understanding the discreteness of nature.

Most modern systematists infer phylogenetic relationships using quantitative methods (Felsenstein 2004). One widely used method, cladistics, involves using shared derived characters, or *synapomorphies*. These characters can be either organismal traits or genes. When two or more species share a synapomorphy relative to an outgroup—a taxon known from independent evidence to be a more distant relative—these species are placed together in a monophyletic group (i.e., a group descended from one ancestral species). Thus, more closely related groups are those sharing more recently evolved synapomorphies.

It is reasonable to suppose that most sympatric species diagnosed by the BSC will be similarly diagnosed as monophyletic groups whose members have synapomorphies. But evolutionary history and reproductive compatibility need not coincide. Perhaps the most common cause of such discordance involves peripatric speciation: colonists originating in only one population of a species invade a new area, and their descendants evolve into a new species. Figure A.1 (A) gives an example of a phylogeny resulting from this scenario. Here, a common ancestor, species A, give rise to three taxa (B_1, B_2, and C). Taxa B_2 and C are the most closely related because they share a derived character or charac-

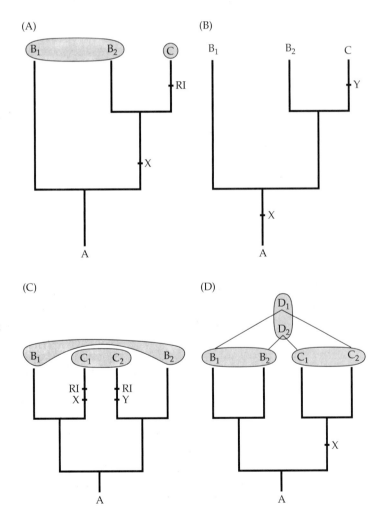

Figure A.1 Phylogenies showing disparities between evolutionary history and reproductive isolation (see text for discussion). (A) Speciation in a peripheral isolate. (B) Species diagnosed by a trait difference using Phylogenetic Species Concept 1 (PSC 1). (C) Parallel speciation yielding a polyphyletic biological species. (D) Allopolyploidy yielding a polyphyletic biological species.

ters, X, that evolved in their own common ancestor and not in the direct ancestor of population B_1. Taxon C, however, may have invaded a new habitat and evolved traits causing reproductive isolation (RI) from the two taxa B_1 and B_2, which themselves can interbreed. The BSC would recognize two species, C and [$B_1 + B_2$], with B_1 and B_2 considered conspecific populations. Most phylogenetic species concepts, however, would recognize a *different* pair of species, B_1 and [$B_2 + C$]. Using cladistics, one would not unite populations B_1 and B_2 based on their reproductive compatibility, for this compatibility is not a shared derived character but a *primitive* character retained from the common ancestor A (a "symplesiomorphy"). In technical terms, biological species [$B_1 + B_2$] is *paraphyletic* with respect to taxon C. That is, within the group [$B_1 + B_2$], members of B_2 are more closely related to members of C than to members of B_1.

In this simple case, taxa B_1 and B_2, although *capable* of interbreeding, do not. But this situation is unrealistic. After all, populations B_1 and B_2 are members of the same biological species and will exchange genes, erasing their distinctness. When this occurs, *some* genes will show the phylogeny depicted in Figure A.1 (A), while other genes will show B_1 and B_2 to be sister taxa, with C an outgroup. If speciation has occurred recently, different loci or traits will not yield congruent phylogenies, and the true history of populations cannot be reconstructed. The discrepancy between the histories of populations and the histories of genes within those populations is the biggest problem afflicting phylogenetic species concepts.

Reproductive isolation and evolutionary history can also conflict when two or more reproductively compatible populations arise independently from different evolutionary lineages, a situation known as "parallel speciation." Limnetic morphs of the threespine stickleback are often cited as an example (see Chapters 4 and 11). Figure A.1 (C) shows a phylogeny resulting from parallel speciation. Here, populations B_1 and B_2 are reproductively compatible, but each has a closer relative, C_1 and C_2, respectively. The latter two populations have evolved different derived traits (X and Y) but are reproductively compatible with each other and reproductively incompatible with B_1 and B_2. The BSC would diagnose two species, [$B_1 + B_2$] and [$C_1 + C_2$]. However, [$C_1 + C_2$] would not be recognized as a single species by most PSCs because it is *polyphyletic* (i.e., a taxon in which different populations or individuals have different common ancestors that reside outside the group). [$B_1 + B_2$] also fails to constitute a phylogenetic species because this entity is paraphyletic. Despite the reproductive relationships among these groups, most PSCs would recognize only a single species comprising the group [$B_1 + B_2 + C_1 + C_2$].

Polyphyly might seem to be rare given the implausibility that two independently evolved species would nevertheless be reproductively compatible with each other. But it may be common in one situation: polyploidy. Figure A.1 (D) shows a phylogeny in which hybridization occurs between two biological species, B and C, eventually producing the allopolyploid species D (see Chapter 9). Two independent hybridizations between different individuals or populations of B (B_1 and B_2) and C (C_1 and C_2) can produce two allotetraploid populations (D_1 and D_2) that are reproductively compatible with each other but incompatible with species B and C. The reproductive compatibility between individuals of species D conceals the fact that this polyploid species includes two groups with independent historical origins.

But recognizing the polyphyly of populations may be as difficult as recognizing the paraphyly of populations. We are not able to directly witness the history of populations, and so must infer it from gene-based phylogenies. In both cases discussed above (Figures A.1 C, D), interbreeding between populations of a biological species can quickly destroy our ability to reconstruct the history of populations, and thus our ability to show that this history is inconsistent with reproductive relationships.

Finally, some systematists dismiss the BSC because they view reproductive isolation as an *apomorphy*—a trait unique to one species—rather than as a synapomorphy that allows cladistic analysis. But reproductive isolation differs from traditional traits used by cladists, for it is not diagnosable in individuals of one taxon. Rather, reproductive isolation is an *interaction*, or joint property of two taxa. Such interactions cannot be incorporated into cladistic studies, although the traits underlying them can.

There are three main versions of the PSC:

1. *PSC1*   A phylogenetic species is an irreducible (basal) cluster of organisms that is diagnosably distinct from other such clusters, and within which there is a parental pattern of ancestry and descent (Cracraft 1989; see also Wheeler and Nixon 1990).

PSC1 is essentially a typological species concept that diagnoses species based on fixed differences in traits. (The term "irreducible" means that a species does not contain other diagnostic groups within it, so that large groups sharing diagnostic traits [e.g., mammals] are not deemed a single species. "Parental pattern of ancestry and descent" is included so that species status is not determined by sex differences or segregating polymorphic traits.)

Although advocates of PSC1 are not explicit on the point, in principle *any* trait can serve to diagnose a new species, even one as trivial as a small difference in color or a single nucleotide difference in DNA sequence. Applying PSC1 would thus tremendously increase the number of named species. *Homo sapiens*, for example, might be divided into several species based on diagnostic differences in morphology, molecules, or a combination of these features. (For diagnostic purposes, combinations of characters can be considered as single "traits.") Applying PSC1 to the birds of paradise, Cracraft (1992) increased the number of named species from about 40 to 90, often diagnosing as a new species an allopatric population having a slight difference in plumage color.

Like all phylogenetic species concepts, PSC1 cannot help us understand why organisms occur in discrete units, whether those units are defined phylogenetically or morphologically. However, its main difficulty is that its use may distort evolutionary history, the very problem it was meant to solve. Such distortion can occur because, under PSC1, species diagnosis is based not on shared derived traits, but on simple diagnostic traits. Figure A.1 (B) gives an example of such distortion in three taxa whose true evolutionary history is shown by the phylogeny. The common ancestor of three taxa, B_1, B_2 and C, evolves a trait X. This state is retained in the descendant species B_1 and B_2. In species C, however, the trait has changed to state Y. Using this trait, the PSC1 would diagnose two species: [$B_1 + B_2$] and C. This distorts the evolutionary history of the group because B_2 and C are more closely related to each other than either is to B_1. Similarly, using novel traits may diagnose a polyploid taxon as a phylogenetic species, even if it had a polyphyletic origin. There is no reason to expect that diagnostic traits will always mirror evolutionary history. Other criticisms of PSC1 are raised by Avise and Ball (1990), Baum (1992), and Baum and Donoghue (1995).

How well does PSC1 handle situations that are problematic for the BSC? In many cases, these two concepts pick out identical species in sympatry, especially when several traits are used.(For an example of this concordance, see Dettman et al.'s [2003] comparison of the BSC with PSC2.) Coordinated *sets* of diagnostic traits cannot be maintained in sympatry without some form of reproductive isolation. Confronting allopatric populations, PSC1 considers them different species if they differ in any trait. Likewise, PSC1 diagnoses each recognizable clone in an asexual group as a different species. Finally, under PSC1, speciation consists of the fixation of a diagnostic character in a lineage, making the process identical to divergent evolution. This type of speciation will occur faster than biological speciation: fixation of one new allele is undoubtedly faster than the evolution of reproductive isolation, which usually requires changes at several loci.

2. *PSC2* A species is the smallest (exclusive) monophyletic group of common ancestry (de Queiroz and Donoghue 1988; see also Rosen 1979, Mishler and Brandon 1987, and Baum and Donoghue 1995).

PSC2 goes back to Ronald Fisher, who suggested that all members of a sexually reproducing species should share "the effective identity of . . . remote ancestry" (1930, p. 124). This concept, updated in light of cladistics by de Queiroz and Donoghue (1988) differs from PSC1 by basing species recognition not on diagnostic characters, but on synapomorphies—shared derived characters that define monophyletic groups.

According to PSC2, a taxon is a species if cladistic analysis shows that it is monophyletic, exclusive (i.e., a group whose members are more closely related to each other than to those of any other group), and includes no other exclusive monophyletic groups within it.

When characterized properly, the units diagnosed by PSC2 will usually be congruent with evolutionary history. The main problem with this concept is operational: how can one *determine* whether a group is monophyletic and exclusive?

The problem arises from population genetics. One wants to know whether populations are exclusive groups sharing a common ancestry, but such a diagnosis can be made only using genes or genetically based traits. Increasingly, systematists rely on gene sequences to reconstruct this ancestry. DNA-based traits have two advantages over traditionally used morphological traits. First, genetic markers are more likely to be selectively neutral and thus to change in a more time-dependent fashion, making them useful for historical reconstruction. Second, gene sequences offer a nearly infinite number of characters, with each nucleotide potentially yielding information about ancestry.

However, the wealth of genetic data also creates a serious problem for the PSC2, because the ancestry of populations must be inferred from the ancestry of genes, and, as has been emphasized many times, *gene trees need not correspond to species trees*. That is, the historical branching pattern of taxa themselves need not coincide with the historical branching pattern of their genes (Avise and Ball 1990; Hey 1994; Avise and Wollenberg 1997).

This problem is demonstrated in Figure A.2 (A), which shows three taxa of haploid organisms, A, B, and C, derived from a common ancestor. The phylogeny of these taxa is represented by the "fat branches" of the tree. The prob-

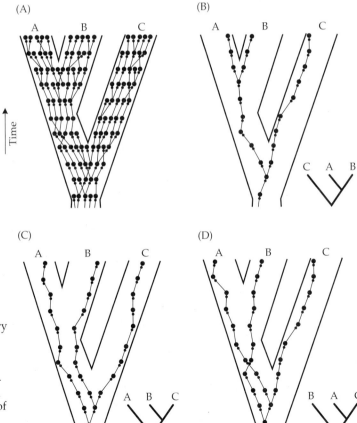

Figure A.2 Gene sorting occurring in a phylogeny whose true population history is shown in (A). Depending on which gene copies are picked for analysis, one can derive all three possible phylogenies, shown by the small diagrams to the lower right of diagrams (B), (C), and (D). (After Hey 1994.)

lem is to use the phylogenies of genes to reconstruct the phylogeny of the fat branches—the "true" population history. To illustrate the problems, we consider a single gene, designating each allele present in an individual with a dot. (The different gene copies do not necessarily differ in sequence, but we assume in this diagram that they can be individually identified.)

At the two successive "fat branching" events, gene flow between lineages is instantaneously prevented by geographic isolation or the rapid evolution of reproductive isolation. At each such split, the two descendant branches contain many different copies of each gene. Through the process of "gene sorting" via drift and selection over generations, these gene copies create their own genealogies—"thin branches"—with some copies leaving no descendants, and others copies leaving varying numbers of descendants. ("Splitting" of a gene phylogeny in Figure A.2 (A), reflects passage of a gene copy to more than one descendant, not new mutations or recombination events, which we ignore.) Gene copies present in the common ancestor can persist in descendants, often for long periods after the populations branch. Eventually, selection and drift will cause all gene copies within a lineage to descend from a single ancestral copy occurring within that lineage (that is, a *coalescence* occurs). When this happens, the gene has become monophyletic within the fat branch.

Until coalescence takes place, however, there can be substantial disparity between the true genealogy of the populations (i.e., A and B are sister groups with respect to the outgroup C), and the genealogy inferred from genes. Figure A.2 (B), (C), and (D) show that, using a single gene, one can obtain all three possible phylogenies between populations, only one of which gives the true population history. Although the populations have become evolutionarily independent taxa at the moment of isolation, in the sense that each now contains a nonoverlapping set of ancestors and descendants, one cannot genetically *demonstrate* that they are monophyletic until considerable time has passed. As noted by Avise and Ball (1990), after two populations become isolated, their genes will go through successive stages of polyphyly and paraphyly before finally becoming *reciprocally monophyletic*—the stage when all gene copies in each population are more closely related to each other than to copies in the other population.

This problem cannot be remedied by using larger samples of alleles or genes, because until reciprocal monophyly is attained, one can obtain conflicting phylogenies using different genes. This can be seen in Figure A.3 for two genes, *zeste* and *YP2*, sampled in four species of *Drosophila* (Hey and Kliman 1993). This group had a common ancestor that existed about 2.5 million years ago. *D. melanogaster* (mel) is an outgroup to the three species *D. simulans* (sim), *D. sechellia* (sec), and *D. mauritiana* (mau). All four species are distinguishable morphologically and show either substantial or complete reproductive isolation. Like *D. melanogaster*, *D. simulans* is cosmopolitan, while *D. sechellia* and *D. mauritiana* are endemic to the islands of the Seychelles and Mauritius, respectively. The endemics presumably arose after colonization of the islands by their common ancestor with *D. simulans*.

Figure A.3 Phylogenies based on sequences of copies of two genes (*zeste* and *Yolk protein 2* [*YP2*]) in four related species of *Drosophila* (sim = *D. simulans*, sec = *D. sechellia*, mau = *D. mauritiana*, and mel = *D. melanogaster*). Both sequences show that *D. simulans* is "paraphyletic," but the paraphyly is almost certainly of these genes and not of the populations themselves. (After Hey and Kliman 1993.)

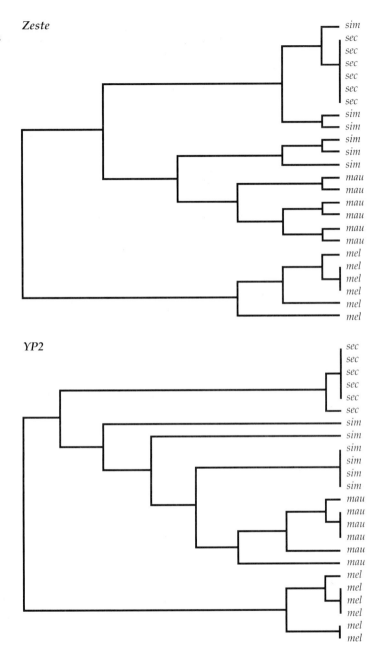

While the phylogeny of *D. melanogaster*, *D. sechellia*, and *D. mauritiana* is resolved under PSC2 using both genes, some sequences from *D. simulans* are more closely related to sequences found in *D. mauritiana* or *D. sechellia* than to other *D. simulans* sequences. That is, *D. simulans* is a paraphyletic species. Some systematists suggest that in such cases the paraphyletic taxon should be called a "metaspecies," so that individuals of *D. simulans* would not be recognized as belonging to *any* species (de Queiroz and Donoghue 1988; Baum and Shaw 1995). A study of 12 additional loci in this group gave similar results, failing to resolve the branching order of the two island colonizations (Kliman et al. 2000).

The most likely explanation for the discordant genealogies among genes is that *D. simulans* is still polymorphic for ancestral alleles that have become monophyletic in its island relatives. If geographic and genetic isolation persists, these species will eventually become monophyletic at all loci, but, as we show in the next section, this may take a very long time. Moreover, if there is any hybridization between the taxa (and there is some evidence for this in the *D. simulans* group), or if balancing selection maintains identical polymorphisms in different species, some genes will *never* become monophyletic within a lineage, and PSC2 status will never be attained.

The PSC2 is thus problematic because it ignores the distinction between monophyly of species and monophyly of genes. The latter is required to diagnose the former, but because of the long period required for different genes to show concordant phylogenies, the PSC2 will fail to diagnose (and resolve the history of) many species that are recognized using other concepts. This is why systematists introduced a third version of the PSC, the *genealogical species concept*.

3. *PSC3* (Also called the "genealogical species concept" or GSC.) A species is a basal, exclusive group of organisms all of whose genes coalesce more recently with each other than with those of any organisms outside the group, and that contains no exclusive group within it (Baum and Donoghue 1995; Shaw 1998).

The GSC was proposed as a way to diagnose the phylogenetic status of populations using genes; it is, in fact, an operational definition of PSC2. Avise and Ball (1990) were the first to consider genetically based monophyly as a way to reconstruct the evolutionary history of taxa, but did not deem it a good way to diagnose species.

Like PSC2, the GSC recognizes species as exclusive groups whose members are more closely related to each other than to individuals of other species. The GSC also recognizes species as basal groups that contain no exclusive subgroups within them. The difference between the two concepts is that the GSC explicitly defines the monophyly of taxa as the monophyly of the genes carried by its members. Although diagnosing a group as a genealogical species (GS) should in principle involve many loci—after all, it is individual organisms and not genes that are members of a species—in practice monophyly is determined using a limited sample of genes.

The main task facing the GSC is specifying *how many loci* must be monophyletic to diagnose a group as a genealogical species. The original formulation by Baum and Donoghue (1995) requires *all* loci to be monophyletic. But this demand is too extreme, as balancing selection that preserves two or more alleles in an ancestral species can keep identical sets of alleles polymorphic in descendants. This is the situation at the MHC locus in humans versus chimps, and in rats versus mice, with both pairs of species having a divergence between 5 and 10 million years (Figueroa et al. 1988; Ayala and Escalante 1996). One also sees ancient polymorphisms of self-incompatibility alleles among species in the genus *Brassica* (Uyenoyama 1995).

Advocates of the GSC recognize the problem with demanding complete monophyly of all genes, but have avoided the question of what proportion of surveyed loci must be monophyletic to allow GS status. Shaw (2001), however, suggests that GS status might be recognized if most loci were monophyletic. Setting this "> 50%" threshold makes judgments about GS status somewhat arbitrary, but no more arbitrary than diagnosing biological species when reproductive isolation is incomplete.

Using coalescent theory, Hudson and Coyne (2002) studied the time to attain GS status when an ancestral species divides into two descendants and the only evolutionary forces operating are mutation and genetic drift. For a single descendant, attaining *complete* monophyly requires a long time, especially if many loci are sampled. To attain a 95% probability of observing monophyly at every sampled gene, where N is the effective size of the population, one requires 1.8 N generations to reach GS status for a single mitochondrial or chloroplast gene, 7.3 N generations for a single nuclear gene, and 26.3 N generations for 11,500 nuclear loci (roughly the number of genealogically independent units within the *Drosophila melanogaster* genome). Attaining *reciprocal* monophyly for both descendant populations takes roughly 10%–30% longer. Directional selection, which speeds the fixation of alleles, will shorten these times; but balancing selection, which retards fixation, will lengthen them. Surprisingly, the number of alleles sampled per gene has little effect on the time required to attain GS status.

Presumably, one uses a sample of loci to infer the GS status of the entire genome. Such a goal requires one to use a large sample of loci and to avoid diagnosing GSs based on single mitochondrial or chloroplast loci. Organelle genes have only one-fourth as many copies as any autosomal locus, and so will become monophyletic well before the rest of the genome. All genes in an organelle are also completely linked, so no additional information about ancestry is gained by using more than one such gene. Nevertheless, genealogical species have been diagnosed on the basis of a single mtDNA haplotype or allozyme locus (e.g., Young and Crother 2001; Leaché and Reeder 2002).

When one relaxes the criteria for GS status, so that only 50% or 95% of sampled loci need be monophyletic, the time to genealogical speciation is reduced. In a sample of 25 nuclear genes, for example, one observes complete reciprocal monophyly with 95% probability after 15.2 N generations. This drops to 4.7 N generations if only 50% of the loci need be monophyletic, and to 11.3 N generations if 95% of the loci need be monophyletic. In the limit, with an infinite sample of loci, one observes complete reciprocal monophyly with 100% probability after 3.8 N generations using the 50% criterion and 8.7 N generations using the 95% criterion.

The conclusion is that one should not adopt a GS criterion requiring *complete* reciprocal monophyly for a large number of loci. Under this extreme view, humans and chimps would be considered one species, and rats and mice another. Moreover, attaining genome-wide monophyly takes so long that, before it occurs, taxa will be recognized as species using nearly every other species concept; indeed, many *additional* branching events might have occurred.

What is the relationship between biological and genealogical speciation? While both processes are accelerated by divergent natural selection and geographic isolation, there is no necessary correspondence between the times when species status is attained under the BSC and the GSC. However, biological speciation is almost certain to precede genealogical speciation if GS status requires complete reciprocal monophyly at many loci. *Drosophila simulans*, for example, is not a genealogical species with respect to *D. mauritiana* under even the "50% monophyly" criterion, and these taxa (although allopatric) have diverged in several morphological traits and show substantial reproductive isolation.

While a "relaxed" version of the GSC seems the most reasonable of all phylogenetic species concepts, we favor the BSC over the GSC for several reasons. First, applying the GSC will often involve designating taxa as metaspecies: large groups of individuals, such as *D. simulans*, will be not be recognized as belonging to any species. Unlike many doubtful cases in the BSC, the term "metaspecies" describes an *ontological* situation (organisms that are not members of any species) rather than an epistemological one (groups that cannot be assigned to recognized species due to a lack of evidence)" (Baum and Shaw 1995, p. 297; our italics). At the moment when an isolated population becomes monophyletic, every individual in every other population instantly loses its status as belonging to any species. It seems odd that, without any change in its own genetic composition, a group can lose species status based on what happens in a remote population. It should be added, however, that systematists disagree on whether the term "metaspecies" should be used, or which entities should be so characterized.

Second, little of biological import occurs at the completion of genealogical speciation. What significance, for example, can one impute to the moment at which the proportion of loci showing exclusivity rises from 50% to 50.1%—the completion of one type of genealogical speciation? In contrast, the completion of biological speciation—the moment when gene flow between sister taxa is no longer possible—corresponds to a biologically meaningful event. It is the moment when taxa become evolutionarily independent (Coyne 1994a). The termination of gene flow also allows genealogies to coalesce without pollution by genes from other taxa. Thus, these reproductive barriers, along with geographical barriers, provide the isolation that *permits* the monophyly required for genealogical speciation. In this sense, reproductive isolation is more fundamental than genetic coalescence.

Finally, genealogical speciation will often be transitory, for the coalescence of genes does not guarantee that geographically isolated populations will remain distinct when they become sympatric. One can envision many small, isolated populations quickly attaining genealogical species status. But range shifts or the disappearance of geographic barriers will quickly eliminate these genealogical species: they will hybridize with other populations and their exclusivity will vanish. In contrast, some forms of reproductive isolation are permanent. It is the permanence of reproductive isolation that guarantees the independence of genealogies among taxa.

The BSC and evolutionary history

Applying the BSC is an exercise not in reconstructing the history of taxa, but in identifying reproductively isolated groups. But implicit in this exercise is the idea that populations of a single biological species are more closely related to each other than to populations of a different species. To justify systematists' assertions that the BSC frequently distorts evolutionary history, one should be able to show many cases in which that history conflicts with reproductive compatibility. The most commonly cited examples are paraphyletic species and polyphyletic species. We have already noted some of the difficulties with using "thin branch" phylogenies of genes to determine whether population phylogenies—"fat branch" phylogenies—are paraphyletic or polyphyletic.

PARAPHYLY. In principle, species with a paraphyletic *origin* should be common. There must be many cases (e.g., peripatric speciation) in which a new biological species, B, originates from only one population of ancestral species A. If species B evolves reproductive isolation from all populations of A, which themselves remain reproductively compatible, then species A is *historically* paraphyletic. That is, if we were present at the moment when the population destined to become species B was geographically isolated, we would see that it derived from only one population of species A.

But we were not present at this moment, and so cannot directly witness the history of taxa. We must rely on the thin branches—the phylogenies of genes—to reconstruct the pattern of fat branches. But gene-based phylogenies can yield false diagnoses of paraphyly for several reasons.

One reason, mentioned above, is that populations of the ancestral species do not remain genetically isolated. Their interbreeding will quickly erase the genetic differences between populations that can be used to diagnose paraphyly. *Recognizable* paraphyly is therefore likely to be a transitory phenomenon. What we expect is that some loci will show paraphyly of alleles in species A relative to those in species B, but that this pattern will not be consistent across all genes. In other words, different traits or loci may yield different patterns of relatedness, with some showing paraphyly and others not. Moreover, as seen in Figure A.2, one expects this discordance even if species A is *not* paraphyletic, for discordance is the expected result when ancestral polymorphisms are sorted into descendant species. Such paraphyly is an ineluctable part of the speciation process and therefore cannot conflict with the BSC.

How many genes, then, must show concordant phylogenies before we are confident that the taxa themselves are paraphyletic? This crucial issue has been almost completely ignored by systematists. In fact, we do not know of any case in which species paraphyly is demonstrated by concordant genealogies of many (or even several) genes. Claims for species paraphyly are almost always based on one or a few loci, usually on the mitochondria (e.g., Melnick et al. 1993; Patton and Smith 1994; Omland 1997; Omland et al. 2000). Such paraphyly tells us little about the evolutionary history of populations because organelle genes may not accurately mirror the rest of the genome.

Indeed, many studies show that mitochondrial and chloroplast DNA introgress between taxa much more readily than does nuclear DNA (e.g., Ferris et al. 1983; Smith 1992; Bernatchez et al. 1995; Howard et al. 1997; Taylor and McPhail 2000; Martinsen et al. 2001; Shaw 2002). The reasons for this are unclear, but may be due to the nature of mitochondrial genes. Most of these genes are constitutively expressed and perform internal metabolic "housekeeping" or protein-synthetic functions, such as producing tRNA or respiratory enzymes. Such functions may be largely divorced from external selective pressures, making mtDNA less responsive than nuclear genes to local environmental differences. This may also be true for cpDNA, which contains genes for photosynthesis, tRNA, and rRNA. Thus, organelle genes, unlike nuclear genes, may function fairly well in the genetic background of a related species. In addition, the spread of adaptive mutations in organelle DNA is not impeded by their linkage to nuclear genes that are divergently adapted between taxa or cause intrinsic postzygotic isolation in hybrids.

The consequence of introgression and linkage is that organelle DNA may appear paraphyletic even when the species themselves are not. Phylogenies based solely on organelle DNA can also distort history in other ways. For example, Shaw (1996b) showed that a mtDNA-based phylogeny of Hawaiian crickets (*Laupala*) was discordant with traditional phylogenies based on morphology and biogeography. The mtDNA phylogenies showed that the most closely related species were sympatric, implying sympatric speciation. However, later phylogenies based on nuclear DNA were concordant with the traditional ones, supporting allopatric speciation following colonization of new islands (Shaw 2002). The most likely reason for this discordance is the introgression of mitochondria between sympatric taxa, which can hybridize. (One would predict that in species having heterogametic females, such as birds and Lepidoptera, mtDNA would introgress less readily. Because Haldane's rule holds in these groups, F_1 hybrid females, which pass on mtDNA, are often sterile.)

Under the "fat branch" approach, the evolutionary history of populations is usually represented by trees with bifurcating branches. However, the techniques used to reconstruct this history involve genes whose diverse genealogies can yield a complicated set of reticulations instead of a definitive phylogeny. Because there is no unitary *genetic* history at the population level, it is almost impossible to recognize true paraphyly among closely related taxa using genetically based phylogenies.

POLYPHYLY.   A polyphyletic species includes individuals or populations having independent evolutionary origins from common ancestors residing outside that species. As noted above, polyphyletic species include independently formed polyploid individuals that interbreed with one another, as well as cases of parallel speciation.

There is strong genetic evidence for a polyphyletic origin of some auto- and allopolyploid plant species (see Chapter 9), and of at least one species of terrestrial snail (Ueshima and Asami 2003). These cases indeed show genuine discordance between the evolutionary history of populations and their reproductive relationships. And because hybridization can form new polyploids repeatedly, this discordance may persist for long periods. In some cases, independently derived polyploids co-occur in nature and interbreed, showing that they are indeed members of the same biological species. But phylogenetic species concepts are unable to deal with such polyphyletic species, as they combine the genes of two ancestral species.

Parallel speciation has a similar effect, except that the independent origins of a single species involve convergent evolution rather than repeated hybridization. There are two possible cases of nonhybrid diploid species having multiple evolutionary origins: limnetic morphs of the threespine stickleback, *Gasterosteus aculeatus* (Rundle et al. 2000), and host races of the stick insect *Timema cristinae* (Nosil et al. 2002). However, in both cases there is ongoing gene flow between sympatric taxa. This can yield inaccurate phylogenies, making speciation events appear independent when they are not.

We conclude that while the BSC may occasionally identify species that are not monophyletic, it is not clear that phylogenetic species concepts are better at dealing with this problem.

References

Abrahamson, W. G. and A. E. Weis. 1997. *Evolutionary Ecology Across Three Trophic Levels: Goldenrods, Gallmakers, and Natural Enemies.* Princeton University Press, Princeton, NJ.

Abrams, P. 1983. The theory of limiting similarity. Annu. Rev. Ecol. Syst. 14: 359–376.

Adam, D., N. Dimitijevic, and M. Schartl. 1993. Tumor suppression in *Xiphophorus* by an accidentally acquired promoter. Science 259: 816–819.

Adams, K. L., R. Cronn, R. Percifield, J. F. Wendel. 2003. Genes duplicated by polyploidy show unequal contributions to the transcriptome and organ-specific reciprocal silencing. Proc. Natl. Acad. Sci. USA 100: 4649–4654.

Alatalo, R. V., D. Erikkson, L. Gustafsson, and A. Lundberg. 1990. Hybridization between pied and collared flycatchers—sexual selection and speciation theory. J. Evol. Biol. 3: 375–389.

Alatalo, R. V., L. Gustafsson, and A. Lundberg. 1994. Male coloration and species recognition in sympatric flycatchers. Proc. R. Soc. Lond. B 256: 113–118.

Albertson, R. C., J. A. Markert, P. D. Danley, and T. D. Kocher. 1999. Phylogeny of a rapidly evolving clade: The cichlid fishes of Lake Malawi, East Africa. Proc. Natl. Acad. Sci. USA 96: 5107–5110.

Albertson, R. C., J. T. Streelman, and T. D. Kocher. 2003. Directional selection has shaped the oral jaws of Lake Malawi cichlid fishes. Proc. Nat. Acad. Sci. USA 100: 5252–5257.

Alexander, R. D. 1968. Life cycle origins, speciation, and related phenomena in crickets. Q. Rev. Biol. 43: 1–41.

Alexander, R. D. and R. S. Bigelow. 1960. Allochronic speciation in field crickets, and a new species *Acheta velitis*. Evolution 14: 334–346.

Alexander, R. D. and T. E. Moore. 1962. The evolutionary relationships of 17-year and 13-year cicadas and three new species (Homoptera, Cicadidae, *Magicicada*). Occas. Pap. Mus. Zool. Univ. Mich. 121: 1–59.

Allen, J. A. 1907. Mutations and the geographic distribution of nearly related species in plants and animals. Am. Nat. 41: 653–655.

Allender, C. J., O. Seehausen, M. E. Knight, G. F. Turner, and N. Maclean. 2003. Divergent selection during speciation of Lake Malawi cichlid fishes inferred from parallel radiations in nuptial coloration. Proc. Natl. Acad. Sci. USA 100: 14074–14079.

Amevigbe, M. D. D., A. Ferrer, S. Champorie, N. Monteny, J. Deunff, and D. Richard-Lenoble. 2000. Isoenzymes of human lice: *Pediculus humanus* and *P. capitis*. Med. Vet. Entomol. 14: 419–425.

Amores, A., A. Force, Y.-L. Yan, L. Joly, C. Amemiya, A. Fritz, R. K. Ho, J. Langeland, V. Prince, Y.-L. Wang, M. Westerfield, M. Ekker, and J. H. Postlethwait. 1998. Zebrafish hox clusters and vertebrate genome evolution. Science 282: 1711–1714.

Anders, F. 1991. Contributions of the Gordon-Kosswig melanoma system to the present concept of neoplasia. Pigment Cell Res. 3: 7–29.

Anderson, E. 1949. *Introgressive Hybridization.* John Wiley and Sons, New York.

Andersson, M. 1982. Female choice selects for extreme tail length in a widowbird. Nature 299:818–820.

Andersson, M. 1994. *Sexual Selection*. Princeton University Press, Princeton, NJ.

Argus, G. W. 1986. The genus *Salix* (Salicaceae) in the southeastern United States. Syst. Bot. Monogr. 9: 1–170.

Arisumi, T. 1985. Rescuing abortive *Impatiens* hybrids through aseptic culture of ovules. J. Am. Soc. Hortic. Sci. 110: 273–276.

Armstrong, R. A. and R. McGehee. 1980. Competitive exclusion. Am. Nat. 115: 151–170.

Arnegard, M. E., and A. S. Kondrashov. 2004. Sympatric speciation by sexual selection alone is unlikely. Evolution 58: 222–237.

Arnold, J. and W. W. Anderson. 1983. Density-regulated selection in a heterogeneous environment. Am. Nat. 121: 656–668.

Arnold, M. L. 1992. Natural hybridization as an evolutionary process. Annu. Rev. Ecol. Syst. 23: 237–261.

Arnold, M. L. 1993. *Iris nelsonii* (Iridaceae): Origin and genetic composition of a homoploid hybrid species. Am. J. Bot. 80: 577–583.

Arnold, M. L. 1997. *Natural Hybridization and Evolution.* Oxford University Press, New York.

Arnold, M. L. and B. D. Bennett. 1993. Natural hybridization in Louisiana Irises: Genetic variation and ecological determinants. Pp. 115–139 *in* R. G. Harrison (ed.) *Hybridization and the Evolutionary Process.* Oxford University Press, New York.

Arnold, M. L. and S. A. Hodges. 1995. Are natural hybrids fit or unfit relative to their parents? Trends Ecol. Evol. 10: 67–71.

Arnold, M. L., J. L. Hamrick, and B. D. Bennett. 1990. Allozyme variation in Louisiana irises: A test for introgression and hybrid speciation. Heredity 65: 297–306.

Arnold, M. L., C. M. Buckner, and J. J. Robinson. 1991. Pollen-mediated introgression and hybrid speciation in Louisiana irises. Proc. Natl. Acad. Sci. USA 88: 1398–1402.

Arnold, S. J., P. A. Verrell, and S. G. Tilley. 1996. The evolution of asymmetry in sexual isolation: a model and a test case. Evolution 50: 1024–1033.

Arnqvist, G. 1998. Comparative evidence for the evolution of genitalia by sexual selection. Nature 393: 784–786.

Arnqvist, G. and I. Danielsson. 1999. Copulatory behavior, genital morphology, and male fertilization success in water striders. Evolution 53: 147–156.

Arnqvist, G., M. Edvardsson, U. Friberg, and T. Nilsson. 2000. Sexual conflict promotes speciation in insects. Proc. Natl. Acad. Sci. USA 97: 10460–10464.

Ashburner, M. 1989. Drosophila: *A Laboratory Handbook.* Cold Spring Harbor Laboratory Press, New York.

Asker, S. E. and L. Jerling. 1992. *Apomixis in Plants.* CRC Press, Boca Raton, FL.

Aspinwall, N. 1974. Genetic analysis of North American populations of the pink salmon, *Onchorhynchus gorbusha,* possible evidence for the neutral mutation-random drift hypothesis. Evolution 28: 295–305.

Atyeo, W. T. and R. M. Windingstad. 1979. Feather mites of the greater sandhill crane. J. Parasitol. 65: 650–658.

Avise, J. C. 1994. *Molecular Markers, Natural History, and Evolution.* Chapman & Hall, New York.

Avise, J. C. 2000. *Phylogeography: The History and Formation of Species.* Harvard University Press, Cambridge, MA.

Avise, J. C. and R. M. Ball. 1990. Principles of genealogical concordance in species concepts and biological taxonomy. Pp. 45–67 *in* D. J. Futuyma and J. Antonovics (eds.) *Oxford Surveys in Evolutionary Biology.* Oxford University Press, Oxford.

Avise, J. C. and K. Wollenberg. 1997. Phylogenetics and the origin of species. Proc. Natl. Acad. Sci. USA 94: 7748–7755.

Avise, J. C., J. Arnold, J. R. M. Ball, E. Bermingham, T. Lamb, J. E. Neigel, C. A. Reeb, and N. C. Saunders. 1987. Intraspecific phylogeography: The mitochondrial DNA bridge between population genetics and systematics. Annu. Rev. Ecol. Syst. 18: 489–522.

Avise, J. C., J. M. Quattro, and R. C. Vrijenhoek. 1992. Molecular clones within organismal clones. Mitochondrial DNA phylogenies and the evolutionary histories of unisexual vertebrates. J. Evol. Biol. 26: 225–246.

Avise, J. C., D. Walker, and G. C. Johns. 1998. Speciation durations and Pleistocene effects on vertebrate phylogeography. Proc. R. Soc. Lond. B 265: 1707–1712.

Ayala, F. J. and A. A. Escalante. 1996. The evolution of human populations—a molecular perspective. Mol. Phylogenet. Evol. 5: 188–201.

Babcock, E. B. and G. L. Stebbins. 1938. The American species of *Crepis.* Their interrelationships and distribution as affected by polyploidy and apomixis. Washington, DC: Pub. 504, Carnegie Institute Washington.

Baird, S. E. 2002. Haldane's rule by sexual transformation in *Caenorhabditis.* Genetics 161: 1349–1353.

Baird, S. E., M. E. Sutherlin, and S. W. Emmons. 1992. Reproductive isolation in Rhabditidae (Nematoda: Secerentea); mechanisms that isolate six species of three genera. Evolution 46: 585–594.

Baker, H. G. 1953. Race formation and reproductive method in flowering plants. Symp. Soc. Exp. Biol. 7: 114–145.

Baker, H. G. 1959. Reproductive methods as factors in speciation in flowering plants. Cold Spring Harb. Symp. Quant. Biol. 24: 177–191.

Baker, H. G. 1955. Self-compatibility and establishment after "long-distance" dispersal. Evolution 9: 347–349.

Baker, M. C. 1983. The behavioral response of female Nuttall's White-Crowned Sparrows to male song of natal and alien dialects. Behav. Ecol. Sociobiol. 12: 309–315.

Baker, M. C. and A. E. M. Baker. 1990. Reproductive behavior of female buntings: Isolating mechanisms in a hybridizing pair of species. Evolution 44: 332–338.

Baker, R. J. and J. W. Bickham. 1986. Speciation by monobrachial centric fusions. Proc. Natl. Acad. Sci. USA 83: 8245–8248.

Baldwin, B. and M. J. Sanderson. 1998. Age and rate of diversification of the Hawaiian silversword alliance. Proc. Natl. Acad. Sci. USA 95: 9402–9406.

Balkau, B. J. and M. W. Feldman. 1973. Selection for migration modification. Genetics 74: 171–174.

Barbash, D. A. and M. Ashburner. 2003. A novel system of fertility rescue in *Drosophila* hybrids reveals a link between hybrid lethality and female sterility. Genetics 163: 217–226.

Barbash, D. A., J. Roote, and M. Ashburner. 2000. The *Drosophila melanogaster Hybrid male rescue* gene causes inviability in male and female species hybrids. Genetics 154: 1747–1771.

Barbash, D. A., D. F. Siino, A. M. Tarone, and J. Roote. 2003. A rapidly evolving MYB–related protein causes species isolation in *Drosophila*. Proc. Natl. Acad. Sci. USA 100: 5302–5307.

Barker, J. S. F. and L. J. Cummins. 1969. The effect of selection for sternopleural bristle number on mating behavior in *Drosophila melanogaster*. Genetics 61: 713–719.

Barlow, G. W. 2000. *The Cichlid Fishes: Nature's Grand Experiment in Evolution*. Perseus, Cambridge, MA.

Barraclough, T. G., and S. Nee. 2001. Phylogenetics and speciation. Trends Ecol. Evol. 16: 391–399.

Barraclough, T. G., and V. Savolainen. 2001. Evolutionary rates and species diversity in flowering plants. Evolution 55: 677–681.

Barraclough, T. G. and A. P. Vogler. 2000. Detecting the geographical pattern of speciation from species-level phylogenies. Am. Nat. 155: 419–434.

Barraclough, T. G. and A. P. Vogler. 2002. Recent diversification rates in North American Tiger Beetles estimated from a dated mtDNA phylogenetic tree. Mol. Biol. Evol. 19: 1706–1716.

Barraclough, T. G., P. H. Harvey, and S. Nee. 1995. Sexual selection and taxonomic diversity in passerine birds. Proc. R. Soc. Lond. B 259: 211–215.

Barraclough, T. G., P. H. Harvey, and S. Nee. 1996. Rate of rbcL gene sequence evolution and species diversification in flowering plants (angiosperms). Proc. R. Soc. Lond. B 263: 589–591.

Barraclough, T. G., S. Nee, and P. H. Harvey. 1998. Sister-group analysis in identifying correlates of diversification. Evol. Ecol. Res. 12: 751–754.

Barraclough, T. G., J. Hogan, and A. P. Vogler. 1999a. Testing whether ecological factors promote cladogenesis in a group of tiger beetles (Coleoptera: Cicindelidae). Proc. R. Soc. Lond. B 266: 1061–1067.

Barraclough, T. G., A. P. Vogler, and P. H. Harvey. 1999b. Revealing the factors that promote speciation. Pp. 202–217 in A. E. Magurran and R. M. May (eds.) *Evolution of Biological Diversity*. Oxford University Press, Oxford.

Barraclough, T. G., C. W. Birky, and A. Burt. 2004. Diversification in sexual and asexual organisms. Evolution: 57: 2166–2172.

Barrett, P. H., P. J. Gautrey, S. Herbert, D. Kohn, and S. Smith. 1987. *Charles Darwin's Notebooks, 1836–1844: Geology, Transmutation of Species, Metaphysical Enquiries*. Cornell University Press, Ithaca, New York.

Barrett, S. J. and P. H. Sneath. 1994. A numerical phenotypic taxonomic study of the genus *Neisseria*. Microbiology 140: 2867–2891.

Barton, N. H. 1980. The fitness of hybrids between two chromosomal races of the grasshopper *Podisma pedestris*. Heredity 45: 47–59.

Barton, N. H. 1988. Speciation. Pp. 185–218 in A. A. Myers and P. S. Giller (eds.) *Analytical Biogeography*. Chapman & Hall, London.

Barton, N. H. 1989. Founder effect speciation. Pp. 229–256 in D. Otte and J. A. Endler (eds.) *Speciation and its Consequences*. Sinauer Associates, Sunderland, MA..

Barton, N. H. 1996. Natural selection and random genetic drift as causes of evolution on islands. Philos. Trans. R. Soc. Lond. B 351: 785–795.

Barton, N. H. 2001. The role of hybridization in evolution. Mol. Ecol. 10: 551–568.

Barton, N. H. and B. Charlesworth. 1984. Genetic revolutions, founder effects, and speciation. Annu. Rev. Ecol. Syst. 15: 133–164.

Barton, N. H. and K. S. Gale. 1993. Genetic analysis of hybrid zones. Pp. 13–45 in R. G. Harrison (ed.) *Hybrid Zones and the Evolutionary Process*. Oxford University Press, New York and Oxford.

Barton, N. H. and G. M. Hewitt. 1981a. Hybrid zones and speciation. Pp. 109–145 in W. R. Atchley and D. S. Woodruf (eds.) *Evolution and Speciation*. Cambridge University Press, Cambridge, UK.

Barton, N. H. and G. M. Hewitt. 1981b. The genetic basis of hybrid inviability between two chromosomal races of the grasshopper *Podisma pedestris*. Heredity 47: 367–383.

Barton, N. H. and G. M. Hewitt. 1985. Analysis of hybrid zones. Annu. Rev. Ecol. Syst. 16: 113–48.

Barton, N. H. and G. M. Hewitt. 1989. Adaptation, speciation, and hybrid zones. Nature 341:497–503.

Basolo, A. L. 1990. Female preference predates the evolution of the sword in swordtail fish. Science 250: 808–810.

Basolo, A. L. 1995. Phylogenetic evidence for the role of a preexisting bias in sexual selection. Proc. R. Soc. Lond. B 259: 307–311.

Bateson, W. 1894. *Materials for the Study of Variation Treated with Especial Regard to Discontinuity in the Origin of Species*. Macmillan and Co., London and New York.

Bateson, W. 1909. Heredity and variation in modern lights. Pp. 85–101 in A. C. Seward (ed.) *Darwin and Modern Science*. Cambridge University Press, Cambridge, UK.

Bateson, W. 1922. Evolutionary faith and modern doubts. Science 55: 55–61.

Baum, D. 1992. Phylogenetic species concepts. Trends Ecol. Evol. 7: 1–2.

Baum, D. A. and M. J. Donoghue. 1995. Choosing among alternative phylogenetic species concepts. Syst. Bot. 20: 560–573.

Baum, D. A. and K. L. Shaw. 1995. Genealogical perspectives on the species problem. Pp. 289–303 *in* P. C. Hoch and A. C. Stephenson (eds.) *Experimental and Molecular Approaches to Plant Biosystematics*. Missouri Botanical Garden, St. Louis, MO.

Baverstock, P. R., M. Adams, R. W. Polkinghorne, and M. Gelder. 1982. A sex-linked enzyme in birds: Z-chromosome conservation but no dosage compensation. Nature 296: 763–767.

Beacham, T. D., R. W. Withler, C. B. Murray, and L. W. Barner. 1988. Variation in body size, morphology, egg size, and biochemical genetics of pink salmon in British Columbia. Trans. Am. Fish. Soc. 117: 109–126.

Bell, M. A. and S. A. Foster. 1994. *The Evolutionary Biology of the Threespine Stickleback*. Pp. 571. Oxford University Press, Oxford.

Bella, J. L., R. K. Butlin, C. Ferris, and G. M. Hewitt. 1992. Asymmetrical homogamy and unequal sex ratio from reciprocal mating-order crosses between *Chorthippus parallelus* subspecies. Heredity 68: 345–352.

Belotte, D., J.-B. Curien, R. C. Maclean, and G. Bell. 2003. An experimental test of local adaptation in soil bacteria. Evolution 57: 27–36.

Berlin, B. 1973. Folk systematics in relation to biological classification and nomenclature. Annu. Rev. Ecol. Syst. 4: 259–271.

Berlin, B. 1992. *Ethnobotanical Classification*. Princeton University Press, Princeton, NJ.

Berlin, B., D. Breedlove, and P. R. Raven. 1974. *Principles of Tzeltal Plant Classification*. Academic Press, New York and London.

Berlocher, S. H. 1999. Host race or species? Allozyme characterization of the 'flowering dogwood fly,' a member of the *Rhagoletis pomonella* complex. Heredity 83: 652–662.

Berlocher, S. H. 2000. Radiation and divergence in the *Rhagoletis pomonella* species group: Inference from allozymes. Evolution 54: 543–557.

Berlocher, S. H. and M. Enquist. 1993. Distribution and host plants of the apple maggot fly, *Rhagoletis pomonella* (Diptera: Tephritidae) in Texas. J. Kans. Entomol. Soc. 66: 145–166.

Berlocher, S. H. and J. L. Feder. 2002. Sympatric speciation in phytophagous insects: Moving beyond controversy? Annu. Rev. Entomol. 47: 773–815.

Berlocher, S. H., and B. A. McPheron. 1996. Population structure of *Rhagoletis pomonella*, the apple maggot fly. Heredity 77: 83–99.

Bernardi, G., L. Findley, and A. Rocha-Olivares. 2003. Vicariance and dispersal across Baja California in disjunct marine fish populations. Evolution 57: 1599–1609.

Bernatchez, L., L. Glemete, C. C. Wilson, and R. G. Danzmann. 1995. Introgression and fixation of Arctic char (*Salvenlinus alpinus*) mitochondrial genome in an allopatric population of brook trout (*Salvelinus fontinalis*). Can. J. Fish. Aquat. Sci. 52: 179–185.

Bernatchez, L., J. A. Vuorinen, R. A. Bodaly, and J. J. Dodson. 1996. Genetic evidence for reproductive isolation and multiple origins of sympatric trophic ecotypes of whitefish (*Coregonus*). Evolution 50: 624–635.

Bernatchez, L., A. Chouinard, and G. Lu. 1999. Integrating molecular genetics and ecology in studies of adaptive radiation: Whitefish, *Coregonus* sp., as a case study. Biol. J. Linn. Soc. Lond. 68: 173–194.

Bernstein, S. and R. Bernstein. 2002. Allometry of male genitalia in a species of soldier-beetle: Support for the one-size-fits-all hypothesis. Evolution 56: 1707–1710.

Bert, T. M., D. M. Hesselman, W. S. Arnold, W. S. Moore, H. Cruz-Lopez, and D. C. Marelli. 1993. High frequency of gonadal neoplasia in a hard clam (*Mercenaria* spp.) hybrid zone. Mar. Biol. 117: 97–104.

Betancourt, A. J., D. C. Presgraves, and W. J. Swanson. 2002. A test for faster X evolution in *Drosophila*. Mol. Biol. Evol. 19: 1816–1819.

Beverley, S. M. and A. C. Wilson. 1995. Ancient origin for Hawaiian Drosophilinae inferred from protein comparisons. Proc. Natl. Acad. Sci. USA 82: 4753–4757.

Biermann, C. H. 1998. The molecular evolution of sperm bindin in six species of sea urchins (Echinoida: Strongylocentrotidae). Mol. Biol. Evol. 15:1761–1771.

Bierne, N., P. David, P. Boudry, and F. Bonhomme. 2002. Assortative fertilization and selection at larval stage in the mussels *Mytilis eduilis* and *M. galloprovincialis*. Evolution 56: 292–298.

Bigelow, R. S. 1965. Hybrid zones and reproductive isolation. Evolution 19: 449–458.

Birkhead, T. R. and A. P. Møller. 1998. *Sperm Competition and Sexual Selection*. Academic Press, London.

Birkhead, T. R. and T. Pizzari. 2002. Postcopulatory sexual selection. Nat. Rev. Genet. 3: 262–273.

Bishop, J. G. and J. A. Hunt. 1988. DNA divergence in and around the Alcohol Dehydrogenase locus in five closely related species of Hawaiian *Drosophila*. Mol. Biol. Evol. 5: 415–341.

Blair, W. F. 1955. Mating call and stage of speciation in the *Microhyla olivacea–M. carolinensis* complex. Evolution 9: 469–480.

Blair, W. F. 1974. Character displacement in frogs. Am. Zool. 14: 1119–1125.

Boake, C. R. B., K. McDonald, S. Maitra, and R. Ganguly. 2003. Forty years of solitude: Life-history divergence and behavioural isolation between laboratory lines of *Drosophila melanogaster*. J. Evol. Biol. 16: 83–90.

Bock, I. R. 1984. Interspecific hybridization in the genus *Drosophila*. Evol. Biol. 18: 41–70.

Bolmgren, K., O. Eriksson, and H. P. Linder. 2003. Contrasting flowering phenology and species richness in abiotically and biotically pollinated angiosperms. Evolution 57: 2001–2011.

Bolnick, D. I. 2004. Waiting for sympatric speciation. Evolution (in press).
Bordenstein, S. R. 2003. Symbiosis and the origin of species. Pp. 283–304 in K. Bourtzis and T. Miller (eds.) *Insect Symbiosis*. CRC Press, Boca Raton, FL.
Bordenstein, S. R. and M. D. Drapeau. 2001. Genotype-by-environment interaction and the Dobzhansky-Muller model of postzygotic isolation. J. Evol. Biol. 14: 490–501.
Bordenstein, S. R., M. D. Drapeau, and J. H. Werren. 2000. Intraspecific variation in sexual isolation in the Jewel wasp *Nasonia*. Evolution 54: 567–573.
Bordenstein, S. R., F. O'Hara, and J. Werren. 2001. *Wolbachia*-induced incompatibility precedes other hybrid incompatibilities in *Nasonia*. Nature 409: 707–710.
Borodin, P. M., M. B. Rogatcheva, A. I. Zhelezova, and S. Oda. 1998. Chromosome pairing in inter-racial hybrids of the house musk shrew (*Suncus murinus*, Insectivora, Soricidae). Genome 41: 79–90.
Boughman, J. W. 2001. Divergent sexual selection enhances reproductive isolation in sticklebacks. Nature 411: 944–948.
Bradshaw, H. D. and D. W. Schemske. 2003. Allelic substitution at a flower colour locus produces a pollinator shift in monkeyflowers. Nature 426: 176–178.
Bradshaw, H. D., S. M. Wilbert, K. G. Otto, and D. W. Schemske. 1995. Genetic mapping of floral traits associated with reproductive isolation in monkeyflowers (*Mimulus*). Nature 376: 762–765.
Bradshaw, H. D., S. M. Wilbert, K. G. Otto, and D. W. Schemske. 1998. Quantitative trait loci affecting differences in floral morphology between two species of monkeyflower (*Mimulus*). Genetics 149: 367–382.
Braig, H. K., H. Guzman, R. B. Tesh, and S. L. O'Neill. 1994. Replacement of the natural *Wolbachia* symbiont of *Drosophila simulans* with a mosquito counterpart. Nature 367: 453–455.
Breen, P. A. and B. E. Adkins. 1980. Spawning in a British Columbia population of northern abalone. Veliger 23: 177–179.
Breeuwer, J. A. J. and G. Jacobs. 1996. *Wolbachia*: Intracellular manipulators of mite reproduction. Exp. Appl. Acarol. 20: 421–434.
Breeuwer, J. A. J. and J. H. Werren. 1990. Microorganisms associated with chromosome destruction and reproductive isolation between two insect species. Nature 346: 558–560.
Breeuwer, J. A. J. and J. H. Werren. 1995. Hybrid breakdown between two haplodiploid species: The role of nuclear and cytoplasmic genes. Evolution 49: 705–717.
Bressac, C. and F. Rousset. 1993. The reproductive incompatibility system in *Drosophila simulans*: DAPI-staining analysis of the *Wolbachia* symbionts in sperm cysts. J. Invertebr. Pathol. 67: 55–64.
Britton-Davidian, J., H. Sonjaya, J. Catalan, G. Cattaneo-Berrebi. 1990. Robertsonian heterozygosity in wild mice: Fertility and transmission rates in Rb (16.17) translocation heterozygotes. Genetica 80: 171–174.
Brochmann, C., L. Borgen, and O. E. Stabbetorp. 2000. Multiple diploid hybrid speciation of the Canary Island endemic *Argyranthemum sundingii* (Asteraceae). Plant Syst. Evol. 220: 77–92.
Bronson, C. L., J. T. C. Grubb, and M. J. Braun. 2003. A test of the endogenous and exogenous selection hypothesis for the maintenance of a narrow avian hybrid zone. Evolution 57:630–637.
Brookfield, J. 2002. Review of *Genes, Categories, and Species* by Jody Hey. Genet. Res. 79:107–108.
Brown, C. W. 1974. Hybridization among the subspecies of the plethodontid salamander *Ensatina eschscholtzii*. Univ. Calif. Pub. Zool. 98: 1–57.
Brown, D. V. and P. E. Eady. 2001. Functional incompatibility between the fertilization systems of two allopatric populations of *Callosobruchus maculatus* (Coleoptera: Bruchidae). Evolution 55: 2257–2262.
Brown, J. H. and M. V. Lomolino. 1998. *Biogeography* (Second Edition). Sinauer Associates, Sunderland, MA.
Brykov, V. A., N. Polyakova, L. A. Skurikhina, and A. D. Kukhlevsky. 1996. Geographical and temporal mitochondrial DNA variability in populations of pink salmon. J. Fish Biol. 48: 899–909.
Buckley, P. A. 1969. Disruption of species-typical behavior patterns in F_1 hybrid *Agapornis* parrots. Zeitschrift fur Tierpsychologie 26: 737–743.
Buerkle, C. A., R. J. Morris, M. A. Asmussen, and L. H. Rieseberg. 2000. The likelihood of homoploid hybrid speciation. Heredity 84: 441–451.
Bulmer, R. N. H. and N. J. Tyler. 1968. Kalam classification of frogs. J. Polyn. Soc. 77:333–385.
Bulmer, R. N. H., J. I. Menzies, and F. Parker. 1975. Kalam classification of reptiles and fishes. J. Polyn. Soc. 84: 267–308.
Bundenberg, d. J. C. M. 1957. Polyploidy in animals. Bibliogr. Genet. 17: 111–228.
Burger, W. C. 1975. The species concept in *Quercus*. Taxon 24: 45–50.
Burgoyne, P. S. 1986. Mammalian X and Y crossover. Nature 319: 258–259.
Burnet, B. and K. Connolly. 1974. Activity and sexual behavior in *Drosophila melanogaster*. Pp. 201–258 in J. H. F. V. Abeelen (ed.) *The Genetics of Behaviour*. North-Holland, Amsterdam.
Burton, R. S. 1990. Hybrid breakdown in developmental time in the copepod *Tigriopus californicus*. Evolution 44: 1814–1822.
Burton, R. S. 1997. Genetic evidence for persistence of marine invertebrate populations in an ephemeral environment. Evolution 51: 993–998.
Bush, G. L. 1969. Sympatric host race formation and speciation in frugivorous flies of the genus *Rhagoletis* (Diptera, Tephritidae). Evolution 23: 237–251.
Bush, G. L. 1993. Host race formation and sympatric speciation in *Rhagoletis* fruit flies (Diptera: Tephritidae). Psyche 99: 335–355.

Bush, G. L., J. L. Feder, S. H. Berlocher, B. A. McPheron, D. C. Smith, and C. A. Chilcote. 1989. Sympatric origins of *R. pomonella*. Nature 339: 346.

Bush, M. K. 1994. Amazonian speciation: A necessarily complex model. J. Biogeogr. 21: 5–17.

Busvine, J. R. 1948. The "head" and "body" races of *Pediculus humanus* L. Parasitology 39: 1–16.

Busvine, J. R. 1978. Evidence from double infestations for the specific status of human head lice and body lice (Anoplura). Syst. Entomol. 3: 1–8.

Butlin, R. K. 1987a. Speciation by reinforcement. Trends Ecol. Evol. 2: 8–13.

Butlin, R. K. 1987b. Species, speciation, and reinforcement. Am. Nat. 130: 461–464.

Butlin, R. K. 1989. Reinforcement of premating isolation. Pp. 158–179 *in* D. Otte and J. A. Endler (eds.) *Speciation and Its Consequences*. Sinauer Associates, Sunderland, MA.

Butlin, R. K. and M. G. Ritchie. 1989. Genetic coupling in mate recognition systems: What is the evidence? Biol. J. Linn. Soc. Lond. 37: 237–246.

Butlin, R. K. and M. G. Ritchie. 1991. Variation in female mate preference across a grasshopper hybrid zone. J. Evol. Biol. 4: 227–240.

Buxton, P. A. 1948. *The Louse. An Account of the Lice Which Infest Man, Their Medical Importance and Control* (Second Edition). Edward Arnold, London.

Cabot, E. L., A. W. Davis, N. A. Johnson, and C.-I. Wu. 1994. Genetics of reproductive isolation in the *Drosophila simulans* clade: complex epistasis underlying hybrid male sterility. Genetics 137: 175–189.

Caillaud, C. M. and S. Via. 2000. Specialized feeding behavior influences both ecological specialization and assortative mating in sympatric host races of pea aphids. Am. Nat. 156: 609–621.

Cain, M. L., V. Andreasen, and D. J. Howard. 1999. Reinforcing selection is effective under a relatively broad set of conditions in a mosaic hybrid zone. Evolution 53: 1343–1353.

Caisse, M. and J. Antonovics. 1978. Evolution in closely adjacent plant populations. IX. Evolution of reproductive isolation in clinal populations. Heredity 40: 371–384.

Callaini, G., R. Dallai, and M. G. Riparbelli. 1997. *Wolbachia*-induced delay of paternal chromosome condensation does not prevent maternal chromosomes from entering anaphase in incompatible crosses of *Drosophila simulans*. J. Cell Sci. 110: 271–280.

Camp, W. H. 1942. The *Crataegus* problem. Castanea 7: 51–55.

Camp, W. H. 1951. Biosystematy. Brittonia 7: 113–127.

Capanna, E. 1982. Robertsonian numerical variation in animal speciation: *Mus musculus*, an emblematic model. Pp. 155–177 *in* C. Barigozzi (ed.) *Mechanisms of Speciation*. Alan R. Liss, New York.

Capanna, E., A. Gropp, H. Winking, G. Noack, and M. C. Civitelli. 1976. Robertsonian metacentrics in the mouse. Chromosoma 58: 341–353.

Cardillo, M. 1999. Latitude and rates of diversification in birds and butterflies. Proc. R. Soc. Lond. B 266: 1221–1225.

Carlquist, S. 1965. *Island Life: A Natural History of the Islands of the World*. Natural History Press, Garden City, New York.

Carney, S. E., S. A. Hodges, and M. L. Arnold. 1996. Effects of differential pollen-tube growth on hybridization in the Louisiana irises. Evolution 50: 1871–1878.

Carr, G. D., B. G. Baldwin, and D. W. Kyhos. 1996. Cytogenetic implications of artificial hybrids between the Hawaiian silversword alliance and North American tarweeds (Asteraceae: Heliantheae—Madiinae). Am. J. Bot. 83: 653–660.

Carson, H. L. 1968. The population flush and its genetic consequences. Pp. 123–138 *in* R. C. Lewontin (ed.) *Population Biology and Evolution*. Syracuse University Press, Syracuse, NY.

Carson, H. L. 1970. Chromosome tracers of the origin of species. Science 168:1414–1418.

Carson, H. L. 1971. Speciation and the founder principle. Stadler Symp. 3: 51–70.

Carson, H. L. 1975. The genetics of speciation at the diploid level. Am. Nat. 109:83–92.

Carson, H. L. 1976. Inference of the time of origin of some *Drosophila* species. Nature 259: 395–396.

Carson, H. L. 1981. Chromosomes and speciation. Chromosomes Today 7: 150–164.

Carson, H. L. 1983. Chromosomal sequences and inter-island colonizations in Hawaiian *Drosophila*. Genetics 103: 465–482.

Carson, H. L. 1989. Sympatric origins of *R. pomonella*. Nature 338: 304.

Carson, H. L. 1992. Inversions in Hawaiian *Drosophila*. Pp. 407–439 *in* C. B. Krimbas and J. R. Powell (eds.) Drosophila *Inversion Polymorphism*. CRC, Boca Raton, FL

Carson, H. L. and D. A. Clague. 1995. Geology and biogeography of the Hawaiian Islands. Pp. 14–29 *in* W. L. Wagner and V. A. Funk (eds.) *Hawaiian Biogeography: Evolution on a Hot-Spot Archipelago*. Smithsonian Institution Press, Washington, DC.

Carson, H. L. and K. Y. Kaneshiro. 1976. *Drosophila* of Hawaii: Systematics and ecological genetics. Annu. Rev. Ecol. Syst. 7: 311–346.

Carson, H. L. and A. R. Templeton. 1984. Genetic revolutions in relation to speciation phenomena: The founding of new populations. Annu. Rev. Ecol. Syst. 15: 97–131.

Carvajal, A. R., M. R. Gandarela, and H. F. Naveira. 1996. A three-locus system of interspecific incompatibility underlies male inviability between *Drosophila buzzatii* and *D. koepferae*. Genetica 98: 1–19.

Caspari, E. and G. S. Watson. 1959. On the evolutionary importance of cytoplasmic sterility in mosquitoes. Evolution 134: 568–570.

Castiglia, R. and E. Capanna. 2000. Contact zone between chromosomal races of *Mus musculus domes-*

ticus. 2. Fertility and segregation in laboratory-reared and wild mice heterozygous for multiple Robertsonian rearrangements. Heredity 85: 147–156.

Castillo-Davis, C. I. and D. L. Hartl. 2002. Genome evolution and developmental constraint in *Caenorhabditis elegans*. Mol. Biol. Evol. 19: 728–735.

Cayouette, J. and P. M. Catling. 1992. Hybridization within the genus *Carex* with special reference to North America. Bot. Rev. 58: 351–438.

Cazemajor, M., C. Landre, and C. Montchamp-Moreau. 1997. The *sex-ratio* trait in *Drosophila simulans*: Genetic analysis of distortion and suppression. Genetics 147: 635–642.

Chang, A. 2004. Conspecific sperm precedence in sister species of *Drosophila* with overlapping ranges. Evolution (in press).

Chapman, T., L. F. Liddle, J. M. Kalb, M. F. Wolfner, and L. Partridge. 1995. Cost of mating in *Drosophila melanogaster* females is mediated by male accessory gland products. Nature 373: 241–244.

Chapman, T., D. M. Neubaum, M. F. Wolfner, and L. Partridge. 2000. The role of male accessory gland protein Acp36DE in sperm competition in *Drosophila melanogaster*. Proc. R. Soc. Lond. B 267: 1097–1105.

Chapman, T., L. A. Herndon, Y. Heifetz, L. Partridge, and M. F. Wolfner. 2001. The Acp26Aa seminal fluid protein is a modulator of early egg-hatchability in *Drosophila melanogaster*. Proc. R. Soc. Lond. B 268: 1647–1654.

Chari, J. and P. Wilson. 2001. Factors limiting hybridization between *Penstemon spectabilis* and *P. contranthifolius*. Can. J. Bot. 79: 1439–1448.

Charlesworth, B., J. A. Coyne, and N. Barton. 1987. The relative rates of evolution of sex chromosomes and autosomes. Am. Nat. 130: 113–146.

Chesser, R. T. and R. M. Zink. 1994. Modes of speciation in birds: A test of Lynch's method. Evolution 48: 490–497.

Chesson, P. 1991. A need for niches? Trends Ecol. Evol. 9: 26–28.

Christie, P. and M. R. Macnair. 1984. Complementary lethal factors in two North American populations of the yellow monkey flower. J. Hered. 75: 510–511.

Christie, P. and M. R. Macnair. 1987. The distribution of postmating reproductive isolating genes in populations of the yellow monkey flower, *Mimulus guttatus*. Evolution 41: 571–578.

Church, S. A. and D. R. Taylor. 2002. The evolution of reproductive isolation in spatially structured populations. Evolution 56: 1859–1862.

Civetta, A. and R. S. Singh. 1995. High divergence of reproductive tract proteins and their association with postzygotic reproductive isolation in *Drosophila melanogaster* and *Drosophila virilis* group species. J. Mol. Evol. 41: 1085–1095.

Claridge, M. F. 1993. Speciation in insect herbivores—the role of acoustic signals in leafhoppers and planthoppers. Pp. 285–297 *in* D. R. Lees and D. Edwards (eds.) *Evolutionary Patterns and Processes*. Academic Press, London.

Claridge, M. F., H. A. Dawah, M. R. Wilson, (eds.) 1997. *Species: The Units of Biodiversity*. Chapman & Hall, London.

Clark, A. G. and A. Civetta. 2000. Protamine wars. Nature 403: 261–262.

Clark, A. G., M. Aguadé, T. Prout, L. G. Harshman, and C. H. Langley. 1995. Variation in sperm displacement and its association with accessory gland protein loci in *Drosophila melanogaster*. Genetics 139: 189–201.

Clark, M. A., N. A. Moran, P. Baumann, and J. J. Wernegreen. 2000. Cospeciation between bacterial endosymbionts (*Buchnera*) and a recent radiation of aphids (*Uroleucon*) and pitfalls of testing for phylogenetic congruence. Evolution 54: 517–525.

Clarke, C. A. and P. M. Sheppard. 1963. Interactions between major genes and polygenes in the determination of the mimetic patterns of *Papilio dardanus*. Evolution 17: 404–413.

Clarke, C. A. and P. M. Sheppard. 1969. Further studies on the genetics of the mimetic butterfly *Papilio memnon*. Philos. Trans. R. Soc. Lond. B 263: 35–70.

Clarke, J. F. G. 1971. The Lepidoptera of Rapa island. Smithsonian Contributions in Zoology 56: 1–282.

Clausen, J., D. D. Keck, and W. M. Hiesey. 1940. Experimental studies on the nature of species. I. Effect of varied environments on western North America plants. Washington, DC: Pub. 520, Carnegie Institute of Washington.

Clausen, J., D. D. Keck, and W. M. Hiesey. 1945. Experimental studies on the nature of species. II. Plant evolution through amphiploidy and autoploidy, with examples from the Madiinae. Washington, DC: Pub. 564, Carnegie Institute of Washington.

Clausen, R. E. and T. H. Goodspeed. 1925. Interspecific hybridization in *Nicotiana*. II. A tetraploid *glutinosa-Tabacum* hybrid, an experimental verification of Winge's hypothesis. Genetics 10: 279–284.

Clay, T. 1949. Some problems in the evolution of a group of ectoparasites. Evolution 3: 279–299.

Clifton, K. E. 1997. Mass spawning by green algae on coral reefs. Science 275: 1116–1118.

Clifton, K. E. and L. M. Clifton. 1999. The phenology of sexual reproduction by green algae (Bryopsidales) on Caribbean coral reefs. J. Phycol. 35: 24–34.

Cohan, F. M. 1984. Can uniform selection retard random genetic divergence between isolated conspecific populations? Evolution 38: 495–504.

Cohan, F. M. 2001. Bacterial species and speciation. Syst. Biol. 50: 513–524.

Cohan, F. M. 2002. What are bacterial species? Annu. Rev. Microbiol. 56: 457–487.

Cohan, F. M. 2004. Periodic selection and ecological diversity in bacteria. (In press) *in* D. Nurminsky (ed.) *Selective Sweeps*. Landes Bioscience, Georgetown, TX.

Colborn, J., R. E. Crabtree, J. B. Shaklee, E. Pfeiler, and B. W. Bowen. 2001. The evolutionary enigma of bonefishes (*Albula* spp.): Cryptic species and ancient separations in a globally distributed shorefish. Evolution 55: 807–820.

Cooley, J. R., C. Simon, D. C. Marshall, K. Slon, and C. Ehrhardt. 2001. Allochronic speciation, secondary contact, and reproductive character displacement in periodical cicadas (Hemiptera: *Magicicada* spp.): Genetic, morphological, and behavioral evidence. Molecular Ecology 10: 661–671.

Cooley, J. R., C. Simon, and D. C. Marshall. 2003. Temporal separation and speciation in periodical cicadas. Bioscience 53: 151–157.

Cordero Rivera, A., J. A. Andrés, A. Cordoba-Aguilar, and C. Utzeri. 2004. Postmating sexual selection: allopatric evolution of sperm competition mechanisms and genital morphology in calopterygid damselflies (Insecta: Odonata). Evolution 58: 349–359.

Cosmides, L. M. and J. Tooby. 1981. Cytoplasmic inheritance and intragenomic conflict. J. Theor. Biol. 89: 83–129.

Coulthart, M. B. and R. S. Singh. 1988. High level of divergence of male reproductive tract proteins between *Drosophila melanogaster* and its sibling species, *D. simulans*. Mol. Biol. Evol. 5: 182–191.

Counterman, B. A., D. Ortiz-Barrientos, and M. A. F. Noor. 2004. Using comparative genomic data to test for faster-X evolution. Evolution (in press).

Cox, R. T., and C. E. Carlton. 2003. A comment on gene introgression versus en masse cycle switching in the evolution of 13-year and 17-year life cycles in periodical cicadas. Evolution 57: 428–432.

Coyne, J. A. 1974. The evolutionary origin of hybrid inviability. Evolution 28: 505–506.

Coyne, J. A. 1984. Genetic basis of male sterility in hybrids between two closely related species of *Drosophila*. Proc. Natl. Acad. Sci. USA 51: 4444–4447.

Coyne, J. A. 1985. The genetic basis of Haldane's rule. Nature 314: 736–738.

Coyne, J. A. 1992a. Genetics and speciation. Nature 355: 511–515.

Coyne, J. A. 1992b. Genetics of sexual isolation in females of the *Drosophila simulans* species complex. Genet. Res. 60: 25–31.

Coyne, J. A. 1993. The genetics of an isolating mechanism between two sibling species of *Drosophila*. Evolution 47: 778–788.

Coyne, J. A. 1994a. Ernst Mayr on the origin of species. Evolution 48: 19–30.

Coyne, J. A. 1994b. Speciation by chromosomes. Trends Ecol. Evol. 9: 76–77.

Coyne, J. A. 1996a. Genetics of differences in pheromonal hydrocarbons between *Drosophila melanogaster* and *D. simulans*. Genetics 143: 353–364.

Coyne, J. A. 1996b. Genetics of sexual isolation in male hybrids of *Drosophila simulans* and *D. mauritiana*. Genet. Res. 68: 211–220.

Coyne, J. A. and B. Charlesworth. 1986. Location of an X-linked factor causing male sterility in hybrids of *Drosophila simulans* and *D. mauritiana*. Heredity 57: 243–246.

Coyne, J. A. and B. Charlesworth. 1989. Genetic analysis of X-linked sterility in hybrids between three sibling species of *Drosophila*. Heredity 62: 97–106.

Coyne, J. and B. Charlesworth. 1997. Genetics of a pheromonal difference affecting sexual isolation between *Drosophila mauritiana* and *D. sechellia*. Genetics 145: 1015–1030.

Coyne, J. A. and B. S. Grant. 1972. Disruptive selection on I-maze activity in *Drosophila melanogaster*. Genetics 71: 185–188.

Coyne, J. A. and M. Kreitman. 1986. Evolutionary genetics of two sibling species, *Drosophila simulans* and *D. sechellia*. Evolution 40: 673–691.

Coyne, J. A. and B. Milstead. 1987. Long-distance migration of *Drosophila*. 3. Dispersal of *D. melanogaster* alleles from a Maryland orchard. Am. Nat. 130: 70–82.

Coyne, J. A. and H. A. Orr. 1989a. Patterns of speciation in *Drosophila*. Evolution 43: 362–381.

Coyne, J. A. and H. A. Orr. 1989b. Two rules of speciation. Pp. 180–207 *in* D. Otte and J. Endler (eds.) *Speciation and Its Consequences*. Sinauer Associates, Sunderland, MA.

Coyne, J. A. and H. A. Orr. 1993. Further evidence against meiotic-drive models of hybrid sterility. Evolution 47: 685–687.

Coyne, J. A. and H. A. Orr. 1997. "Patterns of speciation in *Drosophila*" revisited. Evolution 51:295–303.

Coyne, J. A. and H. A. Orr. 1999. The evolutionary genetics of speciation. Pp. 1–36 *in* A. E. Magurran and R. M. May (eds.) *Evolution of Biological Diversity*. Oxford University Press, Oxford.

Coyne, J. A. and T. D. Price. 2000. Little evidence for sympatric speciation in island birds. Evolution 54: 2166–2171.

Coyne, J. A., H. A. Orr, and D. J. Futuyma. 1988. Do we need a new species concept? Syst. Zool. 37: 190–200.

Coyne, J. A., S. Aulard, and A. Berry. 1991a. Lack of underdominance in a naturally occurring pericentric inversion in *Drosophila melanogaster* and its implications for chromosome evolution. Genetics 129: 791–802.

Coyne, J. A., B. Charlesworth, and H. A. Orr. 1991b. Haldane's rule revisited. Evolution 45: 1710–1714.

Coyne, J. A., W. Meyers, A. P. Crittenden, and P. Sniegowski. 1993. The fertility effects of pericentric inversions in *Drosophila melanogaster*. Genetics 134: 487–496.

Coyne, J. A., K. Mah, and A. Crittenden. 1994. Genetics of a pheromonal difference contributing to reproductive isolation in *Drosophila*. Science 265: 1461–1464.

Coyne, J. A., N. H. Barton, and M. Turelli. 1997. A critique of Sewall Wright's shifting balance theory of evolution. Evolution 51: 643–671.

Coyne, J. A., S. Simeonidis, and P. Rooney. 1998. Relative paucity of genes causing inviability in hybrids between *Drosophila melanogaster* and *D. simulans*. Genetics 150: 1091–1103.

Coyne, J. A., S. Y. Kim, A. S. Chang, D. Lachaise, and S. Elwyn. 2002. Sexual isolation between two species with overlapping ranges: *Drosophila santomea* and *Drosophila yakuba*. Evolution 56: 2424–2434.

Cracraft, J. 1982. Geographic differentiation, cladistics, and vicariance biogeography: Reconstructing the tempo and mode of evolution. Am. Zool. 22: 422–424.

Cracraft, J. 1989. Speciation and its ontology: The empirical consequences of alternative species concepts for understanding patterns and processes of differentiation. Pp. 28–59 *in* D. Otte and J. A. Endler (eds.) *Speciation and Its Consequences*. Sinauer Associates, Sunderland, MA.

Cracraft, J. 1992. The species of the birds-of-paradise (Paradisaeidae): Applying the phylogenetic species concept to a complex pattern of diversification. Cladistics 8: 1–43.

Cracraft, J. and R. O. Prum. 1988. Patterns and processes of diversification: Speciation and historical congruence in some Neotropical birds. Evolution 42: 603–620.

Craft, K. J., M. V. Ashley, and W. D. Koenig. 2002. Limited hybridization between *Quercus lobata* and *Quercus douglasii* (Fagaceae) in a mixed stand in central coastal California. Am. J. Bot. 89: 1792–1798.

Craft, W. A. 1938. The sex ratio in mules and other hybrid mammals. Q. Rev. Biol. 13: 19–40.

Craig, T. P., J. D. Horner, and J. K. Itami. 2001. Genetics, experience, and host-plant preference in *Eurosta solidaginis*: Implication for host shifts and speciation. Evolution 55: 773–782.

Crapon de Caprona, M.-D. and B. Fritzsch. 1984. Interspecific fertile hybrids of haplochromine cichlidae (Teleostei) and their possible importance for speciation. Neth. J. Zool. 34: 503–538.

Cronin, T. M. 1999. *Principles of Paleoclimatology*. Columbia University Press, New York.

Crosby, J. L. 1970. The evolution of genetic discontinuity: Computer models of the selection of barriers to interbreeding between species. Heredity 25: 253–297.

Crossly, S. A. 1974. Changes in mating behavior produced by selection of barriers to interbreeding between species. Evolution 28: 631–647.

Crow, J. F. 1942. Cross fertility and isolating mechanisms in the *Drosophila mulleri* group. Univ. Tex. Pub. 4228: 53–67.

Crowell, K. 1968. Rates of competitive exclusion by the Argentine ant in Bermuda. Ecology 49:551–555.

Cruden, R. W. 1972. Pollination biology of *Nemophilia menziesii* (Hydrophyllaceae) with comments on the evolution of oligolectic bees. Evolution 26: 373–389.

Cruzan, M. B. and M. L. Arnold. 1994. Assortative mating and natural selection in an *Iris* hybrid zone. Evolution 48: 1946–1958.

Cruzan, M. B. and S. C. H. Barrett. 1993. Contribution of cryptic incompatibility to the mating system of *Eichhornia paniculata* (Pontederiaceae). Evolution 47: 925–934.

Cunningham, C. W. and T. M. Collins. 1998. Beyond area relationships: Extinction and recolonization in molecular marine biogeography *in* R. DeSalle and B. Schierwater (eds.) *Molecular Approaches to Ecology and Evolution*. Birkhäuser-Verlag, Basel.

Cunningham, M. A. and M. C. Baker. 1983. Vocal learning in White-crowned sparrows: Sensitive phase and song dialects. Behav. Ecol. Sociobiol. 13: 259–269.

Damuth, J. 1985. Selection among 'species': A formulation in terms of natural functional units. Evolution 39: 407–430.

Damuth, J., and I. L. Heisler. 1988. Alternative formulations of multilevel selection. Biol. Phil. 3: 407–430.

Darlington, C. D. 1937. *Recent Advances in Cytology*. Churchill, London.

Darlington, C. D. 1956. *Chromosome Botany*. George Allen and Unwin Ltd., London.

Darlington, C. D. 1973. *Chromosome Botany and the Origin of Cultivated Plants*. Hafner, New York.

Darwin, C. 1859. *On the Origin of Species by Means of Natural Selection or the Preservation of Favored Races in the Struggle for Life*. J. Murray, London.

Darwin, C. 1871. *The Descent of Man, and Selection in Relation to Sex*. J. Murray, London.

Davies, N., A. Aiello, J. Mallet, A. Pomiankowski, and R. E. Silberglied. 1997. Speciation in two Neotropical butterflies: Extending Haldane's rule. Proc. R. Soc. Lond. B 264: 845–851.

Davies, T. J., V.Savolainen, M. W. Chase, and T. G. Barraclough. 2004. Environmental energy and evolutionary rates in flowering plants. Proc. Roy. Soc. Lond. B (in press).

Davis, A. W. and C.-I. Wu. 1996. The broom of the sorcerer's apprentice: The fine structure of a chromosomal region causing reproductive isolation between two sibling species of *Drosophila*. Genetics 143: 1287–1298.

Davis, A. W., E. G. Noonburg, and C.-I. Wu. 1994. Evidence for complex genic interactions between conspecific chromosomes underlying hybrid female sterility in the *Drosophila simulans* clade. Genetics 137: 191–199.

Davis, A. W., J. Roote, T. Morley, K. Sawamura, S. Herrmann, and M. Ashburner. 1996. Rescue of hybrid sterility in crosses between *D. melanogaster* and *D. simulans*. Nature 380: 157–159.

Davis, J. J. 1915. The pea aphid with relation to forage crops. USDA Bull. 276: 1–67.

Dawkins, R. 1982. Universal darwinism. Pp. 403-425 *in* D. S. Bendall (ed.) *Evolution from Molecules to Men*. Cambridge University Press, Cambridge.

Day, T. 2000. Sexual selection and the evolution of costly female preferences: Spatial effects. Evolution 54: 715–730.

Day, T. and D. Schluter. 1995. The fitness of hybrids. Trends Ecol. Evol. 10: 288.

de Queiroz, A. 1998. Interpreting sister-group tests of key innovation hypotheses. Syst. Biol. 47: 710–718.

de Queiroz, A. 1999. Do image-forming eyes promote evolutionary diversification? Evolution 53: 1654–1664.

de Queiroz, K. and M. J. Donoghue. 1988. Phylogenetic systematics and the species problem. Cladistics 4: 317–338.

de Queiroz, K. and M. J. Donoghue. 1990. Phylogenetic systematics or Nelson's version of cladistics? Cladistics 6: 61–75.

de Villena, F. P.-D. and C. Sapienza. 2001. Female meiosis drives karyotypic evolution in mammals. Genetics 159: 1179–1189.

DeBach, P. and R. A. Sundby. 1963. Competitive displacement among ecological homologues. Hilgardia 34: 105–166.

DeJoode, D. R. and J. F. Wendel. 1992. Genetic diversity and origin of the Hawaiian Islands cotton, *Gossypium tomentosum*. Am. J. Bot. 79: 1311–1319.

DeSalle, R., and D. Grimaldi. 1992. Characters and the systematics of Drosophilidae. J. Heredity 83: 182–188.

del Solar, E. 1966. Sexual isolation caused by selection for positive and negative geotaxis in *Drosophila pseudoobscura*. Proc. Natl. Acad. Sci. USA 56: 484–487.

Delneri, D., I. Colson, S. Grammenoudi, I. N. Roberts, E. J. Louis, and S. G. Oliver. 2003. Engineering evolution to study speciation in yeasts. Nature 422: 68–72.

DeMarais, B. D., T. E. Dowling, M. E. Douglas, W. L. Minckley, and P. C. Marsh. 1992. Origin of *Gila seminuda* (Teleostei: Cyprinidae) through introgressive hybridization: Implications for evolution and conservation. Proc. Natl. Acad. Sci. USA 89: 2747–2751.

Dempster, L. T. and F. Ehrendorfer. 1965. Evolution of the *Galium multiflorum* complex in western North America. II. Critical taxonomic revision. Brittonia 17: 289–334.

de Oliveira, A. K. and A. R. Cordiero. 1980. Adaptation of *Drosophila willistoni* experimental populations to extreme pH medium. Heredity 44: 123–130.

Dermitzakis, E. T., J. P. Masly, H. M. Waldrip, and A. G. Clark. 2000. Non-Mendelian segregation of sex chromosomes in heterospecific *Drosophila* males. Genetics 154: 687–694.

DeSalle, R. 1995. Molecular approaches to biogeographic analysis of Hawaiian Drosophilidae. Pp. 72–89 in W. L. Wagner and V. A. Funk (eds.) *Hawaiian Biogeography: Evolution on a Hot-Spot Archipelago*. Smithsonian Institution Press, Washington, DC.

DeSalle, R. and D. Grimaldi. 1992. Characters and the systematics of Drosophilidae. J. Hered. 83: 182–188.

DeSalle, R. and A. R. Templeton. 1982. Founder effects and the rate of mitochondrial DNA evolution in Hawaiian *Drosophila*. Evolution 42: 1076–1084.

DeSalle, R. L., L. V. Giddings, and A. R. Templeton. 1986. Mitochondrial DNA variability in natural populations of Hawaiian *Drosophila*. II. Genetic and phylogenetic relationships of natural populations of *D. silvestris* and *D. heteroneura*. Heredity 56: 87–96.

Desdevises, Y., S. Morand, and G. Oliver. 2001. Linking specialisation to diversification in the Diplectanidae Bychowsky 1957 (Monogenea, Platyhelminthes). Parasitol. Res. 87: 223–230.

Dettman, J. R., D. J. Jacobson, E. Turner, A. Pringle, and J. W. Taylor. 2003. Reproductive isolation and phylogenetic divergence in Neurospora: Comparing methods of species recognition in a model eukaryote. Evolution 57: 2721–2741.

DeVries, H. 1906. *Species and Varieties, Their Origin by Mutation*, lectures delivered at the University of California. Open Court, Chicago, IL.

Diamond, J. 1966. Zoological classification system of a primitive people. Science 151: 1102–1104.

Diamond, J. M. 1973. Distributional ecology of New Guinea birds. Science 179: 759–769.

Diamond, J. 1986. Evolution of ecological segregation in the New Guinea montane avifauna. Pp. 98–125 in J. Diamond and T. J. Case (eds.) *Community Ecology*. Harper and Rowe, New York.

Diamond, J. M. 1977. Continental and insular speciation in Pacific land birds. Syst. Zool. 26:263–268.

Diamond, J. M. 1992. Horrible plant species. Nature 360: 627–628.

Diaz, A. and M. R. Macnair. 1999. Pollen tube competition as a mechanism of prezygotic reproductive isolation between *Mimulus nasutus* and its presumed progenitor *M. guttatus*. New Phytol. 144: 471–478.

Dickinson, T. A. 1998. Taxonomy of agamic complexes in plants: A role for metapopulation thinking. Folia Geobot. 33: 327–332.

Dieckmann, U. and M. Doebeli. 1999. On the origin of species by sympatric speciation. Nature 400: 354–357.

Diehl, S. R. and G. L. Bush. 1989. The role of habitat preference in adaptation and speciation. Pp. 345–365 in D. Otte and J. Endler (eds.) *Speciation and Its Consequences*. Sinauer Associates, Sunderland, MA.

Dobzhansky, T. 1933. On the sterility of the intrerracial hybrids in *Drosophila pseudoobscura*. Proc. Natl. Acad. Sci. USA 19: 397–403.

Dobzhansky, T. 1934. Studies on hybrid sterility. I. Spermatogenesis in pure and hybrid *Drosophila pseudoobscura*. Z. Zellforch. Microsk. Anat. 21: 169–221.

Dobzhansky, T. 1935. A critique of the species concept in biology. Philos. Sci. 2: 344–355.

Dobzhansky, T. 1936. Studies on hybrid sterility. II. Localization of sterility factors in *Drosophila pseudoobscura* hybrids. Genetics 21: 113–135.

Dobzhansky, T. 1937a. Genetic nature of species differences. Am. Nat. 71: 404–420.

Dobzhansky, T. 1937b. *Genetics and the Origin of Species*. Columbia University Press, New York.

Dobzhansky, T. 1937c. What is a species? Scientia Revista di Scienza 61: 280–286.
Dobzhansky, T. 1940. Speciation as a stage in evolutonary divergence. Am. Nat. 74: 302–321.
Dobzhansky, T. 1946a. Complete reproductive isolation between two morphologically similar species of *Drosophila*. Ecology 27: 205–211.
Dobzhansky, T. 1946b. Genetics of natural populations. XIII. Recombination and variability in populations of *Drosophila pseudoobscura*. Genetics 31: 269–290.
Dobzhansky, T. 1951. *Genetics and the Origin of Species* (Third Edition). Columbia University Press, New York.
Dobzhansky, T. 1970. *Genetics of the Evolutionary Process*. Columbia University Press, New York.
Dobzhansky, T. 1974. Genetic analysis of hybrid sterility within the species *Drosophila pseudoobscura*. Hereditas 77: 81–88.
Dobzhansky, T. 1975. Analysis of incipient reproductive isolation within a species of *Drosophila*. Proc. Natl. Acad. Sci. U.S.A. 72: 3638–3641.
Dobzhansky, T. and G. W. Beadle. 1936. Studies on hybrid sterility. IV. Transplanted testes in *Drosophila pseudoobscura*. Genetics 21: 832–840.
Dobzhansky, T. and R. D. Boche. 1933. Intersterile races of *Drosophila pseudoobscura* Frol. Biol. Zbl. 53: 314–330.
Dobzhansky, T. and C. Epling. 1944. Contributions to the genetics, taxonomy, and ecology of *Drosophila pseudoobscura* and its relatives. Washington, DC: Pub. 54: 1–46, Carnegie Institute Washington.
Dobzhansky, T. and P. C. Koller. 1938. An experimental study of sexual isolation in *Drosophila*. Biol. Zentr. 58: 589–607.
Dobzhansky, T. and O. Pavlovsky. 1966. Spontaneous origin of an incipient species in the *Drosophila paulistorum* complex. Proc. Natl. Acad. Sci. USA 55: 723–733.
Dobzhansky, T. and J. R. Powell. 1975. *Drosophila pseudoobscura* and its American relatives, *Drosophila persimilis* and *Drosophila miranda*. Pp. 537–587 *in* R. C. King (ed.) *Invertebrates of Genetic Interest*. Plenum Press, New York.
Dobzhansky, T., H. Levene, B. Spassky, and N. Spassky. 1959. Release of genetic variability through recombination. III. *Drosophila prosaltans*. Genetics 44: 75–92.
Dod, B., L. S. Jermiin, P. Boursot, V. H. Chapman, J. T. Nielsen, and F. Bonhomme. 1993. Counterselection on sex chromosomes in the Mus *musculus* European hybrid zone. J. Evol. Biol. 6: 529–546.
Dodd, D. M. B. 1989. Reproductive isolation as a consequence of adaptive divergence in *Drosophila pseudoobscura*. Evolution 43: 1308–1311.
Dodd, D. M. B. and J. R. Powell. 1985. Founder-flush speciation: An update of experimental results with *Drosophila*. Evolution 39: 1388–1392.
Dodd, M. E., J. Silvertown, and M. W. Chase. 1999. Phylogenetic analysis of trait evolution and species diversity variation among angiosperm families. Evolution 53: 732–744.
Doebeli, M. 1996. A quantitative genetic competition model for sympatric speciation. J. Evol. Biol. 9: 893–909.
Doebeli, M. and U. Dieckmann. 2003. Speciation along environmental gradients. Nature 421:259–264.
Doherty, J. A. and D. J. Howard. 1996. Lack of preference for conspecific calling songs in female crickets. Anim. Behav. 51: 981–989.
Doherty, P. F., G. Sorci, J. A. Royle, J. E. Hines, J. D. Nichols, and T. Boulinier. 2003. Sexual selection affects local extinction and turnover in bird communities. Proc. Natl. Acad. Sci. USA 100: 5858–5862.
Doi, M., M. Matsuda, M. Tomaru, H. Matsubayashi, and Y. Oguma. 2001. A locus for female discrimination behavior causing sexual isolation in *Drosophila*. Proc. Natl. Acad. Sci. USA 98: 6714–6719.
Dominey, W. J. 1984. Effects of sexual selection and life history on speciation: Species flocks in African cichlids and Hawaiian *Drosophila*. Pp. 231–249 *in* A. A. Echelle and I. Kornfield (eds.) *Evolution of Fish Species Flocks*. University of Maine at Orono Press, Orono, ME.
Dominey, W. J. and A. M. Snyder. 1988. Kleptoparasitism of freshwater crabs by cichlid fishes endemic to Lake Barombi Mbo, Cameroon, West Africa. Env. Biol. Fish 22: 155–160.
Donoghue, M. J. and J. A. Doyle. 2000. Seed plant phylogeny: Demise of the antophyte hypothesis. Curr. Biol. 10: R106–R109.
Dorn, R. D. 1976. A synopsis of American *Salix*. Can. J. Bot. 54: 2769–2789.
Dowling, T. E. and C. L. Secor. 1997. The role of hybridization and introgression in the diversification of animals. Annu. Rev. Ecol. Syst. 28: 593–619.
Doyle, J. A. and M. J. Donoghue. 1986. Seed plant phylogeny and the origin of angiosperms: An experimental cladistic approach. Bot. Rev. 52: 321–431.
Drès, M. and J. Mallet. 2002. Host races in plant-feeding insects and their importance in sympatric speciation. Philos. Trans. R. Soc. Lond. B 357: 471–492.
Dressler, R. L. 1968. Pollination by euglossine bees. Evolution 22: 202–210.
Dressler, R. L. 1981. *The Orchids: Natural History and Classification*. Harvard University Press, Cambridge, MA.
Duda, T. F., and S. R. Palumbi. 1998. Developmental shifts and species selection in gastropods. Proc. Natl. Acad. Sci USA 96: 10272–10277.
Dufour, L. 1844. Anatomie générale des Diptères. Ann. Sciences Naturelles 1: 244–264.
Dufresne, F. and P. D. Hebert. 1994. Hybridization and origins of polyploidy. Proc. R. Soc. Lond. B 258: 141–146.
Durham, W. H. 1991. *Coevolution: Genes, Culture, and Human Diversity*. Stanford University Press, Stanford, CA.

Duvernell, D. D. and B. J. Turner. 1998. Evolution genetics of Death Valley pupfish populations: Mitochondrial DNA sequence variation and population structure. Mol. Ecol. 7: 279–288.

Duvernell, D. D. and B. J. Turner. 1999. Variation and divergence of Death Valley pupfish populations at retrotransposon-defined loci. Mol. Biol. Evol. 16: 363–371.

Dykhuizen, D. E. 1998. Santa Rosalia revisited: Why are there so many species of bacteria? Antonie van Leeuwenhoek J. Microbiol. 73: 25–33.

Eastop, V. F. 1971. Keys for the identification of *Acyrthosiphon* (Hemiptera: Aphididae). Bull. Br. Mus. Nat. Hist. Ent. 26: 1–115.

Eberhard, W. G. 1986. *Sexual Selection and Animal Genitalia*. Harvard University Press. Cambridge, MA.

Eberhard, W. G. 1992. Species isolation, genital mechanics, and the evolution of species-specific genitalia in three species of *Macrodactylus* beetles (Coleoptera, Scarabaeidae, Melolonthinae). Evolution 46: 1774–1783.

Eberhard, W. G. 1993. Evaluating models of sexual selection: Genitalia as a test case. Am. Nat. 142: 564–571.

Eberhard, W. G. 1996. *Female Control: Sexual Selection by Cryptic Female Choice*. Princeton University Press, Princeton, NJ.

Eberhard, W. G. 2001. Species-specific genitalic copulatory courtship in sepsid flies (Diptera, Sepsidae, *Microsepsis*) and theories of genitalic evolution. Evolution 55: 93–102.

Eberhard, W. G., B. A. Huber, F. L. Rodrigues, D. Briceño, I. Salas, and V. Rodriguez. 1998. One size fits all? Relationships between the size and degree of variation in genitalia and other body parts in twenty species of insects and spiders. Evolution 52: 415–431.

Echelle, A. A. and T. E. Dowling. 1992. Mitochondrial DNA variation and evolution of the Death Valley pupfishes (*Cyprinodon*, Cyprinodontidae). Evolution 46: 193–206.

Echelle, A. A.and I. Kornfield, (eds.) 1984. *Evolution of Fish Species Flocks*. University of Maine at Orono Press, Orono, ME.

Edmands, S. 1999. Heterosis and outbreeding depression in interpopulation crosses spanning a wide range of divergence. Evolution 53: 1757–1768.

Edmands, S. 2002. Does parental divergence predict reproductive compatibility? Trends Ecol. Evol. 17: 520–527.

Edmands, S. and R. S. Burton. 1999. Cytochrome c oxidase activity in interpopulation hybrids of a marine copepod: A test for nuclear-nuclear or nuclear-cytoplasmic coadaptation. Evolution 53: 1972–1978.

Ehrendorfer, F. 1961. Evolution of the *Galium multiflorum* complex in western North America. I. Diploids and polyploids in this dioecious group. Madroño 16: 109–122.

Ehrman, L. 1964. Genetic divergence in M. Vetukhiv's experimental populations of *Drosophila pseudoobscura*. Genet. Res. 5: 150–157.

Ehrman, L. 1965. Direct observation of sexual isolation between allopatric and between sympatric strains of the different *Drosophila paulistorum* races. Evolution 19: 459–464.

Ehrman, L. 1969. Genetic divergence in M. Vetukhiv's experimental populations of *Drosophila pseudoobscura*. 5. A further study of rudiments of sexual isolation. Am. Midl. Nat. 82: 272–276.

Ehrman, L., M. M. White, and B. Wallace. 1991. A long-term study involving *Drosophila melanogaster* and toxic media *in* M. K. Hecht, B. Wallace, and R. J. MacIntyre (eds.) *Evolutionary Biology*. Plenum Press, New York.

Eldredge, N. and S. J. Gould. 1972. Punctuated equilibria: An alternative to phyletic gradualism. Pp 82–115 *in* T. J. M. Schopf (ed.) *Models in Paleobiology*. Freeman, Cooper, and Co., San Francisco.

Ellstrand, N. C., R. Whitkus, and L. H. Rieseberg. 1996. Distribution of spontaneous plant hybrids. Proc. Natl. Acad. Sci. USA 193: 5090–5093.

Endler, J. A. 1977. *Geographic Variation, Speciation, and Clines*. Princeton University Press, Princeton, NJ.

Endler, J. A. 1982a. Pleistocene forest refuges: Fact or fancy? Pp. 641–657 *in* G. T. Prance (ed.) *Biological Diversification in the Tropics*. Columbia University Press, New York.

Endler, J. A. 1982b. Problems in distinguishing historical from ecological factors in biogeography. Am. Zool. 22: 441–452.

Endler, J. A. and A. E. Houde. 1995. Geographic variation in female preferences for male traits in *Poecilia reticulata*. Evolution 49: 456–468.

Engels, W. R. and C. R. Preston. 1979. Hybrid dysgenesis in *Drosophila melanogaster*: The biology of male and female sterility. Genetics 92: 161–174.

Ephrussi, B. and G. W. Beadle. 1935. La transplantation des ovaires chez la Drosophile. Bull. Biol. Fr. Belg. 69: 492–502.

Ereshefsky, M. 1992. *The Units of Evolution*. MIT Press, Cambridge, MA.

Etges, W. J. 1998. Premating isolation is determined by larval rearing substrates in cactophilis *Drosophila mojavensis*. IV. Correlated responses in behavioral isolation to artificial selection on a life-history trait. Am. Nat. 152: 129–144.

Evans, M. 2001. Socotra Island Xeric Shrublands, unpublished document, World Wildlife Fund; available at http://www.worldwildlife.org/wildworld/profiles/terrestrial/at/at1318_full.html

Everett, C. A., J. B. Searle, B. M. N. Wallace. 1996. A study of meiotic pairing, nondisjunction and germ cell death in laboratory mice carrying Robertsonian translocations. Genet. Res. 67: 239–247.

Ewens, W. J. 1979. *Mathematical Population Genetics*. Springer-Verlag, Berlin.

Farrell, B. D. 1998. "Inordinate fondness" explained: Why are there so many beetles? Science 281: 555–559.

Farrell, B. D. and C. Mitter. 1993. Phylogenetic determinants of insect/plant community diversity. Pp. 263–266 in R. E. Ricklefs and D. Schluter (eds.) *Species Diversity in Ecological Communities.* University of Chicago Press, Chicago.

Farrell, B. D., D. E. Dussourd, and C. Mitter. 1991. Escalation of plant defense: Do latex and resin canals spur plant diversification? Am. Nat. 138: 881–900.

Farrell, B. D., A. S. Sequiera, B. C. O'Meara, B. B. Normark, J. H. Chung, and B. H. Jordal. 2001. The evolution of agriculture in beetles (Curculionidae: Scolytinae and Platypodinae). Evolution 55: 2011–2027.

Feder, J. L. 1998. The apple maggot fly, *Rhagoletis pomonella:* Flies in the face of conventional wisdom about speciation? Pp. 130–144 in D. J. Howard and S. H. Berlocher (eds.) *Endless Forms: Species and Speciation.* Oxford University Press, New York.

Feder, J. L. and G. L. Bush. 1989. A field test of differential host-plant usage between two sibling species of *Rhagoletis pomonella* fruit flies (Diptera: Tephritidae) and its consequences for sympatric speciation. Evolution 43: 1813–1819.

Feder, J. L., S. Opp, B. Wlazlo, K. Reynolds, W. Go, and S. Spisak. 1994. Host fidelity is an effective pre-mating barrier between sympatric races of the apple maggot fly. Proc. Natl. Acad. Sci. USA 91: 7990–7994.

Feder, J. L., C. A. Chilcote, and G. L. Bush. 1988. Genetic differentiation between sympatric host races of the apple maggot fly *Rhagoletis pomonella.* Nature 336: 61–64.

Feder, J. L., J. B. Roethele, B. Wlazlo, and S. H. Berlocher. 1997a. The selective maintenance of allozyme differences between sympatric host races of the apple maggot fly. Proc. Natl. Acad. Sci. USA 94: 11417–11421.

Feder, J. L., U. Stolz, K. M. Lewis, W. Perry, J. B. Roethele, and A. Rogers. 1997b. The effects of winter length on the genetics of apple and hawthorn races of *Rhagoletis pomonella* (Diptera: Tephritidae). Evolution 51: 1862–1876.

Feder, J. L., S. H. Berlocher, and S. B. Opp. 1998. Sympatric host race formation and speciation in *Rhagoletis* (Diptera: Tephritidae): A tale of two species for Charles Darwin. Pp. 408–441 in S. Mopper and S. Strauss (eds.) *Genetic Structure and Local Adaptation in Natural Insect Populations: Effects of Ecology, Life History, and Behavior.* Chapman and Hall, New York.

Feder, J. L., S. H. Berlocher, J. B. Roethele, H. Dambroski, J. J. Smith, W. L. Perry, V. Gavrilovic, K. E. Filchak, J. Rull, and M. Aluja. 2003a. Allopatric genetic origins for sympatric host-plant shifts and race formation in *Rhagoletis.* Proc. Natl. Acad. Sci. USA 100: 10314–10319.

Feder, J. L., J. B. Roethele, K. Filchak, J. Niedbalski, and J. Romero-Severson. 2003b.. Evidence for inversion polymorphism related to sympatric host race formation in the apple maggot fly, *Rhagoletis pomonella.* Genetics 163: 939–953.

Feil, E. J., M. C. Maiden, M. Achtman, and B. G. Spratt. 1999. The relative contributions of recombination and mutation to the divergence of clones of *Neisseria miningitidis.* Mol. Biol. Evol. 16: 1496–1502.

Feil, E. J., J. M. Smith, M. C. Enright, and B. C. Spratt. 2000. Estimating recombinational parameters in *Streptococcus pneumoniae* from multilocus sequence typing data. Genetics 154: 1439–1450.

Felber, F. 1991. Establishment of a tetraploid cytotype in a diploid population: Effect of relative fitness of the cytotypes. J. Evol. Biol. 4: 195–207.

Felsenstein, J. 1981. Skepticism towards Santa Rosalia, or Why are there so few kinds of animals? Evolution 35: 124–138.

Felsenstein, J. 2004. *Inferring Phylogenies.* Sinauer Associates, Sunderland, MA.

Fenster, C. B. and K. Ritland. 1994. Quantitative genetics of mating system divergence in the yellow monkeyflower species complex. Heredity 73: 422–435.

Fenster, C. B., P. K. Diggle, S. C. H. Barrett, and K. Ritland. 1995. The genetics of floral development differentiating two species of *Mimulus* (Scrophulariaceae). Heredity 74:258–266.

Ferguson, D. and T. Sang. 2001. Speciation through homoploid hybridization between allotetraploids in peonies (*Paeonia*). Proc. Natl. Acad. Sci. USA 98: 3915–3919.

Ferris, S. D. and G. S. Whitt. 1977. Loss of duplicate gene expression after polyploidization. Nature 265: 258–260.

Ferris, S. D., R. D. Sage, C.-M. Huang, J. T. Nielsen, U. Ritte, and A. C. Wilson. 1983. Flow of mitochondrial DNA across a species boundary. Proc. Natl. Acad. Sci. USA 80: 2290–2294.

Ficken, M. S. and R. W. Ficken. 1968a. Courtship of Blue-winged warblers, Golden-winged warblers, and their hybrids. The Wilson Bull. 80: 161–172.

Ficken, M. S. and R. W. Fickin. 1968b. Reproductive isolating mechanisms in the Blue-winged warbler–Golden-winged warbler complex. Evolution 22: 166–179.

Figueroa, F., E. Günter, and J. Klein. 1988. MHC polymorphism pre-dating speciation. Nature 335: 265–267.

Filchak, K. E., J. B. Roethele, and J. L. Feder. 2000. Natural selection and sympatric divergence in the apple maggot *Rhagoletis pomonella.* Nature 407: 739–742.

Fisher, R. A. 1922. On the dominance ratio. Proc. R. Soc. Edinb. Biol. 52: 321–341.

Fisher, R. A. 1930. *The Genetical Theory of Natural Selection.* Clarendon Press, Oxford.

Fishman, L. and J. H. Willis. 2001. Evidence for Dobzhansky–Muller incompatibilities contributing to the sterility of hybrids between *Mimulus guttatus* and *M. nasutus.* Evolution 55: 1932–1942.

Fishman, L., and D. A. Stratton. 2004. The genetics of floral divergence and postzygotic barriers between outcrossing and selfing populations of *Arenaria uniflora* (Caryophyllaceae). Evolution 58: 296–307.

Fishman, L., A. J. Kelly, and J. H. Willis. 2002. Minor QTLs underlie floral traits associated with mating system divergence in *Mimulus*. Evolution 56: 2138–2155.

Fitzpatrick, B. M. 2002. Molecular correlates of reproductive isolation. Evolution 56: 191–198.

Fitzpatrick, B. M. and M. Turelli. 2004. The biogeography of mammalian speciation. Unpublished ms.

Florin, A.-B. and A. Ödeen. 2002. Laboratory environments are not conducive for allopatric speciation. J. Evol. Biol. 15: 10–19.

Forejt, J. and P. Iványi. 1974. Genetic studies on male sterility of hybrids between laboratory and wild mice (*Mus musculus* L.). Genetical Research 24: 189–206.

Forejt, J. 1996. Hybrid sterility in the mouse. Trends in Genetics 12: 412–417.

Foote, D. 2004. Rediscovery of a rare "banana moth" of the Hawaiian *Omiodes* Gueneé (Lepidoptera: Crambidae): A member of a new host-specific complex or an ancient displaced fauna? Occas. Papers Bishop Museum (in press).

Fowler, N. and D. A. Levin. 1984. Ecological constraints on the establishment of a novel polyploid in competition with its diploid progenitor. Am. Nat. 124: 703–711.

Frank, S. H. 1991. Divergence of meiotic drive-suppressors as an explanation for sex-biased hybrid sterility and inviability. Evolution 45: 262–267.

Frith, C. B. and B. M. Beehler. 1998. *The Birds of Paradise*. Oxford University Press, New York.

Fry, J. D. 2003. Multilocus models of sympatric speciation: Bush vs. Rice vs. Felsenstein. Evolution 57: 1735–1746.

Fryer, G. 1997. Biological implications of a suggested Late Pleistocene desiccation of Lake Victoria. Hydrobiologia 354: 177–182.

Fryer, G. 1999. Endemism, speciation, and adaptive radiation in great lakes. Environ. Biol. of Fishes 45: 109–131.

Fryer, G. 2001. On the age and origin of the species flock of haplochromine cichlid fishes of Lake Victoria. Proc. R. Soc. Lond. B 268: 1147–1152.

Fryer, G. and T. D. Iles. 1972. *The Cichlid Fishes of the Great Lakes of Africa*. Oliver and Boyd, Edinburgh.

Fukatami, A. and D. Moriwaki. 1970. Selection for sexual isolation in *Drosophila melanogaster* by a modification of Koopman's method. Jpn. J. Genet. 45: 193–204.

Fulton, M. and S. A. Hodges. 1999. Floral isolation between *Aquilegia formosa* and *A. pubescens*. Proc. R. Soc. Lond. B 266: 2247–2252.

Funk, D. J. 1998. Isolating a role for natural selection in speciation: Host adaptation and sexual isolation in *Neochlamisus bebianea* leaf beetles. Evolution 52: 1744–1759.

Futuyma, D. J. 1983. Speciation [book review]. Science 219: 1059–1060.

Futuyma, D. J. 1987. On the role of species in anagenesis. Am. Nat. 130: 465–473.

Futuyma, D. J. and G. C. Mayer. 1980. Non-allopatric speciation in animals. Syst. Zool. 29: 254–271.

Futuyma, D. J., J. S. Walsh, Jr., T. Morton, D. J. Funk, and M. C. Keese. 1994. Genetic variation in a phylogenetic context: Responses of two specialized leaf beetles (Coleoptera: Chrysomelidae). J. Evol. Biol. 7: 127–146.

Futuyma, D. J., M. C. Keese, and D. J. Funk. 1995. Genetic constraints on macroevolution: The evolution of host affiliation in the leaf beetle genus *Ophraella*. Evolution 49: 797–809.

Fuyama, Y. 1983. Species-specificity of paragonial substances as an isolating mechanism in *Drosophila*. Experientia 39: 190–192.

Gadau, J., R. E. Page, and J. H. Werren. 1999. Mapping of hybrid incompatibility loci in *Nasonia*. Genetics 153: 1731–1741.

Gage, M. J. G., G. A. Parker, S. Nylin, and C. Wicklund. 2002. Sexual selection and speciation in mammals, butterflies, and spiders. Proc. R. Soc. Lond. B 269: 2309–2316.

Galen, C., K. A. Zimmer, and M. E. Newport. 1987. Pollination in floral scent morphs of *Polemonium viscosum*: A mechanism for disruptive selection on flower size. Evolution 41: 599–606.

Galiana, A., A. Moya, and F. J. Ayala. 1993. Founder-flush speciation in *Drosophila pseudoobscura*: A large-scale experiment. Evolution 47: 432–444.

Galindo, B. E., G. W. Moy, W. J. Swanson, and V. D. Vacquier. 2002. Full-length sequence of VERL, the egg vitelline envelope receptor for abalone sperm lysin. Gene 288: 111–117.

Galindo, B. E., V. D. Vacquier, and W. J. Swanson. 2003. Positive selection in the egg receptor for abalone sperm lysin. Proc. Natl. Acad. Sci. USA 100: 4639–4643.

Galis, F. and J. A. J. Metz. 1998. Why are there so many cichlid species? Trends Ecol. Evol. 13:1–3.

Gallez, G. P. and L. D. Gottlieb. 1982. Genetic evidence for the hybrid origin of the diploid plant *Stephanomeria diegensis*. Evolution 36: 1158–1167.

Gardezi, T. and J. da Silva. 1999. Diversity in relation to body size in mammals: A comparative study. Am. Nat. 153: 110–123.

Gaston, K. J. 2000. Global patterns in biodiversity. Nature 405: 220–227.

Gates, R. R. 1924. Polyploidy. Brit. J. Exp. Biol. 1: 153–182.

Gavrilets, S. 1997. Evolution and speciation on holey adaptive landscapes. Trends Ecol. Evol. 12: 307–312.

Gavrilets, S. 1999. A dynamical theory of speciation on holey adaptive landscapes. Am. Nat. 154: 1–22.

Gavrilets, S. 2000a. Rapid evolution of reproductive barriers driven by sexual conflict. Nature 403: 886–889.

Gavrilets, S. 2000b. Waiting time to parapatric speciation. Proc. R. Soc. Lond. B 267: 2483–2492.

Gavrilets, S. 2002. Models of speciation: What have we learned in 40 years? Evolution 57:2197–2215.

Gavrilets, S. and C. R. B. Boake. 1998. On the evolution of premating isolation after a founder event. Am. Nat. 152: 706–716.

Gavrilets, S. and A. Hastings. 1996. Founder-effect speciation: A theoretical reassessment. Am. Nat. 147: 466–491.

Gavrilets, S. and D. Waxman. 2002. Sympatric speciation by sexual conflict. Proc. Natl. Acad. Sci. USA 99: 10533–10538.

Gavrilets, S., H. Li, and M. D. Vose. 1998. Rapid parapatric speciation on holey adaptive landscapes. Proc. R. Soc. Lond. B 265: 1483–1489.

Gavrilets, S., H. Li, and M. D. Vose. 2000. Patterns of parapatric speciation. Evolution 54: 1126–1134.

Genner, M. J., G. F. Turner, S. Barker, and S. J. Hawkins. 1999a. Niche segregation among Lake Malawi cichlid fishes? Evidence from stable isotope signatures. Ecol. Lett. 2:185–190.

Genner, M. J., G. F. Turner, and S. J. Hawkins. 1999b. Foraging of rocky habitat cichlid fishes in Lake Malawi: Coexistence through niche partitioning? Oecologia 121: 283–292.

Gerassimova, H. 1939. Chromosome alterations as a factor of divergence of forms. I. New experimentally produced strains of *C. tectorum* which are physiologically isolated from the original forms owing to reciprocal translocation. Compt. Rend. Acad. Sci. URSS 25: 148–154.

Gerstel, D. U. 1954. A new lethal combination in interspecific cotton hybrids. Genetics 39: 628–639.

Gethmann, R. C. 1988. Crossing over in males of Higher Diptera (Brachycera). J. Hered. 79: 344–350.

Geyer, L. B. and S. R. Palumbi. 2003. Reproductive character displacement and the genetics of gamete recognition in tropical sea urchins. Evolution 57: 1049–1060.

Gharrett, A. J. and W. W. Smoker. 1991. Two generations of hybrids between even- and odd-year pink salmon (*Onchorhynchus gorbuscha*): A test for outbreeding depression? Can. J. Fish. Aquat. Sci. 48: 1744–1749.

Gharrett, A. J., C. Smoot, and A. J. McGregor. 1988. Genetic relationships of even-year northwestern Alaskan pink salmon. Trans. Am. Fish. Soc. 117: 536–545.

Gharrett, A. J., W. W. Smoker, R. R. Reisenbichler, and S. G. Taylor. 1999. Outbreeding depression in hybrids between odd- and even-year pink salmon. Aquaculture 173: 117–129.

Gilbert, D. G. and W. T. Starmer. 1985. Statistics of sexual isolation. Evolution 39: 1380–1383.

Gillespie, J. H. 1991. *The Causes of Molecular Evolution*. Oxford University Press, Oxford.

Gillespie, R. G. and H. B. Croom. 1995. Comparison of speciation mechanisms in web-building and non-web-building groups within a lineage of spiders. Pp. 121–146 in W. L. Wagner and V. A. Funk (eds.) *Hawaiian Biogeography: Evolution on a Hot-Spot Archipelago*. Smithsonian Institution Press, Washington, DC.

Gillespie, R. G. and G. K. Roderick. 2002. Arthropods on islands: Colonization, speciation, and conservation. Annu. Rev. Entomol. 47: 595–632.

Gíslason, D., M. M. Ferguson, S. Skúlason, and S. S. Snorasson. 1999. Rapid and coupled phenotypic differentiation in Icelandic Arctic char (*Salvelinus alpinus*). Can. J. Fish. Aquat. Sci. 56: 2229–2234.

Gittenberger, E. 1988. Sympatric speciation in snails: A largely neglected model. Evolution 42: 826–828.

Gittleman, J. L. and A. Purvis. 1998. Body size and species-richness in carnivores and primates. Proc. R. Soc. Lond. B 265:113–119.

Givnish, T. J. 1998. Adaptive plant evolution on islands: Classical patterns, molecular data, new insights. Pp. 281–304 in P. R. Grant (ed.) *Evolution on Islands*. Oxford University Press, Oxford.

Givnish, T. J., K. J. Sytsma, J. F. Smith, and W. J. Hahn. 1995. Molecular evolution, adaptive radiation, and geographic speciation in *Cyanea* (Campanulaceae, Lobeliodeae). Pp. 259–301 in W. L. Wagner and V. A. Funk (eds.) *Hawaiian Biogeography: Evolution on a Hot-Spot Archipelago*. Smithsonian Institution Press, Washington DC.

Glazner, J. T., B. Devlin, and N. C. Ellstrand. 1998. Biochemical and morphological evidence for host race evolution in desert mistletoe, *Phoradendron californicum* (Viscaceae). Plant Syst. Evol. 161: 13–21.

Goldblatt, P. 1980. Polyploidy in angiosperms: Monocotyledons. Pp. 219–240 in W. H. Lewis (ed.) *Polyploidy: Biological Relevance*. Plenum press, New York.

Goodnight, C. J. 1988. Epistasis and the effect of founder events on the additive genetic variance. Evolution 427: 441–454.

Goodwillie, C. 1999. Multiple origins of self-compatibility in *Linanthus* section *Leptosiphon* (Polemoniaceae): Phylogenetic evidence from internal-transcribed-spacer sequence data. Evolution 53: 1387–1395.

Gottlieb, L. D. 1971. Evolutionary relationships in the outcrossing diploid annual species of *Stephanomeria* (Compositae). Evolution 25: 312–329.

Gottlieb, L. D. 1973. Genetic differentiation, sympatric speciation, and the origin of a diploid species of *Stephanomeria*. Am. J. Bot. 60: 545–553.

Gottlieb, L. D. 1984a. Genetic confirmation of the origin of *Clarkia lingulata*. Evolution 28:244–250.

Gottlieb, L. D. 1984b. Genetics and morphological evolution in plants. Am. Nat. 123: 681–709.

Gottschewski, G. 1940. Eine Analyse Bestimmten *Drosophila pseudoobscura* Rassen–und Artkreuzungen. Z. i. Abst. u. Vererb. 78: 338–398.

Goudet, J. 1999. An improved procedure for testing the effects of key innovations on rate of speciation. Am. Nat. 153: 549–555.

Gould, S. J. 2002. *The Structure of Evolutionary Theory.* Harvard University Press, Cambridge, MA.

Gould, S. J. and N. Eldredge. 1977. Punctuated equilibria: The tempo and mode of evolution reconsidered. Paleobiology 3: 115–151.

Gould, S. J. and E. A. Lloyd. 1999. Individuality and adaptation across levels of selection: How shall we name and generalize the unit of Darwinism? Proc. Natl. Acad. Sci. USA 96:11904–11909.

Goulson, D. and K. Jerrim. 1997. Maintenance of the species boundary between *Silene dioica* and *S. latifolia* (red and white campion). Oikos 79: 115–126.

Grant, B. R. and P. R. Grant. 1993. Evolution of Darwin's finches caused by a rare climatic event. Proc. R. Soc. Lond. B 251: 111–117.

Grant, B. S. and L. E. Mettler. 1969. Disruptive and stabilizing selection on the "escape" behavior of *Drosophila melanogaster.* Genetics 62: 625–637.

Grant, P. R. 1986. *Ecology and Evolution of Darwin's Finches.* Princeton University Press, Princeton, NJ.

Grant, P. R. and B. R. Grant. 1989. Sympatric speciation and Darwin's finches. Pp. 433–457 *in* D. Otte and J. A. Endler (eds.) *Speciation and Its Consequences.* Sinauer Associates, Sunderland, MA.

Grant, P. R. and B. R. Grant. 1992. Hybridization of bird species. Science 256: 193–197.

Grant, P. R. and B. R. Grant. 1996. Speciation and hybridization in island birds. Philos. Trans. R. Soc. Lond. B 351: 765–772.

Grant, P. R. and B. R. Grant. 1997. Genetics and the origin of bird species. Proc. Natl. Acad. Sci. USA 94: 7768–7775.

Grant, P. R. and B. R. Grant. 2002. Adaptive radiation of Darwin's finches. Am. Sci. 90:130–139.

Grant, V. 1949. Pollination systems as isolating mechanisms in angiosperms. Evolution 3: 82–97.

Grant, V. 1957. The plant species in theory and in practice. Pp. 39–80 *in* E. Mayr (ed.) *The Species Problem.* Amer. Assn. Adv. Sci. Pub. 50, Washington, DC, USA.

Grant, V. 1958. The regulation of recombination in plants. Cold Spring Harb. Symp. Quant. Biol. 23: 337–363.

Grant, V. 1964. The genetic structure of races and species in *Gilia.* Advances in Genetics 8: 55–87.

Grant, V. 1966a. Selection for vigor and fertility in the progeny of a highly sterile species hybrid in *Gilia.* Genetics 53: 757–775.

Grant, V. 1966b. The origin of a new species of *Gilia* in a hybridization experiment. Genetics 54: 1189–1199.

Grant, V. 1966c. The selective origin of incompatibility barriers in the plant genus *Gilia.* Am. Nat. 100: 99–118.

Grant, V. 1971. *Plant Speciation.* Columbia University Press, New York.

Grant, V. 1981. *Plant Speciation* (Second Edition). Columbia University Press, New York.

Grant, V. 1992. Floral isolation between ornithophilous and spingophilous species of *Ipomopsis* and *Aquilegia.* Proc. Natl. Acad. Sci. USA 89: 11828–11831.

Grant, V. 1994. Modes and origins of mechanical and ethological isolation in angiosperms. Proc. Natl. Acad. Sci. USA 91: 3–10.

Grant, V. 2002. Frequency of spontaneous amphiploids in *Gilia* (Polemoniaceae) hybrids. Am. J. Botany 89: 1197–1202.

Grant, V. and K. Grant. 1965. *Flower Pollination in the Phlox Family.* Columbia University Press, New York.

Grantham, T. 2001. Hierarchies in evolution. Pp. 190–194 *in* D. E. G. Briggs and P. Crowther (eds.) *Paleobiology II.* Blackwell Scientific Publications, Oxford, UK.

Grantham, T. A. 1995. Hierarchical approaches to macroevolution: Recent work on species selection and the "effect hypothesis". Annu. Rev. Ecol. Syst. 26: 301–321.

Graur, D. and W.-H. Li. 2000. *Fundamentals of Molecular Evolution* (Second Edition). Sinauer Associates, Sunderland, MA.

Greenwood, P. H. 1965. The cichlid fishes of lake Nabugabo, Uganda. Bull. Br. Mus. Nat. Hist. 12: 315–357.

Greenwood, P. H. 1981. *The Haplochromine Fishes of the East African Lakes.* Cornell University Press, Ithaca, NY.

Gregory, P. G. and D. J. Howard. 1993. Laboratory hybridization studies of *Allonemobius fasciatus* and *A. socius* (Orthoptera: Gryllidae). Annals Ent. Soc. Amer. 86: 694–701.

Gregory, P. G. and D. J. Howard. 1994. A postinsemination barrier to fertilization isolates two closely related ground crickets. Evolution 48: 705–710.

Greig, D., R. H. Borts, E. J. Louis and M. Travisano. 2002a. Epistasis and hybrid sterility in *Saccharomyces.* Proc. R. Soc. Lond. B 269: 1167–1171.

Greig, D., E. J. Louis, R. H. Borts, and M. Travisano. 2002b. Hybrid speciation in experimental populations of yeast. Science 298: 1773–1775.

Greig, D., M. Travisano, E. J. Louis, and R. H. Borts. 2003. A role for the mismatch repair system during incipient speciation in *Saccharomyces.* J. Evol. Biol. 16: 429–437.

Griffiths, G. C. D. 1976. The future of Linnaean nomenclature. Syst. Zool. 25: 168–173.

Grimaldi, D., A. C. James, and J. Jaenike. 1992. Systematics and modes of reproductive isolation in the holarctic *Drosophila testacea* species group (Diptera: Drosophilidae). Ann. Entomol. Soc. Am. 85: 671–685.

Gropp, A. and H. Winking. 1981. Robertsonian translocations: Cytology, meiosis, segregation patterns and biological consequences of heterozygosity. Symp. Zool. Soc. Lond. 47: 141–181.

Gross, B. L., A. E. Schwarzbach, and L. H. Rieseberg. 2003. Origin(s) of the diploid hybrid species

Helianthus deserticola (Asteraceae). Am. J. Bot. 90: 1708–1719.

Grudzien, T. A., W. S. Moore, J. R. Cook, and D. Table. 1987. Genetic population structure of the northern flicker (*Colaptes auratus*) hybrid zone. Auk 104: 654–664.

Grula, J. W., J. D. McChesney, and J. O.R. Taylor. 1980. Aphrodisiac pheromones of the sulfur butterflies *Colias eurytheme* and *C. philodice* (Lepidoptera, Pieridae). J. Chem. Ecol. 6:241–256.

Grun, P. 1976. *Cytoplasmic Genetics and Evolution*. Columbia University Press, New York.

Guenet, J.-L., C. Nagamine, D. Simon-Chazottes, X. Montagutelli, and F. Bonhomme. 1990. *Hst-3*: an X-linked hybrid sterility gene. Genetical Research 56: 163–165.

Gupta, J. P., N. Dwivedi, and B. K. Singh. 1980. Natural hybridization in *Drosophila*. Experientia 36: 290.

Gutbrodt, H. and M. Schartl. 1999. Intragenic sex-chromosomal crossovers of *Xmrk* oncogene alleles affect pigment pattern formation and the severity of melanoma in *Xiphophorus*. Genetics 151: 773–783.

Guttman, S. I., and L. A. Weigt. 1989. Electrophoretic evidence of relationships among *Quercus* (oaks) of eastern North America. Can. J. Bot. 67: 339–351.

Haffer, J. 1969. Speciation in Amazonian forest birds. Science 165: 131–137.

Hafner, M. S. and R. D. M. Page. 1995. Molecular phylogenies and host-parasite cospeciation: Gophers and lice as a model system. Philos. Trans. R. Soc. Lond. B 349:77–83.

Hagen, R. H. and J. M. Scriber. 1989. Sex-linked diapause, color, and allozyme loci in *Papilio glaucus*: Linkage analysis and significance in a hybrid zone. J. Hered. 80: 179–185.

Haldane, J. B. S. 1922. Sex ratio and unisexual sterility in animal hybrids. J. Genet. 12: 101–109.

Haldane, J. B. S. 1924. A mathematical theory of natural and artificial selection, Part I. Trans. Camb. Philos. Soc. 23: 19–41.

Haldane, J. B. S. 1926. Can a species concept be justified? Syst. Assoc. Pub. 2: 95–96.

Haldane, J. B. S. 1927. A mathematical theory of natural and artificial selection, Part V: Selection and mutation. Proc. Camb. Philos. Soc. 23: 838–844.

Haldane, J. B. S. 1930. A mathematical theory of natural and artificial selection. Part VI: Isolation. Proc. Camb. Philos. Soc. 26: 220–230.

Haldane, J. B. S. 1932. *The Causes of Evolution*. Longmans, Green & Co, Ltd., New York.

Haldane, J. B. S. 1948. The theory of a cline. J. Genet. 48: 277–284.

Haldane, J. B. S. 1957. The cost of natural selection. J. Genet. 55: 511–524.

Haldane, J. B. S. 1959. Natural selection. Pp. 101–149 *in* P. R. Bell (ed.) *Darwin's Biological Work: Some Aspects Reconsidered*. Camb. Univ. Press, Cambridge, UK.

Hale, D. W. 1986. Heterosynapsis and suppression of chiasmata within heterozygous pericentric inversions of the Sitka deer mouse. Chromosoma 94: 425–432.

Hall, J. P. W. and D. J. Harvey. 2002. The phylogeography of Amazonia revisited: New evidence from riodinid butterflies. Evolution 56: 1489–1497.

Halliburton, R. and G. A. E. Gall. 1981. Disruptive selection and assortative mating in *Tribolium castaneum*. Evolution 35: 829–843.

Hamerton, J. L., N. Canning, M. Ray, and S. Smith. 1975. A cytogenetic survey of 14,069 newborn infants. I. Incidence of chromosome abnormalities. Clin. Genet. 8: 223–243.

Hardin, J. W. 1975. Hybridization and introgression in *Quercus*. J. Arnold Arbor. Harvard University 56: 336–363.

Hardy, D. E. and K. Y. Kaneshiro. 1981. Drosophilidae of Pacific Oceania. Pp. 309–348 *in* M. Ashburner, H. L. Carson, and J. N. Thompson Jr. (eds.) *Genetics and Biology of Drosophila*. Vol. 3a. Academic Press, New York.

Harper, A. A. and D. M. Lambert. 1983. The population genetics of reinforcing selection. Genetica 62: 15–23.

Harrison, R. G. 1985. Barriers to gene exchange between closely related cricket species. II. Life cycle variation and temporal isolation. Evolution 39: 244–259.

Harrison, R. G. 1998. Linking evolutionary pattern and process: The relevance of species concepts for the study of speciation. Pp. 19–31 *in* D. J. Howard and S. H. Berlocher (eds.) *Endless Forms: Species and Speciation*. Oxford University Press, New York.

Harrison, R. G. (ed.) 1992. *Hybrid Zones and the Evolutionary Process*. Oxford University Press, New York.

Harrison, R. G. and S. M. Bogdanowicz. 1995. Mitochondrial-DNA phylogeny of North American field crickets: Perspectives on the evolution of life-cycles, songs, and habitat associations. J. Evol. Biol. 8: 209–232.

Harrison, R. G. and D. M. Rand. 1989. Mosaic hybrid zones and the nature of species boundaries. Pp. 111–133 *in* D. Otte and J. A. Endler (eds.) *Speciation and Its Consequences*. Sinauer Associates, Sunderland, MA.

Harrison, R. H. 1990. Hybrid zones: windows on evolutionary processes. Oxford Surv. Evol. Biol. 7: 69–128.

Haskins, C. P. and E. F. Haskins. 1949. The role of sexual isolation as an isolating mechanism in three species of poeciliid fishes. Evolution 3: 160–169.

Hatfield, T. 1996. Genetic divergence in adaptive characters between sympatric species of sticklebacks. Am. Nat. 149: 1009–1029.

Hatfield, T. and D. Schluter. 1999. Ecological speciation in sticklebacks: Environment-dependent hybrid fitness. Evolution 53: 866–873.

Hauffe, H. C. and J. B. Searle. 1998. Chromosomal heterozygosity and fertility in house mouse (*Mus musculus domesticus*) from northern Italy. Genetics 150: 1143–1154.

Hauschteck-Jungen, E. 1990. Postmating reproductive isolation and modification of the 'sex ratio' trait in *Drosophila subobscura* induced by the sex chromosome gene arrangement $A_{2+3+5+7}$. Genetica 83: 31–44.

Hawthorne, D. J. and S. Via. 2001. Genetic linkage of ecological specialization and reproductive isolation in pea aphids. Nature 412: 904–907.

Hedrick, P. W. 1981. The establishment of chromosomal variants. Evolution 35: 322–332.

Heed, W. B. and R. L. Mangan. 1986. Community ecology of Sonoran desert *Drosophila in* M. Ashburner, H. L. Carson, and J. N. Thompson (eds.) *The Genetics and Biology of Drosophila*, Vol. 3e. Academic Press, New York.

Heikkinen, E. and J. Lumme. 1991. Sterility of male and female hybrids of *Drosophila virilis* and *Drosophila lummei*. Heredity 67: 1–11.

Heilbuth, J. C. 2000. Lower species richness in dioecious clades. Am. Nat. 156:221–241.

Heiser, C. B. 1947. Hybridization between the sunflower species *Helianthus annuus* and *H. petiolaris*. Evolution 1: 249–262.

Heiser, C. B., D. M. Smith, S. Clevenger, and W. C. Martin. 1969. The North American sunflowers *Helianthus*. Mem. Torrey Bot. Club 22: 1–218.

Helbig, A. J. 1991. Inheritance of migratory direction in a bird species: A cross-breeding experiment with SE- and SW-migrating blackcaps (*Sylvia atricapilla*). Behav. Ecol. Sociobiol. 28: 9–12.

Hellberg, M. E. and V. D. Vacquier. 1999. Rapid evolution of fertilization selectivity and lysin cDNA sequences in teguline gastropods. Mol. Biol. Evol. 16: 839–848.

Henikoff, S. and H. S. Malik. 2002. Selfish drivers. Science 417: 227.

Henikoff, S., K. Ahmad, and H. S. Malik. 2001. The centromere paradox: Stable inheritance with rapidly evolving DNA. Science 293: 1098–1102.

Henry, C. S. 1985. Sibling species, call differences, and speciation in green lacewings (Neuroptera: Chrysopidae: *Chrysoperla*). Evolution 39: 965–984.

Henry, C. S., M. Wells, and C. M. Simon. 1999. Convergent evolution of courtship songs among cryptic species of the *Carnea* group of green lacewings (Neuropter: Chrysopidae: *Chrysoperla*). Evolution 53: 1165–1179.

Herndon, L. A. and M. F. Wolfner. 1995. A *Drosophila* seminal fluid protein, Acp26Aa, stimulates egg laying in females for one day after mating. Proc. Natl. Acad. Sci. USA 92:10114–10118.

Heslop-Harrison, J. 1982. Pollen-stigma interaction and cross-incompatibility in the grasses. Science 215: 1358–1364.

Hewitt, G. M. 1989. The division of species by hybrid zones. Pp. 85–110 *in* D. Otte and J. A. Endler (eds.) *Speciation and Its Consequences*. Sinauer Associates, Sunderland, MA.

Hewitt, G. M., P. Mason, and R. A. Nichols. 1989. Sperm precedence and homogamy across a hybrid zone in the alpine grasshopper *Podisma pedestris*. Heredity 62: 343–353.

Hey, J. 1992. Using phylogenetic trees to study speciation and extinction. Evolution 46: 627–640.

Hey, J. H. 1994. Bridging phylogenetics and population genetics with gene tree models. Pp. 435–449 *in* B. Schierwater, B. Streit, G. P. Wagner, and R. DeSalle (eds.) *Molecular Ecology and Evolution: Approaches and Applications*. Birkhäuser-Verlag, Basel.

Hey, J. 2001. *Genes, Categories, and Species*. Oxford University Press, Oxford.

Hey, J. and R. M. Kliman. 1993. Population genetics and phylogenetics of DNA sequence variation at multiple loci within the *Drosophila melanogaster* species complex. Mol. Biol. Evol. 10: 804–822.

Higashi, M., G. Takimoto, and N. Yamamura. 1999. Sympatric speciation by sexual selection. Nature 402: 523–526.

Highton, R. 1991. Taxonomic treatment of genetically differentiated populations. Herpetologica 46: 114–121.

Highton, R. 1998. Is *Ensatina eschscholtzii* a ring species? Herpetologica 54: 254–278.

Hill, G. E. 1994. Geographic variation in male ornamentation and female mate preference in the house finch: A comparative test of models of sexual selection. Behav. Ecol. 5: 64–73.

Hill, W. G. and A. Caballero. 1992. Artificial selection experiments. Annu. Rev. Ecol. Syst. 23:287–310.

Hillis, D. 1988. Systematics of the *Rana pipiens* complex: Puzzle and paradigm. Annu. Rev. Ecol. Syst. 19: 39–63.

Hillis, D. M. 1981. Premating isolation mechanisms among three species of the *Rana pipiens* complex in Texas and southern Oklahoma. Copeia 1981: 312–319.

Hilu, K. W. 1993. Polyploidy and the evolution of domesticated plants. Am. J. Bot. 80: 1494–1499.

Hindar, K. and B. Jonsson. 1993. Ecological polymorphism in Arctic charr. Biol. J. Linn. Soc. Lond. 48: 63–74.

Hirai, H., T. Taguchi, Y. Saitoh, M. Kawanaka, H. Sugiyama, S. Habe, M. Okamoto, M. Hirata, M. Shimada, W. U. Tiu, K. Lai, E. S. Upatham, and T. Agatsuma. 2000. Chromosomal differentiation of the *Schistosoma japonicum* complex. Int. J. Parasitol. 30: 441–452.

Ho, M.-W., and P. Saunders (eds.). 1984. *Beyond Neo-Darwinism*. Academic Press, London.

Hobel, G. and H. C. Gerhardt. 2003. Reproductive character displacement in the acoustic communication system of green tree frogs (*Hyla cinerea*). Evolution 57: 894–904.

Hodges, S. A. 1997. Rapid radiation due to a key innovation in columbines (Rannculaceaea: *Aquilegia*). Pp. 391–405 *in* T. J. Givnish and K. J. Systma (ed.) *Molecular Evolution and Adaptive Radiation*. Cambridge University Press, Cambridge.

Hodges, S. A. and M. L. Arnold. 1994. Floral and ecological isolation between *Aquilegia formosa* and *Aquilegia pubescens*. Proc. Natl. Acad. Sci. USA 91: 2493–2496.

Hoffmann, A. A. and M. Turelli. 1988. Unidirectional incompatibility in *Drosophila simulans*: Inheritance, geographic variation and fitness effects. Genetics 119: 435–444.

Hoffmann, A. A. and M. Turelli. 1997. Cytoplasmic incompatibility in insects *in* S. L. O'Neill, A. A. Hoffmann, and J. H. Werren (eds.) *Influential Passengers: Inherited Microorganisms and Arthropod Reproduction*. Oxford University Press, Oxrod.

Hoffmann, A. A., M. Turelli, and G. M. Simmons. 1986. Unidirectional incompatibility between populations of *Drosophila simulans*. Evolution 40: 692–701.

Hoikkala, A. and J. Lumme. 1984. Genetic control of the difference in male courtship sound between *D. virilis* and *D. lummei*. Behav. Genet. 14: 827–845.

Hoikkala, A., S. Päällysaho, J. Aspi, and J. Lumme. 2000. Localization of genes affecting species differences in male courtship song between *Drosophila virilis* and *D. littoralis*. Genet. Res. 75: 37–45.

Holder, K. K. and J. J. Bull. 2001. Profiles of adaptation in two similar viruses. Genetics 159: 1393–1404.

Holland, B. and W. R. Rice. 1998. Chase-away sexual selection: antagonistic seduction versus resistance. Evolution 52: 1–7.

Holland, B. and W. R. Rice. 1999. Experimental removal of sexual selection reverses intersexual antagonistic coevolution and removes a reproductive load. Proc. Natl. Acad. Sci. USA 96: 5083–5088.

Hollingshead, L. 1930. A lethal factor in *Crepis* effective only in interspecific hybrids. Genetics 15: 114–140.

Hollocher, H. 1996. Island hopping in *Drosophila*: patterns and processes. Philos. Trans. R. Soc. Lond. B 351: 735–743.

Hollocher, H. and C.-I. Wu. 1996. The genetics of reproductive isolation in the *Drosophila simulans* clade: X versus autosomal effects and male vs. female effects. Genetics 143: 1243–1255.

Hollocher, H., K. Agopian, J. Waterbury, R. W. O'Neill, and A. W. Davis. 2000. Characterization of defects in adult germline development and oogenesis of sterile and rescued hybrid females in crosses between *Drosophila simulans* and *Drosophila melanogaster*. J. Exp. Zool. 288: 205–218.

Holman, E. W. 1987. Recognizability of sexual and asexual species of rotifers. Syst. Zool. 36:381–386.

Hosken, D. J., and P. Stockley. 2004. Sexual selection and genital evolution. Trends Ecol. Evol. 19: 87–93.

Howard, D. J. 1993. Reinforcement: Origin, dynamics, and fate of an evolutionary hypothesis *in* R. G. Harrison (ed.) *Hybrid Zones and the Evolutionary Process*. Oxford University Press.

Howard, D. J. 1999. Conspecific sperm and pollen precedence and speciation. Annu. Rev. Ecol. Syst. 30: 109–132.

Howard, D. J. and Berlocher, S. H. (eds). 1998. *Endless Forms: Species and Speciation*. Oxford University Press, Oxford.

Howard, D. J., R. W. Preszler, J. Williams, S. Fenchel, and W. J. Boecklen. 1997. How discrete are oak species? Insights from a hybrid zone between *Quercus grisea* and *Quercus gambelii*. Evolution 51: 747–755.

Howard, D. J., P. G. Gregory, J. M. Chu, and M. L. Cain. 1998a. Conspecific sperm precedence is an effective barrier to hybridization between closely related species. Evolution 52: 511–516.

Howard, D. J., M. Reece, P. G. Gregory, J. Chu, and M. L. Cain. 1998b. The evolution of barriers to fertilization between closely related organisms. Pp. 279–288 *in* D. J. Howard and S. H. Berlocher (eds.) *Endless Forms: Species and Speciation*. Oxford University Press, New York.

Howard, D., J. Marshall, D. Hampton, S. Britch, M. Draney, J. Chu, and R. Cantrell. 2002. The genetics of reproductive isolation: A retrospective and prospective look with comments on ground crickets. Am. Nat. 159: S8–S21.

Hoy, R. R. and R. C. Paul. 1973. Genetic control of song specificity in crickets. Science 180: 82–83.

Huang, Y., G. Orti, M. Sutherlin, A. Duhachek, and A. Zera. 2000. Phylogenetic relationships of North American field crickets inferred from mitochondrial DNA data. Mol. Phylogenet. Evol. 17: 48–57.

Hubbell, S. P. 2001. *The Unified Neutral Theory of Biodiversity and Biogeography*. Princeton University Press, Princeton, NJ.

Hubbell, S. P., and R. B. Foster. 1986. Biology, chance, and history and the structure of tropical rain forest tree communities. Pp. 314-329 *in* J. Diamond and T. J. Case, (eds.) *Community Ecology*. Harper & Row, New York.

Hudson, R. R. and J. A. Coyne. 2002. Mathematical consequences of the genealogical species concept. Evolution 56: 1557–1565.

Huelsenbeck, J. P., B. Rannala, and Z. Yang. 1997. Statistical tests of host-parasite cospeciation. Evolution 51: 410–419.

Hulbert, R. C. 1993. Taxonomic evolution in North American Neogene horses (subfamily Equinae): The rise and fall of an adaptive radiation. Paleobiology 19: 216–234.

Hull, D. L. 1980. Selection and individuality. Annu. Rev. Ecol. Syst. 11: 311–332.

Humphries, J. M., F. L. Bookstein, B. Chernoff, G. R. Smith, R. L. Elder, and S. G. Poss. 1981. Multivariate discrimination by shape in relation to size. Syst. Zool. 30: 291–308.

Hunt, J. A., K. A. Houtchens, L. Brezinsky, F. Shadravan, and J. G. Bishop. 1989. Genomic DNA variation within and between closely related species of Hawaiian *Drosophila*. Pp. 167–180 *in* L. V. Giddings, K. Y. Kaneshiro, and W. W. Anderson (eds.) *Genetics,*

Speciation, and the Founder Principle. Oxford University Press, New York.

Hunter, J. P. 1998. Key innovations and the ecology of macroevolution. Trends Ecol. Evol. 13: 31–36.

Hunter, M. S., S. J. Perlman, and S. E. Kelly. 2003. A bacterial symbiont in the *Bacteroidetes* induces cytoplasmic incompatibility in the parasitoid wasp *Encarsia pergandiella*. Proc. R. Soc. Lond. B 270: 2185–2190.

Hunter, N., S. R. Chambers, E. J. Louis, and R. H. Borts. 1996. The mismatch repair system contributes to meiotic sterility in an interspecific yeast hybrid. EMBO J. 15: 1726–1733.

Hurd, L. E. and R. M. Eisenberg. 1975. Divergent selection for geotactic response and the evolution of reproductive isolation in sympatric and allopatric populations of houseflies. Am. Nat. 109: 353–358.

Hurst, G. D. D. and F. M. Jiggins. 2000. Male-killing bacteria in insects: Mechanisms, incidence, and implications. Emerging Infect. Dis. 6: 329–336.

Hurst, G. D. D. and J. H. Werren. 2001. The role of selfish genetic elements in eukaryotic evolution. Nat. Rev. Genet. 2: 597–606.

Hurst, L. D. and A. Pomiankowski. 1991. Causes of sex ratio bias may account for unisexual sterility in hybrids: A new explanation of Haldane's rule and related phenomena. Genetics 128: 841–858.

Hurst, L. D., W. D. Hamilton, and R. J. Ladle. 1992. Covert sex. Trends Ecol. Evol. 7: 144–145.

Husband, B. C. and H. A. Sabara. 2003. Reproductive isolation between autotetraploids and their diploid progenitors in fireweed, *Chamerion angustifolium* (Onagraceae). New Phytol. 161: 703–713.

Husband, B. C. and D. W. Schemske. 1998. Cytotype distribution at a diploid-tetraploid contact zone in *Chamerion* (*Epilobium*) *angustifolium* (Onagraceae). Am. J. Bot. 85: 1688–1694.

Husband, B. C. and D. W. Schemske. 2000. Ecological mechanisms of reproductive isolation between diploid and tetraploid *Chamerion angustifolium*. J. Ecol. 88: 689–701.

Hutter, P. 1997. Genetics of hybrid inviability in *Drosophila*. Adv. Genet. 36: 157–185.

Hutter, P. and M. Ashburner. 1987. Genetic rescue of inviable hybrids between *Drosophila melanogaster* and its sibling species. Nature 327: 331–333.

Hutter, P., J. Roote, and M. Ashburner. 1990. A genetic basis for the inviability of hybrids between sibling species of *Drosophila*. Genetics 124: 909–920.

Huxley, J. 1942. *Evolution, the Modern Synthesis*. George Allen & Unwin Ltd., London.

Huxley, T. H. 1896–1902. *Collected Essays*. D. Appleton and Company, New York.

Huxley, T. H. 1901. The life and letters of Thomas H. Huxley *in* L. Huxley (ed.) D. Appleton and Company, New York.

Irwin, D. E. 2002. Phylogenetic breaks without geographic barriers to gene flow. Evolution 56: 2383–2394.

Irwin, D. E. and J. H. Irwin. 2002. Circular overlaps: Rare demonstrations of speciation. Auk 119: 596–602.

Irwin, D. E., S. Bensch, and T. D. Price. 2001a. Speciation in a ring. Nature 409: 333–337.

Irwin, D. E., J. H. Irwin, and T. D. Price. 2001b. Ring species as bridges between microevolution and speciation. Genetica 112–113: 223–243.

Isaac, N. J. B., P.-M. Agapow, P. H. Harvey, and A. Purvis. 2003. Phylogenetically nested comparisons for testing continuous correlates of species richness: A simulation study. Evolution 57: 18–26.

Jablonski, D. 1986. Larval ecology and macroevolution of marine invertebrates. Bull. Mar. Sci. 39: 565–587.

Jablonski, D. 1987. Heritability at the species level: Analysis of geographic range of cretaceous mollusks. Science 238: 360–363.

Jablonski, D. 1995. Extinction in the fossil record. Pp. 25–44 *in* R. M. May and J. H. Lawton (eds.) *Extinction Rates*. Oxford University Press, Oxford.

Jablonski, D. 2000. Micro- and macroevolution: Scale and hierarchy in evolutionary biology and paleobiology. Paleobiology, "Deep Time: Paleobiology's Perspective," D. H. Erwin and S. L. Wing (eds.) 26 (Suppl. to No. 4): 15–52.

Jablonski, D. and K. Roy. 2003. Geographical range and speciation in fossil and living molluscs. Proc. R. Soc. Lond. B 270: 401–406.

Jackman, T. R. and D. B. Wake. 1994. Evolutionary and historical analysis of protein variation in the blotched forms of salamanders of the *Ensatina* complex (Amphibia: Plethodontidae). Evolution 48: 876–897.

Jackson, R. C. 1982. Polyploidy and diploidy: New perspectives on chromosome pairing and its evolutionary implications. Am. J. Bot. 69: 1512–1523.

Jaenike, J. 1981. Criteria for ascertaining the existence of host races. Am. Nat. 117: 830–834.

Jaenike, J. 2001. Sex chromosome meiotic drive. Annu. Rev. Ecol. Syst. 32: 25–49.

Jaffe, L. A. and M. Gould. 1985. Polyspermy preventing mechanisms. Pp. 223–250 *in* C. B. Metz and A. Monroy (eds.) *Biology of Fertilization* (Vol. 3). Academic Press, San Diego.

James, T. S., I. Hutchinson, and J. J. Clague. 2002. Improved relative sea-level histories for Victoria and Vancouver, British Columbia, from isolation-basin coring. Geolog. Surv. Can., Curr. Res. A16:1–7.

Jansa, S. A., B. L. Lundrigan, and P. K. Tucker. 2003. Tests for positive selection on immune and reproductive genes in closely related species of the murine genus *Mus*. J. Mol. Evol. 56: 294–307.

Janzen, D. H. 1979. How to be a fig. Annu. Rev. Ecol. Syst. 10: 13–51.

Jiggins, C. D. and J. Mallet. 2000. Bimodal hybrid zones and speciation. Trends Ecol. Evol. 15:250–255.

Jiggins, C. D., M. Linares, R. E. Naisbit, C. Salazar, Z. H. Yang, and J. Mallet. 2001a. Sex-linked hybrid sterility in a butterfly. Evolution 55: 1631–1638.

Jiggins, C. D., R. E. Naisbit, R. L. Coe, and J. Mallet. 2001b. Reproductive isolation caused by colour pattern mimicry. Nature 411: 302–305.

John, B. 1981. Chromosome change and evolutionary change: A critique. Pp. 23–51 *in* W. R. Atchley and D. S. Woodruf (eds.) *Evolution and Speciation: Essays in Honor of M. J. D. White*. Cambridge University Press, Cambridge.

Johnson, M. S. and J. R. G. Turner. 1979. Absence of dosage compensation for a sex-linked enzyme in butterflies. Heredity 43: 71–77.

Johnson, N. A. and M. J. Wade. 1995. Conditions for soft selection favoring the evolution of hybrid inviability. J. Theor. Biol. 176: 493–499.

Johnson, N. A., and C.-I. Wu. 1992. An empirical test of the meiotic drive models of hybrid sterility: Sex ratio data from hybrids between *Drosophila simulans* and *Drosophila sechellia*. Genetics 130: 507–511.

Johnson, N. A., D. E. Perez, E. L. Cabot, H. Hollocher, and C.-I. Wu. 1992. A test of reciprocal X-Y interactions as a cause of hybrid sterility in *Drosophila*. Nature 358: 751–753.

Johnson, N. A., H. Hollocher, E. Noonburg, and C.-I. Wu. 1993. The effects of interspecific Y chromosome replacements on hybrid sterility within the *Drosophila simulans* clade. Genetics 135: 443–453.

Johnson, P. A. and U. Gullberg. 1998. Theory and models of sympatric speciation. Pp. 79–89 *in* D. J. Howard and S. H. Berlocher (eds.) *Endless Forms: Species and Speciation*. Oxford University Press, Oxford.

Johnson, P. A., F. C. Hoppensteadt, J. J. Smith, and G. L. Bush. 1996. Conditions for sympatric speciation—a diploid model incorporating habitat fidelity and non-habitat assortative mating. Evol. Ecol. 10: 187–205.

Johnson, S. D. 1997. Pollination ecotypes of *Satyrium hallackii* (Orchidaceae) in South Africa. Bot. J. Linn. Soc. 123: 225–235.

Johnson, S. D. and K. E. Steiner. 1997. Long-tongued fly pollination and evolution of floral spur length in the *Disa draconis* complex (Orchidaceae). Evolution 51: 45–53.

Johnson, S. D. and K. E. Steiner. 2000. Generalization versus specialization in plant pollination systems. Trends Ecol. Evol. 15: 140–143.

Johnson, T. C., C. A. Scholz, M. R. Talbot, K. Kelts, R. D. Ricketts, G. Ngobi, K. Beuning, I. Ssemmanda, and J. W. McGill. 1996. Late Pleistocene desiccation of Lake Victoria and rapid evolution of cichlid fishes. Science 273: 1091–1093.

Johnson, T. C., K. Kelts, and E. Odada. 2000. The Holocene history of Lake Victoria. Ambio 29:2–11.

Jones, A. G., G. I. Moore, C. Kvarnemo, D. Walker, and J. C. Avise. 2003. Sympatric speciation as a consequence of male pregnancy in seahorses. Proc. Natl. Acad. Sci. USA 100: 6598–6603.

Jones, C. D. 1998. The genetic basis of *Drosophila sechellia's* resistance to a host plant toxin. Genetics 149: 1899–1908.

Jones, C. D. 2001. The genetic basis of larval resistance to a host plant toxin in *Drosophila sechellia*. Genet. Res. 78: 225–233.

Jones, C. D. 2003a. Genetics of egg production in *Drosophila sechellia*. Heredity 92: 249-256.

Jones, C. D. 2004. The genetics of adaptations in *Drosophila sechellia*. Genetica 121(4) (in press).

Jones, E. W. 1959. Biological flora of the British Isles: *Quercus* L. J. Ecology 47: 169–222.

Jones, I. L. and F. M. Hunter. 1993. Mutual sexual selection in a monogamous seabird. Nature 362: 238–239.

Jones, I. L. and F. M. Hunter. 1998. Heterospecific mating preferences for a feather ornament in least auklets. Behav. Ecol. 9: 187–192.

Jordan, D. S. 1905. The origin of species through isolation. Science 22: 545–562.

Jordan, D. S. 1908. The law of geminate species. Am. Nat. 42: 73–80.

Jordan, D. S. and V. L. Kellogg. 1908 (copyright 1907). *Evolution and Animal life: An Elementary Discussion of Facts, Processes, Laws and Theories Relating to the Life and Evolution of Animals*. D. Appleton and Company, New York.

Jordan, K. 1896. On mechanical selection and other problems. Novit. Zool. 3: 426–525.

Jousselin, E., J.-Y. Rasplus, and F. Kjellberg. 2003. Convergence and coevolution in a mutualism: evidence from a molecular phylogeny of *Ficus*. Evolution 57: 1255–1269.

Kandul, N. P., V. A. Lukhtanov, A. V. Dantchenko, J. W. S. Coleman, C. H. Sekercioglu, D. Haig, and N. E. Pierce. 2004. Phylogeny of *Agrodiaetus* Hübner 1822 (Lepidoptera: Lycaenidae) inferred from mtDNA sequences of COI and COII, and nuclear sequences of EF1-a: Karyotype diversification and species radiation. Syst. Biology (in press).

Kaneshiro, K. Y. 1980. Sexual isolation, speciation, and the direction of evolution. Evolution 34:437–444.

Kaneshiro, K. Y. 1989. The dynamics of sexual selection and founder effects in species formation. Pp. 279–296 *in* L. V. Giddings, K. Y. Kaneshiro, and W. W. Anderson (eds.) *Genetics, Speciation and the Founder Principle*. Oxford University Press, New York.

Kaneshiro, K. Y., R. G. Gillespie, and H. L. Carson. 1995. Chromosomes and male genitalia of Hawaiian *Drosophila*: tools for interpreting phylogeny and geography. Pp. 57–71. *Hawaiian Biogeography*. Smithsonian Institution Press, Washington, DC.

Kassen, R., D. Schluter, and J. D. McPhail. 1995. Evolutionary history of threespine sticklebacks (Gasterosteus spp.) in British Columbia: insights from a physiological clock. Canadian J. Zool. 73: 2154–2158.

Katakura, H. 1986. Evidence for the incapacitation of heterospecific sperm in the female genital tract in a pair of closely related ladybirds (Insecta, Coleoptera, Coccinellidae). Zool. Sci. 3: 115–121.

Katakura, H. 1997. Species of *Epilachna* ladybird beetles. Zool. Sci. 14: 869–881.

Katakura, H. and T. Hosogai. 1994. Performance of hybrid ladybird beetles (*Epilachna* spp.) on the host plants of parental species. Entomol. Exp. Appl. 71: 81–85.

Katakura, H., M. Shioi, and Y. Kira. 1989. Reproductive isolation by host specificity in a pair of phytophagous ladybird beetles. Evolution 43: 1045–1053.

Katzourakis, A., A. Purvis, S. Amzeh, R. Rotheray, and F. Gilbert. 2001. Macroevolution of hoverflies (Diptera: Syrphidae): The effect of using higher-level taxa in studies of biodiversity, and correlates of species richness. J. Evol. Biol. 14: 219–227.

Kaufmann, B. P. 1940. The nature of hybrid sterility—abnormal development in eggs of hybrids between *Drosophila miranda* and *Drosophila pseudoobscura*. J. Morphol. 66: 197–212.

Kawecki, T. J. 1996. Sympatric speciation driven by beneficial mutations. Proc. R. Soc. Lond. B 263: 1515–1520.

Kawecki, T. J. 1997. Sympatric speciation via habitat specialization driven by deleterious mutations. Evolution 51: 1751–1763.

Kawecki, T. J. 1998. Red queen meets Santa Rosalia: Arms races and the evolution of host specialization in organisms with parasitic lifestyles. Am. Nat. 152: 635–651.

Kay, K. M. and D. W. Schemske. 2003. Pollinator assemblages and visitation rates for 11 species of Neotropical *Costus* (Costaceae). Biotropica 35: 198–207.

Keeling, M. J., F. M. Jiggins, and J. M. Read. 2003. The invasion and coexistence of competing *Wolbachia* strains. Heredity 91: 382–388.

Keller, M. J. and H. C. Gerhardt. 2001. Polyploidy alters advertisement call structure in gray treefrogs. Proc. R. Soc. Lond. B 268: 341–345.

Kelley, D. B. 1996. Sexual differentiation in *Xenopus laevis*. Pp. 143–176 in R. C. Tinsley and H. R. Kobel (eds.) *The Biology of* Xenopus. Clarendon Press, Oxford.

Kelly, J. K. and M. A. F. Noor. 1996. Speciation by reinforcement: a model derived from studies of *Drosophila*. Genetics 143: 1485–1497.

Kerkis, J. 1933. Development of gonads in hybrids between *Drosophila melanogaster* and *Drosophila simulans*. J. Exp. Zool. 66: 477–509.

Kessel, B. 1998. *Habitat Characteristics of Some Passerine Birds in Western North American Taiga.* University of Alaska Press, Fairbanks, Alaska.

Key, K. H. L. 1968. The concept of stasipatric speciation. Syst. Zool. 17: 14–22.

Khadem, M. and C. B. Krimbas. 1991 Studies of the species barrier between *Drosophila subobscura* and *D. madeirensis*. 1. The genetics of male hybrid sterility. Heredity 67: 157–165.

Kiang, Y. T. and J. L. Hamrick. 1978. Reproductive isolation in the *Mimulus guttatus–M. nasutus* complex. Am. Midl. Nat. 100: 269–276.

Kidwell, M. G. 1983. Intraspecific hybrid sterility. Pp. 125–154 in M. Ashburner, H. L. Carson, and J. J. N. Thompson (eds.) *The Genetics and Biology of* Drosophila, Vol. 3c. Academic Press, London.

Kilias, G., S. N. Alahiotis, and M. Pelecanos. 1980. A multifactorial genetic investigation of speciation theory using *Drosophila melanogaster*. Evolution 34: 730–737.

Kimura, M. 1983. *The Neutral Theory of Molecular Evolution.* Cambridge University Press, Cambridge, UK.

King, M. 1993. *Species Evolution: The Role of Chromosome Change.* Cambridge University Press, Cambridge.

Kinsey, J. D. 1967. Studies of an embryonic lethal hybrid in *Drosophila*. J. Embryol. Exp. Morphol. 17: 405–423.

Kirkpatrick, M. 2001. Reinforcement during ecological speciation. Proc. R. Soc. Lond. B 268: 1–5.

Kirkpatrick, M. and V. Ravigné. 2002. Speciation by natural and sexual selection: models and experiments. Am. Nat. 159: S22–S35.

Kirkpatrick, M. and M. J. Ryan. 1991. The evolution of mating preferences and the paradox of the lek. Nature 350: 33–38.

Kirkpatrick, M. and M. R. Servedio. 1999. The reinforcement of mating preferences on an island. Genetics 151: 865–884.

Kitcher, P. 1984. Species. Philos. Sci. 51: 308–333.

Kittler, R., M. Kayser, and M. Stoneking. 2003. Molecular evolution of *Pediculus humanus* and the origin of clothing. Curr. Biol. 13: 1414–1417.

Klicka, J. and R. M. Zink. 1997. The importance of recent ice ages in speciation: A failed paradigm. Science 277: 1666–1669.

Kliman, R. M., P. Andolfatto, J. A. Coyne, F. Depaulis, M. Kreitman, A. J. Berry, M. McCarter, J. Wakeley, and J. Hey. 2000. The population genetics of the origin and divergence of the *Drosophila simulans* complex species. Genetics 156: 1913–1931.

Knapp, S. and J. Mallet. 2003. Refuting refugia? Science 300: 71–72.

Knight, G. R., A. Robertson, and C. H. Waddington. 1956. Selection for sexual isolation within a species. Evolution 10: 14–22.

Knight, M. E., and G. F. Turner. 2004. Laboratory mating trials indicate incipient speciation by sexual selection among populations of the cichlid fish *Pseudotropheus zebra* from Lake Malawi. Proc. R. Soc. Lond. B 271:675–680.

Knowles, L. L. and T. A. Markow. 2001. Sexually antagonistic coevolution of a postmating-prezygotic isolating character in desert *Drosophila*. Proc. Natl. Acad. Sci. USA 98:8692–8696.

Knowlton, N. 1993. Sibling species in the sea. Annu. Rev. Ecol. Syst. 24: 189–216.

Knowlton, N. and L. A. Weigt. 1998. New dates and new rates for divergence across the Isthmus of Panama. Proc. R. Soc. Lond. B 263: 2257–2263.

Knowlton, N., L. A. Weigt, L. A. Solórzano, D. K. Mills, and E. Bermingham. 1993. Divergence in proteins, mitochondrial DNA, and reproductive compatibility across the Isthmus of Panama. Science 260: 1629–1632.

Knowlton, N., J. L. Maté, H. M. Guzmán, and R. Rowan. 1997. Direct evidence for reproductive isolation among the three species of the *Montrastraea annularis* complex in Central America (Panamá and Honduras). Mar. Biol. 127: 705–711.

Kobel, H. R. 1996. Reproductive capacity of experimental *Xenopus gilli* X × *l. laevis* hybrids *in* R. C. Tinsley and H. R. Kobel (eds.) *The Biology of* Xenopus. Clarendon Press, Oxford.

Kobel, H. R., C. Loumont, and R. C. Tinsley. 1996. The extant species. Pp. 9–34 *in* R. C. Tinsley and H. R. Kobel (eds.) *The Biology of* Xenopus. Clarendon Press, Oxford.

Kocher, T. D. and R. D. Sage. 1986. Further genetic analysis of a hybrid zone between leopard frogs (*Rana pipiens* complex). Evolution 40: 21–33.

Kocher, T. D., J. A. Conroy, K. R. McKaye, J. R. Stauffer, and S. F. Lockwood. 1995. Evolution of NADH dehydrogenase subunit 2 in east African cichlid fish. Mol. Phylogenet. Evol. 4: 420–432.

Kondrashov, A. S. 1986. Multilocus model of sympatric speciation. III. Computer simulations. Theor. Popul. Biol. 29: 1–15.

Kondrashov, A. S. 2003. Accumulation of Dobzhansky–Muller incompatibilities within a spatially structured population. Evolution 57: 151–153.

Kondrashov, A. S. and F. A. Kondrashov. 1999. Interactions among quantitative traits in the course of sympatric speciation. Nature 400: 351–354.

Koopman, K. F. 1950. Natural selection for reproductive isolation between *Drosophila pseudoobscura* and *Drosophila persimilis*. Evolution 4: 135–148.

Koref-Santibañez, S. K. and C. H. Waddington. 1958. The origin of sexual isolation between differerent lines within a species. Evolution 12: 485–493.

Kornfield, I. and P. F. Smith. 2000. African cichlid fishes: Model systems for evolutionary biology. Annu. Rev. Ecol. Syst. 31: 163–196.

Korol, A., E. Rashkovetsky, K. Iliadi, P. Michalak, Y. Ronin, and E. Nevo. 2000. Nonrandom mating in *Drosophila melanogaster* laboratory populations derived from closely adjacent ecologically contrasting slopes at "Evolution Canyon". Proc. Natl. Acad. Sci. USA 97:12637–12642.

Kottler, M. J. 1978. Charles Darwin's biological species concept and theory of geographic speciation: The transmutation notebooks. Ann. Sci. 35: 275–297.

Kresge, N., V. D. Vacquier, and C. D. Stout. 2001. Abalone lysin: The dissolving and evolving sperm protein. BioEssays 23: 95–103.

Kricher, J. 1997. *A Neotropical Companion: An Introduction to the Animals, Plants, and Ecosystems of the New World Tropics*. Princeton University Press, Princeton, NJ.

Krimbas, C. B. 1960. Synthetic sterility in *Drosophila willistoni*. Proc. Natl. Acad. Sci. USA 46: 832–833.

Kruckeberg, A. R. 1957. Variation in fertility of hybrids between isolated populations of the serpentine species, *Streptanthus glandulosus* Hook. Evolution 11:185–211.

Kruckeberg, A. R. 1967. Ecotypic responses to ultramafic soils by some plant species of northwestern United States. Brittonia 19: 133–151.

Kruckeberg, A. R. 1984. California serpentines: Flora, vegetation, geology, soils and management problems. Univ. Cal. Pub. Bot. 78: 1–180.

Kubo, T., and Y. Iwasa. 1995. Inferring the rates of branching and extinction from molecular phylogenies. Evolution 49: 694–704.

Kvist, L., J. Martens, H. Higuchi, A. A. Nazarenko, O. P. Valchuk, and M. Orell. 2003. Evolution and genetic structure of the great tit (*Parus major*) complex. Proc. R. Soc. Lond. B 270: 1447–1454.

Kyriacou, C. P. and J. C. Hall. 1986. Interspecific genetic control of courtship song production and reception in *Drosophila*. Science 232: 494–497.

Lachaise, D., M. Harry, M. Solignac, F. Lemeunier, V. Bénassi, and M.-L. Cariou. 2000. Evolutionary novelties in islands: *Drosophila santomea*, a new *melanogaster* sister species from São Tomé. Proc. R. Soc. Lond. B 267: 1487–1495.

Lambert, D. M. and H. E. H. Paterson. 1984. "On bridging the gap between race and species": the isolation concept and an alternative. Proc. Linn. Soc. NSW 107: 501–514.

Lambert, D. M., B. Michaux, and C. S. White. 1987. Are species self-defining? Syst. Zool. 36:196–205.

Lamnissou, K., M. Loukas, and E. Zouros. 1996. Incompatibilities between Y chromosome and autosomes are responsible for male hybrid sterility in crosses between *Drosophila virilis* and *Drosophila texana*. Heredity 76: 603–609.

Lamont, B. B., T. He, N. J. Enright, S. L. Krauss, and B. P. Miller. 2003. Anthropogenic disturbance promotes hybridization between *Banksia* species by altering their biology. J. Evol. Biol. 16: 551–557.

Lande, R. 1979. Effective deme sizes during long-term evolution estimated from rates of chromosomal rearrangement. Evolution 33: 234–251.

Lande, R. 1980. Genetic variation and phenotypic evolution during allopatric speciation. Am. Nat. 116: 463–479.

Lande, R. 1981. Models of speciation by sexual selection on polygenic traits. Proc. Natl. Acad. Sci. USA 78: 3721–3725.

Lande, R. 1982. Rapid origin of sexual isolation and character divergence in a cline. Evolution 36: 213–223.

Lande, R. 1985a. Expected time for random genetic drift of a population between stable phenotypic states. Proc. Natl. Acad. Sci. USA 82: 7641–7645.

Lande, R. 1985b. The fixation of chromosomal rearrangements in a subdivided population with local extinction and colonization. Heredity 54: 323–332.

Lande, R. and S. J. Arnold. 1983. The measurement of selection on correlated characters. Evolution 37: 1210–1226.

Lande, R., O. Seehausen, and J. M. van Alphen. 2001. Mechanisms or rapid sympatric speciation by sex reversal and sexual selection in cichlid fish. Genetica 112: 435–443.

Langor, D. W., J. R. Spence, and G. R. Pohl. 1990. Host effects on fertility and reproductive success of *Dendroctonus ponderosae* Hopkins (Coleoptera: Scolytidae). Evolution 44: 609–618.

Larson, R. J. 1980. Competition, habitat selection, and the bathymetric segregation of two rockfish (*Sebastes*) species. Ecol. Monogr. 50: 221–239.

Lassy, C. W. and T. L. Karr. 1996. Cytological analysis of fertilization and early embryonic development in incompatible crosses of *Drosophila simulans*. Mech. Dev. 57: 47–58.

Laurie, C. C. 1997. The weaker sex is heterogametic: 75 years of Haldane's rule. Genetics 147: 937–951.

Laven, H. 1951. Crossing experiments with *Culex* strains. Evolution 5: 370–375.

Laven, H. 1957. Vererbung durch Kerngene und das Problem der ausserkaryotischen Vererbung bei *Culex pipiens*. II. Ausserkaryotische Vererbung. Z. Vererbungsl. 88: 478–516.

Lawrence, J. G. and H. Ochman. 1998. Molecular archaeology of the *Escherichia coli* genome. Proc. Natl. Acad. Sci. USA 95: 9413–9417.

Leaché, A. D. and T. W. Reeder. 2002. Molecular systematics of the eastern fence lizard (*Sceloporus undulatus*): a comparison of parsimony, likelihood, and Bayesian approaches. Syst. Biol. 51: 44–68.

Lee, Y. H., T. Ota, and V. D. Vacquier. 1995. Positive selection is a general phenomenon in the evolution of abalone sperm lysin. Mol. Biol. Evol. 12: 231–238.

Legal, L. and M. Plawecki. 1995. Comparative sensitivity of various insects to toxic compounds from *Morinda citrifolia*. Entomol. Problems 26: 155–159.

Legal, L., J. R. David, and J. M. Jallon. 1992. Toxicity and attraction effects produced by *Morinda citrifolia* fruits on the *Drosophila melanogaster* complex of species. Chemoecology 3: 125–129.

Legal, L., B. Chappe, and J.-M. Jallon. 1994. Molecular basis of *Morinda citrifolia* (L.): Toxicity on *Drosophila*. J. Chem. Ecol. 20: 1931–1943.

Lenski, R. E. and M. Travisano. 1994. Dynamics of adaptation and diversification: A 10,000-generation experiment with bacterial populations. Proc. Natl. Acad. Sci. 91: 6808–6814.

Leo, N. P., N. J. H. Campbell, X. Yang, K. Mumcuoglu, and S. C. Barker. 2002. Evidence from mitochondrial DNA that head lice and body lice of humans (Phthiraptera: Pediculidae) are conspecific. J. Med. Entomol. 39: 662–666.

Lessios, H. A. 1984. Possible prezygotic reproductive isolation in sea urchins separated by the Isthmus of Panama. Evolution 38: 1144–1148.

Lessios, H. A. 1998. The first stage of speciation as seen in organisms separated by the Isthmus of Panama. Pp. 186–201 *in* D. H. Howard and S. H. Berlocher, (eds.) *Endless Forms: Species and Speciation*. Oxford University Press, Oxford.

Lessios, H. A., B. D. Kessing, and J. S. Pearse. 2001. Population structure and speciation in tropical seas: Global phylogeography of the sea urchin *Diadema*. Evolution 55: 955–975.

Levene, H. and T. Dobzhansky. 1959. Possible genetic difference between the head louse and the body louse (*Pediculus humanus* L.). Am. Nat. 93: 347–353.

Levin, D. A. 1975. Minority cytotype exclusion in local plant populations. Taxon 24: 35–43.

Levin, D. A. 1976. Consequences of long-term artificial selection, inbreeding, and isolation in *Phlox*: I. The evolution of cross incompatibility. Evolution 30: 335–344.

Levin, D. A. 1978. The origin of isolating mechanisms in flowering plants. Evol. Biol. 11: 185–317.

Levin, D. A. 1983. Polyploidy and novelty in flowering plants. Am. Nat. 122: 1–25.

Levin, D. A. 1985. Reproductive character displacement in *Phlox*. Evolution 39: 1275–1281.

Levin, D. A. 2000. *The Origin, Expansion, and Demise of Plant Species*. Oxford University Press, Oxford.

Levin, D. A. 2002. *The Role of Chromosomal Change in Plant Evolution*. Oxford University Press, Oxford.

Levin, D. A. and A. C. Wilson. 1976. Rates of evolution in seed plants: net increase in diversity of chromosome numbers and species numbers through time. Proc. Natl. Acad. Sci. USA 73: 2086–2090.

Levitan, D. R. 2002. The relationship between conspecific fertilization success and reproductive isolation among three congeneric sea urchins. Evolution 56: 1599–1609.

Levitan, D. R., H. Fukami, J. Jara, D. Kline, T. M. McGovern, K. E. McGhee, C. A. Swanson, and N. Knowlton. 2004. Mechanisms of reproductive isolation among sympatric broadcast-spawning corals of the *Montastraea annularis* species complex. Evolution 58:308–323.

Lewis, D. and L. K. Crowe. 1958. Unilateral incompatibility in flowering plants. Heredity 12:233–256.

Lewis, H. 1962. Catastrophic selection as a factor in speciation. Evolution 16: 257–271.

Lewis, H. 1966. Speciation in flowering plants. Science 152: 167–172.

Lewis, H. 1973. The origin of diploid neospecies in *Clarkia*. Am. Nat. 107: 161–170.

Lewis, H. and M. R. Roberts. 1956. The origin of *Clarkia lingulata*. Evolution 10: 126–138.

Lewontin, R. C. 1961. Evolution and the theory of games. J. Theor. Biol. 1: 382–403.

Lewontin, R. C. 1965. Discussion of paper by Dr. Howard. Pp. 481–485 *in* H. G. Baker and G. L. Stebbins (eds.) *The Genetics of Colonizing Species*. Academic Press, New York.

Lewontin, R. C. 1974. *The Genetic Basis of Evolutionary Change*. Columbia University Press, New York.

Lewontin, R. C. 1983. The organism as the subject and object of evolution. Scientia 118: 63–82.

Li, W.-H. 1997. *Molecular Evolution*. Sinauer Associates, Sunderland, MA.

Liebers, D. de Knijff, P., and A. J. Helbig. 2004. The herring gull complex is not a ring species. Proc. Roy. Soc. Lond. B 271: in press

Liem, K. 1974. Evolutionary strategies and morphological innovations: Cichlid pharyngeal jaws. Syst. Zool. 22: 425–441.

Lifschytz, E. 1987. The developmental program of spermiogenesis in *Drosophila*: A genetic analysis. Int. Rev. Cyt. 109: 211–256.

Liimatainen, J. O. and A. Hoikkala. 1998. Interactions of the males and females of three sympatric *Drosophila virilis*-group species, *D. montana*, *D. littoralis*, and *D. lummei* (Diptera, Drosophilidae) in intra- and interspecific courtships in the wild and in the laboratory. J. Insect Behav. 11: 399–417.

Lijtmaer, D. A., B. Mahler, and P. L. Tubaro. 2003. Hybridization and postzygotic isolation patterns in pigeons and doves. Evolution 57: 1411–1418.

Lin, J.-Z. and K. Ritland. 1997. Quantitative trait loci differentiating the outbreeding *Mimulus guttatus* from the inbreeding *M. platycalyx*. Genetics 146: 1115–1121.

Linder, H. P. and J. Midgley. 1996. Anemophilous plants select pollen from their own species from the air. Oecologia 108: 85–87.

Lindsley, D. L. and K. T. Tokuyasu. 1980. Spermatogenesis. Pp. 225–294 *in* M. Ashburner and T. R. F. Wright (eds.) *The Genetics and Biology of Drosophila*, vol. 2d. Academic Press, New York.

Linn, C., J. L. Feder, S. Nojima, H. R. Dambroski, S. H. Berlocher, and W. Roelofs. 2003. Fruit odor discrimination and sympatric host race formation in *Rhagoletis*. Proc. Natl. Acad. Sci. USA 100: 11490–11493.

Liou, L. W. and T. D. Price. 1994. Speciation by reinforcement of prezygotic isolation. Evolution 48: 1451–1459.

Littlejohn, M. J. 1965. Premating isolation in the *Hyla ewingi* complex (Anura: Hylidae). Evolution 19: 234–243.

Littlejohn, M. J. 1981. Reproductive isolation: A critical review. Pp. 298–334 *in* W. R. Atchley and D. S. Woodruff (eds.) *Evolution and Speciation*. Cambridge University Press, Cambridge, UK.

Littlejohn, M. J. and J. J. Loftus-Hills. 1968. An experimental evaluation of premating isolation in the *Hyla ewingi* complex (Anura: Hylidae). Evolution 22: 659–662.

Lloyd, E. A. and S. J. Gould. 1993. Species selection on variability. Proc. Natl. Acad. Sci. USA 90: 595–599.

Lloyd, M. and H. S. Dybas. 1966a. The periodical cicada problem. I. Population ecology. Evolution 20: 133–149.

Lloyd, M. and H. S. Dybas. 1966b. The periodical cicada problem. II. Evolution. Evolution 20:786–801.

Lofstedt, C., B. S. Hansson, W. Roelofs, and B. O. Bengtsson. 1989. No linkage between genes controlling female pheromone production and male pheromone response in the European corn borer, *Ostrinia nubilalis* Hubner (Lepidoptera: Pyralidae). Genetics 123: 553–556.

Loftus-Hills, J. J. 1975. The evidence for reproductive character displacement between the toads *Bufo americanus* and *Bufo woodhousii fowleri*. Evolution 29: 368–369.

Lorkovic, Z. 1958. Some peculiarities of spatially and sexually restricted gene exchange in the *Erebia tyndarus* group. Cold Spring Harb. Symp. Quant. Biol. 23: 319–325.

Losos, J. B. and R. E. Glor. 2003. Phylogenetic comparative methods and the geography of speciation. Trends Ecol. Evol. 18: 220–227.

Losos, J. B., and D. Schluter. 2000. Analysis of an evolutionary species-area relationship. Nature 408:847–850.

Losos, J. B., D. A. Creer, D. Glossip, R. Goellner, A. Hampton, G. Roberts, N. Haskell, P. Taylor, and J. Ettling. 2000. Evolutionary implications of phenotypic plasticity in the hindlimb of the lizard *Anolis sagrei*. Evolution 54: 301–305.

Louis, J. and J. R. David. 1986. Ecological specialization in the *Drosophila melanogaster* species subgroup: A case study of *D. sechellia*. Acta Oecol. 7: 215–229.

Lovette, I. J. and E. Bermingham. 1999. Explosive speciation in the New World *Dendroica* warblers. Proc. R. Soc. Lond. B 266: 1629–1636.

Lung, O. and M. Wolfner. 1999. *Drosophila* seminal fluid proteins enter the circulatory system of the mated female fly by crossing the posterior vaginal wall. Insect Biochem. Mol. Biol. 29: 1043–1052.

Lynch, J. D. 1978. The distribution of leopard frogs (*Rana blairi* and *Rana pipiens*) (Amphibia, Anura, Ranidae) in Nebraska. J. Herpetol. 12: 157–162.

Lynch, J. D. 1989. The gauge of speciation: On the frequencies of modes of speciation. Pp. 527–556 *in* D. Otte and J. A. Endler (eds.) *Speciation and Its Consequences*. Sinauer Associates, Sunderland, MA.

Lynch, M. 2002. Gene duplication and evolution. Science 297: 945–947.

Lynch, M. and J. S. Conery. 2000. The evolutionary fate and consequences of duplicate genes. Science 290: 1151–1155.

Lynch, M. and A. G. Force. 2000. The origin of interspecific genomic incompatibility via gene duplication. Am. Nat. 156: 590–605.

MacCallum, C. J., B. Nurnberger, N. H. Barton, and J. M. Szymura. 1998. Habitat preference in the *Bombina* hybrid zone in Croatia. Evolution 52: 227–239.

Machado, C. A. and J. Hey. 2003. The causes of phylogenetic conflict in a classic *Drosophila* species group. Proc. R. Soc. Lond. B 270: 1193–1202.

Machado, C. A., E. Jousselin, F. Kjellberg, S. G. Compton, and E. A. Herre. 2001. Phylogenetic relationships, historical biogeography, and character evolution of fig-pollinating wasps. Proc. R. Soc. Lond. B 268: 685–694.

Machado, C. A., R. M. Kliman, J. A. Markert, and J. Hey. 2002. Inferring the history of speciation from multilocus DNA sequence data: The case of *Drosophila pseudoobscura* and close relatives. Mol. Biol. Evol. 19: 472–488.

Macnair, M. R. 1981. Tolerance of higher plants to toxic materials. Pp. 177–207 *in* J. A. Bishop and L. M. Cook (eds.) *Genetic Consequences of Man Made Change*. Academic Press, New York.

Macnair, M. R. 1983. The genetic control of copper tolerance in the yellow monkey flower, *Mimulus guttatus*. Heredity 50: 283–293.

Macnair, M. R. 1987. Heavy metal tolerance in plants: A model evolutionary system. Trends Ecol. Evol. 2: 354–369.

Macnair, M. R. and P. Christie. 1983. Reproductive isolation as a pleiotropic effect of copper tolerance in *Mimulus guttatus*. Heredity 50: 295–302.

Macnair, M. R. and Q. J. Cumbes. 1989. The genetic architecture of interspecific variation in *Mimulus*. Genetics 122: 211–222.

Macnair, M. R. and M. Gardner. 1998. The evolution of edaphic endemics. Pp. 157–171 *in* D. J. Howard and S. H. Berlocher (eds.) *Endless Forms: Species and Speciation*. Oxford University Press, New York.

Magallón, S. and M. J. Sanderson. 2001. Absolute angiosperm diversification rates. Evolution 55: 1762–1780.

Magurran, A. E. and I. W. Ramnarine. 2004. Learned mate recognition and reproductive isolation in guppies. Anim. Behav. (in press).

Majewski, J. and F. M. Cohan. 1998. The effect of mismatch repair and heteroduplex formation on sexual isolation in *Bacillus*. Genetics 148: 13–18.

Majewski, J. and F. M. Cohan. 1999. Adapt globally, act locally: The effect of selective sweeps on bacterial sequence diversity. Genetics 152: 1459–1474.

Majnep, I. S. and R. Bulmer. 1977. *Birds of My Kalam Country*. Aukland University Press/Oxford University Press, Aukland/Oxford.

Malik, H. S. and S. Henikoff. 2001. Adaptive evolution of Cid, a centromere-specific histone in *Drosophila*. Genetics 157: 1293–1298.

Malik, H. S., D. Vermaak, and S. Henikoff. 2002. Recurrent evolution of DNA-binding motifs in the *Drosophila* centromeric histone. Proc. Natl. Acad. Sci. USA 99: 1449–1454.

Malitschek, B., D. Fornzler, and M. Schartl. 1995. Melanoma formation in *Xiphophorus*: A model system for the role of receptor tyrosine kinases in tumorigenesis. BioEssays 17: 1017–1023.

Mallet, J. 1995. A species definition for the Modern Synthesis. Trends Ecol. Evol. 10: 294–299.

Mallet, J. 2001. The speciation revolution. J. Evol. Biol. 14: 887–888.

Mallet, J., W. O. McMillan, and C. D. Jiggins. 1998. Mimicry and warning color at the boundary between races and species. Pp. 390–403 *in* D. Howard and S. H. Berlocher (eds.) *Endless Forms: Species and Speciation*. Oxford University Press, Oxford.

Mampell, K. 1941. Female sterility in interracial hybrids of *Drosophila pseudoobscura*. Proc. Natl. Acad. Sci. USA 27: 337–341.

Mandel, M. J., C. L. Ross, and R. G. Harrison. 2001. Do *Wolbachia* infections play a role in unidirectional incompatibilities in a field cricket hybrid zone? Mol. Ecol. 10: 703–709.

Mantovani, B. and V. Scali. 1992. Hybridogenesis and androgenesis in the stick-insect *Bacillus rossius–grandii benazzii* (Insecta, Phasmatodea). Evolution 46: 783–796.

Markert, J. A., M. E. Arnegard, P. D. Danley, and T. D. Kocher. 1999. Biogeography and population genetics of the Lake Malawi cichlid *Melanochromis auratus*: Habitat transience, philopatry, and speciation. Mol. Ecol. 8: 1013–1026.

Markow, T. A. 1981. Mating preferences are not predictive of the direction of evolution in experimental populations of *Drosophila*. Science 213: 1405–1407.

Markow, T. A. 1991. Sexual isolation among populations of *Drosophila mojavensis*. Evolution 45: 1525–1529.

Marsh, A. L., A. J. Ribbink, and B. A. Marsh. 1981. Sibling species complexes in sympatric populations of *Petrotilapia* Trewavas (Cichlidae, Lake Malawi). Zool. J. Linn. Soc. 71: 253–264.

Marshall, D. C. and J. R. Cooley. 2000. Reproductive character displacement and speciation in periodical cicadas, with description of a new species, 13-year species, *Magicicada neotredecim*. Evolution 54: 1313–1325.

Marshall, D. C., J. R. Cooley, and C. Simon. 2003. Holocene climate shifts, life-cycle plasticity, and speciation in periodical cicadas: A reply to Cox and Carlton. Evolution 57: 433–437.

Marshall, J. L., M. L. Arnold, and D. J. Howard. 2002. Reinforcement: The road not taken. Trends Ecol. Evol. 17: 558–563.

Martens, K., B. Goddeeris, G. Coulter, (eds.) 1994. *Speciation in Ancient Lakes*. Schweizerbart, Stuttgart.

Martin, A. P. and C. M. Simon. 1990. Differing levels of among-population divergence in the mitochondrial DNA of 13- versus 17-year periodical cicadas related to historical biogeography. Evolution 44: 1066–1088.

Martin, F. W. 1967. The genetic control of unilateral incompatibility between two tomato species. Genetics 56: 391–398.

Martínez Wells, M. and C. S. Henry. 1992. The role of courtship songs in reproductive isolation among populations of green lacewings of the genus *Chrysoperla* (Neuroptera: Chrysopidae). Evolution 46: 31–42.

Martinsen, G. D., T. G. Whitham, R. J. Turek, and P. Keim. 2001. Hybrid populations selectively filter gene introgression between species. Evolution 55: 1325–1335.

Masaki, S. 1967. Geographic variation and climatic adaptation in a field cricket (Orthoptera: gryllidae). Evolution 21: 725–741.

Masters, J. C. and H. G. Spencer. 1989. Why we need a new genetic species concept. Syst. Zool. 38: 270–279.

Masters, J. C., R. J. Rayner, I. J. McKay, A. D. Potts, D. Nails, J. W. Ferguson, B. K. Weissenbacher, M. Alsopp, and M. L. Anderson. 1987. The concept of species: Recognition versus isolation. S. Afr. J. Sci. 83: 534–537.

Masterson, J. 1994. Stomatal size in fossil plants: evidence for polyploidy in majority of angiosperms. Science 264: 421–423.

Mather, K. 1943. Polygenic inheritance and natural selection. Biol. Rev. 18: 32–64.

Mather, K. 1949. *Biometrical Genetics*. Dover Publications, Inc., New York.

Matsubara, K., K.-Thidar, and Y. Sano. 2003. A gene block causing cross–incompatibility hidden in wild and cultivated rice. Genetics 165: 343–352.

Maxson, L. R. and R. D. Maxson. 1979. Comparative albumin and biochemical evolution in plethodontid salamanders. Evolution 33: 1057–1062.

Maynard Smith, J. 1966. Sympatric speciation. Am. Nat. 100: 637–650.

Maynard Smith, J. 1978. *The Evolution of Sex*. Cambridge University Press, Cambridge, UK.

Maynard Smith, J. and E. Szathmáry. 1995. *The Major Transitions in Evolution*. W. H. Freeman/Spektrum, Oxford.

Mayr, E. 1942. *Systematics and the Origin of Species*. Columbia University Press, New York.

Mayr, E. 1947. Ecological factors in speciation. Evolution 1: 263–288.

Mayr, E. 1954a. Change of genetic environment and evolution. Pp. 157–180 in J. Huxley, A. C. Hardy, and E. B. Ford (eds.) *Evolution As a Process*. George Allen and Unwin Ltd., London.

Mayr, E. 1954b. Geographic speciation in tropical echinoids. Evolution 8: 1–18.

Mayr, E. 1959. Where are we? Cold Spring Harb. Symp. Quant. Biol. 24: 1–14.

Mayr, E. 1963. *Animal Species and Evolution*. Belknap Press, Cambridge, MA.

Mayr, E. 1969a. Species, speciation, and chromosomes. Pp. 1–7 in K. Benirschke (ed.) *Comparative Mammalian Cytogenetics*. Springer-Verlag, New York.

Mayr, E. 1969b. The biological meaning of species. Biol. J. Linn. Soc. Lond. 1: 311–320.

Mayr, E. 1982. *The Growth of Biological Thought*. Harvard University Press, Cambridge, MA.

Mayr, E. 1988. *Toward a New Philosophy of Biology*. Harvard University Press, Cambridge, MA.

Mayr, E. 1992. A local flora and the biological species concept. Am. J. Bot. 79: 222–238.

Mayr, E. 1995. Species, classification, and evolution. Pp. 3–12 in R. Arai, M. Kato, and Y. Doi (eds.) *Biodiversity and Evolution*. National Science Museum Foundation, Toyko.

Mazer, S. and D. E. Meade. 2000. Geographic variation in flower size in wild radish. Pp. 157–186 in T. A. Mousseau, B. Sinervo and J. Endler (eds.) *Adaptive Genetic Variation in the Wild*. Oxford University Press, New York.

McCarthy, E. M., M. A. Asmussen, and W. W. Anderson. 1995. A theoretical assessment of recombinational speciation. Heredity 74: 502–509.

McCartney, M. A. and H. A. Lessios. 2002. Quantitative analysis of gametic incompatibility between closely related species of Neotropical sea urchins. Biol. Bull. 202: 166–181.

McClure, B. A., R. Cruz-Garcia, B. Beecher, and W. Sulaman. 2000. Factors affecting inter- and intra-specific pollen rejection in *Nicotiana*. Ann. Bot. (Suppl. A) 85: 113–123.

McCune, A. R. 1997. How fast is speciation? Molecular, geological, and phylogenetic evidence from adaptive radiations of fishes. Pp. 585-610 in T. J. Givnish and K. J. Sytsma, eds. Molecular evolution and adaptive radiation. Cambridge University Press, Cambridge.

McCune, A. R. and N. R. Lovejoy. 1998. The relative rate of sympatric and allopatric speciation in fishes. Pp. 172–185 in D. J. Howard and S. H. Berlocher (eds.) *Endless Forms: Species and Speciation*. Oxford University Press, New York.

McDonald, J. H. and M. Kreitman. 1991. Adaptive protein evolution at the *Adh* locus in *Drosophila*. Nature 351: 652–654.

McKaye, K. R. and W. N. Gray. 1984. Extrinsic barriers to gene flow in rock-dwelling cichlids of Lake Malawi: Macrohabitat heterogeneity and reef colonization. Pp. 169–183 in A. A. Echelle and I. Kornfield (eds.) *Evolution of Fish Species Flocks*. University of Maine at Orono Press, Orono, ME.

McKinnon, J. S. and H. D. Rundle. 2002. Speciation in nature: The threespine stickleback model systems. Trends Ecol. Evol. 17: 480–488.

McLain, D. K., M. P. Moulton, and T. P. Redfearn. 1995. Sexual selection and the risk of extinction of introduced birds on oceanic islands. Oikos 74: 27–34.

McNeilly, T. 1968. Evolution in closely adjacent plant populations. III. *Agrostis tenuis* on a small copper mine. Heredity 23: 99–108.

McNeilly, T. and J. Antonovics. 1968. Evolution in closely adjacent plant populations. IV. Barriers to gene flow. Heredity 23: 205–218.

McPeek, M. A. and G. A. Wellborn. 1998. Genetic variation and reproductive isolation among phenotypically divergent amphipod populations. Limnol. Oceanogr. 43: 1162–1169.

McPhail, J. D. 1984. Ecology and evolution of sympatric sticklebacks (Gasterosteus): Morphological and genetic evidence for a species pair in Enos Lake, British Columbia. Canadian J. Zool. 62: 1402–1408.

McPhail, J. D. 1992. Ecology and evolution of sympatric sticklebacks (*Gasterosteus*): Evidence for genetically divergent populations in Paxton Lake, Texada Island, British Columbia. Canadian J. Zool. 70: 361–369.

McPhail, J. D. 1994. Speciation and the evolution of reproductive isolation in the sticklebacks (*Gasterosteus*) of southwestern British Columbia. Pp. 399–437 in M. A. Bell and S. A. Foster (eds.) *The Evolutionary Biology of the Threespine Stickleback*. Oxford University Press, Oxford.

McPheron, B. A., D. C. Smith, and S. H. Berlocher. 1988. Genetic differences between *Rhagoletis pomonella* host races. Nature 336: 64–66.

McQueen, H. A., D. McBride, G. Miele, A. P. Bird, and M. Clinton. 2001. Dosage compensation in birds. Curr. Biol. 11: 253–257.

Meffert, L. M. and E. H. Bryant. 1991. Mating propensity and courtship behavior in serially bottlenecked lines of the housefly. Evolution 45: 293–306.

Melnick, D. J., G. A. Hoelzer, R. Absher, and M. U. Ashley. 1993. mtDNA diversity in Rhesus monkeys reveals overestimates of divergence time and paraphyly with neighboring species. Mol. Biol. Evol. 10: 282–295.

Mendelson, T. C. 2003. Sexual isolation evolves faster than hybrid inviability in a diverse and sexually dimorphic genus of fish (Percidae: *Etheostoma*). Evolution 57: 317–327.

Mendelson, T. C. and K. A. Shaw. 2004. Using AFLPs to reconstruct the history of speciation in an island radiation. Unpublished ms.

Mercot, H., A. Atlan, M. Jacques, and C. Montchamp-Moreau. 1995. Sex-ratio distortion in *Drosophila simulans*: co-occurrence of a meiotic drive and a suppressor of drive. J. Evol. Biol. 8: 283–300.

Metz, E. C. and S. R. Palumbi. 1996. Positive selection and sequence rearrangements generate extensive polymorphism in the gamete recognition protein bindin. Mol. Biol. Evol. 13:397–406.

Metz, E. C., R. E. Kane, H. Yanagimachi, and S. R. Palumbi. 1994. Fertilization between closely related sea urchins is blocked by incompatibilities during sperm-egg attachment and early stages of fusion. Biol. Bull. 187:23–34.

Metz, E. C., G. Gómez-Gutierrez, and V. D. Vacquier. 1998a. Mitochondrial DNA and bindin gene sequence evolution among allopatric species of the sea urchin genus *Arbacia*. Mol. Biol. Evol. 15: 185–195.

Metz, E. C., R. Robles-Sikisaka, and V. D. Vacquier. 1998b. Nonsynonymous substitution in abalone sperm fertilization genes exceeds substitution in introns and mitochondrial DNA. Proc. Natl. Acad. Sci. USA 95: 10676–10681.

Meyer, A. 1993. Phylogenetic relationships and evolutionary processes in east African cichlid fishes. Trends Ecol. Evol. 8: 279–284.

Michaelis, P. 1954. Cytoplasmic inheritance in *Epilobium* and its theoretical significance. Adv. Genet. 6: 287–401.

Michalak, P. and M. A. F. Noor. 2003. Genome-wide patterns of expression in *Drosophila* pure species and hybrid males. Mol. Biol. Evol. 20: 1070–1076.

Michalak, P., I. Minkov, A. Hellin, D. N. Lerman, B. R. Bettencourt, M. E. Feder, A. B. Korol, and E. Nevo. 2001. Genetic evidence for adaptation-driven incipient speciation of *Drosophila melanogaster* along a microclimatic contrast in "Evolution Canyon," Israel. Proc. Natl. Acad. Sci. USA 98: 13195–13200.

Miller, R. B. 1981. Hawkmoths and the geographic patterns of floral variation in *Aquilegia caerulea*. Evolution 35: 763–774.

Miller, R. L. 1997. Specificity of sperm chemotaxis among Great Barrier Reef shallow-water holothurians and ophiuroids. J. Exp. Zool. 279: 189–200.

Miller, R. R. 1950. Speciation in fishes of the genera *Cyprinodon* and *Empetrichthys*, inhabiting the Death Valley region. Evolution 4: 155–163.

Mirov, N. T. 1967. *The genus* Pinus. Ronald Press, New York.

Mishler, B. D. 1990. Reproductive biology and species distinctions in the moss genus *Tortula*, as represented in Mexico. Syst. Bot. 15: 86–97.

Mishler, B. D. and R. N. Brandon. 1987. Individuality, pluralism, and the phylogenetic species concept. Biol. Philos. 2: 397–414.

Mishler, B. D. and A. F. Budd. 1990. Species and evolution in clonal organisms. Syst. Bot. 15: 166–171.

Mishler, B. D. and M. J. Donoghue. 1982. Species concepts: A case for pluralism. Syst. Zool. 31:491–503.

Mitra, S., H. Landel, and S. Pruett-Jones. 1996. Species richness covaries with mating system in birds. Auk 113: 544–551.

Mitrofanov, V. G. and N. V. Sidorova. 1981. Genetics of the sex ratio anomaly in *Drosophila* hybrids of the *virilis* group. Theor. Appl. Genet. 59: 17–22.

Mitter, C., B. Farrell, and B. Wiegmann. 1988. The phylogenetic study of adaptive zones: has phytophagy promoted insect diversity? Am. Nat. 132: 107–128.

Mitter, C., B. Farrell, and D. J. Futuyma. 1991. Phylogenetic studies of insect-plant interactions: insights into the genesis of diversity. Trends Ecol. Evol. 6: 290–293.

Miyatake, T. and T. Shimizu. 1999. Genetic correlations between life-history and behavioral traits can cause reproductive isolation. Evolution 53: 216–224.

Miyatake, T., A. Matsumoto, T. Matsuyama, H.R. Ueda, T. Toyosato, and T. Tanimura. 2002. The period gene and allochronic reproductive isolation in *Bactrocera cucurbitae*. Proc. R. Soc. Lond. B 269: 2467–2472.

Molbo, D., C. A. Machado, J. G. Sevenster, L. Keller, and E. A. Herre. 2003. Cryptic species of fig-pollinating wasps: implications for the evolution of the fig-wasp mutualism, sex allocation, and the precision of adaptation. Proc. Nat. Acad. Sci. USA 100: 5867–5872.

Møller, A. P. and J. J. Cuervo. 1998. Speciation and feather ornamentation in birds. Evolution 52: 859–869.

Montchamp-Moreau, C. and D. Joly. 1997. Abnormal spermiogenesis is associated with the X-linked sex-ratio trait in *Drosophila simulans*. Heredity 79: 24–30.

Montchamp-Moreau, C., J.–F. Ferveur, and M. Jacques. 1991. Geographic distribution and inheritance of three cytoplasmic incompatibility types in *Drosophila simulans*. Genetics 129: 399–407.

Monteiro, L. R. and R. W. Furness. 1997. Speciation through temporal segregation of Madeiran storm petrel (*Oceanodroma castro*) populations in the Azores? Philos. Trans. R. Soc. Lond. B 353: 945–953.

Monti, L., J. Génermont, C. Malosse, and B. Lalanne-Cassou. 1997. A genetic analysis of some components of reproductive isolation between two closely related species, *Spodoptera latifascia* (Walker) and *S. descoinsi* (Lalanne-Cassou and Silvain) (Lepidoptera: Noctuidae). J. Evol. Biol 10: 121–143.

Mooers, A. O., H. D Rundle, and M. C. Whitlock. 1999. The effects of selection and bottlenecks on male mating success in peripheral isolates. Am. Nat. 153: 437–444.

Mooney, H. A. 1966. Influence of soil type on the distribution of two closely related species of *Erigeron*. Ecology 47: 950–958.

Moore, J. A. 1957. An embryologist's view of the species concept. Pp. 325–338 *in* E. Mayr (ed.) *The Species Problem*. Amer. Assn. Adv. Sci., Washington, DC.

Moore, W. S. 1987. Random mating in the northern flicker hybrid zone: implications for the evolution of bright and contrasting plumage patterns in birds. Evolution 41: 539–546.

Moore, W. S. and D. B. Buchanan. 1985. Stability of the northern flicker hybrid zone in historical times: implications for adaptive speciation theory. Evolution 39: 135–151.

Moore, W. S. and J. T. Price. 1993. Nature of selection in the northern flicker hybrid zone and its implications for speciation theory *in* R. G. Harrison (ed.) *Hybrid Zones and the Evolutionary Process*. Oxford University Press, New York.

Moore, W. S., J. H. Graham, and J. T. Price. 1991. Mitochondrial DNA variation in the Northern Flicker (*Colaptes auratus*, Aves). Mol. Biol. Evol. 8: 327–344.

Moran, P., I. Kornfield, and P. N. Reinthal. 1994. Molecular systematics and radiation of the haplochromine cichlids (Teleostei, Perciformes) of Lake Malawi. Copeia 1994: 274–288.

Morell, V. 1999. Ecology returns to speciation studies. Science 284: 2106–2108.

Moritz, C. 1986. The population biology of *Gehyra* (Gekkonidae): Chromosome change and speciation. Syst. Zool. 35: 46–67.

Moritz, C., C. J. Schneider, and D. B. Wake. 1992. Evolutionary relationships within the *Ensatina eschscholtzii* complex confirm the ring species interpretations. Syst. Biol. 41:273–291.

Morrison, D. A., M. McDonald, P. Bankoff, and P. Quirico. 1994. Reproductive isolation mechanisms among four closely related species of *Conospermum* (Proteaceae). Bot. J. Linn. Soc. 116: 13–31.

Mosseler, A. and C. S. Papadopol. 1989. Seasonal isolation as a reproductive barrier among sympatric *Salix* species. Can. J. Bot. 67: 2563–2570.

Moulia, C., J. P. Aussel, F. Bonhomme. P. Boursot, J. T. Nielsen, and F. Renaud. 1991. Wormy mice in a hybrid zone: a genetic control of susceptibility to parasite infection. J. Evol. Biol. 4: 679–687.

Moya, A., A. Galiana, and F. J. Ayala. 1995. Founder-effect speciation theory: failure of experimental corroboration. Proc. Natl. Acad. Sci. USA 92: 3983–3986.

Moyle, L. C., M. S. Olson, and P. Tiffin. 2004. Patterns of reproductive isolation in four angiosperm genera. Evolution (in press).

Muir, G., C. C. Fleming, and S. Schlötterer. 2000. Species status of hybridizing oaks. Nature 405: 1016.

Muller, C. H. 1952. Ecological control of hybridization in *Quercus*: A factor in the mechanism of evolution. Evolution 6: 147–161.

Muller, H. J. 1923. Mutation. Pp. 106–112 *in* C. B. Davenport (ed.) *Eugenics, Genetics and the Family: Proceedings of the Second International Congress of Eugenics, Vol I*. William and Wilkens, Baltimore.

Muller, H. J. 1925. Why polyploidy is rarer in animals than in plants. Am. Nat. 59: 346–353.

Muller, H. J. 1939. Reversibility in evolution considered from the standpoint of genetics. Biol. Rev. Camb. Philos. Soc. 14: 261–280.

Muller, H. J. 1940. Bearing of the *Drosophila* work on systematics. Pp. 185–268 *in* J. S. Huxley (ed.) *The New Systematics*. Clarendon Press, Oxford.

Muller, H. J. 1942. Isolating mechanisms, evolution, and temperature. Biol. Symp. 6: 71–125.

Muller, H. J. and G. Pontecorvo. 1942. Recessive genes causing interspecific sterility and other disharmonies between *Drosophila melanogaster* and *simulans*. Genetics 27: 157.

Muniyamma, M. and J. B. Phipps. 1984. Studies in *Crataegus*: 11. Further cytological evidence for the occurrence of apomixis in North American hawthorns. Can. J. Bot. 62:2316–2324.

Muniyamma, M. and J. B. Phipps. 1985. Studies in *Crataegus*: 12. Cytological evidence for sexuality in some diploid and tetraploid species of North American hawthorns. Can. J. Bot. 63: 1319–1324.

Muntzing, A. 1930a. Outlines to a genetic monograph of the genus *Galeopsis*. Hereditas 13:185–341.

Muntzing, A. 1930b. Uber Chromosomenvermehrung in *Galeopsis*-Kreuzungen und ihre phylogenetische Bedeutung. Hereditas 14: 153–172.

Muntzing, A. 1936. The evolutionary significance of autopolyploidy. Hereditas 21: 263–378.

Nachman, M. W. and P. Myers. 1989. Exceptional chromosomal mutations in a rodent population are not strongly underdominant. Proc. Natl. Acad. Sci. USA 86: 6666–6670.

Nachman, M. W. and J. B. Searle. 1995. Why is the house mouse karyotype so variable? Trends Ecol. Evol. 10: 397–402.

Nagel, L. and D. Schluter. 1998. Body size, natural selection, and speciation in sticklebacks. Evolution 52: 209–218.

Nagl, S., H. Tichy, W. E. Mayer, N. Takezaki, N. Takahata, and J. Klein. 2000. The origin and age of haplochromine fishes in Lake Victoria, East Africa. Proc. R. Soc. Lond. B 267: 1049–1061.

Naisbit, R. E., C. D. Jiggins, and J. L. B. Mallet. 2001. Disruptive sexual selection against hybrids contributes to speciation between *Heliconius cydno* and *H. melpomene*. Proc. R. Soc. Lond. B 268: 1849–1854.

Naisbit, R. E., C. D. Jiggins, and J. Mallet. 2003. Mimicry: developmental genes that contribute to speciation. Evol. Dev. 5: 269–280.

Nason, J. D., N. C. Ellstrand, and M. L. Arnold. 1992. Patterns of hybridization and introgression in populations of oaks, manzanitas, and irises. Am. J. Bot. 79: 101–111.

Nason, J. D., S. B. Heard, and F. R. Williams. 2002. Host-associated genetic differentiation in the goldenrod elliptical-gall moth, *Gnorimoschema gallaesolidaginis* (Lepidoptera: Gelechiidae). Evolution 56: 1475–1488.

Navarro, A. and N. H. Barton. 2003. Accumulating postzygotic isolation genes in parapatry: a new twist on chromosomal speciation. Evolution 57: 447–459.

Naveira, H. and A. Fontdevila. 1986. The evolutionary history of *Drosophila buzzatii*. XII. The genetic basis of sterility in hybrids between *D. buzzattii* and its sibling *D. serido* from Argentina. Genetics 114: 841–857.

Naveira, H. and A. Fontdevila. 1991. The evolutionary history of *Drosophila buzzatii*. XXI. Cumulative action of multiple sterility factors on spermatogenesis in hybrids of *D. buzzatii* and *D. koepferae*. Heredity 67: 57–72.

Naveira, H. F. and X. R. Maside. 1998. The genetics of hybrid male sterility in *Drosophila*. Pp. 330–338 *in* D. J. Howard and S. H. Berlocher (eds.) *Endless Forms: Species and Speciation*. Oxford University Press, Oxford.

Nee, S. 2001. Inferring speciation rates from phylogenies. Evolution 55: 661–668.

Nee, S., E. C. Holmes, R. M. May, and P. H. Harvey. 1994a. Extinction rates can be estimated from molecular phylogenies. Philos. Trans. R. Soc. Lond. B 344: 77–82.

Nee, S., R. M. May, and P. H. Harvey. 1994b. The reconstructed evolutionary process. Philos. Trans. R. Soc. Lond. B 344: 305–311.

Nee, S., T. G. Barraclough, and P. H. Harvey. 1996. Temporal changes in biodiversity: Detecting patterns and identifying causes. Pp. 230–252 *in* K. J. Gaston (ed.) *Biodiversity: A Biology of Numbers and Difference*. Blackwell Scientific Publications, Oxford, UK.

Neff, N. A. and G. R. Smith. 1978. Multivariate analysis of hybrid fishes. Syst. Zool. 28: 176—196.

Nei, M. 1972. Genetic distances between populations. Am. Nat. 106: 282–292.

Nei, M. 1976. Mathematical models of speciation and genetic distance. Pp. 723–766 *in* S. Karlin and E. Nevo (eds.) *Population Genetics and Ecology*. Academic Press, New York.

Nei, M., T. Maruyama, and R. Chakraborty. 1975. The bottleneck effect and genetic variability in populations. Evolution 29: 1–10.

Nei, M., T. Maruyama, and C.-I. Wu. 1983. Models of evolution of reproductive isolation. Genetics 103: 557–579.

Nelson, G. 1989. Species and taxa: Systematics and evolution. Pp. 60–81 *in* D. Otte and J. A. Endler (eds.) *Speciation and Its Consequences*. Sinauer Associates, Sunderland, MA.

Neubaum, D. M. and M. F. Wolfner. 1999. Mated *Drosophila melanogaster* females require a seminal fluid protein, Acp36DE, to store sperm efficiently. Genetics 153: 845–857.

Nevo, E., E. Rashovetsky, T. Pavlicek, and A. Korol. 1998. A complex adaptive syndrome in *Drosophila* caused by microclimatic contrasts. Heredity 80: 9–16.

Newton, I. 2003. *The Speciation and Biogeography of Birds*. Academic Press, Amsterdam.

Niklas, K. J. 1982. Simulated and empiric wind pollination patterns of conifer ovulate cones. Proc. Natl. Acad. Sci. USA 79: 501–514.

Niklas, K. J. 1997. *The Evolutionary Biology of Plants*. University of Chicago Press, Chicago.

Niklas, K. J. and S. L. Buchmann. 1987. The aerodynamics of pollen capture in two sympatric *Ephedra* species. Evolution 41: 104–123.

Niklas, K. J., B. H. Tiffney, and A. H. Knoll. 1983. Patterns in vascular land plant diversification. Nature 303: 314–316.

Nilsson, L. A. 1983. Processes of isolation and introgressive interplay between *Platanthera bifolia* (L.) Rich and *P. chlorantha* (Custer) Reichb. (Orchidaceae). Bot. J. Linn. Soc. 87: 325–350.

Noor, M. A. 1995. Speciation driven by natural selection in *Drosophila*. Nature 375: 674–675.

Noor, M. A. F. 1997a. Genetics of sexual isolation and courtship dysfunction in male hybrids of *Drosophila pseudoobscura* and *Drosophila persimilis*. Evolution 51: 809–815.

Noor, M. A. F. 1997b. How often does sympatry affect sexual isolation in *Drosophila*? Am. Nat. 149: 1156–1163.

Noor, M. A. F. 1999. Reinforcement and other consequences of sympatry. Heredity 83: 503–508.

Noor, M. A. F. 2000. On the evolution of female mating preferences as pleiotropic byproducts of adaptive evolution. Adapt. Behav. 8: 3–12.

Noor, M. A. F., K. L. Grams, L. A. Bertucci, Y. Almendarez, J. Reiland, and K. R. Smith. 2001a. The genetics of reproductive isolation and the potential for gene exchange between *Drosophila pseudoobscura* and *D. persimilis* via backcross hybrid males. Evolution 55:512–521.

Noor, M. A. F., K. L. Grams, L. A. Bertucci, and J. Reiland. 2001b. Chromosomal inversions and the reproductive isolation of species. Proc. Natl. Acad. Sci. USA 98: 12084–12088.

Nosil, P. 2002. Transition rates between specialization and generalization in phytophagous insects. Evolution 56: 1701–1706.

Nosil, P., B. J. Crespi, and C. P. Sandoval. 2002. Host-plant adaptation drives the parallel evolution of reproductive isolation. Nature 417: 440–443.

Nosil, P., B. J. Crespi, and C. P. Sandoval. 2003. Reproductive isolation driven by the combined effects of ecological adaptation and reinforcement. Proc. R. Soc. Lond. B 270: 1911–1918.

Ochman, H., J. G. Lawrence, and E. A. Groisman. 2000. Lateral gene transfer and the nature of bacterial innovation. Nature 405: 299–304.

Ödeen, A. and A. B. Florin. 2000. Effective population size may limit the power of laboratory experiments to demonstrate sympatric and parapatric speciation. Proc. R. Soc. Lond. B 267: 601–616.

Ogden, R. and R. S. Thorpe. 2002. Molecular evidence for ecological speciation in tropical habitats. Proc. Natl. Acad. Sci. USA 99: 13612–13615.

Ohno, S. 1970. *Evolution by Gene Duplication*. Springer-Verlag, Berlin.

Ohta, T. 1993. Amino acid substitution at the *Adh* locus of *Drosophila* is facilitated by small population size. Proc. Natl. Acad. Sci. USA 90: 4548–4551.

Oliver, J. and R. Babcock. 1992. Aspects of the fertilization ecology of broadcast spawning corals: Sperm dilution effects and *in situ* measurements of fertilization. Biol. Bull. 183:409–417.

Ollerton, J. 1996. Reconciling ecological processes with phylogenetic patterns: The apparent paradox of plant-pollinator systems. J. Ecol. 84: 767–769.

Omland, K. E. 1997. Examining two standard assumptions of ancestral reconstructions: Repeated losses of dichromatism in dabbling ducks (*Anatini*). Evolution 51: 1636–1646.

Omland, K. E., C. L. Tarr, W. I. Boarman, J. M. Marzluff, and R. C. Fleischeer. 2000. Cryptic genetic variation and paraphyly in ravens. Proc. R. Soc. Lond. B 267:2475–2482.

O'Neill, S. L. and T. L. Karr. 1990. Bidirectional incompatibility between conspecific populations of *Drosophila simulans*. Nature 348: 178–180.

O'Neill, S. L., R. Giordano, A. M. E. Colbert, T. L. Karr, and H. M. Robertson. 1992. 16S rRNA phylogenetic analysis of the bacterial endosymbionts associated with cytoplasmic incompatibility in insects. Proc. Natl. Acad. Sci. USA 89: 2699–2702.

O'Neill, S. L., A. A. Hoffmann, and J. H. Werren. 1997. *Influential Passengers: Inherited Microorganisms and Arthropod Reproduction*. Oxford University Press, Oxford.

Opler, P. A., H. G. Baker, and G. W. Frankie. 1975. Reproductive biology of some Costa Rican *Cordia* species (Boraginaceae). Biotropica 7: 234–237.

Orr, H. A. 1987. Genetics of male and female sterility in hybrids of *Drosophila pseudoobscura* and *D. persimilis*. Genetics 116: 555–563.

Orr, H. A. 1989a. Does postzygotic isolation result from improper dosage compensation? Genetics 122: 891–894.

Orr, H. A. 1989b. Genetics of sterility in hybrids between two subspecies of *Drosophila*. Evolution 43: 180–189.

Orr, H. A. 1989c. Localization of genes causing postzygotic isolation in two hybridizations involving *Drosophila pseudoobscura*. Heredity 63: 231–237.

Orr, H. A. 1990. "Why polyploidy is rarer in animals than in plants" revisited. Am. Nat. 136: 759–770.

Orr, H. A. 1991a. Genetic basis of postzygotic isolation between *D. melanogaster* and *D. simulans*. Drosoph. Inf. Serv. 70: 161–162.

Orr, H. A. 1991b. Is single-gene speciation possible? Evolution 45: 764–769.

Orr, H. A. 1992. Mapping and characterization of a "speciation gene" in *Drosophila*. Genet. Res. 59: 73–80.

Orr, H. A. 1993a. A mathematical model of Haldane's rule. Evolution 47: 1606–1611.

Orr, H. A. 1993b. Haldane's rule has multiple genetic causes. Nature 361: 532–533.

Orr, H. A. 1995. The population genetics of speciation: The evolution of hybrid incompatibilities. Genetics 139: 1805–1813.

Orr, H. A. 1996. Dobzhansky, Bateson, and the genetics of speciation. Genetics 144: 1331–1335.

Orr, H. A. 1997. Haldane's rule. Annu. Rev. Ecol. Syst. 28: 195–218.

Orr, H. A. 1998a. Testing natural selection vs. genetic drift in phenotypic evolution using quantitative trait locus data. Genetics 149: 2099–2104.

Orr, H. A. 1998b. The population genetics of adaptation: The distribution of factors fixed during adaptive evolution. Evolution 52: 935–949.

Orr, H. A. 1999. Does hybrid lethality depend on sex or genotype? Genetics 152: 1767–1769.

Orr, H. A. 2001. The genetics of species differences. Trends Ecol. Evol. 16: 343–350.

Orr, H. A. and A. Betancourt. 2001. Haldane's sieve and adaptation from the standing genetic variation. Genetics 157: 875–884.

Orr, H. A. and J. A. Coyne. 1989. The genetics of postzygotic isolation in the *Drosophila virilis* group. Genetics 121: 527–537.

Orr, H. A. and J. A. Coyne. 1992. The genetics of adaptation revisited. Am. Nat. 140: 725–742.

Orr, H. A. and S. Irving. 2000. Genetic analysis of the *Hybrid male rescue* locus of *Drosophila*. Genetics 155: 225–231.

Orr, H. A. and S. Irving. 2001. Complex epistasis and the genetic basis of hybrid sterility in the *Drosophila pseudoobscura* Bogota–USA hybridization. Genetics 158: 1089–1100.

Orr, H. A. and S. Irving. 2004. Segregation distortion in hybrids between the Bogota and USA subspecies of *Drosophila pseudoobscura*. Genetics (in press).

Orr, H. A. and L. H. Orr. 1996. Waiting for speciation: The effect of population subdivision on the time to speciation. Evolution 50: 1742–1749.

Orr, H. A. and D. C. Presgraves. 2000. Speciation by postzygotic isolation: Forces, genes and molecules. BioEssays 22: 1085–1094

Orr, H. A. and M. Turelli. 1996. Dominance and Haldane's rule. Genetics 143: 613–616.

Orr, H. A. and M. Turelli. 2001. The evolution of postzygotic isolation: Accumulating Dobzhansky-Muller incompatibilities. Evolution 55: 1085–1094.

Orr, H. A., L. D. Madden, J. A. Coyne, R. Goodwin, and R. S. Hawley. 1997. The developmental genetics of hybrid inviability: A mitotic defect in *Drosophila* hybrids. Genetics 145: 1031–1040.

Osborn, T. C., J. C. Pires, J. A. Birchler, D. L. Auger, Z. J. Chen, H.-S. Lee, L. Comai, A. Madlung, R. W. Doerge, V. Colot, and R. A. Martienssen. 2003. Understanding mechanisms of novel gene expression in polyploids. Trends Genet. 19: 141–147.

Otte, D. 1989. Speciation in Hawaiian crickets. Pp. 483–526 *in* D. Otte and J. Endler (eds.) *Speciation and Its Consequences*. Sinauer Associates, Sunderland, MA.

Otte, D. 1994. *The Crickets of Hawaii: Origin, Systematics, and Evolution*. Orthoptera Society/Academy of Natural Sciences of Philadelphia.

Otte, D. and J. A. Endler (eds.) 1989. *Speciation and Its Consequences*. Sinauer Associates, Sunderland, Mass.

Otto, S. P. and M. C. Whitlock. 1997. The probability of fixation in populations of changing size. Genetics 146: 723–733.

Otto, S. P. and J. Whitton. 2000. Polyploid incidence and evolution. Annu. Rev. Genet. 34: 401–437.

Owen, R. B., R. Crossley, T. C. Johnson, D. Tweddle, I. Kornfield, S. Davison, D. H. Eccles, and D. E. Engstrom. 1990. Major low levels of Lake Malawi and their implications for speciation rates in cichlid fishes. Proc. R. Soc. Lond. B 240: 519–553.

Owens, I. P. F., P. M. Bennett, and P. H. Harvey. 1999. Species richness among birds: body size, life history, sexual selection or ecology? Proc. R. Soc. Lond. B 266: 933–939.

Page, R. D. M. 1994. Parallel phylogenies: Reconstructing the history of host-parasite assemblages. Cladistics 10: 155–173.

Page, R. D. M. 1996. Temporal congruence revisited: Comparison of mitochondrial DNA sequence divergence in cospeciating pocket gophers and their chewing lice. Syst. Biol. 45: 151–167.

Page, R. D. M., P. L. M. Lee, S. A. Becher, R. Griffiths, and D. H. Clayton. 1998. A different tempo of mitochondrial DNA evolution in birds and their parasitic lice. Mol. Phylogenet. Evol. 9: 276–293.

Palmer, E. J. 1948. Hybrid oaks of North America. J. Arnold Arbor. 29: 1–48.

Palopoli, M. F. and C.-I. Wu. 1994. Genetics of hybrid male sterility between *Drosophila* sibling species: a complex web of epistasis is revealed in interspecific studies. Genetics 138: 329–341.

Palumbi, S. R. 1998. Species formation and the evolution of gamete recognition loci. Pp. 271–278 *in* D. J. Howard and S. H. Berlocher (eds.) *Endless Forms: Species and Speciation*. Oxford University Press, New York.

Palumbi, S. R. 1999. All males are not created equal: Fertility differences depend on gamete recognition polymorphisms in sea urchins. Proc. Natl. Acad. Sci. U.S.A. 96: 12632–12637.

Palumbi, S. R. and E. C. Metz. 1991. Strong reproductive isolation between closely related tropical sea urchins (genus *Echinometra*). Mol. Biol. Evol. 8: 227–239.

Palys, T., L. K. Nakamura, and F. M. Cohan. 1997. Discovery and classification of the ecological diversity in the bacterial world: The role of DNA sequence data. Int. J. Syst. Bacteriol. 47: 1145–1156.

Panhuis, T. M., W. J. Swanson, and L. Nunney. 2004. Population genetics of accessory gland proteins and sexual behavior in *Drosophila melanogaster* populations from Evolution Canyon. Evolution 57: 2785—2791.

Pantazidis, A. C. and E. Zouros. 1988. Location of an autosomal factor causing sterility in *Drosophila mojavensis* males carrying the *Drosophila arizonensis* Y chromosome. Heredity 60: 299–304.

Pantazidis, A. C., V. K. Galanopoulos, and E. Zouros. 1993. An autosomal factor from *Drosophila arizonae* restores normal spermatogenesis in *Drosophila mojavensis* males carrying the *D. arizonae* Y chromosome. Genetics 134: 309–318.

Parker, A. and I. Kornfield. 1997. Evolution of the mitochondrial DNA control region in the *mbuna* (Cichlidae) species flock of Lake Malawi, East Africa. J. Mol. Evol. 45: 70–83.

Parks, C. R., N. G. Miller, J. F. Wendel, and K. M. McDougal. 1983. Genetic divergence within the genus *Liriodendron* (Magnoliaceae). Ann. Mo. Bot. Gard. 70: 658–666.

Partridge, L. and G. A. Parker. 1999. Sexual conflict and speciation. Pp. 130–159 *in* A. E. Magurran and R. M. May (eds.) *Evolution of Biological Diversity*. Oxford University Press, Oxford.

Pashley, D. P. and J. A. Martin. 1987. Reproductive incompatibility between host strains of the Fall Armyworm (Lepidoptera: Noctuidae). Ann. Entomol. Soc. Am. 80: 731–733.

Pastuglia, M., D. Roby, C. Dumas, and J. M. Cock. 1997. Rapid induction by wounding and bacterial infection of an S gene family receptor-like kinase gene in *Brassica oleracea*. Plant Cell 9: 49–60.

Paterniani, E. 1969. Selection for reproductive isolation between two populations of maize, *Zea mays* L. Evolution 23: 534–547.

Paterson, A. M., G. P. Wallis, L. J. Wallis, and R. D. Gray. 2000. Seabird and louse coevolution: Complex histories revealed by 12S rRNA sequences and reconciliation analyses. Syst.. Biol. 49: 383–399.

Paterson, H. E. H. 1978. More evidence against speciation by reinforcement. S. Afr. J. Sci. 74: 369–371.

Paterson, H. E. H. 1982. Perspective on speciation by reinforcement. S. Afr. J. Sci. 78: 53–57.

Paterson, H. E. H. 1985. The recognition concept of species. Pp. 21–29 *in* E. S. Vrba (ed.) *Species and Speciation*. Transvaal Museum Monograph No. 4, Pretoria.

Patterson, J. T. 1946. A new type of isolating mechanism in *Drosophila*. Proc. Natl. Acad. Sci. USA 32: 202–208.

Patterson, J. T. and R. K. Griffen. 1944. The genetic mechanism underlying species isolation. Univ. Tex. Pub. 4445: 212–223.

Patterson, J. T. and W. S. Stone 1952. *Evolution in the Genus Drosophila*. Macmillan and Co., New York.

Patton, J. L. and M. F. Smith. 1994. Paraphyly, polyphyly, and the nature of species boundaries in pocket gophers (Genus *Thomomys*). Syst. Biol. 43: 11–26.

Paulay, G. 1985. Adaptive radiation on an isolated oceanic island: The Cryptorhynchinae (Curculionidae) of Rapa revisited. Biol. J. Linn. Soc. 26: 95–187.

Paulson, D. R. 1974. Reproductive isolation in damselflies. Syst. Zool. 23: 40–49.

Paulus, H. F. and C. Gack. 1990. Pollinators as prepollinating isolation factors: Evolution and specialization in *Ophrys* (Orchidaceae). Israel J. Bot. 39: 43–79.

Payne, R. B., L. L. Payne, and J. L. Woods. 1998. Song learning in brood-parasitic indigobirds *Vidua chalybeata*: song mimicry of the host species. Anim. Behav. 55: 1537–1553.

Payne, R. B., L. L. Payne, J. L. Woods, and M. D. Sorenson. 2000. Imprinting and the origin of parasite-host species associations in brood-parasitic, *Vidua chalybeata*. Anim. Behav. 59: 69–81.

Payne, R. J. H. and D. Krakauer. 1997. Sexual selection, space, and speciation. Evolution 51: 1–9.

Payseur, B. A., J. G. Krenz, and M. W. Nachman. 2004. Differential patterns of introgression across the X chromosome in a hybrid zone between two species of house mouse. Unpublished ms.

Peek, A. S., R. A. Feldman, R. A. Lutz, and R. C. Vrijenhoek. 1998. Cospeciation of chemoautotrophic bacteria and deep sea clams. Proc. Natl. Acad. Sci. USA 95: 9962–9966.

Pellmyr, O. 2003. Yuccas, yucca moths, and coevolution: A review. Ann. Mo. Bot. Gard. 90: 35–55.

Perez, D. E. and C.-I. Wu. 1995. Further characterization of the *Odysseus* locus of hybrid sterility in *Drosophila*: One gene is not enough. Genetics 140: 201–206.

Perez, D. E., C.-I. Wu, N. A. Johnson, and M.-L. Wu. 1993. Genetics of reproductive isolation in the *Drosophila simulans* clade: DNA-marker assisted mapping and characterization of a hybrid-male sterility gene, *Odysseus* (*Ods*). Genetics 134: 261–275.

Pernin, P., A. Ataya, and M.-L. Carious. 1992. Genetic structure of natural populations of the free-living amoeba, *Naegleria lovaniensis*. Evidence for sexual reproduction. Heredity 68: 173–181.

Pfennig, K. S. 2003. A test of alternative hypotheses for the evolution of reproductive isolation between spadefoot toads: support for the reinforcement hypothesis. Evolution 57: 2842–2851.

Phillips, P. C. and N. A. Johnson. 1998. The population genetics of synthetic lethals. Genetics 150: 449–458.

Phipps, J. B. and M. Muniyamma. 1980. A taxonomic revision of *Crataegus* (Rosaceae) in Ontario, Canada. Can. J. Bot. 58: 1621–1699.

Piálek, J., H. C. Hauffe, K. M. Rodríguez-Clark and J. B. Searle. 2001. Raciation and speciation in house mice from the Alps: The roles of chromosomes. Mol. Ecol. 10: 613–625.

Plath, M., J. Parzefall, K. E. Korner, and I. Schlupp. 2004. Sexual selection in darkness? Female mating preferences in surface- and cave-dwelling Atlantic mollies, *Poecilia mexicana* (Poeciliidae, Teleostei). Behavioral Ecol. and Sociobiol. 55: 596–601.

Platnick, N. I. and G. H. Nelson. 1978. A method of analysis for historical biogeography. Syst. Zool. 27: 1–16.

Platz, N. E. 1981. Suture zone dynamics: Texas populations of *Rana belandieri* and *Rana blairi*. Copeia 1981: 733–734.

Polechova, J. and N. H. Barton. 2004. "Speciation" in asexual populations. Unpublished ms.

Polyakova, N. E., L. A. Skurikhina, A. D. Kukhlevskii, V. A. Brykov, T. V. Malinina, L. S. Minakhin, and Y. P. Altukhov. 1996. Population genetic structure of pink salmon *Oncorhynchus gorbuscha* (Walbaum) according to restriction analysis of mitochondrial DNA. 2. Comparison of nonoverlapping generations of even and odd years. Genetika 32: 1256–1262.

Pomiankowski, A. 1988. The evolution of female mate preferences for male genetic quality. Oxford Surv. Evol. Biol. 5: 136–184.

Pomiankowski, A. and Y. Iwasa. 1998. Runaway ornament diversity caused by Fisherian sexual selection. Proc. Natl. Acad. Sci. USA 1998: 5106–5111.

Pontecorvo, G. 1943a. Hybrid sterility in artificially produced recombinants between *Drosophila melanogaster* and *D. simulans*. Proc. R. Soc., Edinburgh, B, Biol. 61: 385–397.

Pontecorvo, G. 1943b. Viability interactions between chromosomes of *Drosophila melanogaster* and *Drosophila simulans*. J. Genet. 45: 51–66.

Poulton, E. B. 1908. What is a species? Pp. 46–94 *in Essays on Evolution*. Oxford at the Clarendon Press, Oxford.

Powell, J. A. 1992. Interrelationships of yuccas and yucca moths. Trends Ecol. Evol. 7: 10–15.

Powell, J. R. 1978. The founder-flush speciation theory: An experimental approach. Evolution 32: 465–474.

Powell, J. R. 1983. Interspecific cytoplasmic gene flow in the absence of nuclear gene flow: Evidence from *Drosophila*. Proc. Natl. Acad. Sci. U.S.A. 80: 492–495.

Prager, E. M. and A. C. Wilson. 1975. Slow evolutionary loss of the potential for interspecific hybridization in birds: A manifestation of slow regulatory evolution. Proc. Natl. Acad. Sci. USA 72: 200–204.

Prance, G. T. E. 1982. *Biological Diversification in the Tropics*. Columbia University Press, New York.

Prazmo, W. 1965. Cytogenetic studies on the genus *Aquilegia*. III. Inheritance of the traits distinguishing different complexes in the genus *Aquilegia*. Acta Soc. Bot. Poloniae 34: 403–437.

Presgraves, D. C. 2000. A genetic test of the mechanism of *Wolbachia*-induced cytoplasmic incompatibility in *Drosophila*. Genetics 154: 771–776.

Presgraves, D. C. 2002. Patterns of postzygotic isolation in Lepidoptera. Evolution 56: 1168–1183.

Presgraves, D. C. 2003. A fine-scale genetic analysis of hybrid incompatibilities in *Drosophila*. Genetics 163: 955–972.

Presgraves, D. C. and H. A. Orr. 1998. Haldane's rule in taxa lacking a hemizygous X. Science 282: 952–954.

Presgraves, D. C., L. Balagopalan, S. M. Abmayr, and H. A. Orr. 2003. Adaptive evolution drives divergence of a hybrid inviability gene between two species of *Drosophila*. Nature 423: 715–719.

Price, C. S. C. 1997. Conspecific sperm precedence in *Drosophila*. Nature 388: 663–666.

Price, C. S. C., K. A. Dyer, and J. A. Coyne. 1999. Sperm competition between *Drosophila* males involves both displacement and incapacitation. Nature 400: 449–452.

Price, C. S. C., C. H. Kim, J. Poluszny, and J. A. Coyne. 2000. Mechanisms of conspecific sperm precedence in *Drosophila*. Evolution 54: 2028–2037.

Price, C. S. C., C. H. Kim, C. J. Gronlund, and J. A. Coyne. 2001. Cryptic reproductive isolation in the *Drosophila simulans* clade. Evolution 55: 2028–2037.

Price, T. 1998. Sexual selection and natural selection in bird speciation. Philos. Trans. R. Soc. Lond. B 353: 251–260.

Price, T. 1999. Sexual selection and natural selection in bird speciation. Pp. 93–112 *in* A. Magurran and R. H. May (eds.) *Evolution of Biological Diversity*. Oxford University Press, Oxford.

Price, T. D. and M. M. Bouvier. 2002. The evolution of F_1 post-zygotic incompatibilities in birds. Evolution 56: 2083–2089.

Proctor, H. C. 1991. Courtship in the water mite *Neumania papillator*: Males capitalize on female adaptations for predation. Anim. Behav. 42: 589–598.

Proctor, H. C. 1992. Sensory exploitation and the evolution of male mating behaviour: A cladistic test using water mites (Acari: Parasitengona). Anim. Behav. 44: 745–752.

Prokopy, R. J., S. R. Diehl, and S. S. Cooley. 1988. Behavioral evidence for host races in *Rhagoletis pomonella* flies. Oecologia 76: 138–147.

Promislow, D., R. Montgomerie, and T. E. Martin. 1994. Sexual selection and survival in North American waterfowl. Evolution 48: 2045–2050.

Provine, W. B. 1971. *The Origins of Theoretical Population Genetics*. University of Chicago Press, Chicago.

Provine, W. B. 1989. Founder effects and genetic revolutions in microevolution and speciation: A historical perspective. Pp. 43–76 *in* L. V. Giddings, K. Y. Kaneshiro, and W. W. Anderson (eds.) *Genetics, Speciation, and the Founder Principle*. Oxford University Press, New York.

Prowell, D. P. 1998. Sex linkage and speciation in Lepidoptera. Pp. 309–319 *in* D. J. Howard and S. H. Berlocher (eds.) *Endless Forms: Species and Speciation*. Oxford University Press, Oxford.

Purvis, A., S. Nee, and P. H. Harvey. 1995. Macroevolutionary inferences from primate phylogeny. Proc. R. Soc. Lond. B 260: 329–333.

Qvarnström, A. and E. Forsgren. 1998. Should females prefer dominant males? Trends Ecol. Evol. 13: 498–501.

Rabinowitz, D., J. K. Rapp, V. L. Sork, B. J. Rathcke, G. A. Reese, and J. C. Weaver. 1981. Phenological properties of wind- and insect-pollinated plants. Ecology 62: 49–56.

Ramirez, W. B. 1970. Host specificity of fig wasps. Evolution 24: 680–691.

Ramsey, J. and D. W. Schemske. 1998. Pathways, mechanisms, and rates of polyploid formation in flowering plants. Annu. Rev. Ecol. Syst. 29: 467–501.

Ramsey, J. and D. W. Schemske. 2002. Neopolyploidy in flowering plants. Annu. Rev. Ecol. Syst. 33: 589–639.

Ramsey, J., H. D. Bradshaw, and D. W. Schemske. 2003. Components of reproductive isolation between the monkeyflowers *Mimulus lewisii* and *M. cardinalis* (Scrophulariaceae). Evolution 57: 1520–1534.

Rand, D. M. and R. G. Harrison. 1989. Ecological genetics of a mosaic hybrid zone: Mitochondrial, nuclear, and reproductive differentiation of crickets by soil type. Evolution 43: 432–449.

Randolph, L. F. 1966. *Iris nelsonii*, a new species of Louisiana iris of hybrid origin. Baileya 14: 143–169.

Randolph, L. F., J. Mitra, and I. S. Nelson. 1961. Cytotaxonomic studies of Louisiana irises. Bot. Gaz. 123: 126–133.

Ranz, J. M., C. I. Castillo-Davis, C. D. Meiklejohn, and D. L. Hartl. 2003. Sex-dependent gene expression and evolution of the *Drosophila* transcriptome. Science 300: 1742–1745.

Ranz, J. M., K. Namgyal, G. Gibson, and D. L. Hartl. 2004. Anomalies in the expression profile of interspecific hybrids of *Drosophila melanogaster* and *Drosophila simulans*. Genome Research 14: 373–379.

Räsänen, M. E., A. M. Linna, J. C. R. Santos, and F. R. Negri. 1995. Late Miocene tidal deposits in the Amazonian forest basin. Science 269: 386–390.

Ratcliffe, L. M. and P. R. Grant. 1983a. Species recognition in Darwin's finches (*Geospiza*, Gould). I. Discrimination by morphological cues. Anim. Behav. 31: 1139–1153.

Ratcliffe, L. M. and P. R. Grant. 1983b. Species recognition in Darwin's finches (*Geospiza*, Gould). II. Geographic variation in mate preference. Anim. Behav. 31: 1154–1165.

Rathcke, B. and E. P. Lacey. 1985. Phenological patterns of terrestrial plants. Annu. Rev. Ecol. Syst. 16: 179–214.

Raubenheimer, D. and T. M. Crowe. 1987. The Recognition Species Concept: Is it really an alternative? S. Afr. J. Sci. 83: 530–534.

Raup, D. M. and S. J. Gould. 1974. Stochastic simulation and evolution of morphology: Toward a nomothetic paleontology. Syst. Zool. 23: 305–322.

Raven, P. H. 1976. Systematics and plant population biology. Syst. Bot. 1: 284–316.

Rawson, P. D. and R. S. Burton. 2002. Functional coadaptation between cytochrome c and cytochrome c oxidase within allopatric populations of a marine copepod. Proc. Natl. Acad. Sci. USA 99: 12955–12958.

Read, A. and S. Nee. 1991. Is Haldane's rule significant? Evolution 45: 1707–1709.

Reed, D. L., V. S. Smith, A. R. Rogers, S. L. Hammond, and D. H. 2004. Molecular genetics of human lice supports direct contact between modern and archaic humans. Proc. Natl. Acad. Sci. USA (in press).

Reed, K. M. and J. H. Werren. 1995. Induction of paternal genome loss by the paternal sex ratio chromosome and cytoplasmic incompatibility bacteria (*Wolbachia*): A comparative study of early embryonic events. Mol. Reprod. Dev. 40: 408–418.

Reiland, J. and M. A. F. Noor. 2002. Little qualitative RNA misexpression in sterile male F_1 hybrids of *Drosophila pseudoobscura* and *D. persimilis*. BMC Evol. Biol. 2: 16.

Reed, K. M., J. Sites, J. W., and I. F. Greenbaum. 1992. Synapsis, recombination, and meiotic segregation in the mesquite lizard, *Sceloporus grammicus*, complex. Cytogenet. Cell Genet. 61: 40–45.

Reinhold, K. 1998. Sex linkage among genes controlling sexually selected traits. Behav. Ecol. Sociobiol. 44: 1–7.

Remington, C. L. 1968. Suture-zones of hybrid interaction between recently joined biotas. Evol. Biol. 2: 321–428.

Rhymer, J. M. and D. Simberloff. 1996. Extinction by hybridization and introgression. Annu. Rev. Ecol. Syst. 27: 83–109.

Ribbink, A. J., B. A. Marsh, A. C. Marsh, A. C. Ribbink, and B. J. Sharp. 1983. A preliminary survey of the cichlid fishes of rocky habitats in Lake Malawi. S. Afr. J. Zool. 18:149–310.

Ribera, I., T. G. Barraclough, and A. P. Vogler. 2001. The effect of habitat type on speciation rates and range movements in aquatic beetles: Inferences from species-level phylogenies. Mol. Ecol. 10: 721–725.

Rice, W. R. 1984a. Disruptive selection on habitat preference and the evolution of reproductive isolation: A simulation study. Evolution 38: 1251–1260.

Rice, W. R. 1984b. Sex chromosomes and the evolution of sexual dimorphism. Evolution 38:735–742.

Rice, W. R. 1987. Selection via habitat specialization: The evolution of reproductive isolation as a correlated character. Evolutionary Ecology 1: 1637–1653.

Rice, W. R. 1989. Analyzing tables of statistical tests. Evolution 43:223-225.

Rice, W. R. 1996. Sexually antagonistic male adaptation triggered by experimental arrest of female evolution. Nature 381: 232–234.

Rice, W. R. 1998. Intergenomic conflict, interlocus antagonistic coevolution, and the evolution of reproductive isolation. Pp. 261–270 *in* D. J. Howard and S. H. Berlocher (eds.) *Endless Forms: Species and Speciation*. Oxford University Press, New York.

Rice, W. R. and E. E. Hostert. 1993. Laboratory experiments on speciation: what have we learned in 40 years? Evolution 47: 1637–1653.

Rice, W. R. and G. W. Salt. 1988. Speciation via disruptive selection on habitat preference: Experimental evidence. Am. Nat. 131: 911–917.

Rice, W. R. and G. W. Salt. 1990. The evolution of reproductive isolation as a correlated character under sympatric conditions: Experimental evidence. Evolution 44: 1140–1152.

Richards, A. J. 1972. The *Taraxacum* flora of the British Isles. Watsonia 9 (Suppl.): 1–141.

Richards, A. J. 1973. The origin of *Taraxacum* agamospecies. Bot. J. Linn. Soc. 66: 189-211.

Richards, A. J. 1997. *Plant Breeding Systems*. Chapman & Hall, London.

Richards, O. W. and G. C. Robson. 1926. The species problem and evolution. Nature 117: 345–347, 382–384.

Richman, A. D. and J. R. Kohn. 2000. Evolutionary genetics of self-incompatibility in the Solanaceae. Plant Mol. Biol. 42: 169–179.

Richman, A. D. and T. D. Price. 1992. Evolution of ecological differences in the Old World leaf warblers. Nature 355: 817–821.

Rico, C. and G. F. Turner. 2002. Extreme microallopatric divergence in a cichlid species from Lake Malawi. Mol. Ecol. 11: 1585–1590.

Rico, P., P. Bouteillon, M. J. H. V. Oppen, M. E. Knight, G. M. Hewitt, and G. F. Turner. 2003. No evidence for parallel sympatric speciation in cichlid species of the genus *Pseudotropheus* from northwestern Lake Malawi. J. Evol. Biol. 16: 37–46.

Ridley, M. 1996. *Evolution*. Second Edition. Blackwell Scientific Publications, Oxford, UK.

Rieseberg, L. H. 1991. Homoploid reticulate evolution in *Helianthus* (Asteraceae): Evidence from ribosomal genes. Am. J. Bot. 78: 1218–1237.

Rieseberg, L. H. 1997. Hybrid origins of plant species. Annu. Rev. Ecol. Syst. 28: 359–389.

Rieseberg, L. H. 1998. Genetic mapping as a tool for studying speciation. Pp. 459–487 *in* D. E. Soltis, P. S. Soltis, and J. J. Doyle (eds.) *Molecular Systematics of Plants II: DNA Sequencing*. Chapman and Hall, New York.

Rieseberg, L. H. 2000. Crossing relationships among ancient and experimental sunflower hybrid lineages. Evolution 54: 859–865.

Rieseberg, L. H. 2001. Chromosomal rearrangements and speciation. Trends Ecol. Evol. 16:351–358.

Rieseberg, L. H. and D. Gerber. 1995. Hybridization in the Catalina Island Mountain Mahogany (*Cercocarpus traskiae*): RAPD evidence. Conserv. Biol. 9: 199–203.

Rieseberg, L. H., A. Liston, and D. M. Arias. 1991. Phylogenetic and systematic inferences from chloroplast DNA and isozyme variation in *Helianthus* sect. *Helianthus* (Asteraceae). Syst. Bot. 16: 50–76.

Rieseberg, L. H., A. M. Desrochers, and S. J. Youn. 1995a. Interspecific pollen competition as a reproductive barrier between sympatric species of *Helianthus* (Asteraceae). Am. J. Botany 82: 515–519.

Rieseberg, L. H., C. R. Linder, and G. J. Seiler. 1995b. Chromosomal and genic barriers to introgression in *Helianthus*. Genetics 141: 1163–1171.

Rieseberg, L. H., C. Van Fossen, and A. M. Desrochers. 1995c. Hybrid speciation accompanied by genomic reorganization in wild sunflowers. Nature 375: 313–316.

Rieseberg, L. H., B. Sinervo, C. R. Linder, M. Ungerer, and D. M. Arias. 1996. Role of gene interactions in hybrid speciation: Evidence from ancient and experimental hybrids. Science 272: 741–745.

Rieseberg, L. H., S. J. E. Baird, and A. M. Desrochers. 1998. Patterns of mating in wild sunflower hybrid zones. Evolution 52: 713–726.

Rieseberg, L. H., J. Whitton, and K. Gardner. 1999. Hybrid zones and the genetic architecture of a barrier to gene flow between two sunflower species. Genetics 152: 713–727.

Rieseberg, L. H., D. M. Raymond, Z. Lai, K. Livingstone, T. Nakazato, J. L. Durphy, A. E. Schwarzbach, L. A. Donovan, and C. Lexer. 2003. Major ecological transitions in wild sunflowers facilitated by hybridization. Science 301: 1211–1216.

Riley, H. P. 1938. A character analysis of colonies of *Iris fulva*, *Iris hexagona* var. *giganticaerulea* and natural hybrids. Am. J. Bot. 25: 727–738.

Ringo, J. M. 1977. Why 300 species of Hawaiian *Drosophila*? The sexual selection hypothesis. Evolution 31: 694-696.

Ringo, J., D. Wood, R. Rockwell, and H. Dowse. 1985. An experiment testing two hypotheses of speciation. Am. Nat. 126: 642–661.

Rising, J. D. 1996. *A Guide to the Identification and Natural History of the Sparrows of the United States and Canada*. Academic Press, New York.

Ritchie, M. G. and S. D. F. Phillips. 1998. The genetics of sexual isolation. Pp. 291–308 *in* D.Howard and S. H. Berlocher (eds.) *Endless Forms: Species and Speciation*. Oxford University Press, Oxford.

Ritchie, M. G., R. K. Butlin, and G. M. Hewitt. 1989. Assortative mating across a hybrid zone in *Chorthippus parallelus* (Orthoptera: Acrididae). J. Evol. Biol. 2: 339–352.

Ritchie, M. G., E. J. Halsey, and J. M. Gleason. 1999. *Drosophila* song as a species-specific mating signal and the behavioural importance of Kyriacou and Hall cycles in *D. melanogaster* song. Anim. Behav. 58: 649–657.

Roberts, M. S. and F. M. Cohan. 1995. Recombination and migration rates in natural populations of *Bacillus subtilis* and *Bacillus mojavensis*. Evolution 49: 1081–1094.

Roberts, M. S., K. L. Nakamura, and F. M. Cohan. 1996. *Bacillus vallismortis* sp. nov., a close relative of *Bacillus subtilis*, isolated from soil in Death Valley, California. Int. J. Syst. Bacteriol. 46: 470–475.

Robertson, A. 1967. The nature of quantitative genetic variation. Pp. 265–280 *in* A. Brink (ed.) *Heritage from Mendel*. University of Wisconsin Press, Madison, WI.

Robertson, A. 1968. The spectrum of genetic variation. Pp. 5–16 *in* R. C. Lewontin (ed.) *Population Biology and Evolution*. Syracuse University Press, Syracuse, NY.

Robertson, D. R. 1996. Interspecific competition controls abundance and habitat use of territorial Caribbean damselfishes. Ecology 77: 885–899.

Robertson, F. W. 1966a. A test of sexual isolation in *Drosophila*. Genet. Res. Camb. 8: 181–187.

Robertson, F. W. 1966b. The ecological genetics of growth in *Drosophila*. 8. Adaptation to a new diet. Genet. Res. Camb. 8: 165–179.

Robertson, J. L. and R. Wyatt. 1990. Evidence for pollination ecotypes in the yellow-fringed orchid, *Platanthera ciliaris*. Evolution 44: 121–133.

Robinson, S. K., and J. Terborgh. 1995. Interspecific aggression and habitat selection by Amazonian birds. J. Animal Ecology 64: 1–11.

Robson, G. C. 1928. *The Species Problem*. Oliver and Boyd, London.

Rockwood, E. S., C. G. Kanapi, M. R. Wheeler, and W. S. Stone. 1971. Allozyme changes during the evolution of Hawaiian *Drosophila*. Studies in Genetics VI. Univ. Tex. Pub. 7103: 193–212.

Rodriguez, D. J. 1996. A model for the establishment of polyploidy in plants. Am. Nat. 147: 33–46.

Roelofs, W., T. Glover, X.-H. Tang, I. Sreng, P. Robbins, C. Eckenrode, C. Löfstedt, B. S. Hansson, and B. Bengtson. 1987. Sex pheromone production and perception in European corn borer moths is determined by both autosomal and sex-linked genes. Proc. Natl. Acad. Sci. USA 84: 7585–7589.

Roelofs, W. L. and A. P. Rooney. 2003. Molecular genetics and evolution of pheromone biosynthesis in Lepidoptera. Proc. Natl. Acad. Sci. USA 100: 9179–9184.

Roelofs, W. L., J.-W. Du, X.-H. Tang, P. S. Robbins, and C. J. Eckenrode. 1985. Three European corn borer populations in New York based on sex pheromones and voltinism. J. Chem. Ecol. 11: 829–836.

Roelofs, W. L., W. Liu, G. Hao, H. Jiao, A. P. Rooney, and C. E. Linn. 2002. Evolution of moth sex pheromones via ancestral genes. Proc. Natl. Acad. Sci. USA 99: 13621–13626.

Rolan-Alvarez, E. and A. Caballero. 2000. Estimating sexual selection and sexual isolation effects from mating frequencies. Evolution 54: 30–36.

Rose, M. and W. F. Doolittle. 1983. Molecular biological mechanisms of speciation. Science 220: 157–162.

Rosen, D. E. 1979. Fishes from the uplands and intermontane basins of Guatemala: Revisionary studies and comparative geography. Bull. Am. Mus. Nat. Hist. 162: 267–376.

Ross, C. L. and R. G. Harrison. 2002. A fine-scale spatial analysis of the mosaic hybrid zone between *Gryllus firmus* and *Gryllus pennsylvanicus*. Evolution 56: 2296–2312.

Rousset, F., M. Raymond, and F. Kjellberg. 1991. Cytoplasmic incompatibilities in the mosquito *Culex pipiens*: How to explain a cytotype polymorphism? J. Evol. Biol. 4: 69–81.

Rouyer, F., M.-C. Simmler, C. Johnsson, G. Vergnaud, H. J. Cooke, and J. Weissenbach. 1986. A gradient of sex linkage on the pseudoautosomal region of the human sex chromosomes. Nature 319: 291–295.

Rowntree, V. J. 1996. Feeding, distribution, and reproductive behavior of cyamids (Crustacea: Amphipoda) living on humpback and right whales. Can. J. Zool. 74: 103–109.

Roy, K., J. W. Valentine, D. Jablonski, and S. M. Kidwell. 1996. Scales of climatic variability and time averaging in Pleistocene biotas: Implications for ecology and evolution. Trends Ecol. Evol. 11: 458–463.

Rüber, L., A. Meyer, C. Sturmbauer, and E. Verheyen. 2001. Population structure in two sympatric species of the Lake Tanganyika cichlid tribe Eretmodini: Evidence for introgression. Mol. Ecol. 10: 1207–1225.

Rubinoff, I. 1968. Central American sea-level canal: Possible biological effects. Science 161:857–861.

Rundle, H. D. 2002. A test of ecologically dependent postmating isolation between sympatric sticklebacks. Evolution 56: 322–329.

Rundle, H. D. 2003. Divergent environments and population bottlenecks fail to generate premating isolation in *Drosophila pseudoobscura*. Evolution 57: 2557–2565.

Rundle, H. D. and D. Schluter. 1998. Reinforcement of stickleback mate preferences: Sympatry breeds contempt. Evolution 52: 200–208.

Rundle, H. D., A. O. Mooers, and M. C. Whitlock. 1998. Single founder-flush events and the evolution of reproductive isolation. Evolution 52: 1850–1855.

Rundle, H. D., L. Nagel, J. W. Boughman, and D. Schluter. 2000. Natural selection and parallel speciation in sympatric sticklebacks. Science 287: 306–308.

Russell, S. T. 2003. Evolution of intrinsic post-zygotic reproductive isolation in fish. Ann. Zool. Fennici 40: 321–329.

Ryan, M. J. 1990. Sensory systems, sexual selection, and sensory exploitation. Oxford Surv. Evol. Biol. 7: 157–195.

Ryan, M. J. 1998. Sexual selection, receiver biases, and the evolution of sex differences. Science 281: 1999–2003.

Ryan, M. J. 2001. Seeing red in speciation. Nature 411: 900–901.

Ryan, M. J. and A. S. Rand. 1993a. Sexual selection and signal evolution: The ghost of biases past. Philos. Trans. R. Soc. Lond. B 340: 187–195.

Ryan, M. J. and A. S. Rand. 1993b. Species recognition and sexual selection as a unitary problem in animal communication. Evolution 47: 647–657.

Ryan, M. J. and W. Wagner. 1987. Asymmetries in mating preferences between species: Female swordtails prefer heterospecific mates. Science 236: 595–597.

Rykena, S. 1991. Hybridization experiments as tests for species boundaries in the genus *Lacerta* sensu stricto. Mitt. Zool. Mus. Berl. 67: 55–68.

Saetre, G.-P., T. Moum, S. Bures, M. Kral, M. Adamjan, and J. Moreno. 1997. A sexually selected character displacement in flycatchers reinforces premating isolation. Nature 387: 589–592.

Saetre, G.-P., T. Borge, K. Lindroos, J. Haavbie, B. C. Sheldon, C. Primmer, and A.-C. Syvänen. 2003. Sex chromosome evolution and speciation in flycatchers. Proc. R. Soc. Lond. B 270: 53–59.

Sage, R. D., D. Heyneman, K.-C. Lim, and A. C. Wilson. 1986. Wormy mice in a hybrid zone. Nature 324: 60–63.

Said, K., A. Saad, J. C. Auffray, and J. Britton-Davidian. 1993. Fertility estimates in the Tunisian all-acrocentric and Robertsonian populations of the house mouse and their chromosomal hybrids. Heredity 71: 532–538.

Sainz, A., J. A. Wilder, M. Wolf, and H. Hollocher. 2003. *Drosophila melanogaster* and *D. simulans* rescue strains produce fit offspring, despite divergent centromere-specific histone alleles. Heredity 91: 28–35.

Sanderson, M. J. and M. J. Donoghue. 1994. Shifts in diversification rate with the origin of angiosperms. Science 264: 1590–1593.

Sanderson, M. J. 2002. Estimating absolute rates of molecular evolution and divergence times: A penalized likelihood approach. Molecular Biology and Evolution 19: 101–109.

Sanderson, N. 1989. Can gene flow prevent reinforcement? Evolution 43: 1223–1235.

Sanderson, N., J. M. Szymura, and N. H. Barton. 1992. Variation in mating call across the hybrid zone between the fire-bellied toads *Bombina bombina* and *B. variegata*. Evolution 46: 595–607.

Sandler, L. and E. Novitski. 1957. Meiotic drive as an evolutionary force. Am. Nat. 91: 105–110.

Sanford, W. W. 1968. Distribution of epiphytic orchids in semi-deciduous tropical forest in Southern Nigeria. J. Ecol. 56: 697–705.

Sang, T., D. J. Crawford, and T. F. Stuessy. 1995. Documentation of reticulate evolution in peonies (*Paeonia*) using internal transcribed spacer sequences of nuclear ribosomal DNA: Implications for biogeography and concerted evolution. Proc. Natl. Acad. Sci. USA 92: 6813–6817.

Sasa, M., P. T. Chippendale, and N. A. Johnson. 1998. Patterns of postzygotic isolation in frogs. Evolution 52: 1811–1820.

Sasaki, T., T. Kubo, and H. Ishikawa. 2002. Interspecific transfer of *Wolbachia* between two lepidopteran insects expressing cytoplasmic incompatibility: A *Wolbachia* variant naturally infecting *Cadra cautella* causes male killing in *Ephestia kuehniella*. Genetics 162: 1313–1319.

Sattler, G. D. and M. J. Braun. 2000. Morphometric variation as an indicator of genetic interactions between black-capped and Carolina chickadees at a contact zone in the Appalachian Mountains. Auk 117: 427–444.

Sauer, J. D. 1990. Allopatric speciation: Deduced but not detected. J. Biogeogr. 17: 1–3.

Saumitou-Laprade, P., J. Cuguen, and P. Vernet. 1994. Cytoplasmic male sterility in plants—molecular evidence and the nucleocytoplasmic conflict. Trends Ecol. Evol. 9: 431–435.

Savolainen, R. and K. Vespsäläinen. 2003. Sympatric speciation through intraspecific social parasitism. Proc. Natl. Acad. Sci. USA 100: 7169–7174.

Sawamura, K. and M.-T. Yamamoto. 1993. Cytogenetical localization of *Zygotic hybrid rescue* (*Zhr*), a *Drosophila melanogaster* gene that rescues interspecific hybrids from embryonic lethality. Molec. Gen. Genet. 239: 441–449.

Sawamura, K. and M.-T. Yamamoto. 1997. Characterization of a reproductive isolation gene, *Zygotic hybrid rescue*, of *Drosophila melanogaster* by using minichromosomes. Heredity 79: 97–103.

Sawamura, K., T. Taira, and T. K. Watanabe. 1993a. Hybrid lethal systems in the *Drosophila melanogaster* species complex. I. The *maternal hybrid rescue* (*mhr*) gene of *Drosophila simulans*. Genetics 133: 299–305.

Sawamura, K., T. K. Watanabe, and M.-T. Yamamoto. 1993b. Hybrid lethal systems in the *Drosophila melanogaster* species complex. Genetica 88: 175–185.

Sawamura, K., M.-T. Yamamoto, and T. K. Watanabe. 1993c. Hybrid lethal systems in the *Drosophila melanogaster* species complex. II. The *Zygotic hybrid rescue* (*Zhr*) gene of *D. melanogaster*. Genetics 133: 307–313.

Sawamura, K., A. W. Davis, and C.-I. Wu. 2000. Genetic analysis of speciation by means of introgression into *Drosophila melanogaster*. Proc. Natl. Acad. Sci. USA 97: 2652–2655.

Sawamura, K., J. Roote, C.-I. Wu, and M.-T. Yamamoto. 2004. Genetic complexity underlying hybrid male sterility in Drosophila. Genetics 166: 789–796.

Sawyer, S. A. and D. L. Hartl. 1981. On the evolution of behavioral reproductive isolation: The Wallace effect. Theor. Popul. Biol. 19: 261–273.

Schaeffer, S. W. and E. L. Miller. 1991. Nucleotide sequence analysis of Adh gene estimates the time of geographic isolation of the Bogota population of *Drosophila pseudoobscura*. Proc. Natl. Acad. Sci. USA 1991: 6097–6101.

Schaeffer, S. W. and E. L. Miller. 1992. Estimates of gene flow in *Drosophila pseudooscura* determined from nucleotide sequence analysis of the alcohol dehydrogenase region. Genetics 132: 471–480.

Schäfer, U. 1978. Sterility in *Drosophila hydei* X *D. neohydei* hybrids. Genetica 49: 205–214.

Schäfer, U. 1979. Viability in *Drosophila hydei* X *D. neohydei* hybrids and its regulation by genes located in the sex heterochromatin. Biol. Zent. Bl. 98: 153–161.

Scharloo, W. 1971. Reproductive isolation by disruptive selection: Did it occur? Am. Nat. 105: 83–86.

Schartl, M. 1995. Platyfish and swordtails: A genetic system for the analysis of molecular mechanisms in tumor formation. Trends Genet. 11: 185–189.

Schartl, M., U. Hornung, H. Gutbrod, J.-N. Volff, and J. Wittbrodt. 1999. Melanoma loss-of-function mutants in *Xiphophorus* caused by *Xmrk*-oncogene deletion and gene disruption by a transposable element. Genetics 153: 1385–1394.

Schemske, D. W. 1981. Floral convergence and pollinator sharing in two bee-pollinated tropical herbs. Ecology 62: 946–954.

Schemske, D. W. 2000. Understanding the origin of species. Evolution 54: 1069–1073.

Schemske, D. W. 2002. Ecological and evolutionary perspectives on the origins of tropical diversity. Pp. 163–173 *in* R. L. Chazdon and T. C. Whitmore (eds.) *Foundations of Tropical Forest Biology: Classic Papers with Commentaries*. University of Chicago Press, Chicago, IL.

Schemske, D. W. and H. D. Bradshaw. 1999. Pollinator preference and the evolution of floral traits in monkeyflowers (*Mimulus*). Proc. Natl. Acad. Sci. USA 96: 11910–11915.

Schemske, D. W. and C. Goodwillie. 1996. Morphological and reproductive characteristics of a *Linanthus jepsonii* (Polemoniacea), a newly described, geographically restricted species from northern California. Madroño 43: 453–463.

Schemske, D. W. and C. C. Horvitz. 1984. Variation among floral visitors in pollination ability: A precondition for mutualism specialization. Science 225: 519–521.

Schilthuizen, M. 2000. Dualisms and conflicts in understanding speciation. BioEssays 22: 1134–1141.

Schliewen, U. K., D. Tautz, and S. Pääbo. 1994. Sympatric speciation suggested by monophyly of crater lake cichlids. Nature 368: 629–632.

Schliewen, U., K. Rassmann, M. Markmann, J. Markert, T. Kocher, and D. Tautz. 2001. Genetic and ecological divergence of a monophyletic cichlid species pair under fully sympatric conditions in Lake Ejagham, Cameroon. Mol. Ecol. 10: 1471–1488.

Schlötterer, C. and M. Agis. 2002. Microsatellite analysis of *Drosophila melanogaster* populations along a microclimatic contrast at Lower Nahel Oren Canyon, Mount Carmel. Mol. Biol. Evol. 19: 563–568.

Schluter, D. 1993. Adaptive radiation in sticklebacks: Size, shape, and habitat use efficiency. Ecology 74: 699–709.

Schluter, D. 1995. Adaptive radiation in sticklebacks: Trade-offs in feeding performance and growth. Ecology 76: 82–90.

Schluter, D. 1996a. Ecological causes of adaptive radiation. Am. Nat. 148 (Suppl.): S40–S64.

Schluter, D. 1996b. Ecological speciation in postglacial fishes. Phil. Trans. R. Soc. Lond B 351: 807–814.

Schluter, D. 1998. Ecological causes of speciation. Pp. 114–129 in D. J. Howard and S. H. Berlocher (eds.), *Endless Forms: Species and Speciation*. Oxford University Press, Oxford.

Schluter, D. 2000. *The Ecology of Adaptive Radiation*. Oxford University Press, Oxford.

Schluter, D. and J. D. McPhail. 1992. Ecological character displacement and speciation in sticklebacks. Am. Nat. 140: 85–108.

Schluter, D. M. and L. M. Nagel. 1995. Parallel speciation by natural selection. Am. Nat. 146: 292–301.

Schluter, D. and T. Price. 1993. Honesty, perception, and population divergence in sexually selected traits. Proc. R. Soc. Lond. B 253: 117–122.

Schmitz, U. K. and G. Michaelis. 1988. Dwarfism and male sterility in interspecific hybrids of *Epilobium*. 2. Expression of mitochondrial genes and structure of the mitochondrial DNA. Theor. Appl. Genet. 76: 565–569.

Schnable, P. S. and R. P. Wise. 1998. The molecular basis of cytoplasmic male sterility and fertility restoration. Trends Plant Sci. 3: 175–180.

Schwarzbach, A. E. and L. H. Rieseberg. 2002. Likely multiple origins of a diploid hybrid sunflower species. Mol. Ecol. 11: 1703–1715.

Schwarzbach, A. E., L. A. Donovan, and L. H. Rieseberg. 2001. Transgressive character expression in a hybrid sunflower species. Am. J. Bot. 88: 270–277.

Searle, J. B. 1988. Selection and Robertsonian variation in nature: The case of the common shrew. Pp. 507–531 in A. Daniel (ed.) *The Cytogenetics of Mammalian Autosomal Rearrangements*. Alan R. Liss, New York.

Searle, J. B. 1993. Chromosomal hybrid zones in eutherian mammals. Pp. 309–353 in R. G. Harrison (ed.) *Hybrid Zones and the Evolutionary Process*. Oxford University Press, New York.

Seehausen, O. 1996. *Lake Victoria Rock Cichlids: Taxonomy, Ecology, and Distribution*. Verduijn Cichlids, Zevenhuizen, Germany.

Seehausen, O. 2000. Explosive speciation rates and unusual species richness in haplochromine cichlid fishes: Effects of sexual selection. Adv. Ecol. Res. 31: 237–274.

Seehausen, O. 2002. Patterns in fish radiation are compatible with Pleistocene desiccation of Lake Victoria and its 14600 year history for its cichlid species flock. Proc. R. Soc. Lond. B 267: 491–497.

Seehausen, O. and J. J. M. van Alphen. 1999. Can sympatric speciation by disruptive selection explain rapid evolution of cichlid diversity in Lake Victoria? Ecol. Lett. 2: 262–271.

Seehausen, O., J. J. M. van Alphen, and F. Witte. 1997. Cichlid fish diversity threatened by eutrophication that curbs sexual selection. Science 277: 1808–1811.

Seehausen, O., F. Witte, J. J. M. van Alphen, and N. Bouton. 1998. Direct mate choice maintains diversity among sympatric cichlids in lake Victoria. J. Fish Biol. 53(Suppl. A): 37–55.

Seehausen, O., J. J. M. van Alphen, and R. Lande. 1999. Color polymorphism and sex ratio distortion in a cichlid fish as an incipient stage in sympatric speciation by sexual selection. Ecology Letters 2: 367–378.

Seehausen, O., E. Koetsier, M. V. Schneider, L. J. Chapman, C. A. Chapman, M. E. Knight, G. F. Turner, J. J. M. van Alphen, and R. Bills. 2003. Nuclear markers reveal unexpected genetic variation and a Congolese-Nilotic origin of the Lake Victoria cichlid species flock. Proc. R. Soc. Lond. B 270: 129–137.

Segraves, K. A. and J. N. Thompson. 1999. Plant polyploidy and pollination: Floral traits and insect visits to diploid *Heuchera grossularifolia*. Evolution 53: 1114–1127.

Sene, F. M. and H. L. Carson. 1977. Genetic variation in Hawaiian *Drosophila*. IV. Allozymic similarity between *D. silvestris* and *D. heteroneura* from the island of Hawaii. Genetics 86: 187–198.

Seoighe, C. and K. H. Wolfe. 1998. Extent of genomic rearrangement after genome duplication. Proc. Natl. Acad. Sci. USA 95: 4447–4452.

Sepkoski, J. J. 1999. Rates of speciation in the fossil record. Pp. 260–282 *in* A. E. Magurran and R. M. May (eds.) *Evolution of Biological Diversity*. Oxford University Press, Oxford.

Sepp, S. and J. Paal. 1998. Taxonomic continuum of *Alchemilla* (Rosaceae) in Estonia. Nord. J. Bot. 18: 519–535.

Servedio, M. R. 2000. Reinforcement and the genetics of nonrandom mating. Evolution 54: 21–29.

Servedio, M. R. 2001. Beyond reinforcement: the evolution of premating isolation by direct selection on preferences and postmating, prezygotic incompatibilities. Evolution 55: 1909–1920.

Servedio, M. R. and M. Kirkpatrick. 1997. The effects of gene flow on reinforcement. Evolution 51: 1764–1772.

Servedio, M. R. and M. A. F. Noor. 2003. The role of reinforcement in speciation: Theory and data. Annu. Rev. Ecol. Syst. 34: 339–364.

Shapiro, A. M. and A. H. Porter. 1989. The lock-and-key hypothesis: Evolutionary and biosystematic interpretation of insect genitalia. Annu. Rev. Entomol. 34: 231–245.

Shaw, K. L. 1996. Polygenic inheritance of a behavioral phenotype: Interspecific genetics of song in the Hawaiian cricket genus *Laupala*. Evolution 50: 256–266.

Shaw, K. L. 1998. Species and the diversity of natural groups. Pp. 44–56 *in* D. J. Howard and S. J. Berlocher (eds.) *Endless Forms: Species and Speciation*. Oxford University Press, Oxford.

Shaw, K. L. 2001. The genealogical view of speciation. J. Evol. Biol. 14: 880–882.

Shaw, K. L. 2002. Conflict between nuclear and mitochondrial DNA phylogenies of a recent species radiation: What mitochondrial DNA reveals and conceals about modes of speciation in Hawaiian crickets. Proc. Natl. Acad. Sci. USA 99: 16122–16129.

Shaw, P. W., G. F. Turner, M. R. Idid, R. L. Robinson, and G. R. Carvalho. 2000. Genetic population structure indicates sympatric speciation of Lake Malawi pelagic cichlids. Proc. R. Soc. Lond. B 267: 2273–2280.

Shine, R., R. N. Reed, S. Shetty, M. Lemaster, and R. T. Mason. 2002. Reproductive isolating mechanisms between two sympatric sibling species of sea snakes. Evolution 56: 1655–1662.

Shoemaker, D. D., V. Katju, and J. Jaenike. 1999. *Wolbachia* and the evolution of reproductive isolation between *Drosophila recens* and *Drosophila subquinaria*. Evolution 53: 1157–1164.

Sidow, A. 1996. Gen(om)e duplications in the evolution of early vertebrates. Curr. Opin. Genet. Dev. 6: 715–722.

Silberglied, R. E. and O. R. Taylor. 1978. Ultraviolet reflection and its behavioral role in the courtship of the sulphur butterflies, *Coliqas eurytheme* and *C. philodice* (Lepidoptera: Piridae). Behav. Ecol. Sociobiol. 3: 203–243.

Simmons, L. W. 2001. *Sperm Competition and its Evolutionary Consequences in the Insects*. Princeton University Press, Princeton, NJ.

Simmons, L. W. and M. T. Siva-Jothy. 1998. Sperm competition in insects: Mechanisms and the potential for selection. Pp. 341–434 *in* T. R. Birkhead and A. P. Møller (eds.) *Sperm Competition and Sexual Selection*. Academic Press, London.

Simon, J. C., S. Carre, M. Boutin, N. Prunier-Leterme, B. Sabater-Munoz, A. Latorre, and R. Bournoville. 2003. Host-based divergence in populations of the pea aphid: insights from nuclear markers and the prevalence of facultative symbionts. Proc. Roy. Soc. Lond. B, Biol. Sci. 270: 1703-1712.

Simpson, G. G. 1961. *Principles of Animal Taxonomy*. Columbia University Press, New York.

Sironi, M., C. Badi, L. Sacchi, B. Di Sacco, G. Damiani, and C. Genchi. 1995. Molecular evidence for a close relative of the arthropod endosymbiont *Wolbachia* in a filarial worm. Mol. Biochem. Parasitol. 74: 223–227.

Sites, J. W., Jr. and C. Moritz. 1987. Chromosomal evolution and speciation revisited. Syst. Zool. 36: 153–174.

Skúlason, S. and T. B. Smith. 1996. Resource polymorphisms in vertebrates. Trends Ecol. Evol. 10: 366–370.

Skúlason, S., S. Snorrason, and B. Jónsson. 1999. Sympatric morphs, populations, and speciation in freshwater fish with emphasis on arctic charr. Pp. 70–92 *in* A. E. Magurran and R. M. May (eds.) *Evolution of Biological Diversity*. Oxford University Press, Oxford.

Slatkin, M. 1973. Gene flow and selection in a cline. Genetics 75: 733–756.

Slatkin, M. 1982. Pleiotropy and parapatric speciation. Evolution 36: 263–270.

Slatkin, M. 1985. Gene flow in natural populations. Annu. Rev. Ecol. Syst. 16: 393–430.

Slowinski, J. B. and C. Guyer. 1989. Testing the stochasticity of patterns of organismal diversity: An improved null model. Am. Nat. 134: 907–921.

Slowinski, J. B. and C. Guyer. 1993. Testing whether certain traits have caused amplified diversification: An improved method based on a model of random speciation and extinction. Am. Nat. 142: 1019–1024.

Smith, D. C. 1988. Heritable divergence of *Rhagoletis pomonella* host races by seasonal asynchrony. Nature 336: 66–67.

Smith, G. R. 1992. Introgression in fishes: Significance for paleontology, cladistics, and evolutionary rates. Syst. Biol. 41: 41–57.

Smith, N. H., E. C. Holmens, G. M. Donovan, G. A. Carpenter, and B. G. Spratt. 1999. Networks and groups within the genus *Neisseria*: Analysis of *argF*, *recA*, *rho*, and 16S rRNA sequences from human *Neisseria* species. Mol. Biol. Evol. 16: 773–783.

Smith, T. B. 1987. Bill size polymorphism and interspecific niche utilization in an African finch. Nature 329: 717–719.

Smith, T. B. and S. Skúlason. 1996. Evolutionary significance of resource polymorphisms in fishes, amphibians, and birds. Annu. Rev. Ecol. Syst. 27: 111–133.

Smith, T. B., R. K. Wayne, D. J. Girman, and M. W. Bruford. 1997. A role for ecotones in generating rainforest biodiversity. Science 276: 1855–1857.

Sniegowski, P. 1998. Mismatch repair: origin of species? Curr. Biol. 8: 59–61.

Soans, A. B., D. Pimentel, and J. S. Soans. 1974. Evolution of reproductive isolation in allopatric and sympatric populations. Am. Nat. 108: 117–124.

Sokal, R. R. and T. J. Crovello. 1970. The Biological Species Concept: A critical evaluation. Am. Nat. 104: 127–153.

Soltis, D. E. and P. S. Soltis. 1993. Molecular data and the dynamic nature of polyploidy. Crit. Rev. Plant Sci. 12: 243–275.

Soltis, D. E. and P. S. Soltis. 1999. Polyploidy: recurrent formation and genome evolution. Trends Ecol. Evol. 14: 348–352.

Soltis, P. S. and D. E. Soltis. 2000. The role of genetic and genomic attributes in the success of polyploids. Proc. Natl. Acad. Sci. USA 97: 7051–7057.

Somerson, N. L., L. Ehrman, J. P. Kocka, and F. J. Gottlieb. 1984. Streptococcal L-forms isolated from *Drosophila paulistorum* semispecies cause sterility in male progeny. Proc. Natl. Acad. Sci. USA 81: 282–285.

Song, B. H., X. Q. Wang, X.-R. Wang, L. J. Sun, D. Y. Hong, and P. H. Peng. 2002. Maternal lineages of *Pinus densata*, a diploid hybrid. Mol. Ecol. 11: 1057–1063.

Sorci, G., A. P. Møller, and J. Clobert. 1998. Plumage dichromatism of birds predicts introduction success in New Zealand. J. Anim. Ecol. 67: 263–269.

Sorenson, M. D., K. M. Sefc, and R. B. Payne. 2003. Speciation by host switch in brood parasitic indigobirds. Nature 424: 928–931.

Sota, T. and K. Kubota. 1998. Genital lock-and-key as a selective agent against hybridization. Evolution 52: 1507–1513.

Spencer, H. G., B. H. McArdle, and D. M. Lambert. 1986. A theoretical investigation of speciation by reinforcement. Am. Nat. 128: 241–262.

Spencer, H. G., D. M. Lambert, and B. H. McArdle. 1987. Reinforcement, species, and speciation: A reply to Butlin. Am. Nat. 130: 958–962.

Spiess, E. B. and C. M. Wilke. 1984. Still another attempt to achieve assortative mating by disruptive selection in *Drosophila*. Evolution 38: 505–515.

Spirito, F. 2000. The role of chromosomal rearrangements in speciation. Pp. 320–329 in D. J. Howard and S. H. Berlocher (eds.) *Endless Forms: Species and Speciation*. Oxford University Press, Oxford, UK.

Springer, S. A. and B. J. Crespi. 2004. Rapid evolution of a gamete-recognition protein in a hybrid *Mytilus* population. Unpublished manuscript.

Spurway, H. 1953. Genetics of specific and subspecific differences in European newts. Symp. Soc. Exp. Biol. 7: 200–237.

Stam, E. 2002. Does imbalance in phylogenies reflect only bias? Evolution 56: 1292–1295.

Stanley, S. M. 1998. *Macroevolution: Pattern and Process*. Johns Hopkins University Press, Baltimore, MD.

Stattersfield, A. J., M. J. Crosby, A. J. Long, and D. C. Wege. 1998. *Endemic Bird Areas of the World: Priorities for Biological Conservation*. BirdLife International, Cambridge, UK.

Stauffer R. C. (ed.) 1975. *Charles Darwin's Natural Selection, Being the Second Part of His Big Species Book Written From 1856 to 1858*. Cambridge University Press, Cambridge.

Stebbins, G. L. 1938. Cytological characteristics associated with the different growth habits in the dicotyledons. Am. J. Bot. 25: 189–198.

Stebbins, G. L. 1950. *Variation and Evolution in Plants*. Columbia University Press, New York.

Stebbins, G. L. 1957a. Self-fertilization and population variability in the higher plants. Am. Nat. 91: 337–354.

Stebbins, G. L. 1957b. The hybrid origin of microspecies in the *Elymus glaucus* complex. Cytologia (Suppl.) 36: 336–340.

Stebbins, G. L. 1958. The inviability, weakness, and sterility of interspecific hybrids. Adv. Genet. 9: 147–216.

Stebbins, R. C. 1949. Speciation in salamanders of the plethodontid genus *Ensatina*. Univ. Cal. Pub. Zool. 54: 47–124.

Stephano, J. L. 1992. A study of polyspermy in abalone. Pp. 518–526 in S. A. Shepherd, M. J. Tegner, and S. A. Guzman del Proo (eds.) *Abalone of the World, Biology, Fisheries and Culture*. Fishing News Books (Blackwell), Cambridge, UK.

Stephens, S. G. 1946. The genetics of "Corky." I. The new world alleles and their possible role as an interspecific isolating mechanism. J. Genet. 47: 150–161.

Stephens, S. G. 1950. The genetics of "Corky." II. Further studies on its genetic basis in relation to the general problem of interspecific isolating mechanisms. J. Genet. 50: 9–20.

Sternburg, J. G. and G. P. Waldbauer. 1978. Phenological adaptations in diapause termination by *Cecropia* from different latitudes. Entomol. Exp. Appl. 23: 48–54.

Stewart, W. N. and G. W. Rothwell. 1993. *Paleobotany and the Evolution of Plants* (Second Edition). Cambridge University Press, Cambridge.

Stiassny, M., U. K. Schliewen, and W. Dominey. 1992. A new species flock of cichlid fishes from Lake Bermin, Cameroon, with a description of eight new species of *Tilapia* (Labroidei: Cichlidae). Ichthyol. Explor. Freshwaters 3: 311–346.

Stouthmaer, R., J. A. Breeuwer, and G. D. D. Hurst. 1999. *Wolbachia pipientis*: Microbial manipulator of

arthropod reproduction. Annu. Rev. Microbiol. 53: 71–102.
Stratton, G. E. and G. W. Uetz. 1986. The inheritance of courtship behavior and its role as a reproductive isolating mechanism in two species of *Schizocosa* wolf spiders (Araneae; Lycosidae). Evolution 40: 129–141.
Straw, R. M. 1955. Hybridization, homogamy, and sympatric speciation. Evolution 9: 441–444.
Strong, D. R., J. H. Lawton, and T. R. E. Southwood. 1984. *Insects on Plants: Community Patterns and Mechanisms*. Harvard University Press, Cambridge, MA.
Sturmbauer, C. and A. Meyer. 1993. Mitochondrial phylogeny of the endemic mouthbrooding lineages of cichlid fishes from Lake Tanganyika in Eastern Africa. Mol. Biol. Evol. 10:751–768.
Sturmbauer, C., S. Baric, W. Salzburger, L. Ruber, and E. Verheyen. 2001. Lake level fluctuations synchronize genetic divergences of cichlid fishes in African lakes. Mol. Biol. Evol. 18:144–154.
Sturtevant, A. H. 1920. Genetic studies on *Drosophila simulans*. I. Introduction. Hybrids with *Drosophila melanogaster*. Genetics 5: 488–500.
Sturtevant, A. H. 1938. Essays on evolution. III. On the origin of interspecific sterility. Q. Rev. Biol. 13: 333–335.
Sturtevant, A. H. 1946. Intersexes dependent on a maternal effect in hybrids between *Drosophila repleta* and *D. neorepleta*. Proc. Natl. Acad. Sci. USA 32: 84–87.
Sturtevant, A. H. 1956. A highly specific complementary lethal system in *Drosophila melanogaster*. Genetics 41: 118–123.
Sulloway, F. J. 1979. Geographic isolation in Darwin's thinking: The vicissitudes of a crucial idea. Stud. Hist. Biol. 3: 23–65.
Sultan, S. E. 2000. Phenotypic plasticity for plant development, function, and life history. Trends Plant Sci. 5: 537–542.
Sved, J. A. 1981. A two-sex polygenic model for the evolution of premating isolation. I. Deterministic theory for natural populations. Genetics 97: 197–215.
Swanson, W. J. and V. D. Vacquier. 1997. The abalone egg vitelline envelope receptor for sperm lysin is a giant multivalent molecule. Proc. Natl. Acad. Sci. USA 94: 6724–6729.
Swanson, W. J. and V. D. Vacquier. 1998. Concerted evolution in an egg receptor for a rapidly evolving abalone sperm protein. Science 281: 710–712.
Swanson, W. J. and V. D. Vacquier. 2002a. Reproductive protein evolution. Annu. Rev. Ecol. Syst. 33: 161–179.
Swanson, W. J. and V. D. Vacquier. 2002b. The rapid evolution of reproductive proteins. Nat. Rev. Genet. 3: 137–144.
Swanson, W. J., A. G. Clark, H. M. Waldrip-Dail, M. F. Wolfner, and C. F. Aquadro. 2001a. Evolutionary EST analysis identifies rapidly evolving male reproductive proteins in *Drosophila*. Proc. Natl. Acad. Sci. USA 98: 7375–7379.

Swanson, W. J., Z. Yang, M. F. Wolfner, and C. F. Aquadro. 2001b. Positive Darwinian selection drives the evolution of several female reproductive proteins in mammals. Proc. Natl. Acad. Sci. U.S.A. 98: 2509–2514.
Swanson, W. J., A. Wong, M. F. Wolfner, and C. F. Aquadro. 2004. Evolutionary EST analysis of *Drosophila* female reproductive tracts identifies several genes subjected to positive selection. Unpublished ms.
Szymura, J. M. and N. H. Barton. 1986. Genetic analysis of a hybrid zone between the fire-bellied toads *Bombina bombina* and *B. variegata*, near Cracow in Southern Poland. Evolution 40: 1141–1159.
Szymura, J. M. and N. H. Barton. 1991. The genetic structure of the hybrid zone between the fire-bellied toads *Bombina bombina* and *B. variegata*: comparisons between transects and between loci. Evolution 45: 237–261.

Takahata, N. and M. Slatkin. 1984. Mitochondrial gene flow. Proc. Natl. Acad. Sci. USA 81:1764–1767.
Tal, M. 1980. Physiology of polyploids *in* W. H. Lewis (ed.) *Polyploidy: Biological Relevance*. Plenum Press, New York.
Tao, Y. and D. L. Hartl. 2003. Genetic dissection of hybrid incompatibilities between *Drosophila simulans* and *Drosophila mauritiana*. III. Heterogeneous accumulation of hybrid incompatibilities, degree of dominance and implications for Haldane's rule. Evolution 57: 2580–2598.
Tao, Y., D. L. Hartl, and C. C. Laurie. 2001. Sex-ratio distortion associated with reproductive isolation in *Drosophila*. Proc. Natl. Acad. Sci. USA 98: 13183–13188.
Tao, Y., S. Chen, D. L. Hartl, and C. C. Laurie. 2003a. Genetic dissection of hybrid incompatibilities between *Drosophila simulans* and *D. mauritiana*. I. Differential accumulation of hybrid male sterility effects on the *X* and autosomes. Genetics 164: 1383–1397.
Tao, Y., Z.-B. Zeng, J. Li, D. L. Hartl, and C. C. Laurie. 2003b. Genetic dissection of hybrid incompatibilities between *Drosophila simulans* and *D. mauritiana*. II. Mapping hybrid male sterile loci on the third chromosome. Genetics 164: 1399–1418.
Tauber, C. A. and M. J. Tauber. 1977. Sympatric speciation based on allelic changes at three loci: Evidence from natural populations in two habitats. Science 197: 1298–1299.
Tauber, C. A. and M. J. Tauber. 1989. Sympatric speciation in insects: Perception and perspective. Pp. 307–344 *in* D. Otte and J. A. Endler (eds.) *Speciation and Its Consequences*. Sinauer Associates, Sunderland, MA.
Tauber, C. A., M. J. Tauber, and J. R. Nichols. 1977. Two genes control seasonal isolation in sibling species. Science 197: 592–593.
Taylor, D. J., P. D. N. Hebert, and J. K. Colourne. 1996. Phylogenetics and evolution of the *Daphnia*

longispina group (Crustacea) based on 12S rDNA sequence and allozyme variation. Mol. Phylogenet. Evol. 5: 495–510.
Taylor, E. B. and J. D. McPhail. 1999. Evolutionary history of an adaptive radiation in species pairs of threespine sticklebacks *Gasterosteus*: Insights from mitochondrial DNA. Biol. J. Linn. Soc. 66: 271–291.
Taylor, E. B. and J. D. McPhail. 2000. Historical contingency and ecological determinism interact to prime speciation in sticklebacks, Gasterosteus. Proc. R. Soc. Lond. B 267: 2375–2384.
Taylor, E. B., C. J. Foote, and C. C. Wood. 1996. Molecular genetic evidence for parallel life history evolution within a Pacific salmon (sockeye salmon and kokanee, *Oncorhynchus nerka*). Evolution 50: 401–416.
Taylor, E. B., J. D. McPhail, and D. Schluter. 1997. History of ecological selection in sticklebacks: uniting experimental and phylogenetic approaches. Pp. 511–534 *in* T. J. Givnish and K. J. Systma, (eds.) Molecular Evolution and Adaptive Radiation. Cambridge University Press, Cambridge.
Taylor, J. S., Y. Van de Peer, and A. Meyer. 2001. Genome duplication, divergent resolution and speciation. Trends Genet. 17: 299–301.
Telschow, A., P. Hammerstein, and J. H. Werren. 2002. The effect of *Wolbachia* on genetic divergence between populations: models with two-way migration. Am. Nat. 160: S54–S66.
Templeton, A. R. 1977. Analysis of head shape differences between two interfertile species of Hawaiian *Drosophila*. Evolution 31: 630–641.
Templeton, A. R. 1980a. Modes of speciation and inferences based on genetic distances. Evolution 34: 719–729.
Templeton, A. R. 1980b. The theory of speciation via the founder principle. Genetics 94: 1011–1038.
Templeton, A. R. 1981. Mechanisms of speciation—a population genetics approach. Annu. Rev. Ecol. Syst. 12: 23–48.
Templeton, A. R. 1989. The meaning of species and speciation: A genetic perspective. Pp. 3–27 *in* D. Otte and J. A.Endler (eds.) *Speciation and Its Consequences*. Sinauer Associates, Sunderland, MA.
Templeton, A. R. 1999. Experimental tests of genetic transilience. Evolution 53: 1628–1633.
Thoday, J. M. 1972. Disruptive selection. Proc. R. Soc. Lond. B 182: 109–143.
Thoday, J. M. and J. B. Gibson. 1962. Isolation by disruptive selection. Nature 193: 1164–1166.
Thoday, J. M. and J. B. Gibson. 1970. The probability of isolation by disruptive selection. Am. Nat. 104: 219–230.
Thomas, C. 1878. A list of the species of the tribe Aphidini, family Aphidae, found in the United States, which have been heretofore named with descriptions of some new species. State Lab. Nat. Hist. Bull. 1: 1-16.
Thompson, J. D. and R. Lumaret. 1992. The evolutionary dynamics of polyploid plants: Origins, establishment and persistence. Trends Ecol. Evol. 7: 302–307.
Thompson, V. 1986. Synthetic lethals: a critical review. Evol. Theory 8: 1–13.
Thornton, K. 2002. Rapid divergence of gene duplicates on the *Drosophila melanogaster* X chromosome. Mol. Biol. Evol. 19: 918–925.
Tilley, S. G. and M. J. Mahoney. 1996. Patterns of genetic differentiation in salamanders of the *Desmognathus ochrophaeus* complex (Amphibia: Plethodontidae). Herpetological Monographs 10: 1–42.
Tilley, S. G., P. A. Verrell, and S. J. Arnold. 1990. Correspondence between sexual isolation and allozyme differentiation: A test in the salamander *Desmognathus ochrophaeus*. Proc. Natl. Acad. Sci. USA 87: 2715–2719.
Ting, C.-T., S.-C. Tsaur, M.-L. Wu, and C.-I. Wu. 1998. A rapidly evolving homeobox at the site of a hybrid sterility gene. Science 282: 1501–1504.
Ting, C. T., A. Takahashi, and C.-I. Wu. 2001. Incipient speciation by sexual selection: Concurrent evolution at multiple loci. Proc. Natl. Acad. Sci. USA 98: 6709–6713.
Tomaru, M. and Y. Oguma. 1994. Genetic basis and evolution of species-specific courtship song in the *Drosophila auraria* complex. Genet. Res. 63: 11–17.
Tomaru, M., H. Matsubayashi, and Y. Oguma. 1995. Heterospecific inter-pulse intervals of courtship song elicit female rejection in *Drosophila biauraria*. Anim. Behav. 50: 905–914.
Trewavas, E., J. Green, and S. A. Corbet. 1972. Ecological studies on crater lakes in West Cameroon. Fishes of Barombi Mbo. J. Zool. (Lond.) 167: 41–95.
Trivers, R. L. 1972. Parental investment and sexual selection. Pp. 136–179 *in* B. Campbell (ed.) *Sexual selection and the Descent of Man: 1871–1971*. Aldine Publishing Co., Chicago, IL.
True, J. R., B. S. Weir, and C. C. Laurie. 1996. A genome-wide survey of hybrid incompatibility factors by the introgression of marked segments of *Drosophila mauritiana* chromosomes into *Drosophila simulans*. Genetics 142: 819–837.
Tubaro, P. and D. A. Lijtmaer. 2002. Hybridization patterns and the evolution of reproductive isolation in ducks. Biol. J. Linn. Soc. (Lond.) 77: 193–200.
Tucker, P. K., R. D. Sage, J. Warner, A. C. Wilson, and E. M. Eicher. 1992. Abrupt cline for sex chromosomes in a hybrid zone between two species of mice. Evolution 46: 1146–1163.
Tuomisto, H. and K. Ruokolainen. 1997. The role of ecological knowledge in explaining biogeography and biodiversity in Amazonia. Biodivers. Conserv. 6: 347–357.
Tuomisto, H., K. Ruokolainen, R. Kalliola, A. Linna, W. Danjoy, and Z. Rodriguez. 1995. Dissecting Amazonian biodiversity. Science 269: 63–66.
Turelli, M. 1994. Evolution of incompatibility-inducing microbes and their hosts. Evolution 48: 1500–1513.

Turelli, M. and D. J. Begun. 1997. Haldane's rule and X-chromosome size in *Drosophila*. Genetics 147: 1799–1815.

Turelli, M. and A. A. Hoffmann. 1991. Rapid spread of an inherited incompatibility factor in California *Drosophila*. Nature 353: 440–442.

Turelli, M. and A. A. Hoffmann. 1995. Cytoplasmic incompatibility in *Drosophila simulans*: Dynamics and parameter estimates from natural populations. Genetics 140: 1319–1338.

Turelli, M. and H. A. Orr. 1995. The dominance theory of Haldane's rule. Genetics 140: 389–402.

Turelli, M. and H. A. Orr. 2000. Dominance, epistasis and the genetics of postzygotic isolation. Genetics 154: 1663–1679.

Turelli, M., N. H. Barton, and J. A. Coyne. 2001. Theory and speciation. Trends Ecol. Evol. 16: 330–343.

Turesson, G. 1922. The genotypical response of the plant species to the habitat. Hereditas 3:211–350.

Turesson, G. 1930. The selective effect of climate upon the plant species. Hereditas 14: 99–152.

Turner, B. J. 1974. Genetic divergence of Death Valley pupfish species: biochemical versus morphological evidence. Evolution 28: 281–294.

Turner, G. F. 1994. Speciation in Lake Malawi cichlids: A critical review. Arch. Hydrobiol. 44:139–160.

Turner, G. F. 1999. Explosive speciation of African cichlid fishes. Pp. 113–129 *in* A. E. Magurran and R. M. May (eds.) *Evolution of Biological Diversity*. Oxford University Press, Oxford.

Turner, G. F. and M. T. Burrows. 1995. A model of sympatric speciation by sexual selection. Proc. R. Soc. Lond. B 260: 287–292.

Turner, G. F., O. Seehausen, M. E. Knight, C. J. Allender, and R. L. Robinson. 2001. How many species of cichlid fishes are there in African lakes? Mol. Ecol. 10: 793–806.

Turner, J. R. G. 1985. Fisher's evolutionary faith and the challenge of mimicry. Oxford Surv. Evol. Biol. 2: 159–196.

Ueshima, R. and T. Asami. 2003. Single-gene speciation by left-right reversal. Nature 425:679.

Ungerer, M. C., S. Baird, J. Pan, and L. H. Rieseberg. 1998. Rapid hybrid speciation in wild sunflowers. Proc. Natl. Acad. Sci. USA 95: 11757–11762.

Uyenoyama, M. 1995. A generalized least-squares estimate for the origin of sporophytic self-incompatibility. Genetics 139: 975–992.

Vacquier, V. D. 1998. Evolution of gamete recognition proteins. Science 281: 1996–1999.

Vacquier, V. D. and Y.-H. Lee. 1993. Abalone sperm lysin: Unusual mode of evolution of a gamete recognition protein. Zygote 1: 181–196.

Vacquier, V. D., W. J. Swanson, and Y.-H. Lee. 1997. Positive darwinian selection on two homologous fertilization proteins: What is the selective pressure driving their divergence? J. Mol. Evol. (Suppl. 1): S15–S22.

Val, F. C. 1977. Genetic analysis of the morphological differences between two interfertile species of Hawaiian *Drosophila*. Evolution 31: 611–629.

van Batenburg, F. H. D. and E. Gittenberger. 1996. Ease of fixation of a change in coiling: Computer experiments on chirality in snails. Heredity 76: 278–286.

van Dijken, F. R. and W. Scharloo. 1979. Divergent selection on locomotor activity in *Drosophila melanogaster*. II. Tests for reproductive isolation between selected lines. Behav. Genet. 9: 555–561.

van Doorn, G. S., A. J. Noest, and P. Hogeweg. 1998. Sympatric speciation and extinction driven by environment dependent sexual selection. Proc. R. Soc. Lond. B 265: 1915–1919.

van Oppen, M. J. H., G. F. Turner, C. Rico, J. C. Deutsch, K. M. Ibrahim, R. L. Robinson, and G. M. Hewitt. 1997. Unusually fine-scale genetic structuring found in rapidly speciating Malawi cichlid fishes. Proc. R. Soc. Lond. B 264: 1803–1812.

Van Valen, L. 1976. Ecological species, multispecies, and oaks. Taxon 25: 233–239.

Van Valen, L. 1985. How constant is extinction? Evol. Theory 7:93-106.

Vasek, F. C. and R. H. Sauer. 1971. Seasonal progression of flowering in *Clarkia*. Ecology 52: 1038–1045.

Verdú, M. 2002. Age at maturity and diversification in woody angiosperms. Evolution 56: 1352–1361.

Verheyen, E., W. Salzburger, J. Snooks, and A. Meyer. 2003. Origin of the superflock of cichlid fishes from Lake Victoria, East Africa. Science 300: 325–329.

Via, S. 1991. The genetic structure of host plant adaptation in a spatial patchwork: Demographic variability among reciprocally transplanted pea aphid clones. Evolution 45: 827–852.

Via, S. 1999. Reproductive isolation between sympatric races of pea aphids. I. Gene flow and habitat choice. Evolution 53: 1446–1457.

Via, S. 2001. Sympatric speciation in animals: the ugly ducking grows up. Trends Ecol. Evol. 16: 381–390.

Via, S., A. C. Bouck, and S. Skillman. 2000. Reproductive isolation between sympatric races of pea aphids. II. Selection against migrants and hybrids in the parental environments. Evolution 54: 1626–1637.

Vick, K. W. 1973. Effects of interspecific matings on *Trogoderma glabrum* and *T. inclusum* on oviposition and remating. Ann. Entomol. Soc. Am. 66: 237–239.

Vickery, R. K. 1978. Case studies in evolution of species complexes in *Mimulus*. Evol. Biol. 11:405–507.

Vigneault, G. and E. Zouros. 1986. The genetics of asymmetrical male sterility in *Drosophila mojavensis* and *Drosophila arizonensis* hybrids: interaction between the Y chromosome and autosomes. Evolution 40: 1160–1170.

Vincek, V., C. O'Huigen, Y. Satta, N. Takahata, P. T. Boag, P. R. Grant, B. R. Grant, and J. Klein. 1997. How large was the founding population of Darwin's finches? Proc. R. Soc. Lond. B 264: 111–118.

Vollmer, S. V. and S. R. Palumbi. 2002. Hybridization and the evolution of reef coral diversity. Science 296: 2023–2025.

von Hagen, K. B., and J. W. Kadereit. 2003. The diversification of *Halenia* (Gentianaceae): ecological opportunity versus key innovation. Evolution 57: 2507–2518.

von Siebold, C. T. and H. Stannius. 1854. *Comparative Anatomy*, Vol. 1: *Anatomy of the Invertebrates*, translated from the German, edited with notes and additions, by W. I. Burnett. Gould and Lincoln, Boston, MA.

Vrba, E. 1983. Macroevolutionary trends: New perspectives on the roles of adaptation and incidental effect. Science 221: 387–389.

Vrba, E. 1989. Levels of selection and sorting with special reference to the species level. Evol. Biol. 6: 111–168.

Vrijenhoek, R. C., R. M. Dawley, C. J. Cole, and J. P. Bogart. 1989. A list of the known unisexual vertebrates. NY State Univ. Mus. Bull. 466: 19–23.

Vulic, M., F. Dionisio, F. Taddei, and M. Radman. 1997. Molecular keys to speciation: DNA polymorphism and the control of genetic exchange in enterobacteria. Proc. Natl. Acad. Sci. USA 94: 9763–9767.

Waage, J. K. 1975. Reproductive isolation and the potential for character displacement in the damselflies, *Calopteryx maculata* and *C. aequabilis* (Odonata: Calopterygidae). Syst. Zool. 24: 24–36.

Waage, J. K. 1979. Reproductive character displacement in *Calopteryx* (Odonata: Calopterygidae). Evolution 33: 104–116.

Wade, M. J. and C. J. Goodnight. 1998. The theories of Fisher and Wright in the context of metapopulations: When nature does many small experiments. Evolution 52: 1537–1553.

Wade, M. J., N. A. Johnson, and G. Wardle. 1994a. Analysis of autosomal polygenic variation for the expression of Haldane's rule in flour beetles. Genetics 138: 791–799.

Wade, M. J., H. Patterson, N. W. Chang, and N. Johnson. 1994b. Postcopulatory, prezygotic isolation in flour beetles. Heredity 72: 163–167.

Wagner, A. 2001. Birth and death of duplicated genes in completely sequenced eukaryotes. Trends Genet. 17: 237–239.

Wagner, A., G. P. Wagner, and P. Similion. 1994. Epistasis can facilitate the evolution of reproductive isolation by peak shifts: a two-locus two-allele model. Genetics 138: 533–545.

Wagner, M. 1873. *The Darwinian Theory and the Law of the Migration of Organisms*, translated from the German by James L. Laird. Edward Stanford, London.

Wagner, M. 1889. *Die Entstehung der Arten durch räumliche Sonderung*. Benno Schwalbe, Basel.

Wakabayashi, J. and T. L. Sawyer. 2001. Stream incision, tectonics, uplift, and evolution of topography of the Sierra Nevada, California. J. Geol. 109: 539–562.

Wake, D. B. 1997. Incipient species formation in salamanders of the *Ensatina* complex. Proc. Natl. Acad. Sci. USA 94: 7761–7767.

Wake, D. B. and C. J. Schneider. 1998. Taxonomy of the plethodontid salamander genus *Ensatina*. Herpetologica 54: 279–298.

Wake, D. B. and K. P. Yanev. 1986. Geographic variation in allozymes in a "ring species," the plethodontid salamander *Ensatina eschscholtzii* of western North America. Evolution 40:702–715.

Wake, D. B., K. P. Yanev, and C. W. Brown. 1986. Intraspecific sympatry in a "ring species," the plethodontid salamander *Ensatina eschscholtzii* of southern California. Evolution 40:866–868.

Wake, D. B., K. P. Yanev, and M. M. Frelow. 1989. Sympatry and hybridization in a "ring species": The plethodontid salamander *Ensatina escholtzii*. Pp. 134–157 *in* D. Otte and J. A. Endler (eds.) *Speciation and Its Consequences*. Sinauer Associates, Sunderland, MA.

Walker, T. J. 1974. Character displacement and acoustic insects. Am. Zool. 14: 1137–1150.

Wallace, B. 1959. The influence of genetic systems on geographical distribution. Cold Spring Harb. Symp. Quant. Biol. 20: 16–24.

Wallace, B. 1982. *Drosophila melanogaster* populations selected for resistances to NaCl and $CuSO_4$ in both allopatry and sympatry. J. Hered. 73: 35–42.

Walliker, D., I. A. Quakyi, T. E. Wellems, T. F. McCutchan, A. Szafman, W. T. London, L. M. Corcoran, T. R. Burkot, and R. Carter. 1987. Genetic analysis of the human malaria parasite *Plasmodium falciparium*. Science 236: 1661–1666.

Walsh, B. D. 1867. The apple-worm and the apple-maggot. Amer. J. Hort. 2: 338–343.

Walsh, J. B. 1982. Rate of accumulation of reproductive isolation by chromosome rearrangements. Am. Nat. 120: 510–532.

Walters, S. M. 1986. *Alchemilla*: A challenge to biosystematists. Act. Univ. Upsalla, Symb. Bot. Ups. 27: 193–198.

Wang, H., E. D. McArthur, S. C. Sanderson, J. H. Graham, and D. C. Freeman. 1997. Narrow hybrid zone between two subspecies of big sagebrush (*Artemisia tridentata*: Asteraceae). IV. Reciprocal transplant experiments. Evolution 51: 95–102.

Wang, R. L. and J. Hey. 1996. The speciation history of *Drosophila pseudoobscura* and close relatives: Inferences from DNA sequence variation at the *period* locus. Genetics 144: 1113–1126.

Wang, R. L., J. Wakely, and J. Hey. 1997. Gene flow and natural selection in the origin of *Drosophila pseudoobscura* and close relatives. Genetics 147: 1091–1106.

Wang, X.-R. and A. E. Szmidt. 1994. Hybridization and chloroplast DNA variation in a *Pinus* species complex from Asia. Evolution 48: 1020–1031.

Wang, X.-R., A. E. Szmidt, A. Lewandowski, and Z.-R. Wang. 1990. Evolutionary analysis of *Pinus densata* Masters, a putative Tertiary hybrid. Theor. Appl. Genet. 80: 635–640.

Wang, X.-R., A. E. Szmidt, and O. Savolainen. 2001. Genetic composition and diploid hybrid speciation of a high mountain pine, *Pinus densata*, native to the Tibetan plateau. Genetics 159: 337–346.

Ward, P. S. 1996. A new workerless social parasite in the ant genus *Pseudomyrmex* (Hymenoptera: Formicidae), with a discussion of the origin of social parasitism in ants. Syst. Entomol. 21: 253–263.

Ward, P. S. 2001. Taxonomy, phylogeny and biogeography of the ant genus *Tetraponera* (Hymenoptera: Formicidae) in the Oriental and Australian regions. Invert. Taxon. 15: 589–665.

Ware, A. D. and B. D. Opell. 1989. A test of the mechanical isolation hypothesis in two similar spider species. J. Arachnol. 17: 149–162.

Waser, N. M. 1978. Competition for hummingbird pollination and sequential flowering in two Colorado wildflowers. Ecology 59: 934–944.

Waser, N. M. 1983. Competition for pollination and floral character differences among sympatric plant species: A review of the evidence. Pp. 558 *in* C. E. Jones and R. J. Little (eds.) *Handbook of Experimental Pollination Biology*. Van Nostrand Reinhold, New York.

Waser, N. M. 1993. Population structure, optimal outbreeding and assortative mating in angiosperms. Pp. 173–199 *in* N. M. Thornhill (ed.) *The Natural History of Inbreeding and Outbreeding: Theoretical and Empirical Perspectives*. University of Chicago Press, Chicago, IL.

Waser, N. M. 1998. Pollination, angiosperm speciation, and the nature of species boundaries. Oikos 82: 198–201.

Waser, N. M. 2001. Pollinator behavior and plant speciation: Looking beyond the "ethological isolation" paradigm. Pp. 318–335 *in* L. Chittka and J. D. Thomson (eds.) *Cognitive Ecology of Pollination: Animal Behavior and Floral Evolution*. Cambridge University Press, Cambridge, UK.

Wasserman, M. and H. R. Koepfer. 1977. Character displacement for sexual isolation between *Drosophila mojavensis* and *Drosophila arizonensis*. Evolution 31: 812–823.

Watanabe, T. K. 1979. A gene that rescues the lethal hybrids between *Drosophila melanogaster* and *D. simulans*. Jpn. J. Genet. 54: 325–331.

Watanabe, T. K. and M. Kawanishi. 1979. Mating preference and the direction of evolution in *Drosophila*. Science 205: 906–907.

Waxman, D. and S. Gavrilets. 2004. 20 questions on adaptive dynamics. J. Evol. Biol. (in press).

Weber, K. E. 1990. Increased selection response in larger populations. I. Selection for wing-tip height in *Drosophila melanogaster* at three population sizes. Genetics 125: 579–584.

Weber, K. E. and L. T. Diggins. 1990. Increased selection response in larger populations. II. Selection for ethanol vapor resistance in *Drosophila melanogaster* at two population sizes. Genetics 125: 585–597.

Webster, A. J., R. J. H. Payne, and M. Pagel. 2003. Molecular phylogenies link rates of evolution and speciation. Science 301: 478.

Weeks, A. R., K. T. Reynolds, and A. A. Hoffmann. 2002. *Wolbachia* dynamics and host effects: What has (and has not) been demonstrated? Trends Ecol. Evol. 17: 247–295.

Weiblen, G. D. 2000. Phylogenetic relationships of functionally dioecious *Ficus* (Moraceae) based on ribosomal DNA sequences and morphology. Am. J. Bot. 87: 1342–1357.

Weiblen, G. D. 2001. Phylogenetic relationships of fig wasps pollinating functionally dioecious figs based on mitochondrial DNA sequences and morphology. Syst. Biol. 50: 243–267.

Weiblen, G. D. 2002. How to be a fig wasp. Annu. Rev. Entomol. 47: 299–330.

Weiblen, G. D. and G. L. Bush. 2002. Speciation in fig pollinators and parasites. Mol. Ecol. 11: 1573–1578.

Weis, S. and M. Schartl. 1998. The macromelanophore locus and the melanoma oncogene *Xmrk* are separate genetic entities in the genome of *Xiphophorus*. Genetics 149: 1909–1920.

Welch, A. M., R. D. Semlitsch, and H. C. Gerhardt. 1998. Call duration as an indicator of genetic quality in male gray tree frogs. Science 280: 1928–1930.

Welch, D. M. and M. Meselson. 2000. Evidence for the evolution of bdelloid rotifers without sexual reproduction or genetic exchange. Science 288: 1211–1215.

Welch, M. E. and L. H. Rieseberg. 2002. Patterns of genetic variation suggest a single, ancient origin for the diploid hybrid species *Helianthus paradoxus*. Evolution 56: 2126–2137.

Wells, M. M. and C. S. Henry. 1998. Songs, reproductive isolation, and speciation in cryptic species of insects: A case study using green lacewings. Pp. 217–233 *in* D. J. Howard and S. H. Berlocher (eds.) *Endless Forms: Species and Speciation*. Oxford University Press, New York.

Wen, J. 1999. Origin and evolution of the eastern Asian and eastern North American distinct distributions of flowering plants. Annu. Rev. Ecol. Syst. 30: 421–455.

Wendel, J. F. and G. Percy. 1990. Allozyme diversity and introgression in the Galapagos Islands endemic *Gossypium darwinii* and its relationship to continental *G. barbadense*. Biochem. Syst. Ecol. 18: 517–528.

Werner, E. E. and D. J. Hall. 1977. Competition and habitat shift in two sunfishes (Centrarchideae). Ecology 58: 869–876.

Werner, P. A. and W. J. Platt. 1976. Ecological relationships of co-occurring goldenrods (*Solidago*: Compositae). Am. Nat. 110: 959–971.

Werren, J. H. 1997a. Biology of *Wolbachia*. Annu. Rev. Entomol. 42: 587–609.

Werren, J. H. 1997b. *Wolbachia* and speciation. Pp. 245–260 *in* D. Howard and S. H. Berlocher (eds.) *Endless Forms: Species and Speciation*. Oxford University Press, Oxford, UK.

Werren, J. H. and D. Windsor. 2000. *Wolbachia* infection frequencies in insects: Evidence of a global equilibrium. Proc. R. Soc. Lond. B 267: 1277–1285.

Werren, J. H., D. Windsor, and L. R. Guo. 1995a. Distribution of *Wolbachia* among neotropical arthropods. Proc. R. Soc. Lond. B 262: 187–204.

Werren, J. H., W. Zhang, and L. R. Guo. 1995b. Evolution and phylogeny of *Wolbachia*: Reproductive parasites of arthropods. Proc. R. Soc. Lond. B 261: 55–63.

Werth, C. R. and M. D. Windham. 1991. A model for divergent, allopatric speciation of polyploid pteridophytes resulting from silencing of duplicate-gene expression. Am. Nat. 137: 515–526.

West-Eberhard, M. J. 1983. Sexual selection, social competition, and speciation. Q. Rev. Biol. 58: 155–183.

Westergaard, M. 1940. Studies on cytology and sex determination in polyploid forms of *Melandrium album*. Dan. Bot. Ark. 10: 1–131.

Westergaard, M. 1958. The mechanism of sex determination in dioecious flowering plants. Adv. Genet. 9: 217–289.

Wettstein, F. v. 1927. Die Erscheinung der Heteroploidie, besonders im Pflanzenreich. Ergeb. Biol. 2: 311–356.

Wheeler, Q. D. and R. D. Meier (eds.) 2000. *Species Concepts and Phylogenetic Theory: A Debate*. Columbia University Press, New York.

Wheeler, Q. D. and K. C. Nixon. 1990. Another way of *looking at* the species problem: A reply to de Queiroz and Donoghue. Cladistics 6: 77–81.

White, B. J., C. Crandall, E. S. Raveche, and J. H. Tjio. 1978. Laboratory mice carrying three pairs of Robertsonian translocations: Establishment of a strain and analysis of meiotic segregation. Cytogenet. Cell Genet. 21: 113–138.

White, J. 1973. Viable hybrid young from crossmated periodical cicadas. Ecology 54: 573–580.

White, M. J. D. 1969. Chromosomal rearrangements and speciation in animals. Annu. Rev. Genet. 3: 75–98.

White, M. J. D. 1973. *Animal Cytology and Evolution*. Cambridge University Press, London.

White, M. J. D. 1978. *Modes of Speciation*. W. H. Freeman and Company, San Francisco, CA.

Whitham. T. G. 1989. Plant hybrid zones as sinks for pests. Science 244: 1490–1493.

Whitlock, M. C. and K. Fowler. 1999. The changes in genetic and environmental variance with inbreeding in *Drosophila melanogaster*. Genetics 152: 345–353.

Whitlock, M. C., Phillips, P. C., and K. Fowler. 2002. Persistence of changes in the genetic covariance matrix after a bottleneck. Evolution 56: 1968–1975.

Whittemore, A. T. 1993. Species concepts: A reply to Mayr. Taxon 42: 573–583.

Whittemore, A. T. and B. A. Schaal. 1991. Interspecific gene flow in sympatric oaks. Proc. Natl. Acad. Sci. USA 88: 2540–2544.

Whitton, J. 2004. One down and thousands to go—dissecting polyploid speciation. New Phytol. 161: 607–610.

Wiebes, J. T. 1979. Co-evolution of figs and their insect pollinators. Annu. Rev. Ecol. Syst. 10:1–12.

Wiegmann, B. M., C. Mitter, and B. Farrell. 1993. Diversification of carnivorous parasitic insects: Extraordinary radiation or specialized dead end? Am. Nat. 142:737–754.

Wiernasz, D. C. 1989. Female choice and sexual selection on male wing melanin pattern in *Pieris occidentalis* (Lepidoptera). Evolution 43: 1672–1682.

Wiernasz, D. C. and J. G. Kingsolver. 1992. Wing melanin pattern mediates species recognition in *Pieris occidentalis*. Anim. Behav. 43: 89–94.

Wiley, E. O. 1978. The evolutionary species concept reconsidered. Syst. Zool. 27: 17–26.

Williams, E. G. and J. L. Rouse. 1988. Disparate style lengths contribute to isolation of species in *Rhododendron*. Aust. J. Bot. 36: 183–191.

Williams, E. G. and J. L. Rouse. 1990. Relationships of pollen size, pistil length, and pollen tube growth rates in *Rhododendron* and their influence on hybridization. Sex. Plant Reprod. 3: 7–17.

Williams, J. H., W. J. Boecklen, and D. J. Howard. 2001. Reproductive processes in two oak (*Quercus*) hybrid zones with different levels of hybridization. Heredity 87: 680–690.

Williams, K. S. and C. Simon. 1995. The ecology, behavior, and evolution of periodical cicadas. Annu. Rev. Entomol. 40: 269–295.

Williams, M. A., A. G. Blouin, and M. A. F. Noor. 2001. Courtship songs of *Drosophila pseudoobscura* and *D. persimilis*. II. Genetics of species differences. Heredity 86: 68–77.

Williamson, M. 1981. *Island Populations*. Oxford University Press, Oxford.

Wills, C. J. 1977. A mechanism for rapid allopatric speciation. Am. Nat. 111: 603–605.

Wilson, A. C. 1975. Evolutionary importance of gene regulation. Stadler Symp. 7: 117–133.

Wilson, A. C., L. R. Maxson, and V. M. Sarich. 1974. Two types of molecular evolution: Evidence from studies of interspecific hybridization. Proc. Natl. Acad. Sci. USA 71:2843–2847.

Wilson, D. S. and M. Turelli. 1986. Stable underdominance and the evolutionary invasion of empty niches. Am. Nat. 127: 835–850.

Wilson, E. O. 1965. The challenge from related species. Pp. 7–24 *in* H. G. Baker and G. L. Stebbins (eds.) *The Genetics of Colonizing Species*. Academic Press, New York.

Wilson, E. O. 1971. *The Insect Societies*. Harvard University Press, Cambridge, MA.

Wilson, M. V. H. 1983. Is there a characteristic rate of radiation for the insects? Paleobiology 9:79–85.

Wilson, R. A. 1999. *Species: New Interdisciplinary Essays*. MIT Press, Cambridge MA.

Winge, O. 1917. The chromosomes: Their number and general importance. C. R. Trav. Lab. Carlsberg 13: 131–275.

Wittbrodt, J., D. Adam, B. Malitschek, W. Maueler, F. Raulf, A. Telling, S. M. Robertson, and M. Schartl. 1989. Novel putative receptor tyrosine kinase encoded by the melanoma-inducing *Tu* locus in *Xiphophorus*. Nature 341: 415–421.

Wolfe, A. D., Q.-Y. Xiang, and S. R. Kephart. 1998a. Assessing hybridization in natural populations of *Penstemon* (Scrophulariaceae) using hypervariable inter simple sequence repeat markers. Mol. Ecol. 7: 1101–1125.

Wolfe, A. D., Q.-Y. Xiang, and S. R. Kephart. 1998b. Diploid hybrid speciation in *Penstemon* (Scrophulariaceae). Proc. Natl. Acad. Sci. USA 95: 5112–5115.

Wolfe, K. H. 1997. Molecular evidence for an ancient duplication of the entire yeast genome. Nature 387: 708–713.

Wolfner, M. F. 2002. The gifts that keep on giving: Physiological functions and evolutionary dynamics of male seminal proteins in *Drosophila*. Heredity 88: 85–93.

Wood, C. C. and C. J. Foote. 1996. Evidence for sympatric genetic divergence of anadromous and nonanadromous morphs of sockeye salmon (*Oncorhynchus nerka*). Evolution 50:1265–1279.

Wood, T. K. 1993. Speciation of the *Enchenopa binotata* complex (Insecta: Homoptera: Membracidae). Pp. 300–317 *in* D. R. Lees and D. Edwards (eds.) *Evolutionary Patterns and Processes*. Academic Press, London.

Wood, T. K., K. J. Tilman, A. B. Shantz, C. K. Harris, and J. Pesek. 1999. The role of host-plant fidelity in initiating insect race formation. Evol. Ecol. Res. 1: 317–332.

Wright, D. H. 1983. Species-energy theory: An extension of species-area theory. Oikos 41: 496–506.

Wright, S. 1940a. Breeding structure of populations in relation to speciation. Am. Nat. 74: 232–248.

Wright, S. 1940b. The statistical consequences of Mendelian heredity in relation to speciation. Pp. 161–183 *in* J. S. Huxley (ed.) *The New Systematics*. Oxford University Press, Oxford.

Wright, S. 1948. Evolution, organic. *Encyclopaedia Brittanica*, 14th ed. 8: 915–929.

Wright, S. 1978. *Evolution and the Genetics of Populations*, Vol. 4: *Variability Within and Among Natural Populations*. University of Chicago Press, Chicago, IL.

Wright, S. 1982a. Character change, speciation, and the higher taxa. Evolution 36: 427–443.

Wright, S. 1982b. The shifting balance theory and macroevolution. Annu. Rev. Genet. 16: 1–19.

Wright, S. D., R. D. Gray, and R. C. Gardner. 2003. Energy and the rate of evolution: Inferences from plant rDNA substitution rates in the Western Pacific. Evolution 57: 2893–2898.

Wu, C.-I. 1985. A stochastic simulation study on speciation by sexual selection. Evolution 39:66–82.

Wu, C.-I. 1992. A note on Haldane's rule: Hybrid inviability versus hybrid sterility. Evolution 46: 1584–1587.

Wu, C.-I. and A. T. Beckenbach. 1983. Evidence for extensive genetic differentiation between the sex-ratio and the standard arrangement of *Drosophila pseudoobscura* and *D. persimilis* and identification of hybrid sterility factors. Genetics 105: 71–86.

Wu, C.-I. and A. W. Davis. 1993. Evolution of post-mating reproductive isolation—the composite nature of Haldane's rule and its genetic bases. Am. Nat. 142: 187–212.

Wu, C.-I. and M. F. Palopoli. 1994. Genetics of postmating reproductive isolation in animals. Annu. Rev. Genet. 27: 283–308.

Wu, C.-I. and C. T. Ting. 2004. Genes and speciation. Nature Reviews Genetics 5: 114–122.

Wu, C.-I., H. Hollocher, D. J. Begun, C. F. Aquadro, Y. Xu, and M.-L. Wu. 1995. Sexual isolation in *Drosophila melanogaster*: A possible case of incipient speciation. Proc. Natl. Acad. Sci. USA 92: 2519–2523.

Wu, C.I., N. A. Johnson, and M. F. Palopoli. 1996. Haldane's rule and its legacy: Why are there so many sterile males? Trends Ecol. Evol. 11: 281–284.

Wullschleger, E. B., J. Wiehn, and J. Jokela. 2002. Reproductive character displacement between the closely related freshwater snails *Lymnaea peregra* and *L. ovata*. Evol. Ecol. Res. 4: 247–257.

Wyckoff, G. J., W. Wang, and C.-I. Wu. 2000. Rapid evolution of male reproductive genes in the descent of man. Nature 403: 304–309.

Xiang, Q.-Y., D. E. Soltis, and P. S. Soltis. 1998. The eastern Asian and eastern and western North American floristic disjunction: Congruent phylogenetic patterns in seven diverse genera. Mol. Phylogenet. Evol. 10: 178–190.

Xiang, Q.-Y., D. E. Soltis, P. S. Soltis, S. R. Manchester, and D. J. Crawford. 2000. Timing the eastern Asian-eastern North American floristic disjunction: Molecular clocks confirm paleontological estimates. Mol. Phylogenet. Evol. 15: 462–472.

Yamada, H., M. Matsuda, and Y. Oguma. 2002. Genetics of sexual isolation based on courtship song between two sympatric species: *Drosophila ananassae* and *D. pallidosa*. Genetica 116: 225–237.

Yang, Z., W. J. Swanson, and V. D. Vacquier. 2000. Maximum-likelihood analysis of molecular adaptation in abalone sperm lysin reveals variable selective pressures among lineages and sites. Mol. Biol. Evol. 17: 1446–1455.

Yen, J. H. and A. R. Bar. 1971. New hypothesis of the cause of cytoplasmic incompatibility in *Culex pipiens*. Nature 232: 657–658.

Yen, J. H. and A. R. Bar. 1973. The etiological agent of cytoplasmic incompatibility in *Culex pipiens* L. J. Invertebr. Pathol. 22: 242–250.

Yoo, B. H. 1980. Long-term selection for a quantitative character in large replicate populations of *Drosophila melanogaster*. II. Lethals and visible mutants with large effects. Genet. Res. 35: 19–31.

Young, J. E. and B. I. Crother. 2001. Allozyme evidence for the separation of *Rana areolata* and *Rana capita* and for the resurrection of *Rana sevosa*. Copeia 2001: 382–388.

Yule, G. U. 1925. A mathematical theory of evolution, based on the conclusions of Dr. J. C. Willis, F. R. S. Philos. Trans. R. Soc. Lond. B 213: 21–87.

Zeng, L.-W. 1996. Resurrecting Muller's theory of Haldane's rule. Genetics 143: 603–607.

Zeng, L.-W. and R. S. Singh. 1993. The genetic basis of Haldane's rule and the nature of asymmetric hybrid male sterility among *Drosophila simulans*, *Drosophila mauritiana* and *Drosophila sechellia*. Genetics 134: 251–260.

Zeng, Z. B., J. J. Liu, C. H. Kao, J. M. Mercer, and C. C. Laurie. 2000. Genetic architecture of a morphological shape difference between two *Drosophila* species. Genetics 154: 299–310.

Zhang, Z., T. M. Hambuch, and J. Parsch. 2004. Molecular evolution of sex-biased genes in *Drosophila*. Unpublished manuscript.

Zhou, Z. and S. Zheng. 2003. The missing link in *Gingkgo* evolution. Nature 423: 821–822.

Zimmerman, E. C. 1960. Possible evidence of rapid evolution in Hawaiian moths. Evolution 14:137–138.

Zouros, E. 1973. Genic differentiation associated with the early stages of speciation in the mulleri subgroup of *Drosophila*. Evolution 27: 601–621.

Zouros, E. 1981. The chromosomal basis of sexual isolation in two sibling species of *Drosophila*: *D. arizonensis* and *D. mohavensis*. Genetics 97: 703–718.

Author Index

A

Abrahamson, W. G., 164
Abrams, P., 35
Adam, D., 315
Adams, K. L., 332
Adkins, B. E., 245
Agis, M., 123
Alatalo, R. V., 360
Albertson, R. C., 149, 150
Alexander, R. D., 166
Allen, J. A., 91
Allender, C. J., 149, 152
Amevigbe, M. D. D., 157
Amores, A., 325, 327
Anders, F., 315
Anderson, E., 37, 71, 343, 347
Andersson, M., 218, 220, 223
Angus, G. W., 203
Antonovics, J., 122, 207, 361, 371, 456
Arisumi, T., 66
Armstrong, R. A., 35
Arnold, M. L., 71, 191, 197, 202, 342, 343, 347, 348, 350
Arnold, S. J., 226
Arnqvist, G., 231, 240, 407, 436
Asami, T., 111, 410, 472
Ashburner, M., 260, 310, 311, 317, 318
Asker, S. E., 20
Aspinwall, N., 202, 209
Atyeo, W. T., 155, 157
Avise, J. C., 15, 18, 100, 101, 421, 463–465
Awadalla, P., 318
Ayala, F. J., 468

B

Babcock, E. B., 19
Babcock, R., 204
Baird, S. E., 286
Baker, A. E. M., 215
Baker, H. G., 20, 22, 212
Baker, M. C., 215, 221
Baker, R. J., 262, 263
Baldwin, B., 420
Balkau, B. J., 113, 366
Ball, R. M., 101, 463–465
Bar, A. R., 278, 279
Barbash, D. A., 307, 317–319
Barlow, G. W., 147
Barraclough, T. G., 51, 52, 111, 122, 169–172, 177, 188, 407, 418, 419, 423, 434, 436, 440–442
Barrett, P. H., 11
Barrett, S. C. H., 212
Barrett, S. J., 23, 24
Barton, N. H., 31, 33, 42, 67, 97, 112, 114–117, 119, 120, 135, 177, 251, 257, 262, 303, 306, 309, 343, 369, 379, 386, 394–398, 400, 403, 412, 451
Basolo, A. L., 214
Bateson, W., 2, 26, 248, 269
Baum, D. A., 27, 459, 463, 467, 469
Baverstock, P. R., 336
Beacham, T. D., 209
Beadle, G. W., 311, 312
Beckenbach, A. T., 274
Beehler, B. M., 219, 227, 428

Begun, D. J., 290
Bell, M. A., 146
Bella, J. L., 236
Belotte, D., 52
Bennett, B. D., 71, 347, 350
Berlin, B., 13, 17
Berlocher, S. H., 1, 158–161, 164, 449
Bermingham, E., 420
Bernardi, G., 101
Bernatchez, L., 146, 471
Bernstein, R., 231
Bernstein, S., 231
Bert, T. M., 253
Betancourt, A., 295, 380
Beverley, S. M., 108
Bickham, J. W., 262, 263
Biermann, C. H., 242
Bierne, N., 237
Bigelow, R. S., 167, 371
Birkhead, T. R., 232, 244
Bishop, J. G., 403
Blair, W. F., 353, 359, 360
Boake, C. R. B., 376
Boche, R. D., 311
Bock, I. R., 253
Bogdanowicz, S. M., 167
Bolmgren, K., 439
Bolnick, D. I., 135, 415
Bordenstein, S. R., 226, 268, 276, 278–280
Borodin, P., 264
Boughman, J. W., 146, 222
Bouvier, M. M., 42, 65, 67, 73, 77–79, 285, 286, 293, 421, 425
Bradsaw, H. D., 192, 196, 200–202, 408

Braig, H. K., 277
Braun, M. J., 42
Breen, P. A., 245
Breeuwer, J. A. J., 260, 268, 277, 279, 290
Bressac, C., 277
Britton-Davidian, J., 264
Brochmann, C., 350
Bronson, C. L., 42, 119, 251
Brookfield, J., 25, 450
Brown, C. W., 103
Brown, D. V., 240
Brown, J. H., 92, 124
Bryant, E. H., 400
Brykov, V. A., 209
Buchanan, D. B., 99
Buchmann, S. L., 229
Buckley, P. A., 254, 255
Buerkle, C. A., 341, 342, 350, 415
Bull, J. J., 394
Bulmer, R. N. H., 13
Bundenberg, d. J. C. M., 335–336
Burger, W. C., 43
Burgoyne, P. S., 262
Burrows, M. T., 416
Burton, R. S., 310, 319
Bush, G. L., 84, 95, 97, 126, 131, 132, 155, 156, 159, 161, 176, 187, 196
Busvine, J. R., 156
Butlin, R. K., 226, 354, 362, 452
Buxton, P. A., 156

C

Caballero, A., 214, 394
Cabot, E. L., 276, 303, 307, 312
Caillaud, C. M., 162, 163
Cain, M. L., 371
Caisse, M., 371
Callaini, 277
Camp, W. H., 161
Capanna, E., 262–264
Cardillo, M, 436, 437, 439
Carlquist, S., 107
Carlton, C. E., 168
Carney, S. E., 233, 237
Carr, G. D., 330
Carson, H. L., 73, 75, 107–109, 161, 180, 260, 388, 391, 392, 394, 396, 399, 402, 408, 421

Carvajal, A. R., 300, 303, 307
Caspari, E., 277, 279
Castiglia, J., 264
Castillo-Davis, C. I., 309
Catling, P. M., 42
Cayouette, J., 42
Cazemajor, M., 296
Change, A., 236
Chapman, T., 238, 244
Chari, J., 62
Charlesworth, B., 225, 226, 290, 292, 294, 303, 315, 394–396, 400, 403, 412
Chesser, R. T., 169
Chesson, P., 35
Christie, P., 189, 192, 268, 272, 303, 385, 409
Church, S. A., 276
Civetta, A., 239, 294
Clague, D. A., 108
Claridge, M. F., 25, 39
Clark, A. G., 238, 239
Clarke, C. A., 34, 449
Clausen, J., 185–187, 189, 206, 324
Clausen, R. E., 322
Clay, T., 155
Clifton, K. E., 202, 204
Clifton, L. M., 202, 204
Cohan, F. M, 23, 24, 51, 454
Colborn, J., 94, 101
Collins, T. M., 101
Conery, J. S., 312, 313, 325
Cooley, J. F., 359
Cooley, J. R., 167, 168
Cordero Rivera, A., 240
Cosmides, L. M., 294
Coulthart, M. B., 294
Counterman, B. A., 295
Cox, R. T., 168
Coyne, J. A., 40, 48, 61–64, 73–78, 80, 81, 99, 107, 123, 138, 145, 169, 181, 191, 214–216, 225, 227, 229, 232, 261, 262, 268, 278, 280, 285–288, 290, 292, 294, 296–298, 300, 303, 304, 307, 311, 315, 361–366, 372, 374–376, 402, 410, 415, 421, 422, 427, 452, 468, 469
Cracraft, J., 27, 46, 92, 97, 122, 462
Craft, K. J., 44

Craft, W. A., 285
Craig, T. P., 164
Crapon de Caprona, M.-D., 149
Crespi, B. J., 244, 359
Cronin, T. M., 142
Croom, H. B., 108
Crosby, J. L., 367
Crossley, S. A., 355
Crother, B. I., 468
Crovello, T. J., 36, 448
Crow, J. F., 268, 271, 303, 310
Crowe, L. K., 241
Crowe, T. M., 452
Crowell, K., 458
Cruden, R. W., 198
Cruzan, M. B., 212
Cuervo, J. J., 436, 442
Cumbes, Q. J., 212
Cunningham, C. W., 101
Cunningham, M. A., 221

D

da Silva, J., 436, 441
Damuth, J., 443, 444
Danielsson, I., 231
Darlington, C. D., 267, 324, 325
Darwin, C., 1, 9, 10, 11, 37, 83, 87, 106, 112, 125, 247
David, J. R., 185
Davies, N., 254
Davies, T. J., 441
Davis, A. W., 285, 288, 290, 292, 307, 310, 317
Davis, J. J., 162
Dawkins, R., 49
Day, T., 343, 378
de Queiroz, A., 433
de Queiroz, K., 27, 463, 467
De Villena, F. P.-D., 259, 297
DeBach, P., 35
DeJoode, D. R., 330
del Solar, E., 90
Delneri, D., 259
DeMarais, B. D., 350
Dempster, L. T., 335
Dermitzakis, E. T., 296
DeSalle, R., 108, 391, 402, 403, 422
Desdevises, Y., 441
Dettman, J. R., 361
DeVries, H., 2

Diamond, J. M., 13, 16, 145, 169, 187, 188
Diaz, A., 233, 237, 242
Dickinson, T. A., 21
Dieckmann, U., 32, 52, 53, 99, 114, 134, 135, 137, 191, 201, 415, 416
Diehl, S. R., 131, 132
Diggins, L. T., 394, 450
Dobzhansky, T., 12, 17, 26, 28, 29, 32, 33, 36, 37, 41, 49, 50, 51, 86, 87, 156, 157, 180, 181, 253, 255, 261, 267–269, 271, 300, 301, 303, 307, 311, 312, 314, 333, 342, 353, 358
Dod, B., 290
Dodd, D. M. B., 90, 106, 398, 399
Dodd, M. E., 198, 434, 436, 438, 439, 440
Doebeli, M., 32, 52, 53, 99, 114, 131, 134, 135, 137, 191, 201, 415, 416
Doherty, J. A., 361
Doherty, P. F., 438
Doi, M., 225
Dominey, W. J., 70, 148, 153
Donoghue, M. J., 11, 14, 27, 46, 432, 433, 463, 467
Doolittle, W. F., 314, 319
Dorn, R. D., 203
Dowling, T. E., 93, 334
Doyle, J. A., 432
Drapeau, M. D., 268
Drès, M., 158, 159, 164, 209
Dressler, R. L., 64, 194, 198, 200
Duda, T. F., 445
Dufour, L., 228
Dufresne, F., 334
Durham, W. H., 14
Duvernell, D. D., 93
Dybas, H. S., 167, 168, 202
Dykhuizen, D. E., 24

E

Eady, P. E., 240
Eastop, V. F., 162, 164
Eberhard, W. G., 227, 229–231. 450
Echelle, A. A., 93, 147
Edmands, S., 80, 102, 310

Ehrendorfer, F., 335
Ehrman, L., 356, 358
Eisenberg, R. M., 89, 138
Eldredge, N., 410, 442
Ellstrand, N. C., 16, 41, 42, 43
Endler, J. A., 1, 95, 113, 114, 118, 121, 222
Engels, W. R., 314, 319
Enquist, M., 161
Ephrussi, B., 312
Epling, C., 450
Ereshefsky, M., 25, 39
Escalante, A. A., 468
Etges, W. J., 90
Evans, M., 107
Everett, C. A., 264
Ewens, W. J., 398

F

Farrell, B. D., 405, 436–440
Feder, J. L., 31, 64, 158–162, 164, 187, 450
Feil, E. J., 23
Felber, F., 331
Feldman, M. W., 113, 366
Felsenstein, J., 115, 127, 131, 357, 359, 373, 374, 459
Fenster, C. B., 213
Ferguson, D., 350
Ferris, S. D., 325, 471
Ficken, M. S., 252
Ficken, R. W., 252
Figueroa, F., 468
Filchak, K. E., 161
Firtzsch, B., 149
Fisher, R. A., 3, 112, 113, 219, 366–367, 386
Fishman, L., 211, 213, 268
Fitzpatrick, B. M., 122, 172, 174, 177, 403
Florin, A.-B., 90, 141, 399
Fontdevila, A., 303
Foote, C. J., 145
Foote, D., 426
Force, A. G., 275, 312, 313
Forejt, J., 260, 264, 307
Forsgren, E., 220
Foster, S. A., 146
Fowler, K., 394
Fowler, N., 331
Frank, S. H., 295, 296
Frith, C. B., 219, 428

Frith, C. M., 227
Fry, J. D., 132
Fryer, G., 147, 148, 150, 419
Fukatami, A., 356
Fulton, M., 194, 197
Funk, D. J., 189, 405
Furness, R. W., 209
Futuyma, D. J., 68, 85, 117, 121, 126, 160, 183, 258–260, 271, 402
Fuyama, Y., 233, 235, 236

G

Gack, C., 64, 194, 198
Gadau, J., 268, 274, 298, 303, 307
Gage, M. J. G., 438
Gale, K. S., 119, 306
Galen, C., 200
Galiana, A., 400
Galindo, B. E., 242, 244, 245
Galis, F., 154
Gall, G. A. E., 138
Gallez, G. P., 349
Gardezi, T., 436, 441
Gardner, M., 21, 120, 189, 203
Gaston, K. J., 440
Gates, R. B., 333
Gavrilets, S., 115, 135, 137, 244, 376, 396, 397, 414, 416
Genner, M. J., 70, 148, 154
Gerassimova, H., 257, 343
Gerber, D., 43
Gerhardt, H. C., 252, 329, 332, 360
Gerstel, D. U., 268
Gethmann, R. C., 262
Geyer, L. B., 243, 359
Gharrett, A. J., 209
Gibson, J. B., 138, 356–357
Gilbert, D. G., 214
Gillespie, J. H., 272, 403
Gillespie, R. G., 108, 426
Gíslason, D., 145, 176
Gittenberger, E., 111, 410
Gittleman, J. L., 434, 441
Givnish, T. J., 108, 109
Glazner, J. T., 165
Glor, R. E., 169, 175
Goldblatt, P., 326
Goodnight, C. J., 56, 394
Goodspeed, T. H., 322

Goodwillie, C., 21, 63
Gottlieb, L. D., 110, 212, 266, 349
Gottschewski, G., 307
Goudet, J., 437
Gould, M., 244, 245
Gould, S. J., 2, 49, 83, 410, 431, 442–445
Goulson, D., 195
Grant, B. R., 202, 252, 420
Grant, R. S., 138
Grant, K., 15, 41, 42, 65, 109, 110, 145, 200
Grant, P. R., 202, 215, 220, 252, 360, 420
Grant, V., 1, 4, 6, 15, 19–21, 33, 40, 41, 42, 63, 65, 81, 109, 110, 145, 193, 194, 198, 200, 266, 267, 311, 323, 324, 326–328, 337, 338, 340, 341, 343, 344, 350, 351, 361, 365
Grantham, T. A., 443
Graur, D., 72
Gray, W. N., 151, 152
Greenwood, P. H., 147, 149
Gregory, P. G., 61, 233–236
Greig, D., 265, 266, 344
Griffen, R. K., 303, 310
Griffiths, G. C. D., 16
Grimaldi, D., 233
Gropp, A., 264
Gross, B. L., 347
Grudzien, T. A., 97
Grula, J. W., 215
Grun, P., 295
Guenet, J.-L., 290, 307
Gullberg, U., 131
Gupta, J. P., 15, 41
Gutbrodt, H., 315
Guttman, S. I., 44
Guyer, C., 432, 434, 436–438

H

Hafner, M. S., 157, 189, 190
Hagen, R. H., 290
Haldane, J. B. S., 3, 86, 113, 284, 322, 386, 390
Hale, D. W., 262
Hall, D. J., 187
Hall, J. C., 227
Hall, J. P. W., 95–97, 121
Halliburton, R., 138

Hamerton, J. L., 259
Hamrick, J. L., 212
Haney, T., 157
Hardin, J. W., 43
Hardy, D. E., 391
Harper, A. A., 356, 370
Harrison, R. G., 47, 97, 120, 167, 183, 206–207, 450, 455
Hartl, D. L., 275, 292, 293, 295, 297, 303, 304, 306, 309, 367
Harvey, D. J., 95–97, 121
Haskins, C. P., 3, 217
Haskins, E. F., 3, 217
Hastings, A., 396, 397, 414
Hatfield, T., 145, 146, 250, 255
Hauffe, H. C., 262, 264
Hauschteck-Jungen, E., 296
Hawthorne, D. J., 163, 192, 193
Hebert, P. D., 334
Hedrick, P. W., 259
Heed, W. B., 214
Heikkinen, E., 303
Heilbuth, J. C., 436, 441
Heiser, C. B., 345, 346
Heisler, I. L., 444
Helbig, A. J., 251
Hellberg, M. E., 235
Henikoff, S., 259, 295, 297
Henry, C. S., 166, 215
Herndon, L. A., 236
Heslop-Harrison, J., 233, 235
Hewitt, G. M., 31, 33, 67, 97, 112, 115–117, 119, 120, 236, 303, 306, 369, 379, 451
Hey, J., 25, 41, 305, 309, 418, 419, 420, 423
Highton, R., 40, 103
Hill, G. E., 214
Hill, W. G., 394
Hillis, D. M., 42, 77, 207, 208, 360
Hilu, K. W., 326
Hindar, K., 145
Hirai, H., 350
Ho, M.-W., 49
Hobel, G., 252, 360
Hodges, S. A., 191, 194, 197, 198, 202, 342, 431, 436, 439
Hoffmann, A. A., 277–279, 426
Hoikkala, A., 225
Holder, K. K., 394
Holland, B., 218, 239, 244
Hollingshead, L., 298

Hollocher, H., 290, 292, 295, 303, 311, 312
Holman, E. W., 22
Horvitz, C. C., 195
Hosken, D. J., 231
Hosogai, T., 70, 187
Hostert, E. E., 87, 89, 250, 354, 355, 357
Houde, A. E., 222
Howard, D. J., 1, 31, 43–45, 61, 70, 183, 233–236, 242, 244, 271, 353, 358, 361, 363, 371, 372, 375, 471
Hoy, R. R., 215
Huang, Y., 167
Hubbell, S. P., 35
Hudson, R. R., 48, 468
Huelsenbeck, J. P., 189
Hulbert, R. C., 420–422
Hull, D. L., 444
Humphries, J. M., 15
Hunt, J. A., 403
Hunter, F. M., 214
Hunter, J. P., 431
Hunter, N., 266, 276
Hurd, L. E., 89, 138
Hurst, G. D. D., 280, 311
Hurst, L. D., 18, 295, 296
Husband, B. C., 63, 201, 209, 329, 331, 332
Hutter, P., 317
Huxley, J., 55, 247

I

Iles, T. D., 147, 148, 419
Irving, S., 296, 303, 305, 307, 317, 318
Irwin, D. E., 103, 104, 215, 219
Irwin, J. H., 103
Isaac, N. J. B, 434
Iványi, P., 264, 307
Iwasa, Y., 87, 219, 414, 423

J

Jablonski, D., 142, 421, 435, 443, 444
Jackman, T. R., 103
Jackson, R. C., 324
Jacobs, G., 277
Jaenike, J., 158, 295

Jaffe, L. A., 244, 245
James, T. S., 146
Jansa, S. A., 239
Janzen, D. H., 195
Jerling, L., 20
Jerrim, K., 195
Jiggins, C. D., 65–67, 121, 122, 221, 249, 381, 409
Jiggins, F. M., 280
John, B., 262
Johnson, M. S., 335
Johnson, N. A., 268, 271, 274, 296, 312, 365
Johnson, P. A., 131, 134, 147
Johnson, S. D., 198–200
Johnson, T. C., 150
Joly, D., 296
Jones, A. G., 146
Jones, C. D., 107, 192, 193
Jones, E. W., 43
Jones, I. L., 214
Jonsson, B., 145
Jordan, D. S., 2, 84, 86, 91, 92, 168
Jordan, K., 2
Jousselin, E., 200

K

Kadereit, J. W., 431, 439
Kaliszewska, Z., 157
Kandul, N. P., 419
Kaneshiro, K. Y., 226, 391
Karr, T. L., 277–279, 426
Kassen, R., 146
Katakura, H., 70, 187, 234
Katzourakis, A., 436, 438
Kaufmann, B. P., 303, 310
Kawanishi, M., 64, 226
Kawecki, T. J., 132–133
Kay, K. M., 361
Keeling, M. J., 279, 280
Keller, M. J., 329, 332
Kelley, D. B., 293
Kellogg, C. L., 91
Kelly, J. K., 221, 374, 415
Kerkis, J., 312
Kessel, B., 184
Key, K. H. L., 258
Khadem, M., 303
Kiang, Y. T., 212
Kidwell, M. G., 314, 319
Kilias, G., 90, 399

Kimura, M., 272
King, M., 257, 259, 263, 264, 268
Kingsolver, J. G., 214, 222
Kinsey, J. D., 310
Kirkpatrick, M., 65, 85, 217, 218, 223, 371, 374, 375, 379, 404
Kitcher, P., 26
Kittler, R., 157
Klicka, J., 81, 97, 421, 425
Kliman, R. M., 47, 64, 467
Knapp, S., 95
Knight, G. R., 355
Knight, M. E., 149, 152
Knowles, L. L., 240
Knowlton, N., 37, 92, 93, 202, 204
Kobel, H. R., 293, 297
Kocher, T. D., 42, 150, 207
Koepfer, H. R., 359
Kohn, J. R., 238
Koller, P. C., 358
Kondrashov, A. S., 32, 53, 131, 135, 137, 191, 275, 416
Kondrashov, F. A., 32, 53, 135, 137, 191, 416
Koopman, K. F., 138, 355
Kornfield, I., 147–149, 153, 154
Korol, A., 123
Kottler, M. J., 11
Kreitman, M., 303, 403
Kresge, N., 234, 235
Kricher, J., 95
Krimbas, C. B., 271, 303
Kruckeberg, A. R., 101, 120, 189
Kubo, T., 423
Kubota, K., 228, 230
Kvist, L., 104
Kyriacou, C. P., 227

L

Lacey, E. P., 205
Lachaise, D., 41, 185
Lambert, D. M., 356, 370, 451
Lamnissou, K., 303
Lamont, B. B., 71, 186, 206
Lande, R., 113, 217, 219, 258, 263, 395, 412–414, 416
Langor, D. W., 405
Larson, R. J., 187
Lassy, C. W., 277
Laurie, C. C., 285

Laven, H., 278
Lawrence, J. G., 23
Leaché, A. D., 468
Lee, Y. H., 235, 242, 243, 359
Legal, L., 61, 107, 185
Lenski, R. E., 394
Leo, N. P., 157
Lessios, H. A., 61, 92, 94, 226
Levene, H., 156, 157
Levin, D. A., 81, 181, 186, 204, 211, 233, 234, 266, 324–326, 329–333, 361, 420, 421
Levitan, D. R., 204, 233
Lewis, D., 241
Lewis, H., 110, 257
Lewontin, R. C., 4, 272, 305, 314, 353, 394, 458
Li, W.-H., 72, 242
Liebers, D., 104
Liem, K., 148
Lifschytz, E., 312
Liimatainen, J. O., 225
Lijtmaer, D. A., 79, 80, 275, 285, 286, 421, 425
Lin, J.-Z., 213
Linder, H. P., 229
Lindsley, D. L., 260, 312
Linn, C., 160
Liou, L. W., 66, 221, 372, 373, 415
Littlejohn, M. J., 353, 359–360
Lloyd, E. A., 443, 444
Lloyd, M., 167, 168, 202
Lofstedt, C., 225
Loftus-Hills, J. F., 360, 361
Lomolino, M. V., 92, 124
Lorkovic, Z., 229
Losos, J. B., 99, 145, 169, 175
Louis, J., 185
Lovejoy, N. R., 147, 150, 414, 419, 427
Lovette, I. J., 420
Lumaret, R., 325, 328, 329, 424
Lumme, J., 225, 303
Lung, O., 246
Lynch, J. D., 168, 170, 187
Lynch, M., 275, 312, 313, 325

M

MacCallum, C. J., 187, 191
Machado, C. A., 34, 42, 68, 196, 305, 309

Macnair, M. R., 21, 120, 189, 192, 193, 203, 212, 233, 237, 242, 268, 272, 303, 385, 409
Magallón, S., 418, 420, 423
Magurran, A. E., 360
Mahoney, M. J., 101
Majewski, J., 23, 24
Majnep, I. S., 13
Malik, H. S., 295, 297
Malitschek, B., 315
Mallet, J., 11, 27, 30, 32, 66, 96, 121, 122, 126, 158, 159, 164, 209, 447–449, 451
Mampell, K., 307
Mandel, M. J., 278
Mangan, R. L., 214
Mantovani, B., 286
Markert, J. A., 152
Markow, T. A., 240, 405
Marsh, A. L., 148
Marshall, D. C., 168
Marshall, J. L., 353, 354, 361, 369, 380
Martens, K., 147
Martin, A. P., 168
Martin, F. W., 241
Martin, J. A., 254
Martinsen, G. D., 44, 471
Masaki, S., 206
Maside, X. R., 300
Masly, J. P., 271, 306, 312
Masters, J. C., 451, 452
Masterson, J., 326–327
Mather, K., 300
Matsubara, K., 268
Maxson, L. R., 77
Maxson, R. D., 77
Mayer, G. C., 85, 117, 126, 160, 258–260, 402
Maynard Smith, J., 49, 126, 131, 165, 442
Mayr, E., 1–6, 10, 13, 15, 21, 25, 27, 29, 30, 32, 33, 35, 38, 39, 45, 55, 68, 69, 83, 84, 86, 91, 92, 102, 104, 107, 126, 136, 147, 154, 156, 157, 166, 168, 180, 182, 205, 209, 214, 216, 230, 257, 298, 314, 319, 389–391, 394, 397, 426, 447
Mazer, S., 200
McCarthy, E. M., 340, 341, 350
McCartney, M. A., 226
McClure, B. A., 241
McCune, A. R., 147, 150, 414, 419, 427
McDonald, J. H., 403
McGehee, R., 35
McKaye, K. R., 151
McKinnon, J. S., 145, 146
McLain, D. K., 438
McNeilly, T., 122, 207, 361, 456
McPeek, M. A., 406
McPhail, J. D., 34, 146, 184, 250, 406, 407, 471
McPheron, B. A., 160, 161
McQueen, H. A., 336
Meade, D. E., 200
Meffert, L. M., 400
Meier, R. D., 25, 39
Melnick, D. J., 471
Mendelson, T. C., 76, 77, 79, 108
Mercot, H., 296
Meselson, M., 18
Metz, E. C., 63, 70, 233, 235, 243, 359
Metz, J. A. J., 154
Meyer, A., 147, 152
Meyer, S., 150
Michaelis, P., 311
Michalak, P., 123, 293
Midgley, J., 229
Miller, E. L., 305, 376
Miller, R. B., 200
Miller, R. L., 233, 234
Miller, R. R., 93
Milstead, B., 123
Mirov, N. T., 349
Mishler, B. D., 11, 14, 20, 46
Mitra, S., 407, 438, 441
Mitrofanov, V. G., 303, 310
Mitter, C., 404–405, 434, 436, 439, 440
Miyatake, T., 90
Molbo, D., 155, 196
Moller, A. P., 232, 436, 442
Montchamp-Moreau, C., 278, 296
Monteiro, L. R., 209
Monti, L., 225
Mooers, A. O., 400
Mooney, H. A., 186
Moore, J. A., 371
Moore, T. E., 167
Moore, W. S., 97, 99, 121
Moran, P., 150
Morell, V., 179
Moritz, C., 103, 261, 263, 264, 402
Moriwaki, D., 356
Morrison, D., 63
Mosseler, A., 203
Moulia, C., 253
Moya, A., 400
Moyle, L. C., 79–81, 364
Muir, G., 44
Muller, H. J., 38, 43, 67, 68, 258, 267–271, 274, 287, 290, 300, 307, 312, 333, 334, 372, 384, 386
Muniyamma, M., 161
Muntzing, A., 322, 326
Myers, P., 261, 402

N

Nachman, M. W., 261, 263, 264, 402
Nagel, L., 68, 131, 146, 184, 221, 409
Nagl, S., 150
Naisbit, R. E., 66, 67, 249, 381, 409
Nason, J. D., 44
Navarro, A., 117, 177, 257, 309
Naveira, H. F., 300, 303
Nee, S., 286, 418, 423, 434, 438
Neff, N. A., 15
Nei, M., 3, 73, 87, 172, 220, 271, 272, 387, 388, 395, 414, 421
Nelson, G., 11, 97, 459
Neubaum, D. M., 238
Nevo, E., 123
Newton, I., 99
Niklas, K. J., 229, 233, 234, 420
Nilsson, L. A., 64, 194, 252
Nixon, K. C., 459, 462
Noor, M. A. F., 68, 71, 117, 162, 177, 207, 221, 225, 227, 254, 257, 293, 303, 309, 353, 354, 359, 364, 365, 369, 374–379, 415
Nosil, P., 102, 359, 406, 407, 445, 472
Novitski, E., 295

O

O'Hara, P., 254
O'Neill, S. L., 276–279, 426
Ochman, H., 23
Ödeen, A., 90, 141, 399
Ogden, R., 121
Oguma, Y., 225
Ohno, S., 313
Ohta, T., 402
Oliver, J., 204
Ollerton, J., 198
Omland, K. E., 471
Opell, B. D., 230
Opler, P. A., 202, 204
Orr, H. A., 38, 40, 61–63, 73–78,
    80, 81, 87, 107, 111, 181, 191,
    216, 232, 268–276, 280,
    285–294, 296–298, 300, 303,
    305, 307, 308, 310, 311, 317,
    318, 335, 362–364, 366, 372,
    374–376, 397, 410, 414, 415,
    421, 422, 427
Orr, L. H., 276
Osborn, T. C., 332
Otte, D., 1, 108, 359, 377
Otto, S. P., 326, 327, 329, 331,
    332, 335, 336, 396
Owen, R. B., 149, 152, 426
Owens, I. P. F., 436, 437, 439,
    441

P

Page, R. D. M., 157, 189, 190
Palmer, E. J., 43
Palopoli, M. F., 274, 307, 312
Palumbi, S. R., 42, 63, 70,
    233–235, 243, 359, 445
Palys, T., 52
Panhuis, T. M., 123
Pantazidis, A. Z., 268, 303
Papadopol, C. S., 203
Parker, A., 149
Parker, G. A., 380
Partridge, L., 380
Pashley, D. P., 254
Pastuglia, M., 241, 242
Paterniani, E., 355
Paterson, A. M., 190
Paterson, H. E. H., 370,
    451–453

Patterson, J. T., 27, 232, 233,
    234, 271, 303, 310, 361
Patton, J. L., 471
Paul, J., 20, 21
Paul, R. C., 215
Paulson, D. R., 228
Paulus, H. F., 64, 194, 198
Pavlovsky, O., 87
Payne, R. B., 222
Payseur, B. A., 290
Pellmyr, O., 195, 196
Percy, G., 330
Perez, D. E., 260, 307, 312, 315
Pernin, P., 18
Pfenning, K. S., 360
Phillips, P. C., 271
Phillips, S. D. F., 224, 226
Phipps, J. B., 161
Piálek, J., 263, 264
Pizzari, T., 244
Plath, M., 215
Platnick, N. I., 97
Platt, W. J., 182, 183
Platz, N. E., 207
Plawecki, M., 61
Polechova, J., 114, 135
Polyakova, N. E., 209
Pomiankowski, A., 87, 219,
    295, 296, 414
Pontecorvo, G., 268, 271, 290,
    303, 312
Porter, A. H., 230
Poulton, E. B., 2, 269
Powell, J. A., 195
Powell, J. R., 17, 34, 41, 106,
    311, 399
Prager, E. M., 73
Prance, G. T. E., 95
Prazmo, W., 192
Presgraves, D. C., 31, 42, 57,
    60, 63, 67, 77, 78, 268, 275,
    277, 285, 286, 290, 293, 298,
    303, 304–308, 316– 317, 319,
    334, 421
Preston, C. R., 314, 319
Price, C. S. C, 64
Price, J. T., 97, 99, 121
Price, T. D., 42, 65–67, 73,
    77–79, 81, 99, 107, 145, 169,
    188, 214, 219, 221, 233–237,
    240, 285, 286, 293, 372, 373,
    415, 421, 425, 437
Proctor, H. C., 218

Prokopy, R. J., 160, 161
Promislow, D., 438
Provine, W. B., 2, 388
Prowell, D. P., 224, 226
Purvis, A., 420, 434, 441

Q

Qvarnström, A., 220

R

Rabinowitz, D., 205
Ramirez, W. B., 200
Ramnarine, I. W., 360
Ramsey, J., 57, 58, 62, 63, 69,
    181, 186, 196, 265, 321,
    323–325, 328, 329, 331, 332,
    415
Rand, A. S., 214, 215
Rand, D. M., 120, 183
Randolph, L. F., 347, 348
Ranz, J. M., 293, 294
Räsänen, M. E., 97
Ratcliffe, L. M., 215, 220, 360
Rathcke, B., 205
Raubenheimer, D., 452
Raup, D. M., 49
Raven, P. H., 11, 16
Ravigné, V., 65, 85, 375, 379,
    404
Rawson, P. D., 310, 319
Read, A., 286
Reed, D. L., 157
Reed, K. M., 262, 277
Reeder, T. W., 468
Reiland, J., 293
Reinhold, K., 226, 292
Remington, C. L., 97, 99
Rhymer, J. M., 37
Ribbink, A. J., 148
Ribera, I., 171
Ricci, C., 22
Rice, W. R., 32, 87, 89, 131, 133,
    134, 138–141, 176, 191, 218,
    226, 239, 244, 250, 354, 355,
    357, 416
Richards, A. J., 19, 20
Richards, O. W., 2
Richman, A. D., 188, 238
Rico, C., 152

Rieseberg, L. H., 43, 68, 71, 117, 162, 177, 233, 237, 257, 266, 267, 303, 307, 308, 327, 337–339, 345–348, 350, 409
Riley, H. P., 37, 71, 347, 348
Ringo, J. D., 399
Ringo, J. M., 87, 89, 90, 217
Rising, J. D., 184
Ritchie, M. G., 214, 224–226, 362
Ritland, K., 213
Roberts, M. R., 110
Roberts, M. S., 23, 24
Robertson, A., 300
Robertson, F. W., 90, 187, 356
Robertson, J. L., 200
Robinson, S. K., 187
Robson, G. C., 2
Rockwood, E. S., 402
Roderick, G. K., 426
Rodriguez, D. J., 331
Roelofs, W., 215, 225, 226
Rolan-Alvarez, E., 214
Rooney, A. P., 226
Rose, M., 314, 319
Ross, C. L., 120
Rothwell, G. W., 425
Rouse, J. L., 233–235, 242
Rousset, F., 277, 279
Rouyer, F., 262
Rowntree, V. J., 157
Roy, K., 142
Rüber, L., 149
Rubinoff, I., 92
Rundle, H. D., 145, 146, 221, 250, 360, 377, 400, 401, 405, 406, 472
Ruokolainen, K., 97
Russell, S. T., 66, 77, 79
Ryan, M. J., 214, 215, 217–219, 223
Rykena, S., 286

S

Sabara, H. A., 63, 331, 332
Saetre, G.-P., 290, 360
Sage, R. D., 42, 207, 253
Said, K., 264
Sainz, A., 298
Salt, G. W., 133, 138–141, 176
Sanderson, M. J., 172, 418, 420, 423, 432, 433
Sanderson, N., 371, 372
Sandler, L., 295
Sanford, W. W., 64
Sang, T., 349, 350
Sapienza, C., 259, 297
Sasa, M., 77, 78, 421
Sasaki, T., 277
Sattler, G. D., 42
Sauer, J. D., 84
Sauer, R. H., 203
Saumitou-Laprade, P., 311
Saunders, P., 49
Savolainen, R., 154
Savolainen, V., 436, 440
Sawamura, K., 290, 293, 306, 307, 317, 318
Sawyer, S. A., 367
Sawyer, T. L., 104
Scali, V., 286
Schaal, B. A., 42, 43, 44
Schaeffer, S. W., 305, 376
Schäfer, U., 303
Scharloo, W., 357
Schartl, M., 315
Schemske, D. W., 34, 62, 63, 69, 183, 192, 195, 196, 200–202, 209, 265, 321, 323–325, 327–329, 331, 332, 361, 408, 415, 430, 439
Schilthuizen, M., 30, 31, 126
Schliewen, U. K., 147, 150, 152, 153, 176, 419
Schlötterer, C., 123
Schluter, D., 35, 63, 68, 99, 131, 145, 146, 184, 185, 188, 219, 221, 250, 255, 329, 343, 360, 377, 385, 386, 404, 406, 409, 411, 428, 430, 440
Schmitz, U. K., 311
Schnable, P. S., 311
Schneider, C. J., 103
Schwarzbach, A. E., 345
Scriber, J. M., 290
Searle, J. B., 262–264
Secor, C. L., 334
Seehausen, O., 37, 70, 71, 147–151, 154, 186, 216
Seger, J., 157
Segraves, K. A., 331, 333
Sene, F. M., 402
Seoighe, C., 326
Sepkoski, J. J., 420, 422
Sepp, S., 20, 21
Servedio, M. R., 354, 371, 374
Shapiro, A. M., 230
Shaw, K. A., 108
Shaw, K. L., 27, 48, 108, 151, 153, 225, 467–469, 471
Sheppard, P. M., 34, 449
Shimizu, T., 90
Shine, R., 215, 226
Shoemaker, D. D., 279
Sidorova, N. V., 303, 310
Sidow, A., 326
Silberglied, R. E., 215
Simerloff, D., 37
Simmons, L. W., 232, 236, 244
Simon, C., 167, 168
Simon, J.C., 164
Simpson, G. G., 27, 456
Singh, R. S., 268, 286, 294
Sironi, M., 276
Sites, J. W., 261, 264, 402
Siva-Jothy, M. T., 236
Skúlason, S., 145
Slatkin, M., 85, 113, 143, 357
Slowinski, J. B., 432, 434, 436–438
Smith, D. C., 160, 161
Smith, G. R., 15, 471
Smith, M. F., 471
Smith, N. H., 23
Smith, P. F., 147, 148, 153, 154
Smith, T. B., 121, 145, 449
Smoker, W. W., 209
Sneath, P. H., 23, 24
Sniegowski, P., 266
Snyder, A. M., 153
Soans, A. B., 117, 138
Sokal, R. R., 36, 448
Soltis, D. E., 326, 328, 329
Soltis, P. S., 326, 328, 329
Somerson, N. L., 280
Song, B., 349
Sorci, G., 438
Sorenson, M. D., 222
Sota, T., 228, 230
Spencer, H. G., 370–373, 452
Spiess, E. B., 138
Spirito, F., 257
Springer, S. A., 244, 359
Spurway, H., 286
Stam, E., 433
Stanley, S. M., 419–421, 425, 431, 443
Stannius, H., 230

Starmer, W. T., 214
Stattersfield, A. J., 107
Stauffer, R. C., 83, 112
Stebbins, G. L., 3, 16, 19, 20, 43, 66, 71, 103, 180, 181, 205, 212, 232, 265, 266, 271, 314, 323, 326, 328, 331, 334–336, 338, 340, 344, 350
Steiner, K. E., 198, 200
Stephano, J. L., 245
Stephens, S. G., 365
Sternburg, J. G., 206
Stewart, W. N., 425
Stiassny, M., 147, 153
Stockey, P., 231
Stone, W. S., 233, 234, 271, 303, 310, 361
Stouthmaer, R., 276, 277
Stratton, D. A., 211, 213
Stratton, G. E., 252, 255
Straw, R. M., 201
Strong, D. R., 404
Sturmbauer, C., 152
Sturtevant, A. H., 253, 258, 286, 303, 310, 317
Sulloway, F. J., 83, 247
Sultan, S. E., 43
Sundby, R. A., 35
Sved, J. L., 368–370, 380
Swanson, W. J., 235, 238, 239, 241, 242, 245, 294
Szathmáry, E., 49
Szmidt, A. E., 349
Szymura, J. M., 42, 119, 303, 306

T

Takahata, N., 143
Tal, M., 333
Tao, Y., 275, 290, 292, 293, 295–297, 303, 304, 306, 307, 311
Tauber, C. A., 165, 192, 210
Tauber, M. J., 165, 192
Tautz, D., 147
Taylor, D. H., 350
Taylor, D. J., 146
Taylor, D. R., 276
Taylor, E. B., 406, 407, 471
Taylor, J. S., 312
Taylor, O. R., 215
Telschow, A. P., 279

Templeton, A. R., 27, 52, 180, 225, 368, 370, 376, 391–394, 396, 399, 400, 402, 452–454
Terborgh, J., 187
Thoday, J. M., 138, 356–357
Thomas, C., 162
Thompson, J. D., 325, 328, 329, 424
Thompson, J. N., 331, 333
Thompson, V., 271
Thornton, K., 313
Thorpe, R. S., 121
Tilley, S. G., 101
Ting, C.-T., 224–225, 306, 315
Tokuyasu, K. T., 260, 312
Tomaru, M., 214, 225
Tooby, J., 294
Travisano, M., 394
Trewavas, E., 153
Trivers, R. L., 380
True, J. R., 275, 290, 292, 295, 303, 310, 311
Tubaro, P. L., 285, 286
Tucker, P. K., 290
Tuomisto, H., 97
Turelli, M., 87, 122, 136, 172, 174, 177, 275, 277–279, 288–291, 308, 310, 375, 397, 398, 414, 426, 449
Turesson, G., 189, 206
Turner, B. J., 93
Turner, G. F., 40, 147–149, 152, 416, 419
Turner, J. R. G., 300, 335
Tyler, N. J., 13

U

Ueshima, R., 111, 410, 472
Uetz, G. W., 252, 255
Ungerer, M. C., 345, 346, 426
Uyenoyama, M., 468

V

Vacquier, V. D., 233–235, 238, 241, 242
Val, F. C., 225
van Alphen, J. M., 70, 148, 149
van Batenburg, F. H. D., 111, 410
van Oppen, M. J. H., 152

Van Valen, L., 27, 35, 43, 419, 420, 457–459
Vasek, F. C., 203
Verdú, M., 437, 441
Verheyen, E., 150
Verpäläinen, K., 154
Via, S., 162, 163, 177, 192, 193
Vick, K. W., 233, 234
Vickery, R. K., 81
Vigneault, G., 274, 303
Vincek, V., 403
Vogler, A. P., 111, 122, 169–172, 177, 419
Vollmer, S. V., 42
von Hagen, K. B., 431, 439
von Siebold, C. T., 230
Vrba, E., 443
Vrijenhoek, R. C., 18
Vulic, M., 23

W

Waage, J. K., 359
Wade, M. J., 56, 233, 236, 271, 365
Wagner, A., 325, 396
Wagner, M., 1, 84, 91
Wagner, W., 219
Wakabayashi, J., 104
Wake, D. B., 103, 104
Waldbauer, G. P., 206
Walker, T. J., 361, 364
Wallace, B., 257, 356
Walliker, D., 18
Walsh, B. D., 159, 258, 259
Walsh, J. B., 414
Walters, S. M., 20
Wang, H., 186, 250–251, 305, 309
Wang, X.-R., 349
Ward, P., 419
Ward, P. S., 46, 154
Ware, A. D., 230
Waser, N. M., 102, 195, 198, 205, 207
Wasserman, M., 359
Watanabe, T. K., 64, 226, 317
Watson, G. S., 277, 279
Waxman, D., 135, 137
Weber, K. E., 394
Webster, A. J., 440
Weeks, A. R., 277, 279

Weiblen, G.D., 155, 156, 176, 195, 196, 200
Weigt, L. A., 44, 93
Weis, A. E., 164
Weis, S., 315
Welch, A. M., 223
Welch, D. M., 18
Welch, M. E., 347
Wellborn, G. A., 406
Wells, M. M., 166, 215
Wen, J., 93, 123
Wendel, J. F., 330
Werner, E. E., 187
Werner, P. A., 182, 183
Werren, J. H., 254, 260, 268, 276–279, 290, 295, 311
Werth, C. R., 312
West-Eberhard, M. J., 3, 217
Westergaard, M., 335
Wettstein, F. v., 334
Wheeler, Q. D., 25, 39, 459, 462
White, B. J., 154
White, J., 167, 262
White, M. J. D., 1, 84, 112, 116, 143–144, 156, 157, 180, 257, 258, 260, 285, 334
Whitham, T. G., 253
Whitlock, M. C., 394, 396
Whitt, G. S., 325
Whittemore, A. T., 15, 16, 42–44
Whitton, M. C., 326, 327, 329, 331–333, 335, 336
Wiebes, J. T., 194, 195
Wiegmann, B. M., 441

Wiernasz, D. C., 214, 222
Wiley, E. O., 27, 456, 458
Wilke, C. M., 138
Williams, E. G., 233–235, 242
Williams, J. H., 45, 225
Williams, K. S., 167
Williamson, M., 107
Willis, J. H., 268
Wills, C. J., 87
Wilson, A. C., 25, 39, 72, 73, 108, 326, 421
Wilson, D. S., 449
Wilson, E. O., 154–155, 367, 368
Wilson, M. V. H., 419
Wilson, P., 62
Windham, M. D., 312
Windingstad, R. M., 155, 157
Windsor, D., 277
Winge, O., 322
Winking, H., 264
Wise, R. P., 311
Wittbrodt, J., 268, 274, 303, 307, 314
Wolfe, A. D., 201, 350
Wolfe, K. H., 326
Wolfner, M. F., 236, 238, 246
Wood, C. C., 145
Wood, T. K., 165, 203
Wright, D. H., 440
Wright, S., 34, 35, 56, 300
Wu, C.-I., 60, 64, 75, 77, 221, 224–225, 241, 274, 285, 288, 290, 292, 295, 296, 298–300, 303, 307, 310–312, 315, 319, 388

Wullschleger, E. B., 359
Wyatt, R., 200
Wyckoff, G. J., 239

X

Xiang, Q.-Y., 93, 94

Y

Yamada, H., 225
Yamamoto, M.-T., 318
Yanev, K. P., 103
Yang, Z., 242, 244
Yen, J. H., 278, 279
Yoo, B. H., 394
Young, J. E., 468
Yule, G. U., 417

Z

Zeng, L.-W., 268, 286, 287
Zeng, S., 425
Zeng, Z. B., 223, 225, 232
Zhang, Z., 294
Zhou, Z., 425
Zimmerman, E. C., 426
Zink, R. M., 81, 97, 169, 421, 425
Zouros, E., 225, 268, 274, 303

Subject Index

A

abalones
   gametic isolation in, 234–235, 241–242
Acyrthosiphon pisum
   as evidence of sympatric speciation, 162–164
Acyrthosiphon pisum pisum, 162, 192, 193
Aedes, 276, 293
Aegilops, 265
Aethia, 214
agamic complexes, 18–19, 21, 23, 53
agamospecies, 20
Agaoinidea, 195
Agapornis personata fischeri, 254
Agapornis roseicollis, 254
Agrodiaetus, 419
Agrostis, 361
Agrostis tenuis, 122, 207, 456
Albula, 94
Albula vulpes, 94
Alchemilla, 19, 20, 21
Alcidae, 171
allochronic isolation. *See* Temporal isolation
allochronic speciation
   and sympatric speciation, 166–168
Allonemobius, 61
Allonemobius fasciatus, 70, 234, 236, 361
Allonemobius socius, 70, 234, 236, 361

allo-parapatric speciation, 112
allopatric speciation, 85–111
   basic premise of, 123
   in cichlids, 151–152
   definition of, 83
   evolution of habitat isolation in, 189–191
   evolution of temporal isolation in, 206–207
   and habitat isolation, 185–186
   laboratory experiments of, 88–89
   modes of, 86
   as null hypothesis in experiments, 142
   rate of speciation in, 414–415, 427–428
   versus sympatric and parapatric speciation, 84
   See also Peripatric speciation; Vicariant speciation
allopatric taxa
   and biological species concept, 39–40
   habitat isolation between, 60–61
allopolyploids, 143, 323
   frequency of, 328–330
allo-sympatric speciation, 112
   Rhagoletis pomonella as possible example of, 162
Alpheus, 93
Ammodramus maritimus, 100
anagenesis, 55

Anartia, 254
Andropadus virens, 121
angiosperms
   polyploidy in, 326, 327
Anolis, 99, 144
Anolis roquet, 121
Anopheles, 293
antagonistic sexual selection, 218
Anthoxanthum, 361
Anthoxanthum odoratum, 122, 207
Aphytis chrysomphali, 35
Aphytis lingnanensis, 35
Apocryptophagus, 155–156, 176
apomixis, 211–212
apparent speciation interval, 413
   calculation of, 423–424
apparent speciation rate, 413
Aquilegia, 192, 198, 202
Aquilegia caerulea, 200
Aquilegia formosa, 191, 194, 197
Aquilegia pubescens, 191, 194, 197
Arbacia, 243, 359
Arctic charr
   as evidence of sympatric speciation, 176–177, 191
area cladogram, 95
Arenaria uniflora, 213
Argyranthemum brousonetii, 350
Argyranthemum frutescens, 350
Argyranthemum sundingii, 350
Armeria maritima, 22

Artemisia tridentata, 250
Artemisia tridentata tridentata, 186
Artemisia tridentata vaseyana, 186
assortative mating, 130–131
  models of, 131
  and niche preference, 133–134
auditory isolation, 215
autoallopolyploids, 323
autogamy, 211–213
autopolyploids, 322–323, 324
  frequency of, 328–330

B

Bacillus, 23
bacteria, reproductive isolation in, 23
bacterial speciation, 23–24, 51–52
Bacteriodetes, 276
Bactrocera cucurbitae, 89, 90
balanced phylogeny, 417
Banksia, 71, 205
barriers
  isolating (*See* Reproductive isolating barriers)
  multiple, 63
  nongenetic, 36
Bateson, William
  and problem of hybrid sterility, 248, 269
bdelloid rotifers, 18, 22
behavioral hybrid sterility, 29, 251–252, 254
behavioral isolation, 28, 64, 213–226
  detecting and measuring, 213–214
  drift as cause of, 217, 221
  effect on gene flow, 216
  evolution of, 216–223
  examples of, 214–215
  genetics of, 223–227
  nongenetic forms of, 221–222
  relative importance in speciation, 215–216
  and sexual selection, 217–221
  and sympatry, 81

traits involved in, 214
Betula, 405
between-lineage epistasis, 56
biogeography
  effect on divergence times, 427–428
  use in ascertaining mode of speciation, 168–175
biological features
  and diversification rates, 81
biological speciation interval, 412–413
  calculation of, 424
biological speciation rate, 412–413
biological species concept, 26, 247
  advantages of, 38–39
  and allopatric taxa, 39–40
  alternatives to, 27, 447–469
  and ecological differentiation, 35, 49–50
  and evolutionary history, 46–48, 470–472
  versus genotypic cluster species concept, 31–32
  oak species as test of, 43–45
  and paraphyly, 46–48, 470–472
  and polyphyly, 472
  problems with, 39–48
  relativistic aspect of, 55
birds
  sympatric speciation in, 143
Bombina, 187, 306
Bombina bombina, 119, 120, 191, 302, 306, 361
Bombina variegata, 119, 120, 191, 302, 306, 361
Bombus, 201
bottleneck, 106, 390, 395
Bovidae, 420
Brassica, 468
Bufo americanus, 361
Bufo woodhousii, 361

C

Caenorhabditis, 286
Caenorhabditis elegans, 325
Calathea ovandensis, 195
Callosobruchus maculatus, 240
Calopteryx, 359

Calopteryx haemorrhoidalis, 240
Campsis, 94
Carabus, 230
Carabus iwakianus, 228
Carabus maiyasanus, 228
Carex, 20, 42
Carpodacus mexicanus, 214
Catasetum discolor, 194
Catasetum saccatum, 194
Caulophyllum, 94
Cebidae, 420
centric fusions
  as cause of hybrid sterility, 262–265
centromeric DNA sequences
  role in Haldane's rule, 297–298
Cerasolen, 196
Cervus elaphus, 39
Chaetodipus, 174
Chamerion, 331, 332
Chamerion angustifolium, 63, 201, 210, 331
Charis, 96
Charis cleonis, 95
chemical-based isolation, 215
Chorthippus parallelus, 362
chromosomal model of parapatric speciation, 116–117
chromosomal speciation
  in animals, 259–265
  in plants, 265–267
  in plants *versus* animals, 266–267
  theory of, 116–117, 256–259, 309
chromosome doubling test, 265
chromosome rearrangements
  and isolating barriers, 71–72
  role in speciation, 116–117, 256–267, 308–309, 401–402
Chrysanthemum, 323, 326
Chrysoperla carnea, 165
Chrysoperla downesi, 165, 166, 192, 210
Chrysoperla plorabunda, 165, 166, 192, 210, 215
cichlids
  as evidence for sympatric speciation, 147–154, 176–177, 191

haplochromine, 71
Cicindela, 419
Cicindela dorsalis, 100
cladistics, 459–461
cladogenesis, 55
Clarkia, 203, 257, 420
Clarkia biloba, 110
Clarkia lingulata, 110–111
clinal isolation, 101–102
clinal speciation, 111, 113–115
cluster analysis
 in bacteria, 23, 24
 on sympatric individuals, 20–21
clustering
 in bacteria, 51–52
 explanations for, 51–54
 statistical identification of, 15–16
 in uniparental *versus* sexually reproducing eukaryotes, 22
coexistence
 as problem in sympatric speciation, 127, 129–130
 and reproductive isolation, 36
cohesion species concept, 27, 452–455
Colaptes auratus auratus, 99
Colaptes auratus cafer, 99
Coleoptera, 144, 419
Colias, 215
Colias eurytheme, 215
Colias philodice, 215
Collinsia, 344
Collocaliinae, 190
colonization event
 and species formation, 108, 109
Colubridae, 419
comb phylogeny, 417
competitive gametic isolation, 29, 232, 233, 236–238
complex conspecific epistasis, 307
congruent clines
 and parapatric speciation, 119
Conospermum, 63
conspecific pollen precedence (CPP), 236, 237

conspecific sperm precedence (CSP), 236–237
continuous-resource model of disruptive natural selection, 134–136
copulatory behavioral isolation, 29
Cordia, 204
Coregonus, 15
Coregonus clupeaformis, 146
Cornus, 94
cospeciation
 in habitat isolation, 189–190
 in pollinator isolation, 200
Costus allenii, 361
Costus guanaiensis, 361
Costus laevis, 361
Crassostrea virginica, 100
Crataegus, 19, 159, 161
Cratogeomys, 190
Crepis, 268, 271, 299
Crepis tectorum, 343
Culex, 278, 279
Culex pipiens, 279
Cyanea, 106, 108, 109
Cyprinidae, 419
Cyprinodon, 15, 93
cytoplasmic incompatibility (CI), 276–280

D

Daphnia galeata mendotae, 350
Darwin, Charles
 principle of divergence, 125
 and problem of hybrid sterility, 247–248
 theory of speciation, 1–2, 10–11
 view of geographic isolation, 83
 view of sympatric speciation, 125–126
Decumaria, 94
Delphinium gypsophilum, 323
Delphinium nelsonii, 207
Dendroica, 420
Dendrosicyos socotrana, 107
Dennyus, 190
Deronectes, 171
Desmognathus ochrophaeus, 101
destroy-the-hybrid experiments, 355–356

Diadema, 94
Diadema antillarum, 61
Diadema mexicanum, 61
Dichelopa, 144
Dicurus admilis, 182
Dicurus ludwigii, 182
differential fusion, 376
Dipodomys, 173, 174
Dipodomys microps, 172, 174
Dipodomys ordii, 172, 174
Diptera, 419
directional speciation, 445
direct selection
 on preferences, 217, 218–219
 role in speciation, 383–384
 on traits, 217, 219–220
Disa draconis, 200
discrete-habitat model
 of disruptive natural selection, 130–134
disruptive natural selection, 130–136
 continuous-resource model of, 134–136
 discrete-habitat model of, 130–134
disruptive selection, 30–31, 191, 356–357
disruptive sexual selection, 128–130
divergent selection
 effect on reproductive isolation, 87, 88–89, 90–91
diversification rates
 and biological features, 81–82
Dobzhansky, Theodosius
 theory of speciation, 2–3
Dobzhansky–Muller model, 269–272, 299, 300–301, 397
 as cause of Haldane's rule, 287
 mathematical treatment of, 272–276
dominance theory
 as cause of Haldane's rule, 287–292, 298
drift. *See* Genetic drift
Drosophila, 15, 41, 60, 61–64, 68, 73–81, 89, 106, 107, 109, 181, 213–216, 223–227, 232, 234–239, 241, 244–246, 257, 260, 261, 266, 271, 277–280,

283, 285, 286, 288, 290, 293, 294, 298, 299, 315–318, 326, 335, 343, 355, 357, 358, 361–364, 366, 375, 376, 385, 391, 393, 401, 403, 415, 420–422, 424, 427–429, 465, 466
   Hawaiian, 391, 393, 402, 403, 419, 422
Drosophila aldrichi-2, 302
Drosophila ananassae, 224–225
Drosophila arizonae, 224–225, 240, 268, 359
Drosophila arizonensis, 302, 359
Drosophila auraria, 224–225
Drosophila biauraria, 224–225
Drosophila buzatti, 302
Drosophila heteroneura, 224–225, 402, 403
Drosophila hydei, 302
Drosophila koepferae, 302
Drosophila limpiensis, 361
Drosophila littoralis, 224–225
Drosophila lummei, 224–225, 302
Drosophila madeirensis, 302
Drosophila mauritiana, 47, 64, 185, 224–225, 229, 232, 234, 236, 237, 288, 292, 293, 295, 302, 304, 305–308, 315–317, 465, 466, 469
Drosophila melanogaster, 57, 88, 107, 123, 138–139, 224–225, 238, 239, 244, 253, 268, 271, 288, 293–295, 302, 304–306, 308, 310–312, 314, 316–318, 325, 355, 356, 399–401, 420, 423, 465, 466, 468
Drosophila microspina, 361
Drosophila miranda, 17, 302, 358
Drosophila mojavensis, 89, 224–225, 240, 268, 302, 359, 405
Drosophila montana, 302, 310
Drosophila mulleri, 302
Drosophila neohydei, 302
Drosophila neorepleta, 302
Drosophila obscura, 17
Drosophila pallidosa, 224–225
Drosophila paulistorum, 87, 359
Drosophila persimilis, 17, 34, 224–225, 254–255, 261, 274, 291, 293, 301, 302, 307, 309, 311, 355, 358, 359, 376

Drosophila pseudoobscura, 17, 34, 88, 90, 106, 224–225, 254, 255, 261, 268, 274, 291, 293, 296, 301, 302, 305, 307, 309, 311, 355, 358, 359, 376, 398–400
Drosophila pulchrella, 236
Drosophila quechua, 302, 307
Drosophila recens, 280
Drosophila repleta, 302
Drosophila santomea, 236, 278, 302, 361
Drosophila sechellia, 47, 60, 107, 185, 192, 193, 224–225, 233, 288, 292, 296, 302, 306, 317, 465, 466
Drosophila silvestris, 224–225, 402–403
Drosophila simulans, 47, 57, 64, 88, 192, 224–225, 229, 232–234, 236, 237, 253, 268, 277, 278, 288, 292–296, 302, 304, 305–308, 311, 312, 314–318, 399, 465–467, 469
Drosophila subobscura, 296, 302
Drosophila subquiniaria, 280
Drosophila suzukii, 236
Drosophila teissieri, 288
Drosophila texana, 268, 302, 310
Drosophila virilis, 224–225, 268, 302
Drosophila willistoni, 88, 302, 307
Drosophila yakuba, 236, 278, 302, 361
Drosophilidae, 391
ducks
   and behavioral isolation, 65
duplicate gene hypothesis
   of intrinsic postzygotic isolation, 312–313

E

Echinometra, 63, 70, 235, 243, 359
Echinometra oblonga, 243
ecological character displacement, 377–378
ecological differentiation
   and biological species concept, 35
   in polyploidy, 332–333

ecological disturbance
   hybridization following, 70–71
ecological inviability, 29, 249–251
ecological isolation, 28, 64, 179–210
   in early speciation studies, 179–181
   See also Habitat isolation; Pollinator isolation; Temporal isolation
ecological niches
   as explanation of species, 49–54
ecological species concept, 27, 457–459
ecotones
   definition of, 95
   discontinuities at, 121
effect hypothesis, 443
Ellipsoptera, 171, 188, 407
Elymus, 344
Enchenopa, 166, 203
Enchenopa binotata, 165, 203
enhanced isolation. *See* Reinforcement
Ensatina, 104
Ensatina eschscholtzii, 103–104
Ephedra, 229
Epilachna, 187
Epilachna niponica, 70
Epilachna yasutomii, 70
Epilobium, 311
epistasis
   between-lineage, 56
   complex conspecific, 307
   as conservative evolutionary force, 389
   interspecific, 56
   intraspecific, 56
Equinae, 420, 424
Equus, 420
Erebia cassiodes, 229
Erebia nivalis, 229
Erigeron, 186
Escherichia coli, 23
Eschrichtius robustus, 157
Etheostoma, 76
ethological isolation. *See* Behavioral isolation
ethological pollinator isolation, 193–194

Eubalena australis, 157
Eubalena glacilis, 157
Euhadra aomoriensis, 111
eukaryotic taxa
   sexually reproducing, 12–17
   uniparentally reproducing, 18–22, 45
Eulaema cingulata, 194
Eurosta solidaginsi, 164–165
Eutamias, 174
evolutionary history
   and biological species concept, 46–48, 470–472
evolutionary species concept, 27, 456–457
extinction
   key factors influencing, 435, 438–439
   and net diversification intervals, 422
   and reinforcement, 370–371
extrinsic hybrid inviability, 31, 249–251
extrinsic postzygotic isolation, 29, 62, 66, 67, 249–253
   behavioral sterility, 29, 249, 251–252
   classification of, 249
   definition of, 248
   ecological inviability, 249–251
   evolution of, 255–256
   frequency of, 255
   versus intrinsic postzygotic isolation, 252–253

F

faster-heterogametic theory
   role in Haldane's rule, 297
faster-male theory
   as cause of Haldane's rule, 292–294, 298
faster-X theory
   as cause of Haldane's rule, 294–295
ferns
   polyploidy in, 326, 327
Ficedula, 360
Ficus, 155–156, 195, 196
fig wasps
   as evidence of sympatric speciation, 155–156, 176–177
   mutualism in, 188, 191, 195–196
fish
   as evidence of sympatric speciation, 145–147
Flexamia, 171
Floral isolation. *See* Pollinator isolation
flowering asynchrony
   role in prezygotic isolation, 331
folk species
   and Linnaean species, 12–14
Formicidae, 419
fossils
   delineating species in, 45–46
founder effect
   criticism of theories of, 394–396
   definition of, 106
   speciation, 106, 111
      experimental evidence, 399–401
      molecular data for, 402–403
founder event
   definition of, 105
   in models of speciation, 390–395, 397
   and species formation, 109
founder-flush-crash model, 391–392

G

Galax aphylla, 323
Gallus gallus, 79
Gambusia, 419
gametic isolation, 29, 232–245
   competitive, 29, 232, 233, 236–238
   definition of, 232
   evolution of, 241–245
   examples of, 233–238
   noncompetitive, 29, 232, 233–236
   relative importance in speciation, 238–240
   role of genetic drift in, 241–242
   role of sexual selection in, 238, 244–245
Gasterosteus aculeatus, 68, 146, 223, 249, 360, 472
Gastrophryne carolinensis, 359
Gastrophryne olivacea, 359
Gazella, 173
geminate species pairs, 92–93
   estimated divergence times in, 94
genealogical species concept, 27, 467–470
gene flow
   as conservative evolutionary force, 389
   and habitat isolation, 62, 63, 183, 184
   and reproductive barriers, 57–61
   and species concept, 34
   and temporal isolation, 202, 209
genetic distance
   and hybrid inviability, 77–81
   and hybrid sterility, 77–81
   and postzygotic isolation, 73–75, 76, 77–81
   and prezygotic isolation, 73–75, 76, 77–81
genetic drift
   as cause of behavioral isolation, 217, 221
   as cause of gametic isolation, 241–242
   role in speciation, 5, 387–393
      case studies, 409–410
      chromosomal data, 401–402
      experimental evidence, 399–401
      molecular data, 402–404
      for species recognition, 220–221
   on traits, 219–220
genetic revolution model, 389–391
genetic transilience model, 392–393
genic speciation, 267–276

between-locus incompatibilities, 268–269
  evolution of, 269–272
  mathematical models of, 272–276
  within-locus incompatibilities, 267
genomic allopolyploids, 323
genotypic cluster species concept (GCSC), 27, 30, 31–32, 447–451
Geomydoecus, 190
Geomys, 190
Geospiza, 403
Geospiza difficilis, 360
Geospiza fortis, 202, 215, 220, 252
Geospiza fuliginosa, 252, 360
Geospiza scandans, 202, 215, 220
Geukensia demissa, 100
Gila, 33
Gilia, 81, 265, 311, 328, 344, 351, 361, 365
Gilia malior, 344
Gilia modocensis, 344
Gilia seminuda, 350
Gilia spendens, 200
Ginkgoaceae, 425
Ginkgo biloba, 425
Glycine, 79, 80, 364
good-genes models, 223, 244
Gossypium, 365
Gossypium darwinii, 330
Gossypium hirsutum, 323
Gossypium tomemtosum, 330
Gruidae, 171, 419, 420
Grus canadensis tabida, 155
Gryllus, 120, 278
Gryllus firmus, 120, 183, 206
Gryllus pennsylvanicus, 120, 167, 183, 207
Gryllus veletis, 167
gymnosperms
  polyploidy in, 326

H

habitat
  versus niche, 182
habitat isolation, 28, 60–61, 182–193
Hadenoecini, 419, 420

Haldane's rule, 5, 75, 78–80, 254, 260, 280, 284–299
  causes of, 286–298
  definition of, 284
  frequency of, 285
Haliotis, 234, 243, 359
Haliotis tuberculata, 243
haplochromine cichlids, 71
Hawaiian *Drosophila*, 391, 393, 402
Hedylepta, 426
Helianthus, 68, 237, 342, 350
  recombinational speciation in, 345–347
Helianthus annuus, 302, 308–309, 343, 345–347, 409
Helianthus anomalus, 345–348, 409, 426
Helianthus deserticola, 347, 348
Helianthus paradoxus, 347, 348
Helianthus petiolaris, 302, 308–309, 343, 345–347, 409
Helianthus tuberosus, 323
Heliconius, 66, 67, 381
Heliconius cydno, 66, 221, 249, 409
Heliconius melpomene, 66, 221, 249, 409
Heliconius melpomene melpomene, 66
heterogamety
  and postzygotic isolation, 286
Heuchera grossularifolia, 331
Hieraceum, 20
Hieracium umbellatum, 189
Hipparionini, 420, 424
Hmr (Hybrid male rescue) mutation, 317, 318
homoploid hybrid speciation. *See* Recombinational speciation
Homo sapiens, 462
Horkelia californica, 187
Horkelia cuneata, 187
Horkelia fusca, 187
host races
  definition of, 158
  as evidence for sympatric speciation, 157–166
host-specific parasites
  as evidence for sympatric speciation, 154–157

host-specific species
  as evidence for sympatric speciation, 165–166
Huxley, T. H.
  and problem of hybrid sterility, 248
Hyalella azteca, 406
Hyalophora cecropia, 206
hybrid breakdown, 254
hybrid incompatibilities
  complexity of, 307
  probability of, 308
hybrid inviability, 29, 253, 280
  absolute strength of, 58
  compared with hybrid sterility, 60, 75, 79–80
  in *D. melanogaster-D. simulans* hybrids, 316
  developmental biology of, 310–311
  difficulty of reversing, 68
  extrinsic, 31, 249–251
  and gene flow, 77–81
  and mammals, 73
  number of genes causing, 302–303, 306
  and snowball effect, 274–275
  in *Xiphophorus* fish, 314–315
hybridization, 16, 18, 33, 38
  and biological species concept, 40–43
  following ecological disturbance, 70–71, 186
  and recombinational speciation, 342–343
Hybrid male rescue (*Hmr*) mutation, 317, 318
hybrid rescue mutations, 317–318
hybrid speciation. *See* Recombinational speciation
hybrid sterility, 29, 253–254, 280
  absolute strength of, 58
  behavioral, 29, 251–252, 254
  centric fusions as cause of, 262–265
  chromosomal speciation and, 259–267
  compared with hybrid inviability, 60, 75, 79–80

in *D. simulans–D. mauritiana* hybrids, 315–316
developmental biology of, 311–312
difficulty of reversing, 68
Dobzhansky–Muller model of, 269–272
and gene flow, 77–81
meiotic drive as cause of, 296
number of genes causing, 302–303, 306
physiological, 29, 253–254
and postzygotic isolation, 66–67
problem of, 247–248
and snowball effect, 274–275
hybrid zones
mosaic, 120
and parapatric speciation, 118–121
primary, 118
secondary, 118
statistical analysis of, 306
and vicariant speciation, 97–99
Hyla, 359
Hyla chrysoscelis, 332
Hyla cinerea, 252, 360
Hyla ewingi, 359
Hyla gratiosa, 252, 360
Hyla verreauxi, 359
Hyla versicolor, 332
Hymenoptera, 419

I

Icterus bullockii, 99
Icterus galbula, 99
Ilybius, 171
imprinting
as cause of behavioral isolation, 221–222
indirect selection
on preferences, 217, 219
role in speciation, 384–385
inquilism, 154–155
interbreeding, 17
interspecific epistasis, 56
intraspecific epistasis, 56
intrinsic gametic incompatibility, 234–235

intrinsic postzygotic isolation, 29, 56, 67, 80, 253–254
classification of, 249
definition of, 248–249
developmental basis of, 309–312
distribution of genes involved in, 308–309
duplicate genes in, 312–313
evolution of, 256
versus extrinsic postzygotic isolation, 252–253
frequency of, 255
genes causing, 313–319
genes suppressing, 317–319
genetic basis of, 299–319
genetic modes of, 256–280
Haldane's rule in, 284–299
number of genes involved in, 299–307
See also Hybrid inviability; Hybrid sterility
introgression, 34
analysis of, 223, 304
and biological species concept, 40–43
inviability
ecological, 29
hybrid (*See* Hybrid inviability)
Ipomopsis, 198
Ipomopsis aggregata, 207
Iris, 37, 237
recombinational speciation in, 347–348, 350
Iris brevicaulis, 343, 348
Iris fulva, 38, 71, 186, 343, 347, 348
Iris hexagona, 38, 71, 186, 347
Iris versicolor, 323
island species
as evidence for sympatric speciation, 143–145
isolation
auditory, 215
behavioral (*See* Behavioral isolation)
chemical-based, 215
clinal, 101–102
ecological, 28, 64, 179–210
(*See also* Habitat isolation; Pollinator isolation; Temporal isolation)

in early speciation studies, 179–181
enhanced (*See* Reinforcement)
gametic (*See* Gametic isolation)
habitat (*See* habitat isolation)
mating system, 28, 211–213
mechanical (*See* Mechanical isolation)
pollinator (*See* Pollinator isolation)
structural, 227–229
tactile, 215, 227–229, 230
temporal (*See* Temporal isolation)

J

Jordan's law, 91

K

key factors
affecting speciation rate, 429–442
tests of, 431–441
multiple comparisons, 433–435
single comparisons, 432–433

L

Lacerta, 286
land snails
peripatric speciation in, 111
large X-effect, 226
role in Haldane's rule, 290–292, 294
Larus argentatus, 104
Laticauda, 215
Laupala, 108, 359, 377, 471
Laupala kohalensis, 224–225
Laupala paranigra, 224–225
Lepidoptera, 60, 67, 77, 78, 144, 195, 226, 280, 284–286, 293, 299, 335, 421, 424, 471
Lepomis, 187
Lepomis cyanellus, 15
Lepomis macrochirus, 15

Lethal hybrid rescue *(Lhr)* mutation, 317
limiting similarity principle, 35
Limulus polyphemus, 100
Linanthus, 21, 63
Linepithema humile, 458
Linnaean species, and folk species, 12–14
Liriodendron, 94
Liriodendron chinese, 93
Liriodendron tulipifera, 93
lock-and-key hypothesis, 228, 230
Lycopersicum, 265

M

Macrodactylus, 229
Macrophiothrix, 234
Mactridae, 419
Magicicada, 167, 168, 359
Magicicada cassini, 167
Magicicada septendecim, 167, 168
Magicicada septendecula, 167
Magicicada tredecassini, 167
Magicicada tredecim, 167, 168
Magicicada tredecula, 167
Malaclemys terrapin, 100
Malurus, 171
Malus pumila, 159
mammals
 hybrid inviability in, 73
maternal hybrid rescue *(mhr)* mutation, 317
mating system isolation, 28, 211–213
Maylandia zebra, 152
Mayr, Ernst, 3–6
 analysis of sympatric speciation, 126, 136
 biological species concept of, 3, 10, 15, 28
 model of speciation, 389–391
mechanical isolation, 28, 227–232
 evolution of, 230–231
 examples of, 228–229
 genetics of, 231–232
 relative importance on selection, 229–230
 structural, 227–228
 tactile, 215, 227–228, 230
mechanical pollinator isolation, 193, 194
meiotic drive, 259
 as cause of Haldane's rule, 295–298
Menispermum, 94
Menispermum canadensis, 93
Menispermum dauricum, 93
Mercenaria, 253
Microcryptorhynchus, 144
Microsepsis armillata, 229
Microsepsis eberhardi, 229
microspecies, 20
Mimulus, 21, 61, 63, 81, 198, 212, 237, 242, 385
Mimulus cardinalis, 57, 58, 181, 186, 192, 196, 200, 201, 408
Mimulus guttatus, 192, 193, 268, 272, 302, 409
Mimulus lewisii, 57, 58, 181, 186, 192, 196, 200, 201, 408
Mimulus nasutus, 268
minority cytotype exclusion, 331
Mitchella, 94
Modern Synthesis, 25, 26, 32, 33, 37, 84, 268, 384
 behavioral isolation in, 213, 216
 ecology in, 179–180
 theory of speciation, 2–4
molecular systematics
 in comparative studies, 168
monocots
 polyploidy in, 326
monogonont rotifers, 22
Montastraea annularis, 204
Montastraea faveolata, 204
Montastraea franksi, 204
Morinda citrifolia, 61, 185, 193
Mormoopidae, 174
mtDNA studies, 100–101, 153, 171–175
multilocus model
 of assortative mating, 131
multiple barriers, 63
Muridae, 420
Musca domestica, 89, 117, 138
Mus musculus domesticus, 252, 263, 264, 290, 402
Mus musculus musculus, 252, 264
Mus spretus, 264, 290
mutationalist school of speciation, 2
mutualism
 obligatory, 188, 195
Myotis, 174
Myrmica, 154
Mytilus, 359
Mytilus galloprovincalis, 244
Mytilus trossulus, 244

N

Nasonia, 260, 268, 278, 307, 356
Nasonia giraulti, 254, 278, 302
Nasonia longicornis, 278
Nasonia vitripennis, 254, 278, 302
natural selection
 as cause of gametic isolation, 242
 disruptive, 130–136
  continuous-resource model of, 134–136
  discrete-habitat model of, 130–134
 role in speciation, 5, 385–386
  case studies, 408–409
  comparative data, 404–407
  experimental evidence, 398–399
  molecular data, 402–403
Necoclamisus bebbianae, 189
Neisseria, 23, 24
Neisseria meningitidis, 23
Neochlamisus bebbianae, 405
net diversification interval, 413
 calculation of, 417–423
net diversification rate, 413
 calculation of, 417–423
Neurospora, 361
niche
 ecological, 49–50
 versus habitat, 182
niche adaptation, 130
 and assortative mating, 131
 and niche preference, 131–133
niche preference, 130
 and assortative mating, 133–134

and niche adaptation, 131–133
Nicotiana, 322, 344
Nicotiana tabacum, 323
no-gene model
　of assortative mating, 131, 209
noncompetitive gametic isolation, 29, 232, 233–236
nongenetic barriers, 36
Notostraca, 425
Notropis spliopterus, 15
Notropis whippeli, 15
Numida meleagris, 79
Nup96 gene
　as cause of intrinsic isolation, 315–316

O

oak species
　as test of biological species concept, 43–45
Oceanodroma castro, 209
Ochotona, 172
OdsH (Odysseus-H) gene
　as cause of intrinsic isolation, 307, 315–316
Omiodes, 426
Oncorhynchus gorbuscha, 209
Oncorhynchus nerka, 145, 146
one-locus, one-allele model
　of assortative mating, 131
one-locus, two-allele model
　of assortative mating, 131
Ophraella, 183
Ophrys, 194
orchids, 199, 200
　and pollinator isolation, 64, 194, 198
Origin of Species, 1, 9, 10, 11, 83, 86–87
Orthogeomys, 172, 173, 190
Orthoptera, 144
Ostrinia furnacalis, 226
Ostrinia nubilalis, 215, 224–226

P

Pachysandra, 94
Paeonia emodi, 349
Paeonia lactiflora, 349
Paeonia officinalis, 350
Paeonia veitchu, 349
Paeonia xinjiangensis, 349
Panthera leo, 40
Panthera pardus, 40
Papilio dardanus, 34
Papilio memnon, 449
Pappogeomys, 190
para-allopatric speciation, 95, 112
　models of, 115–116
parallel speciation, 405–407
parapatric speciation, 111–123
　versus allopatric speciation, 84
　chromosomal model of, 116–117
　definition of, 83, 111
　evidence from nature, 118–123
　evolution of habitat isolation in, 189–191
　evolution of pollinator isolation in, 198–200
　evolution of temporal isolation in, 206–207
　experimental evidence for, 117
　historical observations of, 122–123
　models of, 111–112
　non-ecological models of, 115
　rate of speciation in, 415–416
paraphyly
　and biological species concept, 46–48, 470–472
parasites
　as evidence for sympatric speciation, 154–157
parasitic fig wasps
　as evidence of sympatric speciation, 155–156, 176–177
　mutualism in, 188, 191, 195–196
Passerina amoena, 99, 215
Passerina cyanea, 99, 215
peak shift models, 388–393
　criticism of, 394–395
　recent, 396–398
Pediculus humanus capitis, 156
Pediculus humanus corporis, 156
Penstemon, 350
Penstemon centranthifolius, 201
Penstemon clevelandii, 350
Penstemon grinnellii, 201
Penstemon spectabilis, 201
Penthorum, 94
peripatric speciation, 105–111
　on archipelagos, 107–110
　evidence from nature, 106–111
　experimental evidence for, 106
　in land snails, 111
　in peripheral isolates, 110–111
　on single oceanic islands, 106–107
　theory of, 105–106
　and vicariant speciation, 105
Perognathus, 174
Peromyscus boylii, 172, 173
Petrobium arboreum, 107
Pheidole megacephala, 458
pheromone-based isolation, 215
Pheucticus ludovicianus, 99
Pheucticus melanocephalus, 99
Phlox, 81
Phoradendron californicum, 165
Phryma, 94
Phylloscopus, 105, 188
Phylloscopus trochiloides, 104, 215
phylogenetic species concepts, 27, 39, 459–470
　version 1, 462–463
　version 2, 463–467
　version 3, 467–470
phylogeography
　definition of, 100
Phylox, 361
Physalaemus, 215
Pieris occidentalis, 214–215, 222
Pieris protodice, 214–215, 222
Pinaroloxias inornata, 107
Pinus, 350
Pinus densata, 349
Pinus tabulaeformis, 349
Pinus yunnanensis, 349
Platanaceae, 423
Platanthera bifolia, 64, 252

Platanthera chlorantha, 64, 252
Platanthera ciliaris, 200
pleiotropy
    as feature of speciation, 56
Podisma, 251, 306
Podisma pedestris, 120, 302, 306
Podophyllum, 94
Poecile, 251
Poecile atricapilla, 42
Poecile carolinensis, 42
Poecilia, 360
Poecilia mexicana, 215
Poecilia reticulata, 222
Poeciliopsis monacha-lucida, 18
Polemonium viscosum, 200
pollinator isolation, 28, 71, 193–202, 332
    absolute strength of, 58
    between allopatric taxa, 60–61
    cospeciation in, 189–190
    definition of, 182
    detecting and measuring, 184–185, 194–195
    effect on gene flow, 62, 63, 183, 184, 206
    ethological, 193–194
    evolution of, 188–191, 198–201
        in allopatry and parapatry, 189–191, 198–200
        in sympatry, 191, 200–201
    examples of, 186–188, 195–197
    genetics of, 191–193, 201–202
    macrospatial, 183
    mechanical, 193, 194
    microspatial, 182–183
    and orchids, 64
    problem of allopatry in, 185–186
    relative importance in speciation, 188, 197–198
pollinator specificity
    role in prezygotic isolation, 331
polyphyly, 472
polyploidy, 60, 69–70, 256, 321–337
    classification of, 322–324
    definition of, 321

discovery of, 322
ecology and persistence of, 330–333
frequency of auto- *versus* allopolyploidy, 328–330
genetic origins of, 324–325
    nonreduction during meiosis, 324–325
    polyspermy, 325
    somatic doubling, 324
incidence of, 326–328
rate in animals *versus* plants, 333–337
rate of speciation by, 415
and reproductive isolation, 60
polyspermy, 244–245, 325
population bottleneck, 106, 390, 395
Populus, 44
postmating prezygotic isolation, 29, 56, 232
    See also Gametic isolation
postzygotic isolation, 29, 56, 59–60, 181, 247–281
    classification of, 249
        problems with, 252–253
    effect on gene flow, 62, 63
    extrinsic (*See* Extrinsic postzygotic isolation)
    and genetic distance, 73–75, 76, 77–81
    and heterogamety, 286
    importance of, 66–69
    intrinsic (*See* Intrinsic postzygotic isolation)
    versus prezygotic isolation, 65–69, 372
    reinforcement of, 365–366
    role of selection in, 409
    via hybrid sterility, 66–67
    See also Extrinsic postzygotic isolation; Intrinsic postzygotic isolation
Potentilla glandulosa, 189
premating isolation, 28, 56, 66
prezygotic isolation, 29, 59–60, 181
    and genetic distance, 73–75, 76, 77–81
    in polyploidy, 330–332
    postmating, 29, 56, 232 (*See also* Gametic isolation)

    versus postzygotic isolation, 65–69, 372
    role of selection in, 408–409
    in sympatry, 358–362
        versus allopatry, 362–365
Procamallanus, 154
prokaryotes, 22–24
Pseudotropheus zebra, 149
punctuated equilibrium, 442
Pundamilia nyerei, 216
Pundamilia pundamilis, 216
pure birth model, 417
Pyrenestes, 449

Q

Quercus, 43
Quercus douglasii, 44
Quercus gambelii, 44, 45, 183
Quercus grisea, 44, 45, 183
Quercus kelloggii, 44
Quercus lobata, 44
Quercus petraea, 44
Quercus robus, 44
Quercus wislizenii, 44

R

Rana, 207, 360
Rana berlandieri, 207, 208
Rana blairi, 187, 208
Rana pipiens, 187
Rana sphenocephala, 187, 208
Raphanus sativus, 200
recognition species concept, 27, 451–452
recombinational speciation, 337–351
    definition of, 337–338
    evidence from nature, 344–350, 351
    experimental work on, 343–344
    rate of speciation in, 415
    theories of, 338–342
        early, 338–340
        evaluation of, 350–351
        recent, 340–342
refugia theory, 95, 97
reinforcement, 353–381
    A. R. Wallace's theory of, 353

alternatives to, 375–381
definition of, 353
evidence from nature, 357–365
in evolution of temporal isolation, 207–208
experimental evidence for, 355–357
and extinction, 370–371
of postzygotic isolation, 365–366
rate of speciation in, 415
theory of, 366–375
    early history, 366–368
    objections to, 369–372
    revival of, 372–375
reproductive character displacement
    in evolution of temporal isolation, 207–208
reproductive isolating barriers
    absolute strength of, 62
    associated with chromosome rearrangements, 71–72
    classification of, 28–29
    comparative studies of, 72–82
    determining those important in speciation, 69–72
    early-acting *vs.* later-acting, 58–59
    and gene flow, 57–61
    importance of in speciation, 31, 32
    postzygotic (*See* Postzygotic isolation)
    premating, 28, 56
    prezygotic (*See* Prezygotic isolation)
    relative strength of, 63–65
    in sympatric species, 57
reproductive isolation
    and coexistence, 36
    effect of divergent selection on, 87, 88–89, 90–91
    evolutionary rate of origin, 72–81
    as explanation for species, 50–54
    and gene exchange, 34
    identifying and measuring, 61–72

    parapatric origin of, 122–123
    and polyploidy, 60
    relative rates of, 73–75
    reversibility of, 37
    and speciation, 32–33
    time course of, 72–81
    traits promoting the evolution of, 81–82
reproductive proteins
    evolutionary rate of, 238–240
rescue mutations
    hybrid, 317–318
Rhagoletis, 171
Rhagoletis mendex, 187
Rhagoletis pomonella, 31, 64, 132, 158, 187, 205, 210, 426, 449, 450
    as evidence of sympatric speciation, 159–162, 164, 166, 176
Rhinichthys, 15
Rhododendron, 234, 235, 242
Richardonsius, 15
ring species
    and vicariant speciation, 102–105
Rivulus, 419
Rosaceae, 423
rotifers
    bdelloid, 18, 22
    monogonont, 22
Rubus, 20
runaway sexual selection, 219, 378

S

Saccharomyces, 259, 344
Saccharomyces cerevisiae, 259, 266, 325, 326, 344
Saccharomyces paradoxus, 259, 266, 344
Salix, 203, 405
Saltugilia grinnellii, 200
Salvelinus alpinus, 145
Satyrium hallackii, 199
Schistosoma sinensium, 350
Schizocosa ocreata, 251, 252
Schizocosa rovneri, 251, 252
sea urchins
    gametic isolation in, 235

Sebastes, 187
segmental allopolyploids, 323
selection
    direct
        on preferences, 217–218–219
        role in speciation, 383–384
        on traits, 217, 218–219
    disruptive, 30–31, 191, 356–357
    divergent
        effect on reproductive isolation, 87, 88–89, 90–91
        experiments, 355–357
        on preferences, 218–219
        role in speciation, 383–387
    See also Natural selection; Sexual selection
selfing, 211–213
    effects of, 212
Senecio, 20
sensory drive, 218
sensory exploitation, 218
sexually reproducing eukaryotic taxa, 12–17
    versus asexual taxa, 22
sexual selection
    antagonistic, 218, 244–245
    as cause of behavioral isolation, 217–221
    as cause of gametic isolation, 238, 244–245
    disruptive, 128–130
    effect on extinction rates, 438
    on genitalia, 230–231
    role in speciation, 3, 386–387, 435, 438
        comparative data, 407–408
    runaway, 219, 231
Silene, 79, 80, 198, 364
Silene dioica, 195
Silene latifolia, 195
silversword alliance, 330
single bottleneck design, 106
sister-group comparisons, 436–442
snowball effect, 274–275, 300
Solidago, 164–165, 182–183
Solidago altissima, 165

Solidago canadensis, 183
Solidago gigantea, 183
Solidago graminifolia, 183
Solidago missouriensis, 183
Solidago nemoralis, 183
Solidago speciosa, 183
Sorex, 173
Sorex araneus, 263
Spea, 360
speciation
  allochronic, 166–168
  allo-parapatric, 112
  allopatric (*See* Allopatric speciation)
  allo-sympatric, 112
  bacterial, 23–24, 51–52
  biogeography of, 168–175
  central problem of, 57
  chromosomal (*See* Chromosomal speciation)
  clinal, 111, 113–115
  controversies over types of, 84
  directional, 445
  genetics of, 313–314
  history of theories of, 1–6
  hybrid models of, 112
  para-allopatric, 95, 112
    models of, 115–116
  parapatric (*See* Parapatric speciation)
  peak shift models of, 388–393
  peripatric (*See* Peripatric speciation)
  rapid, 426–427, 428
  resulting from few isolating barriers, 69–70
  role of drift, 387–393
  role of selection, 383–387
  slow, 425
  stasipatric, 112, 116–117
  stepping-stone, 115–116
  sympatric (*See* Sympatric speciation)
  vicariant (*See* Vicariant speciation)
speciation rate, 411–416
  apparent, 413
  biological, 412–413
  calculation of, 416–425
  definitions of, 412–413
  effect of biogeography on, 427–428
  and extinction rates, 435
  extreme, 425–427
  key factors affecting, 429–442
    sister-group comparisons of, 436–441
    tests of, 431–435
  in models of speciation, 413–416
species
  characterization of, 30
  delineating in fossils, 45–46
  explanations for, 49
  folk, 12–14
  reality of, 10–25
    and groups with little or no sexual reproduction, 17–25
    and higher taxa, 16–17
    and sexually reproducing eukaryotic taxa, 12–17
species concepts, 25–48
  alternative, 27, 48
  and gene flow, 34
  goal of, 26
  history of, 25
species drift, 444
species selection, 431, 442–445
species sorting, 443
Spermophilus, 172, 173
Spodoptera, 225, 254
Spodoptera descoinsi, 224–225
Spodoptera latifascia, 224–225
stasipatric model, 257
stasipatric speciation, 112, 116–117
Stegastes, 187
Stephanomeria, 350
Stephanomeria diegensis, 349
Stephanomeria exigua, 349
Stephanomeria virgata, 349
stepping-stone speciation, 115–116
sterility. *See* Hybrid sterility
Streptanthus, 79
Strepthanthus glandulosus, 101
Streptococcus pneumoniae, 23
Strongylocentrotus, 243
structural isolation, 227–228
structuralist school of biology, 49
Suncus murinus, 264
suture zones
  and vicariant speciation, 97–99
swamping effect, 371
Sylvia, 171
Sylvia atricapilla, 171
sympatric speciation, 31, 125–178
  absence of in birds, 99–100
  allochronic speciation in, 166–168
  versus allopatric speciation, 84
  as alternative to reinforcement, 378–379
  criteria for ascertaining, 142
  Darwin's views on, 125–126
  definition of, 126
  evidence from nature, 141–175
    biogeography of speciation, 168–175
    from habitat islands, 143–157
    from host races and host-specific species, 157–166
    temporal isolation, 166–168
  evolution of habitat isolation in, 191
  evolution of pollinator isolation in, 200–201
  evolution of temporal isolation in, 208–210
  experimental evidence for, 138–141
  history of work on, 126
  versus parapatric speciation, 126
  prezygotic isolation in, 358–362
    versus allopatry, 362–365
  rate of speciation in, 415–416, 427–428
  theory of, 127–137
    assessment of, 136–137
    disruptive natural selection, 127, 130–136

SUBJECT INDEX

disruptive sexual selection, 127, 128–130
sympatric taxa
  isolating barriers in, 57–60
synonymy, 22

T

tactile isolation, 215, 227–229, 230
Tamias cinericollis, 174–175
Tamias dorsalis, 174–175
Tanysiptera, 394
Tanysiptera galatea, 389
Taraxacum, 19, 20
Teleogryllus commodus, 215
Teleogryllus emma, 206
Teleogryllus oceanicu, 215
temporal isolation, 28, 71, 202–210
  detecting and measuring, 203–204
  effect on gene flow, 202, 209
  evolution of, 206–210
    in allopatry and parapatry, 206–207
    reinforcement in, 207–208
    reproductive character displacement in, 207–208
    in sympatry, 208–210
  examples of, 204–205
  genetics of, 210
  relative importance in speciation, 205–206
  role in sympatric speciation, 166–168, 203
  unusual features of, 203
tension zones
  and parapatric speciation, 119, 120
Teredinidae, 419
Tetragnatha, 108, 419
Thomomydoecus, 190
Thomomys, 190
Tigriopus californicus, 102, 310
Tilapia, 153
Timema cristinae, 102, 359, 406, 472
Tragopogon, 265
transition time for biological speciation, 412

Tribolium castaneum, 138, 271
Tribolium freemani, 271
Triticum, 265
Triturus, 286

U

unbalanced female test, 288, 311
uniparentally reproducing eukaryotic taxa, 18–22, 45
  versus sexual taxa, 22
uniparental reproduction, 18

V

Vandiemenella, 257
Veneridae, 419
Vermivora, 252
Viburnum, 203
vicariance biogeography, 92
vicariant speciation
  clinal isolation and, 101–102
  definition of, 86
  evidence from nature, 91–105
  experimental evidence for, 87–91
  and peripatric speciation, 105
  ring species and, 102–105
  theory of, 86–87

W

Wallace, A. R.
  theory of reinforcement, 353
wasps
  as evidence of sympatric speciation, 155–156, 176–177
  and figs
    mutualism in, 188, 191, 195–196
Wolbachia, 80, 426
  biology of, 277–278
  as cause of intrinsic postzygotic isolation, 276–280
  role in speciation, 278–280

X

X chromosome
  effect on hybrid male sterility, 260
X-effects
  on behavioral isolation, 226
  large, 290–292, 294
Xiphophorus, 171, 314, 315, 419
Xiphophorus helleri, 214, 268, 302, 314
Xiphophorus maculatus, 214, 268, 302, 314
Xiphophorus nigrensis, 219
Xiphophorus pygmaeus, 219
Xmrk-2 gene
  as cause of intrinsic isolation, 314–315

Z

Zea mays, 355
Zonotrichia altricapilla, 184
Zonotrichia leucophrys, 184
Zonotrichia leucophrys nutalli, 221
Zygogeomys, 190
Zygotic hybrid rescue *(Zhr)* mutation, 318

Ollscoil na hÉireann, Gaillimh